## Variational Bayesian Learning Theory

Variational Bayesian learning is one of the most popular methods in machine learning. Designed for researchers and graduate students in machine learning, this book summarizes recent developments in the nonasymptotic and asymptotic theory of variational Bayesian learning and suggests how this theory can be applied in practice.

The authors begin by developing a basic framework with a focus on conjugacy, which enables the reader to derive tractable algorithms. Next, it summarizes nonasymptotic theory, which, although limited in application to bilinear models, precisely describes the behavior of the variational Bayesian solution and reveals its sparsity-inducing mechanism. Finally, the text summarizes asymptotic theory, which reveals phase transition phenomena depending on the prior setting, thus providing suggestions on how to set hyperparameters for particular purposes. Detailed derivations allow readers to follow along without prior knowledge of the mathematical techniques specific to Bayesian learning.

SHINICHI NAKAJIMA is a senior researcher at Technische Universität Berlin. His research interests include the theory and applications of machine learning, and he has published papers at numerous conferences and in journals such as *The Journal of Machine Learning Research, The Machine Learning Journal, Neural Computation*, and *IEEE Transactions on Signal Processing*. He currently serves as an area chair for Neural Information Processing Systems (NIPS) and an action editor for Digital Signal Processing.

KAZUHO WATANABE is an associate professor at Toyohashi University of Technology. His research interests include statistical machine learning and information theory, and he has published papers at numerous conferences and in journals such as *The Journal of Machine Learning Research, The Machine Learning Journal, IEEE Transactions on Information Theory*, and *IEEE Transactions on Neural Networks and Learning Systems*.

MASASHI SUGIYAMA is the director of the RIKEN Center for Advanced Intelligence Project and professor of Complexity Science and Engineering at the University of Tokyo. His research interests include the theory, algorithms, and applications of machine learning. He has written several books on machine learning, including *Density Ratio Estimation in Machine Learning*. He served as program cochair and general cochair of the NIPS conference in 2015 and 2016, respectively, and received the Japan Academy Medal in 2017.

# Variational Bayesian Learning Theory

SHINICHI NAKAJIMA
*Technische Universität Berlin*

KAZUHO WATANABE
*Toyohashi University of Technology*

MASASHI SUGIYAMA
*University of Tokyo*

CAMBRIDGE
UNIVERSITY PRESS

CAMBRIDGE
UNIVERSITY PRESS

Shaftesbury Road, Cambridge CB2 8EA, United Kingdom

One Liberty Plaza, 20th Floor, New York, NY 10006, USA

477 Williamstown Road, Port Melbourne, VIC 3207, Australia

314–321, 3rd Floor, Plot 3, Splendor Forum, Jasola District Centre, New Delhi – 110025, India

103 Penang Road, #05–06/07, Visioncrest Commercial, Singapore 238467

Cambridge University Press is part of Cambridge University Press & Assessment, a department of the University of Cambridge.

We share the University's mission to contribute to society through the pursuit of education, learning and research at the highest international levels of excellence.

www.cambridge.org
Information on this title: www.cambridge.org/9781107430761

DOI: 10.1017/9781139879354



First published 2019
First paperback edition 2025

*A catalogue record for this publication is available from the British Library*

*Library of Congress Cataloging-in-Publication data*
Names: Nakajima, Shinichi, author. | Watanabe, Kazuho, author. | Sugiyama, Masashi, 1974- author.
Title: Variational Bayesian learning theory / Shinichi Nakajima (Technische Universität Berlin), Kazuho Watanabe (Toyohashi University of Technology), Masashi Sugiyama (University of Tokyo).
Description: Cambridge ; New York, NY : Cambridge University Press, 2019. | Includes bibliographical references and index.
Identifiers: LCCN 2019005983| ISBN 9781107076150 (hardback : alk. paper) | ISBN 9781107430761 (pbk. : alk. paper)
Subjects: LCSH: Bayesian field theory. | Probabilities.
Classification: LCC QC174.85.B38 N35 2019 | DDC 519.2/33–dc23
LC record available at https://lccn.loc.gov/2019005983

ISBN 978-1-107-07615-0 Hardback
ISBN 978-1-107-43076-1 Paperback

# Contents

# Preface

Bayesian learning is a statistical inference method that provides estimators and other quantities computed from the *posterior distribution*—the conditional distribution of unknown variables given observed variables. Compared with *point estimation* methods such as maximum likelihood (ML) estimation and maximum a posteriori (MAP) learning, Bayesian learning has the following advantages:

- Theoretically optimal.
  The posterior distribution is what we can obtain best about the unknown variables from observation. Therefore, Bayesian learning provides most accurate predictions, provided that the assumed model is appropriate.
- Uncertainty information is available.
  Sharpness of the posterior distribution indicates the reliability of estimators. The credible interval, which can be computed from the posterior distribution, provides probabilistic bounds of unknown variables.
- Model selection and hyperparameter estimation can be performed in a single framework.
  The marginal likelihood can be used as a criterion to evaluate how well a statistical model (which is typically a combination of model and prior distributions) fits the observed data, taking account of the flexibility of the model as a penalty.
- Less prone to overfitting.
  It was theoretically proven that Bayesian learning overfits the observation noise less than MAP learning.

On the other hand, Bayesian learning has a critical drawback—computing the posterior distribution is computationally hard in many practical models. This is because Bayesian learning requires *expectation* operations or integral computations, which cannot be analytically performed except for simple cases.

Accordingly, various approximation methods, including deterministic and sampling methods, have been proposed.

Variational Bayesian (VB) learning is one of the most popular deterministic approximation methods to Bayesian learning. VB learning aims to find the closest distribution to the Bayes posterior under some constraints, which are designed so that the expectation operation is tractable. The simplest and most popular approach is the *mean field approximation* where the approximate posterior is sought in the space of *decomposable* distributions, i.e., groups of unknown variables are forced to be independent of each other. In many practical models, Bayesian learning is intractable *jointly* for all unknown parameters, while it is tractable if the dependence between groups of parameters is ignored. Such a case often happens because many practical models have been constructed by combining simple models in which Bayesian learning is analytically tractable. This property is called *conditional conjugacy*, and makes VB learning computationally tractable.

Since its development, VB learning has shown good performance in many applications. Its good aspects and downsides have been empirically observed and qualitatively discussed. Some of those aspects seem inherited from full Bayesian learning, while some others seem to be artifacts by forced independence constraints. We have dedicated ourselves to theoretically clarifying the behavior of VB learning quantitatively, which is the main topic of this book.

This book starts from the formulation of Bayesian learning methods. In Part I, we introduce Bayesian learning and VB learning, emphasizing how conjugacy and conditional conjugacy make the computation tractable. We also briefly introduce other approximation methods and relate them to VB learning. In Part II, we derive algorithms of VB learning for popular statistical models, on which theoretical analysis will be conducted in the subsequent parts.

We categorize the theory of VB learning into two parts, and exhibit them separately. Part III focuses on *nonasymptotic* theory, where we do not assume the availability of a large number of samples. This analysis so far has been applied only to a class of *bilinear* models, but we can make detailed discussions including analytic forms of global solutions and theoretical performance guarantees. On the other hand, Part IV focuses on asymptotic theory, where the number of observed samples is assumed to be large. This approach has been applied to a broad range of statistical models, and successfully elucidated the *phase transition* phenomenon of VB learning. As a practical outcome, this analysis provides a guideline on how to set hyperparameters for different purposes.

Recently, a lot of variations of VB learning have been proposed, e.g., more accurate inference methods beyond the mean field approximation, stochastic gradient optimization for big data analysis, and sampling based update rules for automatic (black-box) inference to cope with general nonconjugate likelihoods including deep neural networks. Although we briefly introduce some of those recent works in Part I, they are not in the central scope of this book. We rather focus on the simplest mean field approximation, of which the behavior has been clarified quantitatively by theory.

This book was completed under the support by many people. Shinichi Nakajima deeply thanks Professor Klaus-Robert Müller and the members in Machine Learning Group in Technische Universität Berlin for their direct and indirect support during the period of book writing. Special thanks go to Sergej Dogadov, Hannah Marienwald, Ludwig Winkler, Dr. Nico Gönitz, and Dr. Pan Kessel, who reviewed chapters of earlier versions, found errors and typos, provided suggestions to improve the presentation, and kept encouraging him in proceeding book writing. The authors also thank Lauren Cowles and her team in Cambridge University Press, as well as all other staff members who contributed to the book production process, for their help, as well as their patience on the delays in our manuscript preparation. Lauren Cowles, Clare Dennison, Adam Kratoska, and Amy He have coordinated the project since its proposal, and Harsha Vardhanan in SPi Global has managed the copy-editing process with Andy Saff.

The book writing project was partially supported by the following organizations: the German Research Foundation (GRK 1589/1) by the Federal Ministry of Education and Research (BMBF) under the Berlin Big Data Center project (Phase 1: FKZ 01IS14013A and Phase 2: FKz 01IS18025A), the Japan Society for the Promotion of Science (15K16050), and the International Research Center for Neurointelligence (WPI-IRCN) at The University of Tokyo Institutes for Advanced Study.

# Nomenclature

| | |
|---|---|
| $a, b, c, \ldots, A, B, C, \ldots$ | : Scalars. |
| $\boldsymbol{a}, \boldsymbol{b}, \boldsymbol{c}, \ldots$ (bold-faced small letters) | : Vectors. |
| $\boldsymbol{A}, \boldsymbol{B}, \boldsymbol{C}, \ldots$ (bold-faced capital letters) | : Matrices. |
| $\mathcal{A}, \mathcal{B}, \mathcal{C}, \ldots$ (calligraphic capital letters) | : Tensors or sets. |

$(\cdot)_{l,m}$ : $(l, m)$th element of a matrix.

$\top$ : Transpose of a matrix or vector.

$\mathrm{tr}(\cdot)$ : Trace of a matrix.

$\det(\cdot)$ : Determinant of a matrix.

$\odot$ : Hadamard (elementwise) product.

$\otimes$ : Kronecker product.

$\times_n$ : $n$-mode tensor product.

$|\cdot|$ : Absolute value of a scalar. It applies element-wise for a vector or matrix.

$\mathrm{sign}(\cdot)$ : Sign operator such that $\mathrm{sign}(x) = \begin{cases} 1 & \text{if } x \geq 0, \\ -1 & \text{otherwise.} \end{cases}$ It applies elementwise for a vector or matrix.

$\{\cdots\}$ : Set consisting of specified entities.

$\{\cdots\}^D$ : $D$fold Cartesian product, i.e.,

$$\mathbb{X}^D \equiv \{(x_1, \ldots, x_D)^\top; x_d \in \mathbb{X} \text{ for } d = 1, \ldots, D\}.$$

$\#(\cdot)$ : Cardinality (the number of entities) of a set.

$\mathbb{R}$ : The set of all real numbers.

$\mathbb{R}_+$ : The set of all nonnegative real numbers.

$\mathbb{R}_{++}$ : The set of all positive real numbers.

$\mathbb{R}^D$ : The set of all $D$-dimensional real (column) vectors.

$[\cdot,\cdot]$ : The set of real numbers in a range, i.e.,
$$[l,u] = \{x \in \mathbb{R}; l \leq x \leq u\}.$$

$[\cdot,\cdot]^D$ : The set of $D$-dimensional real vectors whose entries are in a range, i.e., $[l,u]^D \equiv \{\boldsymbol{x} \in \mathbb{R}^D; l \leq x_d \leq u \text{ for } d = 1,\ldots,D\}$.

$\mathbb{R}^{L\times M}$ : The set of all $L \times M$ real matrices.

$\mathbb{R}^{M_1\times M_2\times\cdots\times M_N}$ : The set of all $M_1 \times M_2 \times \cdots \times M_N$ real tensors.

$\mathbb{I}$ : The set of all integers.

$\mathbb{I}_{++}$ : The set of all positive integers.

$\mathbb{C}$ : The set of all complex numbers.

$\mathbb{S}^D$ : The set of all $D \times D$ symmetric matrices.

$\mathbb{S}^D_+$ : The set of all $D \times D$ positive semidefinite matrices.

$\mathbb{S}^D_{++}$ : The set of all $D \times D$ positive definite matrices.

$\mathbb{D}^D$ : The set of all $D \times D$ diagonal matrices.

$\mathbb{D}^D_+$ : The set of all $D \times D$ positive semidefinite diagonal matrices.

$\mathbb{D}^D_{++}$ : The set of all $D \times D$ positive definite diagonal matrices.

$\mathbb{H}^{K-1}_N$ : The set of all possible histograms for $N$ samples and $K$ categories, i.e., $\mathbb{H}^{K-1}_N \equiv \{\boldsymbol{x} \in \{0,\ldots,N\}^K; \sum_{k=1}^K x_k = N\}$.

$\Delta^{K-1}$ : The standard $(K-1)$-simplex, i.e.,
$$\Delta^{K-1} \equiv \{\boldsymbol{\theta} \in [0,1]^K; \textstyle\sum_{k=1}^K \theta_k = 1\}).$$

$(\boldsymbol{a}_1,\ldots,\boldsymbol{a}_M)$ : Column vectors of $\boldsymbol{A}$, i.e., $\boldsymbol{A} = (\boldsymbol{a}_1,\ldots,\boldsymbol{a}_M) \in \mathbb{R}^{L\times M}$.

$(\widetilde{\boldsymbol{a}}_1,\ldots,\widetilde{\boldsymbol{a}}_L)$ : Row vectors of $\boldsymbol{A}$, i.e., $\boldsymbol{A} = (\widetilde{\boldsymbol{a}}_1,\ldots,\widetilde{\boldsymbol{a}}_L)^\top \in \mathbb{R}^{L\times M}$.

$\mathbf{Diag}(\cdot)$ : Diagonal matrix with specified diagonal elements, i.e.,
$$(\mathbf{Diag}(\boldsymbol{x}))_{l,m} = \begin{cases} x_l & \text{if } l = m, \\ 0 & \text{otherwise.} \end{cases}$$

$\mathbf{diag}(\cdot)$ : Column vector consisting of the diagonal entries of a matrix, i.e.,
$$(\mathbf{diag}(\boldsymbol{X}))_l = X_{l,l}.$$

$\mathbf{vec}(\cdot)$ : Vectorization operator concatenating all column vectors of a matrix into a long column vector, i.e., $\mathbf{vec}(\boldsymbol{A}) = (\boldsymbol{a}_1^\top,\ldots,\boldsymbol{a}_M^\top)^\top \in \mathbb{R}^{LM}$ for a matrix $\boldsymbol{A} = (\boldsymbol{a}_1,\ldots,\boldsymbol{a}_M) \in \mathbb{R}^{L\times M}$.

$\boldsymbol{I}_D$ : $D$-dimensional $(D \times D)$ identity matrix.

$\boldsymbol{\Gamma}$ : A diagonal matrix.

$\boldsymbol{\Omega}$ : An orthogonal matrix.

$\boldsymbol{e}_k$ : One of $K$ expression, i.e., $\boldsymbol{e}_k = (\underbrace{0,\ldots,0,\overbrace{\ 1\ }^{k\text{th}},0,\ldots,0}_{K})^\top \in \{0,1\}^K$.

$\mathbf{1}_K$ : $K$-dimensional vector with all elements equal to one, i.e.,
$$\boldsymbol{e}_k = (\underbrace{1,\ldots,1}_{K})^\top.$$

$\mathrm{Gauss}_D(\boldsymbol{\mu}, \boldsymbol{\Sigma})$ : $D$-dimensional Gaussian distribution with mean $\boldsymbol{\mu}$ and covariance $\boldsymbol{\Sigma}$.

$\mathrm{MGauss}_{D_1,D_2}(\boldsymbol{M}, \boldsymbol{\Sigma} \otimes \boldsymbol{\Psi})$ : $D_1 \times D_2$ dimensional matrix variate Gaussian distribution with mean $\boldsymbol{M}$ and covariance $\boldsymbol{\Sigma} \otimes \boldsymbol{\Psi}$.

$\mathrm{Gamma}(\alpha, \beta)$ : Gamma distribution with shape parameter $\alpha$ and scale parameter $\beta$.

$\mathrm{InvGamma}(\alpha, \beta)$ : Inverse-Gamma distribution with shape parameter $\alpha$ and scale parameter $\beta$.

$\mathrm{Wishart}_D(\boldsymbol{V}, \nu)$ : $D$-dimensional Wishart distribution with scale matrix $\boldsymbol{V}$ and degree of freedom $\nu$.

$\mathrm{InvWishart}_D(\boldsymbol{V}, \nu)$ : $D$-dimensional inverse-Wishart distribution with scale matrix $\boldsymbol{V}$ and degree of freedom $\nu$.

$\mathrm{Multinomial}(\boldsymbol{\theta}, N)$ : Multinomial distribution with event probabilities $\boldsymbol{\theta}$ and number of trials $N$.

$\mathrm{Dirichlet}(\boldsymbol{\phi})$ : Dirichlet distribution with concentration parameters $\boldsymbol{\phi}$.

$\mathrm{Prob}(\cdot)$ : Probability of an event.

$p(\cdot), q(\cdot)$ : Probability distribution (probability mass function for discrete random variables, and probability density function for continuous random variables). Typically $p$ is used for a model distribution and $q$ is used for the true distribution.

$r(\cdot)$ : A trial distribution (a variable of a functional) for approximation.

$\langle f(\boldsymbol{x}) \rangle_{p(\boldsymbol{x})}$ : Expectation value of $f(\boldsymbol{x})$ over distribution $p(\boldsymbol{x})$, i.e.,

$$\langle f(\boldsymbol{x}) \rangle_{p(\boldsymbol{x})} \equiv \int f(\boldsymbol{x}) p(\boldsymbol{x}) d\boldsymbol{x}.$$

$\widehat{\cdot}$ : Estimator for an unknown variable, e.g., $\widehat{\boldsymbol{x}}$ and $\widehat{\boldsymbol{A}}$ are estimators for a vector $\boldsymbol{x}$ and a matrix $\boldsymbol{A}$, respectively.

$\mathbf{Mean}(\cdot)$ : Mean of a random variable.

$\mathrm{Var}(\cdot)$ : Variance of a random variable.

$\mathbf{Cov}(\cdot)$ : Covariance of a random variable.

$\mathrm{KL}(\cdot||\cdot)$ : Kullbuck–Leibler divergence between distributions, i.e.,

$$\mathrm{KL}(p||q) \equiv \left\langle \log \frac{p(\boldsymbol{x})}{q(\boldsymbol{x})} \right\rangle_{p(\boldsymbol{x})}.$$

$\delta(\boldsymbol{\mu}; \widehat{\boldsymbol{\mu}})$ : Dirac delta function located at $\widehat{\boldsymbol{\mu}}$. It also denotes its approximation (called Pseudo-delta function) with its entropy finite.

GE : Generalization error.

TE : Training error.

$F$ : Free energy.

| | |
|---|---|
| $O(f(N))$ | : A function such that $\limsup_{N\to\infty} \lvert O(f(N))/f(N)\rvert < \infty$. |
| $o(f(N))$ | : A function such that $\lim_{N\to\infty} o(f(N))/f(N) = 0$. |
| $\Omega(f(N))$ | : A function such that $\liminf_{N\to\infty} \lvert\Omega(f(N))/f(N)\rvert > 0$ |
| $\omega(f(N))$ | : A function such that $\lim_{N\to\infty} \lvert\omega(f(N))/f(N)\rvert = \infty$. |
| $\Theta(f(N))$ | : A function such that $\limsup_{N\to\infty} \lvert\Theta(f(N))/f(N)\rvert < \infty$ and $\liminf_{N\to\infty} \lvert\Theta(f(N))/f(N)\rvert > 0$. |
| $O_{\mathrm{p}}(f(N))$ | : A function such that $\limsup_{N\to\infty} \left\lvert O_{\mathrm{p}}(f(N))/f(N)\right\rvert < \infty$ in probability. |
| $o_{\mathrm{p}}(f(N))$ | : A function such that $\lim_{N\to\infty} o_{\mathrm{p}}(f(N))/f(N) = 0$ in probability. |
| $\Omega_{\mathrm{p}}(f(N))$ | : A function such that $\liminf_{N\to\infty} \left\lvert\Omega_{\mathrm{p}}(f(N))/f(N)\right\rvert > 0$ in probability |
| $\omega_{\mathrm{p}}(f(N))$ | : A function such that $\lim_{N\to\infty} \left\lvert\omega_{\mathrm{p}}(f(N))/f(N)\right\rvert = \infty$ in probability. |
| $\Theta_{\mathrm{p}}(f(N))$ | : A function such that $\limsup_{N\to\infty} \left\lvert\Theta_{\mathrm{p}}(f(N))/f(N)\right\rvert < \infty$ and $\liminf_{N\to\infty} \left\lvert\Theta_{\mathrm{p}}(f(N))/f(N)\right\rvert > 0$ in probability. |

# Part I

# Formulation

# 1
# Bayesian Learning

Bayesian learning is an inference method based on the fundamental law of probability, called the Bayes theorem. In this first chapter, we introduce the framework of Bayesian learning with simple examples where Bayesian learning can be performed analytically.

## 1.1 Framework

*Bayesian learning* considers the following situation. We have observed a set $\mathcal{D}$ of data, which are subject to a *conditional distribution* $p(\mathcal{D}|\boldsymbol{w})$, called the *model distribution*, of the data given unknown *model parameter* $\boldsymbol{w}$. Although the value of $\boldsymbol{w}$ is unknown, vague information on $\boldsymbol{w}$ is provided as a *prior distribution* $p(\boldsymbol{w})$. The conditional distribution $p(\mathcal{D}|\boldsymbol{w})$ is also called the *model likelihood* when it is seen as a function of the unknown parameter $\boldsymbol{w}$.

### 1.1.1 Bayes Theorem and Bayes Posterior

Bayesian learning is based on the following basic factorization property of the *joint distribution* $p(\mathcal{D}, \boldsymbol{w})$:

$$\underbrace{p(\boldsymbol{w}|\mathcal{D})}_{\text{posterior}} \underbrace{p(\mathcal{D})}_{\text{marginal}} = \underbrace{p(\mathcal{D}, \boldsymbol{w})}_{\text{joint}} = \underbrace{p(\mathcal{D}|\boldsymbol{w})}_{\text{likelihood}} \underbrace{p(\boldsymbol{w})}_{\text{prior}}, \tag{1.1}$$

where the marginal distribution is given by

$$p(\mathcal{D}) = \int_{\mathcal{W}} p(\mathcal{D}, \boldsymbol{w}) d\boldsymbol{w} = \int_{\mathcal{W}} p(\mathcal{D}|\boldsymbol{w}) p(\boldsymbol{w}) d\boldsymbol{w}. \tag{1.2}$$

Here, the integration is performed in the domain $\mathcal{W}$ of the parameter $\boldsymbol{w}$. Note that, if the domain $\mathcal{W}$ is discrete, integration should be replaced with summation, i.e., for any function $f(\boldsymbol{w})$,

$$\int_{\mathcal{W}} f(\boldsymbol{w})d\boldsymbol{w} \to \sum_{\boldsymbol{w}' \in \mathcal{W}} f(\boldsymbol{w}').$$

The *posterior distribution*, the distribution of the unknown parameter $\boldsymbol{w}$ given the observed data set $\mathcal{D}$, is derived by dividing both sides of Eq. (1.1) by the marginal distribution $p(\mathcal{D})$:

$$p(\boldsymbol{w}|\mathcal{D}) = \frac{p(\mathcal{D}, \boldsymbol{w})}{p(\mathcal{D})} \propto p(\mathcal{D}, \boldsymbol{w}). \tag{1.3}$$

Here, we emphasized that the posterior distribution is proportional to the joint distribution $p(\mathcal{D}, \boldsymbol{w})$ because the marginal distribution $p(\mathcal{D})$ is a constant (as a function of $\boldsymbol{w}$). In other words, the joint distribution is an *unnormalized posterior distribution*. Eq. (1.3) is called the *Bayes theorem*, and the posterior distribution computed exactly by Eq. (1.3) is called the *Bayes posterior* when we distinguish it from its approximations.

**Example 1.1** (Parametric density estimation) Assume that the observed data $\mathcal{D} = \{\boldsymbol{x}^{(1)}, \ldots, \boldsymbol{x}^{(N)}\}$ consist of $N$ *independent and identically distributed (i.i.d.)* samples from the model distribution $p(\boldsymbol{x}|\boldsymbol{w})$. Then, the model likelihood is given by $p(\mathcal{D}|\boldsymbol{w}) = \prod_{n=1}^{N} p(\boldsymbol{x}^{(n)}|\boldsymbol{w})$, and therefore, the posterior distribution is given by

$$p(\boldsymbol{w}|\mathcal{D}) = \frac{\prod_{n=1}^{N} p(\boldsymbol{x}^{(n)}|\boldsymbol{w})p(\boldsymbol{w})}{\int \prod_{n=1}^{N} p(\boldsymbol{x}^{(n)}|\boldsymbol{w})p(\boldsymbol{w})d\boldsymbol{w}} \propto \prod_{n=1}^{N} p(\boldsymbol{x}^{(n)}|\boldsymbol{w})p(\boldsymbol{w}).$$

**Example 1.2** (Parametric regression) Assume that the observed data $\mathcal{D} = \{(\boldsymbol{x}^{(1)}, \boldsymbol{y}^{(1)}), \ldots, (\boldsymbol{x}^{(N)}, \boldsymbol{y}^{(N)})\}$ consist of $N$ i.i.d. input–output pairs from the model distribution $p(\boldsymbol{x}, \boldsymbol{y}|\boldsymbol{w}) = p(\boldsymbol{y}|\boldsymbol{x}, \boldsymbol{w})p(\boldsymbol{x})$. Then, the likelihood function is given by $p(\mathcal{D}|\boldsymbol{w}) = \prod_{n=1}^{N} p(\boldsymbol{y}^{(n)}|\boldsymbol{x}^{(n)}, \boldsymbol{w})p(\boldsymbol{x}^{(n)})$, and therefore, the posterior distribution is given by

$$p(\boldsymbol{w}|\mathcal{D}) = \frac{\prod_{n=1}^{N} p(\boldsymbol{y}^{(n)}|\boldsymbol{x}^{(n)}, \boldsymbol{w})p(\boldsymbol{w})}{\int \prod_{n=1}^{N} p(\boldsymbol{y}^{(n)}|\boldsymbol{x}^{(n)}, \boldsymbol{w})p(\boldsymbol{w})d\boldsymbol{w}} \propto \prod_{n=1}^{N} p(\boldsymbol{y}^{(n)}|\boldsymbol{x}^{(n)}, \boldsymbol{w})p(\boldsymbol{w}).$$

Note that the input distribution $p(\boldsymbol{x})$ does not affect the posterior, and accordingly is often ignored in practice.

### 1.1.2 Maximum A Posteriori Learning

Since the joint distribution $p(\mathcal{D}, \boldsymbol{w})$ is just the product of the likelihood function and the prior distribution (see Eq. (1.1)), it is usually easy to compute. Therefore, it is relatively easy to perform *maximum a posteriori (MAP) learning*, where the parameters are point-estimated so that the posterior probability is maximized, i.e.,

$$\widehat{\boldsymbol{w}}^{\mathrm{MAP}} = \underset{\boldsymbol{w}}{\operatorname{argmax}}\, p(\boldsymbol{w}|\mathcal{D}) = \underset{\boldsymbol{w}}{\operatorname{argmax}}\, p(\mathcal{D}, \boldsymbol{w}). \tag{1.4}$$

MAP learning includes *maximum likelihood (ML) learning*,

$$\widehat{\boldsymbol{w}}^{\mathrm{ML}} = \underset{\boldsymbol{w}}{\operatorname{argmax}}\, p(\mathcal{D}|\boldsymbol{w}), \tag{1.5}$$

as a special case with the flat prior $p(\boldsymbol{w}) \propto 1$.

### 1.1.3 Bayesian Learning

On the other hand, *Bayesian learning* requires integration of the joint distribution with respect to the parameter $\boldsymbol{w}$, which is often computationally hard. More specifically, performing Bayesian learning means computing at least one of the following quantities:

***Marginal likelihood*** **(zeroth moment)**

$$p(\mathcal{D}) = \int p(\mathcal{D}, \boldsymbol{w}) d\boldsymbol{w}. \tag{1.6}$$

This quantity has been already introduced in Eq. (1.2) as the normalization factor of the posterior distribution. As seen in Section 1.1.5 and subsequent sections, the marginal likelihood plays an important role in model selection and hyperparameter estimation.

***Posterior mean*** **(first moment)**

$$\widehat{\boldsymbol{w}} = \langle \boldsymbol{w} \rangle_{p(\boldsymbol{w}|\mathcal{D})} = \frac{1}{p(\mathcal{D})} \int \boldsymbol{w} \cdot p(\mathcal{D}, \boldsymbol{w}) d\boldsymbol{w}, \tag{1.7}$$

where $\langle \cdot \rangle_p$ denotes the expectation value over the distribution $p$, i.e., $\langle \cdot \rangle_{p(\boldsymbol{w})} = \int \cdot p(\boldsymbol{w}) d\boldsymbol{w}$. This quantity is also called the *Bayesian estimator*. The Bayesian estimator or the model distribution with the Bayesian estimator plugged in (see the plug-in predictive distribution (1.10)) can be the final output of Bayesian learning.

***Posterior covariance*** **(second moment)**

$$\widehat{\boldsymbol{\Sigma}}_{w} = \left\langle (\boldsymbol{w} - \widehat{\boldsymbol{w}})(\boldsymbol{w} - \widehat{\boldsymbol{w}})^{\top} \right\rangle_{p(\boldsymbol{w}|\mathcal{D})} = \frac{1}{p(\mathcal{D})} \int (\boldsymbol{w} - \widehat{\boldsymbol{w}})(\boldsymbol{w} - \widehat{\boldsymbol{w}})^{\top} p(\mathcal{D}, \boldsymbol{w}) d\boldsymbol{w}, \tag{1.8}$$

where $\top$ denotes the transpose of a matrix or vector. This quantity provides the credibility information, and is used to assess the confidence level of the Bayesian estimator.

***Predictive distribution*** **(expectation of model distribution)**

$$p(\mathcal{D}^{\text{new}}|\mathcal{D}) = \langle p(\mathcal{D}^{\text{new}}|\boldsymbol{w}) \rangle_{p(\boldsymbol{w}|\mathcal{D})} = \frac{1}{p(\mathcal{D})} \int p(\mathcal{D}^{\text{new}}|\boldsymbol{w}) p(\mathcal{D}, \boldsymbol{w}) d\boldsymbol{w}, \tag{1.9}$$

where $p(\mathcal{D}^{\text{new}}|\boldsymbol{w})$ denotes the model distribution on *unobserved* new data $\mathcal{D}^{\text{new}}$. In the i.i.d. case such as Examples 1.1 and 1.2, it is sufficient to compute the predictive distribution for a single new sample $\mathcal{D}^{\text{new}} = \{\boldsymbol{x}\}$.

Note that each of the four quantities (1.6) through (1.9) requires to compute the expectation of some function $f(\boldsymbol{w})$ over the unnormalized posterior distribution $p(\mathcal{D}, \boldsymbol{w})$ on $\boldsymbol{w}$, i.e., $\int f(\boldsymbol{w}) p(\mathcal{D}, \boldsymbol{w}) d\boldsymbol{w}$. Specifically, the marginal likelihood, the posterior mean, and the posterior covariance are the zeroth, the first, and the second moments of the unnormalized posterior distribution, respectively. The expectation is analytically intractable except for some simple cases, and numerical computation is also hard when the dimensionality of the unknown parameter $\boldsymbol{w}$ is high. This is the main bottleneck of Bayesian learning, with which many approximation methods have been developed to cope.

It hardly happens that the first moment (1.7) or the second moment (1.8) are computationally tractable but the zeroth moment (1.6) is not. Accordingly, we can say that performing Bayesian learning on the parameter $\boldsymbol{w}$ amounts to obtaining the *normalized* posterior distribution $p(\boldsymbol{w}|\mathcal{D})$. It sometimes happens that computing the predictive distribution (1.9) is still intractable even if the zeroth, the first, and the second moments can be computed based on some approximation. In such a case, the model distribution with the Bayesian estimator plugged in, called the *plug-in predictive distribution*,

$$p(\mathcal{D}^{\text{new}}|\widehat{\boldsymbol{w}}), \tag{1.10}$$

is used for prediction in practice.

### 1.1.4 Latent Variables

So far, we introduced the observed data set $\mathcal{D}$ as a known variable, and the model parameter $\boldsymbol{w}$ as an unknown variable. In practice, more varieties of known and unknown variables can be involved.

Some probabilistic models have *latent variables* (or *hidden variables*) $\boldsymbol{z}$, which can be involved in the original model, or additionally introduced for computational reasons. They are typically attributed to each of the observed samples, and therefore have large degrees of freedom. However, they are just additional unknown variables, and there is no reason in inference to distinguish them from the model parameters $\boldsymbol{w}$.[1] The joint posterior over the parameters and the latent variables is given by Eq. (1.3) with $\boldsymbol{w}$ and $p(\boldsymbol{w})$ replaced with $\overline{\boldsymbol{w}} = (\boldsymbol{w}, \boldsymbol{z})$ and $p(\overline{\boldsymbol{w}}) = p(\boldsymbol{z}|\boldsymbol{w})p(\boldsymbol{w})$, respectively.

**Example 1.3** (Mixture models) A mixture model is often used for parametric density estimation (Example 1.1). The model distribution is given by

$$p(\boldsymbol{x}|\boldsymbol{w}) = \sum_{k=1}^{K} \alpha_k p(\boldsymbol{x}|\boldsymbol{\tau}_k), \tag{1.11}$$

where $\boldsymbol{w} = \{\alpha_k, \boldsymbol{\tau}_k; \alpha_k \geq 0, \sum_{k=1}^{K} \alpha_k = 1\}_{k=1}^{K}$ is the unknown parameters. The mixture model (1.11) is the weighted sum of $K$ distributions, each of which is parameterized by the component parameter $\boldsymbol{\tau}_k$. The domain of the *mixing weights* $\boldsymbol{\alpha} = (\alpha_1, \ldots, \alpha_K)^\top$, also called as the *mixture coefficients*, forms the *standard* $(K-1)$*-simplex*, denoted by $\Delta^{K-1} \equiv \{\boldsymbol{\alpha} \in \mathbb{R}_+^K; \sum_{k=1}^{K} \alpha_k = 1\}$ (see Figure 1.1). Figure 1.2 shows an example of the mixture model with three one-dimensional Gaussian components.

The likelihood,

$$\begin{aligned} p(\mathcal{D}|\boldsymbol{w}) &= \prod_{n=1}^{N} p(\boldsymbol{x}^{(n)}|\boldsymbol{w}), \\ &= \prod_{n=1}^{N} \left( \sum_{k=1}^{K} \alpha_k p(\boldsymbol{x}|\boldsymbol{\tau}_k) \right), \end{aligned} \tag{1.12}$$

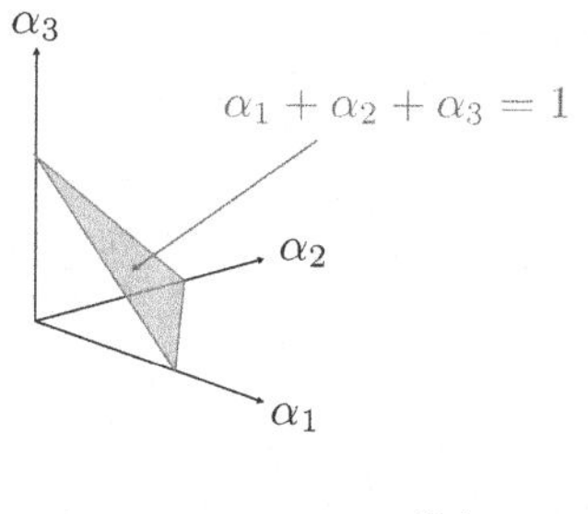


Figure 1.1 $(K-1)$-simplex, $\Delta^{K-1}$, for $K = 3$.

[1] For this reason, the latent variables $\boldsymbol{z}$ and the model parameters $\boldsymbol{w}$ are also called *local latent variables* and *global latent variables*, respectively.

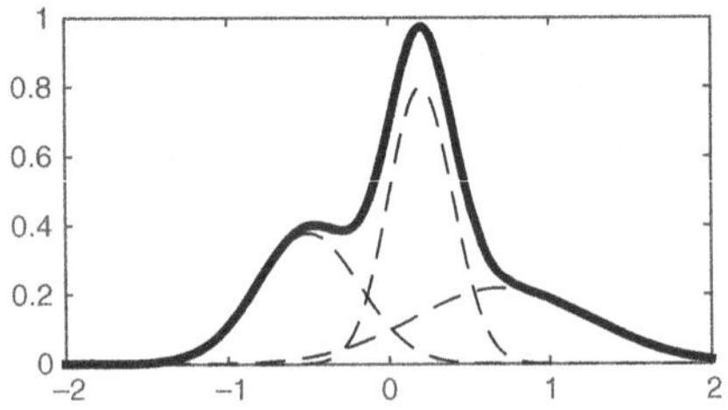


Figure 1.2 Gaussian mixture.

for $N$ observed i.i.d. samples $\mathcal{D} = \{\boldsymbol{x}^{(1)}, \ldots, \boldsymbol{x}^{(N)}\}$ has $O(K^N)$ terms, which makes even ML learning intractable. This intractability arises from the summation inside the multiplication in Eq. (1.12). By introducing latent variables, we can turn this summation into a multiplication, and make Eq. (1.12) tractable.

Assume that each sample $\boldsymbol{x}$ belongs to a single component $k$, and is drawn from $p(\boldsymbol{x}|\boldsymbol{\tau}_k)$. To describe the assignment, we introduce a latent variable $z \in \mathcal{Z} \equiv \{\boldsymbol{e}_k\}_{k=1}^K$ associated with each observed sample $\boldsymbol{x}$, where $\boldsymbol{e}_k \in \{0, 1\}^K$ is the $K$-dimensional binary vector, called the *one-of-K representation*, with one at the $k$th entry and zeros at the other entries:

$$\boldsymbol{e}_k = \underbrace{(0, \ldots, 0, \overbrace{\;1\;}^{k\text{th}}, 0, \ldots, 0)}_{K}^{\top}.$$

Then, we have the following model:

$$p(\boldsymbol{x}, z|\boldsymbol{w}) = p(\boldsymbol{x}|z, \boldsymbol{w})p(z|\boldsymbol{w}), \tag{1.13}$$

$$\text{where} \qquad p(\boldsymbol{x}|z, \boldsymbol{w}) = \prod_{k=1}^{K} \{p(\boldsymbol{x}|\boldsymbol{\tau}_k)\}^{z_k}, \qquad p(z|\boldsymbol{w}) = \prod_{k=1}^{K} \alpha_k^{z_k}.$$

The conditional distribution (1.13) on the observed variable $\boldsymbol{x}$ and the latent variable $z$ given the parameter $\boldsymbol{w}$ is called the *complete likelihood*.

Note that marginalizing the complete likelihood over the latent variable recovers the original mixture model:

$$p(\boldsymbol{x}|\boldsymbol{w}) = \int_{\mathcal{Z}} p(\boldsymbol{x}, z|\boldsymbol{w})dz = \sum_{z \in \{\boldsymbol{e}_k\}_{k=1}^K} \prod_{k=1}^{K} \{\alpha_k p(\boldsymbol{x}|\boldsymbol{\tau}_k)\}^{z_k} = \sum_{k=1}^{K} \alpha_k p(\boldsymbol{x}|\boldsymbol{\tau}_k).$$

This means that, if samples are generated from the model distribution (1.13), and only $\boldsymbol{x}$ is recorded, the observed data follow the original mixture model (1.11).

In the literature, latent variables tend to be marginalized out even in MAP learning. For example, the *expectation-maximization (EM) algorithm* (Dempster et al., 1977), a popular MAP solver for latent variable models, seeks a (local) maximizer of the posterior distribution with the latent variables marginalized out, i.e.,

$$\widehat{\boldsymbol{w}}^{\mathrm{EM}} = \operatorname*{argmax}_{\boldsymbol{w}} p(\boldsymbol{w}|\mathcal{D}) = \operatorname*{argmax}_{\boldsymbol{w}} \int_{\mathcal{Z}} p(\mathcal{D}, \boldsymbol{w}, \boldsymbol{z}) d\boldsymbol{z}. \tag{1.14}$$

However, we can also maximize the posterior jointly over the parameters and the latent variables, i.e.,

$$(\widehat{\boldsymbol{w}}^{\mathrm{MAP-hard}}, \widehat{\boldsymbol{z}}^{\mathrm{MAP-hard}}) = \operatorname*{argmax}_{\boldsymbol{w},\boldsymbol{z}} p(\boldsymbol{w}, \boldsymbol{z}|\mathcal{D}) = \operatorname*{argmax}_{\boldsymbol{w},\boldsymbol{z}} p(\mathcal{D}, \boldsymbol{w}, \boldsymbol{z}). \tag{1.15}$$

For clustering based on the mixture model in Example 1.3, the EM algorithm (1.14) gives a *soft assignment*, where the expectation value $\widehat{\boldsymbol{z}}^{\mathrm{EM}} \in \Delta^{K-1} \subset [0, 1]^K$ is substituted into the joint distribution $p(\mathcal{D}, \boldsymbol{w}, \boldsymbol{z})$, while the joint maximization (1.15) gives the *hard assignment*, where the optimal assignment $\widehat{\boldsymbol{z}}^{\mathrm{MAP-hard}} \in \{\boldsymbol{e}_k\}_{k=1}^K \subset \{0, 1\}^K$ is looked for in the binary domain.

### 1.1.5 Empirical Bayesian Learning

In many practical cases, it is reasonable to use a prior distribution parameterized by *hyperparameters* $\boldsymbol{\kappa}$. The hyperparameters can be tuned by hand or based on some criterion outside the Bayesian framework. A popular method of the latter is the *cross validation*, where the hyperparameters are tuned so that an (preferably unbiased) estimator of the performance criterion is optimized. In such cases, the hyperparameters should be treated as *known* variables when Bayesian learning is performed.

On the other hand, the hyperparameters can be estimated within the Bayesian framework. In this case, there is again no reason to distinguish the hyperparameters from the other unknown variables $(\boldsymbol{w}, \boldsymbol{z})$. The joint posterior over all unknown variables is given by Eq. (1.3) with $\boldsymbol{w}$ and $p(\boldsymbol{w})$ replaced with $\overline{\boldsymbol{w}} = (\boldsymbol{w}, \boldsymbol{\kappa}, \boldsymbol{z})$ and $p(\overline{\boldsymbol{w}}) = p(\boldsymbol{z}|\boldsymbol{w})p(\boldsymbol{w}|\boldsymbol{\kappa})p(\boldsymbol{\kappa})$, respectively, where $p(\boldsymbol{\kappa})$ is called a *hyperprior*. A popular approach, called *empirical Bayesian (EBayes) learning* (Efron and Morris, 1973), applies Bayesian learning on $\boldsymbol{w}$ (and $\boldsymbol{z}$) and point-estimate $\boldsymbol{\kappa}$, i.e.,

$$\widehat{\boldsymbol{\kappa}}^{\mathrm{EBayes}} = \operatorname*{argmax}_{\boldsymbol{\kappa}} p(\mathcal{D}, \boldsymbol{\kappa}) = \operatorname*{argmax}_{\boldsymbol{\kappa}} p(\mathcal{D}|\boldsymbol{\kappa})p(\boldsymbol{\kappa}),$$

$$\text{where} \quad p(\mathcal{D}|\boldsymbol{\kappa}) = \int p(\mathcal{D}, \boldsymbol{w}, \boldsymbol{z}|\boldsymbol{\kappa}) d\boldsymbol{w} d\boldsymbol{z}.$$

Here the marginal likelihood $p(\mathcal{D}|\boldsymbol{\kappa})$ is seen as the likelihood of the hyperparameter $\boldsymbol{\kappa}$, and MAP learning is performed by maximizing the joint distribution $p(\mathcal{D}, \boldsymbol{\kappa})$ of the observed data $\mathcal{D}$ and the hyperparameter $\boldsymbol{\kappa}$, which can be seen as an *unnormalized posterior distribution* of the hyperparameter. The hyperprior is often assumed to be flat: $p(\boldsymbol{\kappa}) \propto 1$.

With an appropriate design of priors, empirical Bayesian learning combined with approximate Bayesian learning is often used for *automatic relevance determination (ARD)*, where irrelevant degrees of freedom of the statistical model are automatically pruned out. Explaining the ARD property of approximate Bayesian learning is one of the main topics of theoretical analysis in Parts III and IV.

## 1.2 Computation

Now, let us explain how Bayesian learning is performed in simple cases. We start from introducing *conjugacy*, an important notion in performing Bayesian learning.

### 1.2.1 Popular Distributions

Table 1.1 summarizes several distributions that are frequently used as a model distribution (or likelihood function) $p(\mathcal{D}|\boldsymbol{w})$ or a prior distribution $p(\boldsymbol{w})$ in Bayesian learning. The domain $\mathcal{X}$ of the random variable $\boldsymbol{x}$ and the domain $\mathcal{W}$ of the parameters $\boldsymbol{w}$ are shown in the table.

Some of the distributions in Table 1.1 have complicated function forms, involving Beta or Gamma functions. However, such complications are mostly in the *normalization constant*, and can often be ignored when it is sufficient to find the *shape* of a function. In Table 1.1, the normalization constant is separated by a dot, so that one can find the simple main part. As will be seen shortly, we often refer to the normalization constant when we need to perform integration of a function, which is in the same form as the main part of a popular distribution.

Below we summarize abbreviations of distributions:

$$\mathrm{Gauss}_M(\boldsymbol{x}; \boldsymbol{\mu}, \boldsymbol{\Sigma}) \equiv \frac{1}{(2\pi)^{M/2} \det(\boldsymbol{\Sigma})^{1/2}} \cdot \exp\left( -\frac{1}{2} (\boldsymbol{x} - \boldsymbol{\mu})^\top \boldsymbol{\Sigma}^{-1} (\boldsymbol{x} - \boldsymbol{\mu}) \right), \tag{1.16}$$

$$\mathrm{Gamma}(x; \alpha, \beta) \equiv \frac{\beta^\alpha}{\Gamma(\alpha)} \cdot x^{\alpha-1} \exp(-\beta x), \tag{1.17}$$

Table 1.1 *Popular distributions. The following notation is used:* $\mathbb{R}$ *: The set of all real numbers,* $\mathbb{R}_{++}$ *: The set of all positive real numbers,* $\mathbb{I}_{++}$ *: The set of all positive integers,* $\mathbb{S}_{++}^{M}$ *: The set of all* $M \times M$ *positive definite matrices,* $\mathbb{H}_N^{K-1} \equiv \{\boldsymbol{x} \in \{0, \ldots, N\}^K; \sum_{k=1}^{K} x_k = N\}$ *: The set of all possible histograms for N samples and K categories,* $\Delta^{K-1} \equiv \{\boldsymbol{\theta} \in [0,1]^K; \sum_{k=1}^{K} \theta_k = 1\}$ *: The standard* $(K-1)$*-simplex,* $\det(\cdot)$ *:Determinant of matrix,* $\mathcal{B}(y, z) \equiv \int_0^1 t^{y-1}(1-t)^{z-1} dt$ *: Beta function,* $\Gamma(y) \equiv \int_0^\infty t^{y-1}\exp(-t)dt$ *: Gamma function, and* $\Gamma_M(y) \equiv \int_{\boldsymbol{T} \in \mathbb{S}_{++}^M} \det(\boldsymbol{T})^{y-(M+1)/2} \exp(-\mathrm{tr}(\boldsymbol{T}))d\boldsymbol{T}$ *: Multivariate Gamma function.*

| Probability distribution | $p(\boldsymbol{x}\vert\boldsymbol{w})$ | $\boldsymbol{x} \in \mathcal{X}$ | $\boldsymbol{w} \in \mathcal{W}$ |
|---|---|---|---|
| Isotropic Gaussian | $\mathrm{Gauss}_M(\boldsymbol{x}; \boldsymbol{\mu}, \sigma^2 \boldsymbol{I}_M) \equiv \frac{1}{(2\pi\sigma^2)^{M/2}} \cdot \exp\left(-\frac{1}{2\sigma^2}\Vert\boldsymbol{x}-\boldsymbol{\mu}\Vert^2\right)$ | $\boldsymbol{x} \in \mathbb{R}^M$ | $\boldsymbol{\mu} \in \mathbb{R}^M, \sigma^2 > 0$ |
| Gaussian | $\mathrm{Gauss}_M(\boldsymbol{x}; \boldsymbol{\mu}, \boldsymbol{\Sigma}) \equiv \frac{1}{(2\pi)^{M/2}\det(\boldsymbol{\Sigma})^{1/2}} \cdot \exp\left(-\frac{1}{2}(\boldsymbol{x}-\boldsymbol{\mu})^\top \boldsymbol{\Sigma}^{-1}(\boldsymbol{x}-\boldsymbol{\mu})\right)$ | $\boldsymbol{x} \in \mathbb{R}^M$ | $\boldsymbol{\mu} \in \mathbb{R}^M, \boldsymbol{\Sigma} \in \mathbb{S}_{++}^M$ |
| Gamma | $\mathrm{Gamma}(x; \alpha, \beta) \equiv \frac{\beta^\alpha}{\Gamma(\alpha)} \cdot x^{\alpha-1}\exp(-\beta x)$ | $x > 0$ | $\alpha > 0, \beta > 0$ |
| Wishart | $\mathrm{Wishart}_M(\boldsymbol{X}; \boldsymbol{V}, \nu) \equiv \frac{1}{(2^\nu \vert\boldsymbol{V}\vert)^{M/2}\Gamma_M(\frac{\nu}{2})} \cdot \det(\boldsymbol{X})^{\frac{\nu-M-1}{2}} \exp\left(-\frac{\mathrm{tr}(\boldsymbol{V}^{-1}\boldsymbol{X})}{2}\right)$ | $\boldsymbol{X} \in \mathbb{S}_{++}^M$ | $\boldsymbol{V} \in \mathbb{S}_{++}^M, \nu > M-1$ |
| Bernoulli | $\mathrm{Binomial}_1(x; \theta) \equiv \theta^x(1-\theta)^{1-x}$ | $x \in \{0, 1\}$ | $\theta \in [0, 1]$ |
| Binomial | $\mathrm{Binomial}_N(x; \theta) \equiv \binom{N}{x}\theta^x(1-\theta)^{N-x}$ | $x \in \{0, \ldots, N\}$ | $\theta \in [0, 1]$ |
| Multinomial | $\mathrm{Multinomial}_{K,N}(\boldsymbol{x}; \boldsymbol{\theta}) \equiv N! \cdot \prod_{k=1}^{K}(x_k!)^{-1}\theta_k^{x_k}$ | $\boldsymbol{x} \in \mathbb{H}_N^{K-1}$ | $\boldsymbol{\theta} \in \Delta^{K-1}$ |
| Beta | $\mathrm{Beta}(x; a, b) \equiv \frac{1}{\mathcal{B}(a,b)} \cdot x^{a-1}(1-x)^{b-1}$ | $x \in [0, 1]$ | $a > 0, b > 0$ |
| Dirichlet | $\mathrm{Dirichlet}_K(\boldsymbol{x}; \boldsymbol{\phi}) \equiv \frac{\Gamma(\sum_{k=1}^{K}\phi_k)}{\prod_{k=1}^{K}\Gamma(\phi_k)} \cdot \prod_{k=1}^{K} x_k^{\phi_k - 1}$ | $\boldsymbol{x} \in \Delta^{K-1}$ | $\boldsymbol{\phi} \in \mathbb{R}_{++}^K$ |

$$\mathrm{Wishart}_M(\boldsymbol{X}; \boldsymbol{V}, \nu) \equiv \frac{1}{(2^{\nu}|\boldsymbol{V}|)^{M/2}\Gamma_M\left(\frac{\nu}{2}\right)} \cdot \det(\boldsymbol{X})^{\frac{\nu-M-1}{2}} \exp\left(-\frac{\mathrm{tr}(\boldsymbol{V}^{-1}\boldsymbol{X})}{2}\right), \tag{1.18}$$

$$\mathrm{Binomial}_N(x; \theta) \equiv \binom{N}{x} \cdot \theta^x (1-\theta)^{N-x}, \tag{1.19}$$

$$\mathrm{Multinomial}_{K,N}(\boldsymbol{x}; \boldsymbol{\theta}) \equiv N! \cdot \prod_{k=1}^{K} (x_k!)^{-1} \theta_k^{x_k}, \tag{1.20}$$

$$\mathrm{Beta}(x; a, b) \equiv \frac{1}{\mathcal{B}(a,b)} \cdot x^{a-1}(1-x)^{b-1}, \tag{1.21}$$

$$\mathrm{Dirichlet}_K(\boldsymbol{x}; \boldsymbol{\phi}) \equiv \frac{\Gamma(\sum_{k=1}^{K} \phi_k)}{\prod_{k=1}^{K} \Gamma(\phi_k)} \cdot \prod_{k=1}^{K} x_k^{\phi_k - 1}. \tag{1.22}$$

The distributions in Table 1.1 are categorized into four groups, which are separated by dashed lines. In each group, an upper distribution family is a special case of a lower distribution family. Note that the following hold:

$$\mathrm{Gamma}(x; \alpha, \beta) = \mathrm{Wishart}_1\left(x; \frac{1}{2\beta}, 2\alpha\right),$$

$$\mathrm{Binomial}_N(x; \theta) = \mathrm{Multinomial}_{2,N}\left((x, N-x)^\top; (\theta, 1-\theta)^\top\right),$$

$$\mathrm{Beta}(x; a, b) = \mathrm{Dirichlet}_2\left((x, 1-x)^\top; (a, b)^\top\right).$$

### 1.2.2 Conjugacy

Let us think about the *function form* of the posterior (1.3):

$$p(\boldsymbol{w}|\mathcal{D}) = \frac{p(\mathcal{D}|\boldsymbol{w})p(\boldsymbol{w})}{p(\mathcal{D})} \propto p(\mathcal{D}|\boldsymbol{w})p(\boldsymbol{w}),$$

which is determined by the function form of the product of the model likelihood $p(\mathcal{D}|\boldsymbol{w})$ and the prior $p(\boldsymbol{w})$. Note that we here call the conditional $p(\mathcal{D}|\boldsymbol{w})$ NOT the *model distribution* but the *model likelihood*, since we are interested in the function form of the posterior, a distribution of the parameter $\boldsymbol{w}$.

*Conjugacy* is defined as the relation between the likelihood $p(\mathcal{D}|\boldsymbol{w})$ and the prior $p(\boldsymbol{w})$.

**Definition 1.4** (Conjugate prior) A prior $p(\boldsymbol{w})$ is called *conjugate* with a likelihood $p(\mathcal{D}|\boldsymbol{w})$, if the posterior $p(\boldsymbol{w}|\mathcal{D})$ is in the same distribution family as the prior.

### 1.2.3 Posterior Distribution

Here, we introduce computation of the posterior distribution in simple cases where a conjugate prior exists and is adopted.

#### Isotropic Gaussian Model

Let us compute the posterior distribution for the isotropic Gaussian model:

$$p(\boldsymbol{x}|\boldsymbol{w}) = \mathrm{Gauss}_M(\boldsymbol{x}; \boldsymbol{\mu}, \sigma^2 \boldsymbol{I}_M) = \frac{1}{(2\pi\sigma^2)^{M/2}} \cdot \exp\left(-\frac{1}{2\sigma^2}\|\boldsymbol{x} - \boldsymbol{\mu}\|^2\right). \tag{1.23}$$

The likelihood for $N$ i.i.d. samples $\mathcal{D} = \{\boldsymbol{x}^{(1)}, \ldots, \boldsymbol{x}^{(N)}\}$ is written as

$$p(\mathcal{D}|\boldsymbol{w}) = \prod_{n=1}^{N} p(\boldsymbol{x}^{(n)}|\boldsymbol{w}) = \frac{\exp\left(-\frac{1}{2\sigma^2}\sum_{n=1}^{N}\|\boldsymbol{x}^{(n)} - \boldsymbol{\mu}\|^2\right)}{(2\pi\sigma^2)^{MN/2}}. \tag{1.24}$$

**Gaussian Likelihood** As noted in Section 1.2.2, we should see Eq. (1.24), which is the distribution of observed data $\mathcal{D}$, as a function of the parameter $\boldsymbol{w}$. Naturally, the function form depends on which parameters are estimated in the *Bayesian way*. The isotropic Gaussian has two parameters $\boldsymbol{w} = (\boldsymbol{\mu}, \sigma^2)$, and we first consider the case where the variance parameter $\sigma^2$ is known, and the posterior of the mean parameter $\boldsymbol{\mu}$ is estimated, i.e., we set $\boldsymbol{w} = \boldsymbol{\mu}$. This case contains the case where $\sigma^2$ is unknown but point-estimated in the empirical Bayesian procedure or tuned outside the Bayesian framework, e.g., by performing cross-validation (we set $\boldsymbol{w} = \boldsymbol{\mu}, \kappa = \sigma^2$ in the latter case).

Omitting the constant (with respect to $\boldsymbol{\mu}$), the likelihood (1.24) can be written as

$$\begin{aligned} p(\mathcal{D}|\boldsymbol{\mu}) &\propto \exp\left(-\frac{1}{2\sigma^2}\sum_{n=1}^{N}\|\boldsymbol{x}^{(n)} - \boldsymbol{\mu}\|^2\right) \\ &\propto \exp\left(-\frac{1}{2\sigma^2}\sum_{n=1}^{N}\|(\boldsymbol{x}^{(n)} - \overline{\boldsymbol{x}}) + (\overline{\boldsymbol{x}} - \boldsymbol{\mu})\|^2\right) \\ &= \exp\left(-\frac{1}{2\sigma^2}\left(\sum_{n=1}^{N}\|\boldsymbol{x}^{(n)} - \overline{\boldsymbol{x}}\|^2 + N\|\overline{\boldsymbol{x}} - \boldsymbol{\mu}\|^2\right)\right) \\ &\propto \exp\left(-\frac{N}{2\sigma^2}\|\boldsymbol{\mu} - \overline{\boldsymbol{x}}\|^2\right) \\ &\propto \mathrm{Gauss}_M\left(\boldsymbol{\mu}; \overline{\boldsymbol{x}}, \frac{\sigma^2}{N}\boldsymbol{I}_M\right), \end{aligned} \tag{1.25}$$

where $\overline{\boldsymbol{x}} = \frac{1}{N}\sum_{n=1}^{N}\boldsymbol{x}^{(n)}$ is the *sample mean*. Note that we omitted the factor $\exp\left(-\frac{1}{2\sigma^2}\sum_{n=1}^{N}\|\boldsymbol{x}^{(n)} - \overline{\boldsymbol{x}}\|^2\right)$ as a constant in the fourth equation.

The last equation (1.25) implies that, as a function of the mean parameter $\boldsymbol{\mu}$, the model likelihood $p(\mathcal{D}|\boldsymbol{\mu})$ has the same form as the isotropic Gaussian with mean $\overline{\boldsymbol{x}}$ and variance $\frac{\sigma^2}{N}$. Eq. (1.25) also implies that the ML estimator for the mean parameter is given by

$$\widehat{\boldsymbol{\mu}}^{\mathrm{ML}} = \overline{\boldsymbol{x}}.$$

Thus, we found that the likelihood function for the mean parameter of the isotropic Gaussian is in the Gaussian form. This comes from the following facts:

- The isotropic Gaussian model for a single sample $\boldsymbol{x}$ is in the Gaussian form also as a function of the mean parameter, i.e., $\mathrm{Gauss}_M(\boldsymbol{x}; \boldsymbol{\mu}, \sigma^2 \boldsymbol{I}_M) \propto \mathrm{Gauss}_M(\boldsymbol{\mu}; \boldsymbol{x}, \sigma^2 \boldsymbol{I}_M)$.
- The isotropic Gaussians are *multiplicatively closed*, i.e., the product of isotropic Gaussians with different means is a Gaussian: $p(\mathcal{D}|\boldsymbol{\mu}) \propto \prod_{n=1}^{N} \mathrm{Gauss}_M(\boldsymbol{\mu}; \boldsymbol{x}^{(n)}, \sigma^2 \boldsymbol{I}_M) \propto \mathrm{Gauss}_M\left(\boldsymbol{\mu}; \overline{\boldsymbol{x}}, \frac{\sigma^2}{N} \boldsymbol{I}_M\right)$.

Since the isotropic Gaussian is multiplicatively closed and the likelihood (1.25) is in the Gaussian form, the isotropic Gaussian prior must be conjugate. Let us choose the isotropic Gaussian prior,

$$p(\boldsymbol{\mu}|\boldsymbol{\mu}_0, \sigma_0^2) = \mathrm{Gauss}_M(\boldsymbol{\mu}; \boldsymbol{\mu}_0, \sigma_0^2 \boldsymbol{I}_M) \propto \exp\left(-\frac{1}{2\sigma_0^2}\|\boldsymbol{\mu} - \boldsymbol{\mu}_0\|^2\right),$$

for hyperparameters $\boldsymbol{\kappa} = (\boldsymbol{\mu}_0, \sigma_0^2)$. Then, the function form of the posterior is given by

$$\begin{aligned}
p(\boldsymbol{\mu}|\mathcal{D}, \boldsymbol{\mu}_0, \sigma_0^2) &\propto p(\mathcal{D}|\boldsymbol{\mu}) p(\boldsymbol{\mu}|\boldsymbol{\mu}_0, \sigma_0^2) \\
&\propto \mathrm{Gauss}_M\left(\boldsymbol{\mu}; \overline{\boldsymbol{x}}, \frac{\sigma^2}{N}\right) \mathrm{Gauss}_M(\boldsymbol{\mu}; \boldsymbol{\mu}_0, \sigma_0^2) \\
&\propto \exp\left(-\frac{N}{2\sigma^2}\left\|\boldsymbol{\mu} - \overline{\boldsymbol{x}}\right\|^2 - \frac{1}{2\sigma_0^2}\left\|\boldsymbol{\mu} - \boldsymbol{\mu}_0\right\|^2\right) \\
&\propto \exp\left(-\frac{N\sigma^{-2} + \sigma_0^{-2}}{2}\left\|\boldsymbol{\mu} - \frac{N\sigma^{-2}\overline{\boldsymbol{x}} + \sigma_0^{-2}\boldsymbol{\mu}_0}{N\sigma^{-2} + \sigma_0^{-2}}\right\|^2\right) \\
&\propto \mathrm{Gauss}_M\left(\boldsymbol{\mu}; \frac{N\sigma^{-2}\overline{\boldsymbol{x}} + \sigma_0^{-2}\boldsymbol{\mu}_0}{N\sigma^{-2} + \sigma_0^{-2}}, \frac{1}{N\sigma^{-2} + \sigma_0^{-2}}\right).
\end{aligned}$$

Therefore, the posterior is

$$p(\boldsymbol{\mu}|\mathcal{D}, \boldsymbol{\mu}_0, \sigma_0^2) = \mathrm{Gauss}_M\left(\boldsymbol{\mu}; \frac{N\sigma^{-2}\overline{\boldsymbol{x}} + \sigma_0^{-2}\boldsymbol{\mu}_0}{N\sigma^{-2} + \sigma_0^{-2}}, \frac{1}{N\sigma^{-2} + \sigma_0^{-2}}\right). \tag{1.26}$$

Note that the *equality* holds in Eq. (1.26). We omitted constant factors in the preceding derivation. But once the function form of the posterior is found, the normalization factor is unique. If the function form coincides with one of the well-known distributions (e.g., ones given in Table 1.1), one can find the normalization constant (from the table) without any further computation.

Multiplicative closedness of a function family of the model likelihood is essential in performing Bayesian learning. Such families are called the *exponential family*:

**Definition 1.5** (Exponential families) A family of distributions is called the exponential family if it is written as

$$p(\boldsymbol{x}|\boldsymbol{w}) = p(\boldsymbol{t}|\boldsymbol{\eta}) = \exp\left(\boldsymbol{\eta}^\top \boldsymbol{t} - A(\boldsymbol{\eta}) + B(\boldsymbol{t})\right), \tag{1.27}$$

where $\boldsymbol{t} = \boldsymbol{t}(\boldsymbol{x})$ is a function, called *sufficient statistics*, of the random variable $\boldsymbol{x}$, and $\boldsymbol{\eta} = \boldsymbol{\eta}(\boldsymbol{w})$ is a function, called *natural parameters*, of the parameter $\boldsymbol{w}$.

The essential property of the exponential family is that the interaction between the random variable and the parameter occurs only in the log linear form, i.e., $\exp(\boldsymbol{\eta}^\top \boldsymbol{t})$. Note that, although $A(\cdot)$ and $B(\cdot)$ are arbitrary functions, $A(\cdot)$ does not depend on $\boldsymbol{t}$, and $B(\cdot)$ does not depend on $\boldsymbol{\eta}$.

Assume that $N$ observed samples $\mathcal{D} = (\boldsymbol{t}^{(1)}, \ldots, \boldsymbol{t}^{(N)}) = (\boldsymbol{t}(\boldsymbol{x}^{(1)}), \ldots, \boldsymbol{t}(\boldsymbol{x}^{(N)}))$ are drawn from the exponential family distribution (1.27). If we use the exponential family prior $p(\boldsymbol{\eta}) = \exp\left(\boldsymbol{\eta}^\top \boldsymbol{t}^{(0)} - A_0(\boldsymbol{\eta}) + B_0(\boldsymbol{t}^{(0)})\right)$, then the posterior is given as an exponential family distribution with the same set of natural parameters $\boldsymbol{\eta}$:

$$p(\boldsymbol{\eta}|\mathcal{D}) = \exp\left(\boldsymbol{\eta}^\top \sum_{n=0}^{N} \boldsymbol{t}^{(n)} - A'(\boldsymbol{\eta}) + B'(\mathcal{D})\right),$$

where $A'(\boldsymbol{\eta})$ and $B'(\mathcal{D})$ are a function of $\boldsymbol{\eta}$ and a function of $\mathcal{D}$, respectively. Therefore, the conjugate prior for the exponential family distribution is the exponential family with the same natural parameters $\boldsymbol{\eta}$.

All distributions given in Table 1.1 are exponential families. For example, the sufficient statistics and the natural parameters for the univariate Gaussian are given by $\boldsymbol{\eta} = (\frac{\mu}{\sigma^2}, -\frac{1}{2\sigma^2})^\top$ and $\boldsymbol{t} = (x, x^2)^\top$, respectively. The mixture model (1.11) is a common *non*exponential family distribution.

**Gamma Likelihood** Next we consider the posterior distribution of the variance parameter $\sigma^2$ with the mean parameter regarded as a constant, i.e., $w = \sigma^2$.

Omitting the constants (with respect to $\sigma^2$) of the model likelihood (1.24), we have

$$p(\mathcal{D}|\sigma^2) \propto (\sigma^2)^{-MN/2} \exp\left(-\frac{1}{2\sigma^2}\sum_{n=1}^{N}\|\boldsymbol{x}^{(n)} - \boldsymbol{\mu}\|^2\right).$$

If we see the likelihood as a function of the *inverse* of $\sigma^2$, we find that it is proportional to the *Gamma distribution*:

$$\begin{aligned} p(\mathcal{D}|\sigma^{-2}) &\propto (\sigma^{-2})^{MN/2} \exp\left(-\left(\frac{1}{2}\sum_{n=1}^{N}\|\boldsymbol{x}^{(n)} - \boldsymbol{\mu}\|^2\right)\sigma^{-2}\right) \\ &\propto \mathrm{Gamma}\left(\sigma^{-2}; \frac{MN}{2} + 1, \frac{1}{2}\sum_{n=1}^{N}\|\boldsymbol{x}^{(n)} - \boldsymbol{\mu}\|^2\right). \end{aligned} \tag{1.28}$$

Since the mode of the Gamma distribution is known as $\mathrm{argmax}_x \mathrm{Gamma}(x; \alpha, \beta) = \frac{\alpha-1}{\beta}$, Eq. (1.28) implies that the ML estimator for the variance parameter is given by

$$\widehat{\sigma}^{2\,\mathrm{ML}} = \frac{1}{\widehat{\sigma}^{-2\,\mathrm{ML}}} = \frac{\frac{1}{2}\sum_{n=1}^{N}\|\boldsymbol{x}^{(n)} - \boldsymbol{\mu}\|^2}{\frac{MN}{2} + 1 - 1} = \frac{1}{MN}\sum_{n=1}^{N}\|\boldsymbol{x}^{(n)} - \boldsymbol{\mu}\|^2.$$

Now we found that the model likelihood of the isotropic Gaussian is in the Gamma form as a function of the inverse variance $\sigma^{-2}$. Since the Gamma distribution is in the exponential family and multiplicatively closed, the Gamma prior is conjugate.

If we use the Gamma prior

$$p(\sigma^{-2}|\alpha_0, \beta_0) = \mathrm{Gamma}(\sigma^{-2}; \alpha_0, \beta_0) \propto (\sigma^{-2})^{\alpha_0 - 1} \exp(-\beta_0 \sigma^{-2})$$

with hyperparameters $\boldsymbol{\kappa} = (\alpha_0, \beta_0)$, the posterior can be written as

$$\begin{aligned} p(\sigma^{-2}|\mathcal{D}, \alpha_0, \beta_0) &\propto p(\mathcal{D}|\sigma^{-2})p(\sigma^{-2}|\alpha_0, \beta_0) \\ &\propto \mathrm{Gamma}\left(\sigma^{-2}; \frac{MN}{2} + 1, \frac{1}{2}\sum_{n=1}^{N}\|\boldsymbol{x}^{(n)} - \boldsymbol{\mu}\|^2\right)\mathrm{Gamma}(\sigma^{-2}; \alpha_0, \beta_0) \\ &\propto (\sigma^{-2})^{MN/2 + \alpha_0 - 1} \exp\left(-\left(\frac{1}{2}\sum_{n=1}^{N}\|\boldsymbol{x}^{(n)} - \boldsymbol{\mu}\|^2 + \beta_0\right)\sigma^{-2}\right), \end{aligned}$$

and therefore

$$p(\sigma^{-2}|\mathcal{D}, \alpha_0, \beta_0) = \mathrm{Gamma}\left(\sigma^{-2}; \frac{MN}{2} + \alpha_0, \frac{1}{2}\sum_{n=1}^{N}\|\boldsymbol{x}^{(n)} - \boldsymbol{\mu}\|^2 + \beta_0\right). \tag{1.29}$$

**Isotropic Gauss-Gamma Likelihood** Finally, we consider the general case where both the mean and variance parameters are unknown, i.e., $\boldsymbol{w} = (\boldsymbol{\mu}, \sigma^{-2})$. The likelihood is written as

$$p(\mathcal{D}|\boldsymbol{\mu}, \sigma^{-2}) \propto (\sigma^{-2})^{MN/2} \exp\left(-\left(\frac{1}{2}\sum_{n=1}^{N} \|\boldsymbol{x}^{(n)} - \boldsymbol{\mu}\|^2\right)\sigma^{-2}\right)$$
$$= (\sigma^{-2})^{MN/2} \exp\left(-\left(\frac{N\|\boldsymbol{\mu} - \overline{\boldsymbol{x}}\|^2}{2} + \frac{\sum_{n=1}^{N} \|\boldsymbol{x}^{(n)} - \overline{\boldsymbol{x}}\|^2}{2}\right)\sigma^{-2}\right)$$
$$\propto \text{GaussGamma}_M\left(\boldsymbol{\mu}, \sigma^{-2}\middle|\overline{\boldsymbol{x}}, N\boldsymbol{I}_M, \frac{M(N-1)}{2} + 1, \frac{\sum_{n=1}^{N} \|\boldsymbol{x}^{(n)} - \overline{\boldsymbol{x}}\|^2}{2}\right),$$

where

$$\text{GaussGamma}_M(\boldsymbol{x}, \tau|\boldsymbol{\mu}, \lambda\boldsymbol{I}_M, \alpha, \beta)$$
$$\equiv \text{Gauss}_M(\boldsymbol{x}|\boldsymbol{\mu}, (\tau\lambda)^{-1}\boldsymbol{I}_M) \cdot \text{Gamma}(\tau|\alpha, \beta)$$
$$= \frac{\exp\left(-\frac{\tau\lambda}{2}\|\boldsymbol{x} - \boldsymbol{\mu}\|^2\right)}{(2\pi(\tau\lambda)^{-1})^{M/2}} \cdot \frac{\beta^\alpha}{\Gamma(\alpha)} \tau^{\alpha-1} \exp(-\beta\tau)$$
$$= \frac{\beta^\alpha}{(2\pi/\lambda)^{M/2}\Gamma(\alpha)} \tau^{\alpha+\frac{M}{2}-1} \exp\left(-\left(\frac{\lambda\|\boldsymbol{x} - \boldsymbol{\mu}\|^2}{2} + \beta\right)\tau\right)$$

is the *isotropic Gauss-Gamma distribution* on the random variable $\boldsymbol{x} \in \mathbb{R}^M$, $\tau > 0$ with parameters $\boldsymbol{\mu} \in \mathbb{R}^M, \lambda > 0, \alpha > 0, \beta > 0$.

Note that, although the isotropic Gauss-Gamma distribution is the product of an isotropic Gaussian distribution and a Gamma distribution, the random variables $\boldsymbol{x}$ and $\tau$ are not independent of each other. This is because the isotropic Gauss-Gamma distribution is a *hierarchical model* $p(\boldsymbol{x}|\tau)p(\tau)$, where the variance parameter $\sigma^2 = (\tau\lambda)^{-1}$ for the isotropic Gaussian depends on the random variable $\tau$ of the Gamma distribution.

Since the isotropic Gauss-Gamma distribution is multiplicatively closed, it is a conjugate prior. Choosing the isotropic Gauss-Gamma prior

$$p(\boldsymbol{\mu}, \sigma^{-2}|\boldsymbol{\mu}_0, \lambda_0, \alpha_0, \beta_0) = \text{GaussGamma}_M(\boldsymbol{\mu}, \sigma^{-2}|\boldsymbol{\mu}_0, \lambda_0\boldsymbol{I}_M, \alpha_0, \beta)$$
$$\propto (\sigma^{-2})^{\alpha_0+\frac{M}{2}-1} \exp\left(-\left(\frac{\lambda_0\|\boldsymbol{\mu} - \boldsymbol{\mu}_0\|^2}{2} + \beta_0\right)\sigma^{-2}\right)$$

with hyperparameters $\boldsymbol{\kappa} = (\boldsymbol{\mu}_0, \lambda_0, \alpha_0, \beta_0)$, the posterior is given by

$$p(\boldsymbol{\mu}, \sigma^{-2}|\mathcal{D}, \boldsymbol{\kappa}) \propto p(\mathcal{D}|\boldsymbol{\mu}, \sigma^{-2})p(\boldsymbol{\mu}, \sigma^{-2}|\boldsymbol{\kappa})$$
$$\propto \text{GaussGamma}_M\left(\boldsymbol{\mu}, \sigma^{-2}\middle|\overline{\boldsymbol{x}}, N\boldsymbol{I}_M, \frac{M(N-1)}{2} + 1, \frac{\sum_{n=1}^{N} \|\boldsymbol{x}^{(n)} - \overline{\boldsymbol{x}}\|^2}{2}\right)$$
$$\cdot \text{GaussGamma}_M(\boldsymbol{\mu}, \sigma^{-2}|\boldsymbol{\mu}_0, \lambda_0\boldsymbol{I}_M, \alpha_0, \beta)$$

$$
\begin{aligned}
&\propto (\sigma^{-2})^{MN/2} \exp\left(-\left(\frac{N\|\boldsymbol{\mu}-\overline{\boldsymbol{x}}\|^2}{2} + \frac{\sum_{n=1}^{N}\|\boldsymbol{x}^{(n)}-\overline{\boldsymbol{x}}\|^2}{2}\right)\sigma^{-2}\right) \\
&\qquad \cdot (\sigma^{-2})^{\alpha_0+\frac{M}{2}-1} \exp\left(-\left(\frac{\lambda_0\|\boldsymbol{\mu}-\boldsymbol{\mu}_0\|^2}{2} + \beta_0\right)\sigma^{-2}\right) \\
&\propto (\sigma^{-2})^{M(N+1)/2+\alpha_0-1} \\
&\qquad \cdot \exp\left(-\left(\frac{N\|\boldsymbol{\mu}-\overline{\boldsymbol{x}}\|^2+\lambda_0\|\boldsymbol{\mu}-\boldsymbol{\mu}_0\|^2}{2} + \frac{\sum_{n=1}^{N}\|\boldsymbol{x}^{(n)}-\overline{\boldsymbol{x}}\|^2}{2} + \beta_0\right)\sigma^{-2}\right) \\
&\propto (\sigma^{-2})^{\widehat{\alpha}+\frac{M}{2}-1} \exp\left(-\left(\frac{\widehat{\lambda}\left\|\boldsymbol{\mu}-\widehat{\boldsymbol{\mu}}\right\|^2}{2} + \widehat{\beta}\right)\sigma^{-2}\right),
\end{aligned}
$$

where

$$
\begin{aligned}
\widehat{\boldsymbol{\mu}} &= \frac{N\overline{\boldsymbol{x}} + \lambda_0\boldsymbol{\mu}_0}{N+\lambda_0}, \\
\widehat{\lambda} &= N + \lambda_0, \\
\widehat{\alpha} &= \frac{MN}{2} + \alpha_0, \\
\widehat{\beta} &= \frac{\sum_{n=1}^{N}\|\boldsymbol{x}^{(n)}-\overline{\boldsymbol{x}}\|^2}{2} + \frac{N\lambda_0\|\overline{\boldsymbol{x}}-\boldsymbol{\mu}_0\|^2}{2(N+\lambda_0)} + \beta_0.
\end{aligned}
$$

Thus, the posterior is obtained as

$$
p(\boldsymbol{\mu}, \sigma^{-2}|\mathcal{D}, \boldsymbol{\kappa}) = \mathrm{GaussGamma}_M(\boldsymbol{\mu}, \sigma^{-2}|\widehat{\boldsymbol{\mu}}, \widehat{\lambda}\boldsymbol{I}_M, \widehat{\alpha}, \widehat{\beta}). \tag{1.30}
$$

Although the Gauss-Gamma distribution seems a bit more complicated than the ones in Table 1.1, its moments are known. Therefore, Bayesian learning with a conjugate prior can be analytically performed also when both parameters $\boldsymbol{w} = (\boldsymbol{\mu}, \sigma^{-2})$ are estimated.

**Gaussian Model**

Bayesian learning can be performed for a general Gaussian model in a similar fashion to the isotropic case. Consider the $M$-dimensional Gaussian distribution,

$$
p(\boldsymbol{x}|\boldsymbol{w}) = \mathrm{Gauss}_M(\boldsymbol{x}; \boldsymbol{\mu}, \boldsymbol{\Sigma}) \equiv \frac{1}{(2\pi)^{M/2}\det(\boldsymbol{\Sigma})^{1/2}} \cdot \exp\left(-\frac{1}{2}(\boldsymbol{x}-\boldsymbol{\mu})^\top\boldsymbol{\Sigma}^{-1}(\boldsymbol{x}-\boldsymbol{\mu})\right) \tag{1.31}
$$

with mean and covariance parameters $\boldsymbol{w} = (\boldsymbol{\mu}, \boldsymbol{\Sigma})$. The likelihood for $N$ i.i.d. samples $\mathcal{D} = \{\boldsymbol{x}^{(1)}, \ldots, \boldsymbol{x}^{(N)}\}$ is written as

$$
p(\mathcal{D}|\boldsymbol{w}) = \prod_{n=1}^{N} p(\boldsymbol{x}^{(n)}|\boldsymbol{w}) = \frac{\exp\left(-\frac{1}{2}\sum_{n=1}^{N}(\boldsymbol{x}^{(n)}-\boldsymbol{\mu})^\top\boldsymbol{\Sigma}^{-1}(\boldsymbol{x}^{(n)}-\boldsymbol{\mu})\right)}{(2\pi)^{NM/2}\det(\boldsymbol{\Sigma})^{N/2}}. \tag{1.32}
$$

**Gaussian Likelihood** Let us first compute the posterior distribution on the mean parameter $\boldsymbol{\mu}$, with the covariance parameter regarded as a known constant. In this case, the likelihood can be written as

$$
\begin{aligned}
p(\mathcal{D}|\boldsymbol{\mu}) &\propto \exp\left(-\frac{1}{2}\sum_{n=1}^{N}(\boldsymbol{x}^{(n)}-\boldsymbol{\mu})^{\top}\boldsymbol{\Sigma}^{-1}(\boldsymbol{x}^{(n)}-\boldsymbol{\mu})\right) \\
&\propto \exp\left(-\frac{1}{2}\sum_{n=1}^{N}\left((\boldsymbol{x}^{(n)}-\overline{\boldsymbol{x}})+(\overline{\boldsymbol{x}}-\boldsymbol{\mu})\right)^{\top}\cdot\boldsymbol{\Sigma}^{-1}\left((\boldsymbol{x}^{(n)}-\overline{\boldsymbol{x}})+(\overline{\boldsymbol{x}}-\boldsymbol{\mu})\right)\right) \\
&= \exp\left(-\frac{1}{2\sigma^2}\left(\sum_{n=1}^{N}(\boldsymbol{x}^{(n)}-\overline{\boldsymbol{x}})^{\top}\boldsymbol{\Sigma}^{-1}(\boldsymbol{x}^{(n)}-\overline{\boldsymbol{x}})+N(\overline{\boldsymbol{x}}-\boldsymbol{\mu})^{\top}\boldsymbol{\Sigma}^{-1}(\overline{\boldsymbol{x}}-\boldsymbol{\mu})\right)\right) \\
&\propto \exp\left(-\frac{N}{2}(\boldsymbol{\mu}-\overline{\boldsymbol{x}})^{\top}\boldsymbol{\Sigma}^{-1}(\boldsymbol{\mu}-\overline{\boldsymbol{x}})\right) \\
&\propto \mathrm{Gauss}_M\left(\boldsymbol{\mu};\overline{\boldsymbol{x}},\frac{1}{N}\boldsymbol{\Sigma}\right). \qquad (1.33)
\end{aligned}
$$

Therefore, with the conjugate Gaussian prior

$$
p(\boldsymbol{\mu}|\boldsymbol{\mu}_0,\boldsymbol{\Sigma}_0) = \mathrm{Gauss}_M(\boldsymbol{\mu};\boldsymbol{\mu}_0,\boldsymbol{\Sigma}_0) \propto \exp\left(-\frac{1}{2}(\boldsymbol{\mu}-\boldsymbol{\mu}_0)^{\top}\boldsymbol{\Sigma}_0^{-1}(\boldsymbol{\mu}-\boldsymbol{\mu}_0)\right),
$$

with hyperparameters $\boldsymbol{\kappa} = (\boldsymbol{\mu}_0,\boldsymbol{\Sigma}_0)$, the posterior is written as

$$
\begin{aligned}
p(\boldsymbol{\mu}|\mathcal{D},\boldsymbol{\mu}_0,\boldsymbol{\Sigma}_0) &\propto p(\mathcal{D}|\boldsymbol{\mu})p(\boldsymbol{\mu}|\boldsymbol{\mu}_0,\boldsymbol{\Sigma}_0) \\
&\propto \mathrm{Gauss}_M\left(\boldsymbol{\mu};\overline{\boldsymbol{x}},\frac{1}{N}\boldsymbol{\Sigma}\right)\mathrm{Gauss}_M(\boldsymbol{\mu};\boldsymbol{\mu}_0,\boldsymbol{\Sigma}_0) \\
&\propto \exp\left(-\frac{N(\boldsymbol{\mu}-\overline{\boldsymbol{x}})^{\top}\boldsymbol{\Sigma}^{-1}(\boldsymbol{\mu}-\overline{\boldsymbol{x}})+(\boldsymbol{\mu}-\boldsymbol{\mu}_0)^{\top}\boldsymbol{\Sigma}_0^{-1}(\boldsymbol{\mu}-\boldsymbol{\mu}_0)}{2}\right) \\
&\propto \exp\left(-\frac{(\boldsymbol{\mu}-\widehat{\boldsymbol{\mu}})^{\top}\widehat{\boldsymbol{\Sigma}}^{-1}(\boldsymbol{\mu}-\widehat{\boldsymbol{\mu}})}{2}\right),
\end{aligned}
$$

where

$$
\begin{aligned}
\widehat{\boldsymbol{\mu}} &= \left(N\boldsymbol{\Sigma}^{-1}+\boldsymbol{\Sigma}_0^{-1}\right)^{-1}\left(N\boldsymbol{\Sigma}^{-1}\overline{\boldsymbol{x}}+\boldsymbol{\Sigma}_0^{-1}\boldsymbol{\mu}_0\right), \\
\widehat{\boldsymbol{\Sigma}} &= \left(N\boldsymbol{\Sigma}^{-1}+\boldsymbol{\Sigma}_0^{-1}\right)^{-1}.
\end{aligned}
$$

Thus, we have

$$
p(\boldsymbol{\mu}|\mathcal{D},\boldsymbol{\mu}_0,\boldsymbol{\Sigma}_0) = \mathrm{Gauss}_M\left(\boldsymbol{\mu};\widehat{\boldsymbol{\mu}},\widehat{\boldsymbol{\Sigma}}\right). \qquad (1.34)
$$

**Wishart Likelihood** If we see the mean parameter $\boldsymbol{\mu}$ as a given constant, the model likelihood (1.32) can be written as follows, as a function of the covariance parameter $\boldsymbol{\Sigma}$:

$$
\begin{aligned}
p(\mathcal{D}|\boldsymbol{\Sigma}^{-1}) &\propto \det\left(\boldsymbol{\Sigma}^{-1}\right)^{N/2} \exp\left(-\frac{\sum_{n=1}^{N}(\boldsymbol{x}^{(n)}-\boldsymbol{\mu})^\top\boldsymbol{\Sigma}^{-1}(\boldsymbol{x}^{(n)}-\boldsymbol{\mu})}{2}\right) \\
&\propto \det\left(\boldsymbol{\Sigma}^{-1}\right)^{N/2} \exp\left(-\frac{\mathrm{tr}\left(\sum_{n=1}^{N}(\boldsymbol{x}^{(n)}-\boldsymbol{\mu})(\boldsymbol{x}^{(n)}-\boldsymbol{\mu})^\top\boldsymbol{\Sigma}^{-1}\right)}{2}\right) \\
&\propto \mathrm{Wishart}_M\left(\boldsymbol{\Sigma}^{-1}; \left(\sum_{n=1}^{N}(\boldsymbol{x}^{(n)}-\boldsymbol{\mu})(\boldsymbol{x}^{(n)}-\boldsymbol{\mu})^\top\right)^{-1}, M+N+1\right).
\end{aligned}
$$

Here, as in the isotropic Gaussian case, we computed the distribution on the *inverse* $\boldsymbol{\Sigma}^{-1}$ of the covariance parameter. With the *Wishart distribution*

$$
\begin{aligned}
p(\boldsymbol{\Sigma}^{-1}|\boldsymbol{V}_0,\nu_0) &= \mathrm{Wishart}_M(\boldsymbol{\Sigma}^{-1}; \boldsymbol{V}_0,\nu_0) \\
&= \frac{1}{\left(2^{\nu_0}\det(\boldsymbol{V}_0)\right)^{M/2}\Gamma_M\left(\frac{\nu_0}{2}\right)} \cdot \det\left(\boldsymbol{\Sigma}^{-1}\right)^{\frac{\nu_0-M-1}{2}} \exp\left(-\frac{\mathrm{tr}(\boldsymbol{V}_0^{-1}\boldsymbol{\Sigma}^{-1})}{2}\right)
\end{aligned}
$$

for hyperparameters $\boldsymbol{\kappa} = (\boldsymbol{V}_0, \nu_0)$ as a conjugate prior, the posterior is computed as

$$
\begin{aligned}
p(\boldsymbol{\Sigma}^{-1}|\mathcal{D},\boldsymbol{V}_0,\nu_0) &\propto p(\mathcal{D}|\boldsymbol{\Sigma}^{-1})p(\boldsymbol{\Sigma}^{-1}|\boldsymbol{V}_0,\nu_0) \\
&\propto \mathrm{Wishart}_M\left(\boldsymbol{\Sigma}^{-1}; \left(\sum_{n=1}^{N}(\boldsymbol{x}^{(n)}-\boldsymbol{\mu})(\boldsymbol{x}^{(n)}-\boldsymbol{\mu})^\top\right)^{-1}, M+N+1\right) \\
&\qquad \cdot \mathrm{Wishart}_M(\boldsymbol{\Sigma}^{-1}; \boldsymbol{V}_0,\nu_0) \\
&\propto \det\left(\boldsymbol{\Sigma}^{-1}\right)^{\frac{N}{2}} \exp\left(-\frac{\mathrm{tr}\left(\left(\sum_{n=1}^{N}(\boldsymbol{x}^{(n)}-\boldsymbol{\mu})(\boldsymbol{x}^{(n)}-\boldsymbol{\mu})^\top\right)\boldsymbol{\Sigma}^{-1}\right)}{2}\right) \\
&\qquad \cdot \det\left(\boldsymbol{\Sigma}^{-1}\right)^{\frac{\nu_0-M-1}{2}} \exp\left(-\frac{\mathrm{tr}\left(\boldsymbol{V}_0^{-1}\boldsymbol{\Sigma}^{-1}\right)}{2}\right) \\
&\propto \det\left(\boldsymbol{\Sigma}^{-1}\right)^{\frac{\nu_0-M+N-1}{2}} \exp\left(-\frac{\mathrm{tr}\left(\left(\sum_{n=1}^{N}(\boldsymbol{x}^{(n)}-\boldsymbol{\mu})(\boldsymbol{x}^{(n)}-\boldsymbol{\mu})^\top+\boldsymbol{V}_0^{-1}\right)\boldsymbol{\Sigma}^{-1}\right)}{2}\right).
\end{aligned}
$$

Thus we have

$$
\begin{aligned}
&p(\boldsymbol{\Sigma}^{-1}|\mathcal{D},\boldsymbol{V}_0,\nu_0) \\
&\quad = \mathrm{Wishart}_M\left(\boldsymbol{\Sigma}^{-1}; \left(\sum_{n=1}^{N}(\boldsymbol{x}^{(n)}-\boldsymbol{\mu})(\boldsymbol{x}^{(n)}-\boldsymbol{\mu})^\top+\boldsymbol{V}_0^{-1}\right)^{-1}, N+\nu_0\right). \qquad (1.35)
\end{aligned}
$$

Note that the Wishart distribution can be seen as a multivariate extension of the Gamma distribution and is reduced to the Gamma distribution for $M = 1$:

$$\mathrm{Wishart}_1(x; V, \nu) = \mathrm{Gamma}(x; \nu/2, 1/(2V)).$$

**Gauss-Wishart Likelihood** When both parameters $\boldsymbol{w} = (\boldsymbol{\mu}, \boldsymbol{\Sigma}^{-1})$ are unknown, the model likelihood (1.32) is seen as

$$\begin{aligned}
p(\mathcal{D}|\boldsymbol{\mu}, \boldsymbol{\Sigma}^{-1}) &\propto \det\left(\boldsymbol{\Sigma}^{-1}\right)^{N/2} \exp\left(-\frac{\sum_{n=1}^{N}(\boldsymbol{x}^{(n)}-\boldsymbol{\mu})^\top \boldsymbol{\Sigma}^{-1}(\boldsymbol{x}^{(n)}-\boldsymbol{\mu})}{2}\right) \\
&\propto \det\left(\boldsymbol{\Sigma}^{-1}\right)^{N/2} \exp\left(-\frac{\mathrm{tr}\left(\sum_{n=1}^{N}(\boldsymbol{x}^{(n)}-\boldsymbol{\mu})(\boldsymbol{x}^{(n)}-\boldsymbol{\mu})^\top \boldsymbol{\Sigma}^{-1}\right)}{2}\right) \\
&\propto \det\left(\boldsymbol{\Sigma}^{-1}\right)^{N/2} \exp\left(-\frac{\mathrm{tr}\left(\sum_{n=1}^{N}\left((\boldsymbol{x}^{(n)}-\overline{\boldsymbol{x}})+(\overline{\boldsymbol{x}}-\boldsymbol{\mu})\right)\left((\boldsymbol{x}^{(n)}-\overline{\boldsymbol{x}})+(\overline{\boldsymbol{x}}-\boldsymbol{\mu})\right)^\top \boldsymbol{\Sigma}^{-1}\right)}{2}\right) \\
&\propto \det\left(\boldsymbol{\Sigma}^{-1}\right)^{N/2} \exp\left(-\frac{\mathrm{tr}\left(N(\boldsymbol{\mu}-\overline{\boldsymbol{x}})(\boldsymbol{\mu}-\overline{\boldsymbol{x}})^\top + \sum_{n=1}^{N}(\boldsymbol{x}^{(n)}-\overline{\boldsymbol{x}})(\boldsymbol{x}^{(n)}-\overline{\boldsymbol{x}})^\top)\boldsymbol{\Sigma}^{-1}\right)}{2}\right) \\
&\propto \mathrm{GaussWishart}_M\left(\boldsymbol{\mu}, \boldsymbol{\Sigma}^{-1}; \overline{\boldsymbol{x}}, N, \left(\textstyle\sum_{n=1}^{N}(\boldsymbol{x}^{(n)} - \overline{\boldsymbol{x}})(\boldsymbol{x}^{(n)} - \overline{\boldsymbol{x}})^\top\right)^{-1}, M + N\right),
\end{aligned}$$

where

$$\begin{aligned}
&\mathrm{GaussWishart}_M(\boldsymbol{x}, \boldsymbol{\Lambda}|\boldsymbol{\mu}, \lambda, \boldsymbol{V}, \nu) \\
&\quad \equiv \mathrm{Gauss}_M(\boldsymbol{x}|\boldsymbol{\mu}, (\lambda\boldsymbol{\Lambda})^{-1})\mathrm{Wishart}_M(\boldsymbol{\Lambda}|\boldsymbol{V}, \nu) \\
&\quad = \frac{\exp\left(-\frac{\lambda}{2}(\boldsymbol{x}-\boldsymbol{\mu})^\top \boldsymbol{\Lambda}(\boldsymbol{x}-\boldsymbol{\mu})\right)}{(2\pi)^{M/2}\det(\lambda\boldsymbol{\Lambda})^{-1/2}} \cdot \frac{\det(\boldsymbol{\Lambda})^{\frac{\nu-M-1}{2}} \exp\left(-\frac{\mathrm{tr}(\boldsymbol{V}^{-1}\boldsymbol{\Lambda})}{2}\right)}{(2^{\nu}\det(\boldsymbol{V}))^{M/2}\Gamma_M\left(\frac{\nu}{2}\right)} \\
&\quad = \frac{\lambda^{M/2}}{(2^{\nu+1}\pi\det(\boldsymbol{V}))^{M/2}\Gamma_M\left(\frac{\nu}{2}\right)} \det(\boldsymbol{\Lambda})^{\frac{\nu-M}{2}} \exp\left(-\frac{\mathrm{tr}\left((\lambda(\boldsymbol{x}-\boldsymbol{\mu})(\boldsymbol{x}-\boldsymbol{\mu})^\top + \boldsymbol{V}^{-1})\boldsymbol{\Lambda}\right)}{2}\right)
\end{aligned}$$

is the *Gauss–Wishart distribution* on the random variables $\boldsymbol{x} \in \mathbb{R}^M, \boldsymbol{\Lambda} \in \mathbb{S}_{++}^M$ with parameters $\boldsymbol{\mu} \in \mathbb{R}^M, \lambda > 0, \boldsymbol{V} \in \mathbb{S}_{++}^M, \nu > M - 1$.

With the conjugate Gauss–Wishart prior,

$$\begin{aligned}
p(\boldsymbol{\mu}, \boldsymbol{\Sigma}^{-1}|\boldsymbol{\mu}_0, \lambda_0, \alpha_0, \beta_0) &= \mathrm{GaussWishart}_M(\boldsymbol{\mu}, \boldsymbol{\Sigma}^{-1}|\boldsymbol{\mu}_0, \lambda_0, \boldsymbol{V}_0, \nu_0) \\
&\propto \det\left(\boldsymbol{\Sigma}^{-1}\right)^{\frac{\nu-M}{2}} \exp\left(-\frac{\mathrm{tr}\left((\lambda_0(\boldsymbol{\mu}-\boldsymbol{\mu}_0)(\boldsymbol{\mu}-\boldsymbol{\mu}_0)^\top + \boldsymbol{V}_0^{-1})\boldsymbol{\Sigma}^{-1}\right)}{2}\right)
\end{aligned}$$

with hyperparameters $\boldsymbol{\kappa} = (\boldsymbol{\mu}_0, \lambda_0, \boldsymbol{V}_0, \nu_0)$, the posterior is written as

$$\begin{aligned}
p(\boldsymbol{\mu}, \boldsymbol{\Sigma}^{-1}|\mathcal{D}, \boldsymbol{\kappa}) &\propto p(\mathcal{D}|\boldsymbol{\mu}, \boldsymbol{\Sigma}^{-1})p(\boldsymbol{\mu}, \boldsymbol{\Sigma}^{-1}|\boldsymbol{\kappa}) \\
&\propto \mathrm{GaussWishart}_M\left(\boldsymbol{\mu}, \boldsymbol{\Sigma}^{-1}; \overline{\boldsymbol{x}}, N, \left(\textstyle\sum_{n=1}^{N}(\boldsymbol{x}^{(n)} - \overline{\boldsymbol{x}})(\boldsymbol{x}^{(n)} - \overline{\boldsymbol{x}})^\top\right)^{-1}, M + N\right) \\
&\qquad \cdot \mathrm{GaussWishart}_M(\boldsymbol{\mu}, \boldsymbol{\Sigma}^{-1}|\boldsymbol{\mu}_0, \lambda_0, \boldsymbol{V}_0, \nu_0)
\end{aligned}$$

$$\propto \det\left(\boldsymbol{\Sigma}^{-1}\right)^{N/2} \exp\left(-\frac{\mathrm{tr}\left(N(\boldsymbol{\mu}-\overline{\boldsymbol{x}})(\boldsymbol{\mu}-\overline{\boldsymbol{x}})^\top+\sum_{n=1}^N(\boldsymbol{x}^{(n)}-\overline{\boldsymbol{x}})(\boldsymbol{x}^{(n)}-\overline{\boldsymbol{x}})^\top)\boldsymbol{\Sigma}^{-1}\right)}{2}\right)$$

$$\cdot \det\left(\boldsymbol{\Sigma}^{-1}\right)^{\frac{\nu_0-M}{2}} \exp\left(-\frac{\mathrm{tr}\left((\lambda_0(\boldsymbol{\mu}-\boldsymbol{\mu}_0)(\boldsymbol{\mu}-\boldsymbol{\mu}_0)^\top+\boldsymbol{V}_0^{-1})\boldsymbol{\Sigma}^{-1}\right)}{2}\right)$$

$$\propto \det\left(\boldsymbol{\Sigma}^{-1}\right)^{\frac{\widehat{\nu}-M}{2}} \exp\left(-\mathrm{tr}\left(\frac{\left(\widehat{\lambda}(\boldsymbol{\mu}-\widehat{\boldsymbol{\mu}})(\boldsymbol{\mu}-\widehat{\boldsymbol{\mu}})^\top \widehat{\boldsymbol{V}}^{-1}\right)\boldsymbol{\Sigma}^{-1}}{2}\right)\right),$$

where

$$\widehat{\boldsymbol{\mu}} = \frac{N\overline{\boldsymbol{x}} + \lambda_0\boldsymbol{\mu}_0}{N+\lambda_0},$$

$$\widehat{\lambda} = N + \lambda_0,$$

$$\widehat{\boldsymbol{V}} = \left(\sum_{n=1}^N(\boldsymbol{x}^{(n)}-\overline{\boldsymbol{x}})(\boldsymbol{x}^{(n)}-\overline{\boldsymbol{x}})^\top + \tfrac{N\lambda_0}{N+\lambda_0}(\overline{\boldsymbol{x}}-\boldsymbol{\mu}_0)(\overline{\boldsymbol{x}}-\boldsymbol{\mu}_0)^\top + \boldsymbol{V}_0^{-1}\right)^{-1},$$

$$\widehat{\nu} = N + \nu_0.$$

Thus, we have the posterior distribution as the Gauss–Wishart distribution:

$$p(\boldsymbol{\mu}, \boldsymbol{\Sigma}^{-1}|\mathcal{D}, \boldsymbol{\kappa}) = \mathrm{GaussWishart}_M\left(\boldsymbol{\mu}, \boldsymbol{\Sigma}^{-1}|\widehat{\boldsymbol{\mu}}, \widehat{\lambda}, \widehat{\boldsymbol{V}}, \widehat{\nu}\right). \tag{1.36}$$

**Linear Regression Model**

Consider the *linear regression model*, where an input variable $\boldsymbol{x} \in \mathbb{R}^M$ and an output variable $y \in \mathbb{R}$ are assumed to satisfy the following probabilistic relation:

$$y = \boldsymbol{a}^\top\boldsymbol{x} + \varepsilon, \tag{1.37}$$

$$p(\varepsilon|\sigma^2) = \mathrm{Gauss}_1(\varepsilon; 0, \sigma^2) = \frac{1}{\sqrt{2\pi\sigma^2}} \cdot \exp\left(-\frac{\varepsilon^2}{2\sigma^2}\right). \tag{1.38}$$

Here $\boldsymbol{a}$ and $\sigma^2$ are called the *regression parameter* and the *noise variance parameter*, respectively. By substituting $\varepsilon = y - \boldsymbol{a}^\top\boldsymbol{x}$, which is obtained from Eq. (1.37), into Eq. (1.38), we have

$$p(y|\boldsymbol{x}, \boldsymbol{w}) = \mathrm{Gauss}_1(y; \boldsymbol{a}^\top\boldsymbol{x}, \sigma^2) = \frac{1}{\sqrt{2\pi\sigma^2}} \cdot \exp\left(-\frac{(y-\boldsymbol{a}^\top\boldsymbol{x})^2}{2\sigma^2}\right).$$

The likelihood function for $N$ observed i.i.d.[2] samples,

$$\mathcal{D} = (\boldsymbol{y}, \boldsymbol{X}),$$

[2] In the context of regression, i.i.d. usually means that the observation noise $\varepsilon^{(n)} = y^{(n)} - \boldsymbol{a}^\top\boldsymbol{x}^{(n)}$ is independent for different samples, i.e., $p(\{\varepsilon^{(n)}\}_{n=1}^N) = \prod_{n=1}^N p(\varepsilon^{(n)})$, and the independence between the input $(\boldsymbol{x}^{(1)}, \ldots, \boldsymbol{x}^{(N)})$, i.e., $p(\{\boldsymbol{x}^{(n)}\}_{n=1}^N) = \prod_{n=1}^N p(\boldsymbol{x}^{(n)})$, is not required.

is given by

$$p(\mathcal{D}|\boldsymbol{w}) = \frac{1}{(2\pi\sigma^2)^{N/2}} \cdot \exp\left(-\frac{\|\boldsymbol{y} - \boldsymbol{X}\boldsymbol{a}\|^2}{2\sigma^2}\right), \tag{1.39}$$

where we defined

$$\boldsymbol{y} = (y^{(1)}, \ldots, y^{(N)})^\top \in \mathbb{R}^N, \quad \boldsymbol{X} = (\boldsymbol{x}^{(1)}, \ldots, \boldsymbol{x}^{(N)})^\top \in \mathbb{R}^{N \times M}.$$

**Gaussian Likelihood** The computation of the posterior is similar to the isotropic Gaussian case. As in Section 1.2.3, we first consider the case where only the regression parameter $\boldsymbol{a}$ is estimated, with the noise variance parameter $\sigma^2$ regarded as a known constant.

One can guess that the likelihood (1.39) is Gaussian as a function of $\boldsymbol{a}$, since it is an exponential of a concave quadratic function. Indeed, by expanding the exponent and completing the square for $\boldsymbol{a}$, we obtain

$$\begin{aligned} p(\mathcal{D}|\boldsymbol{a}) &\propto \exp\left(-\frac{\|\boldsymbol{y} - \boldsymbol{X}\boldsymbol{a}\|^2}{2\sigma^2}\right) \\ &\propto \exp\left(-\frac{\left(\boldsymbol{a}-(\boldsymbol{X}^\top\boldsymbol{X})^{-1}\boldsymbol{X}^\top\boldsymbol{y}\right)^\top \boldsymbol{X}^\top\boldsymbol{X}\left(\boldsymbol{a}-(\boldsymbol{X}^\top\boldsymbol{X})^{-1}\boldsymbol{X}^\top\boldsymbol{y}\right)}{2\sigma^2}\right) \\ &\propto \mathrm{Gauss}_M\left(\boldsymbol{a}; (\boldsymbol{X}^\top\boldsymbol{X})^{-1}\boldsymbol{X}^\top\boldsymbol{y}, \sigma^2(\boldsymbol{X}^\top\boldsymbol{X})^{-1}\right). \end{aligned} \tag{1.40}$$

Eq. (1.40) implies that, when $\boldsymbol{X}^\top\boldsymbol{X}$ is *nonsingular* (i.e., its inverse exists), the ML estimator for $\boldsymbol{a}$ is given by

$$\widehat{\boldsymbol{a}}^{\mathrm{ML}} = (\boldsymbol{X}^\top\boldsymbol{X})^{-1}\boldsymbol{X}^\top\boldsymbol{y}. \tag{1.41}$$

Therefore, with the conjugate Gaussian prior

$$p(\boldsymbol{a}|\boldsymbol{a}_0, \boldsymbol{\Sigma}_0) = \mathrm{Gauss}_M(\boldsymbol{a}; \boldsymbol{a}_0, \boldsymbol{\Sigma}_0) \propto \exp\left(-\frac{1}{2}(\boldsymbol{a} - \boldsymbol{a}_0)^\top \boldsymbol{\Sigma}_0^{-1}(\boldsymbol{a} - \boldsymbol{a}_0)\right)$$

for hyperparameters $\boldsymbol{\kappa} = (\boldsymbol{a}_0, \boldsymbol{\Sigma}_0)$, the posterior is Gaussian:

$$\begin{aligned} p(\boldsymbol{a}|\mathcal{D}, \boldsymbol{a}_0, \boldsymbol{\Sigma}_0) &\propto p(\mathcal{D}|\boldsymbol{a})p(\boldsymbol{a}|\boldsymbol{a}_0, \boldsymbol{\Sigma}_0) \\ &\propto \mathrm{Gauss}_M\left(\boldsymbol{a}; \boldsymbol{a}_0, \tfrac{1}{N}\sigma^2(\boldsymbol{X}^\top\boldsymbol{X})^{-1}\right) \mathrm{Gauss}_M(\boldsymbol{a}; \boldsymbol{a}_0, \boldsymbol{\Sigma}_0) \\ &\propto \exp\left(-\frac{\frac{\left(\boldsymbol{a}-(\boldsymbol{X}^\top\boldsymbol{X})^{-1}\boldsymbol{X}^\top\boldsymbol{y}\right)^\top \boldsymbol{X}^\top\boldsymbol{X}\left(\boldsymbol{a}-(\boldsymbol{X}^\top\boldsymbol{X})^{-1}\boldsymbol{X}^\top\boldsymbol{y}\right)}{\sigma^2} + (\boldsymbol{a}-\boldsymbol{a}_0)^\top\boldsymbol{\Sigma}_0^{-1}(\boldsymbol{a}-\boldsymbol{a}_0)}{2}\right) \\ &\propto \exp\left(-\frac{(\boldsymbol{a} - \widehat{\boldsymbol{a}})^\top \widehat{\boldsymbol{\Sigma}}_{\boldsymbol{a}}^{-1} (\boldsymbol{a} - \widehat{\boldsymbol{a}})}{2}\right), \end{aligned}$$

where

$$\widehat{\boldsymbol{a}} = \left(\frac{\boldsymbol{X}^\top \boldsymbol{X}}{\sigma^2} + \boldsymbol{\Sigma}_0^{-1}\right)^{-1} \left(\frac{\boldsymbol{X}^\top \boldsymbol{y}}{\sigma^2} + \boldsymbol{\Sigma}_0^{-1}\boldsymbol{a}_0\right),$$

$$\widehat{\boldsymbol{\Sigma}}_a = \left(\frac{\boldsymbol{X}^\top \boldsymbol{X}}{\sigma^2} + \boldsymbol{\Sigma}_0^{-1}\right)^{-1}.$$

Thus we have

$$p(\boldsymbol{a}|\mathcal{D}, \boldsymbol{a}_0, \boldsymbol{\Sigma}_0) = \mathrm{Gauss}_M\left(\boldsymbol{a}; \widehat{\boldsymbol{a}}, \widehat{\boldsymbol{\Sigma}}_a\right). \tag{1.42}$$

**Gamma Likelihood** When only the noise variance parameter $\sigma^2$ is unknown, the model likelihood (1.39) is in the Gamma form, as a function of the inverse $\sigma^{-2}$:

$$\begin{aligned} p(\mathcal{D}|\sigma^{-2}) &\propto (\sigma^{-2})^{NM/2} \exp\left(-\frac{\|\boldsymbol{y} - \boldsymbol{X}\boldsymbol{a}\|^2}{2}\sigma^{-2}\right) \\ &\propto \mathrm{Gamma}\left(\sigma^{-2}; \frac{NM}{2} + 1, \frac{\|\boldsymbol{y} - \boldsymbol{X}\boldsymbol{a}\|^2}{2}\right), \end{aligned} \tag{1.43}$$

which implies that the ML estimator is

$$\widehat{\sigma}^{2\,\mathrm{ML}} = \frac{1}{\widehat{\sigma}^{-2\,\mathrm{ML}}} = \frac{1}{MN}\sum_{n=1}^{N} \|\boldsymbol{y} - \boldsymbol{X}\boldsymbol{a}\|^2.$$

With the conjugate Gamma prior

$$p(\sigma^{-2}|\alpha_0, \beta_0) = \mathrm{Gamma}(\sigma^{-2}; \alpha_0, \beta_0) \propto (\sigma^{-2})^{\alpha_0 - 1}\exp(-\beta_0\sigma^{-2})$$

with hyperparameters $\boldsymbol{\kappa} = (\alpha_0, \beta_0)$, the posterior is computed as

$$\begin{aligned} p(\sigma^{-2}|\mathcal{D}, \alpha_0, \beta_0) &\propto p(\mathcal{D}|\sigma^{-2})p(\sigma^{-2}|\alpha_0, \beta_0) \\ &\propto \mathrm{Gamma}\left(\sigma^{-2}; \frac{MN}{2} + 1, \frac{1}{2}\|\boldsymbol{y} - \boldsymbol{X}\boldsymbol{a}\|^2\right)\mathrm{Gamma}(\sigma^{-2}; \alpha_0, \beta_0) \\ &\propto (\sigma^{-2})^{MN/2 + \alpha_0 - 1}\exp\left(-\left(\frac{1}{2}\|\boldsymbol{y} - \boldsymbol{X}\boldsymbol{a}\|^2 + \beta_0\right)\sigma^{-2}\right). \end{aligned}$$

Therefore,

$$p(\sigma^{-2}|\mathcal{D}, \alpha_0, \beta_0) = \mathrm{Gamma}\left(\sigma^{-2}; \frac{MN}{2} + \alpha_0, \frac{1}{2}\|\boldsymbol{y} - \boldsymbol{X}\boldsymbol{a}\|^2 + \beta_0\right). \tag{1.44}$$

**Gauss-Gamma Likelihood** When we estimate both parameters $\boldsymbol{w} = (\boldsymbol{a}, \sigma^{-2})$, the likelihood (1.39) is written as

$$
\begin{aligned}
p(\mathcal{D}|\boldsymbol{a},\sigma^{-2}) &\propto (\sigma^{-2})^{NM/2}\exp\left(-\frac{\|\boldsymbol{y}-\boldsymbol{X}\boldsymbol{a}\|^2}{2}\sigma^{-2}\right)\\
&\propto (\sigma^{-2})^{NM/2}\exp\left(-\frac{\left(\boldsymbol{a}-\widehat{\boldsymbol{a}}^{\mathrm{ML}}\right)^\top\boldsymbol{X}^\top\boldsymbol{X}\left(\boldsymbol{a}-\widehat{\boldsymbol{a}}^{\mathrm{ML}}\right)+\|\boldsymbol{y}-\boldsymbol{X}\widehat{\boldsymbol{a}}^{\mathrm{ML}}\|^2}{2}\sigma^{-2}\right)\\
&\propto \mathrm{GaussGamma}_M\left(\boldsymbol{a},\sigma^{-2};\widehat{\boldsymbol{a}}^{\mathrm{ML}},\boldsymbol{X}^\top\boldsymbol{X},\frac{M(N-1)}{2}+1,\frac{\|\boldsymbol{y}-\boldsymbol{X}\widehat{\boldsymbol{a}}^{\mathrm{ML}}\|^2}{2}\right),
\end{aligned}
$$

where $\widehat{\boldsymbol{a}}^{\mathrm{ML}}$ is the ML estimator, given by Eq. (1.41), for the regression parameter, and

$$
\begin{aligned}
&\mathrm{GaussGamma}_M(\boldsymbol{x},\tau|\boldsymbol{\mu},\boldsymbol{\Lambda},\alpha,\beta)\\
&\quad\equiv \mathrm{Gauss}_M(\boldsymbol{x}|\boldsymbol{\mu},(\tau\boldsymbol{\Lambda})^{-1})\cdot\mathrm{Gamma}(\tau|\alpha,\beta)\\
&\quad= \frac{\exp\left(-\frac{\tau}{2}(\boldsymbol{x}-\boldsymbol{\mu})^\top\boldsymbol{\Lambda}(\boldsymbol{x}-\boldsymbol{\mu})\right)}{(2\pi\tau^{-1})^{M/2}\det(\boldsymbol{\Lambda})^{-1/2}}\cdot\frac{\beta^\alpha}{\Gamma(\alpha)}\tau^{\alpha-1}\exp(-\beta\tau)\\
&\quad= \frac{\beta^\alpha}{(2\pi)^{M/2}\det(\boldsymbol{\Lambda})^{-1/2}\Gamma(\alpha)}\tau^{\alpha+\frac{M}{2}-1}\exp\left(-\left(\frac{(\boldsymbol{x}-\boldsymbol{\mu})^\top\boldsymbol{\Lambda}(\boldsymbol{x}-\boldsymbol{\mu})}{2}+\beta\right)\tau\right)
\end{aligned}
$$

is the (general) Gauss-Gamma distribution on the random variable $\boldsymbol{x}\in\mathbb{R}^M$, $\tau>0$ with parameters $\boldsymbol{\mu}\in\mathbb{R}^M,\boldsymbol{\Lambda}\in\mathbb{S}^M_{++},\alpha>0,\beta>0$. With the conjugate Gauss-Gamma prior

$$
\begin{aligned}
p(\boldsymbol{a},\sigma^{-2}|\boldsymbol{\kappa}) &= \mathrm{GaussGamma}_M(\boldsymbol{a},\sigma^{-2}|\boldsymbol{\mu}_0,\boldsymbol{\Lambda}_0,\alpha_0,\beta_0)\\
&\propto (\sigma^{-2})^{\alpha_0+\frac{M}{2}-1}\exp\left(-\left(\frac{(\boldsymbol{a}-\boldsymbol{\mu}_0)^\top\boldsymbol{\Lambda}_0(\boldsymbol{a}-\boldsymbol{\mu}_0)}{2}+\beta_0\right)\sigma^{-2}\right)
\end{aligned}
$$

for hyperparameters $\boldsymbol{\kappa}=(\boldsymbol{\mu}_0,\boldsymbol{\Lambda}_0,\alpha_0,\beta_0)$, the posterior is computed as

$$
\begin{aligned}
p(\boldsymbol{a},\sigma^{-2}|\mathcal{D},\boldsymbol{\kappa}) &\propto p(\mathcal{D}|\boldsymbol{a},\sigma^{-2})p(\boldsymbol{a},\sigma^{-2}|\boldsymbol{\kappa})\\
&\propto \mathrm{GaussGamma}_M\left(\boldsymbol{a},\sigma^{-2};\widehat{\boldsymbol{a}}^{\mathrm{ML}},\boldsymbol{X}^\top\boldsymbol{X},\frac{M(N-1)}{2}+1,\frac{\|\boldsymbol{y}-\boldsymbol{X}\widehat{\boldsymbol{a}}^{\mathrm{ML}}\|^2}{2}\right)\\
&\qquad\cdot\mathrm{GaussGamma}_M(\boldsymbol{a},\sigma^{-2}|\boldsymbol{\mu}_0,\boldsymbol{\Lambda}_0,\alpha_0,\beta_0)\\
&\propto (\sigma^{-2})^{NM/2}\exp\left(-\frac{\left(\boldsymbol{a}-\widehat{\boldsymbol{a}}^{\mathrm{ML}}\right)^\top\boldsymbol{X}^\top\boldsymbol{X}\left(\boldsymbol{a}-\widehat{\boldsymbol{a}}^{\mathrm{ML}}\right)+\|\boldsymbol{y}-\boldsymbol{X}\widehat{\boldsymbol{a}}^{\mathrm{ML}}\|^2}{2}\sigma^{-2}\right)\\
&\qquad\cdot(\sigma^{-2})^{\alpha_0+\frac{M}{2}-1}\exp\left(-\left(\frac{(\boldsymbol{a}-\boldsymbol{\mu}_0)^\top\boldsymbol{\Lambda}_0(\boldsymbol{a}-\boldsymbol{\mu}_0)}{2}+\beta_0\right)\sigma^{-2}\right)\\
&\propto (\sigma^{-2})^{\widehat{\alpha}+\frac{M}{2}-1}\exp\left(-\left(\frac{(\boldsymbol{a}-\widehat{\boldsymbol{\mu}})^\top\widehat{\boldsymbol{\Lambda}}(\boldsymbol{a}-\widehat{\boldsymbol{\mu}})}{2}+\widehat{\beta}\right)\sigma^{-2}\right),
\end{aligned}
$$

where

$$
\begin{aligned}
\widehat{\boldsymbol{\mu}} &= (\boldsymbol{X}^\top\boldsymbol{X}+\boldsymbol{\Lambda}_0)^{-1}\left(\boldsymbol{X}^\top\boldsymbol{X}\widehat{\boldsymbol{a}}^{\mathrm{ML}}+\boldsymbol{\Lambda}_0\boldsymbol{\mu}_0\right),\\
\widehat{\boldsymbol{\Lambda}} &= \boldsymbol{X}^\top\boldsymbol{X}+\boldsymbol{\Lambda}_0,\\
\widehat{\alpha} &= \frac{NM}{2}+\alpha_0,\\
\widehat{\beta} &= \frac{\|\boldsymbol{y}-\boldsymbol{X}\widehat{\boldsymbol{a}}^{\mathrm{ML}}\|^2}{2}+\frac{(\widehat{\boldsymbol{a}}^{\mathrm{ML}}-\boldsymbol{\mu}_0)^\top\boldsymbol{\Lambda}_0(\boldsymbol{X}^\top\boldsymbol{X}+\boldsymbol{\Lambda}_0)^{-1}\boldsymbol{X}^\top\boldsymbol{X}(\widehat{\boldsymbol{a}}^{\mathrm{ML}}-\boldsymbol{\mu}_0)}{2}+\beta_0.
\end{aligned}
$$

Thus, we obtain

$$p(\boldsymbol{a}, \sigma^{-2}|\mathcal{D}, \boldsymbol{\kappa}) = \text{GaussGamma}_M(\boldsymbol{a}, \sigma^{-2}|\widehat{\boldsymbol{\mu}}, \widehat{\boldsymbol{\Lambda}}, \widehat{\alpha}, \widehat{\beta}). \tag{1.45}$$

**Multinomial Model**

The *multinomial distribution*, which expresses a distribution over the *histograms* of independent events, is another frequently used basic component in Bayesian modeling. For example, it appears in *mixture models* and *latent Dirichlet allocation*.

Assume that exclusive $K$ events occur with the probability

$$\boldsymbol{\theta} = (\theta_1, \ldots, \theta_K) \in \Delta^{K-1} \equiv \left\{ \boldsymbol{\theta} \in \mathbb{R}^K; 0 \leq \theta_k \leq 1, \sum_{k=1}^{K} \theta_k = 1 \right\}.$$

Then, the histogram

$$\boldsymbol{x} = (x_1, \ldots, x_K) \in \mathbb{H}_N^{K-1} \equiv \left\{ \boldsymbol{x} \in \mathbb{I}^K; 0 \leq x_k \leq N; \sum_{k=1}^{K} x_k = N \right\}$$

of events after $N$ iterations follows the *multinomial distribution*, defined as

$$p(\boldsymbol{x}|\boldsymbol{\theta}) = \text{Multinomial}_{K,N}(\boldsymbol{x}; \boldsymbol{\theta}) \equiv N! \cdot \prod_{k=1}^{K} \frac{\theta_k^{x_k}}{x_k!}. \tag{1.46}$$

$\boldsymbol{\theta}$ is called the *multinomial parameter*.

As seen shortly, calculation of the posterior with its conjugate prior is surprisingly easy.

**Dirichlet Likelihood** As a function of the multinomial parameter $\boldsymbol{w} = \boldsymbol{\theta}$, it is easy to find that the likelihood (1.46) is in the form of the *Dirichlet distribution*:

$$p(\boldsymbol{x}|\boldsymbol{\theta}) \propto \text{Dirichlet}_K(\boldsymbol{\theta}; \boldsymbol{x} + \mathbf{1}_K),$$

where $\mathbf{1}_K$ is the $K$-dimensional vector with all elements equal to 1. Since the Dirichlet distribution is an exponential family and hence multiplicatively closed, it is conjugate for the multinomial parameter. With the conjugate Dirichlet prior

$$p(\boldsymbol{\theta}|\boldsymbol{\phi}) = \text{Dirichlet}_K(\boldsymbol{\theta}; \boldsymbol{\phi}) \propto \prod_{k=1}^{K} \theta_k^{\phi-1}$$

with hyperparameters $\boldsymbol{\kappa} = \boldsymbol{\phi}$, the posterior is computed as

$$
\begin{aligned}
p(\boldsymbol{\theta}|\boldsymbol{x}, \boldsymbol{\phi}) &\propto p(\boldsymbol{x}|\boldsymbol{\theta})p(\boldsymbol{\theta}|\boldsymbol{\phi}) \\
&\propto \text{Dirichlet}_K(\boldsymbol{\theta}; \boldsymbol{x} + \mathbf{1}_K) \cdot \text{Dirichlet}_K(\boldsymbol{\theta}; \boldsymbol{\phi}) \\
&\propto \prod_{k=1}^{K} \theta_k^{x_k} \cdot \theta_k^{\phi_k - 1} \\
&\propto \prod_{k=1}^{K} \theta_k^{x_k + \phi_k - 1}.
\end{aligned}
$$

Thus we have

$$
p(\boldsymbol{\theta}|\boldsymbol{x}, \boldsymbol{\phi}) = \text{Dirichlet}_K(\boldsymbol{\theta}; \boldsymbol{x} + \boldsymbol{\phi}). \tag{1.47}
$$

**Special Cases** For $K = 2$, the multinomial distribution is reduced to the *binomial distribution*:

$$
\begin{aligned}
p(x_1|\theta_1) &= \text{Multinomial}_{2,N}\left((x_1, N - x_1)^\top; (\theta_1, 1 - \theta_1)^\top\right) \\
&= \text{Binomial}_N(x_1; \theta_1) \\
&= \binom{N}{x_1} \cdot \theta_1^{x_1}(1 - \theta_1)^{N - x_1}.
\end{aligned}
$$

Furthermore, it is reduced to the *Bernoulli distribution* for $K = 2$ and $N = 1$:

$$
\begin{aligned}
p(x_1|\theta_1) &= \text{Binomial}_1(x_1; \theta_1) \\
&= \theta_1^{x_1}(1 - \theta_1)^{1 - x_1}.
\end{aligned}
$$

Similarly, its conjugate Dirichlet distribution for $K = 2$ is reduced to the *Beta distribution*:

$$
\begin{aligned}
p(\theta_1|\phi_1, \phi_2) &= \text{Dirichlet}_2\left((\theta_1, 1 - \theta_1)^\top; (\phi_1, \phi_2)^\top\right) \\
&= \text{Beta}(\theta_1; \phi_1, \phi_2) \\
&= \frac{1}{\mathcal{B}(\phi_1, \phi_2)} \cdot \theta_1^{\phi_1 - 1}(1 - \theta_1)^{\phi_2 - 1},
\end{aligned}
$$

where $\mathcal{B}(\phi_1, \phi_2) = \frac{\Gamma(\phi_1)\Gamma(\phi_2)}{\Gamma(\phi_1 + \phi_2)}$ is the *Beta function*. Naturally, the Beta distribution is conjugate to the binomial and the Bernoulli distributions, and the posterior can be computed as easily as for the multinomial case.

With a conjugate prior in the form of a popular distribution, the four quantities introduced in Section 1.1.3, i.e., the marginal likelihood, the posterior mean, the posterior covariance, and the predictive distribution, can be obtained analytically. In the following subsections, we show how they are obtained.

Table 1.2 *First and second moments of common distributions.* $\mathbf{Mean}(\boldsymbol{x}) = \langle \boldsymbol{x} \rangle_{p(\boldsymbol{x}|\boldsymbol{w})}$, $\mathrm{Var}(x) = \left\langle (x - \mathrm{Mean}(x))^2 \right\rangle_{p(x|\boldsymbol{w})}$, $\mathbf{Cov}(\boldsymbol{x}) = \langle (\boldsymbol{x} - \mathbf{Mean}(\boldsymbol{x}))(\boldsymbol{x} - \mathbf{Mean}(\boldsymbol{x}))^\top \rangle_{p(\boldsymbol{x}|\boldsymbol{w})}$, $\Psi(z) \equiv \frac{d}{dz} \log \Gamma(z)$ : *Digamma function, and* $\Psi_m(z) \equiv \frac{d^m}{dz^m} \Psi(z)$: *Polygamma function of order m.*

| $p(\boldsymbol{x}\|\boldsymbol{w})$ | First moment | Second moment |
|---|---|---|
| $\mathrm{Gauss}_M(\boldsymbol{x}; \boldsymbol{\mu}, \boldsymbol{\Sigma})$ | $\mathbf{Mean}(\boldsymbol{x}) = \boldsymbol{\mu}$ | $\mathbf{Cov}(\boldsymbol{x}) = \boldsymbol{\Sigma}$ |
| $\mathrm{Gamma}(x; \alpha, \beta)$ | $\mathrm{Mean}(x) = \frac{\alpha}{\beta}$<br>$\mathrm{Mean}(\log x)$<br>$= \Psi(\alpha) - \log \beta$ | $\mathrm{Var}(x) = \frac{\alpha}{\beta^2}$<br>$\mathrm{Var}(\log x) = \Psi_1(\alpha)$ |
| $\mathrm{Wishart}_M(\boldsymbol{X}; \boldsymbol{V}, \nu)$ | $\mathbf{Mean}(\boldsymbol{X}) = \nu \boldsymbol{V}$ | $\mathrm{Var}(x_{m,m'}) = \nu(V_{m,m'}^2 + V_{m,m} V_{m',m'})$ |
| $\mathrm{Multinomial}_{K,N}(\boldsymbol{x}; \boldsymbol{\theta})$ | $\mathbf{Mean}(\boldsymbol{x}) = N\boldsymbol{\theta}$ | $(\mathbf{Cov}(\boldsymbol{x}))_{k,k'} = \begin{cases} N\theta_k(1-\theta_k) & (k = k') \\ -N\theta_k\theta_{k'} & (k \neq k') \end{cases}$ |
| $\mathrm{Dirichlet}_K(\boldsymbol{x}; \boldsymbol{\phi})$ | $\mathbf{Mean}(\boldsymbol{x}) = \frac{1}{\sum_{k=1}^K \phi_k} \boldsymbol{\phi}$<br>$\mathrm{Mean}(\log x_k)$<br>$= \Psi(\phi_k) - \Psi(\sum_{k'=1}^K \phi_{k'})$ | $(\mathbf{Cov}(\boldsymbol{x}))_{k,k'} = \begin{cases} \frac{\phi_k(\tau-\phi_k)}{\tau^2(\tau+1)} & (k = k') \\ -\frac{\phi_k\phi_{k'}}{\tau^2(\tau+1)} & (k \neq k') \end{cases}$<br>where $\tau = \sum_{k=1}^K \phi_k$ |

### 1.2.4 Posterior Mean and Covariance

As seen in Section 1.2.3, by adopting a conjugate prior having a form of one of the common family distributions, such as the one in Table 1.1, we can have the posterior distribution in the same common family.[3] In such cases, we can simply use the known form of moments, which are summarized in Table 1.2. For example, the posterior (1.42) for the regression parameter $\boldsymbol{a}$ (when the noise variance $\sigma^2$ is treated as a known constant) is the following Gaussian distribution:

$$p(\boldsymbol{a}|\mathcal{D}, \boldsymbol{a}_0, \boldsymbol{\Sigma}_0) = \mathrm{Gauss}_M\left(\boldsymbol{a}; \widehat{\boldsymbol{a}}, \widehat{\boldsymbol{\Sigma}}_a\right),$$

$$\text{where} \quad \widehat{\boldsymbol{a}} = \left(\frac{\boldsymbol{X}^\top \boldsymbol{X}}{\sigma^2} + \boldsymbol{\Sigma}_0^{-1}\right)^{-1} \left(\frac{\boldsymbol{X}^\top \boldsymbol{y}}{\sigma^2} + \boldsymbol{\Sigma}_0^{-1} \boldsymbol{a}_0\right),$$

$$\widehat{\boldsymbol{\Sigma}}_a = \left(\frac{\boldsymbol{X}^\top \boldsymbol{X}}{\sigma^2} + \boldsymbol{\Sigma}_0^{-1}\right)^{-1}.$$

[3] If we would say that the prior is in the family that contains all possible distributions, this family would be the conjugate prior for any likelihood function, which is however useless. Usually, the notion of the conjugate prior implicitly requires that moments (at least the normalization constant and the first moment) of any family member can be computed analytically.

Therefore, the posterior mean and the posterior covariance are simply given by

$$\langle \boldsymbol{a} \rangle_{p(\boldsymbol{a}|\mathcal{D},\boldsymbol{a}_0,\boldsymbol{\Sigma}_0)} = \widehat{\boldsymbol{a}},$$
$$\left\langle (\boldsymbol{a} - \langle \boldsymbol{a} \rangle)(\boldsymbol{a} - \langle \boldsymbol{a} \rangle)^\top \right\rangle_{p(\boldsymbol{a}|\mathcal{D},\boldsymbol{a}_0,\boldsymbol{\Sigma}_0)} = \widehat{\boldsymbol{\Sigma}}_{\boldsymbol{a}},$$

respectively. The posterior (1.29) of the (inverse) variance parameter $\sigma^{-2}$ of the isotropic Gaussian distribution (when the mean parameter $\boldsymbol{\mu}$ is treated as a known constant) is the following Gamma distribution:

$$p(\sigma^{-2}|\mathcal{D}, \alpha_0, \beta_0) = \mathrm{Gamma}\left(\sigma^{-2}; \frac{MN}{2} + \alpha_0, \frac{1}{2}\sum_{n=1}^{N} \|\boldsymbol{x}^{(n)} - \boldsymbol{\mu}\|^2 + \beta_0\right).$$

Therefore, the posterior mean and the posterior variance are given by

$$\left\langle \sigma^{-2} \right\rangle_{p(\sigma^{-2}|\mathcal{D},\alpha_0,\beta_0)} = \frac{\frac{MN}{2} + \alpha_0}{\frac{1}{2}\sum_{n=1}^{N} \|\boldsymbol{x}^{(n)} - \boldsymbol{\mu}\|^2 + \beta_0},$$
$$\left\langle \left(\sigma^{-2} - \left\langle \sigma^{-2} \right\rangle\right)^2 \right\rangle_{p(\sigma^{-2}|\mathcal{D},\alpha_0,\beta_0)} = \frac{\frac{MN}{2} + \alpha_0}{(\frac{1}{2}\sum_{n=1}^{N} \|\boldsymbol{x}^{(n)} - \boldsymbol{\mu}\|^2 + \beta_0)^2},$$

respectively.

Also in other cases, the posterior mean and the posterior covariances can be easily computed by using Table 1.2, if the form of the posterior distribution is in the table.

### 1.2.5 Predictive Distribution

The predictive distribution (1.9) for a new data set $\mathcal{D}^{\mathrm{new}}$ can be computed analytically, if the posterior distribution is in the exponential family, and hence multiplicatively closed. In this section, we show two examplary cases, the linear regression model and the multinomial model.

#### Linear Regression Model

Consider the linear regression model:

$$p(y|\boldsymbol{x}, \boldsymbol{a}) = \mathrm{Gauss}_1(y; \boldsymbol{a}^\top \boldsymbol{x}, \sigma^2) = \frac{1}{\sqrt{2\pi\sigma^2}} \cdot \exp\left(-\frac{(y - \boldsymbol{a}^\top \boldsymbol{x})^2}{2\sigma^2}\right), \tag{1.48}$$

where only the regression parameter is unknown, i.e., $\boldsymbol{w} = \boldsymbol{a} \in \mathbb{R}^M$, and the noise variance parameter $\sigma^2$ is treated as a known constant. We choose the zero-mean Gaussian as a conjugate prior:

$$p(\boldsymbol{a}|\boldsymbol{C}) = \mathrm{Gauss}_M(\boldsymbol{a}; \boldsymbol{0}, \boldsymbol{C}) = \frac{\exp\left(-\frac{1}{2}\boldsymbol{a}^\top \boldsymbol{C}^{-1} \boldsymbol{a}\right)}{(2\pi)^{M/2} \det(\boldsymbol{C})^{1/2}}, \tag{1.49}$$

where $\boldsymbol{C}$ is the prior covariance.

When $N$ i.i.d. samples $\mathcal{D} = (\boldsymbol{X}, \boldsymbol{y})$, where

$$\boldsymbol{y} = (y^{(1)}, \ldots, y^{(N)})^\top \in \mathbb{R}^N, \qquad \boldsymbol{X} = (\boldsymbol{x}^{(1)}, \ldots, \boldsymbol{x}^{(N)})^\top \in \mathbb{R}^{N \times M},$$

are observed, the posterior is given by

$$p(\boldsymbol{a}|\boldsymbol{y}, \boldsymbol{X}, \boldsymbol{C}) = \mathrm{Gauss}_M\left(\boldsymbol{a}; \widehat{\boldsymbol{a}}, \widehat{\boldsymbol{\Sigma}}_a\right)$$
$$= \frac{1}{(2\pi)^{M/2} \det\left(\widehat{\boldsymbol{\Sigma}}_a\right)^{1/2}} \cdot \exp\left(-\frac{(\boldsymbol{a} - \widehat{\boldsymbol{a}})^\top \widehat{\boldsymbol{\Sigma}}_a^{-1} (\boldsymbol{a} - \widehat{\boldsymbol{a}})}{2}\right), \tag{1.50}$$

where

$$\widehat{\boldsymbol{a}} = \left(\frac{\boldsymbol{X}^\top \boldsymbol{X}}{\sigma^2} + \boldsymbol{C}^{-1}\right)^{-1} \frac{\boldsymbol{X}^\top \boldsymbol{y}}{\sigma^2} = \widehat{\boldsymbol{\Sigma}}_a \frac{\boldsymbol{X}^\top \boldsymbol{y}}{\sigma^2}, \tag{1.51}$$
$$\widehat{\boldsymbol{\Sigma}}_a = \left(\frac{\boldsymbol{X}^\top \boldsymbol{X}}{\sigma^2} + \boldsymbol{C}^{-1}\right)^{-1}. \tag{1.52}$$

This is just a special case of the posterior (1.42) for the linear regression model with the most general Gaussian prior.

Now, let us compute the predictive distribution on the output $y^*$ for a new given input $\boldsymbol{x}^*$. As defined in Eq. (1.9), the predictive distribution is the expectation value of the model distribution (1.48) (for a new input–output pair) over the posterior distribution (1.50):

$$\begin{aligned}
p(y^*|\boldsymbol{x}^*, \boldsymbol{y}, \boldsymbol{X}, \boldsymbol{C}) &= \langle p(y^*|\boldsymbol{x}^*, \boldsymbol{a})\rangle_{p(\boldsymbol{a}|\boldsymbol{y}, \boldsymbol{X}, \boldsymbol{C})} \\
&= \int p(y^*|\boldsymbol{x}^*, \boldsymbol{a}) p(\boldsymbol{a}|\boldsymbol{y}, \boldsymbol{X}, \boldsymbol{C}) d\boldsymbol{a} \\
&= \int \mathrm{Gauss}_1(y^*; \boldsymbol{a}^\top \boldsymbol{x}^*, \sigma^2) \mathrm{Gauss}_M\left(\boldsymbol{a}; \widehat{\boldsymbol{a}}, \widehat{\boldsymbol{\Sigma}}_a\right) d\boldsymbol{a} \\
&\propto \int \exp\left(-\frac{(y^* - \boldsymbol{a}^\top \boldsymbol{x}^*)^2}{2\sigma^2} - \frac{(\boldsymbol{a} - \widehat{\boldsymbol{a}})^\top \widehat{\boldsymbol{\Sigma}}_a^{-1} (\boldsymbol{a} - \widehat{\boldsymbol{a}})}{2}\right) d\boldsymbol{a} \\
&\propto \exp\left(-\frac{y^{*2}}{2\sigma^2}\right) \int \exp\left(-\frac{\boldsymbol{a}^\top \left(\widehat{\boldsymbol{\Sigma}}_a^{-1} + \frac{\boldsymbol{x}^* \boldsymbol{x}^{*\top}}{\sigma^2}\right) \boldsymbol{a} - 2\boldsymbol{a}^\top \left(\widehat{\boldsymbol{\Sigma}}_a^{-1} \widehat{\boldsymbol{a}} + \frac{\boldsymbol{x}^* y^*}{\sigma^2}\right)}{2}\right) d\boldsymbol{a} \\
&\propto \exp\left(-\frac{\sigma^{-2} y^{*2} - \left(\widehat{\boldsymbol{\Sigma}}_a^{-1} \widehat{\boldsymbol{a}} + \frac{\boldsymbol{x}^* y^*}{\sigma^2}\right)^\top \left(\widehat{\boldsymbol{\Sigma}}_a^{-1} + \frac{\boldsymbol{x}^* \boldsymbol{x}^{*\top}}{\sigma^2}\right)^{-1} \left(\widehat{\boldsymbol{\Sigma}}_a^{-1} \widehat{\boldsymbol{a}} + \frac{\boldsymbol{x}^* y^*}{\sigma^2}\right)}{2}\right) \\
&\quad \cdot \int \exp\left(-\frac{(\boldsymbol{a} - \breve{\boldsymbol{a}})^\top \left(\widehat{\boldsymbol{\Sigma}}_a^{-1} + \frac{\boldsymbol{x}^* \boldsymbol{x}^{*\top}}{\sigma^2}\right) (\boldsymbol{a} - \breve{\boldsymbol{a}})}{2}\right) d\boldsymbol{a},
\end{aligned} \tag{1.53}$$

where

$$\breve{\boldsymbol{a}} = \left(\widehat{\boldsymbol{\Sigma}}_a^{-1} + \frac{\boldsymbol{x}^* \boldsymbol{x}^{*\top}}{\sigma^2}\right)^{-1} \left(\widehat{\boldsymbol{\Sigma}}_a^{-1} \widehat{\boldsymbol{a}} + \frac{\boldsymbol{x}^* y^*}{\sigma^2}\right).$$

Note that, although the preceding computation is similar to the one for the posterior distribution in Section 1.2.3, any factor that depends on $y^*$ cannot be ignored even if it does not depend on $\boldsymbol{a}$, since the goal is to obtain the distribution on $y^*$.

The integrand in Eq. (1.53) coincides with the main part of

$$\mathrm{Gauss}_M\left(\boldsymbol{a}; \breve{\boldsymbol{a}}, \left(\widehat{\boldsymbol{\Sigma}}_{\boldsymbol{a}}^{-1} + \frac{\boldsymbol{x}^*\boldsymbol{x}^{*\top}}{\sigma^2}\right)^{-1}\right)$$

without the normalization factor. Therefore, the integral is the inverse of the normalization factor, i.e.,

$$\int \exp\left(-\frac{(\boldsymbol{a}-\breve{\boldsymbol{a}})^\top\left(\widehat{\boldsymbol{\Sigma}}_{\boldsymbol{a}}^{-1} + \frac{\boldsymbol{x}^*\boldsymbol{x}^{*\top}}{\sigma^2}\right)(\boldsymbol{a}-\breve{\boldsymbol{a}})}{2}\right)d\boldsymbol{a} = (2\pi)^{M/2}\det\left(\widehat{\boldsymbol{\Sigma}}_{\boldsymbol{a}}^{-1} + \frac{\boldsymbol{x}^*\boldsymbol{x}^{*\top}}{\sigma^2}\right)^{-1/2},$$

which is a constant with respect to $y^*$. Therefore, by using Eqs. (1.51) and (1.52), we have

$$\begin{aligned}
&p(y^*|\boldsymbol{x}^*, \boldsymbol{y}, \boldsymbol{X}, \boldsymbol{C}) \\
&\propto \exp\left(-\frac{\sigma^{-2}y^{*2} - \left(\widehat{\boldsymbol{\Sigma}}_{\boldsymbol{a}}^{-1}\widehat{\boldsymbol{a}} + \frac{\boldsymbol{x}^*y^*}{\sigma^2}\right)^\top\left(\widehat{\boldsymbol{\Sigma}}_{\boldsymbol{a}}^{-1} + \frac{\boldsymbol{x}^*\boldsymbol{x}^{*\top}}{\sigma^2}\right)^{-1}\left(\widehat{\boldsymbol{\Sigma}}_{\boldsymbol{a}}^{-1}\widehat{\boldsymbol{a}} + \frac{\boldsymbol{x}^*y^*}{\sigma^2}\right)}{2}\right) \\
&\propto \exp\left(-\frac{y^{*2} - (\boldsymbol{X}^\top\boldsymbol{y} + \boldsymbol{x}^*y^*)^\top(\boldsymbol{X}^\top\boldsymbol{X} + \boldsymbol{x}^*\boldsymbol{x}^{*\top} + \sigma^2\boldsymbol{C}^{-1})^{-1}(\boldsymbol{X}^\top\boldsymbol{y} + \boldsymbol{x}^*y^*)}{2\sigma^2}\right) \\
&\propto \exp\left(-\frac{1}{2\sigma^2}\left\{y^{*2}\left(1 - \boldsymbol{x}^{*\top}\left(\boldsymbol{X}^\top\boldsymbol{X} + \boldsymbol{x}^*\boldsymbol{x}^{*\top} + \sigma^2\boldsymbol{C}^{-1}\right)^{-1}\boldsymbol{x}^*\right)\right.\right. \\
&\qquad \left.\left. - 2y^*\boldsymbol{x}^{*\top}\left(\boldsymbol{X}^\top\boldsymbol{X} + \boldsymbol{x}^*\boldsymbol{x}^{*\top} + \sigma^2\boldsymbol{C}^{-1}\right)^{-1}\boldsymbol{X}^\top\boldsymbol{y}\right\}\right) \\
&\propto \exp\left(-\frac{1 - \boldsymbol{x}^{*\top}(\boldsymbol{X}^\top\boldsymbol{X} + \boldsymbol{x}^*\boldsymbol{x}^{*\top} + \sigma^2\boldsymbol{C}^{-1})^{-1}\boldsymbol{x}^*}{2\sigma^2}\right. \\
&\qquad \left. \cdot\left(y^* - \frac{\boldsymbol{x}^{*\top}(\boldsymbol{X}^\top\boldsymbol{X} + \boldsymbol{x}^*\boldsymbol{x}^{*\top} + \sigma^2\boldsymbol{C}^{-1})^{-1}\boldsymbol{X}^\top\boldsymbol{y}}{1 - \boldsymbol{x}^{*\top}(\boldsymbol{X}^\top\boldsymbol{X} + \boldsymbol{x}^*\boldsymbol{x}^{*\top} + \sigma^2\boldsymbol{C}^{-1})^{-1}\boldsymbol{x}^*}\right)^2\right) \\
&\propto \exp\left(-\frac{(y^* - \widehat{y})^2}{2\widehat{\sigma}_y^2}\right),
\end{aligned}$$

where

$$\widehat{y} = \frac{\boldsymbol{x}^{*\top}\left(\boldsymbol{X}^\top\boldsymbol{X} + \boldsymbol{x}^*\boldsymbol{x}^{*\top} + \sigma^2\boldsymbol{C}^{-1}\right)^{-1}\boldsymbol{X}^\top\boldsymbol{y}}{1 - \boldsymbol{x}^{*\top}\left(\boldsymbol{X}^\top\boldsymbol{X} + \boldsymbol{x}^*\boldsymbol{x}^{*\top} + \sigma^2\boldsymbol{C}^{-1}\right)^{-1}\boldsymbol{x}^*},$$

$$\widehat{\sigma}_y^2 = \frac{\sigma^2}{1 - \boldsymbol{x}^{*\top}\left(\boldsymbol{X}^\top\boldsymbol{X} + \boldsymbol{x}^*\boldsymbol{x}^{*\top} + \sigma^2\boldsymbol{C}^{-1}\right)^{-1}\boldsymbol{x}^*}.$$

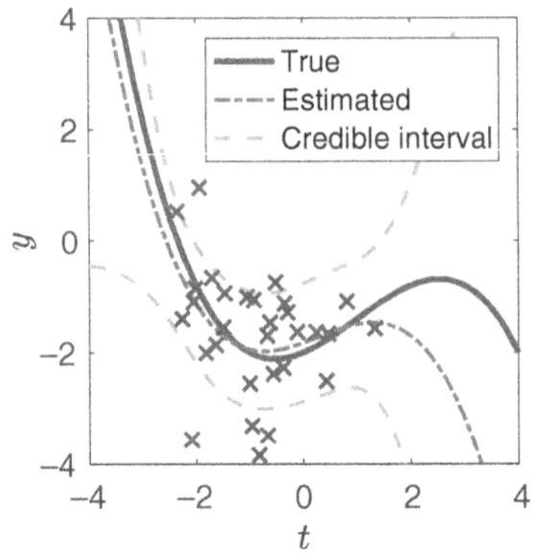


Figure 1.3 Predictive distribution of the linear regression model.

Thus, the predictive distribution has been analytically obtained:

$$p(y^*|\boldsymbol{x}^*, \boldsymbol{y}, \boldsymbol{X}, \boldsymbol{C}) = \mathrm{Gauss}_1\left(y^*; \widehat{y}, \widehat{\sigma}_y^2\right). \tag{1.54}$$

Figure 1.3 shows an example of the predictive distribution of the linear regression model. The curve labeled as "True" indicates the mean $y = \boldsymbol{a}^{*}\boldsymbol{x}$ of the true regression model $y = \boldsymbol{a}^{*}\boldsymbol{x} + \varepsilon$, where $\boldsymbol{a}^{*} = (-2, 0.4, 0.3, -0.1)^{\top}$, $\boldsymbol{x} = (1, t, t^2, t^3)^{\top}$, and $\varepsilon \sim \mathrm{Gauss}_1(0, 1^2)$. The crosses are $N = 30$ i.i.d. observed samples generated from the true regression model and the input distribution $t \sim \mathrm{Uniform}(-2.4, 1.6)$, where $\mathrm{Uniform}(l, u)$ denotes the uniform distribution on $[l, u]$. The regression model (1.48) with the prior (1.49) for the hyperparameters $\boldsymbol{C} = 10000 \cdot \boldsymbol{I}_M, \sigma^2 = 1$ was trained with the observed samples. The curve labeled as "Estimated" and the pair of curves labeled as "Credible interval" show the mean $\widehat{y}$ and the *credible interval* $\widehat{y} \pm \widehat{\sigma}_y$ of the predictive distribution (1.54), respectively.

Reflecting the fact that the samples are observed only in the middle region ($t \in [-2.4, 1.6]$), the credible interval is large in outer regions. The larger interval implies that the "Estimated" function is less reliable, and we see that the gap from the "True" function is indeed large. Since the true function is unknown in practical situations, the variance of the predictive distribution is important information on the reliability of the estimated result.

### Multinomial Model

Let us compute the predictive distribution of the multinomial model:

$$p(\boldsymbol{x}|\boldsymbol{\theta}) = \mathrm{Multinomial}_{K,N}(\boldsymbol{x}; \boldsymbol{\theta}) \propto \prod_{k=1}^{K} \frac{\theta_k^{x_k}}{x_k!},$$

$$p(\boldsymbol{\theta}|\boldsymbol{\phi}) = \mathrm{Dirichlet}_K(\boldsymbol{\theta}; \boldsymbol{\phi}) \propto \prod_{k=1}^{K} \theta_k^{\phi_k - 1},$$

with the observed data $\mathcal{D} = \boldsymbol{x} = (x_1, \ldots, x_K) \in \mathbb{H}_N^{K-1}$ and the unknown parameter $\boldsymbol{w} = \boldsymbol{\theta} = (\theta_1, \ldots, \theta_K) \in \Delta^{K-1}$.

The posterior was derived in Eq. (1.47):

$$p(\boldsymbol{\theta}|\boldsymbol{x}, \boldsymbol{\phi}) = \mathrm{Dirichlet}_K(\boldsymbol{\theta}; \boldsymbol{x} + \boldsymbol{\phi}) \propto \prod_{k=1}^{K} \theta_k^{x_k + \phi_k - 1}.$$

Therefore, the predictive distribution for a new single sample $\boldsymbol{x}^* \in \mathbb{H}_1^{K-1}$ is given by

$$\begin{aligned} p(\boldsymbol{x}^*|\boldsymbol{x}, \boldsymbol{\phi}) &= \langle p(\boldsymbol{x}^*|\boldsymbol{\theta}) \rangle_{p(\boldsymbol{\theta}|\boldsymbol{x}, \boldsymbol{\phi})} \\ &= \int p(\boldsymbol{x}^*|\boldsymbol{\theta}) p(\boldsymbol{\theta}|\boldsymbol{x}, \boldsymbol{\phi}) d\boldsymbol{\theta} \\ &= \int \mathrm{Multinomial}_{K,1}(\boldsymbol{x}^*; \boldsymbol{\theta}) \mathrm{Dirichlet}_K(\boldsymbol{\theta}; \boldsymbol{x} + \boldsymbol{\phi}) d\boldsymbol{\theta} \\ &\propto \int \prod_{k=1}^{K} \theta_k^{x_k^*} \cdot \theta_k^{x_k + \phi_k - 1} d\boldsymbol{\theta} \\ &= \int \prod_{k=1}^{K} \theta_k^{x_k^* + x_k + \phi_k - 1} d\boldsymbol{\theta}. \end{aligned} \tag{1.55}$$

In the fourth equation, we ignored the factors that depend neither on $\boldsymbol{x}^*$ nor on $\boldsymbol{\theta}$.

The integrand in Eq. (1.55) is the main part of $\mathrm{Dirichlet}_K(\boldsymbol{\theta}; \boldsymbol{x}^* + \boldsymbol{x} + \boldsymbol{\phi})$, and therefore, the integral is equal to the inverse of its normalization factor:

$$\begin{aligned} \int \prod_{k=1}^{K} \theta_k^{x_k^* + x_k + \phi_k - 1} d\boldsymbol{\theta} &= \frac{\prod_{k=1}^{K} \Gamma(x_k^* + x_k + \phi_k)}{\Gamma(\sum_{k=1}^{K} x_k^* + x_k + \phi_k)} \\ &= \frac{\prod_{k=1}^{K} \Gamma(x_k^* + x_k + \phi_k)}{\Gamma(N + \sum_{k=1}^{K} \phi_k + 1)}. \end{aligned}$$

Thus, by using the identity $\Gamma(x+1) = x\Gamma(x)$ for the Gamma function, we have

$$\begin{aligned} p(\boldsymbol{x}^*|\boldsymbol{x}, \boldsymbol{\phi}) &\propto \prod_{k=1}^{K} \Gamma(x_k^* + x_k + \phi_k) \\ &\propto \prod_{k=1}^{K} (x_k + \phi_k)^{x_k^*} \Gamma(x_k + \phi_k) \end{aligned}$$

$$
\begin{aligned}
&\propto \prod_{k=1}^{K} (x_k + \phi_k)^{x_k^*} \\
&\propto \prod_{k=1}^{K} \left( \frac{x_k + \phi_k}{\sum_{k'=1}^{K} x_{k'} + \phi_{k'}} \right)^{x_k^*} \\
&= \mathrm{Multinomial}_{K,1}(\boldsymbol{x}^*; \widehat{\boldsymbol{\theta}}),
\end{aligned} \tag{1.56}
$$

where

$$
\widehat{\theta}_k = \frac{x_k + \phi_k}{\sum_{k'=1}^{K} x_{k'} + \phi_{k'}}. \tag{1.57}
$$

From Eq. (1.47) and Table 1.2, we can easily see that the predictive mean $\widehat{\boldsymbol{\theta}}$, specified by Eq. (1.57), coincides with the posterior mean, i.e., the Bayesian estimator:

$$
\widehat{\boldsymbol{\theta}} = \langle \boldsymbol{\theta} \rangle_{\mathrm{Dirichlet}_K(\boldsymbol{\theta}; \boldsymbol{x}+\boldsymbol{\phi})} .
$$

Therefore, in the multinomial model, the predictive distribution coincides with the model distribution with the Bayesian estimator plugged in.

In the preceding derivation, we performed the integral computation and derived the form of the predictive distribution. However, the necessary information to determine the predictive distribution is the probability table on the events $\boldsymbol{x}^* \in \mathbb{H}_1^{K-1} = \{\boldsymbol{e}_k\}_{k=1}^K$, of which the degree of freedom is only $K$. Therefore, the following simple calculation gives the same result:

$$
\begin{aligned}
\mathrm{Prob}(\boldsymbol{x}^* = \boldsymbol{e}_k | \boldsymbol{x}, \boldsymbol{\phi}) &= \langle \mathrm{Multinomial}_{K,1}(\boldsymbol{e}_k; \boldsymbol{\theta}) \rangle_{\mathrm{Dirichlet}_K(\boldsymbol{\theta}; \boldsymbol{x}+\boldsymbol{\phi})} \\
&= \langle \theta_k \rangle_{\mathrm{Dirichlet}_K(\boldsymbol{\theta}; \boldsymbol{x}+\boldsymbol{\phi})} \\
&= \widehat{\theta}_k,
\end{aligned}
$$

which specifies the function form of the predictive distribution, given by Eq. (1.56).

### 1.2.6 Marginal Likelihood

Let us compute the marginal likelihood of the linear regression model, defined by Eqs. (1.48) and (1.49):

$$
\begin{aligned}
p(\mathcal{D}|\boldsymbol{C}) &= p(\boldsymbol{y}|\boldsymbol{X}, \boldsymbol{C}) \\
&= \langle p(\boldsymbol{y}|\boldsymbol{X}, \boldsymbol{a}) \rangle_{p(\boldsymbol{a}|\boldsymbol{C})} \\
&= \int p(\boldsymbol{y}|\boldsymbol{X}, \boldsymbol{a}) p(\boldsymbol{a}|\boldsymbol{C}) d\boldsymbol{a}
\end{aligned}
$$

$$
\begin{aligned}
&= \int \mathrm{Gauss}_N(\boldsymbol{y}; \boldsymbol{X}\boldsymbol{a}, \sigma^2 \boldsymbol{I}_N)\mathrm{Gauss}_M(\boldsymbol{a}; \boldsymbol{0}, \boldsymbol{C}) d\boldsymbol{a} \\
&= \int \frac{\exp\left(-\frac{\|\boldsymbol{y}-\boldsymbol{X}\boldsymbol{a}\|^2}{2\sigma^2}\right)}{(2\pi\sigma^2)^{N/2}} \cdot \frac{\exp\left(-\frac{1}{2}\boldsymbol{a}^\top \boldsymbol{C}^{-1}\boldsymbol{a}\right)}{(2\pi)^{M/2}\det(\boldsymbol{C})^{1/2}} d\boldsymbol{a} \\
&= \frac{\exp\left(-\frac{\|\boldsymbol{y}\|^2}{2\sigma^2}\right)}{(2\pi\sigma^2)^{N/2}(2\pi)^{M/2}\det(\boldsymbol{C})^{1/2}} \\
&\quad \cdot \int \exp\left(-\frac{-2\boldsymbol{a}^\top \frac{\boldsymbol{X}^\top \boldsymbol{y}}{\sigma^2} + \boldsymbol{a}^\top \left(\frac{\boldsymbol{X}^\top \boldsymbol{X}}{\sigma^2} + \boldsymbol{C}^{-1}\right)\boldsymbol{a}}{2}\right) d\boldsymbol{a} \\
&= \frac{\exp\left(-\frac{1}{2}\left(\frac{\|\boldsymbol{y}\|^2}{\sigma^2} - \widehat{\boldsymbol{a}}^\top \widehat{\boldsymbol{\Sigma}}_a^{-1} \widehat{\boldsymbol{a}}\right)\right)}{(2\pi\sigma^2)^{N/2}(2\pi)^{M/2}\det(\boldsymbol{C})^{1/2}} \\
&\quad \cdot \int \exp\left(-\frac{(\boldsymbol{a}-\widehat{\boldsymbol{a}})^\top \widehat{\boldsymbol{\Sigma}}_a^{-1} (\boldsymbol{a}-\widehat{\boldsymbol{a}})}{2}\right) d\boldsymbol{a},
\end{aligned} \tag{1.58}
$$

where $\widehat{\boldsymbol{a}}$ and $\widehat{\boldsymbol{\Sigma}}_a$ are, respectively, the posterior mean and the posterior covariance, given by Eqs. (1.51) and (1.52).

By using

$$
\int \exp\left(-\frac{(\boldsymbol{a}-\widehat{\boldsymbol{a}})^\top \widehat{\boldsymbol{\Sigma}}_a^{-1} (\boldsymbol{a}-\widehat{\boldsymbol{a}})}{2}\right) d\boldsymbol{a} = \sqrt{(2\pi)^M \det\left(\widehat{\boldsymbol{\Sigma}}_a\right)},
$$

and Eq. (1.58), we have

$$
\begin{aligned}
p(\boldsymbol{y}|\boldsymbol{X}, \boldsymbol{C}) &= \frac{\exp\left(-\frac{1}{2}\left(\frac{\|\boldsymbol{y}\|^2}{\sigma^2} - \frac{\boldsymbol{y}^\top \boldsymbol{X}\widehat{\boldsymbol{\Sigma}}_a \boldsymbol{X}^\top \boldsymbol{y}}{\sigma^4}\right)\right)}{(2\pi\sigma^2)^{N/2}(2\pi)^{M/2}\det(\boldsymbol{C})^{1/2}} \sqrt{(2\pi)^M \det\left(\widehat{\boldsymbol{\Sigma}}_a\right)} \\
&= \frac{\exp\left(-\frac{\|\boldsymbol{y}\|^2 - \boldsymbol{y}^\top \boldsymbol{X}(\boldsymbol{X}^\top \boldsymbol{X} + \sigma^2 \boldsymbol{C}^{-1})^{-1} \boldsymbol{X}^\top \boldsymbol{y}}{2\sigma^2}\right)}{(2\pi\sigma^2)^{N/2}\det(\boldsymbol{C}\boldsymbol{X}^\top \boldsymbol{X} + \sigma^2 \boldsymbol{I}_M)^{1/2}},
\end{aligned} \tag{1.59}
$$

where we also used Eqs. (1.51) and (1.52).

Eq. (1.59) is an explicit expression of the marginal likelihood as a function of the hyperparameter $\boldsymbol{\kappa} = \boldsymbol{C}$. Based on it, we perform EBayes learning in Section 1.2.7.

### 1.2.7 Empirical Bayesian Learning

In empirical Bayesian (EBayes) learning, the hyperparameter $\boldsymbol{\kappa}$ is estimated by maximizing the marginal likelihood $p(\mathcal{D}|\boldsymbol{\kappa})$. The negative logarithm of the marginal likelihood,

$$F^{\mathrm{Bayes}} = -\log p(\mathcal{D}|\boldsymbol{\kappa}), \tag{1.60}$$

is called the *Bayes free energy* or *stochastic complexity*.[4] Since $\log(\cdot)$ is a monotonic function, maximizing the marginal likelihood is equivalent to minimizing the Bayes free energy.

Eq. (1.59) implies that the Bayes free energy of the linear regression model is given by

$$\begin{aligned} 2F^{\mathrm{Bayes}} &= -2\log p(\boldsymbol{y}|\boldsymbol{X}, \boldsymbol{C}) \\ &= N\log(2\pi\sigma^2) + \log\det(\boldsymbol{C}\boldsymbol{X}^\top\boldsymbol{X} + \sigma^2\boldsymbol{I}_M) \\ &\quad + \frac{\|\boldsymbol{y}\|^2 - \boldsymbol{y}^\top\boldsymbol{X}\left(\boldsymbol{X}^\top\boldsymbol{X} + \sigma^2\boldsymbol{C}^{-1}\right)^{-1}\boldsymbol{X}^\top\boldsymbol{y}}{\sigma^2}. \end{aligned} \tag{1.61}$$

Let us restrict the prior covariance to be diagonal:

$$\boldsymbol{C} = \mathbf{Diag}(c_1^2, \ldots, c_M^2) \in \mathbb{D}^M. \tag{1.62}$$

The prior (1.49) with diagonal covariance (1.62) is called the *automatic relevance determination (ARD)* prior, which is known to make the EBayes estimator sparse (Neal, 1996). In the following example, we see this effect by setting the design matrix to identity, $\boldsymbol{X} = \boldsymbol{I}_M$, which enables us to derive the EBayes solution analytically.

Under the identity design matrix, the Bayes free energy (1.61) can be decomposed as

$$\begin{aligned} 2F^{\mathrm{Bayes}} &= N\log(2\pi\sigma^2) + \log\det(\boldsymbol{C} + \sigma^2\boldsymbol{I}_M) + \frac{\|\boldsymbol{y}\|^2 - \boldsymbol{y}^\top\left(\boldsymbol{I}_M + \sigma^2\boldsymbol{C}^{-1}\right)^{-1}\boldsymbol{y}}{\sigma^2} \\ &= N\log(2\pi\sigma^2) + \frac{\|\boldsymbol{y}\|^2}{\sigma^2} + \sum_{m=1}^{M}\left(\log(c_m^2 + \sigma^2) - \frac{y_m^2}{\sigma^2\left(1 + \sigma^2 c_m^{-2}\right)}\right) \\ &= \sum_{m=1}^{M} 2F_m^* + \text{const.}, \end{aligned} \tag{1.63}$$

where

$$2F_m^* = \log\left(1 + \frac{c_m^2}{\sigma^2}\right) - \frac{y_m^2}{\sigma^2}\left(1 + \frac{\sigma^2}{c_m^2}\right)^{-1}. \tag{1.64}$$

In Eq. (1.63), we omitted the constant factors with respect to the hyperparameter $\boldsymbol{C}$. As the remaining terms are decomposed into each component $m$, we can independently minimize $F_m^*$ with respect to $c_m^2$.

[4] The logarithm of the marginal likelihood $\log p(\mathcal{D}|\boldsymbol{\kappa})$ is called the *log marginal likelihood* or *evidence*.

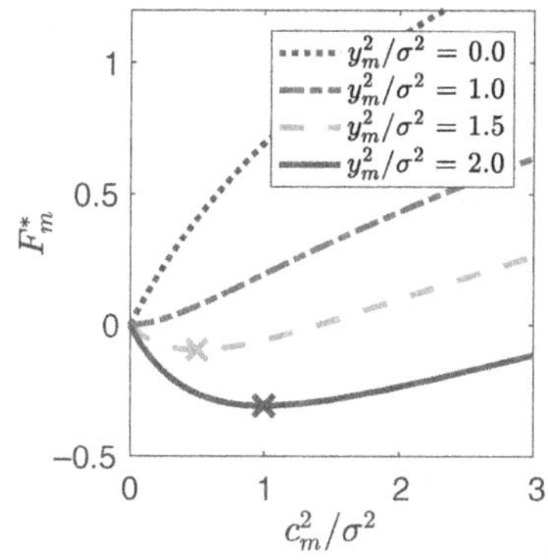


Figure 1.4 The (componentwise) Bayes free energy (1.64) of linear regression model with the ARD prior. The minimizer is shown as a cross if it lies in the positive region of $c_m^2/\sigma^2$.

The derivative of Eq. (1.64) with respect to $c_m^2$ is

$$\begin{aligned} 2\frac{\partial F_m^*}{\partial c_m^2} &= \frac{1}{c_m^2 + \sigma^2} - \frac{y_m^2}{(1 + \sigma^2 c_m^{-2})^2 c_m^4} \\ &= \frac{1}{c_m^2 + \sigma^2} - \frac{y_m^2}{(c_m^2 + \sigma^2)^2} \\ &= \frac{c_m^2 - (y_m^2 - \sigma^2)}{(c_m^2 + \sigma^2)^2}. \end{aligned} \tag{1.65}$$

Eq. (1.65) implies that $F_m^*$ is monotonically increasing over all domain $c_m^2 > 0$ when $y_m^2 \leq \sigma^2$, and has the unique minimizer in the region $c_m^2 > 0$ when $y_m^2 > \sigma^2$. Specifically, the minimizer is given by

$$\widehat{c}_m^2 = \begin{cases} y_m^2 - \sigma^2 & \text{if } y_m^2 > \sigma^2, \\ +0 & \text{otherwise.} \end{cases} \tag{1.66}$$

Figure 1.4 shows the (componentwise) Bayes free energy (1.64) for different observations, $y_m^2 = 0, \sigma^2, 1.5\sigma^2, 2\sigma^2$. The minimizer is in the positive region of $c_m^2$ if and only if $y_m^2 > \sigma^2$.

If the EBayes estimator is given by $\widehat{c}_m^2 \to +0$, it means that the *prior* distribution for the $m$th component $a_m$ of the regression parameter is the *Dirac delta function* located at the origin.[5] This formally means that we *a priori*

[5] When $y_m^2 \leq \sigma^2$, the Bayes free energy (1.64) decreases as $c_m^2$ approaches to 0. However, the domain of $c_m^2$ is restricted to be positive, and therefore, $\widehat{c}_m^2 = 0$ is not the solution. We express this solution as $\widehat{c}_m^2 \to +0$.

*knew* that $a_m = 0$, i.e., we choose a model that does not contain the *m*th component.

By substituting Eq. (1.66) into the Bayes posterior mean (1.51), we obtain the EBayes estimator:

$$\widehat{a}_m^{\mathrm{EBayes}} = \widehat{c}_m^2 \left(\widehat{c}_m^2 + \sigma^2\right)^{-1} y_m$$
$$= \begin{cases} \left(1 - \frac{\sigma^2}{y_m^2}\right) y_m & \text{if } y_m^2 > \sigma^2, \\ 0 & \text{otherwise.} \end{cases} \quad (1.67)$$

The form of the estimator (1.67) is called the *James–Stein (JS) estimator* having interesting properties including the *domination* over the ML estimator (Stein, 1956; James and Stein, 1961; Efron and Morris, 1973) (see Appendix A).

Note that the assumption that $\boldsymbol{X} = \boldsymbol{I}_M$ is not practical. For a general design matrix $\boldsymbol{X}$, the Bayes free energy is not decomposable into each component. Consequently, the prior variances $\{c_m^2\}_{m=1}^M$ that minimize the Bayes free energy (1.61) interact with each other. Therefore, the preceding simple mechanism is not applied. However, it is empirically observed that many prior variances tend to go to $\widehat{c}_m^2 \to +0$, so that the EBayes estimator $\widehat{\boldsymbol{a}}^{\mathrm{EBayes}}$ is sparse.

# 2
# Variational Bayesian Learning

In Chapter 1, we saw examples where the model likelihood has a conjugate prior, with which Bayesian learning can be performed analytically. However, many practical models do not have conjugate priors. Even in such cases, the notion of conjugacy is still useful. Specifically, we can make use of the *conditional conjugacy*, which comes from the fact that many practical models are built by combining basic distributions. In this chapter, we introduce *variational Bayesian (VB) learning*, which makes use of the conditional conjugacy, and approximates the Bayes posterior by solving a constrained minimization problem.

## 2.1 Framework

VB learning is derived by casting Bayesian learning as an optimization problem with respect to the posterior distribution (Hinton and van Camp, 1993; MacKay, 1995; Opper and Winther, 1996; Attias, 1999; Jordan et al., 1999; Jaakkola and Jordan, 2000; Ghahramani and Beal, 2001; Bishop, 2006; Wainwright and Jordan, 2008).

### 2.1.1 Free Energy Minimization

Let $r(\boldsymbol{w})$, or $r$ for short, be an arbitrary distribution, which we call a *trial distribution*, on the parameter $\boldsymbol{w}$, and consider the *Kullback–Leibler (KL) divergence* from the trial distribution $r(\boldsymbol{w})$ to the Bayes posterior $p(\boldsymbol{w}|\mathcal{D})$:

$$\mathrm{KL}\left(r(\boldsymbol{w})\|p(\boldsymbol{w}|\mathcal{D})\right) = \int r(\boldsymbol{w}) \log \frac{r(\boldsymbol{w})}{p(\boldsymbol{w}|\mathcal{D})} d\boldsymbol{w} = \left\langle \log \frac{r(\boldsymbol{w})}{p(\boldsymbol{w}|\mathcal{D})} \right\rangle_{r(\boldsymbol{w})}. \tag{2.1}$$

Since the KL divergence is equal to zero if and only if the two distributions coincide with each other, the minimizer of Eq. (2.1) is the Bayes posterior, i.e.,

$$p(\boldsymbol{w}|\mathcal{D}) = \underset{r}{\operatorname{argmin}} \, \mathrm{KL}\left(r(\boldsymbol{w})\|p(\boldsymbol{w}|\mathcal{D})\right). \tag{2.2}$$

The problem (2.2) is equivalent to the following problem:

$$p(\boldsymbol{w}|\mathcal{D}) = \underset{r}{\operatorname{argmin}} \, F(r), \tag{2.3}$$

where the functional of $r$,

$$F(r) = \int r(\boldsymbol{w}) \log \frac{r(\boldsymbol{w})}{p(\boldsymbol{w}, \mathcal{D})} d\boldsymbol{w} = \left\langle \log \frac{r(\boldsymbol{w})}{p(\boldsymbol{w}, \mathcal{D})} \right\rangle_{r(\boldsymbol{w})} \tag{2.4}$$

$$= \mathrm{KL}\left(r(\boldsymbol{w})\|p(\boldsymbol{w}|\mathcal{D})\right) - \log p(\mathcal{D}), \tag{2.5}$$

is called the *free energy*. Intuitively, we replaced the posterior distribution $p(\boldsymbol{w}|\mathcal{D})$ in the KL divergence (2.1) with its unnormalized version—the joint distribution $p(\mathcal{D}, \boldsymbol{w}) = p(\boldsymbol{w}|\mathcal{D})p(\mathcal{D})$—in the free energy (2.4). The equivalence holds because the normalization factor $p(\mathcal{D})$ does not depend on $\boldsymbol{w}$, and therefore $\langle \log p(\mathcal{D}) \rangle_{r(\boldsymbol{w})} = \log p(\mathcal{D})$ does not depend on $r$. Note that the free energy (2.4) is a generalization of the Bayes free energy, defined by Eq. (1.60): The free energy (2.4) is a functional of an arbitrary distribution $r$, and equal to the Bayes free energy (1.60) for the Bayes poterior $r(\boldsymbol{w}) = p(\boldsymbol{w}|\mathcal{D})$. Since the KL divergence is nonnegative, Eq. (2.5) implies that the free energy $F(r)$ is an upper-bound of the Bayes free energy $-\log p(\mathcal{D})$ for any distribution $r$. Since the log marginal likelihood $\log p(\mathcal{D})$ is called the evidence, $-F(r)$ is also called the *evidence lower-bound (ELBO)*.

As mentioned in Section 1.1.2, the joint distribution is easy to compute in general. However, the minimization problem in Eq. (2.3) can still be computationally intractable, because the objective functional (2.4) involves the *expectation* over the distribution $r(\boldsymbol{w})$. Actually, it can be hard to even evaluate the objective functional for most of the possible distributions. To make the evaluation of the objective functional tractable *for optimal* $r(\boldsymbol{w})$, we restrict the search space to $\mathcal{G}$. Namely, we solve the following problem:

$$\min_{r} F(r) \quad \text{s.t.} \quad r \in \mathcal{G}, \tag{2.6}$$

where s.t. is an abbreviation for "subject to."

We can choose a tractable distribution class directly for $\mathcal{G}$, e.g., Gaussian, such that the expectation for evaluating the free energy is tractable for any $r \in \mathcal{G}$. However, in many practical models, a weaker constraint restricts the *optimal distribution* to be in a tractable class, thanks to *conditional conjugacy*.

### 2.1.2 Conditional Conjugacy

Let us consider a few examples where the model likelihood has no conjugate prior. The likelihood of the matrix factorization model (which will be discussed in detail in Section 3.1) is given by

$$p(V|A, B) = \frac{\exp\left(-\frac{1}{2\sigma^2}\left\|V - BA^\top\right\|_{\mathrm{Fro}}^2\right)}{(2\pi\sigma^2)^{LM/2}}, \tag{2.7}$$

where $V \in \mathbb{R}^{L\times M}$ is an observed random variable, and $A \in \mathbb{R}^{M\times H}$ and $B \in \mathbb{R}^{L\times H}$ are the parameters to be estimated. Although $\sigma^2 \in \mathbb{R}_{++}$ can also be unknown, let us treat it as a hyperparameter, i.e., a constant when computing the posterior distribution.

If we see Eq. (2.7) as a function of the parameters $w = (A, B)$, its function form is the exponential of a polynomial including a fourth-order term $\left\|BA^\top\right\|_{\mathrm{Fro}}^2 = \mathrm{tr}(BA^\top AB^\top)$. Therefore, no conjugate prior exists for this likelihood with respect to the parameters $w = (A, B)$.[1]

The next example is a mixture of Gaussians (which will be discussed in detail in Section 4.1.1):

$$p(\mathcal{D}, \mathcal{H}|w) = \prod_{n=1}^{N}\prod_{k=1}^{K}\left\{\alpha_k \frac{\exp\left(-\frac{\|x^{(n)}-\mu_k\|^2}{2\sigma^2}\right)}{(2\pi\sigma^2)^{M/2}}\right\}^{z_k^{(n)}}, \tag{2.8}$$

where $\mathcal{D} = \{x^{(n)}\}_{n=1}^N$ are observed data, $\mathcal{H} = \{z^{(n)}\}_{n=1}^N$ are hidden variables, and $w = (\alpha, \{\mu_k\}_{k=1}^K)$ are parameters. For simplicity, we here assume that all Gaussian components have the same variance $\sigma^2$, which is treated as a hyperparameter, i.e., we compute the joint posterior distribution of the hidden variables $\{z^{(n)}\}_{n=1}^N$ and the parameters $w$, regarding the hyperparameter $\sigma^2$ as a constant.

If we see Eq. (2.8) as a function of $(\{z^{(n)}\}_{n=1}^N, \alpha, \{\mu_k\}_{k=1}^K)$, no conjugate prior exists. More specifically, it has a factor $\prod_{n=1}^N \prod_{k=1}^K \alpha_k^{z_k^{(n)}}$, and we cannot compute

$$\sum_{z_k^{(n)} \in \{e_k\}_{k=1}^K} \int \prod_{n=1}^{N}\prod_{k=1}^{K} \alpha_k^{z_k^{(n)}} d\alpha_k$$

analytically for general $N$, which is required when evaluating moments.

[1] Here, "no conjugate prior" means that there is no *useful* and *nonconditional* conjugate prior, such that the posterior is in the same distribution family with computable moments. We might say that the exponential function of fourth-order polynomials is conjugate to the likelihood (2.7), since the posterior is within the same family. However, this statement is useless in practice because we cannot compute moments of the distribution analytically.

The same difficulty happens in the latent Dirichlet allocation model (which will be discussed in detail in Section 4.2.4). The likelihood is written as

$$p(\mathcal{D}, \mathcal{H}|\boldsymbol{w}) = \prod_{m=1}^{M} \prod_{n=1}^{N^{(m)}} \prod_{h=1}^{H} \left\{ \Theta_{m,h} \prod_{l=1}^{L} B_{l,h}{}^{w_l^{(n,m)}} \right\}^{z_h^{(n,m)}}, \tag{2.9}$$

where $\mathcal{D} = \{\{\boldsymbol{w}^{(n,m)}\}_{n=1}^{N^{(m)}}\}_{m=1}^{M}$ are observed data, $\mathcal{H} = \{\{\boldsymbol{z}^{(n,m)}\}_{n=1}^{N^{(m)}}\}_{m=1}^{M}$ are hidden variables, and $\boldsymbol{w} = (\boldsymbol{\Theta}, \boldsymbol{B})$ are parameters to be estimated. Computing the sum (over the hidden variables $\mathcal{H}$) of the integral (over the parameters $\boldsymbol{w}$) is intractable for practical problem sizes.

Readers might find that Eqs. (2.7), (2.8), and (2.9) are not much more complicated than the conjugate cases: Eq. (2.7) is similar to the Gaussian form, and Eqs. (2.8) and (2.9) are in the form of the multinomial or Dirichlet distribution, where we have unknowns both in the base and in the exponent. Indeed, they are in a known form if we regard a part of unknowns as fixed constants.

The likelihood (2.7) of the matrix factorization model is in the Gaussian form of $\boldsymbol{A}$ if we see $\boldsymbol{B}$ as a constant, or vice versa. The likelihood (2.8) of a mixture of Gaussians is in the multinomial form of the hidden variables $\mathcal{H} = \{\boldsymbol{z}^{(n)}\}_{n=1}^{N}$ if we see the parameters $\boldsymbol{w} = (\boldsymbol{\alpha}, \{\boldsymbol{\mu}_k\}_{k=1}^{K})$ as constants, and it is the (independent) product of the Dirichlet form of $\boldsymbol{\alpha}$ and the Gaussian form of $\{\boldsymbol{\mu}_k\}_{k=1}^{K}$ if we see the hidden variables $\mathcal{H} = \{\boldsymbol{z}^{(n)}\}_{n=1}^{N}$ as constants. Similarly, the likelihood (2.9) of the latent Dirichlet allocation model is in the multinomial form of the hidden variables $\mathcal{H} = \{\{\boldsymbol{z}^{(n,m)}\}_{n=1}^{N^{(m)}}\}_{m=1}^{M}$ if we see the parameters $\boldsymbol{w} = (\boldsymbol{\Theta}, \boldsymbol{B})$ as constants, and it is the product of the Dirichlet form of the row vectors $\{\widetilde{\boldsymbol{\theta}}_m\}_{m=1}^{M}$ of $\boldsymbol{\Theta}$ and the Dirichlet form of the column vectors $\{\boldsymbol{\beta}_h\}_{h=1}^{H}$ of $\boldsymbol{B}$ if we see the hidden variables $\mathcal{H} = \{\{\boldsymbol{z}^{(n,m)}\}_{n=1}^{N^{(m)}}\}_{m=1}^{M}$ as constants.

Since the likelihoods in the Gaussian, multinomial, and Dirichlet forms have conjugate priors, the aforementioned properties can be described with the notion of *conditional conjugacy*, which is defined as follows:

**Definition 2.1** (Conditionally conjugate prior) Let us divide the unknown parameters $\boldsymbol{w}$ (or more generally all unknown variables including hidden variables) into two parts $\boldsymbol{w} = (\boldsymbol{w}_1, \boldsymbol{w}_2)$. If the posterior of $\boldsymbol{w}_1$,

$$p(\boldsymbol{w}_1|\boldsymbol{w}_2, \mathcal{D}) \propto p(\mathcal{D}|\boldsymbol{w}_1, \boldsymbol{w}_2)p(\boldsymbol{w}_1), \tag{2.10}$$

is in the same distribution family as the prior $p(\boldsymbol{w}_1)$ (where $\boldsymbol{w}_2$ is regarded as a given constant or condition), the prior $p(\boldsymbol{w}_1)$ is called a *conditionally conjugate prior* of the model likelihood $p(\mathcal{D}|\boldsymbol{w})$ with respect to the parameter $\boldsymbol{w}_1$, given the fixed parameter $\boldsymbol{w}_2$.

### 2.1.3 Constraint Design

Once conditional conjugacy for all unknowns is found, designing tractable VB learning is straightforward.

Let us divide the unknown parameters $\boldsymbol{w}$ into $S$ groups, i.e., $\boldsymbol{w} = (\boldsymbol{w}_1, \ldots, \boldsymbol{w}_S)$, such that, for each $s = 1, \ldots, S$, the model likelihood $p(\mathcal{D}|\boldsymbol{w}) = p(\mathcal{D}|\boldsymbol{w}_s, \{\boldsymbol{w}_{s'}\}_{s' \neq s})$ has a conditionally conjugate prior $p(\boldsymbol{w}_s)$ with respect to $\boldsymbol{w}_s$, given $\{\boldsymbol{w}_{s'}\}_{s' \neq s}$ as fixed constants. Then, if we use the prior

$$p(\boldsymbol{w}) = \prod_{s=1}^{S} p(\boldsymbol{w}_s), \tag{2.11}$$

the posterior distribution

$$p(\boldsymbol{w}|\mathcal{D}) \propto p(\mathcal{D}|\boldsymbol{w})p(\boldsymbol{w})$$

is, *as a function of* $\boldsymbol{w}_s$, in the same distribution family as the prior $p(\boldsymbol{w}_s)$. Therefore, moments of the posterior distribution are tractable, if the other parameters $\{\boldsymbol{w}_{s'}\}_{s' \neq s}$ are given.

To make use of this property, we impose on the approximate posterior the independence constraint between the parameter groups,

$$r(\boldsymbol{w}) = \prod_{s=1}^{S} r_s(\boldsymbol{w}_s), \tag{2.12}$$

which allows us to compute moments with respect to $\boldsymbol{w}_s$ independently from the other parameters $\{\boldsymbol{w}_{s'}\}_{s' \neq s}$. In VB learning, we solve the minimization problem (2.6) under the constraint (2.12). This makes the expectation computation, which is required in evaluating the free energy (2.4), tractable (on any stationary points for $r$). Namely, we define the *VB posterior* as

$$\widehat{r} = \underset{r}{\mathrm{argmin}} F(r) \quad \text{s.t.} \quad r(\boldsymbol{w}) = \prod_{s=1}^{S} r_s(\boldsymbol{w}_s). \tag{2.13}$$

Note that it is not guaranteed that the free energy $F(r) = \left\langle \log \frac{r(\boldsymbol{w})}{p(\mathcal{D}|\boldsymbol{w})p(\boldsymbol{w})} \right\rangle_{r(\boldsymbol{w})}$ is tractable for any $r$ satisfying the constraint (2.12). However, the constraint allows us to optimize each factor $\{r_s\}_{s=1}^{S}$ separately. To optimize each factor, we rely on *calculus of variations*, which will be explained in Section 2.1.4. By applying calculus of variations, the free energy is expressed as an explicit function with a *finite* number of unknown parameters.

### 2.1.4 Calculus of Variations

*Calculus of variations* is a method, developed in physics, to derive conditions that any optimal function minimizing a (smooth) functional should satisfy (Courant and Hilbert, 1953). Specifically, it gives (infinitely many) stationary conditions of the functional with respect to the variable.

The change of the functional $F(r)$ with respect to an *infinitesimal change* of the variable $r$ (which is a function of $\boldsymbol{w}$) is called a *variation* and written as $\delta I$. For $r$ to be a stationary point of the functional, the variation must be equal to zero for all possible values of $\boldsymbol{w}$. Since the free energy (2.4) does not depend on the derivatives of $r(\boldsymbol{w})$, the variation $\delta I$ is simply the derivative with respect to $r$. Therefore, the stationary conditions are given by

$$\delta I = \frac{\partial F}{\partial r} = 0, \quad \forall \boldsymbol{w} \in \mathcal{W}, \tag{2.14}$$

which is a special case of the *Euler–Lagrange equation*. If we see the function $r(\boldsymbol{w})$ as a (possibly) infinite-dimensional vector with the parameter value $\boldsymbol{w}$ as its index, the variation $\delta I = \delta I(\boldsymbol{w})$ can be interpreted as the gradient of the functional $F(r)$ in the $|\mathcal{W}|$-dimensional space. As the stationary conditions in a finite-dimensional space require that all entries of the gradient equal to zero, the optimal function $r(\boldsymbol{w})$ should satisfy Eq. (2.14) for any parameter values $\boldsymbol{w} \in \mathcal{W}$.

In Section 2.1.5, we see that, by applying the stationary conditions (2.14) to the free energy minimization problem (2.13) with the independence constraint taken into account, we can find that each factor $r_s(\boldsymbol{w}_s)$ of the approximate posterior is in the same distribution family as the corresponding prior $p_s(\boldsymbol{w}_s)$, thanks to the conditional conjugacy.

### 2.1.5 Variational Bayesian Learning

Let us solve the problem (2.13) to get the VB posterior

$$\widehat{r} = \underset{r}{\operatorname{argmin}} F(r) \quad \text{s.t.} \quad r(\boldsymbol{w}) = \prod_{s=1}^{S} r_s(\boldsymbol{w}_s).$$

We use the decomposable conditionally conjugate prior (2.11):

$$p(\boldsymbol{w}) = \prod_{s=1}^{S} p(\boldsymbol{w}_s),$$

which means that, for each $s = 1, \ldots, S$, the posterior $p(\boldsymbol{w}_s | \{\boldsymbol{w}_{s'}\}_{s' \neq s}, \mathcal{D})$ for $\boldsymbol{w}_s$ is in the same form as the corresponding prior $p(\boldsymbol{w}_s)$, given $\{\boldsymbol{w}_{s'}\}_{s' \neq s}$ as fixed constants.

Now we apply the calculus of variations, and compute the stationary conditions (2.14). The free energy can be written as

$$F(r) = \int \left( \prod_{s=1}^{S} r_s(\boldsymbol{w}_s) \right) \left( \log \frac{\prod_{s=1}^{S} r_s(\boldsymbol{w}_s)}{p(\mathcal{D}|\boldsymbol{w}) \prod_{s=1}^{S} p(\boldsymbol{w}_s)} \right) d\boldsymbol{w}. \tag{2.15}$$

Taking the derivative of Eq. (2.15) with respect to $r_s(\boldsymbol{w}_s)$ for any $s = 1, \ldots, S$ and $\boldsymbol{w}_s \in \mathcal{W}$, we obtain the following stationary conditions:

$$\begin{aligned}
0 = \frac{\partial F}{\partial r_s} &= \int \left( \prod_{s' \neq s} r_{s'}(\boldsymbol{w}_{s'}) \right) \left( \log \frac{\prod_{s'=1}^{S} r_{s'}(\boldsymbol{w}_{s'})}{p(\mathcal{D}|\boldsymbol{w}) \prod_{s'=1}^{S} p(\boldsymbol{w}_{s'})} + 1 \right) d\boldsymbol{w} \\
&= \left\langle \log \frac{\prod_{s'=1}^{S} r_{s'}(\boldsymbol{w}_{s'})}{p(\mathcal{D}|\boldsymbol{w}) \prod_{s'=1}^{S} p(\boldsymbol{w}_{s'})} \right\rangle_{\prod_{s' \neq s} r_{s'}(\boldsymbol{w}_{s'})} + 1 \\
&= \left\langle \log \frac{\prod_{s' \neq s} r_{s'}(\boldsymbol{w}_{s'})}{p(\mathcal{D}|\boldsymbol{w}) \prod_{s' \neq s} p(\boldsymbol{w}_{s'})} \right\rangle_{\prod_{s' \neq s} r_{s'}(\boldsymbol{w}_{s'})} + \log \frac{r_s(\boldsymbol{w}_s)}{p(\boldsymbol{w}_s)} + 1 \\
&= \left\langle \log \frac{1}{p(\mathcal{D}|\boldsymbol{w})} \right\rangle_{\prod_{s' \neq s} r_{s'}(\boldsymbol{w}_{s'})} + \log \frac{r_s(\boldsymbol{w}_s)}{p(\boldsymbol{w}_s)} + \text{const.}
\end{aligned} \tag{2.16}$$

Note the following on Eq. (2.16):

- The right-hand side is a function of $\boldsymbol{w}_s$ ($\boldsymbol{w}_{s'}$ for $s' \neq s$ are integrated out).
- For each $s$, Eq. (2.16) must hold for any possible value of $\boldsymbol{w}_s$, which can fully specify the function form of the posterior $r_s(\boldsymbol{w}_s)$.
- To make Eq. (2.16) satisfied for any $\boldsymbol{w}_s$, it is necessary that

  $$-\langle \log p(\mathcal{D}|\boldsymbol{w}) \rangle_{\prod_{s' \neq s} r_{s'}(\boldsymbol{w}_{s'})} + \log \frac{r_s(\boldsymbol{w}_s)}{p(\boldsymbol{w}_s)}$$

  is a constant.

The last note leads to the following relation:

$$r_s(\boldsymbol{w}_s) \propto p(\boldsymbol{w}_s) \exp \langle \log p(\mathcal{D}|\boldsymbol{w}) \rangle_{\prod_{s' \neq s} r_{s'}(\boldsymbol{w}_{s'})}. \tag{2.17}$$

As a function of $\boldsymbol{w}_s$, Eq. (2.17) can be written as

$$\begin{aligned}
r_s(\boldsymbol{w}_s) &\propto \exp \langle \log p(\mathcal{D}|\boldsymbol{w}) p(\boldsymbol{w}_s) \rangle_{\prod_{s' \neq s} r_{s'}(\boldsymbol{w}_{s'})} \\
&\propto \exp \langle \log p(\boldsymbol{w}_s | \{\boldsymbol{w}_{s'}\}_{s' \neq s}, \mathcal{D}) \rangle_{\prod_{s' \neq s} r_{s'}(\boldsymbol{w}_{s'})} \\
&\propto \exp \int \log p(\boldsymbol{w}_s | \{\boldsymbol{w}_{s'}\}_{s' \neq s}, \mathcal{D}) \prod_{s' \neq s} r_{s'}(\boldsymbol{w}_{s'}) d\boldsymbol{w}_{s'}.
\end{aligned} \tag{2.18}$$

Due to the conditional conjugacy, $p(\boldsymbol{w}_s | \{\boldsymbol{w}_{s'}\}_{s' \neq s}, \mathcal{D})$ is in the same form as the prior $p(\boldsymbol{w}_s)$. As the intergral operator $g(x) = \int f(x; \alpha) d\alpha$ can be interpreted as an infinite number of additions of parametric functions $f(x; \alpha)$ over all possible values of $\alpha$, the operator $h(x) = \exp \int \log f(x; \alpha) d\alpha$ corresponds to an infinite number of multiplications of $f(x;\alpha)$ over all possible values of $\alpha$. Therefore, Eq. (2.18) implies that the VB posterior $r_s(\boldsymbol{w}_s)$ is in the same form as the prior $p(\boldsymbol{w}_s)$, if the distribution family is multiplicatively closed.

Assume that the prior $p(\boldsymbol{w}_s)$ for each group of parameters is in a multiplicatively closed distribution family. Then, we may express the corresponding VB posterior $r_s(\boldsymbol{w}_s)$ in a parametric form, of which the parameters are called *variational parameters*, without any loss of accuracy or optimality. The last question is whether we can compute the expectation value of the log-likelihood $\log p(\mathcal{D}|\boldsymbol{w})$ for each factor $r_s(\boldsymbol{w}_s)$ of the approximate posterior. In many cases, this expectation can be computed analytically, which allows us to express the stationary conditions (2.17) as a finite number of equations in explicit forms of the variational parameters.

Typically, the obtained stationary conditions are used to update the variational parameters in an iterative algorithm, which gives a local minimizer $\widehat{r}$ of the free energy (2.4). We call the minimizer $\widehat{r}$ the *VB posterior*, and its mean

$$\widehat{\boldsymbol{w}} = \langle \boldsymbol{w} \rangle_{\widehat{r}(\boldsymbol{w})} \tag{2.19}$$

the *VB estimator*.

The computation of predictive distribution

$$p(\mathcal{D}^{\mathrm{new}}|\mathcal{D}) = \langle p(\mathcal{D}^{\mathrm{new}}|\boldsymbol{w}) \rangle_{\widehat{r}(\boldsymbol{w})}$$

can be hard even after finding the VB posterior $\widehat{r}(\boldsymbol{w})$. This is natural because we need approximation for the function form of the likelihood $p(\mathcal{D}|\boldsymbol{w})$, and now we need to compute the integral with the integrand involving the same function form. In many practical cases, the *plug-in predictive distribution* $p(\mathcal{D}^{\mathrm{new}}|\widehat{\boldsymbol{w}})$, i.e., the model distribution with the VB estimator plugged in, is substituted for the predictive distribution.

### 2.1.6 Empirical Variational Bayesian Learning

When the model involves hyperparameters $\boldsymbol{\kappa}$ in the likelihood and/or in the prior, the joint distribution is dependent on $\boldsymbol{\kappa}$, i.e.,

$$p(\mathcal{D}, \boldsymbol{w}|\boldsymbol{\kappa}) = p(\boldsymbol{w}|\boldsymbol{\kappa})p(\mathcal{D}|\boldsymbol{w}, \boldsymbol{\kappa}),$$

and so is the free energy:

$$\begin{aligned} F(r, \boldsymbol{\kappa}) &= \int r(\boldsymbol{w}) \log \frac{r(\boldsymbol{w})}{p(\mathcal{D}, \boldsymbol{w}|\boldsymbol{\kappa})} d\boldsymbol{w} \\ &= \left\langle \log \frac{r(\boldsymbol{w})}{p(\boldsymbol{w}, \mathcal{D}|\boldsymbol{\kappa})} \right\rangle_{r(\boldsymbol{w})} \qquad (2.20) \\ &= \mathrm{KL}\left(r(\boldsymbol{w}) \| p(\boldsymbol{w}|\mathcal{D}, \boldsymbol{\kappa})\right) - \log p(\mathcal{D}|\boldsymbol{\kappa}). \qquad (2.21) \end{aligned}$$

Similarly to the empirical Bayesian learning, the hyperparameters can be estimated from observation by minimizing the free energy simultaneously with respect to $r$ and $\boldsymbol{\kappa}$:

$$(\widehat{r}, \widehat{\boldsymbol{\kappa}}) = \underset{r, \boldsymbol{\kappa}}{\operatorname{argmin}} F(r, \boldsymbol{\kappa}).$$

This approach is called the *empirical VB (EVB) learning.*

EVB learning amounts to minimizing the sum of the KL divergence to the Bayes posterior and the marginal likelihood (see Eq. (2.21)). Conceptually, minimizing any weighted sum of those two terms is reasonable to find the VB posterior and the hyperparameters at the same time. But only the unweighted sum makes the objective tractable—under this choice, the objective is written with the joint distribution as in Eq. (2.20), while any other choice requires explicitly accessing the Bayes posterior and the marginal likelihood separately.

### 2.1.7 Techniques for Nonconjugate Models

In Sections 2.1.2 through 2.1.5, we saw how to design tractable VB learning by making use of the conditional conjugacy. However, there are also many cases where a reasonable model does not have a conditionally conjugate prior. A frequent and important example is the case where the likelihood involves the *sigmoid function*,

$$\sigma(\boldsymbol{x}; \boldsymbol{w}) = \frac{1}{1 + e^{-\boldsymbol{w}^\top \boldsymbol{x}}}, \tag{2.22}$$

or a function with a similar shape, e.g., the *error function*, the hyperbolic tangent, and the *rectified linear unit (ReLU)*. We face such cases, for example, in solving *classification* problems and in adopting *neural network* structure with a nonlinear *activation function*.

To maintain the tractability in such cases, we need to explicitly restrict the function form of the approximate posterior $r(\boldsymbol{w}; \widehat{\boldsymbol{\lambda}})$, and optimize its variational parameters $\widehat{\boldsymbol{\lambda}}$ by free energy minimization. Namely, we solve the VB learning problem (2.6) with the search space $\mathcal{G}$ set to the function space of a simple distribution family, e.g., the Gaussian distribution $r(\boldsymbol{w}; \widehat{\boldsymbol{\lambda}}) = \mathrm{Gauss}_D(\boldsymbol{w}; \widehat{\boldsymbol{w}}, \widehat{\boldsymbol{\Sigma}})$ parameterized with the variational parameters $\widehat{\boldsymbol{\lambda}} = (\widehat{\boldsymbol{w}}, \widehat{\boldsymbol{\Sigma}})$ consisting of the mean and the covariance parameters. Then, the VB learning problem (2.6) is reduced to the following unconstrained minimization problem,

$$\min_{\widehat{\boldsymbol{\lambda}}} F(\widehat{\boldsymbol{\lambda}}), \tag{2.23}$$

of the free energy

$$F(\widehat{\lambda}) = \int r(w;\widehat{\lambda}) \log \frac{r(w;\widehat{\lambda})}{p(w,\mathcal{D})} dw = \left\langle \log \frac{r(w;\widehat{\lambda})}{p(w,\mathcal{D})} \right\rangle_{r(w;\widehat{\lambda})}, \tag{2.24}$$

which is a function of the variational parameters $\widehat{\lambda}$.

It is often the case that the free energy (2.24) is still intractable in computing the expectation value of the log joint probability, $\log p(w,\mathcal{D}) = \log p(\mathcal{D}|w)p(w)$, over the approximate posterior $r(w;\widehat{\lambda})$ (because of the intractable function form of the likelihood $p(\mathcal{D}|w)$ or the prior $p(w)$). In this section, we introduce a few techniques developed for coping with such intractable functions.

**Local Variational Approximation**

The first method is to bound the joint distribution $p(w,\mathcal{D})$ with a simple function, of which the expectation value over the approximate distribution $r(w;\widehat{\lambda})$ is tractable.

As seen in Section 2.1.1, the free energy (2.24) is an upper-bound of the Bayes free energy, $F^{\mathrm{Bayes}} \equiv -\log p(\mathcal{D})$, for any $\widehat{\lambda}$. Consider further upper-bounding the free energy as

$$F(\widehat{\lambda}) \leq \overline{F}(\widehat{\lambda},\xi) \equiv \int r(w;\widehat{\lambda}) \log \frac{r(w;\widehat{\lambda})}{\underline{p}(w;\xi)} dw \tag{2.25}$$

by replacing the joint distribution $p(w,\mathcal{D})$ with its parametric lower-bound $\underline{p}(w;\xi)$ such that

$$0 \leq \underline{p}(w;\xi) \leq p(w,\mathcal{D}) \tag{2.26}$$

for any $w \in \mathcal{W}$ and $\xi \in \Xi$. Here, we introduced another set of variational parameters $\xi$ with its domain $\Xi$. Let us choose a lower-bound $\underline{p}(w;\xi)$ such that its function form with respect to $w$ is the same as the approximate posterior $r(w;\widehat{\lambda})$. More specifically, we assume that, for any given $\xi$, there exists $\widehat{\lambda}$ such that

$$\underline{p}(w;\xi) \propto r(w;\widehat{\lambda}) \tag{2.27}$$

as a function of $w$.[2] Since the direct minimization of $F(\widehat{\lambda})$ is intractable, we instead minimize its upper-bound $\overline{F}(\widehat{\lambda},\xi)$ jointly over $\widehat{\lambda}$ and $\xi$. Namely, we solve the problem

$$\min_{\widehat{\lambda},\xi} \overline{F}(\widehat{\lambda},\xi), \tag{2.28}$$

[2] The parameterization, i.e., the function form with respect to the variational parameters, can be different between $\underline{p}(w;\xi)$ and $r(w;\widehat{\lambda})$.

to find the approximate posterior $r(w;\widehat{\lambda})$ such that $\overline{F}(\widehat{\lambda},\xi)$ ($\geq F(\widehat{\lambda})$) is closest to the Bayes free energy $F^{\mathrm{Bayes}}$ (when $\xi$ is also optimized).

Let

$$q(w;\xi) = \frac{\underline{p}(w;\xi)}{\underline{Z}(\xi)} \tag{2.29}$$

be the distribution created by normalizing the lower-bound with its normalization factor

$$\underline{Z}(\xi) = \int \underline{p}(w;\xi)dw. \tag{2.30}$$

Note that the normalization factor (2.30) is trivially a lower-bound of the marginal likelihood, i.e.,

$$\underline{Z}(\xi) \leq \int p(w, \mathcal{D})dw = p(\mathcal{D}),$$

and is tractable because of the assumption (2.27) that $\underline{p}$ is in the same simple function form as $r$.

With Eq. (2.29), the upper-bound (2.25) is expressed as

$$\begin{aligned}\overline{F}(\widehat{\lambda},\xi) &= \int r(w;\widehat{\lambda}) \log \frac{r(w;\widehat{\lambda})}{q(w;\xi)} dw - \log \underline{Z}(\xi)\\ &= \mathrm{KL}\left(r(w;\widehat{\lambda}) \| q(w;\xi)\right) - \log \underline{Z}(\xi),\end{aligned} \tag{2.31}$$

which implies that the optimal $\widehat{\lambda}$ is attained when

$$r(w;\widehat{\lambda}) = q(w;\xi) \tag{2.32}$$

for any $\xi \in \Xi$ (the assumption (2.27) guarantees the attainability). Thus, by putting this back into Eq. (2.31), the problem (2.28) is reduced to

$$\max_{\xi} \underline{Z}(\xi), \tag{2.33}$$

which amounts to maximizing the lower-bound (2.30) of the marginal likelihood $p(\mathcal{D})$. Once the maximizer $\widehat{\xi}$ is obtained, Eq. (2.32) gives the optimal approximate posterior.

Such an approximation scheme for nonconjugate models is called *local variational approximation* or *direct site bounding* (Jaakkola and Jordan, 2000; Girolami, 2001; Bishop, 2006; Seeger, 2008, 2009), which will be discussed further with concrete examples in Chapter 5. Existing nonconjugate models applied with the local variational approximation form the bound in Eq. (2.26) based on the convexity of a function. In such a case, the gap between $\overline{F}(\xi)$ and $F$ turns out to be the expected Bregman divergence associated with the convex function (see Section 5.3.1). A similar approach can be applied to *expectation propagation*, another approximation method introduced in Section 2.2.3. There, by upper-bounding the joint probability $p(\boldsymbol{w}, \mathcal{D})$, we minimize an upper-bound of $\mathrm{KL}\left(p(\boldsymbol{w}|\mathcal{D}) \| r(\boldsymbol{w}; \widehat{\boldsymbol{\lambda}})\right)$ (see Section 2.2.3).

**Black Box Variational Inference**

As the available data size increases, and the benefit of using big data has been proven, for example, by the breakthrough in deep learning (Krizhevsky et al., 2012), scalable training algorithms have been intensively developed, to enable big data analysis on billions of data samples. The *stochastic gradient descent* (Robbins and Monro, 1951; Spall, 2003), where a noisy gradient of the objective function is cheaply computed from a subset of the whole data in each iteration, has become popular, and has been adopted for VB learning (Hoffman et al., 2013; Khan et al., 2016).

The *black-box variational inference* was proposed as a general method to compute a noisy gradient of the free energy in nonconjugate models (Ranganath et al., 2013; Wingate and Weber, 2013; Kingma and Welling, 2014). As a function of the variational parameters $\widehat{\boldsymbol{\lambda}}$, the gradient of the free energy (2.24) can be evaluated by

$$\begin{aligned}
\frac{\partial F}{\partial \widehat{\boldsymbol{\lambda}}} &= \frac{\partial}{\partial \widehat{\boldsymbol{\lambda}}} \int r(\boldsymbol{w}; \widehat{\boldsymbol{\lambda}}) \log \frac{r(\boldsymbol{w}; \widehat{\boldsymbol{\lambda}})}{p(\mathcal{D}, \boldsymbol{w})} d\boldsymbol{w} \\
&= \int \frac{\partial r(\boldsymbol{w}; \widehat{\boldsymbol{\lambda}})}{\partial \widehat{\boldsymbol{\lambda}}} \log \frac{r(\boldsymbol{w}; \widehat{\boldsymbol{\lambda}})}{p(\mathcal{D}, \boldsymbol{w})} d\boldsymbol{w} + \int r(\boldsymbol{w}; \widehat{\boldsymbol{\lambda}}) \frac{\partial}{\partial \widehat{\boldsymbol{\lambda}}} \left( \log r(\boldsymbol{w}; \widehat{\boldsymbol{\lambda}}) \right) d\boldsymbol{w} \\
&= \int r(\boldsymbol{w}; \widehat{\boldsymbol{\lambda}}) \frac{\partial \log r(\boldsymbol{w}; \widehat{\boldsymbol{\lambda}})}{\partial \widehat{\boldsymbol{\lambda}}} \log \frac{r(\boldsymbol{w}; \widehat{\boldsymbol{\lambda}})}{p(\mathcal{D}, \boldsymbol{w})} d\boldsymbol{w} + \frac{\partial}{\partial \widehat{\boldsymbol{\lambda}}} \int r(\boldsymbol{w}; \widehat{\boldsymbol{\lambda}}) d\boldsymbol{w} \\
&= \left\langle \frac{\partial \log r(\boldsymbol{w}; \widehat{\boldsymbol{\lambda}})}{\partial \widehat{\boldsymbol{\lambda}}} \log \frac{r(\boldsymbol{w}; \widehat{\boldsymbol{\lambda}})}{p(\mathcal{D}, \boldsymbol{w})} \right\rangle_{r(\boldsymbol{w}; \widehat{\boldsymbol{\lambda}})}.
\end{aligned} \tag{2.34}$$

Assume that we restrict the approximate posterior $r(\boldsymbol{w}; \widehat{\boldsymbol{\lambda}})$ to be in a simple distribution family, from which samples can be easily drawn, and its *score function*, the gradient of the log probability, is easily computed, e.g., an analytic form is available. Then, Eq. (2.34) can be easily computed by drawing samples from $r(\boldsymbol{w}; \widehat{\boldsymbol{\lambda}})$, and computing the sample average. With some variance reduction techniques, the stochastic gradient with the black box gradient estimator (2.34) has shown to be useful for VB learning in general nonconjugate models. A notable advantage is that it does not require any model specific analysis to implement the gradient estimation, since Eq. (2.34) can be evaluated as long as the log joint probability $p(\mathcal{D}, \boldsymbol{w}) = p(\mathcal{D}|\boldsymbol{w})p(\boldsymbol{w})$ of the model can be evaluated for drawn samples of $\boldsymbol{w}$.

## 2.2 Other Approximation Methods

There are several other methods for approximate Bayesian learning, which are briefly introduced in this section.

### 2.2.1 Laplace Approximation

In the *Laplace approximation*, the posterior is approximated by a Gaussian:

$$r(\boldsymbol{w}) = \mathrm{Gauss}_D(\boldsymbol{w}; \widehat{\boldsymbol{w}}, \widehat{\boldsymbol{\Sigma}}).$$

VB learning finds the variational parameters $\widehat{\boldsymbol{\lambda}} = (\widehat{\boldsymbol{w}}, \widehat{\boldsymbol{\Sigma}})$ by minimizing the free energy (2.4), i.e., solving the problem (2.6) with the search space $\mathcal{G}$ restricted to the Gaussian distributions. Instead, the Laplace approximation estimates the mean and the covariance by

$$\widehat{\boldsymbol{w}}^{\mathrm{LA}} (= \widehat{\boldsymbol{w}}^{\mathrm{MAP}}) = \operatorname*{argmax}_{\boldsymbol{w}} p(\mathcal{D}|\boldsymbol{w}) p(\boldsymbol{w}), \tag{2.35}$$

$$\widehat{\boldsymbol{\Sigma}}^{\mathrm{LA}} = \widehat{\boldsymbol{F}}^{-1}, \tag{2.36}$$

where the entries of $\widehat{\boldsymbol{F}} \in \mathbb{S}_{++}^D$ are given by

$$\widehat{F}_{i,j} = - \left. \frac{\partial^2 \log p(\mathcal{D}|\boldsymbol{w}) p(\boldsymbol{w})}{\partial w_i \partial w_j} \right|_{\boldsymbol{w} = \widehat{\boldsymbol{w}}^{\mathrm{LA}}}. \tag{2.37}$$

Namely, the Laplace approximation first finds the MAP estimator for the mean, and then computes Eq.(2.37) at $\boldsymbol{w} = \widehat{\boldsymbol{w}}^{\mathrm{LA}}$ to estimate the inverse covariance, which corresponds to the second-order *Taylor approximation* to $\log p(\mathcal{D}|\boldsymbol{w}) p(\boldsymbol{w})$. Note that, for the flat prior $p(\boldsymbol{w}) \propto 1$, Eq. (2.37) is reduced to the *Fisher information*:

$$F_{i,j} = \left\langle \frac{\partial \log p(\mathcal{D}|\boldsymbol{w})}{\partial w_i} \frac{\partial \log p(\mathcal{D}|\boldsymbol{w})}{\partial w_j} \right\rangle_{p(\mathcal{D}|\boldsymbol{w})} = - \left\langle \frac{\partial^2 \log p(\mathcal{D}|\boldsymbol{w})}{\partial w_i \partial w_j} \right\rangle_{p(\mathcal{D}|\boldsymbol{w})}.$$

In general, the Laplace approximation is computationally less demanding than VB learning, since no integral computation is involved, and the inverse covariance estimation (2.36) is performed only once after the MAP mean estimator (2.35) is found.

### 2.2.2 Partially Bayesian Learning

*Partially Bayesian (PB) learning* is MAP learning after some of the unknown parameters are integrated out. This approach can be described in the free energy minimization framework (2.6) with a strnger constraint than VB learning.

Let us split the unknown parameters $\boldsymbol{w}$ into two parts $\boldsymbol{w} = (\boldsymbol{w}_1, \boldsymbol{w}_2)$, and assume that we integrate $\boldsymbol{w}_1$ out and point-estimate $\boldsymbol{w}_2$. Integrating $\boldsymbol{w}_1$ out means that we consider the exact posterior on $\boldsymbol{w}_1$, and MAP estimating $\boldsymbol{w}_2$ means that we approximate the posterior $\boldsymbol{w}_2$ with the delta function. Namely, PB learning solves the following problem:

$$\min_r F(r) \qquad \text{s.t.} \qquad r(\boldsymbol{w}) = r_1(\boldsymbol{w}_1) \cdot \delta(\boldsymbol{w}_2; \widehat{\boldsymbol{w}}_2), \tag{2.38}$$

where the free energy $F(r)$ is defined by Eq. (2.4), and $\delta(\boldsymbol{w}; \widehat{\boldsymbol{w}})$ is the *Dirac delta function* located at $\widehat{\boldsymbol{w}}$.

Using the constraint in Eq. (2.38), under which the variables to be optimized are $r_1$ and $\widehat{\boldsymbol{w}}_2$, we can express the free energy as

$$\begin{aligned} F(r_1, \widehat{\boldsymbol{w}}_2) &= \left\langle \log \frac{r_1(\boldsymbol{w}_1) \cdot \delta(\boldsymbol{w}_2; \widehat{\boldsymbol{w}}_2)}{p(\mathcal{D}|\boldsymbol{w}_1, \boldsymbol{w}_2) p(\boldsymbol{w}_1) p(\boldsymbol{w}_2)} \right\rangle_{r_1(\boldsymbol{w}_1) \cdot \delta(\boldsymbol{w}_2; \widehat{\boldsymbol{w}}_2)} \\ &= \left\langle \log \frac{r_1(\boldsymbol{w}_1)}{p(\mathcal{D}|\boldsymbol{w}_1, \widehat{\boldsymbol{w}}_2) p(\boldsymbol{w}_1) p(\widehat{\boldsymbol{w}}_2)} \right\rangle_{r_1(\boldsymbol{w}_1)} + \left\langle \log \delta(\boldsymbol{w}_2; \widehat{\boldsymbol{w}}_2) \right\rangle_{\delta(\boldsymbol{w}_2; \widehat{\boldsymbol{w}}_2)} \\ &= \left\langle \log \frac{r_1(\boldsymbol{w}_1)}{p(\boldsymbol{w}_1|\widehat{\boldsymbol{w}}_2, \mathcal{D})} \right\rangle_{r_1(\boldsymbol{w}_1)} - \log p(\mathcal{D}|\widehat{\boldsymbol{w}}_2) p(\widehat{\boldsymbol{w}}_2) + \left\langle \log \delta(\boldsymbol{w}_2; \widehat{\boldsymbol{w}}_2) \right\rangle_{\delta(\boldsymbol{w}_2; \widehat{\boldsymbol{w}}_2)}, \end{aligned} \tag{2.39}$$

where

$$p(\mathcal{D}|\widehat{\boldsymbol{w}}_2) = \left\langle p(\mathcal{D}|\boldsymbol{w}_1, \widehat{\boldsymbol{w}}_2) \right\rangle_{p(\boldsymbol{w}_1)} = \int p(\mathcal{D}|\boldsymbol{w}_1, \widehat{\boldsymbol{w}}_2) p(\boldsymbol{w}_1) d\boldsymbol{w}_1. \tag{2.40}$$

The free energy (2.39) depends on $r_1$ only through the first term, which is the KL divergence, $\mathrm{KL}\left(r_1(\boldsymbol{w}_1) \| p(\boldsymbol{w}_1|\widehat{\boldsymbol{w}}_2, \mathcal{D})\right)$, from the trial distribution to the Bayes posterior (conditioned on $\widehat{\boldsymbol{w}}_2$). Therefore, the minimizer for $r_1$ is trivially the conditional Bayes posterior

$$r_1(\boldsymbol{w}_1) = p(\boldsymbol{w}_1|\widehat{\boldsymbol{w}}_2, \mathcal{D}), \tag{2.41}$$

with which the first term in Eq. (2.39) vanishes. The third term in Eq. (2.39) is the entropy of the delta function, which diverges to infinity but is independent of $\widehat{\boldsymbol{w}}_2$. By regarding the delta function as a distribution with its width narrow enough to express a point estimate, while its entropy is finite (although it is very large), we can ignore the third term. Thus, the free energy minimization problem (2.38) can be written as

$$\min_{\widehat{\boldsymbol{w}}_2} - \log p(\mathcal{D}|\widehat{\boldsymbol{w}}_2) p(\widehat{\boldsymbol{w}}_2), \tag{2.42}$$

which amounts to MAP learning for $\boldsymbol{w}_2$ after $\boldsymbol{w}_1$ is marginalized out.

This method is computationally beneficial when the likelihood $p(\mathcal{D}|\boldsymbol{w}) = p(\mathcal{D}|\boldsymbol{w}_1, \boldsymbol{w}_2)$ is conditionally conjugate to the prior $p(\boldsymbol{w}_1)$ with respect to $\boldsymbol{w}_1$, given $\boldsymbol{w}_2$. Thanks to the conditional conjugacy, the posterior (2.41) of $\boldsymbol{w}_1$ is in a known form, and its normalization factor (2.40), which is required when evaluating the objective in Eq. (2.42), can be obtained analytically.

PB learning was applied in many previous works. For example, in the *expectation-maximization (EM) algorithm* (Dempster et al., 1977), latent variables are integrated out and parameters are point-estimated. In the first probabilisitic interpretation of principal component analysis (PCA) (Tipping and Bishop, 1999), one factor of the matrix factorization was called a latent variable and integrated out, while the other factor was called a parameter and point-estimated.

The same idea has been adopted for Gibbs sampling and VB learning, where some of the unknown parameters are integrated out based on the conditional conjugacy, and the other parameters are estimated by the corresponding learning method. Those methods are called *collapsed Gibbs sampling* (Griffiths and Steyvers, 2004) and *collapsed VB learning* (Kurihara et al., 2007; Teh et al., 2007; Sato et al., 2012), respectively. Following this terminology, PB learning may be also called *collapsed MAP learning*. The collapsed version is in general more accurate and more computationally efficient than the uncollapsed counterpart, since it imposes a weaker constraint and applies a nonexact numerical estimation to a smaller number of unknowns.

### 2.2.3 Expectation Propagation

As explained in Section 2.1.1, VB learning amounts to minimizing the KL divergence $\mathrm{KL}(r(\boldsymbol{w})\|p(\boldsymbol{w}|\mathcal{D}))$ from the approximate posterior to the Bayes posterior. *Expectation propagation (EP)* is an alternative deterministic approximation scheme, which minimizes the KL divergence *from the Bayes posterior to the approximate posterior* (Minka, 2001b), i.e.,

$$\min_{r} \mathrm{KL}(p(\boldsymbol{w}|\mathcal{D})\|r(\boldsymbol{w})) \qquad \text{s.t.} \qquad r \in \mathcal{G}. \tag{2.43}$$

Clearly from its definition, the KL divergence,

$$\mathrm{KL}(q(\boldsymbol{x})\|p(\boldsymbol{x})) = \int q(\boldsymbol{x}) \log \frac{q(\boldsymbol{x})}{p(\boldsymbol{x})} d\boldsymbol{x},$$

diverges to $+\infty$ if the support of $q(\boldsymbol{x})$ is not covered by the support of $p(\boldsymbol{x})$, while it remains finite if the support of $p(\boldsymbol{x})$ is not covered by the support of $q(\boldsymbol{x})$. Due to this asymmetric property of the KL divergence, VB learning and EP can provide drastically different approximate posteriors—VB learning,

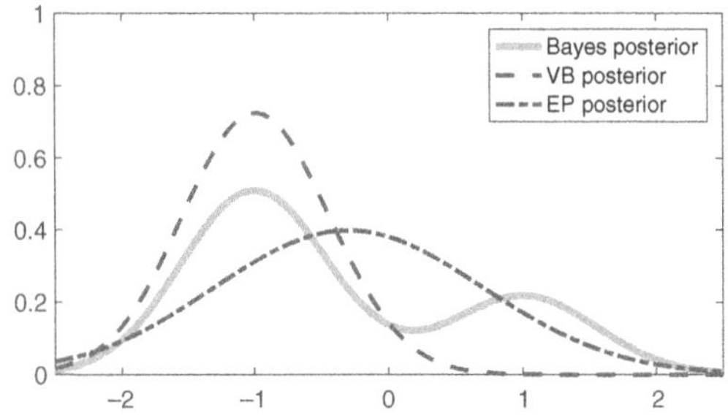


Figure 2.1 Bayes posterior, VB posterior, and EP posterior.

minimizing KL $(r(\boldsymbol{w})\|p(\boldsymbol{w}|\mathcal{D}))$, tends to provide a posterior that approximates a single mode of the Bayes posterior, while EP, minimizing KL $(p(\boldsymbol{w}|\mathcal{D})\|r(\boldsymbol{w}))$, tends to provide a posterior with a broad support covering all modes of the Bayes posterior (see the illustration in Figure 2.1).

**Moment Matching Algorithm**

The EP problem (2.43) is typically solved by *moment matching*. It starts with expressing the posterior distribution by the product of factors,

$$p(\boldsymbol{w}|\mathcal{D}) = \frac{1}{Z}\prod_n t_n(\boldsymbol{w}),$$

where $Z = p(\mathcal{D})$ is the marginal likelihood. For example, in the parametric density estimation (Example 1.1) with i.i.d. samples $\mathcal{D} = \{\boldsymbol{x}^{(1)}, \ldots, \boldsymbol{x}^{(N)}\}$, the factor can be set to $t_n(\boldsymbol{w}) = p(\boldsymbol{x}^{(n)}|\boldsymbol{w})$ and $t_0(\boldsymbol{w}) = p(\boldsymbol{w})$. In EP, the approximating posterior is also assumed to have the same form,

$$r(\boldsymbol{w}) = \frac{1}{\widetilde{Z}}\prod_n \widetilde{t}_n(\boldsymbol{w}), \tag{2.44}$$

where $\widetilde{Z}$ is the normalization constant and becomes an approximation of the marginal likelihood Z. Note that the factorization is not over the elements of $\boldsymbol{w}$.

EP tries to minimize the KL divergence,

$$\mathrm{KL}(p\|r) = \mathrm{KL}\left(\frac{1}{Z}\prod_n t_n(\boldsymbol{w}) \,\middle\|\, \frac{1}{\widetilde{Z}}\prod_n \widetilde{t}_n(\boldsymbol{w})\right),$$

which is approximately carried out by refining each factor while the other factors are fixed, and cycling through all the factors. To refine the factor $\widetilde{t}_n(\boldsymbol{w})$, we define the unnormalized distribution,

$$r^{\neg n}(\boldsymbol{w}) = \frac{r(\boldsymbol{w})}{\widetilde{t}_n(\boldsymbol{w})},$$

and the following distribution is used as an estimator of the true posterior:

$$\widehat{p}_n(\boldsymbol{w}) = \frac{t_n(\boldsymbol{w})r^{\neg n}(\boldsymbol{w})}{\widetilde{Z}_n},$$

where $\widetilde{Z}_n = \int t_n(\boldsymbol{w})r^{\neg n}(\boldsymbol{w})d\boldsymbol{w}$ is the normalization constant. That is, the new approximating posterior $r^{\mathrm{new}}(\boldsymbol{w})$ is computed so that it minimizes $\mathrm{KL}(\widehat{p}_n \| r^{\mathrm{new}})$. Usually, the approximating posterior is assumed to be a member of the exponential family. In that case, the minimization of $\mathrm{KL}(\widehat{p}_n \| r^{\mathrm{new}})$ is reduced to the moment matching between $\widehat{p}_n$ and $r^{\mathrm{new}}$. Namely, the parameter of $r^{\mathrm{new}}$ is determined so that its moments are matched with those of $\widehat{p}_n$.

The new approximating posterior $r^{\mathrm{new}}$ yields the refinement of the factor $\widetilde{t}_n(\boldsymbol{w})$,

$$\widetilde{t}_n(\boldsymbol{w}) = \widetilde{Z}_n \frac{r^{\mathrm{new}}(\boldsymbol{w})}{r^{\neg n}(\boldsymbol{w})},$$

where the multiplication of $\widetilde{Z}_n$ is derived from the zeroth-order moment matching between $\widehat{p}_n$ and $r^{\mathrm{new}}$, $\int t_n(\boldsymbol{w})r^{\neg n}(\boldsymbol{w})d\boldsymbol{w} = \int \widetilde{t}_n(\boldsymbol{w})r^{\neg n}(\boldsymbol{w})d\boldsymbol{w}$.

After several passes through all the factors, if the factors converge, then the posterior is approximated by Eq. (2.44), and the marginal likelihood is approximated by $\widetilde{Z} = \int \prod_n \widetilde{t}_n(\boldsymbol{w})d\boldsymbol{w}$ or alternatively by updating it as $\widetilde{Z} \leftarrow \widetilde{Z}\widetilde{Z}_n$ whenever the factor $\widetilde{t}_n(\boldsymbol{w})$ is refined. Although the convergence of EP is not guaranteed, it is known that if EP converges, the resulting approximating posterior is a stationary point of a certain energy function (Minka, 2001b).

**Local Variational Approximation for EP**

In Section 2.1.1, we saw that VB learning *minimizes an upper-bound* (the free energy (2.4)) of the Bayes free energy $F^{\mathrm{Bayes}} \equiv -\log p(\mathcal{D})$ (or equivalently maximizing the ELBO). We can say that EP does the opposite. Namely, the EP problem (2.43) *maximizes a lowerbound* of the Bayes free energy:

$$\max_r E(r) \quad \text{s.t.} \quad r \in \mathcal{G}, \tag{2.45}$$

$$\text{where} \quad E(r) = -\int \frac{p(\boldsymbol{w},\mathcal{D})}{p(\mathcal{D})} \log \frac{p(\boldsymbol{w},\mathcal{D})}{r(\boldsymbol{w})} d\boldsymbol{w} \tag{2.46}$$

$$= -\int p(\boldsymbol{w}|\mathcal{D}) \log \frac{p(\boldsymbol{w}|\mathcal{D})}{r(\boldsymbol{w})} d\boldsymbol{w} - \log p(\mathcal{D})$$

$$= -\mathrm{KL}\left(p(\boldsymbol{w}|\mathcal{D}) \| r(\boldsymbol{w})\right) - \log p(\mathcal{D}). \tag{2.47}$$

The maximization form (2.45) of the EP problem can be solved by local variational approximation, which is akin to the local variational approximation for VB learning (Section 2.1.7). Let us restrict the search space $\mathcal{G}$ for the approximate posterior $r(\boldsymbol{w};\widehat{\boldsymbol{\nu}})$ to the function space of a simple distribution family, e.g., Gaussian, parameterized with variational parameters $\widehat{\nu}$. Then, the EP problem (2.45) is reduced to the following unconstrained maximization problem,

$$\max_{\widehat{\nu}} E(\widehat{\nu}), \tag{2.48}$$

of the objective function written as

$$E(\widehat{\nu}) = -\int \frac{p(w, \mathcal{D})}{p(\mathcal{D})} \log \frac{p(w, \mathcal{D})}{p(\mathcal{D}) r(w; \widehat{\nu})} dw - \log p(\mathcal{D}). \tag{2.49}$$

Consider lower-bounding the objective (2.49) as

$$E(\widehat{\nu}) \geq \underline{E}(\widehat{\nu}, \eta) \equiv -\int \frac{\overline{p}(w; \eta)}{p(\mathcal{D})} \max\left\{0, \log \frac{\overline{p}(w; \eta)}{p(\mathcal{D}) r(w; \widehat{\nu})}\right\} dw - \log p(\mathcal{D}) \tag{2.50}$$

by using a parametric upper-bound $\overline{p}(w; \eta)$ of the joint distribution such that

$$\overline{p}(w; \eta) \geq p(w, \mathcal{D}) \tag{2.51}$$

for any $w \in \mathcal{W}$ and $\eta \in H$, where $\eta$ is another set of variational parameters with its domain $H$.[3] Let us choose an upper-bound $\overline{p}(w; \eta)$ such that its function form with respect to $w$ is the same as the approximate posterior $r(w; \widehat{\nu})$. More specifically, we assume that, for any given $\eta$, there exists $\widehat{\nu}$ such that

$$\overline{p}(w; \eta) \propto r(w; \widehat{\nu}) \tag{2.52}$$

as a function of $w$.

Since the direct maximization of $E(\widehat{\nu})$ is intractable, we instead maximize its lower-bound $\underline{E}(\widehat{\nu}, \eta)$ jointly over $\widehat{\nu}$ and $\eta$. Namely, we solve the problem,

$$\max_{\widehat{\nu}, \eta} \underline{E}(\widehat{\nu}, \eta), \tag{2.53}$$

to find the approximate posterior $r(w; \widehat{\nu})$ such that $\underline{E}(\widehat{\nu}, \eta)$ $(\leq E(\widehat{\nu}))$ is closest to the Bayes free energy $F^{\mathrm{Bayes}}$ (when $\eta$ is also optimized).

Let

$$q(w; \eta) = \frac{\overline{p}(w; \eta)}{\overline{Z}(\eta)} \tag{2.54}$$

be the distribution created by normalizing the upper-bound with its normalization factor

$$\overline{Z}(\eta) = \int \overline{p}(w; \eta) dw. \tag{2.55}$$

[3] The two sets, $\widehat{\nu}$ and $\eta$, of variational parameters play the same roles as $\widehat{\lambda}$ and $\xi$, respectively, in the local variational approximation for VB learning.

Note that the normalization factor (2.55) is trivially an upper-bound of the marginal likelihood, i.e.,

$$\overline{Z}(\boldsymbol{\eta}) \geq \int p(\boldsymbol{w}, \mathcal{D})d\boldsymbol{w} = p(\mathcal{D}),$$

and is tractable because of the assumption (2.52) that $\overline{p}$ is in the same simple function form as $r$.

With Eq. (2.54), the lower-bound (2.50) is expressed as

$$\begin{aligned}\underline{E}(\widehat{\boldsymbol{\nu}}, \boldsymbol{\eta}) &= -\int \frac{\overline{Z}(\boldsymbol{\eta})q(\boldsymbol{w};\boldsymbol{\eta})}{p(\mathcal{D})} \max\left\{0, \log \frac{\overline{Z}(\boldsymbol{\eta})q(\boldsymbol{w};\boldsymbol{\eta})}{p(\mathcal{D})r(\boldsymbol{w};\widehat{\boldsymbol{\nu}})}\right\} d\boldsymbol{w} - \log p(\mathcal{D}) \\ &= -\frac{\overline{Z}(\boldsymbol{\eta})}{p(\mathcal{D})} \int q(\boldsymbol{w};\boldsymbol{\eta}) \max\left\{0, \log \frac{q(\boldsymbol{w};\boldsymbol{\eta})}{r(\boldsymbol{w};\widehat{\boldsymbol{\nu}})} + \log \frac{\overline{Z}(\boldsymbol{\eta})}{p(\mathcal{D})}\right\} d\boldsymbol{w} - \log p(\mathcal{D}).\end{aligned} \tag{2.56}$$

Eq. (2.56) is upper-bounded by

$$\begin{aligned}&-\frac{\overline{Z}(\boldsymbol{\eta})}{p(\mathcal{D})} \int q(\boldsymbol{w};\boldsymbol{\eta}) \left(\log \frac{q(\boldsymbol{w};\boldsymbol{\eta})}{r(\boldsymbol{w};\widehat{\boldsymbol{\nu}})} + \log \frac{\overline{Z}(\boldsymbol{\eta})}{p(\mathcal{D})}\right) d\boldsymbol{w} - \log p(\mathcal{D}) \\ &= -\frac{\overline{Z}(\boldsymbol{\eta})}{p(\mathcal{D})} \mathrm{KL}\left(q(\boldsymbol{w};\boldsymbol{\eta}) \| r(\boldsymbol{w};\widehat{\boldsymbol{\nu}})\right) - \frac{\overline{Z}(\boldsymbol{\eta})}{p(\mathcal{D})} \log \frac{\overline{Z}(\boldsymbol{\eta})}{p(\mathcal{D})} - \log p(\mathcal{D}),\end{aligned} \tag{2.57}$$

which, for any $\boldsymbol{\eta} \in \boldsymbol{H}$, is maximized when $\widehat{\boldsymbol{\nu}}$ is such that

$$r(\boldsymbol{w};\widehat{\boldsymbol{\nu}}) = q(\boldsymbol{w};\boldsymbol{\eta}) \tag{2.58}$$

(the assumption (2.52) guarantees the attainability). With this optimal $\widehat{\boldsymbol{\nu}}$, Eq. (2.57) coincides with Eq. (2.56). Thus, after optimization with respect to $\widehat{\boldsymbol{\nu}}$, the lower-bound (2.56) is given as

$$\max_{\widehat{\boldsymbol{\nu}}} \underline{E}(\widehat{\boldsymbol{\nu}}, \boldsymbol{\eta}) = -\frac{\overline{Z}(\boldsymbol{\eta})}{p(\mathcal{D})} \log \frac{\overline{Z}(\boldsymbol{\eta})}{p(\mathcal{D})} - \log p(\mathcal{D}). \tag{2.59}$$

Since $x \log x$ for $x \geq 1$ is monotonically increasing, maximizing the lower-bound (2.59) is achieved by solving

$$\min_{\boldsymbol{\eta}} \overline{Z}(\boldsymbol{\eta}). \tag{2.60}$$

Once the minimizer $\widehat{\boldsymbol{\eta}}$ is obtained, Eq. (2.58) gives the optimal approximate posterior.

The problem (2.60) amounts to minimizing an upper-bound of the marginal likelihood. This is in contrast to the local variational approximation for VB learning, where a lower-bound of the marginal likelihood is maximized in the end (compare Eq. (2.33) and Eq. (2.60)).

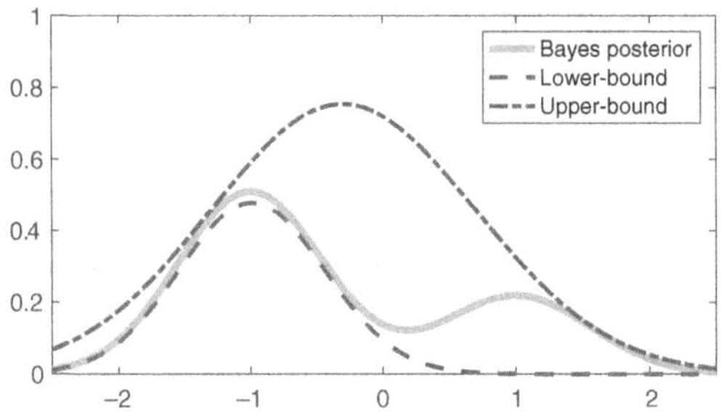


Figure 2.2 Bayes posterior and its tightest lower- and upper-bounds, formed by a Gaussian.

Remembering that the joint distribution is proportional to the Bayes posterior, i.e., $p(\boldsymbol{w}|\mathcal{D}) = p(\boldsymbol{w}, \mathcal{D})/p(\mathcal{D})$, we can say that the VB posterior is the normalized version of the tightest (in terms of the total mass) lower-bound of the Bayes posterior, while the EP posterior is the normalized version of the tightest upper-bound of the Bayes posterior. Figure 2.2 illustrates the tightest upper-bound and the tightest lower-bound of the Bayes posterior, which correspond to unnormalized versions of the VB posterior and the EP posterior, respectively (compare Figures 2.1 and 2.2). This view also explains the tendency of VB learning and EP—a lower-bound (the VB posterior) must be zero wherever the Bayes posterior is zero, while an upper-bound (the EP posterior) must be positive wherever the Bayes posterior is positive.

### 2.2.4 Metropolis–Hastings Sampling

If a sufficient number of samples $\{\boldsymbol{w}^{(1)}, \ldots, \boldsymbol{w}^{(L)}\}$ from the posterior distribution (1.3) are obtained, the expectation $\int f(\boldsymbol{w})p(\boldsymbol{w}|\mathcal{D})d\boldsymbol{w}$ required for computing the quantities such as Eqs. (1.6) through (1.9) can be approximated by

$$\frac{1}{L}\sum_{l=1}^{L} f(\boldsymbol{w}^{(l)}).$$

The *Metropolis–Hastings sampling* and the Gibbs sampling are most popular methods to sample from the (unnormalized) posterior distribution in the framework of *Markov chain Monte Carlo (MCMC)*.

In the Metropolis–Hastings sampling, we draw samples from a simple distribution $q(\boldsymbol{w}|\boldsymbol{w}^{(t)})$ called a proposal distribution, which is conditioned on the current state $\boldsymbol{w}^{(t)}$ of the parameter (or latent variables) $\boldsymbol{w}$. The proposal distribution is chosen to be a simple distribution such as a Gaussian centered at $\boldsymbol{w}^{(t)}$ if $\boldsymbol{w}$ is continuous or the uniform distribution in a certain neighborhood of $\boldsymbol{w}^{(t)}$ if $\boldsymbol{w}$ is discrete. At each cycle of the algorithm, we draw a candidate sample $\boldsymbol{w}^*$ from the proposal distribution $q(\boldsymbol{w}|\boldsymbol{w}^{(t)})$, and we accept it with probability

$$\min\left(1, \frac{p(\boldsymbol{w}^*, \mathcal{D})}{p(\boldsymbol{w}^{(t)}, \mathcal{D})} \frac{q(\boldsymbol{w}^{(t)}|\boldsymbol{w}^*)}{q(\boldsymbol{w}^*|\boldsymbol{w}^{(t)})}\right).$$

If $\boldsymbol{w}^*$ is accepted, then the next state $\boldsymbol{w}^{(t+1)}$ is moved to $\boldsymbol{w}^*$, $\boldsymbol{w}^{(t+1)} = \boldsymbol{w}^*$; otherwise, it stays at the current state, $\boldsymbol{w}^{(t+1)} = \boldsymbol{w}^{(t)}$. We repeat this procedure until a sufficiently long sequence of states is obtained. Note that if the proposal distribution is symmetric, i.e., $q(\boldsymbol{w}|\boldsymbol{w}') = q(\boldsymbol{w}'|\boldsymbol{w})$ for any $\boldsymbol{w}$ and $\boldsymbol{w}'$, in which case the algorithm is called the Metropolis algorithm, the probability of acceptance depends on the ratio of the posteriors,

$$\frac{p(\boldsymbol{w}^*, \mathcal{D})}{p(\boldsymbol{w}^{(t)}, \mathcal{D})} = \frac{p(\boldsymbol{w}^*, \mathcal{D})/Z}{p(\boldsymbol{w}^{(t)}, \mathcal{D})/Z} = \frac{p(\boldsymbol{w}^*|\mathcal{D})}{p(\boldsymbol{w}^{(t)}|\mathcal{D})},$$

and if $\boldsymbol{w}^*$ has higher posterior probability (density) than $\boldsymbol{w}^{(t)}$, it is accepted with probability 1.

To guarantee that the distribution of the sampled sequence converges to the posterior distribution, we discard a first part of the sequence, which is called *burn-in*. Usually, after the burn-in period, we retain only every $M$th sample and discard the other samples so that the retained samples can be considered as independent if $M$ is sufficiently large.

### 2.2.5 Gibbs Sampling

Another popular MCMC method is *Gibbs sampling*, which makes use of the conditional conjugacy. More specifically, it is applicable when we can compute and draw samples from the conditional distribution of a variable of $\boldsymbol{w} \in \mathbb{R}^J$,

$$p(w_j|w_1, \ldots, w_{j-1}, w_{j+1}, \ldots, w_J, \mathcal{D}) \equiv p(w_j|\boldsymbol{w}_{\neg j}, \mathcal{D}),$$

conditioned on the rest of the variables of $\boldsymbol{w}$.

Assuming that $\boldsymbol{w}^{(t)}$ is obtained at the $t$th cycle of the Gibbs sampling algorithm, the next sample of each variable is drawn from the conditional distribution,

$$p(w_j^{(t+1)}|\boldsymbol{w}_{\neg j}^{(t)}, \mathcal{D}),$$

where

$$\boldsymbol{w}_{\neg j}^{(t)} = (w_1^{(t+1)}, \ldots, w_{j-1}^{(t+1)}, w_{j+1}^{(t)}, \ldots, w_J^{(t)})$$

from $j = 1$ to $J$ in turn.

This sampling procedure can be viewed as a special case of the Metropolis–Hastings algorithm. If the proposal distribution $q(\boldsymbol{w}|\boldsymbol{w}^{(t)})$ is chosen to be

$$p(w_j|w_{\neg j}^{(t)}, \mathcal{D})\delta(w_{\neg j} - w_{\neg j}^{(t)}),$$

then the probability that the candidate $w^*$ is accepted is 1 since $w_{\neg j}^* = w_{\neg j}^{(t)}$ implies that

$$\begin{aligned}\frac{p(w^*, \mathcal{D})}{p(w^{(t)}, \mathcal{D})} \frac{q(w^{(t)}|w^*)}{q(w^*|w^{(t)})} &= \frac{p(w^*|\mathcal{D})}{p(w^{(t)}|\mathcal{D})} \frac{p(w_j^{(t)}|w_{\neg j}^{(t)}, \mathcal{D})}{p(w_j^*|w_{\neg j}^{(t)}, \mathcal{D})} \\ &= \frac{p(w_{\neg j}^*|\mathcal{D})p(w_j^*|w_{\neg j}^*, \mathcal{D})}{p(w_{\neg j}^{(t)}|\mathcal{D})p(w_j^{(t)}|w_{\neg j}^{(t)}, \mathcal{D})} \frac{p(w_j^{(t)}|w_{\neg j}^{(t)}, \mathcal{D})}{p(w_j^*|w_{\neg j}^{(t)}, \mathcal{D})} = 1.\end{aligned}$$

As we have seen in the Metropolis–Hastings and Gibbs sampling algorithms, MCMC methods do not require the knowledge of the normalization constant $Z = \int p(w, \mathcal{D})dw$. Note that, however, even if we have samples from the posterior, we need additional steps to compute $Z$ with the samples. A simple way is to calculate the expectation of the inverse of the likelihood by the sample average,

$$\left\langle \frac{1}{p(\mathcal{D}|w)} \right\rangle_{p(w|\mathcal{D})} \simeq \frac{1}{L} \sum_{l=1}^{L} \frac{1}{p(\mathcal{D}|w^{(l)})}.$$

It provides an estimate of the inverse of $Z$ because

$$\left\langle \frac{1}{p(\mathcal{D}|w)} \right\rangle_{p(w|\mathcal{D})} = \int \frac{1}{p(\mathcal{D}|w)} \frac{p(\mathcal{D}|w)p(w)}{Z} dw = \frac{1}{Z} \int p(w)dw = \frac{1}{Z}.$$

However, this estimator is known to have high variance. A more sophisticated sampling method to compute $Z$ was developed by Chib (1995), while it requires multiple runs of MCMC sampling.

A new efficient method to compute the marginal likelihood was recently proposed and named a *widely applicable Bayesian information criterion (WBIC)*, which requires only a single run of MCMC sampling from a generalized posterior distribution (Watanabe, 2013). This method computes the expectation of the negative log-likelihood,

$$\langle -\log p(\mathcal{D}|w) \rangle_{p^{(\beta)}(w|\mathcal{D})},$$

over the $\beta$-generalized posterior distribution defined as

$$p^{(\beta)}(w|\mathcal{D}) \propto p(\mathcal{D}|w)^{\beta} p(w),$$

with $\beta = 1/\log N$, where $N$ is the number of i.i.d. samples. The computed (approximated) expectation is proved to have the same leading terms as those of the asymptotic expansion of $-\log Z$ as $N \to \infty$.

# Part II

# Algorithm

# 3
# VB Algorithm for Multilinear Models

In this chapter, we derive iterative VB algorithms for multilinear models with Gaussian noise, where we can rely on the conditional conjugacy with respect to each linear factor. The models introduced in this chapter will be further analyzed in Part III, where the global solution or its approximation is analytically derived, and the behavior of the VB solution is investigated in detail.

## 3.1 Matrix Factorization

Assume that we observe a matrix $\boldsymbol{V} \in \mathbb{R}^{L\times M}$, which is the sum of a target matrix $\boldsymbol{U} \in \mathbb{R}^{L\times M}$ and a noise matrix $\boldsymbol{\mathcal{E}} \in \mathbb{R}^{L\times M}$:

$$\boldsymbol{V} = \boldsymbol{U} + \boldsymbol{\mathcal{E}}.$$

In the *matrix factorization (MF)* model (Srebro and Jaakkola, 2003; Srebro et al., 2005; Lim and Teh, 2007; Salakhutdinov and Mnih, 2008; Ilin and Raiko, 2010) or the *probabilistic principal component analysis (probabilistic PCA)* (Tipping and Bishop, 1999; Bishop, 1999b), the target matrix is assumed to be low rank, and therefore can be factorized as

$$\boldsymbol{U} = \boldsymbol{B}\boldsymbol{A}^{\top},$$

where $\boldsymbol{A} \in \mathbb{R}^{M\times H}$, $\boldsymbol{B} \in \mathbb{R}^{L\times H}$ for $H \leq \min(L, M)$ are unknown parameters to be estimated, and $\top$ denotes the transpose of a matrix or vector. Here, the rank of $\boldsymbol{U}$ is upper-bounded by $H$. We denote a column vector of a matrix by a bold lowercase letter, and a row vector by a bold lowercase letter with a tilde, namely,

$$\boldsymbol{A} = (\boldsymbol{a}_1, \ldots, \boldsymbol{a}_H) = (\widetilde{\boldsymbol{a}}_1, \ldots, \widetilde{\boldsymbol{a}}_M)^{\top} \in \mathbb{R}^{M\times H},$$
$$\boldsymbol{B} = (\boldsymbol{b}_1, \ldots, \boldsymbol{b}_H) = \left(\widetilde{\boldsymbol{b}}_1, \ldots, \widetilde{\boldsymbol{b}}_L\right)^{\top} \in \mathbb{R}^{L\times H}.$$

### 3.1.1 VB Learning for MF

Assume that the observation noise $\mathcal{E}$ is independent Gaussian:

$$p(\boldsymbol{V}|\boldsymbol{A},\boldsymbol{B}) \propto \exp\left(-\frac{1}{2\sigma^2}\left\|\boldsymbol{V}-\boldsymbol{B}\boldsymbol{A}^\top\right\|_{\mathrm{Fro}}^2\right), \tag{3.1}$$

where $\|\cdot\|_{\mathrm{Fro}}$ denotes the Frobenius norm.

**Conditional Conjugacy**

If we treat $\boldsymbol{B}$ as a constant, the likelihood (3.1) is in the Gaussian form of $\boldsymbol{A}$. Similarly, if we treat $\boldsymbol{A}$ as a constant, the likelihood (3.1) is in the Gaussian form of $\boldsymbol{B}$. Therefore, conditional conjugacy with respect to $\boldsymbol{A}$ given $\boldsymbol{B}$, as well as with respect to $\boldsymbol{B}$ given $\boldsymbol{A}$, holds if we adopt Gaussian priors:

$$p(\boldsymbol{A}) \propto \exp\left(-\frac{1}{2}\mathrm{tr}\left(\boldsymbol{A}\boldsymbol{C}_A^{-1}\boldsymbol{A}^\top\right)\right), \tag{3.2}$$

$$p(\boldsymbol{B}) \propto \exp\left(-\frac{1}{2}\mathrm{tr}\left(\boldsymbol{B}\boldsymbol{C}_B^{-1}\boldsymbol{B}^\top\right)\right), \tag{3.3}$$

where $\mathrm{tr}(\cdot)$ denotes the trace of a matrix.

Typically, the prior covariance matrices $\boldsymbol{C}_A$ and $\boldsymbol{C}_B$ are restricted to be diagonal, which induces low-rankness (we discuss this mechanism in Chapter 7):

$$\boldsymbol{C}_A = \mathbf{Diag}(c_{a_1}^2,\ldots,c_{a_H}^2),$$
$$\boldsymbol{C}_B = \mathbf{Diag}(c_{b_1}^2,\ldots,c_{b_H}^2),$$

for $c_{a_h}, c_{b_h} > 0, h = 1,\ldots,H$.

**Variational Bayesian Algorithm**

Thanks to the conditional conjugacy, the following independence constraint makes the approximate posterior Gaussian:

$$r(\boldsymbol{A},\boldsymbol{B}) = r_A(\boldsymbol{A})r_B(\boldsymbol{B}). \tag{3.4}$$

The VB learning problem (2.13) is then reduced to

$$\widehat{r} = \underset{r}{\mathrm{argmin}} F(r) \quad \text{s.t.} \quad r(\boldsymbol{A},\boldsymbol{B}) = r_A(\boldsymbol{A})r_B(\boldsymbol{B}). \tag{3.5}$$

Under the constraint (3.4), the free energy is written as

$$\begin{aligned} F(r) &= \left\langle \log \frac{r_A(\boldsymbol{A})r_B(\boldsymbol{B})}{p(\boldsymbol{V}|\boldsymbol{A},\boldsymbol{B})p(\boldsymbol{A})p(\boldsymbol{B})}\right\rangle_{r_A(\boldsymbol{A})r_B(\boldsymbol{B})} \\ &= \int r_A(\boldsymbol{A})r_B(\boldsymbol{B}) \log \frac{r_A(\boldsymbol{A})r_B(\boldsymbol{B})}{p(\boldsymbol{V}|\boldsymbol{A},\boldsymbol{B})p(\boldsymbol{A})p(\boldsymbol{B})} d\boldsymbol{A}d\boldsymbol{B}. \end{aligned} \tag{3.6}$$

Following the recipe described in Section 2.1.5, we take the derivatives of the free energy (3.6) with respect to $r_A(\boldsymbol{A})$ and $r_B(\boldsymbol{B})$, respectively. Thus, we obtain the following stationary conditions:

$$r_A(\boldsymbol{A}) \propto p(\boldsymbol{A}) \exp \langle \log p(\boldsymbol{V}|\boldsymbol{A}, \boldsymbol{B}) \rangle_{r_B(\boldsymbol{B})}, \tag{3.7}$$

$$r_B(\boldsymbol{B}) \propto p(\boldsymbol{B}) \exp \langle \log p(\boldsymbol{V}|\boldsymbol{A}, \boldsymbol{B}) \rangle_{r_A(\boldsymbol{A})}. \tag{3.8}$$

By substituting the likelihood (3.1) and the prior (3.2) into Eq. (3.7), we obtain

$$\begin{aligned} r_A(\boldsymbol{A}) &\propto \exp\left(-\frac{1}{2}\mathrm{tr}\left(\boldsymbol{A}\boldsymbol{C}_A^{-1}\boldsymbol{A}^\top\right) - \frac{1}{2\sigma^2}\left\langle \|\boldsymbol{V} - \boldsymbol{B}\boldsymbol{A}^\top\|_{\mathrm{Fro}}^2 \right\rangle_{r_B(\boldsymbol{B})}\right) \\ &\propto \exp\left(-\frac{1}{2}\mathrm{tr}\left(\boldsymbol{A}\boldsymbol{C}_A^{-1}\boldsymbol{A}^\top + \sigma^{-2}\left\langle -2\boldsymbol{V}^\top\boldsymbol{B}\boldsymbol{A}^\top + \boldsymbol{A}\boldsymbol{B}^\top\boldsymbol{B}\boldsymbol{A}^\top \right\rangle_{r_B(\boldsymbol{B})}\right)\right) \\ &\propto \exp\left(-\frac{\mathrm{tr}\left((\boldsymbol{A} - \widehat{\boldsymbol{A}})\widehat{\boldsymbol{\Sigma}}_A^{-1}(\boldsymbol{A} - \widehat{\boldsymbol{A}})^\top\right)}{2}\right), \end{aligned} \tag{3.9}$$

where

$$\widehat{\boldsymbol{A}} = \sigma^{-2}\boldsymbol{V}^\top \langle \boldsymbol{B} \rangle_{r_B(\boldsymbol{B})} \widehat{\boldsymbol{\Sigma}}_A, \tag{3.10}$$

$$\widehat{\boldsymbol{\Sigma}}_A = \sigma^2 \left( \left\langle \boldsymbol{B}^\top \boldsymbol{B} \right\rangle_{r_B(\boldsymbol{B})} + \sigma^2 \boldsymbol{C}_A^{-1} \right)^{-1}. \tag{3.11}$$

Similarly, by substituting the likelihood (3.1) and the prior (3.3) into Eq. (3.8), we obtain

$$\begin{aligned} r_B(\boldsymbol{B}) &\propto \exp\left(-\frac{1}{2}\mathrm{tr}\left(\boldsymbol{B}\boldsymbol{C}_B^{-1}\boldsymbol{B}^\top\right) - \frac{1}{2\sigma^2}\left\langle \|\boldsymbol{V} - \boldsymbol{B}\boldsymbol{A}^\top\|_{\mathrm{Fro}}^2 \right\rangle_{r_A(\boldsymbol{A})}\right) \\ &\propto \exp\left(-\frac{1}{2}\mathrm{tr}\left(\boldsymbol{B}\boldsymbol{C}_B^{-1}\boldsymbol{B}^\top + \sigma^{-2}\left\langle -2\boldsymbol{V}\boldsymbol{A}\boldsymbol{B}^\top + \boldsymbol{B}\boldsymbol{A}^\top\boldsymbol{A}\boldsymbol{B}^\top \right\rangle_{r_A(\boldsymbol{A})}\right)\right) \\ &\propto \exp\left(-\frac{\mathrm{tr}\left((\boldsymbol{B} - \widehat{\boldsymbol{B}})\widehat{\boldsymbol{\Sigma}}_B^{-1}(\boldsymbol{B} - \widehat{\boldsymbol{B}})^\top\right)}{2}\right), \end{aligned} \tag{3.12}$$

where

$$\widehat{\boldsymbol{B}} = \sigma^{-2}\boldsymbol{V} \langle \boldsymbol{A} \rangle_{r_A(\boldsymbol{A})} \widehat{\boldsymbol{\Sigma}}_B, \tag{3.13}$$

$$\widehat{\boldsymbol{\Sigma}}_B = \sigma^2 \left( \left\langle \boldsymbol{A}^\top \boldsymbol{A} \right\rangle_{r_A(\boldsymbol{A})} + \sigma^2 \boldsymbol{C}_B^{-1} \right)^{-1}. \tag{3.14}$$

Eqs. (3.9) and (3.12) imply that the posteriors are Gaussian. More specifically, they can be written as

$$r_A(\boldsymbol{A}) = \mathrm{MGauss}_{M,H}(\boldsymbol{A}; \widehat{\boldsymbol{A}}, \boldsymbol{I}_M \otimes \widehat{\boldsymbol{\Sigma}}_A), \tag{3.15}$$

$$r_B(\boldsymbol{B}) = \mathrm{MGauss}_{L,H}(\boldsymbol{B}; \widehat{\boldsymbol{B}}, \boldsymbol{I}_L \otimes \widehat{\boldsymbol{\Sigma}}_B), \tag{3.16}$$

where $\otimes$ denotes the *Kronecker product*, and

$$\mathrm{MGauss}_{D_1,D_2}(\boldsymbol{X}; \boldsymbol{M}, \breve{\boldsymbol{\Sigma}}) \equiv \mathrm{Gauss}_{D_1 \cdot D_2}(\mathbf{vec}(\boldsymbol{X}^\top); \mathbf{vec}(\boldsymbol{M}^\top), \breve{\boldsymbol{\Sigma}}) \tag{3.17}$$

denotes the *matrix variate Gaussian* distribution (Gupta and Nagar, 1999). Here, $\mathbf{vec} : \mathbb{R}^{D_2 \times D_1} \mapsto \mathbb{R}^{D_2 D_1}$ is the *vectorization operator*, which concatenates all column vectors of a matrix into a long column vector. Note that, if the covariance has a specific structure expressed as $\breve{\boldsymbol{\Sigma}} = \boldsymbol{\Sigma} \otimes \boldsymbol{\Psi} \in \mathbb{R}^{D_2 D_1 \times D_2 D_1}$, such as Eqs. (3.15) and (3.16), the matrix variate Gaussian distribution can be written as

$$\mathrm{MGauss}_{D_1,D_2}(\boldsymbol{X}; \boldsymbol{M}, \boldsymbol{\Sigma} \otimes \boldsymbol{\Psi}) \equiv \frac{1}{(2\pi)^{D_1 D_2/2} \det(\boldsymbol{\Sigma})^{D_2/2} \det(\boldsymbol{\Psi})^{D_1/2}} \cdot \exp\left(-\frac{1}{2}\mathrm{tr}\left(\boldsymbol{\Sigma}^{-1}(\boldsymbol{X} - \boldsymbol{M})\boldsymbol{\Psi}^{-1}(\boldsymbol{X} - \boldsymbol{M})^\top\right)\right). \tag{3.18}$$

The fact that the posterior is Gaussian is a consequence of the forced independence between $\boldsymbol{A}$ and $\boldsymbol{B}$ and conditional conjugacy. The parameters, $\left\{\widehat{\boldsymbol{A}}, \widehat{\boldsymbol{B}}, \widehat{\boldsymbol{\Sigma}}_A, \widehat{\boldsymbol{\Sigma}}_B\right\}$, defining the VB posterior (3.15) and (3.16), are the *variational parameters*.

Since $r_A(\boldsymbol{A})$ and $r_B(\boldsymbol{B})$ are Gaussian, the first and the (noncenterized) second moments can be expressed with variational parameters as follows:

$$\begin{aligned}
\langle \boldsymbol{A} \rangle_{r_A(\boldsymbol{A})} &= \widehat{\boldsymbol{A}}, \\
\left\langle \boldsymbol{A}^\top \boldsymbol{A} \right\rangle_{r_A(\boldsymbol{A})} &= \widehat{\boldsymbol{A}}^\top \widehat{\boldsymbol{A}} + M\widehat{\boldsymbol{\Sigma}}_A, \\
\langle \boldsymbol{B} \rangle_{r_B(\boldsymbol{B})} &= \widehat{\boldsymbol{B}}, \\
\left\langle \boldsymbol{B}^\top \boldsymbol{B} \right\rangle_{r_B(\boldsymbol{B})} &= \widehat{\boldsymbol{B}}^\top \widehat{\boldsymbol{B}} + L\widehat{\boldsymbol{\Sigma}}_B.
\end{aligned}$$

By substituting the preceding into Eqs. (3.10), (3.11), (3.13), and (3.14), we have the following relations among the variational parameters:

$$\widehat{\boldsymbol{A}} = \sigma^{-2}\boldsymbol{V}^\top \widehat{\boldsymbol{B}}\widehat{\boldsymbol{\Sigma}}_A, \tag{3.19}$$

$$\widehat{\boldsymbol{\Sigma}}_A = \sigma^2\left(\widehat{\boldsymbol{B}}^\top \widehat{\boldsymbol{B}} + L\widehat{\boldsymbol{\Sigma}}_B + \sigma^2 \boldsymbol{C}_A^{-1}\right)^{-1}, \tag{3.20}$$

$$\widehat{\boldsymbol{B}} = \sigma^{-2}\boldsymbol{V}\widehat{\boldsymbol{A}}\widehat{\boldsymbol{\Sigma}}_B, \tag{3.21}$$

$$\widehat{\boldsymbol{\Sigma}}_B = \sigma^2\left(\widehat{\boldsymbol{A}}^\top \widehat{\boldsymbol{A}} + M\widehat{\boldsymbol{\Sigma}}_A + \sigma^2 \boldsymbol{C}_B^{-1}\right)^{-1}. \tag{3.22}$$

As we see shortly, Eqs. (3.19) through (3.22) are stationary conditions for variational parameters, which can be used as update rules for *coordinate descent* local search (Bishop, 1999b).

**Free Energy as a Function of Variational Parameters**

By substituting Eqs. (3.15) and (3.16) into Eq. (3.6), we can explicitly write down the free energy as (not a functional but) a function of the unknown variational parameters $\left\{\widehat{\boldsymbol{A}}, \widehat{\boldsymbol{B}}, \widehat{\boldsymbol{\Sigma}}_A, \widehat{\boldsymbol{\Sigma}}_B\right\}$:

$$\begin{aligned}
2F &= 2\left\langle \log \frac{r_A(\boldsymbol{A})r_B(\boldsymbol{B})}{p(\boldsymbol{V}|\boldsymbol{A},\boldsymbol{B})p(\boldsymbol{A})p(\boldsymbol{B})} \right\rangle_{r_A(\boldsymbol{A})r_B(\boldsymbol{B})} \\
&= 2\left\langle \log \frac{r_A(\boldsymbol{A})r_B(\boldsymbol{B})}{p(\boldsymbol{A})p(\boldsymbol{B})} \right\rangle_{r_A(\boldsymbol{A})r_B(\boldsymbol{B})} - 2\left\langle \log p(\boldsymbol{V}|\boldsymbol{A},\boldsymbol{B}) \right\rangle_{r_A(\boldsymbol{A})r_B(\boldsymbol{B})} \\
&= \Bigg\langle M \log \frac{\det(\boldsymbol{C}_A)}{\det\left(\widehat{\boldsymbol{\Sigma}}_A\right)} + L \log \frac{\det(\boldsymbol{C}_B)}{\det\left(\widehat{\boldsymbol{\Sigma}}_B\right)} + \mathrm{tr}\left(\boldsymbol{C}_A^{-1}\boldsymbol{A}^\top\boldsymbol{A} + \boldsymbol{C}_B^{-1}\boldsymbol{B}^\top\boldsymbol{B}\right) \\
&\quad - \mathrm{tr}\left(\widehat{\boldsymbol{\Sigma}}_A^{-1}(\boldsymbol{A}-\widehat{\boldsymbol{A}})^\top(\boldsymbol{A}-\widehat{\boldsymbol{A}}) + \widehat{\boldsymbol{\Sigma}}_B^{-1}(\boldsymbol{B}-\widehat{\boldsymbol{B}})^\top(\boldsymbol{B}-\widehat{\boldsymbol{B}})\right) \\
&\quad + LM\log(2\pi\sigma^2) + \frac{\|\boldsymbol{V}-\boldsymbol{B}\boldsymbol{A}^\top\|_{\mathrm{Fro}}^2}{\sigma^2} \Bigg\rangle_{r_A(\boldsymbol{A})r_B(\boldsymbol{B})} \\
&= M \log \frac{\det(\boldsymbol{C}_A)}{\det\left(\widehat{\boldsymbol{\Sigma}}_A\right)} + L \log \frac{\det(\boldsymbol{C}_B)}{\det\left(\widehat{\boldsymbol{\Sigma}}_B\right)} - \mathrm{tr}\left(M\widehat{\boldsymbol{\Sigma}}_A^{-1}\widehat{\boldsymbol{\Sigma}}_A + L\widehat{\boldsymbol{\Sigma}}_B^{-1}\widehat{\boldsymbol{\Sigma}}_B\right) \\
&\quad + \mathrm{tr}\left(\boldsymbol{C}_A^{-1}\left(\widehat{\boldsymbol{A}}^\top\widehat{\boldsymbol{A}} + M\widehat{\boldsymbol{\Sigma}}_A\right) + \boldsymbol{C}_B^{-1}\left(\widehat{\boldsymbol{B}}^\top\widehat{\boldsymbol{B}} + L\widehat{\boldsymbol{\Sigma}}_B\right)\right) \\
&\quad + LM\log(2\pi\sigma^2) + \left\langle \frac{\|(\boldsymbol{V}-\widehat{\boldsymbol{B}}\widehat{\boldsymbol{A}}^\top)+(\widehat{\boldsymbol{B}}\widehat{\boldsymbol{A}}^\top-\boldsymbol{B}\boldsymbol{A}^\top)\|_{\mathrm{Fro}}^2}{\sigma^2} \right\rangle_{r_A(\boldsymbol{A})r_B(\boldsymbol{B})} \\
&= M \log \frac{\det(\boldsymbol{C}_A)}{\det\left(\widehat{\boldsymbol{\Sigma}}_A\right)} + L \log \frac{\det(\boldsymbol{C}_B)}{\det\left(\widehat{\boldsymbol{\Sigma}}_B\right)} - (L+M)H \\
&\quad + \mathrm{tr}\left(\boldsymbol{C}_A^{-1}\left(\widehat{\boldsymbol{A}}^\top\widehat{\boldsymbol{A}} + M\widehat{\boldsymbol{\Sigma}}_A\right) + \boldsymbol{C}_B^{-1}\left(\widehat{\boldsymbol{B}}^\top\widehat{\boldsymbol{B}} + L\widehat{\boldsymbol{\Sigma}}_B\right)\right) \\
&\quad + LM\log(2\pi\sigma^2) + \frac{\|\boldsymbol{V}-\widehat{\boldsymbol{B}}\widehat{\boldsymbol{A}}^\top\|_{\mathrm{Fro}}^2}{\sigma^2} + \left\langle \frac{\|\widehat{\boldsymbol{B}}\widehat{\boldsymbol{A}}^\top-\boldsymbol{B}\boldsymbol{A}^\top\|_{\mathrm{Fro}}^2}{\sigma^2} \right\rangle_{r_A(\boldsymbol{A})r_B(\boldsymbol{B})} \\
&= LM\log(2\pi\sigma^2) + \frac{\left\|\boldsymbol{V}-\widehat{\boldsymbol{B}}\widehat{\boldsymbol{A}}^\top\right\|_{\mathrm{Fro}}^2}{\sigma^2} + M \log \frac{\det(\boldsymbol{C}_A)}{\det\left(\widehat{\boldsymbol{\Sigma}}_A\right)} + L \log \frac{\det(\boldsymbol{C}_B)}{\det\left(\widehat{\boldsymbol{\Sigma}}_B\right)} \\
&\quad - (L+M)H + \mathrm{tr}\Big\{\boldsymbol{C}_A^{-1}\left(\widehat{\boldsymbol{A}}^\top\widehat{\boldsymbol{A}} + M\widehat{\boldsymbol{\Sigma}}_A\right) + \boldsymbol{C}_B^{-1}\left(\widehat{\boldsymbol{B}}^\top\widehat{\boldsymbol{B}} + L\widehat{\boldsymbol{\Sigma}}_B\right) \\
&\quad + \sigma^{-2}\left(-\widehat{\boldsymbol{A}}^\top\widehat{\boldsymbol{A}}\widehat{\boldsymbol{B}}^\top\widehat{\boldsymbol{B}} + \left(\widehat{\boldsymbol{A}}^\top\widehat{\boldsymbol{A}} + M\widehat{\boldsymbol{\Sigma}}_A\right)\left(\widehat{\boldsymbol{B}}^\top\widehat{\boldsymbol{B}} + L\widehat{\boldsymbol{\Sigma}}_B\right)\right)\Big\}. \qquad (3.23)
\end{aligned}$$

Now, the VB learning problem is reduced from the function *optimization* (3.5) to the following *variable optimization*:

$$\begin{aligned} \text{Given} \quad & \boldsymbol{C}_A, \boldsymbol{C}_A \in \mathbb{D}^H_{++}, \quad \sigma^2 \in \mathbb{R}_{++}, \\ \min_{\widehat{\boldsymbol{A}}, \widehat{\boldsymbol{B}}, \widehat{\boldsymbol{\Sigma}}_A, \widehat{\boldsymbol{\Sigma}}_B} \quad & F, \\ \text{s.t.} \quad & \widehat{\boldsymbol{A}} \in \mathbb{R}^{M \times H}, \widehat{\boldsymbol{B}} \in \mathbb{R}^{L \times H}, \quad \widehat{\boldsymbol{\Sigma}}_A, \widehat{\boldsymbol{\Sigma}}_B \in \mathbb{S}^H_{++}, \end{aligned} \tag{3.24}$$

where $\mathbb{R}_{++}$ is the set of positive real numbers, $\mathbb{S}^D_{++}$ is the set of $D \times D$ (symmetric) positive definite matrices, and $\mathbb{D}^D_{++}$ is the set of $D \times D$ positive definite diagonal matrices.

We note the following:

- Once the solution $\left\{\widehat{\boldsymbol{A}}, \widehat{\boldsymbol{B}}, \widehat{\boldsymbol{\Sigma}}_A, \widehat{\boldsymbol{\Sigma}}_B\right\}$ of the problem (3.24) is obtained, Eqs. (3.15) and (3.16) specify the VB posterior $\widehat{r}(\boldsymbol{A}, \boldsymbol{B}) = r_A(\boldsymbol{A}) r_B(\boldsymbol{B})$.
- We treated the prior covariances $\boldsymbol{C}_A$ and $\boldsymbol{C}_B$ and the noise variance $\sigma^2$ as hyperparameters, and therefore assumed to be given when the VB problem was solved. However, they can be estimated through the empirical Bayesian procedure, which is explained shortly. They can also be treated as random variables, and their VB posterior can be computed by adopting conjugate Gamma priors and minimizing the free energy under an appropriate independence constraint.
- Eqs. (3.19) through (3.22) coincide with the stationary conditions of the free energy (3.23), which are derived from the derivatives with respect to $\widehat{\boldsymbol{A}}, \widehat{\boldsymbol{\Sigma}}_A, \widehat{\boldsymbol{B}}$, and $\widehat{\boldsymbol{\Sigma}}_B$, respectively. Therefore, iterating Eqs. (3.19) through (3.22) gives a local solution to the problem (3.24).

### Empirical Variational Bayesian Algorithm

The empirical variational Bayesian (EVB) procedure can be performed by minimizing the free energy also with respect to the hyperparameters:

$$\begin{aligned} \min_{\widehat{\boldsymbol{A}}, \widehat{\boldsymbol{B}}, \widehat{\boldsymbol{\Sigma}}_A, \widehat{\boldsymbol{\Sigma}}_B, \boldsymbol{C}_A, \boldsymbol{C}_A, \sigma^2} \quad & F, \\ \text{s.t.} \quad & \widehat{\boldsymbol{A}} \in \mathbb{R}^{M \times H}, \widehat{\boldsymbol{B}} \in \mathbb{R}^{L \times H}, \quad \widehat{\boldsymbol{\Sigma}}_A, \widehat{\boldsymbol{\Sigma}}_B \in \mathbb{S}^H_{++}, \\ & \boldsymbol{C}_A, \boldsymbol{C}_A \in \mathbb{D}^H_{++}, \quad \sigma^2 \in \mathbb{R}_{++}. \end{aligned} \tag{3.25}$$

By differentiating the free energy (3.23) with respect to each entry of $\boldsymbol{C}_A$ and $\boldsymbol{C}_B$, we have, for $h = 1, \ldots, H$,

$$c_{a_h}^2 = \left\|\widehat{\boldsymbol{a}}_h\right\|^2 / M + \left(\widehat{\boldsymbol{\Sigma}}_A\right)_{h,h}, \tag{3.26}$$

$$c_{b_h}^2 = \left\|\widehat{\boldsymbol{b}}_h\right\|^2 / L + \left(\widehat{\boldsymbol{\Sigma}}_B\right)_{h,h}. \tag{3.27}$$

**Algorithm 1** EVB learning for matrix factorization.

1: Initialize the variational parameters $(\widehat{A}, \widehat{\Sigma}_A, \widehat{B}, \widehat{\Sigma}_B)$, and the hyperparameters $(C_A, C_B, \sigma^2)$, for example, $\widehat{A}_{m,h}, \widehat{B}_{l,h} \sim \text{Gauss}_1(0, \tau)$, $\widehat{\Sigma}_A = \widehat{\Sigma}_B = C_A = C_B = \tau I_H$, and $\sigma^2 = \tau^2$ for $\tau^2 = \|V\|^2_{\text{Fro}}/(LM)$.
2: Apply (substitute the right-hand side into the left-hand side) Eqs. (3.20), (3.19), (3.22), and (3.21) to update $\widehat{\Sigma}_A, \widehat{A}, \widehat{\Sigma}_B$, and $\widehat{B}$, respectively.
3: Apply Eqs. (3.26) and (3.27) for all $h = 1, \ldots, H$, and Eq. (3.28) to update $C_A, C_B$, and $\sigma^2$, respectively.
4: Prune the $h$th component if $c^2_{a_h} c^2_{b_h} < \varepsilon$, where $\varepsilon > 0$ is a small threshold, e.g., set to $\varepsilon = 10^{-4}$.
5: Evaluate the free energy (3.23).
6: Iterate Steps 2 through 5 until convergence (until the energy decrease becomes smaller than a threshold).

Similarly, by differentiating the free energy (3.23) with respect to $\sigma^2$, we have

$$\sigma^2 = \frac{\|V\|^2_{\text{Fro}} - \text{tr}\left(2V^\top \widehat{B}\widehat{A}^\top\right) + \text{tr}\left((\widehat{A}^\top \widehat{A} + M\widehat{\Sigma}_A)(\widehat{B}^\top \widehat{B} + L\widehat{\Sigma}_B)\right)}{LM}. \tag{3.28}$$

Eqs. (3.26)–(3.28) are used as update rules for the prior covariances $C_A, C_B$, and the noise variance $\sigma^2$, respectively.

Starting from some initial value, iterating Eqs. (3.19) through (3.22) and Eqs. (3.26) through (3.28) gives a local solution for EVB learning. Algorithm 1 summarizes this iterative procedure. If we appropriately set the hyperparameters $(C_A, C_B, \sigma^2)$ in Step 1 and skip Steps 3 and 4, Algorithm 1 is reduced to (nonempirical) VB learning.

We note the following for implementation:

- Due to the automatic relevance determination (ARD) effect in EVB learning (see Chapter 7), $c^2_{a_h} c^2_{b_h}$ converges to zero for some $h$. For this reason, "pruning" in Step 4 is necessary for numerical stability ($\log\det(C)$ diverges if $C$ is singular). If the $h$th component is pruned, the corresponding $h$th column of $\widehat{A}$ and $\widehat{B}$ and the $h$th column and row of $\widehat{\Sigma}_A, \widehat{\Sigma}_B, C_A, C_B$ should be removed, and the rank $H$ should be reduced accordingly.
- In principle, the update rules never increase the free energy. However, pruning can slightly increase it.
- When computing the free energy by Eq. (3.23), $\log\det(\cdot)$ should be computed as twice the sum of the log of the diagonals of the Cholesky decomposition, i.e.,

$$\log\det(\boldsymbol{C}) = 2\sum_{h=1}^{H}(\log(\mathbf{Chol}(\boldsymbol{C}))_{h,h}).$$

Otherwise, $\det(\cdot)$ can be huge for practical size of $H$, causing numerical instability.

**Simple Variational Bayesian Learning (with Columnwise Independence)**
The updates (3.19) through (3.22) require inversion of an $H \times H$ matrix. One can derive a faster VB learning algorithm by using a stronger constraint for the VB learning. More specifically, instead of the matrixwise independence (3.4), we assume the independence between the column vectors of $\boldsymbol{A} = (\boldsymbol{a}_1, \ldots, \boldsymbol{a}_H)$ and $\boldsymbol{B} = (\boldsymbol{b}_1, \ldots, \boldsymbol{b}_H)$ (Ilin and Raiko, 2010; Nakajima and Sugiyama, 2011; Kim and Choi, 2014):

$$r(\boldsymbol{A}, \boldsymbol{B}) = \prod_{h=1}^{H} r_{a_h}(\boldsymbol{a}_h) \prod_{h=1}^{H} r_{b_h}(\boldsymbol{b}_h). \tag{3.29}$$

By applying the same procedure as that with the matrixwise independence constraint, we can derive the solution to

$$\widehat{r} = \underset{r}{\mathrm{argmin}} F(r) \quad \text{s.t.} \quad r(\boldsymbol{A}, \boldsymbol{B}) = \prod_{h=1}^{H} r_{a_h}(\boldsymbol{a}_h) \prod_{h=1}^{H} r_{b_h}(\boldsymbol{b}_h), \tag{3.30}$$

which is in the form of the matrix variate Gaussian:

$$r_A(\boldsymbol{A}) = \mathrm{MGauss}_{M,H}(\boldsymbol{A}; \widehat{\boldsymbol{A}}, \boldsymbol{I}_M \otimes \widehat{\boldsymbol{\Sigma}}_A) = \prod_{h=1}^{H} \mathrm{Gauss}_M(\boldsymbol{a}_h; \widehat{\boldsymbol{a}}_h, \widehat{\sigma}^2_{a_h} \boldsymbol{I}_M),$$

$$r_B(\boldsymbol{B}) = \mathrm{MGauss}_{L,H}(\boldsymbol{B}; \widehat{\boldsymbol{B}}, \boldsymbol{I}_L \otimes \widehat{\boldsymbol{\Sigma}}_B) = \prod_{h=1}^{H} \mathrm{Gauss}_L(\boldsymbol{b}_h; \widehat{\boldsymbol{b}}_h, \widehat{\sigma}^2_{b_h} \boldsymbol{I}_L),$$

with the variational parameters,

$$\begin{aligned}
\widehat{\boldsymbol{A}} &= (\widehat{\boldsymbol{a}}_1, \ldots, \widehat{\boldsymbol{a}}_H), \\
\widehat{\boldsymbol{B}} &= (\widehat{\boldsymbol{b}}_1, \ldots, \widehat{\boldsymbol{b}}_H), \\
\widehat{\boldsymbol{\Sigma}}_A &= \mathbf{Diag}(\widehat{\sigma}^2_{a_1}, \ldots, \widehat{\sigma}^2_{a_H}), \\
\widehat{\boldsymbol{\Sigma}}_B &= \mathbf{Diag}(\widehat{\sigma}^2_{b_1}, \ldots, \widehat{\sigma}^2_{b_H}).
\end{aligned}$$

Here $\mathbf{Diag}(\cdots)$ denotes the diagonal matrix with the specified diagonal entries. The stationary conditions are given as follows: for all $h = 1, \ldots, H$,

$$\widehat{\boldsymbol{a}}_h = \frac{\widehat{\sigma}^2_{a_h}}{\sigma^2}\left(\boldsymbol{V} - \sum_{h' \neq h} \widehat{\boldsymbol{b}}_{h'} \widehat{\boldsymbol{a}}_{h'}^{\top}\right)^{\top} \widehat{\boldsymbol{b}}_h, \tag{3.31}$$

$$\widehat{\sigma}^2_{a_h} = \sigma^2 \left( \left\| \widehat{\boldsymbol{b}}_h \right\|^2 + L\widehat{\sigma}^2_{b_h} + \frac{\sigma^2}{c^2_{a_h}} \right)^{-1}, \tag{3.32}$$

$$\widehat{\boldsymbol{b}}_h = \frac{\widehat{\sigma}^2_{b_h}}{\sigma^2} \left( \boldsymbol{V} - \sum_{h' \neq h} \widehat{\boldsymbol{b}}_{h'} \widehat{\boldsymbol{a}}_{h'}^\top \right) \widehat{\boldsymbol{a}}_h, \tag{3.33}$$

$$\widehat{\sigma}^2_{b_h} = \sigma^2 \left( \left\| \widehat{\boldsymbol{a}}_h \right\|^2 + M\widehat{\sigma}^2_{a_h} + \frac{\sigma^2}{c^2_{b_h}} \right)^{-1}. \tag{3.34}$$

The free energy is given by Eq. (3.23) with the posterior covariances $\widehat{\boldsymbol{\Sigma}}_A$ and $\widehat{\boldsymbol{\Sigma}}_B$ restricted to be diagonal. The stationary conditions for the hyperparameters are unchanged, and given by Eqs. (3.26) through (3.28). Therefore, Algorithm 1 with Eqs. (3.31) through (3.34), substituted for Eqs. (3.19) through (3.22), gives a local solution to the VB problem (3.30) with the columnwise independence constraint.

We call this variant *simple VB (SimpleVB) learning*. In Chapter 6, it will be shown that, in the fully observed MF model, the SimpleVB problem (3.30) with columnwise independence and the original VB problem (3.5) with matrixwise independence actually give the equivalent solution.

### 3.1.2 Special Cases

Probabilistic principal component analysis and reduced rank regression are special cases of matrix factorization. Therefore, they can be trained by Algorithm 1 with or without small modifications.

**Probabilistic Principal Component Analysis** *Probabilistic principal component analysis* (Tipping and Bishop, 1999; Bishop, 1999b) is a probabilisitic model of which the ML estimation corresponds to the classical principal component analysis (PCA) (Hotelling, 1933). The observation $\boldsymbol{v} \in \mathbb{R}^L$ is assumed to be driven by a latent vector $\widetilde{\boldsymbol{a}} \in \mathbb{R}^H$ in the following form:

$$\boldsymbol{v} = \boldsymbol{B}\widetilde{\boldsymbol{a}} + \boldsymbol{\varepsilon}.$$

Here, $\boldsymbol{B} \in \mathbb{R}^{L \times H}$ specifies the linear relationship between $\widetilde{\boldsymbol{a}}$ and $\boldsymbol{v}$, and $\boldsymbol{\varepsilon} \in \mathbb{R}^L$ is a Gaussian noise subject to $\mathrm{Gauss}_L(\boldsymbol{0}, \sigma^2 \boldsymbol{I}_L)$.

Suppose that we are given $M$ observed samples $\boldsymbol{V} = (\boldsymbol{v}_1, \ldots, \boldsymbol{v}_M)$ generated from the latent vectors $\boldsymbol{A}^\top = (\widetilde{\boldsymbol{a}}_1, \ldots, \widetilde{\boldsymbol{a}}_M)$, and each latent vector is subject to $\widetilde{\boldsymbol{a}} \sim \mathrm{Gauss}_H(\boldsymbol{0}, \boldsymbol{I}_H)$. Then, the probabilistic PCA model is written as Eqs. (3.1) and (3.2) with $\boldsymbol{C}_A = \boldsymbol{I}_H$. Having the prior (3.2) on $\boldsymbol{B}$, it is equivalent to the MF model.

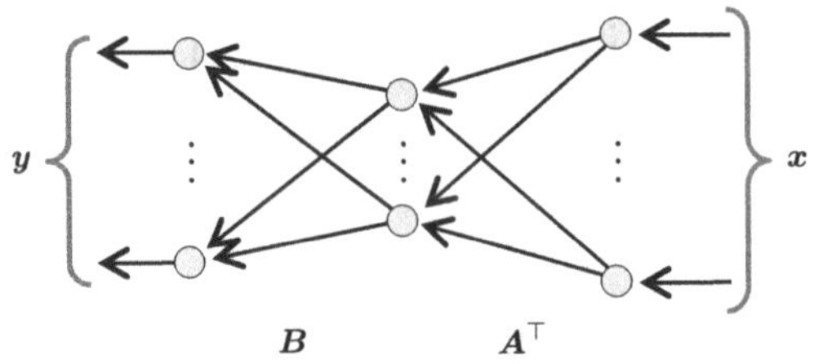


Figure 3.1 Reduced rank regression model.

If we apply VB or EVB learning, the intrinsic dimension $H$ is automatically selected without additional procedure (Bishop, 1999b). This useful property is caused by the ARD (Neal, 1996), which makes the estimators for the irrelevant column vectors of $\boldsymbol{A}$ and $\boldsymbol{B}$ zero. In Chapter 7, this phenomenon is explained in terms of *model-induced regularization (MIR)*, while in Chapter 8, a theoretical guarantee of the dimensionality estimation is given.

**Reduced Rank Regression** *Reduced rank regression (RRR)* (Baldi and Hornik, 1995; Reinsel and Velu, 1998) is aimed at learning a relation between an input vector $\boldsymbol{x} \in \mathbb{R}^M$ and an output vector $\boldsymbol{y} \in \mathbb{R}^L$ by using the following linear model:

$$\boldsymbol{y} = \boldsymbol{B}\boldsymbol{A}^\top \boldsymbol{x} + \boldsymbol{\varepsilon}, \tag{3.35}$$

where $\boldsymbol{A} \in \mathbb{R}^{M \times H}$ and $\boldsymbol{B} \in \mathbb{R}^{L \times H}$ are parameters to be estimated, and $\boldsymbol{\varepsilon} \sim \mathrm{Gauss}_L(\boldsymbol{0}, \sigma'^2 \boldsymbol{I}_L)$ is a Gaussian noise. RRR can be seen as a linear neural network (Figure 3.1), of which the model distribution is given by

$$p(\boldsymbol{y}|\boldsymbol{x}, \boldsymbol{A}, \boldsymbol{B}) = \left(2\pi\sigma'^2\right)^{-L/2} \exp\left(-\frac{1}{2\sigma'^2} \left\|\boldsymbol{y} - \boldsymbol{B}\boldsymbol{A}^\top \boldsymbol{x}\right\|^2\right). \tag{3.36}$$

Thus, we can interpret this model as first projecting the input vector $\boldsymbol{x}$ onto a lower-dimensional latent subspace by $\boldsymbol{A}^\top$ and then performing linear prediction by $\boldsymbol{B}$.

Suppose we are given $N$ pairs of input and output vectors:

$$\mathcal{D} = \left\{(\boldsymbol{x}^{(n)}, \boldsymbol{y}^{(n)}) | \boldsymbol{x}^{(n)} \in \mathbb{R}^M, \boldsymbol{y}^{(n)} \in \mathbb{R}^L, n = 1, \ldots, N\right\}. \tag{3.37}$$

Then, the likelihood of the RRR model (3.36) is expressed as

$$\begin{aligned} p(\mathcal{D}|\boldsymbol{A}, \boldsymbol{B}) &= \prod_{n=1}^{N} p(\boldsymbol{y}^{(n)}|\boldsymbol{x}^{(n)}, \boldsymbol{A}, \boldsymbol{B}) p(\boldsymbol{x}^{(n)}) \\ &\propto \exp\left(-\frac{1}{2\sigma'^2} \sum_{n=1}^{N} \left\|\boldsymbol{y}^{(n)} - \boldsymbol{B}\boldsymbol{A}^\top \boldsymbol{x}^{(n)}\right\|^2\right). \end{aligned} \tag{3.38}$$

Note that we here ignored the input distributions $\prod_{n=1}^{N} p(\boldsymbol{x}^{(n)})$ as constants (see Example 1.2 in Section 1.1.1). Let us assume that the samples are *centered*:

$$\frac{1}{N}\sum_{n=1}^{N} \boldsymbol{x}^{(n)} = \boldsymbol{0} \quad \text{and} \quad \frac{1}{N}\sum_{n=1}^{N} \boldsymbol{y}^{(n)} = \boldsymbol{0}.$$

Furthermore, let us assume that the input samples are *prewhitened* (Hyvärinen et al., 2001), i.e., they satisfy

$$\frac{1}{N}\sum_{n=1}^{N} \boldsymbol{x}^{(n)}\boldsymbol{x}^{(n)\top} = \boldsymbol{I}_M.$$

Let

$$\boldsymbol{V} = \boldsymbol{\Sigma}_{XY} = \frac{1}{N}\sum_{n=1}^{N} \boldsymbol{y}^{(n)}\boldsymbol{x}^{(n)\top} \tag{3.39}$$

be the sample *cross-covariance* matrix, and

$$\sigma^2 = \frac{\sigma'^2}{N} \tag{3.40}$$

be a rescaled noise variance. Then the exponent of the likelihood (3.38) can be written as

$$\begin{aligned}
&-\frac{1}{2\sigma'^2}\sum_{n=1}^{N} \left\|\boldsymbol{y}^{(n)} - \boldsymbol{B}\boldsymbol{A}^\top\boldsymbol{x}^{(n)}\right\|^2 \\
&= -\frac{1}{2\sigma'^2}\sum_{n=1}^{N}\left\{\left\|\boldsymbol{y}^{(n)}\right\|^2 - 2\mathrm{tr}\left(\boldsymbol{y}^{(n)}\boldsymbol{x}^{(n)\top}\boldsymbol{A}\boldsymbol{B}^\top\right) + \mathrm{tr}\left(\boldsymbol{B}\boldsymbol{A}^\top\boldsymbol{x}^{(n)}\boldsymbol{x}^{(n)\top}\boldsymbol{A}\boldsymbol{B}^\top\right)\right\} \\
&= -\frac{1}{2\sigma'^2}\left\{\sum_{n=1}^{N}\left\|\boldsymbol{y}^{(n)}\right\|^2 - 2N\mathrm{tr}\left(\boldsymbol{V}\boldsymbol{A}\boldsymbol{B}^\top\right) + N\mathrm{tr}\left(\boldsymbol{A}\boldsymbol{B}^\top\boldsymbol{B}\boldsymbol{A}^\top\right)\right\} \\
&= -\frac{1}{2\sigma'^2}\left(N\left\|\boldsymbol{V} - \boldsymbol{B}\boldsymbol{A}^\top\right\|_{\mathrm{Fro}}^2 + \sum_{n=1}^{N}\left\|\boldsymbol{y}^{(n)}\right\|^2 - N\left\|\boldsymbol{V}\right\|_{\mathrm{Fro}}^2\right) \\
&= -\frac{1}{2\sigma^2}\left\|\boldsymbol{V} - \boldsymbol{B}\boldsymbol{A}^\top\right\|_{\mathrm{Fro}}^2 - \frac{1}{2\sigma^2}\left(\frac{1}{N}\sum_{n=1}^{N}\left\|\boldsymbol{y}^{(n)}\right\|^2 - \left\|\boldsymbol{V}\right\|_{\mathrm{Fro}}^2\right).
\end{aligned} \tag{3.41}$$

The first term in Eq. (3.41) coincides with the log-likelihood of the MF model (3.1), and the second term is constant with respect to $\boldsymbol{A}$ and $\boldsymbol{B}$. Thus, RRR is reduced to MF, as far as the posteriors for $\boldsymbol{A}$ and $\boldsymbol{B}$ are concerned.

However, the second term depends on the rescaled noise variance $\sigma^2$, and therefore, should be considered when $\sigma^2$ is estimated based on the free energy minimization principle. Furthermore, the normalization constant of the likelihood (3.38) differs from that of the MF model. Taking these differences into account, the VB free energy of the RRR model (3.38) with the priors (3.2) and (3.3) is given by

$$
\begin{aligned}
2F^{\mathrm{RRR}} &= NL\log(2\pi N\sigma^2) + \frac{\frac{1}{N}\sum_{n=1}^{N}\left\|\boldsymbol{y}^{(n)}\right\|^2 - \|\boldsymbol{V}\|_{\mathrm{Fro}}^2}{\sigma^2} \\
&\quad + \frac{\left\|\boldsymbol{V} - \widehat{\boldsymbol{B}}\widehat{\boldsymbol{A}}^\top\right\|_{\mathrm{Fro}}^2}{\sigma^2} + M\log\frac{\det(\boldsymbol{C}_A)}{\det\left(\widehat{\boldsymbol{\Sigma}}_A\right)} + L\log\frac{\det(\boldsymbol{C}_B)}{\det\left(\widehat{\boldsymbol{\Sigma}}_B\right)} \\
&\quad - (L+M)H + \mathrm{tr}\left\{\boldsymbol{C}_A^{-1}\left(\widehat{\boldsymbol{A}}^\top\widehat{\boldsymbol{A}} + M\widehat{\boldsymbol{\Sigma}}_A\right) + \boldsymbol{C}_B^{-1}\left(\widehat{\boldsymbol{B}}^\top\widehat{\boldsymbol{B}} + L\widehat{\boldsymbol{\Sigma}}_B\right)\right. \\
&\quad \left. + \sigma^{-2}\left(-\widehat{\boldsymbol{A}}^\top\widehat{\boldsymbol{A}}\widehat{\boldsymbol{B}}^\top\widehat{\boldsymbol{B}} + \left(\widehat{\boldsymbol{A}}^\top\widehat{\boldsymbol{A}} + M\widehat{\boldsymbol{\Sigma}}_A\right)\left(\widehat{\boldsymbol{B}}^\top\widehat{\boldsymbol{B}} + L\widehat{\boldsymbol{\Sigma}}_B\right)\right)\right\}. \qquad (3.42)
\end{aligned}
$$

Note that the difference from Eq. (3.23) is only in the first two terms. Accordingly, the stationary conditions for the variational parameters $\widehat{\boldsymbol{A}}, \widehat{\boldsymbol{B}}, \widehat{\boldsymbol{\Sigma}}_A$, and $\widehat{\boldsymbol{\Sigma}}_B$, and those for the prior covariances $\boldsymbol{C}_A$ and $\boldsymbol{C}_B$ (in EVB learning) are the same, i.e., the update rules given by Eqs. (3.19) through (3.22), (3.26), and (3.27) are valid for the RRR model. The update rule for the rescaled noise variance is different from Eq. (3.28), and given by

$$
(\sigma^2)^{\mathrm{RRR}} = \frac{\frac{1}{N}\sum_{n=1}^{N}\left\|\boldsymbol{y}^{(n)}\right\|^2 - \mathrm{tr}\left(2\boldsymbol{V}^\top\widehat{\boldsymbol{B}}\widehat{\boldsymbol{A}}^\top\right) + \mathrm{tr}\left((\widehat{\boldsymbol{A}}^\top\widehat{\boldsymbol{A}} + M\widehat{\boldsymbol{\Sigma}}_A)(\widehat{\boldsymbol{B}}^\top\widehat{\boldsymbol{B}} + L\widehat{\boldsymbol{\Sigma}}_B)\right)}{NL}, \qquad (3.43)
$$

which was obtained from the derivative of Eq. (3.42), instead of Eq. (3.23), with respect to $\sigma^2$.

Once the rescaled noise variance $\sigma^2$ is estimated, Eq. (3.40) gives the original noise variance $\sigma'^2$ of the RRR model (3.38).

## 3.2 Matrix Factorization with Missing Entries

One of the major applications of MF is *collaborative filtering (CF)*, where only a part of the entries in $\boldsymbol{V}$ are observed, and the task is to predict missing entries. We can derive a VB algorithm for this scenario, similarly to the fully observed case.

### 3.2.1 VB Learning for MF with Missing Entries

To express missing entris, the likelihood (3.1) should be replaced with

$$
p(\boldsymbol{V}|\boldsymbol{A},\boldsymbol{B}) \propto \exp\left(-\frac{1}{2\sigma^2}\left\|\mathcal{P}_\Lambda(\boldsymbol{V}) - \mathcal{P}_\Lambda\left(\boldsymbol{B}\boldsymbol{A}^\top\right)\right\|_{\mathrm{Fro}}^2\right), \qquad (3.44)
$$

where $\Lambda$ denotes the set of observed indices, and $\mathcal{P}_\Lambda(\boldsymbol{V})$ denotes the matrix of the same size as $\boldsymbol{V}$ with its entries given by

$$(\mathcal{P}_\Lambda(\boldsymbol{V}))_{l,m} = \begin{cases} V_{l,m} & \text{if } (l,m) \in \Lambda, \\ 0 & \text{otherwise.} \end{cases}$$

**Conditional Conjugacy**

Since the likelihood (3.44) is still in a Gaussian form of $\boldsymbol{A}$ if $\boldsymbol{B}$ is regarded as a constant, or vise versa, the conditional conjugacy with respect to $\boldsymbol{A}$ and $\boldsymbol{B}$, respectively, still holds if we adopt the Gaussian priors (3.2) and (3.3):

$$p(\boldsymbol{A}) \propto \exp\left(-\frac{1}{2}\mathrm{tr}\left(\boldsymbol{A}\boldsymbol{C}_A^{-1}\boldsymbol{A}^\top\right)\right),$$

$$p(\boldsymbol{B}) \propto \exp\left(-\frac{1}{2}\mathrm{tr}\left(\boldsymbol{B}\boldsymbol{C}_B^{-1}\boldsymbol{B}^\top\right)\right).$$

The posterior will be still Gaussian, but in a broader class than the fully observed case, as will be seen shortly.

**Variational Bayesian Algorithm**

With the missing entries, the stationary condition (3.7) becomes

$$\begin{aligned} r_A(\boldsymbol{A}) &\propto \exp\left(-\frac{1}{2}\mathrm{tr}\left(\boldsymbol{A}\boldsymbol{C}_A^{-1}\boldsymbol{A}^\top\right) - \frac{1}{2\sigma^2}\left\langle\left\|\mathcal{P}_\Lambda(\boldsymbol{V}) - \mathcal{P}_\Lambda\left(\boldsymbol{B}\boldsymbol{A}^\top\right)\right\|_{\mathrm{Fro}}^2\right\rangle_{r_B(\boldsymbol{B})}\right) \\ &\propto \exp\left(-\frac{1}{2}\mathrm{tr}\left(\boldsymbol{A}\boldsymbol{C}_A^{-1}\boldsymbol{A}^\top\right)\right. \\ &\qquad \left. + \sigma^{-2}\sum_{(l,m)\in\Lambda}\left\langle -2V_{l,m}\sum_{h=1}^{H} B_{l,h}A_{m,h} + \sum_{h=1}^{H}\sum_{h'=1}^{H} B_{l,h}B_{l,h'}A_{m,h}A_{m,h'}\right\rangle_{r_B(\boldsymbol{B})}\right) \\ &\propto \exp\left(-\frac{\sum_{m=1}^{M}\left((\widetilde{\boldsymbol{a}}_m - \widehat{\widetilde{\boldsymbol{a}}}_m)^\top \widehat{\boldsymbol{\Sigma}}_{A,m}^{-1}(\widetilde{\boldsymbol{a}}_m - \widehat{\widetilde{\boldsymbol{a}}}_m)\right)}{2}\right), \end{aligned} \tag{3.45}$$

where

$$\widehat{\widetilde{\boldsymbol{a}}}_m = \sigma^{-2}\widehat{\boldsymbol{\Sigma}}_{A,m}\sum_{l:(l,m)\in\Lambda} V_{l,m}\left\langle\widetilde{\boldsymbol{b}}_l\right\rangle_{r_B(\boldsymbol{B})}, \tag{3.46}$$

$$\widehat{\boldsymbol{\Sigma}}_{A,m} = \sigma^2\left(\sum_{l:(l,m)\in\Lambda}\left\langle\widetilde{\boldsymbol{b}}_l\widetilde{\boldsymbol{b}}_l^\top\right\rangle_{r_B(\boldsymbol{B})} + \sigma^2\boldsymbol{C}_A^{-1}\right)^{-1}. \tag{3.47}$$

Here, $\sum_{(l,m)\in\Lambda}$ denotes the sum over $l$ and $m$ such that $(l,m) \in \Lambda$, and $\sum_{l:(l,m)\in\Lambda}$ denotes the sum over $l$ such that $(l,m) \in \Lambda$ for given $m$.

Similarly, we have

$$r_B(\boldsymbol{B}) \propto \exp\left(-\frac{1}{2}\mathrm{tr}\left(\boldsymbol{B}\boldsymbol{C}_B^{-1}\boldsymbol{B}^\top\right) - \frac{1}{2\sigma^2}\left\langle\left\|\mathcal{P}_\Lambda(\boldsymbol{V}) - \mathcal{P}_\Lambda\left(\boldsymbol{B}\boldsymbol{A}^\top\right)\right\|_{\mathrm{Fro}}^2\right\rangle_{r_A(\boldsymbol{A})}\right)$$

$$\propto \exp\left(-\frac{\sum_{l=1}^{L}\left((\widetilde{\boldsymbol{b}}_m - \widehat{\widetilde{\boldsymbol{b}}}_l)^\top \widehat{\boldsymbol{\Sigma}}_{B,l}^{-1}(\widetilde{\boldsymbol{b}}_l - \widehat{\widetilde{\boldsymbol{b}}}_l)\right)}{2}\right), \tag{3.48}$$

where

$$\widehat{\widetilde{\boldsymbol{b}}}_l = \sigma^{-2}\widehat{\boldsymbol{\Sigma}}_{B,l}\sum_{m:(l,m)\in\Lambda} V_{l,m}\left\langle\widetilde{\boldsymbol{a}}_m\right\rangle_{r_A(\boldsymbol{A})}, \tag{3.49}$$

$$\widehat{\boldsymbol{\Sigma}}_{B,l} = \sigma^2\left(\sum_{m:(l,m)\in\Lambda}\left\langle\widetilde{\boldsymbol{a}}_m\widetilde{\boldsymbol{a}}_m^\top\right\rangle_{r_A(\boldsymbol{A})} + \sigma^2\boldsymbol{C}_B^{-1}\right)^{-1}. \tag{3.50}$$

Eqs. (3.45) and (3.48) imply that $\boldsymbol{A}$ and $\boldsymbol{B}$ are Gaussian in the following form:

$$r_A(\boldsymbol{A}) = \mathrm{MGauss}_{M,H}(\boldsymbol{A}; \widehat{\boldsymbol{A}}, \breve{\boldsymbol{\Sigma}}_A) = \prod_{m=1}^{M}\mathrm{Gauss}_H(\widetilde{\boldsymbol{a}}_m; \widehat{\widetilde{\boldsymbol{a}}}_m, \widehat{\boldsymbol{\Sigma}}_{A,m}),$$

$$r_B(\boldsymbol{B}) = \mathrm{MGauss}_{L,H}(\boldsymbol{B}; \widehat{\boldsymbol{B}}, \breve{\boldsymbol{\Sigma}}_B) = \prod_{l=1}^{L}\mathrm{Gauss}_H(\widetilde{\boldsymbol{b}}_m; \widehat{\widetilde{\boldsymbol{b}}}_l, \widehat{\boldsymbol{\Sigma}}_{B,l}),$$

where

$$\breve{\boldsymbol{\Sigma}}_A = \begin{pmatrix} \widehat{\boldsymbol{\Sigma}}_{A,1} & \boldsymbol{0} & \cdots & \boldsymbol{0} \\ \boldsymbol{0} & \widehat{\boldsymbol{\Sigma}}_{A,2} & & \vdots \\ \vdots & & \ddots & \boldsymbol{0} \\ \boldsymbol{0} & \cdots & \boldsymbol{0} & \widehat{\boldsymbol{\Sigma}}_{A,M} \end{pmatrix}, \quad \breve{\boldsymbol{\Sigma}}_B = \begin{pmatrix} \widehat{\boldsymbol{\Sigma}}_{B,1} & \boldsymbol{0} & \cdots & \boldsymbol{0} \\ \boldsymbol{0} & \widehat{\boldsymbol{\Sigma}}_{B,2} & & \vdots \\ \vdots & & \ddots & \boldsymbol{0} \\ \boldsymbol{0} & \cdots & \boldsymbol{0} & \widehat{\boldsymbol{\Sigma}}_{B,L} \end{pmatrix}.$$

Note that the posterior covariances cannot be expressed with a Kronecker product, unlike the fully observed case. However, the posteriors are Gaussian, and moments are given by

$$\left\langle\widetilde{\boldsymbol{a}}_m\right\rangle_{r_A(\boldsymbol{A})} = \widehat{\widetilde{\boldsymbol{a}}}_m,$$

$$\left\langle\widetilde{\boldsymbol{a}}_m\widetilde{\boldsymbol{a}}_m^\top\right\rangle_{r_A(\boldsymbol{A})} = \widehat{\widetilde{\boldsymbol{a}}}_m\widehat{\widetilde{\boldsymbol{a}}}_m^\top + \widehat{\boldsymbol{\Sigma}}_{A,m},$$

$$\left\langle\widetilde{\boldsymbol{b}}_l\right\rangle_{r_B(\boldsymbol{B})} = \widehat{\widetilde{\boldsymbol{b}}}_l,$$

$$\left\langle\widetilde{\boldsymbol{b}}_l\widetilde{\boldsymbol{b}}_l^\top\right\rangle_{r_B(\boldsymbol{B})} = \widehat{\widetilde{\boldsymbol{b}}}_l\widehat{\widetilde{\boldsymbol{b}}}_l^\top + \widehat{\boldsymbol{\Sigma}}_{B,l}.$$

Thus, Eqs. (3.46), (3.47), (3.49), and (3.50) lead to

$$\widehat{\overline{\boldsymbol{a}}}_m = \sigma^{-2}\widehat{\boldsymbol{\Sigma}}_{A,m} \sum_{l:(l,m)\in\Lambda} V_{l,m}\widehat{\overline{\boldsymbol{b}}}_l, \tag{3.51}$$

$$\widehat{\boldsymbol{\Sigma}}_{A,m} = \sigma^2 \left( \sum_{l:(l,m)\in\Lambda} \left( \widehat{\overline{\boldsymbol{b}}}_l \widehat{\overline{\boldsymbol{b}}}_l^\top + \widehat{\boldsymbol{\Sigma}}_{B,l} \right) + \sigma^2 \boldsymbol{C}_A^{-1} \right)^{-1}, \tag{3.52}$$

$$\widehat{\overline{\boldsymbol{b}}}_l = \sigma^{-2}\widehat{\boldsymbol{\Sigma}}_{B,l} \sum_{m:(l,m)\in\Lambda} V_{l,m}\widehat{\overline{\boldsymbol{a}}}_m, \tag{3.53}$$

$$\widehat{\boldsymbol{\Sigma}}_{B,l} = \sigma^2 \left( \sum_{m:(l,m)\in\Lambda} \left( \widehat{\overline{\boldsymbol{a}}}_m \widehat{\overline{\boldsymbol{a}}}_m^\top + \widehat{\boldsymbol{\Sigma}}_{A,m} \right) + \sigma^2 \boldsymbol{C}_B^{-1} \right)^{-1}, \tag{3.54}$$

which are used as update rules for local search (Lim and Teh, 2007).

### Free Energy as a Function of Variational Parameters

An explicit form of the free energy can be obtained in a similar fashion to the fully observed case:

$$\begin{aligned}
2F &= \#(\Lambda)\cdot\log(2\pi\sigma^2) + M\log\det(\boldsymbol{C}_A) + L\log\det(\boldsymbol{C}_B) \\
&\quad - \sum_{m=1}^{M}\log\det\left(\widehat{\boldsymbol{\Sigma}}_{A,m}\right) - \sum_{l=1}^{L}\log\det\left(\widehat{\boldsymbol{\Sigma}}_{B,l}\right) - (L+M)H \\
&\quad + \operatorname{tr}\left\{ \boldsymbol{C}_A^{-1}\left( \widehat{\boldsymbol{A}}^\top\widehat{\boldsymbol{A}} + \sum_{m=1}^{M}\widehat{\boldsymbol{\Sigma}}_{A,m} \right) + \boldsymbol{C}_B^{-1}\left( \widehat{\boldsymbol{B}}^\top\widehat{\boldsymbol{B}} + \sum_{l=1}^{L}\widehat{\boldsymbol{\Sigma}}_{B,l} \right) \right\} \\
&\quad + \sigma^{-2}\sum_{(l,m)\in\Lambda}\left( V_{l,m}^2 - 2V_{l,m}\widehat{\overline{\boldsymbol{a}}}_m^\top\widehat{\overline{\boldsymbol{b}}}_l + \operatorname{tr}\left\{ \left( \widehat{\overline{\boldsymbol{a}}}_m\widehat{\overline{\boldsymbol{a}}}_m^\top + \widehat{\boldsymbol{\Sigma}}_{A,m} \right)\left( \widehat{\overline{\boldsymbol{b}}}_l\widehat{\overline{\boldsymbol{b}}}_l^\top + \widehat{\boldsymbol{\Sigma}}_{B,l} \right) \right\} \right),
\end{aligned} \tag{3.55}$$

where $\#(\Lambda)$ denotes the number of observed entries.

### Empirical Variational Bayesian Algorithm

By taking derivatives of the free energy (3.55), we can derive update rules for the hyperparameters:

$$c_{a_h}^2 = \frac{\left\|\widehat{\boldsymbol{a}}_h\right\|^2 + \left(\sum_{m=1}^{M}\widehat{\boldsymbol{\Sigma}}_{A,m}\right)_{h,h}}{M}, \tag{3.56}$$

$$c_{b_h}^2 = \frac{\left\|\widehat{\boldsymbol{b}}_h\right\|^2 + \left(\sum_{l=1}^{L}\widehat{\boldsymbol{\Sigma}}_{B,l}\right)_{h,h}}{L}, \tag{3.57}$$

**Algorithm 2** EVB learning for matrix factorization with missing entries.

1: Initialize the variational parameters $(\widehat{\boldsymbol{A}}, \{\widehat{\boldsymbol{\Sigma}}_{A,m}\}_{m=1}^{M}, \widehat{\boldsymbol{B}}, \{\widehat{\boldsymbol{\Sigma}}_{B,l}\}_{l=1}^{L})$, and the hyperparameters $(\boldsymbol{C}_A, \boldsymbol{C}_B, \sigma^2)$, for example, $\widehat{A}_{m,h}, \widehat{B}_{l,h} \sim \mathrm{Gauss}_1(0, \tau)$, $\widehat{\boldsymbol{\Sigma}}_{A,m} = \widehat{\boldsymbol{\Sigma}}_{B,l} = \boldsymbol{C}_A = \boldsymbol{C}_B = \tau \boldsymbol{I}_H$, and $\sigma^2 = \tau^2$ for $\tau^2 = \sum_{(l,m)\in\Lambda} V_{l,m}^2 / \#(\Lambda)$.
2: Apply (substitute the right-hand side into the left-hand side) Eqs. (3.52), (3.51), (3.54), and (3.53) to update $\{\widehat{\boldsymbol{\Sigma}}_{A,m}\}_{m=1}^{M}, \widehat{\boldsymbol{A}}, \{\widehat{\boldsymbol{\Sigma}}_{B,l}\}_{l=1}^{L}$, and $\widehat{\boldsymbol{B}}$, respectively.
3: Apply Eqs. (3.56) and (3.57) for all $h = 1, \ldots, H$, and Eq. (3.58) to update $\boldsymbol{C}_A, \boldsymbol{C}_B$, and $\sigma^2$, respectively.
4: Prune the $h$th component if $c_{a_h}^2 c_{b_h}^2 < \varepsilon$, where $\varepsilon > 0$ is a threshold, e.g., set to $\varepsilon = 10^{-4}$.
5: Evaluate the free energy (3.55).
6: Iterate Steps 2 through 5 until convergence (until the energy decrease becomes smaller than a threshold).

$$\sigma^2 = \frac{\sum_{(l,m)\in\Lambda} \left( V_{l,m} - 2V_{l,m} \widehat{\widetilde{\boldsymbol{a}}}_m^{\top} \widehat{\widetilde{\boldsymbol{b}}}_l + \mathrm{tr}\left\{ \left( \widehat{\widetilde{\boldsymbol{a}}}_m \widehat{\widetilde{\boldsymbol{a}}}_m^{\top} + \widehat{\boldsymbol{\Sigma}}_{A,m} \right) \left( \widehat{\widetilde{\boldsymbol{b}}}_l \widehat{\widetilde{\boldsymbol{b}}}_l^{\top} + \widehat{\boldsymbol{\Sigma}}_{B,l} \right) \right\} \right)}{\#(\Lambda)}. \quad (3.58)$$

Algorithm 2 summarizes the EVB algorithm for MF with missing entries. Again, if we appropriately set the hyperparameters $(\boldsymbol{C}_A, \boldsymbol{C}_B, \sigma^2)$ in Step 1 and skip Steps 3 and 4, Algorithm 2 is reduced to (nonempirical) VB learning.

**Simple Variational Bayesian Learning (with Columnwise Independence)**
Similarly to the fully observed case, we can reduce the computational burden and the memory requirement of VB learning by adopting the columnwise independence (Ilin and Raiko, 2010):

$$r(\boldsymbol{A}, \boldsymbol{B}) = \prod_{h=1}^{H} r_{a_h}(\boldsymbol{a}_h) \prod_{h=1}^{H} r_{b_h}(\boldsymbol{b}_h). \quad (3.59)$$

By applying the same procedure as the matrixwise independence case, we can derive the solution to

$$\widehat{r} = \underset{r}{\mathrm{argmin}} F(r) \quad \text{s.t.} \quad r(\boldsymbol{A}, \boldsymbol{B}) = \prod_{h=1}^{H} r_{a_h}(\boldsymbol{a}_h) \prod_{h=1}^{H} r_{b_h}(\boldsymbol{b}_h), \quad (3.60)$$

which is in the form of the matrix variate Gaussian,

$$r_A(\boldsymbol{A}) = \mathrm{MGauss}_{M,H}(\boldsymbol{A}; \widehat{\boldsymbol{A}}, \breve{\boldsymbol{\Sigma}}_A) = \prod_{m=1}^{M}\prod_{h=1}^{H} \mathrm{Gauss}_1(A_{m,h}; \widehat{A}_{m,h}, \widehat{\sigma}^2_{A_{m,h}}),$$

$$r_B(\boldsymbol{B}) = \mathrm{MGauss}_{L,H}(\boldsymbol{B}; \widehat{\boldsymbol{B}}, \breve{\boldsymbol{\Sigma}}_B) = \prod_{l=1}^{L}\prod_{h=1}^{H} \mathrm{Gauss}_1(B_{l,h}; \widehat{B}_{l,h}, \widehat{\sigma}^2_{B_{l,h}}),$$

with diagonal posterior covariances, i.e.,

$$\breve{\boldsymbol{\Sigma}}_A = \begin{pmatrix} \widehat{\boldsymbol{\Sigma}}_{A,1} & \mathbf{0} & \cdots & \mathbf{0} \\ \mathbf{0} & \widehat{\boldsymbol{\Sigma}}_{A,2} & & \vdots \\ \vdots & & \ddots & \mathbf{0} \\ \mathbf{0} & \cdots & \mathbf{0} & \widehat{\boldsymbol{\Sigma}}_{A,M} \end{pmatrix}, \quad \breve{\boldsymbol{\Sigma}}_B = \begin{pmatrix} \widehat{\boldsymbol{\Sigma}}_{B,1} & \mathbf{0} & \cdots & \mathbf{0} \\ \mathbf{0} & \widehat{\boldsymbol{\Sigma}}_{B,2} & & \vdots \\ \vdots & & \ddots & \mathbf{0} \\ \mathbf{0} & \cdots & \mathbf{0} & \widehat{\boldsymbol{\Sigma}}_{B,L} \end{pmatrix},$$

for

$$\widehat{\boldsymbol{\Sigma}}_{A,m} = \mathbf{Diag}(\widehat{\sigma}^2_{A_{m,1}}, \ldots, \widehat{\sigma}^2_{A_{m,H}}),$$
$$\widehat{\boldsymbol{\Sigma}}_{B,l} = \mathbf{Diag}(\widehat{\sigma}^2_{B_{l,1}}, \ldots, \widehat{\sigma}^2_{B_{l,H}}).$$

The stationary conditions are given as follows: for all $l = 1, \ldots, L$, $m = 1, \ldots, M$, and $h = 1, \ldots, H$,

$$\widehat{A}_{m,h} = \frac{\widehat{\sigma}^2_{A_{m,h}}}{\sigma^2} \sum_{l;(l,m)\in\Lambda} \left( V_{l,m} - \sum_{h'\neq h} \widehat{B}_{l,h'}\widehat{A}_{m,h'} \right) \widehat{B}_{l,h}, \tag{3.61}$$

$$\widehat{\sigma}^2_{A_{m,h}} = \sigma^2 \left( \sum_{l;(l,m)\in\Lambda} \left( \widehat{B}^2_{l,h} + \widehat{\sigma}^2_{B_{l,h}} \right) + \frac{\sigma^2}{c^2_{a_h}} \right)^{-1}, \tag{3.62}$$

$$\widehat{B}_{l,h} = \frac{\widehat{\sigma}^2_{B_{l,h}}}{\sigma^2} \sum_{m;(l,m)\in\Lambda} \left( V_{l,m} - \sum_{h'\neq h} \widehat{A}_{m,h'}\widehat{B}_{l,h'} \right) \widehat{A}_{m,h}, \tag{3.63}$$

$$\widehat{\sigma}^2_{B_{l,h}} = \sigma^2 \left( \sum_{m;(l,m)\in\Lambda} \left( \widehat{A}^2_{m,h} + \widehat{\sigma}^2_{A_{m,h}} \right) + \frac{\sigma^2}{c^2_{b_h}} \right)^{-1}. \tag{3.64}$$

The free energy is given by Eq. (3.55) with the posterior covariances $\{\widehat{\boldsymbol{\Sigma}}_{A,m}, \widehat{\boldsymbol{\Sigma}}_{B,l}\}$ restricted to be diagonal. The stationary conditions for the hyperparameters are unchanged, and given by Eqs. (3.56) through (3.58). Therefore, Algorithm 2 with Eqs. (3.61) through (3.64) substituted for Eqs. (3.51) through (3.54) gives a local solution to the VB problem (3.60) with the columnwise independence constraint.

SimpleVB learning is much more practical when missing entries exist. In the fully observed case, the posterior covariances $\widehat{\boldsymbol{\Sigma}}_A$ and $\widehat{\boldsymbol{\Sigma}}_B$ are common to all rows of $\boldsymbol{A}$ and to all rows of $\boldsymbol{B}$, respectively, while in the partially observed case, we need to store the posterior covariances $\widehat{\boldsymbol{\Sigma}}_{A,m}$ and $\widehat{\boldsymbol{\Sigma}}_{B,l}$ for all $m = 1, \ldots, M$ and $l = 1, \ldots, L$. Since $L$ and $M$ can be huge, e.g., in collaborative filtering applications, the required memory size is significantly reduced by restricting the covariances to be diagonal.

## 3.3 Tensor Factorization

A matrix is a two-dimensional array of numbers. We can extend this notion to an $N$-dimensional array, which is called an *N-mode tensor*. Namely, a tensor $\mathcal{V} \in \mathbb{R}^{M^{(1)} \times \cdots \times M^{(N)}}$ consists of $\prod_{n=1}^{N} M^{(n)}$ entries lying in an $N$-dimensional array, where $M^{(n)}$ denotes the length in *mode n*. In this section, we derive VB learning for tensor factorization.

### 3.3.1 Tucker Factorization

Similarly to the rank of a matrix, we can control the degree of freedom of a tensor by controlling its *tensor rank*. Although there are a few different definitions of the tensor rank and corresponding ways of factorization, we here focus on *Tucker factorization (TF)* (Tucker, 1996; Kolda and Bader, 2009) defined as follows:

$$\mathcal{V} = \mathcal{G} \times_1 \boldsymbol{A}^{(1)} \cdots \times_N \boldsymbol{A}^{(N)} + \mathcal{E},$$

where $\mathcal{V} \in \mathbb{R}^{M^{(1)} \times \cdots \times M^{(N)}}$, $\mathcal{G} \in \mathbb{R}^{H^{(1)} \times \cdots \times H^{(N)}}$, and $\{\boldsymbol{A}^{(n)} \in \mathbb{R}^{M^{(n)} \times H^{(n)}}\}$ are an observed tensor, a *core tensor*, and *factor matrices*, respectively. $\mathcal{E} \in \mathbb{R}^{M^{(1)} \times \cdots \times M^{(N)}}$ is noise and $\times_n$ denotes the *n-mode tensor product*. *Parafac* (Harshman, 1970), another popular way of tensor factorization, can be seen as a special case of Tucker factorization where the core tensor is *superdiagonal*, i.e., only the entries $G_{h^{(1)},\ldots,h^{(N)}}$ for $h^{(1)} = h^{(2)} = \cdots = h^{(N)}$ are nonzero.

### 3.3.2 VB Learning for TF

The probabilistic model for MF is straightforwardly extended to TF (Chu and Ghahramani, 2009; Mørup and Hansen, 2009). Assume Gaussian noise and Gaussian priors:

$$p(\mathcal{V}|\mathcal{G}, \{\boldsymbol{A}^{(n)}\}) \propto \exp\left( -\frac{\left\| \mathcal{V} - \mathcal{G} \times_1 \boldsymbol{A}^{(1)} \cdots \times_N \boldsymbol{A}^{(N)} \right\|^2}{2\sigma^2} \right), \tag{3.65}$$

$$p(\mathcal{G}) \propto \exp\left(-\frac{\mathbf{vec}(\mathcal{G})^\top \left(\boldsymbol{C}_{G^{(N)}} \otimes \cdots \otimes \boldsymbol{C}_{G^{(1)}}\right)^{-1} \mathbf{vec}(\mathcal{G})}{2}\right), \tag{3.66}$$

$$p(\{\boldsymbol{A}^{(n)}\}) \propto \exp\left(-\frac{\sum_{n=1}^{N} \mathrm{tr}(\boldsymbol{A}^{(n)} \boldsymbol{C}_{A^{(n)}}^{-1} \boldsymbol{A}^{(n)\top})}{2}\right), \tag{3.67}$$

where $\otimes$ and $\mathbf{vec}(\cdot)$ denote the *Kronecker product* and the *vectorization operator*, respectively. $\{\boldsymbol{C}_{G^{(n)}}\}$ and $\{\boldsymbol{C}_{A^{(n)}}\}$ are the prior covariances restricted to be diagonal, i.e.,

$$\boldsymbol{C}_{G^{(n)}} = \mathbf{Diag}\left(c^2_{g_1^{(n)}}, \ldots, c^2_{g_{H^{(n)}}^{(n)}}\right),$$

$$\boldsymbol{C}_{A^{(n)}} = \mathbf{Diag}\left(c^2_{a_1^{(n)}}, \ldots, c^2_{a_{H^{(n)}}^{(n)}}\right).$$

We denote $\breve{\boldsymbol{C}}_G = \boldsymbol{C}_{G^{(N)}} \otimes \cdots \otimes \boldsymbol{C}_{G^{(1)}}$.

### Conditional Conjugacy

Since the TF model is multilinear, the likelihood (3.65) is in the Gaussian form of the core tensor $\mathcal{G}$ and of each of the factor matrices $\{\boldsymbol{A}^{(n)}\}$, if the others are fixed as constants. Therefore, the Gaussian priors (3.66) and (3.67) are conditionally conjugate for each parameter, and the posterior will be Gaussian.

### Variational Bayesian Algorithm

Based on the conditional conjugacy, we impose the following constraint on the VB posterior:

$$r(\mathcal{G}, \{\boldsymbol{A}^{(n)}\}) = r(\mathcal{G}) \prod_{n=1}^{N} r(\boldsymbol{A}^{(n)}).$$

Then, the free energy can be written as

$$F(r) = \int r(\mathcal{G}) \left(\prod_{n=1}^{N} r(\boldsymbol{A}^{(n)})\right) \left(\log p(\mathcal{V}, \mathcal{G}, \{\boldsymbol{A}^{(n)}\}) - \log r(\mathcal{G}) - \sum_{n=1}^{N} \log r(\boldsymbol{A}^{(n)})\right) \cdot d\mathcal{G} \left(\prod_{n=1}^{N} d\boldsymbol{A}^{(n)}\right). \tag{3.68}$$

Using the variational method, we obtain

$$0 = \int \left(\prod_{n=1}^{N} r(\boldsymbol{A}^{(n)})\right) \left(\log p(\mathcal{V}, \mathcal{G}, \{\boldsymbol{A}^{(n)}\}) - \log r(\mathcal{G}) - \sum_{n=1}^{N} \log r(\boldsymbol{A}^{(n)}) - 1\right) \cdot \left(\prod_{n=1}^{N} d\boldsymbol{A}^{(n)}\right),$$

and therefore

$$r(\mathcal{G}) \propto \exp\langle \log p(\mathcal{V}, \mathcal{G}, \{A^{(n)}\}) \rangle_{r(\{A^{(n)}\})}$$
$$\propto p(\mathcal{G}) \exp\langle \log p(\mathcal{V}|\mathcal{G}, \{A^{(n)}\}) \rangle_{r(\{A^{(n)}\})}. \tag{3.69}$$

Similarly, we can also obtain

$$0 = \int r(\mathcal{G}) \left( \prod_{n' \neq n} r(A^{(n)}) \right) \left( \log p(\mathcal{V}, \mathcal{G}, \{A^{(n)}\}) - \log r(\mathcal{G}) - \sum_{n=1}^{N} \log r(A^{(n)}) - 1 \right)$$
$$\cdot \left( \prod_{n' \neq n} dA^{(n)} \right),$$

and therefore

$$r(A^{(n)}) \propto \exp\langle \log p(\mathcal{V}, \mathcal{G}, \{A^{(n)}\}) \rangle_{r(\mathcal{G}) r(\{A^{(n')}\}_{n' \neq n})}$$
$$\propto p(A^{(n)}) \exp\langle \log p(\mathcal{V}|\mathcal{G}, \{A^{(n)}\}) \rangle_{r(\mathcal{G}) r(\{A^{(n')}\}_{n' \neq n})}. \tag{3.70}$$

Eqs. (3.69) and (3.70) imply that the VB posteriors are Gaussian. The expectation in Eq. (3.69) can be calculated as follows:

$$\langle \log p(\mathcal{V}|\mathcal{G}, \{A^{(n)}\}) \rangle_{r(\{A^{(n)}\})}$$
$$= -\frac{1}{2\sigma^2} \left\langle \left\| \mathcal{V} - \mathcal{G}(\times_1 A^{(1)}) \cdots (\times_N A^{(N)}) \right\|^2 \right\rangle_{r(\{A^{(n)}\})} + \text{const.}$$
$$= -\frac{1}{2\sigma^2} \Big\langle -2\mathbf{vec}(\mathcal{V})^\top (A^{(N)} \otimes \cdots \otimes A^{(1)}) \mathbf{vec}(\mathcal{G})$$
$$+ \mathbf{vec}(\mathcal{G})^\top (A^{(N)\top} A^{(N)} \otimes \cdots \otimes A^{(1)\top} A^{(1)}) \mathbf{vec}(\mathcal{G}) \Big\rangle_{r(\{A^{(n)}\})} + \text{const.}$$

Substituting the preceding calculation and the prior (3.66) into Eq. (3.69) gives

$$\log r(\mathcal{G}) = \log p(\mathcal{G}) \langle \log p(\mathcal{V}|\mathcal{G}, \{A^{(n)}\}) \rangle_{r(\{A^{(n)}\})} + \text{const.}$$
$$= -\frac{1}{2} (\breve{g} - \widehat{\breve{g}})^\top \widehat{\breve{\Sigma}}_G^{-1} (\breve{g} - \widehat{\breve{g}}) + \text{const.},$$

where

$$\breve{g} = \mathbf{vec}(\mathcal{G}),$$
$$\breve{v} = \mathbf{vec}(\mathcal{V}),$$
$$\widehat{\breve{g}} = \frac{\widehat{\breve{\Sigma}}_G}{\sigma^2} \left( \widehat{A}^{(N)} \otimes \cdots \otimes \widehat{A}^{(1)} \right)^\top \breve{v}, \tag{3.71}$$
$$\widehat{\breve{\Sigma}}_G = \sigma^2 \left( \left\langle A^{(N)\top} A^{(N)} \otimes \cdots \otimes A^{(1)\top} A^{(1)} \right\rangle_{r(\{A^{(n)}\})} + \sigma^2 \breve{C}_G^{-1} \right)^{-1}. \tag{3.72}$$

Similarly, the expectation in Eq. (3.70) can be calculated as follows:

$$\begin{aligned}
&\langle \log p(\mathcal{V}|\mathcal{G}, \boldsymbol{A}^{(n)})\rangle_{r(\mathcal{G})r(\{\boldsymbol{A}^{(n')}\}_{n'\neq n})} \\
&= -\frac{1}{2\sigma^2}\left\langle \left\| \mathcal{V} - \mathcal{G}(\times_1 \boldsymbol{A}^{(1)})\cdots(\times_N \boldsymbol{A}^{(N)}) \right\|^2 \right\rangle_{r(\mathcal{G})r(\{\boldsymbol{A}^{(n')}\}_{n'\neq n})} + \text{const.} \\
&= -\frac{1}{2\sigma^2}\Bigg\{ \mathrm{tr}\left( -2\boldsymbol{V}_{(n)}^\top \boldsymbol{A}^{(n)} \widehat{\boldsymbol{G}}_{(n)} (\widehat{\boldsymbol{A}}^{(N)} \otimes \cdots \otimes \widehat{\boldsymbol{A}}^{(n+1)} \otimes \widehat{\boldsymbol{A}}^{(n-1)} \cdots \otimes \widehat{\boldsymbol{A}}^{(1)})^\top \right) \\
&\quad + \mathrm{tr}\Big( \boldsymbol{A}^{(n)} \Big\langle \boldsymbol{G}_{(n)} (\boldsymbol{A}^{(N)\top}\boldsymbol{A}^{(N)} \otimes \cdots \otimes \boldsymbol{A}^{(n+1)\top}\boldsymbol{A}^{(n+1)} \\
&\qquad\qquad \otimes \boldsymbol{A}^{(n-1)\top}\boldsymbol{A}^{(n-1)} \cdots \otimes \boldsymbol{A}^{(1)\top}\boldsymbol{A}^{(1)}) \boldsymbol{G}_{(n)}^\top \Big\rangle_{r(\mathcal{G})r(\{\boldsymbol{A}^{(n')}\}_{n'\neq n})} \boldsymbol{A}^{(n)\top} \Big) \Bigg\}.
\end{aligned}$$

Substituting the preceding calculation and the prior (3.67) into Eq. (3.70) gives

$$\begin{aligned}
\log r(\boldsymbol{A}^{(n)}) &= \log p(\boldsymbol{A}^{(n)}) \exp\langle \log p(\mathcal{V}|\mathcal{G}, \{\boldsymbol{A}^{(n)}\})\rangle_{r(\mathcal{G})r(\{\boldsymbol{A}^{(n')}\}_{n'\neq n})} + \text{const.} \\
&= -\frac{1}{2}\mathrm{tr}\left( (\boldsymbol{A}^{(n)} - \widehat{\boldsymbol{A}}^{(n)}) \widehat{\boldsymbol{\Sigma}}_{\boldsymbol{A}^{(n)}}^{-1} (\boldsymbol{A}^{(n)} - \widehat{\boldsymbol{A}}^{(n)})^\top \right),
\end{aligned}$$

where

$$\widehat{\boldsymbol{A}}^{(n)} = \frac{1}{\sigma^2}\boldsymbol{V}_{(n)}(\widehat{\boldsymbol{A}}^{(N)} \otimes \cdots \otimes \widehat{\boldsymbol{A}}^{(n+1)} \otimes \widehat{\boldsymbol{A}}^{(n-1)} \cdots \otimes \widehat{\boldsymbol{A}}^{(1)})\widehat{\boldsymbol{G}}_{(n)}^\top \widehat{\boldsymbol{\Sigma}}_{\boldsymbol{A}^{(n)}}, \tag{3.73}$$

$$\begin{aligned}
\widehat{\boldsymbol{\Sigma}}_{\boldsymbol{A}^{(n)}} = \sigma^2 \Big( \Big\langle \boldsymbol{G}_{(n)} (\boldsymbol{A}^{(N)\top}\boldsymbol{A}^{(N)} \otimes \cdots \otimes \boldsymbol{A}^{(n+1)\top}\boldsymbol{A}^{(n+1)} \\
\otimes \boldsymbol{A}^{(n-1)\top}\boldsymbol{A}^{(n-1)} \otimes \cdots \otimes \boldsymbol{A}^{(1)\top}\boldsymbol{A}^{(1)}) \boldsymbol{G}_{(n)}^\top \Big\rangle_{r(\mathcal{G})r(\{\boldsymbol{A}^{(n')}\}_{n'\neq n})} + \sigma^2 \boldsymbol{C}_{A^{(n)}}^{-1} \Big)^{-1}.
\end{aligned} \tag{3.74}$$

Thus, the VB posterior is given by

$$\begin{aligned}
r(\mathcal{G}, \{\boldsymbol{A}^{(n)}\}) = \mathrm{Gauss}_{\prod_{n=1}^N H^{(n)}} \left( \breve{\boldsymbol{g}}; \widehat{\breve{\boldsymbol{g}}}, \widehat{\breve{\boldsymbol{\Sigma}}}_G \right) \\
\cdot \prod_{n=1}^N \mathrm{MGauss}_{M^{(n)}, H^{(n)}} \left( \boldsymbol{A}^{(n)}; \widehat{\boldsymbol{A}}^{(n)}, \boldsymbol{I}_{M^{(n)}} \otimes \widehat{\boldsymbol{\Sigma}}_{\boldsymbol{A}^{(n)}} \right),
\end{aligned} \tag{3.75}$$

where the means and the covariances satisfy

$$\widehat{\breve{\boldsymbol{g}}} = \frac{\widehat{\breve{\boldsymbol{\Sigma}}}_G}{\sigma^2} \left( \widehat{\boldsymbol{A}}^{(N)} \otimes \cdots \otimes \widehat{\boldsymbol{A}}^{(1)} \right)^\top \breve{\boldsymbol{v}}, \tag{3.76}$$

$$\begin{aligned}
\widehat{\breve{\boldsymbol{\Sigma}}}_G = \sigma^2 \Bigg( \left( \widehat{\boldsymbol{A}}^{(N)\top}\widehat{\boldsymbol{A}}^{(N)} + M^{(N)}\widehat{\boldsymbol{\Sigma}}_{\boldsymbol{A}^{(N)}} \right) \otimes \cdots \otimes \left( \widehat{\boldsymbol{A}}^{(1)\top}\widehat{\boldsymbol{A}}^{(1)} + M^{(1)}\Sigma_{\boldsymbol{A}^{(1)}} \right) \\
+ \sigma^2 \breve{\boldsymbol{C}}_G^{-1} \Bigg)^{-1},
\end{aligned} \tag{3.77}$$

$$\widehat{A}^{(n)} = \frac{1}{\sigma^2} V_{(n)} (\widehat{A}^{(N)} \otimes \cdots \otimes \widehat{A}^{(n+1)} \otimes \widehat{A}^{(n-1)} \cdots \otimes \widehat{A}^{(1)}) \widehat{G}_{(n)}^{\top} \widehat{\Sigma}_{A^{(n)}}, \tag{3.78}$$

$$\widehat{\Sigma}_{A^{(n)}} = \sigma^2 \Bigg( \Bigg\langle G_{(n)} \Big( (\widehat{A}^{(N)\top} \widehat{A}^{(N)} + M^{(N)} \widehat{\Sigma}_{A^{(N)}}) \otimes \cdots \otimes (\widehat{A}^{(n+1)\top} \widehat{A}^{(n+1)} + M^{(n+1)} \widehat{\Sigma}_{A^{(n+1)}}) \otimes (\widehat{A}^{(n-1)\top} \widehat{A}^{(n-1)} + M^{(n-1)} \widehat{\Sigma}_{A^{(n-1)}}) \otimes \cdots \otimes (\widehat{A}^{(1)\top} \widehat{A}^{(1)} + M^{(1)} \widehat{\Sigma}_{A^{(1)}}) \Big) G_{(n)}^{\top} \Bigg\rangle_{r(\mathcal{G})} + \sigma^2 C_{A^{(n)}}^{-1} \Bigg)^{-1}. \tag{3.79}$$

The expectation in Eqs. (3.79) is explicitly given by

$$\begin{aligned} &\Bigg\langle G_{(n)} \Big( (\widehat{A}^{(N)\top} \widehat{A}^{(N)} + M^{(N)} \widehat{\Sigma}_{A^{(N)}}) \otimes \cdots \otimes (\widehat{A}^{(n+1)\top} \widehat{A}^{(n+1)} + M^{(n+1)} \widehat{\Sigma}_{A^{(n+1)}}) \\ &\quad \otimes (\widehat{A}^{(n-1)\top} \widehat{A}^{(n-1)} + M^{(n-1)} \widehat{\Sigma}_{A^{(n-1)}}) \otimes \cdots \otimes (\widehat{A}^{(1)\top} \widehat{A}^{(1)} + M^{(1)} \widehat{\Sigma}_{A^{(1)}}) \Big) G_{(n)}^{\top} \Bigg\rangle_{r(\mathcal{G})} \\ &= \widehat{G}_{(n)} \Big( (\widehat{A}^{(N)\top} \widehat{A}^{(N)} + M^{(N)} \widehat{\Sigma}_{A^{(N)}}) \otimes \cdots \otimes (\widehat{A}^{(n+1)\top} \widehat{A}^{(n+1)} + M^{(n+1)} \widehat{\Sigma}_{A^{(n+1)}}) \\ &\quad \otimes (\widehat{A}^{(n-1)\top} \widehat{A}^{(n-1)} + M^{(n-1)} \widehat{\Sigma}_{A^{(n-1)}}) \otimes \cdots \otimes (\widehat{A}^{(1)\top} \widehat{A}^{(1)} + M^{(1)} \widehat{\Sigma}_{A^{(1)}}) \Big) \widehat{G}_{(n)}^{\top} + \Xi^{(n)}, \end{aligned} \tag{3.80}$$

where the entries of $\Xi^{(n)} \in \mathbb{R}^{H^{(n)} \times H^{(n)}}$ are specified as

$$\begin{aligned} \Xi^{(n)}_{h^{(n)}, h'^{(n)}} = &\sum_{(h^{(1)}, h'^{(1)}), \ldots, (h^{(n-1)}, h'^{(n-1)}), (h^{(n+1)}, h'^{(n+1)}), \ldots, (h^{(N)}, h'^{(N)})} (\widehat{A}^{(N)\top} \widehat{A}^{(N)} + M^{(N)} \widehat{\Sigma}_{A^{(N)}})_{h^{(N)}, h'^{(N)}} \\ &\cdots (\widehat{A}^{(n+1)\top} \widehat{A}^{(n+1)} + M^{(n+1)} \widehat{\Sigma}_{A^{(n+1)}})_{h^{(n+1)}, h'^{(n+1)}} (\widehat{A}^{(n-1)\top} \widehat{A}^{(n-1)} + M^{(n-1)} \widehat{\Sigma}_{A^{(n-1)}})_{h^{(n-1)}, h'^{(n-1)}} \\ &\cdots (\widehat{A}^{(1)\top} \widehat{A}^{(1)} + M^{(1)} \widehat{\Sigma}_{A^{(1)}})_{h^{(1)}, h'^{(1)}} (\Sigma_{\mathcal{G}})_{(h^{(1)}, h'^{(1)}), \ldots, (h^{(N)}, h'^{(N)})}. \end{aligned}$$

Here, we used the tensor expression of $\Sigma_{\mathcal{G}} \in \mathbb{R}^{\prod_{n=1}^{N} 2H^{(n)}}$ for the core posterior covariance $\widehat{\breve{\Sigma}}_{G}$.

### Free Energy as a Function of Variational Parameters

Substituting Eq. (3.75) into Eq. (3.68), we have

$$2F = 2 \Bigg\langle \log r(\mathcal{G}) + \sum_{n=1}^{N} \log r(A^{(n)}) - \log p(\mathcal{V} | \mathcal{G}, \{A^{(n)}\}) p(\mathcal{G}) \prod_{n=1}^{N} p(A^{(n)}) \Bigg\rangle_{r(\mathcal{G}) (\prod_{n=1}^{N} r(A^{(n)}))}$$

$$
\begin{aligned}
&= \left(\prod_{n=1}^{N} M^{(n)}\right)\log(2\pi\sigma^2) + \log\frac{\det\left(\breve{\boldsymbol{C}}_G\right)}{\det\left(\widehat{\breve{\boldsymbol{\Sigma}}}_G\right)} + \sum_{n=1}^{N} M^{(n)}\log\frac{\det\left(\boldsymbol{C}_{A^{(n)}}\right)}{\det\left(\widehat{\boldsymbol{\Sigma}}_{A^{(n)}}\right)} \\
&\quad + \frac{\|\mathcal{V}\|^2}{\sigma^2} - \prod_{n=1}^{N} H^{(n)} - \prod_{n=1}^{N}(M^{(n)}H^{(n)}) \\
&\quad + \mathrm{tr}\left(\breve{\boldsymbol{C}}_G^{-1}(\widehat{\breve{\boldsymbol{g}}\breve{\boldsymbol{g}}^\top} + \widehat{\breve{\boldsymbol{\Sigma}}}_G)\right) + \sum_{n=1}^{N}\mathrm{tr}\left(\boldsymbol{C}_{A^{(n)}}^{-1}(\widehat{\boldsymbol{A}}^{(n)\top}\widehat{\boldsymbol{A}}^{(n)} + M^{(n)}\widehat{\boldsymbol{\Sigma}}_{A^{(n)}})\right) \\
&\quad - \frac{2}{\sigma^2}\breve{\boldsymbol{v}}^\top(\widehat{\boldsymbol{A}}^{(N)} \otimes \cdots \otimes \widehat{\boldsymbol{A}}^{(1)})\widehat{\breve{\boldsymbol{g}}} \\
&\quad + \frac{1}{\sigma^2}\mathrm{tr}\left\{\left((\widehat{\boldsymbol{A}}^{(N)\top}\widehat{\boldsymbol{A}}^{(N)} + M^{(N)}\widehat{\boldsymbol{\Sigma}}_{A^{(N)}}) \otimes \cdots \otimes (\widehat{\boldsymbol{A}}^{(1)\top}\widehat{\boldsymbol{A}}^{(1)} + M^{(1)}\widehat{\boldsymbol{\Sigma}}_{A^{(1)}})\right)\right. \\
&\qquad \left.\cdot(\widehat{\breve{\boldsymbol{g}}\breve{\boldsymbol{g}}^\top} + \widehat{\breve{\boldsymbol{\Sigma}}}_G)\right\}. \qquad (3.81)
\end{aligned}
$$

**Empirical Variational Bayesian Algorithm**

The derivative of the free energy (3.81) with respect to $\breve{\boldsymbol{C}}_G$ gives

$$2\frac{\partial F}{\partial \breve{\boldsymbol{C}}_G} = M^{(n)}\left(\breve{\boldsymbol{C}}_G^{-1} - \breve{\boldsymbol{C}}_G^{-2}\left(\breve{\boldsymbol{g}}\breve{\boldsymbol{g}}^\top + \widehat{\breve{\boldsymbol{\Sigma}}}_G\right)\right).$$

Since it holds that

$$\frac{\partial \breve{\boldsymbol{C}}_G}{\partial(\boldsymbol{C}_{G^{(n)}})_{h,h}} = \boldsymbol{C}_{G^{(N)}} \otimes \cdots \otimes \boldsymbol{C}_{G^{(n+1)}} \otimes \boldsymbol{E}_{(H^{(n)},h,h)} \otimes \boldsymbol{C}_{G^{(n-1)}} \otimes \cdots \otimes \boldsymbol{C}_{G^{(1)}},$$

where $\boldsymbol{E}_{(H,h,h')} \in \mathbb{R}^{H\times H}$ is the matrix with the $(h,h')$th entry equal to one and the others equal to zero, we have

$$
\begin{aligned}
&2\frac{\partial F}{\partial(\boldsymbol{C}_{G^{(n)}})_{h,h}} = 2\mathrm{tr}\left(\frac{\partial F}{\partial \breve{\boldsymbol{C}}_G}\frac{\partial \breve{\boldsymbol{C}}_G}{\partial(\boldsymbol{C}_{G^{(n)}})_{h,h}}\right) \\
&= M^{(n)}\left\|\mathbf{vec}\left(\boldsymbol{I}_{H^{(N)}} \otimes \cdots \otimes \boldsymbol{I}_{H^{(n+1)}} \otimes (\boldsymbol{C}_{G^{(n)}})_{h,h}^{-1}\boldsymbol{E}_{(H^{(n)},h,h)} \otimes \boldsymbol{I}_{H^{(n-1)}} \otimes \cdots \otimes \boldsymbol{I}_{H^{(1)}}\right.\right. \\
&\quad \left.\left.-\boldsymbol{C}_{G^{(N)}}^{-1} \otimes \cdots \otimes \boldsymbol{C}_{G^{(n+1)}}^{-1} \otimes (\boldsymbol{C}_{G^{(n)}})_{h,h}^{-2}\boldsymbol{E}_{(H^{(n)},h,h)} \otimes \boldsymbol{C}_{G^{(n-1)}}^{-1} \otimes \cdots \otimes \boldsymbol{C}_{G^{(1)}}^{-1}\left(\breve{\boldsymbol{g}}\breve{\boldsymbol{g}}^\top + \widehat{\breve{\boldsymbol{\Sigma}}}_G\right)\right)\right\|_1 \\
&= M^{(n)}(\boldsymbol{C}_{G^{(n)}})_{h,h}^{-2}\left\|\mathbf{vec}\left(\boldsymbol{I}_{H^{(N)}} \otimes \cdots \otimes \boldsymbol{I}_{H^{(n+1)}} \otimes (\boldsymbol{C}_{G^{(n)}})_{h,h}\boldsymbol{E}_{(H^{(n)},h,h)} \otimes \boldsymbol{I}_{H^{(n-1)}} \otimes \cdots \otimes \boldsymbol{I}_{H^{(1)}}\right.\right. \\
&\quad \left.\left.-\boldsymbol{C}_{G^{(N)}}^{-1} \otimes \cdots \otimes \boldsymbol{C}_{G^{(n+1)}}^{-1} \otimes \boldsymbol{E}_{(H^{(n)},h,h)} \otimes \boldsymbol{C}_{G^{(n-1)}}^{-1} \otimes \cdots \otimes \boldsymbol{C}_{G^{(1)}}^{-1}\left(\breve{\boldsymbol{g}}\breve{\boldsymbol{g}}^\top + \widehat{\breve{\boldsymbol{\Sigma}}}_G\right)\right)\right\|_1 \\
&= M^{(n)}(\boldsymbol{C}_{G^{(n)}})_{h,h}^{-2}\left(\left(\textstyle\prod_{n'\neq n} H^{(n')}\right)(\boldsymbol{C}_{G^{(n)}})_{h,h} - \mathbf{diag}\left(\breve{\boldsymbol{g}}\breve{\boldsymbol{g}}^\top + \widehat{\breve{\boldsymbol{\Sigma}}}_G\right)^\top\right. \\
&\quad \left.\mathbf{diag}\left(\boldsymbol{C}_{G^{(N)}}^{-1} \otimes \cdots \otimes \boldsymbol{C}_{G^{(n+1)}}^{-1} \otimes \boldsymbol{E}_{(H^{(n)},h,h)} \otimes \boldsymbol{C}_{G^{(n-1)}}^{-1} \otimes \cdots \otimes \boldsymbol{C}_{G^{(1)}}^{-1}\right)\right), \qquad (3.82)
\end{aligned}
$$

**Algorithm 3** EVB learning for Tucker factorization.

1: Initialize the variational parameters $(\widehat{\breve{\boldsymbol{g}}}, \widehat{\breve{\boldsymbol{\Sigma}}}_G, \{\widehat{\boldsymbol{A}}^{(n)}\}, \{\widehat{\boldsymbol{\Sigma}}_{A^{(n)}}\})$, and the hyperparameters $(\{\boldsymbol{C}_{G^{(n)}}\}, \{\boldsymbol{C}_{A^{(n)}}\}, \sigma^2)$, for example, $\widehat{\breve{g}}_h \sim \mathrm{Gauss}_1(0, \tau)$, $\widehat{\boldsymbol{A}}^{(n)}_{m,h} \sim \mathrm{Gauss}_1\left(0, \tau^{1/N}\right)$, $\widehat{\breve{\boldsymbol{\Sigma}}}_G = \tau \boldsymbol{I}_{\prod_{n=1}^{N} H^{(n)}}$, $\boldsymbol{C}_{G^{(n)}} = \tau \boldsymbol{I}_{H^{(n)}}$, $\widehat{\boldsymbol{\Sigma}}_{A^{(n)}} = \boldsymbol{C}_{A^{(n)}} = \tau^{1/N} \boldsymbol{I}_{H^{(n)}}$, and $\sigma^2 = \tau^2$ for $\tau^2 = \|\mathcal{V}\|^2 / \prod_{n=1}^{N} M^{(n)}$.
2: Apply (substitute the right-hand side into the left-hand side) Eqs. (3.77), (3.76), (3.79), and (3.78) to update $\widehat{\breve{\boldsymbol{\Sigma}}}_G, \widehat{\breve{\boldsymbol{g}}}, \{\widehat{\boldsymbol{\Sigma}}_{A^{(n)}}\}$, and $\{\widehat{\boldsymbol{A}}^{(n)}\}$, respectively.
3: Apply Eqs. (3.83) through (3.85) to update $\{\boldsymbol{C}_{G^{(n)}}\}, \{\boldsymbol{C}_{A^{(n)}}\}$, and $\sigma^2$, respectively.
4: Prune the $h$th component if $c^2_{g_h^{(n)}} c^2_{a_h^{(n)}} < \varepsilon$, where $\varepsilon > 0$ is a threshold, e.g., set to $\varepsilon = 10^{-4}$.
5: Evaluate the free energy (3.81).
6: Iterate Steps 2 through 5 until convergence (until the energy decrease becomes smaller than a threshold).

where $\|\cdot\|_1$ denotes the $\ell_1$-norm, and $\mathbf{diag}(\cdot)$ denotes the column vector consisting of the diagonal entries of a matrix. Thus, the prior covariance for the core tensor can be updated by

$$c^2_{g_h^{(n)}} = \frac{\mathbf{diag}\left(\breve{\boldsymbol{g}}\breve{\boldsymbol{g}}^\top + \widehat{\breve{\boldsymbol{\Sigma}}}_G\right)^\top \mathbf{diag}\left(\boldsymbol{C}^{-1}_{G(N)} \otimes \cdots \otimes \boldsymbol{C}^{-1}_{G(n+1)} \otimes \boldsymbol{E}_{(H^{(n)},h,h)} \otimes \boldsymbol{C}^{-1}_{G(n-1)} \otimes \cdots \otimes \boldsymbol{C}^{-1}_{G(1)}\right)}{\prod_{n' \neq n} H^{(n')}}. \tag{3.83}$$

The derivative of the free energy (3.81) with respect to $\boldsymbol{C}_{A^{(n)}}$ gives

$$2\frac{\partial F}{\partial c^2_{a_h^{(n)}}} = M^{(n)}\left(c^{-2}_{a_h^{(n)}} - c^{-4}_{a_h^{(n)}}\left(\frac{\|\widehat{\boldsymbol{a}}_h^{(n)}\|^2}{M^{(n)}} + (\widehat{\boldsymbol{\Sigma}}_{A^{(n)}})_{h,h}\right)\right),$$

which leads to the following update rule for the prior covariance for the factor matrices:

$$c^2_{a_h^{(n)}} = \frac{\|\widehat{\boldsymbol{a}}_h^{(n)}\|^2}{M^{(n)}} + (\widehat{\boldsymbol{\Sigma}}_{A^{(n)}})_{h,h}. \tag{3.84}$$

Finally, the derivative of the free energy (3.81) with respect to $\sigma^2$ gives

$$2\frac{\partial F}{\partial \sigma^2} = \frac{\prod_{n=1}^{N} M^{(n)}}{\sigma^2} - \frac{1}{\sigma^4}\left(\|\mathcal{V}\|^2 - 2\breve{\boldsymbol{v}}^\top(\widehat{\boldsymbol{A}}^{(N)} \otimes \cdots \otimes \widehat{\boldsymbol{A}}^{(1)})\widehat{\breve{\boldsymbol{g}}}\right.$$
$$\left. + \mathrm{tr}\left\{\left((\widehat{\boldsymbol{A}}^{(N)\top}\widehat{\boldsymbol{A}}^{(N)} + M^{(N)}\widehat{\boldsymbol{\Sigma}}_{A^{(N)}}) \otimes \cdots \otimes (\widehat{\boldsymbol{A}}^{(1)\top}\widehat{\boldsymbol{A}}^{(1)} + M^{(1)}\widehat{\boldsymbol{\Sigma}}_{A^{(1)}})\right)(\widehat{\breve{\boldsymbol{g}}}\widehat{\breve{\boldsymbol{g}}}^\top + \widehat{\breve{\boldsymbol{\Sigma}}}_G)\right\}\right),$$

which leads to the update rule for the noise variance as follows:

$$\widehat{\sigma}^2 = \frac{1}{\prod_{n=1}^{N} M^{(n)}} \Bigg( \|\mathcal{V}\|^2 - 2\breve{\boldsymbol{v}}^\top (\widehat{\boldsymbol{A}}^{(N)} \otimes \cdots \otimes \widehat{\boldsymbol{A}}^{(1)}) \widehat{\breve{\boldsymbol{g}}}$$
$$+ \mathrm{tr}\left\{ \left( (\widehat{\boldsymbol{A}}^{(N)\top} \widehat{\boldsymbol{A}}^{(N)} + M^{(N)} \widehat{\boldsymbol{\Sigma}}_{A^{(N)}}) \otimes \cdots \otimes (\widehat{\boldsymbol{A}}^{(1)\top} \widehat{\boldsymbol{A}}^{(1)} + M^{(1)} \widehat{\boldsymbol{\Sigma}}_{A^{(1)}}) \right) (\widehat{\breve{\boldsymbol{g}}\breve{\boldsymbol{g}}^\top} + \widehat{\breve{\boldsymbol{\Sigma}}}_G) \right\} \Bigg). \tag{3.85}$$

Algorithm 3 summarizes the EVB algorithm for Tucker factorization. If we appropriately set the hyperparameters $(\{\boldsymbol{C}_{G^{(n)}}\}, \{\boldsymbol{C}_{A^{(n)}}\}, \sigma^2)$ in Step 1 and skip Steps 3 and 4, Algorithm 3 is reduced to (nonempirical) VB learning.

## 3.4 Low-Rank Subspace Clustering

PCA *globally* embeds data points into a low-dimensional subspace. As more flexible tools, *subspace clustering* methods, which *locally* embed the data into the union of subspaces, have been developed. In this section, we derive VB learning for subspace clustering.

### 3.4.1 Subspace Clustering Methods

Most clustering methods, such as $k$-means (MacQueen, 1967; Lloyd, 1982) and spectral clustering (Shi and Malik, 2000), (explicitly or implicitly) assume that there are sparse areas between dense areas, and separate the dense areas as clusters (Figure 3.2 left). On the other hand, there are some data, e.g., projected trajectories of points on a rigid body in 3D space, where data points can be assumed to lie in a union of small dimensional subspaces (Figure 3.2 right). Note that a point lying in a subspace is not necessarily far from a point lying in another subspace if those subspaces intersect each other. Subspace clustering methods have been developed to analyze this kind of data.

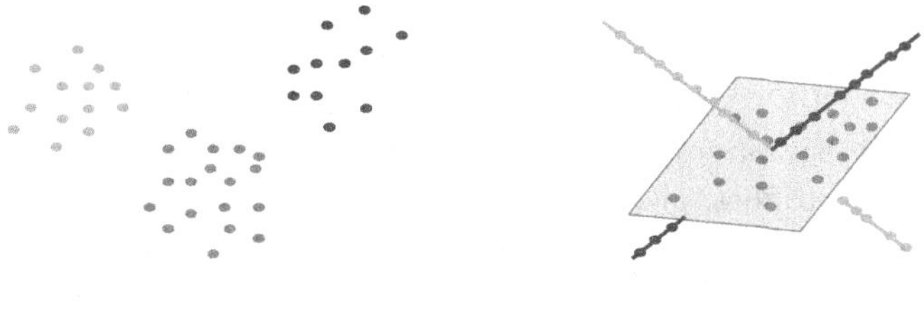

Figure 3.2 Clustering (left) and subspace clustering (right).

Let $\boldsymbol{V} = (\boldsymbol{v}_1, \ldots, \boldsymbol{v}_M) \in \mathbb{R}^{L \times M}$ be $L$-dimensional observed samples of size $M$. We assume that each $\boldsymbol{v}_m$ is approximately expressed as a linear combination of $M'$ *words* in a dictionary, $\boldsymbol{D} = (\boldsymbol{d}_1, \ldots, \boldsymbol{d}_{M'}) \in \mathbb{R}^{L \times M'}$, i.e.,

$$\boldsymbol{V} = \boldsymbol{D}\boldsymbol{U} + \boldsymbol{\mathcal{E}},$$

where $\boldsymbol{U} \in \mathbb{R}^{M' \times M}$ is unknown coefficients, and $\boldsymbol{\mathcal{E}} \in \mathbb{R}^{L \times M}$ is noise. In subspace clustering, the observed matrix $\boldsymbol{V}$ itself is often used as a dictionary $\boldsymbol{D}$. Then, a convex formulation of the *sparse subspace clustering (SSC)* (Soltanolkotabi and Candès, 2011; Elhamifar and Vidal, 2013) is given by

$$\min_{\boldsymbol{U}} \|\boldsymbol{V} - \boldsymbol{V}\boldsymbol{U}\|_{\mathrm{Fro}}^2 + \lambda \|\boldsymbol{U}\|_1 \,, \quad \text{s.t. } \mathbf{diag}(\boldsymbol{U}) = \boldsymbol{0}, \tag{3.86}$$

where $\boldsymbol{U} \in \mathbb{R}^{M \times M}$ is a parameter to be estimated, $\lambda > 0$ is a regularization coefficient. $\|\cdot\|_1$ is the $\ell_1$-norm of a matrix. The first term in Eq. (3.86) together with the constraint requires that each data point $\boldsymbol{v}_m$ is accurately expressed as a linear combination of *other* data points, $\{\boldsymbol{v}_{m'}\}$ for $m' \neq m$. The second term, which is the $\ell_1$-*regularizer*, enforces that the number of samples contributing to the linear combination should be small, which leads to low-dimensionality of each obtained subspace. After the solution $\widehat{\boldsymbol{U}}$ to the problem (3.86) is obtained, the matrix $\mathrm{abs}(\widehat{\boldsymbol{U}}) + \mathrm{abs}(\widehat{\boldsymbol{U}}^\top)$, where $\mathrm{abs}(\cdot)$ takes the absolute value elementwise, is regarded as an affinity matrix, and a *spectral clustering algorithm*, such as the *normalized cuts* (Shi and Malik, 2000), is applied to obtain clusters.

Another popular method for subspace clustering is *low-rank subspace clustering (LRSC)* or *low-rank representation* (Liu et al., 2010; Liu and Yan, 2011; Liu et al., 2012; Vidal and Favaro, 2014), where low-dimensional subspaces are sought by enforcing the low-rankness of $\boldsymbol{U}$ with the *trace norm*:

$$\min_{\boldsymbol{U}} \|\boldsymbol{V} - \boldsymbol{V}\boldsymbol{U}\|_{\mathrm{Fro}}^2 + \lambda \|\boldsymbol{U}\|_{\mathrm{tr}} \,. \tag{3.87}$$

Since LRSC enforces the low-rankness of $\boldsymbol{U}$, the constraint $\mathbf{diag}(\boldsymbol{U}) = \boldsymbol{0}$ is not necessary, which makes its optimization problem (3.87) significantly simpler than the optimization problem (3.86) for SSC. Thanks to this simplicity, the global solution of Eq. (3.87) has been analytically obtained (Vidal and Favaro, 2014).

Good properties of SSC and LRSC have been theoretically shown (Liu et al., 2010, 2012; Soltanolkotabi and Candès, 2011; Elhamifar and Vidal, 2013; Vidal and Favaro, 2014). It is observed that they behave differently in different situations, and each of SSC and LRSC shows advantages and disadvantages over the other, i.e., neither SSC nor LRSC is known to dominate the other in the general situations. In the rest of this section, we focus on LRSC and derive its VB learning algorithm.

### 3.4.2 VB Learning for LRSC

We start with the following probabilistic model, of which the maximum a posteriori (MAP) estimator coincides with the solution to the convex formulation (3.87) under a certain hyperparameter setting:

$$p(\boldsymbol{V}|\boldsymbol{A},\boldsymbol{B}) \propto \exp\left(-\frac{1}{2\sigma^2}\left\|\boldsymbol{V}-\boldsymbol{V}\boldsymbol{B}\boldsymbol{A}^\top\right\|_{\mathrm{Fro}}^2\right), \tag{3.88}$$

$$p(\boldsymbol{A}) \propto \exp\left(-\frac{1}{2}\mathrm{tr}(\boldsymbol{A}\boldsymbol{C}_A^{-1}\boldsymbol{A}^\top)\right), \tag{3.89}$$

$$p(\boldsymbol{B}) \propto \exp\left(-\frac{1}{2}\mathrm{tr}(\boldsymbol{B}\boldsymbol{C}_B^{-1}\boldsymbol{B}^\top)\right). \tag{3.90}$$

Here, we factorized $\boldsymbol{U}$ as $\boldsymbol{U} = \boldsymbol{B}\boldsymbol{A}^\top$, where $\boldsymbol{A} \in \mathbb{R}^{M\times H}$ and $\boldsymbol{B} \in \mathbb{R}^{M\times H}$ for $H \leq \min(L, M)$ are the parameters to be estimated (Babacan et al., 2012a). This factorization is known to induce low-rankness through the MIR mechanism, which will be discussed in Chapter 7. We assume that the prior covariances are diagonal:

$$\boldsymbol{C}_A = \mathbf{Diag}(c_{a_1}^2,\ldots,c_{a_H}^2), \qquad \boldsymbol{C}_B = \mathbf{Diag}(c_{b_1}^2,\ldots,c_{b_H}^2).$$

**Conditional Conjugacy**

The model likelihood (3.88) of LRSC is similar to the model likelihood (3.1) of MF, and it is in the Gaussian form with respect to $\boldsymbol{A}$ if $\boldsymbol{B}$ is regarded as a constant, or vice versa. Therefore, the priors (3.89) and (3.90) are conditionally conjugate for $\boldsymbol{A}$ and $\boldsymbol{B}$, respectively.

**Variational Bayesian Algorithm**

The conditional conjugacy implies that the following independence constraint on the approximate posterior leads to a tractable algorithm:

$$r(\boldsymbol{A},\boldsymbol{B}) = r(\boldsymbol{A})r(\boldsymbol{B}).$$

In the same way as MF, we can show that the VB posterior is Gaussian in the following form:

$$r(\boldsymbol{A}) \propto \exp\left(-\frac{\mathrm{tr}\left((\boldsymbol{A}-\widehat{\boldsymbol{A}})\widehat{\boldsymbol{\Sigma}}_A^{-1}(\boldsymbol{A}-\widehat{\boldsymbol{A}})^\top\right)}{2}\right),$$

$$r(\boldsymbol{B}) \propto \exp\left(-\frac{(\breve{\boldsymbol{b}}-\widehat{\breve{\boldsymbol{b}}})^\top\widehat{\breve{\boldsymbol{\Sigma}}}_B^{-1}(\breve{\boldsymbol{b}}-\widehat{\breve{\boldsymbol{b}}})}{2}\right), \tag{3.91}$$

where $\breve{b} = \mathbf{vec}(B) \in \mathbb{R}^{MH}$. The variational parameters satisfy the following stationary conditions:

$$\widehat{A} = \frac{1}{\sigma^2} V^\top V \widehat{B} \widehat{\Sigma}_A, \tag{3.92}$$

$$\widehat{\Sigma}_A = \sigma^2 \left( \left\langle B^\top V^\top V B \right\rangle_{r(B)} + \sigma^2 C_A^{-1} \right)^{-1}, \tag{3.93}$$

$$\widehat{\breve{b}} = \frac{\widehat{\breve{\Sigma}}_B}{\sigma^2} \mathbf{vec}\left( V^\top V \widehat{A} \right), \tag{3.94}$$

$$\widehat{\breve{\Sigma}}_B = \sigma^2 \left( (\widehat{A}^\top \widehat{A} + M \widehat{\Sigma}_A) \otimes V^\top V + \sigma^2 (C_B^{-1} \otimes I_M) \right)^{-1}, \tag{3.95}$$

where the entries of $\langle B^\top V^\top V B \rangle_{r(B)}$ in Eq. (3.93) are explicitly given by

$$\left( \left\langle B^\top V^\top V B \right\rangle_{r(B)} \right)_{h,h'} = \left( \widehat{B}^\top V^\top V \widehat{B} \right)_{h,h'} + \mathrm{tr}\left( V^\top V \widehat{\Sigma}_B^{(h,h')} \right). \tag{3.96}$$

Here $\widehat{\Sigma}_B^{(h,h')} \in \mathbb{R}^{M \times M}$ is the $(h, h')$th block matrix of $\widehat{\breve{\Sigma}}_B \in \mathbb{R}^{MH \times MH}$, i.e.,

$$\widehat{\breve{\Sigma}}_B = \begin{pmatrix} \widehat{\Sigma}_B^{(1,1)} & \cdots & \widehat{\Sigma}_B^{(1,H)} \\ \vdots & \ddots & \vdots \\ \widehat{\Sigma}_B^{(H,1)} & \cdots & \widehat{\Sigma}_B^{(H,H)} \end{pmatrix}.$$

**Free Energy as a Function of Variational Parameters**

The free energy can be explicitly written as

$$2F = LM \log(2\pi\sigma^2) + \frac{\left\| V - V\widehat{B}\widehat{A}^\top \right\|_{\mathrm{Fro}}^2}{\sigma^2} + M \log \frac{\det(C_A)}{\det(\widehat{\Sigma}_A)} + \log \frac{\det(C_B \otimes I_M)}{\det(\widehat{\breve{\Sigma}}_B)}$$
$$- 2MH + \mathrm{tr}\left\{ C_A^{-1} \left( \widehat{A}^\top \widehat{A} + M \widehat{\Sigma}_A \right) \right\} + \mathrm{tr}\left\{ C_B^{-1} \widehat{B}^\top \widehat{B} \right\} + \mathrm{tr}\left\{ (C_B^{-1} \otimes I_M) \widehat{\breve{\Sigma}}_B \right\}$$
$$+ \mathrm{tr}\left\{ \sigma^{-2} V^\top V \left( -\widehat{B}\widehat{A}^\top \widehat{A}\widehat{B}^\top + \left\langle B(\widehat{A}^\top \widehat{A} + M \widehat{\Sigma}_A) B^\top \right\rangle_{r(B)} \right) \right\}, \tag{3.97}$$

where the expectation in the last term is given by

$$\left( \left\langle B(\widehat{A}^\top \widehat{A} + M \widehat{\Sigma}_A) B \right\rangle_{r(B)}^\top \right)_{m,m'} = \left( \widehat{B}(\widehat{A}^\top \widehat{A} + M \widehat{\Sigma}_A) \widehat{B}^\top \right)_{m,m'}$$
$$+ \mathrm{tr}\left( (\widehat{A}^\top \widehat{A} + M \widehat{\Sigma}_A) \widehat{\Sigma}_B^{(m,m')} \right). \tag{3.98}$$

**Algorithm 4** EVB learning for low-rank subspace clustering.

1: Initialize the variational parameters $(\widehat{\boldsymbol{A}}, \widehat{\boldsymbol{\Sigma}}_A, \widehat{\boldsymbol{B}}, \breve{\widehat{\boldsymbol{\Sigma}}}_B)$, and the hyperparameters $(\boldsymbol{C}_A, \boldsymbol{C}_B, \sigma^2)$, for example, $\widehat{A}_{m,h}, \widehat{B}_{m,h} \sim \text{Gauss}_1(0, 1^2)$, $\widehat{\boldsymbol{\Sigma}}_A = \boldsymbol{C}_A = \boldsymbol{C}_B = \boldsymbol{I}_H$, $\breve{\widehat{\boldsymbol{\Sigma}}}_B = \boldsymbol{I}_{MH}$, and $\sigma^2 = \|\boldsymbol{V}\|^2/(LM)$.
2: Apply (substitute the right-hand side into the left-hand side) Eqs. (3.93), (3.92), (3.95), and (3.94) to update $\widehat{\boldsymbol{\Sigma}}_A, \widehat{\boldsymbol{A}}, \breve{\widehat{\boldsymbol{\Sigma}}}_B$, and $\widehat{\boldsymbol{B}}$, respectively.
3: Apply Eqs. (3.99) through (3.101) to update $\boldsymbol{C}_A, \boldsymbol{C}_B$, and $\sigma^2$, respectively.
4: Prune the $h$th component if $c_{a_h}^2 c_{b_h}^2 < \varepsilon$, where $\varepsilon > 0$ is a threshold, e.g., set to $\varepsilon = 10^{-4}$.
5: Evaluate the free energy (3.97).
6: Iterate Steps 2 through 5 until convergence (until the energy decrease becomes smaller than a threshold).

Here, $\widehat{\boldsymbol{\Sigma}}_B^{(m,m')} \in \mathbb{R}^{H \times H}$ is defined as

$$\widehat{\boldsymbol{\Sigma}}_B^{(m,m')} = \begin{pmatrix} \left(\widehat{\boldsymbol{\Sigma}}_B^{(1,1)}\right)_{m,m'} & \cdots & \left(\widehat{\boldsymbol{\Sigma}}_B^{(1,H)}\right)_{m,m'} \\ \vdots & \ddots & \vdots \\ \left(\widehat{\boldsymbol{\Sigma}}_B^{(H,1)}\right)_{m,m'} & \cdots & \left(\widehat{\boldsymbol{\Sigma}}_B^{(H,H)}\right)_{m,m'} \end{pmatrix}.$$

**Empirical Variational Bayesian Algorithm**

By differentiating the free energy (3.97) with respect to the hyperparameters, we can obtain the stationary conditions for the hyperparameters:

$$c_{a_h}^2 = \left\|\widehat{\boldsymbol{a}}_h\right\|^2 / M + \left(\widehat{\boldsymbol{\Sigma}}_A\right)_{h,h}, \tag{3.99}$$

$$c_{b_h}^2 = \left(\left\|\widehat{\boldsymbol{b}}_h\right\|^2 + \operatorname{tr}\left(\widehat{\boldsymbol{\Sigma}}_B^{(h,h)}\right)\right) / M, \tag{3.100}$$

$$\widehat{\sigma}^2 = \frac{\operatorname{tr}\left(\boldsymbol{V}^\top \boldsymbol{V}\left(\boldsymbol{I}_M - 2\widehat{\boldsymbol{B}}\widehat{\boldsymbol{A}}^\top + \left\langle \boldsymbol{B}(\widehat{\boldsymbol{A}}^\top \widehat{\boldsymbol{A}} + M\widehat{\boldsymbol{\Sigma}}_A)\boldsymbol{B}^\top \right\rangle_{r(\boldsymbol{B})}\right)\right)}{LM}. \tag{3.101}$$

Algorithm 4 summarizes the EVB algorithm. If we appropriately set the hyperparameters $(\boldsymbol{C}_A, \boldsymbol{C}_B, \sigma^2)$ in Step 1 and skip Steps 3 and 4, Algorithm 4 is reduced to (nonempirical) VB learning.

**Variational Bayesian Algorithm under the Kronecker Product Covariance Constraint**

The standard VB algorithm, given in Algorithm 4, for LRSC requires the inversion of an $MH \times MH$ matrix, which is prohibitively huge in practical applications. As a remedy, Babacan et al. (2012a) proposed to restrict the posterior $r(\boldsymbol{B})$ for $\boldsymbol{B}$ to be the matrix variate Gaussian with the Kronecker product covariance (KPC) structure, as Eq. (3.18). Namely, we restrict the approximate posterior to be in the following form:

$$\begin{aligned} r(\boldsymbol{B}) &= \mathrm{MGauss}_{M,H}(\boldsymbol{B}; \widehat{\boldsymbol{B}}, \widehat{\boldsymbol{\Psi}}_{\boldsymbol{B}} \otimes \widehat{\boldsymbol{\Sigma}}_{B}) \\ &\propto \exp\left(-\frac{1}{2}\mathrm{tr}\left(\widehat{\boldsymbol{\Psi}}_{\boldsymbol{B}}^{-1}(\boldsymbol{B} - \widehat{\boldsymbol{B}})\widehat{\boldsymbol{\Sigma}}_{\boldsymbol{B}}^{-1}(\boldsymbol{B} - \widehat{\boldsymbol{B}})^{\top}\right)\right). \end{aligned} \tag{3.102}$$

Under this additional constraint, the free energy is written as

$$\begin{aligned} 2F^{\mathrm{KPC}} &= LM\log(2\pi\sigma^2) + \frac{\left\|\boldsymbol{V} - \boldsymbol{V}\widehat{\boldsymbol{B}}\widehat{\boldsymbol{A}}^{\top}\right\|^2}{\sigma^2} + M\log\frac{\det(\boldsymbol{C}_A)}{\det\left(\widehat{\boldsymbol{\Sigma}}_A\right)} + M\log\frac{\det(\boldsymbol{C}_B)}{\det\left(\widehat{\boldsymbol{\Sigma}}_{\boldsymbol{B}}\right)} \\ &\quad + H\log\frac{1}{\det\left(\widehat{\boldsymbol{\Psi}}_{\boldsymbol{B}}\right)} - 2MH \\ &\quad + \mathrm{tr}\left\{\boldsymbol{C}_A^{-1}\left(\widehat{\boldsymbol{A}}^{\top}\widehat{\boldsymbol{A}} + M\widehat{\boldsymbol{\Sigma}}_A\right)\right\} + \mathrm{tr}\left\{\boldsymbol{C}_B^{-1}\left(\widehat{\boldsymbol{B}}^{\top}\widehat{\boldsymbol{B}} + \mathrm{tr}(\widehat{\boldsymbol{\Psi}}_{\boldsymbol{B}})\widehat{\boldsymbol{\Sigma}}_{\boldsymbol{B}}\right)\right\} \\ &\quad + \mathrm{tr}\left\{\sigma^{-2}\boldsymbol{V}^{\top}\boldsymbol{V}\left(M\widehat{\boldsymbol{B}}\widehat{\boldsymbol{\Sigma}}_{\boldsymbol{A}}\widehat{\boldsymbol{B}}^{\top} + \mathrm{tr}\left((\widehat{\boldsymbol{A}}^{\top}\widehat{\boldsymbol{A}} + M\widehat{\boldsymbol{\Sigma}}_A)\widehat{\boldsymbol{\Sigma}}_{\boldsymbol{B}}\right)\widehat{\boldsymbol{\Psi}}_{\boldsymbol{B}}\right)\right\}. \end{aligned} \tag{3.103}$$

By differentiating the free energy (3.103) with respect to each variational parameter, we obtain the following update rules:

$$\widehat{\boldsymbol{A}} = \frac{1}{\sigma^2}\boldsymbol{V}^{\top}\boldsymbol{V}\widehat{\boldsymbol{B}}\widehat{\boldsymbol{\Sigma}}_A, \tag{3.104}$$

$$\widehat{\boldsymbol{\Sigma}}_A = \sigma^2\left(\widehat{\boldsymbol{B}}^{\top}\boldsymbol{V}^{\top}\boldsymbol{V}\widehat{\boldsymbol{B}} + \mathrm{tr}\left(\boldsymbol{V}^{\top}\boldsymbol{V}\widehat{\boldsymbol{\Psi}}_{\boldsymbol{B}}\right)\widehat{\boldsymbol{\Sigma}}_{\boldsymbol{B}} + \sigma^2\boldsymbol{C}_A^{-1}\right)^{-1}, \tag{3.105}$$

$$\widehat{\boldsymbol{B}}^{\mathrm{new}} = \widehat{\boldsymbol{B}}^{\mathrm{old}} - \alpha\left(\widehat{\boldsymbol{B}}^{\mathrm{old}}\boldsymbol{C}_B^{-1} + \frac{1}{\sigma^2}\boldsymbol{V}^{\top}\boldsymbol{V}\left(-\widehat{\boldsymbol{A}} + \widehat{\boldsymbol{B}}^{\mathrm{old}}(\widehat{\boldsymbol{A}}^{\top}\widehat{\boldsymbol{A}} + M\widehat{\boldsymbol{\Sigma}}_A)\right)\right), \tag{3.106}$$

$$\widehat{\boldsymbol{\Sigma}}_{\boldsymbol{B}} = \sigma^2\left(\frac{\mathrm{tr}\left(\boldsymbol{V}^{\top}\boldsymbol{V}\widehat{\boldsymbol{\Psi}}_{\boldsymbol{B}}\right)}{M}(\widehat{\boldsymbol{A}}^{\top}\widehat{\boldsymbol{A}} + M\widehat{\boldsymbol{\Sigma}}_A) + \frac{\sigma^2\mathrm{tr}(\widehat{\boldsymbol{\Psi}}_{\boldsymbol{B}})}{M}\boldsymbol{C}_B^{-1}\right)^{-1}, \tag{3.107}$$

$$\widehat{\boldsymbol{\Psi}}_{\boldsymbol{B}} = \sigma^2\left(\frac{\mathrm{tr}\left((\widehat{\boldsymbol{A}}^{\top}\widehat{\boldsymbol{A}} + M\widehat{\boldsymbol{\Sigma}}_A)\widehat{\boldsymbol{\Sigma}}_{\boldsymbol{B}}\right)}{H}\boldsymbol{V}^{\top}\boldsymbol{V} + \frac{\sigma^2\mathrm{tr}(\boldsymbol{C}_B^{-1}\widehat{\boldsymbol{\Sigma}}_B)}{H}\boldsymbol{I}_M\right)^{-1}, \tag{3.108}$$

$$c_{a_h}^2 = (\widehat{\boldsymbol{A}}^{\top}\widehat{\boldsymbol{A}} + M\widehat{\boldsymbol{\Sigma}}_A)_{h,h}/M, \tag{3.109}$$

$$c_{b_h}^2 = \left(\widehat{\boldsymbol{B}}^{\top}\widehat{\boldsymbol{B}} + \mathrm{tr}(\widehat{\boldsymbol{\Psi}}_{\boldsymbol{B}})\widehat{\boldsymbol{\Sigma}}_{\boldsymbol{B}}\right)_{h,h}/M, \tag{3.110}$$

$$\begin{aligned} \widehat{\sigma}^2 &= \frac{1}{LM}\mathrm{tr}\left(\boldsymbol{V}^{\top}\boldsymbol{V}\left(\boldsymbol{I}_M - 2\widehat{\boldsymbol{B}}\widehat{\boldsymbol{A}}^{\top} + \mathrm{tr}\left((\widehat{\boldsymbol{A}}^{\top}\widehat{\boldsymbol{A}} + M\widehat{\boldsymbol{\Sigma}}_A)\widehat{\boldsymbol{\Sigma}}_{\boldsymbol{B}}\right)\widehat{\boldsymbol{\Psi}}_{\boldsymbol{B}}\right.\right. \\ &\quad \left.\left. + \widehat{\boldsymbol{B}}(\widehat{\boldsymbol{A}}^{\top}\widehat{\boldsymbol{A}} + M\widehat{\boldsymbol{\Sigma}}_A)\widehat{\boldsymbol{B}}^{\top}\right)\right). \end{aligned} \tag{3.111}$$

**Algorithm 5** EVB learning for low-rank subspace clustering under the Kronecker product covariance constraint (3.102).

1: Initialize the variational parameters $(\widehat{\boldsymbol{A}}, \widehat{\boldsymbol{\Sigma}}_A, \widehat{\boldsymbol{B}}, \widehat{\boldsymbol{\Sigma}}_B, \widehat{\boldsymbol{\Psi}}_B)$, and the hyperparameters $(\boldsymbol{C}_A, \boldsymbol{C}_B, \sigma^2)$, for example, $\widehat{A}_{m,h}, \widehat{B}_{m,h} \sim \mathrm{Gauss}_1(0, 1^2)$, $\widehat{\boldsymbol{\Sigma}}_A = \widehat{\boldsymbol{\Sigma}}_B = \boldsymbol{C}_A = \boldsymbol{C}_B = \boldsymbol{I}_H$, $\widehat{\boldsymbol{\Psi}}_B = \boldsymbol{I}_M$, and $\sigma^2 = \|V\|^2/(LM)$.
2: Apply (substitute the right-hand side into the left-hand side) Eqs. (3.105), (3.104), (3.107), and (3.108) to update $\widehat{\boldsymbol{\Sigma}}_A, \widehat{\boldsymbol{A}}, \widehat{\boldsymbol{\Sigma}}_B$, and $\widehat{\boldsymbol{\Psi}}_B$, respectively.
3: Apply Eq. (3.106) $T$ times (e.g., $T = 20$) to update $\widehat{\boldsymbol{B}}$.
4: Apply Eqs. (3.109) through (3.111) to update $\boldsymbol{C}_A, \boldsymbol{C}_B$, and $\sigma^2$, respectively.
5: Prune the $h$th component if $c_{a_h}^2 c_{b_h}^2 < \varepsilon$, where $\varepsilon > 0$ is a threshold, e.g., set to $\varepsilon = 10^{-4}$.
6: Evaluate the free energy (3.103).
7: Iterate Steps 2 through 6 until convergence (until the energy decrease becomes smaller than a threshold).

Note that Eq. (3.106) is the gradient descent algorithm for $\widehat{\boldsymbol{B}}$ with the step size $\alpha > 0$.

Algorithm 5 summarizes the EVB algorithm under the KPC constraint, which we call *KPC approximation (KPCA)*. If we appropriately set the hyperparameters $(\boldsymbol{C}_A, \boldsymbol{C}_B, \sigma^2)$ in Step 1 and skip Steps 4 and 5, Algorithm 5 is reduced to (nonempirical) VB learning.

## 3.5 Sparse Additive Matrix Factorization

PCA is known to be sensitive to outliers in data and generally fails in their presence. To cope with outliers, robust PCA, where spiky noise is captured by an elementwise sparse term, was proposed (Candès et al., 2011). In this section, we introduce a generalization of robust PCA, called *sparse additive matrix factorization (SAMF)* (Nakajima et al., 2013b) and derive its VB learning algorithm.

### 3.5.1 Robust PCA and Matrix Factorization

In *robust PCA*, the observed matrix $\boldsymbol{V} \in \mathbb{R}^{L \times M}$ is modeled as follows:

$$\boldsymbol{V} = \boldsymbol{U}^{\text{low-rank}} + \boldsymbol{U}^{\text{element}} + \boldsymbol{\mathcal{E}}, \tag{3.112}$$

where $\boldsymbol{U}^{\text{low-rank}} \in \mathbb{R}^{L \times M}$ is a low-rank matrix, $\boldsymbol{U}^{\text{element}} \in \mathbb{R}^{L \times M}$ is an elementwise sparse matrix, and $\boldsymbol{\mathcal{E}} \in \mathbb{R}^{L \times M}$ is a (typically dense) noise matrix.

Given the observed matrix $V$, one can infer each term in the right-hand side of Eq. (3.112) by solving the following convex problem (Candès et al., 2011):

$$\min_{U^{\text{low-rank}}, U^{\text{element}}} \left\| V - U^{\text{low-rank}} - U^{\text{element}} \right\|_{\text{Fro}}^2 + \lambda_1 \left\| U^{\text{low-rank}} \right\|_{\text{tr}} + \lambda_2 \left\| U^{\text{element}} \right\|_1 ,$$

where the *trace norm* $\|\cdot\|_{\text{tr}}$ induces low-rank sparsity, and the $\ell_1$-*norm* $\|\cdot\|_1$ induces elementwise sparsity. The regularization coefficients $\lambda_1$ and $\lambda_2$ control the strength of sparsity.

In Bayesian modeling, the low-rank matrix is commonly expressed as the product of two matrices, $A \in \mathbb{R}^{M \times H}$ and $B \in \mathbb{R}^{L \times H}$:

$$U^{\text{low-rank}} = BA^\top = \sum_{h=1}^{H} b_h a_h^\top . \tag{3.113}$$

Trivially, low-rankness is forced if $H$ is set to a small value. However, when VB learning is applied, the estimator can be low-rank even if we adopt the full-rank model, i.e., $H = \min(L, M)$. This phenomenon is caused by MIR, which will be discussed in Chapter 7.

### 3.5.2 Sparse Matrix Factorization Terms

SAMF (Nakajima et al., 2013b) was proposed as a generalization of robust PCA, where various types of sparsity are induced by combining different types of factorization. For example, the following factorization *implicitly* induces rowwise, columnwise, and elementwise sparsity, respectively:

$$U^{\text{row}} = \Gamma_E D = (\gamma_1^e \widetilde{d}_1, \ldots, \gamma_L^e \widetilde{d}_L)^\top , \tag{3.114}$$

$$U^{\text{column}} = E \Gamma_D = (\gamma_1^d e_1, \ldots, \gamma_M^d e_M), \tag{3.115}$$

$$U^{\text{element}} = E \odot D, \tag{3.116}$$

where $\Gamma_D = \mathbf{Diag}(\gamma_1^d, \ldots, \gamma_M^d) \in \mathbb{R}^{M \times M}$ and $\Gamma_E = \mathbf{Diag}(\gamma_1^e, \ldots, \gamma_L^e) \in \mathbb{R}^{L \times L}$ are diagonal matrices, and $D, E \in \mathbb{R}^{L \times M}$. $\odot$ denotes the Hadamard product, i.e., $(E \odot D)_{l,m} = E_{l,m} D_{l,m}$. The reason why the factorizations (3.114) through (3.116) induce the corresponding types of sparsity is explained in Section 7.5.

As a general expression of sparsity inducing factorizations, we define a sparse matrix factorization (SMF) term with a mapping $G$ consisting of partitioning, rearrangement, and factorization:

$$U = G(\{U'^{(k)}\}_{k=1}^K; \mathcal{X}), \text{ where } U'^{(k)} = B^{(k)} A^{(k)\top}. \tag{3.117}$$

Here, $\{A^{(k)}, B^{(k)}\}_{k=1}^K$ are parameters to be estimated, and $G(\cdot; \mathcal{X}) : \mathbb{R}^{\prod_{k=1}^K (L'^{(k)} \times M'^{(k)})} \mapsto \mathbb{R}^{L \times M}$ is a designed function associated with an index mapping $\mathcal{X}$ (explained shortly).

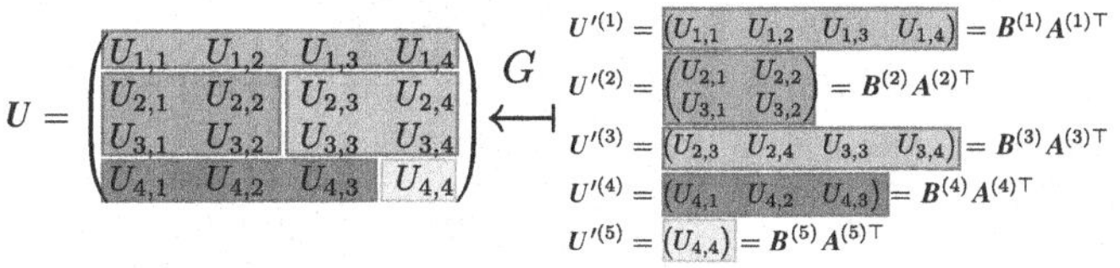


Figure 3.3 An example of SMF-term construction. $\boldsymbol{G}(\cdot;\mathcal{X})$ with $\mathcal{X}:(k,l',m')\mapsto(l,m)$ maps the set $\{\boldsymbol{U}'^{(k)}\}_{k=1}^{K}$ of the PR matrices to the target matrix $\boldsymbol{U}$, so that $U'^{(k)}_{l',m'} = U_{\mathcal{X}(k,l',m')} = U_{l,m}$.

Figure 3.3 illustrates how to construct an SMF term. First, we partition the entries of $\boldsymbol{U}$ into $K$ parts. Then, by rearranging the entries in each part, we form *partitioned-and-rearranged (PR) matrices* $\boldsymbol{U}'^{(k)} \in \mathbb{R}^{L'^{(k)}\times M'^{(k)}}$ for $k = 1,\ldots,K$. Finally, each of $\boldsymbol{U}'^{(k)}$ is decomposed into the product of $\boldsymbol{A}^{(k)} \in \mathbb{R}^{M'^{(k)}\times H'^{(k)}}$ and $\boldsymbol{B}^{(k)} \in \mathbb{R}^{L'^{(k)}\times H'^{(k)}}$, where $H'^{(k)} \leq \min(L'^{(k)}, M'^{(k)})$.

In Eq. (3.117), the function $\boldsymbol{G}(\cdot;\mathcal{X})$ is responsible for partitioning and rearrangement: it maps the set $\{\boldsymbol{U}'^{(k)}\}_{k=1}^{K}$ of the PR matrices to the target matrix $\boldsymbol{U} \in \mathbb{R}^{L\times M}$, based on the one-to-one map $\mathcal{X}:(k,l',m')\mapsto(l,m)$ from the indices of the entries in $\{\boldsymbol{U}'^{(k)}\}_{k=1}^{K}$ to the indices of the entries in $\boldsymbol{U}$ such that

$$\left(\boldsymbol{G}(\{\boldsymbol{U}'^{(k)}\}_{k=1}^{K};\mathcal{X})\right)_{l,m} = U_{l,m} = U_{\mathcal{X}(k,l',m')} = U'^{(k)}_{l',m'}. \tag{3.118}$$

When VB learning is applied, the SMF-term expression (3.117) induces partitionwise sparsity and low-rank sparsity in each partition. Accordingly, partitioning, rearrangement, and factorization should be designed in the following way. Suppose that we are given a required sparsity structure on a matrix (examples of possible side information that suggests particular sparsity structures are given in Section 3.5.3). We first partition the matrix, according to the required sparsity. Some partitions can be submatrices. We rearrange each of the submatrices on which we do not want to impose low-rank sparsity into a long vector ($\boldsymbol{U}'^{(3)}$ in the example in Figure 3.3). We leave the other submatrices which we want to be low-rank ($\boldsymbol{U}'^{(2)}$) and the original vectors ($\boldsymbol{U}'^{(1)}$ and $\boldsymbol{U}'^{(4)}$) and scalars ($\boldsymbol{U}'^{(5)}$) as they are. Finally, we factorize each of the PR matrices to induce sparsity.

Let us, for example, assume that rowwise sparsity is required. We first make the rowwise partition, i.e., separate $\boldsymbol{U} \in \mathbb{R}^{L\times M}$ into $L$ pieces of $M$-dimensional row vectors $\boldsymbol{U}'^{(l)} = \widetilde{\boldsymbol{u}}_l^{\top} \in \mathbb{R}^{1\times M}$. Then, we factorize each partition as $\boldsymbol{U}'^{(l)} = \boldsymbol{B}^{(l)}\boldsymbol{A}^{(l)\top}$ (see the top illustration in Figure 3.4). Thus, we obtain the rowwise sparse term (3.114). Here, $\mathcal{X}(k,1,m') = (k,m')$ makes the following connection between Eqs. (3.114) and (3.117): $\gamma_l^e = \boldsymbol{B}^{(k)} \in \mathbb{R}, \widetilde{\boldsymbol{d}}_l = \boldsymbol{A}^{(k)} \in \mathbb{R}^{M\times 1}$ for $k = l$. Similarly, requiring columnwise and elementwise sparsity leads to

Table 3.1 *Examples of SMF terms.*

| Factorization | Induced sparsity | $K$ | $(L'^{(k)}, M'^{(k)})$ | $\mathcal{X} : (k, l', m') \mapsto (l, m)$ |
|---|---|---|---|---|
| $U = BA^\top$ | low-rank | 1 | $(L, M)$ | $\mathcal{X}(1, l', m') = (l', m')$ |
| $U = \Gamma_E D$ | rowwise | $L$ | $(1, M)$ | $\mathcal{X}(k, 1, m') = (k, m')$ |
| $U = E\Gamma_D$ | columnwise | $M$ | $(L, 1)$ | $\mathcal{X}(k, l', 1) = (l', k)$ |
| $U = E \odot D$ | elementwise | $L \times M$ | $(1, 1)$ | $\mathcal{X}(k, 1, 1) = \textit{vec-order}(k)$ |

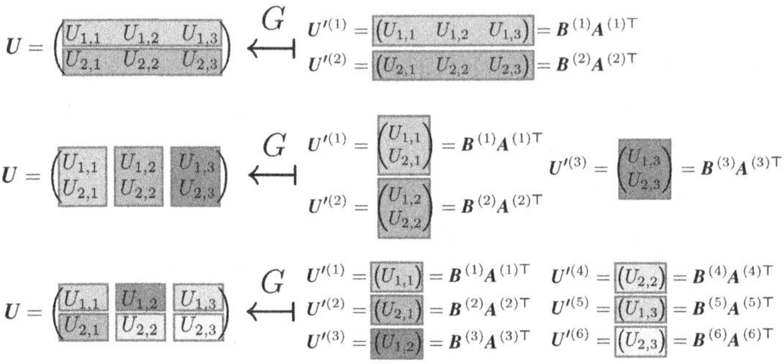

Figure 3.4 SMF-term construction for the rowwise (top), the columnwise (middle), and the elementwise (bottom) sparse terms.

Eqs. (3.115) and (3.116), respectively (see the bottom two illustrations in Figure 3.4). Table 3.1 summarizes how to design these SMF terms, where $\textit{vec-order}(k) = (1 + ((k-1) \bmod L), \lceil k/L \rceil)$ goes along the columns one after another in the same way as the *vec* operator forming a vector by stacking the columns of a matrix (in other words, $(U'^{(1)}, \ldots, U'^{(K)})^\top = \textit{vec}(U)$).

Now we define the SAMF model as the sum of SMF terms (3.117):

$$V = \sum_{s=1}^{S} U^{(s)} + \mathcal{E}, \tag{3.119}$$

$$\text{where} \qquad U^{(s)} = G(\{B^{(k,s)} A^{(k,s)\top}\}_{k=1}^{K^{(s)}}; \mathcal{X}^{(s)}). \tag{3.120}$$

### 3.5.3 Examples of SMF Terms

In practice, SMF terms should be designed based on side information. Suppose that $V \in \mathbb{R}^{L \times M}$ consists of $M$ samples of $L$-dimensional sensor outputs. In robust PCA (3.112), we add an elementwise sparse term (3.116) to the low-rank term (3.113), assuming that the low-rank signal is expected to be

Figure 3.5 Foreground/background video separation task.

contaminated with spiky noise when observed. Here, we can say that the existence of spiky noise is used as side information.

Similarly, if we expect that a small number of sensors can be broken, and their outputs are unreliable over all $M$ samples, we should add the rowwise sparse term (3.114) to separate the low-rank signal from rowwise noise:

$$\boldsymbol{V} = \boldsymbol{U}^{\text{low-rank}} + \boldsymbol{U}^{\text{row}} + \boldsymbol{\mathcal{E}}.$$

If we expect some accidental disturbances occurred during the observation, but do not know their exact locations (i.e., which samples are affected), the columnwise sparse term (3.115) can effectively capture such disturbances.

The SMF expression (3.117) enables us to use side information in a more flexible way, and its advantage has been shown in a *foreground/background video separation* problem (Nakajima et al., 2013b). The top image in Figure 3.5 is a frame of a video available from the *Caviar Project* website,[1] and the task is to separate *moving* objects (bottom-right) from the background (bottom-left). Previous approaches (Candès et al., 2011; Ding et al., 2011; Babacan et al., 2012b) first constructed the observation matrix $\boldsymbol{V}$ by stacking all pixels in each frame into each column (Figure 3.6), and then fitted it by the robust PCA model (3.112). Here, the low-rank term and the elementwise

[1] The European Commission (EC)-funded CAVIAR project/IST 2001 37540, found at URL: http://homepages.inf.ed.ac.uk/rbf/CAVIAR/.

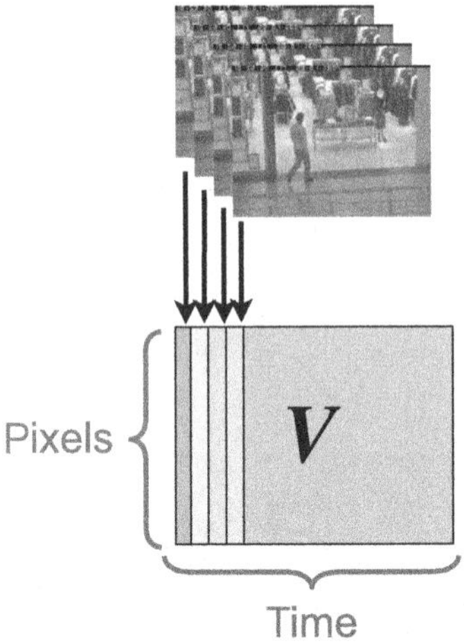


Figure 3.6 The observation matrix $V$ is constructed by stacking all pixels in each frame into each column.

sparse term are expected to capture the static background and the moving foreground, respectively. However, we can also rely on the natural assumption that the pixels in a segment sharing similar intensities tend to belong to the same object. Under this assumption as side information, we can adopt a segmentwise sparse term, for which the PR matrix is constructed based on a precomputed oversegmented image (Figure 3.7). The segmentwise sparse term has been shown to capture the foreground more accurately than the elementwise sparse term in this application. Details will be discussed in Chapter 11.

### 3.5.4 VB Learning for SAMF

Let us summarize the parameters of the SAMF model (3.119) as follows:

$$\boldsymbol{\Theta} = \{\boldsymbol{\Theta}_{\boldsymbol{A}}^{(s)}, \boldsymbol{\Theta}_{\boldsymbol{B}}^{(s)}\}_{s=1}^{S}, \qquad \text{where} \qquad \boldsymbol{\Theta}_{\boldsymbol{A}}^{(s)} = \{\boldsymbol{A}^{(k,s)}\}_{k=1}^{K^{(s)}}, \quad \boldsymbol{\Theta}_{\boldsymbol{B}}^{(s)} = \{\boldsymbol{B}^{(k,s)}\}_{k=1}^{K^{(s)}}.$$

As in the MF model, we assume independent Gaussian noise and priors. Then, the likelihood and the priors are given by

$$p(\boldsymbol{V}|\boldsymbol{\Theta}) \propto \exp\left(-\frac{1}{2\sigma^2}\left\|\boldsymbol{V} - \sum_{s=1}^{S}\boldsymbol{U}^{(s)}\right\|_{\text{Fro}}^2\right), \tag{3.121}$$

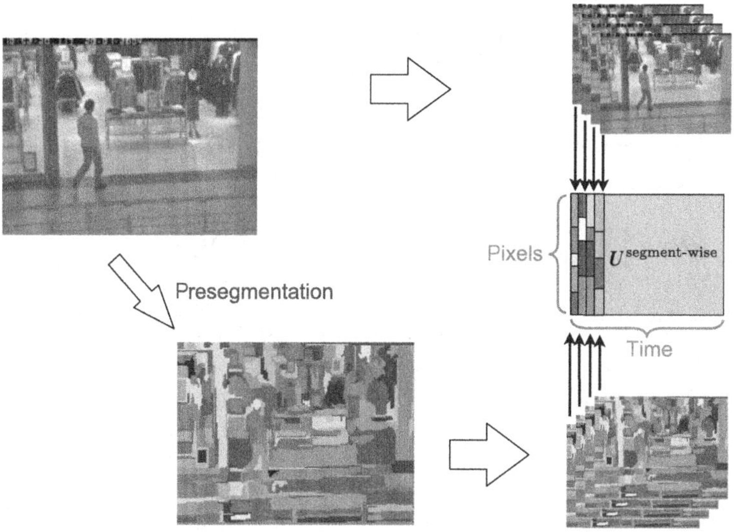


Figure 3.7 Construction of a segmentwise sparse term. The original frame is presegmented, based on which the segmentwise sparse term is constructed as an SMF term.

$$p(\{\boldsymbol{\Theta}_{A}^{(s)}\}_{s=1}^{S}) \propto \exp\left( -\frac{1}{2} \sum_{s=1}^{S} \sum_{k=1}^{K^{(s)}} \mathrm{tr}\left( \boldsymbol{A}^{(k,s)} \boldsymbol{C}_{A}^{(k,s)-1} \boldsymbol{A}^{(k,s)\top} \right) \right), \tag{3.122}$$

$$p(\{\boldsymbol{\Theta}_{B}^{(s)}\}_{s=1}^{S}) \propto \exp\left( -\frac{1}{2} \sum_{s=1}^{S} \sum_{k=1}^{K^{(s)}} \mathrm{tr}\left( \boldsymbol{B}^{(k,s)} \boldsymbol{C}_{B}^{(k,s)-1} \boldsymbol{B}^{(k,s)\top} \right) \right). \tag{3.123}$$

We assume that the prior covariances of $\boldsymbol{A}^{(k,s)}$ and $\boldsymbol{B}^{(k,s)}$ are diagonal:

$$\boldsymbol{C}_{A}^{(k,s)} = \mathbf{Diag}(c_{a_1}^{(k,s)2}, \ldots, c_{a_H}^{(k,s)2}),$$
$$\boldsymbol{C}_{B}^{(k,s)} = \mathbf{Diag}(c_{b_1}^{(k,s)2}, \ldots, c_{b_H}^{(k,s)2}).$$

**Conditional Conjugacy**

As seen in Eq. (3.121), the SAMF model is the MF model for the parameters $(\boldsymbol{\Theta}_{A}^{(s)}, \boldsymbol{\Theta}_{B}^{(s)})$ in the $s$th SMF term, if the other parameters $\{\boldsymbol{\Theta}_{A}^{(s')}, \boldsymbol{\Theta}_{B}^{(s')}\}_{s' \neq s}$ are regarded as constants. Therefore, the Gaussian priors (3.122) and (3.123) are conditionally conjugate for each of $\boldsymbol{\Theta}_{A}^{(s)}$ and $\boldsymbol{\Theta}_{B}^{(s)}$ in each of the SMF terms.

**Variational Bayesian Algorithm**

Based on the conditional conjugacy, we solve the VB learning problem under the following independence constraint (Babacan et al., 2012b):

$$r(\boldsymbol{\Theta}) = \prod_{s=1}^{S} r_{\boldsymbol{A}}^{(s)}(\boldsymbol{\Theta}_{\boldsymbol{A}}^{(s)}) r_{\boldsymbol{B}}^{(s)}(\boldsymbol{\Theta}_{\boldsymbol{B}}^{(s)}). \tag{3.124}$$

Following the standard procedure described in Section 2.1.5, we can find that the VB posterior, which minimizes the free energy (2.15), is in the following form:

$$\begin{aligned} r(\boldsymbol{\Theta}) &= \prod_{s=1}^{S} \prod_{k=1}^{K^{(s)}} \Big( \mathrm{MGauss}_{M'^{(k,s)},H'^{(k,s)}}(\boldsymbol{A}^{(k,s)}; \widehat{\boldsymbol{A}}^{(k,s)}, \widehat{\boldsymbol{\Sigma}}_{\boldsymbol{A}}^{(k,s)}) \\ &\qquad \cdot \mathrm{MGauss}_{L'^{(k,s)},H'^{(k,s)}}(\boldsymbol{B}^{(k,s)}; \widehat{\boldsymbol{B}}^{(k,s)}, \widehat{\boldsymbol{\Sigma}}_{\boldsymbol{B}}^{(k,s)}) \Big) \\ &= \prod_{s=1}^{S} \prod_{k=1}^{K^{(s)}} \Big( \prod_{m'=1}^{M'^{(k,s)}} \mathrm{Gauss}_{H'^{(k,s)}}(\widetilde{\boldsymbol{a}}_{m'}^{(k,s)}; \widehat{\widetilde{\boldsymbol{a}}}_{m'}^{(k,s)}, \widehat{\boldsymbol{\Sigma}}_{\boldsymbol{A}}^{(k,s)}) \\ &\qquad \cdot \prod_{l'=1}^{L'^{(k,s)}} \mathrm{Gauss}_{H'^{(k,s)}}(\widetilde{\boldsymbol{b}}_{l'}^{(k,s)}; \widehat{\widetilde{\boldsymbol{b}}}_{l'}^{(k,s)}, \widehat{\boldsymbol{\Sigma}}_{\boldsymbol{B}}^{(k,s)}) \Big) \end{aligned} \tag{3.125}$$

with the variational parameters satisfying the stationary conditions given by

$$\widehat{\boldsymbol{A}}^{(k,s)} = \sigma^{-2} \boldsymbol{Z}'^{(k,s)\top} \widehat{\boldsymbol{B}}^{(k,s)} \widehat{\boldsymbol{\Sigma}}_{\boldsymbol{A}}^{(k,s)}, \tag{3.126}$$

$$\widehat{\boldsymbol{\Sigma}}_{\boldsymbol{A}}^{(k,s)} = \sigma^2 \left( \widehat{\boldsymbol{B}}^{(k,s)\top} \widehat{\boldsymbol{B}}^{(k,s)} + L'^{(k,s)} \widehat{\boldsymbol{\Sigma}}_{\boldsymbol{B}}^{(k,s)} + \sigma^2 \boldsymbol{C}_A^{(k,s)-1} \right)^{-1}, \tag{3.127}$$

$$\widehat{\boldsymbol{B}}^{(k,s)} = \sigma^{-2} \boldsymbol{Z}'^{(k,s)} \widehat{\boldsymbol{A}}^{(k,s)} \widehat{\boldsymbol{\Sigma}}_{\boldsymbol{B}}^{(k,s)}, \tag{3.128}$$

$$\widehat{\boldsymbol{\Sigma}}_{\boldsymbol{B}}^{(k,s)} = \sigma^2 \left( \widehat{\boldsymbol{A}}^{(k,s)\top} \widehat{\boldsymbol{A}}^{(k,s)} + M'^{(k,s)} \widehat{\boldsymbol{\Sigma}}_{\boldsymbol{A}}^{(k,s)} + \sigma^2 \boldsymbol{C}_B^{(k,s)-1} \right)^{-1}. \tag{3.129}$$

Here, $\boldsymbol{Z}'^{(k,s)} \in \mathbb{R}^{L'^{(k,s)} \times M'^{(k,s)}}$ is defined as

$$Z'^{(k,s)}_{l',m'} = Z^{(s)}_{\mathcal{X}^{(s)}(k,l',m')}, \qquad \text{where} \quad \boldsymbol{Z}^{(s)} = \boldsymbol{V} - \sum_{s' \neq s} \widehat{\boldsymbol{U}}^{(s)}. \tag{3.130}$$

**Free Energy as a Function of Variational Parameters**

The free energy can be explicitly written as

$$\begin{aligned} 2F &= LM \log(2\pi\sigma^2) + \frac{\|\boldsymbol{V}\|_{\mathrm{Fro}}^2}{\sigma^2} \\ &+ \sum_{s=1}^{S} \sum_{k=1}^{K^{(s)}} \left( M'^{(k,s)} \log \frac{\det\left(\boldsymbol{C}_A^{(k,s)}\right)}{\det\left(\widehat{\boldsymbol{\Sigma}}_A^{(k,s)}\right)} + L'^{(k,s)} \log \frac{\det\left(\boldsymbol{C}_B^{(k,s)}\right)}{\det\left(\widehat{\boldsymbol{\Sigma}}_B^{(k,s)}\right)} \right) \\ &+ \sum_{s=1}^{S} \sum_{k=1}^{K^{(S)}} \mathrm{tr} \Big\{ \boldsymbol{C}_A^{(k,s)-1} (\widehat{\boldsymbol{A}}^{(k,s)\top} \widehat{\boldsymbol{A}}^{(k,s)} + M'^{(k,s)} \widehat{\boldsymbol{\Sigma}}_A^{(k,s)}) \\ &\qquad + \boldsymbol{C}_B^{(k,s)-1} (\widehat{\boldsymbol{B}}^{(k,s)\top} \widehat{\boldsymbol{B}}^{(k,s)} + L'^{(k,s)} \widehat{\boldsymbol{\Sigma}}_B^{(k,s)}) \Big\} \end{aligned}$$

$$+\frac{1}{\sigma^2}\mathrm{tr}\left\{-2\boldsymbol{V}^\top\left(\sum_{s=1}^{S}\boldsymbol{G}(\{\widehat{\boldsymbol{B}}^{(k,s)}\widehat{\boldsymbol{A}}^{(k,s)\top}\}_{k=1}^{K^{(s)}};\mathcal{X}^{(s)})\right)\right.$$
$$\left.+2\sum_{s=1}^{S}\sum_{s'=s+1}^{S}\boldsymbol{G}^\top(\{\widehat{\boldsymbol{B}}^{(k,s)}\widehat{\boldsymbol{A}}^{(k,s)\top}\}_{k=1}^{K^{(s)}};\mathcal{X}^{(s)})\boldsymbol{G}(\{\widehat{\boldsymbol{B}}^{(k,s')}\widehat{\boldsymbol{A}}^{(k,s')\top}\}_{k=1}^{K^{(s')}};\mathcal{X}^{(s')})\right\}$$
$$+\frac{1}{\sigma^2}\sum_{s=1}^{S}\sum_{k=1}^{K^{(S)}}\mathrm{tr}\left\{(\widehat{\boldsymbol{A}}^{(k,s)\top}\widehat{\boldsymbol{A}}^{(k,s)}+M'^{(k,s)}\widehat{\boldsymbol{\Sigma}}_A^{(k,s)})(\widehat{\boldsymbol{B}}^{(k,s)\top}\widehat{\boldsymbol{B}}^{(k,s)}+L'^{(k,s)}\widehat{\boldsymbol{\Sigma}}_B^{(k,s)})\right\}$$
$$-\sum_{s=1}^{S}\sum_{k=1}^{K^{(S)}}(L'^{(k,s)}+M'^{(k,s)})H'^{(k,s)}. \tag{3.131}$$

### Empirical Variational Bayesian Algorithm

The following stationary conditions for the hyperparameters can be obtained from the derivatives of the free energy (3.131):

$$c_{a_h}^{(k,s)2}=\left\|\widehat{\boldsymbol{a}}_h^{(k,s)}\right\|^2/M'^{(k,s)}+(\widehat{\boldsymbol{\Sigma}}_A^{(k,s)})_{hh}, \tag{3.132}$$
$$c_{b_h}^{(k,s)2}=\left\|\widehat{\boldsymbol{b}}_h^{(k,s)}\right\|^2/L'^{(k,s)}+(\widehat{\boldsymbol{\Sigma}}_B^{(k,s)})_{hh}, \tag{3.133}$$

---

**Algorithm 6** EVB learning for sparse additive matrix factorization.

---

1: Initialize the variational parameters $\{\widehat{\boldsymbol{A}}^{(k,s)},\widehat{\boldsymbol{\Sigma}}_A^{(k,s)},\widehat{\boldsymbol{B}}^{(k,s)},\widehat{\boldsymbol{\Sigma}}_B^{(k,s)}\}_{k=1,s=1}^{K^{(s)}\ S}$, and the hyperparameters $\{\boldsymbol{C}_A^{(k,s)},\boldsymbol{C}_B^{(k,s)}\}_{k=1,s=1}^{K^{(s)}\ S},\sigma^2$, for example, $\widehat{A}_{m,h}^{(k,s)},\widehat{B}_{l,h}^{(k,s)}\sim\mathrm{Gauss}_1(0,\tau)$, $\widehat{\boldsymbol{\Sigma}}_A^{(k,s)}=\widehat{\boldsymbol{\Sigma}}_B^{(k,s)}=\boldsymbol{C}_A^{(k,s)}=\boldsymbol{C}_B^{(k,s)}=\tau\boldsymbol{I}_{H'^{(k,s)}}$, and $\sigma^2=\tau^2$ for $\tau^2=\|\boldsymbol{V}\|_{\mathrm{Fro}}^2/(LM)$.
2: Apply (substitute the right-hand side into the left-hand side) Eqs. (3.127), (3.126), (3.129), and (3.128) for each $k$ and $s$ to update $\widehat{\boldsymbol{\Sigma}}_A^{(k,s)},\widehat{\boldsymbol{A}}^{(k,s)},\widehat{\boldsymbol{\Sigma}}_B^{(k,s)}$, and $\widehat{\boldsymbol{B}}^{(k,s)}$, respectively.
3: Apply Eqs. (3.132) and (3.133) for all $h'=1,\ldots,H'^{(k,s)}$, $k$ and $s$, and Eq. (3.134) to update $\boldsymbol{C}_A^{(k,s)},\boldsymbol{C}_B^{(k,s)}$, and $\sigma^2$, respectively.
4: Prune the $h$th component if $c_{a_h}^{(k,s)2}c_{b_h}^{(k,s)2}<\varepsilon$, where $\varepsilon>0$ is a threshold, e.g., set to $\varepsilon=10^{-4}$.
5: Evaluate the free energy (3.131).
6: Iterate Steps 2 through 5 until convergence (until the energy decrease becomes smaller than a threshold).

---

$$\sigma^2 = \frac{1}{LM} \Bigg\{ \|V\|_{\mathrm{Fro}}^2 - 2 \sum_{s=1}^{S} \mathrm{tr}\left( \widehat{U}^{(s)\top} \left( V - \sum_{s'=s+1}^{S} \widehat{U}^{(s')} \right) \right)$$
$$+ \sum_{s=1}^{S} \sum_{k=1}^{K^{(s)}} \mathrm{tr}\left( \left( \widehat{B}^{(k,s)\top} \widehat{B}^{(k,s)} + L'^{(k,s)} \widehat{\Sigma}_{\mathrm{B}}^{(k,s)} \right) \cdot \left( \widehat{A}^{(k,s)\top} \widehat{A}^{(k,s)} + M'^{(k,s)} \widehat{\Sigma}_{\mathrm{A}}^{(k,s)} \right) \right) \Bigg\}. \tag{3.134}$$

Algorithm 6 summarizes the EVB algorithm for SAMF. If we appropriately set the hyperparameters $\{C_A^{(k,s)}, C_B^{(k,s)}\}_{k=1,s=1}^{K^{(s)}\ S}, \sigma^2$ in Step 1 and skip Steps 3 and 4, Algorithm 6 is reduced to (nonempirical) VB learning.

# 4

# VB Algorithm for Latent Variable Models

In this chapter, we discuss VB learning for *latent variable models*. Starting with finite mixture models as the simplest example, we overview the VB learning algorithms for more complex latent variable models such as Bayesian networks and hidden Markov models.

Let $\mathcal{H}$ denote the set of *(local) latent variables* and $\boldsymbol{w}$ denote a model parameter vector (or the set of *global latent variables*). In this chapter, we consider the latent variable model for training data $\mathcal{D}$:

$$p(\mathcal{D}|\boldsymbol{w}) = \sum_{\mathcal{H}} p(\mathcal{D}, \mathcal{H}|\boldsymbol{w}).$$

Let us employ the following factorized model to approximate the posterior distribution for $\boldsymbol{w}$ and $\mathcal{H}$:

$$r(\boldsymbol{w}, \mathcal{H}) = r_w(\boldsymbol{w}) r_{\mathcal{H}}(\mathcal{H}). \tag{4.1}$$

Applying the general VB framework explained in Section 2.1.5 to the preceding model leads to the following update rules for $\boldsymbol{w}$ and $\mathcal{H}$:

$$r_w(\boldsymbol{w}) = \frac{1}{C_w} p(\boldsymbol{w}) \exp \left\langle \log p(\mathcal{D}, \mathcal{H}|\boldsymbol{w}) \right\rangle_{r_{\mathcal{H}}(\mathcal{H})}, \tag{4.2}$$

$$r_{\mathcal{H}}(\mathcal{H}) = \frac{1}{C_{\mathcal{H}}} \exp \left\langle \log p(\mathcal{D}, \mathcal{H}|\boldsymbol{w}) \right\rangle_{r_w(\boldsymbol{w})}. \tag{4.3}$$

In the following sections, we discuss these update rules for some specific examples of latent variable models.

## 4.1 Finite Mixture Models

A *finite mixture model* $p(\boldsymbol{x}|\boldsymbol{w})$ of an $L$-dimensional input $\boldsymbol{x} \in \mathbb{R}^L$ with a parameter vector $\boldsymbol{w} \in \mathbb{R}^M$ is defined by

$$p(\boldsymbol{x}|\boldsymbol{w}) = \sum_{k=1}^{K} \alpha_k p(\boldsymbol{x}|\boldsymbol{\tau}_k), \tag{4.4}$$

where integer $K$ is the number of components and $\boldsymbol{\alpha} = (\alpha_1, \ldots, \alpha_K)^\top \in \Delta^{K-1}$ is the set of mixing weights (Example 1.3). The parameter $\boldsymbol{w}$ of the model is $\boldsymbol{w} = \{\alpha_k, \boldsymbol{\tau}_k\}_{k=1}^{K}$.

The finite mixture model can be rewritten as follows by using a hidden variable $\boldsymbol{z} = (z_1, \ldots, z_K)^\top \in \{\boldsymbol{e}_1, \ldots, \boldsymbol{e}_K\}$,

$$p(\boldsymbol{x}, \boldsymbol{z}|\boldsymbol{w}) = \prod_{k=1}^{K} \left[\alpha_k p(\boldsymbol{x}|\boldsymbol{\tau}_k)\right]^{z_k}. \tag{4.5}$$

Here $\boldsymbol{e}_k \in \{0, 1\}^K$ is the $K$-dimensional binary vector, called the *one-of-K representation*, with one at the $k$th entry and zeros at the other entries:

$$\boldsymbol{e}_k = (\underbrace{0, \ldots, 0, \overbrace{1}^{k\text{-th}}, 0, \ldots, 0}_{K})^\top.$$

The hidden variable $\boldsymbol{z}$ is not observed and is representing the component from which the data sample $\boldsymbol{x}$ is generated. If the data sample $\boldsymbol{x}$ is from the $k$th component, then $z_k = 1$, otherwise, $z_k = 0$. Then

$$\sum_{\boldsymbol{z}} p(\boldsymbol{x}, \boldsymbol{z}|\boldsymbol{w}) = p(\boldsymbol{x}|\boldsymbol{w})$$

holds where the sum over $\boldsymbol{z}$ goes through all possible values of the hidden variable.

### 4.1.1 Mixture of Gaussians

If the component distribution in Eq. (4.4) is chosen to be a Gaussian distribution,

$$p(\boldsymbol{x}|\boldsymbol{\tau}) = \mathrm{Gauss}_L(\boldsymbol{x}; \boldsymbol{\mu}, \boldsymbol{\Sigma}),$$

the finite mixture model is called the *mixture of Gaussians* or the *Gaussian mixture model (GMM)*.

In some applications, the parameters are restricted to the means of each component, and it is assumed that there is no correlation between each input dimension. In this case, since $L = M$, the model is written by

$$p(\boldsymbol{x}|\boldsymbol{w}) = \sum_{k=1}^{K} \frac{\alpha_k}{(2\pi\sigma^2)^{M/2}} \exp\left(-\frac{\|\boldsymbol{x} - \boldsymbol{\mu}_k\|^2}{2\sigma^2}\right), \tag{4.6}$$

where $\sigma > 0$ is a constant.

In this subsection, the uncorrelated GMM (4.6) is considered in the VB framework by further assuming that $\sigma^2 = 1$ for simplicity. The joint model for the observed and hidden variables (4.5) is given by the product of the following two distributions:

$$p(z|\alpha) = \mathrm{Multinomial}_{K,1}(z;\alpha), \tag{4.7}$$

$$p(\boldsymbol{x}|z,\{\boldsymbol{\mu}_k\}_{k=1}^K) = \prod_{k=1}^K \{\mathrm{Gauss}_M(\boldsymbol{x};\boldsymbol{\mu}_k,\boldsymbol{I}_M)\}^{z_k}. \tag{4.8}$$

Thus, for the set of hidden variables $\mathcal{H} = \{z^{(n)}\}_{n=1}^N$ and the complete data set $\{\mathcal{D},\mathcal{H}\} = \{\boldsymbol{x}^{(n)}, z^{(n)}\}_{n=1}^N$, the complete likelihood is given by

$$p(\mathcal{D},\mathcal{H}|\alpha,\{\boldsymbol{\mu}_k\}_{k=1}^K) = \prod_{n=1}^N \prod_{k=1}^K \left\{\alpha_k \mathrm{Gauss}_M(\boldsymbol{x}^{(n)};\boldsymbol{\mu}_k,\boldsymbol{I}_M)\right\}^{z_k^{(n)}}. \tag{4.9}$$

*ML learning* of the GMM is carried out by the expectation-maximization (EM) algorithm (Dempster et al., 1977), which corresponds to a clustering algorithm called the soft K-means (MacKay, 2003, ch. 22).

Because of the conditional conjugacy (Section 2.1.2) of the parameters $\boldsymbol{\alpha} = (\alpha_1,\ldots,\alpha_K)^\top \in \Delta^{K-1}$ and $\{\boldsymbol{\mu}_k\}_{k=1}^K$, we assume that the prior of the parameters is the product of the following two distributions:

$$p(\alpha|\phi) = \mathrm{Dirichlet}_K(\alpha;(\phi,\ldots,\phi)^\top), \tag{4.10}$$

$$p(\{\boldsymbol{\mu}_k\}_{k=1}^K|\boldsymbol{\mu}_0,\xi) = \prod_{k=1}^K \mathrm{Gauss}_M(\boldsymbol{\mu}_k|\boldsymbol{\mu}_0,(1/\xi)\boldsymbol{I}_M), \tag{4.11}$$

where $\xi > 0$, $\boldsymbol{\mu}_0 \in \mathbb{R}^M$ and $\phi > 0$ are the hyperparameters.

### VB Posterior for the Gaussian Mixture Model

Let

$$\overline{N}_k = \sum_{n=1}^N \left\langle z_k^{(n)} \right\rangle_{r_{\mathcal{H}}(\mathcal{H})} \tag{4.12}$$

and

$$\overline{\boldsymbol{x}}_k = \frac{1}{\overline{N}_k} \sum_{n=1}^N \left\langle z_k^{(n)} \right\rangle_{r_{\mathcal{H}}(\mathcal{H})} \boldsymbol{x}^{(n)}, \tag{4.13}$$

where $z_k^{(n)} = 1$ if the $n$th data sample $\boldsymbol{x}^{(n)}$ is from the $k$th component; otherwise, $z_k^{(n)} = 0$. The variable $\overline{N}_k$ is the expected number of times data come from the $k$th component, and $\overline{\boldsymbol{x}}_k$ is the mean of them. Note that the variables $\overline{N}_k$ and $\overline{\boldsymbol{x}}_k$ satisfy the constraints $\sum_{k=1}^K \overline{N}_k = N$ and $\sum_{k=1}^K \overline{N}_k \overline{\boldsymbol{x}}_k = \sum_{n=1}^N \boldsymbol{x}^{(n)}$. From

(4.2) and the respective priors (4.10) and (4.11), the VB posterior $r_w(\boldsymbol{w}) = r_\alpha(\boldsymbol{\alpha})r_\mu(\{\boldsymbol{\mu}_k\}_{k=1}^K)$ is obtained as the product of the following two distributions:

$$r_\alpha(\boldsymbol{\alpha}) = \mathrm{Dirichlet}_K\left(\boldsymbol{\alpha}; (\widehat{\phi}_1, \ldots, \widehat{\phi}_K)^\top\right), \tag{4.14}$$

$$r_\mu(\{\boldsymbol{\mu}_k\}_{k=1}^K) = \prod_{k=1}^K \mathrm{Gauss}_M\left(\boldsymbol{\mu}_k; \widehat{\boldsymbol{\mu}}_k, \widehat{\sigma}_k^2 \boldsymbol{I}_M\right), \tag{4.15}$$

where

$$\widehat{\phi}_k = \overline{N}_k + \phi, \tag{4.16}$$

$$\widehat{\sigma}_k^2 = \frac{1}{\overline{N}_k + \xi}, \tag{4.17}$$

$$\widehat{\boldsymbol{\mu}}_k = \frac{\overline{N}_k \overline{\boldsymbol{x}}_k + \xi \boldsymbol{\mu}_0}{\overline{N}_k + \xi}. \tag{4.18}$$

From Eq. (4.3), the VB posterior $r_{\mathcal{H}}(\mathcal{H})$ is given by

$$r_{\mathcal{H}}(\mathcal{H}) = \frac{1}{C_{\mathcal{H}}} \prod_{n=1}^N \exp\left(z_k^{(n)} \left\{ \Psi(\widehat{\phi}_k) - \Psi\left(\sum_{k'=1}^K \widehat{\phi}_{k'}\right) - \frac{\|\boldsymbol{x}^{(n)} - \widehat{\boldsymbol{\mu}}_k\|^2}{2} - \frac{M}{2}\left(\log 2\pi + \frac{1}{\overline{N}_k + \xi}\right)\right\}\right),$$

where $\Psi$ is the di-gamma (psi) function, and we used

$$\langle \log \alpha_k \rangle_{r_\alpha(\boldsymbol{\alpha})} = \Psi(\widehat{\phi}_k) - \Psi\left(\sum_{k'=1}^K \widehat{\phi}_{k'}\right). \tag{4.19}$$

That is, $r_{\mathcal{H}}(\mathcal{H}) = r_z(\{\boldsymbol{z}^{(n)}\}_{n=1}^N)$ is the multinomial distribution:

$$r_z(\{\boldsymbol{z}^{(n)}\}_{n=1}^N) = \prod_{n=1}^N r_z(\boldsymbol{z}^{(n)}) = \prod_{n=1}^N \mathrm{Multinomial}_{K,1}\left(\boldsymbol{z}^{(n)}; \widehat{\boldsymbol{z}}^{(n)}\right),$$

where $\widehat{\boldsymbol{z}}^{(n)} \in \Delta^{K-1}$ is

$$\widehat{z}_k^{(n)} = \left\langle z_k^{(n)} \right\rangle_{r_{\mathcal{H}}(\mathcal{H})} = \frac{\overline{z}_k^{(n)}}{\sum_{k'=1}^K \overline{z}_{k'}^{(n)}}, \tag{4.20}$$

for

$$\overline{z}_k^{(n)} = \exp\left(\Psi(\widehat{\phi}_k) - \Psi\left(\sum_{k'=1}^K \widehat{\phi}_{k'}\right) - \frac{1}{2}\left\|\boldsymbol{x}^{(n)} - \widehat{\boldsymbol{\mu}}_k\right\|^2 + M\widehat{\sigma}_k^2\right). \tag{4.21}$$

The free energy as a function of variational parameters is expressed as follows:

$$
\begin{aligned}
F &= \left\langle \log \frac{r_{\mathcal{H}}(\mathcal{H}) r_w(\boldsymbol{w})}{p(\boldsymbol{w})} \right\rangle_{r_{\mathcal{H}}(\mathcal{H}) r_w(\boldsymbol{w})} - \langle \log p(\mathcal{D}, \mathcal{H} | \boldsymbol{w}) \rangle_{r_{\mathcal{H}}(\mathcal{H}) r_w(\boldsymbol{w})} \\
&= \left\langle \log \frac{(\widehat{z}_k^{(n)})^{z_k^{(n)}} \frac{\Gamma(\sum_{k=1}^K \widehat{\phi}_k)}{\prod_{k=1}^K \Gamma(\widehat{\phi}_k)} \prod_{k=1}^K \alpha_k^{\widehat{\phi}_k - 1} \frac{\exp\left(-\frac{\|\mu_k - \widehat{\mu}_k\|^2}{2\widehat{\sigma}_k^2}\right)}{(2\pi\widehat{\sigma}_k^2)^{M/2}}}{\frac{\Gamma(K\phi)}{(\Gamma(\phi))^K} \prod_{k=1}^K \alpha_k^{\phi - 1} \left(\frac{\xi}{2\pi}\right)^{M/2} \exp\left(-\frac{\xi \|\mu_k - \mu_0\|^2}{2}\right)} \right\rangle_{r_{\mathcal{H}}(\mathcal{H}) r_w(\boldsymbol{w})} \\
&\quad - \left\langle \log \prod_{n=1}^N \prod_{k=1}^K \left\{ \alpha_k \frac{\exp\left(-\frac{\|\boldsymbol{x}^{(n)} - \mu_k\|^2}{2}\right)}{(2\pi)^{M/2}} \right\}^{z_k^{(n)}} \right\rangle_{r_{\mathcal{H}}(\mathcal{H}) r_w(\boldsymbol{w})} \\
&= \log \left( \frac{\Gamma(\sum_{k=1}^K \widehat{\phi}_k)}{\prod_{k=1}^K \Gamma(\widehat{\phi}_k)} \right) - \log \left( \frac{\Gamma(K\phi)}{(\Gamma(\phi))^K} \right) - \frac{M}{2} \sum_{k=1}^K \log \left( \xi \widehat{\sigma}_k^2 \right) - \frac{KM}{2} \\
&\quad + \sum_{n=1}^N \sum_{k=1}^K \widehat{z}_k^{(n)} \log \widehat{z}_k^{(n)} + \sum_{k=1}^K \left( \widehat{\phi}_k - \phi - \overline{N}_k \right) \left( \Psi(\widehat{\phi}_k) - \Psi(\textstyle\sum_{k'=1}^K \widehat{\phi}_{k'}) \right) \\
&\quad + \sum_{k=1}^K \frac{\xi \left( \|\widehat{\mu}_k - \mu_0\|^2 + M\widehat{\sigma}_k^2 \right)}{2} + \sum_{k=1}^K \frac{\overline{N}_k \left( M \log(2\pi) + M\widehat{\sigma}_k^2 \right)}{2} \\
&\quad + \sum_{k=1}^K \frac{\overline{N}_k \|\overline{\boldsymbol{x}}_k - \widehat{\mu}_k\|^2 + \sum_{n=1}^N \widehat{z}_k^{(n)} \|\boldsymbol{x}^{(n)} - \overline{\boldsymbol{x}}_k\|^2}{2}. \qquad (4.22)
\end{aligned}
$$

The prior hyperparameters, $(\phi, \mu_0, \xi)$, can be estimated by the EVB learning (Section 2.1.6). Computing the partial derivatives, we have

$$\frac{\partial F}{\partial \phi} = K\left(\Psi(\phi) - \Psi(K\phi)\right) - \sum_{k=1}^K \left( \Psi(\widehat{\phi}_k) - \Psi\left(\textstyle\sum_{k'=1}^K \widehat{\phi}_{k'}\right) \right), \qquad (4.23)$$

$$\frac{\partial F}{\partial \mu_0} = \xi \sum_{k=1}^K (\mu_0 - \widehat{\mu}_k), \qquad (4.24)$$

$$\frac{\partial F}{\partial \xi} = -\frac{M}{2} \left( \frac{K}{\xi} - \sum_{k=1}^K \left( \frac{\|\widehat{\mu}_k - \mu_0\|^2}{M} + \widehat{\sigma}_k^2 \right) \right). \qquad (4.25)$$

The stationary conditions $\frac{\partial F}{\partial \mu_0} = \mathbf{0}$ and $\frac{\partial F}{\partial \xi} = 0$ yield the following update rules:

$$\mu_0 = \frac{1}{K} \sum_{k=1}^K \widehat{\mu}_k, \qquad (4.26)$$

**Algorithm 7** EVB learning for the Gaussian mixture model.

1: Initialize the variational parameters $(\{\widehat{z}^{(n)}\}_{n=1}^{N}, \{\widehat{\phi}_k\}_{k=1}^{K}, \{\widehat{\mu}_k, \widehat{\sigma}_k^2\}_{k=1}^{K})$, and the hyperparameters $(\phi, \mu_0, \xi)$.
2: Apply (substitute the right-hand side into the left-hand side) Eqs. (4.21), (4.20), (4.12), (4.13), (4.16), (4.17), and (4.18) to update $\{\widehat{z}^{(n)}\}_{n=1}^{N}$, $\{\widehat{\phi}_k\}_{k=1}^{K}$, and $\{\widehat{\mu}_k, \widehat{\sigma}_k^2\}_{k=1}^{K}$.
3: Apply Eqs. (4.29), (4.26), and (4.27) to update $\phi$, $\mu_0$ and $\xi$, respectively.
4: Evaluate the free energy (4.22).
5: Iterate Steps 2 through 4 until convergence (until the energy decrease becomes smaller than a threshold).

$$\xi = \left\{ \frac{1}{K} \sum_{k=1}^{K} \left( \frac{\|\widehat{\mu}_k - \mu_0\|^2}{M} + \widehat{\sigma}_k^2 \right) \right\}^{-1}. \tag{4.27}$$

Since the stationary condition $\frac{\partial F}{\partial \phi} = 0$ is not explicitly solved for $\phi$, the *Newton–Raphson* step is usually used for updating $\phi$. With the second derivative,

$$\frac{\partial^2 F}{\partial \phi^2} = K \left( \Psi^{(1)}(\phi) - K \Psi^{(1)}(K\phi) \right), \tag{4.28}$$

the update rule is obtained as follows:

$$\begin{aligned} \phi^{\mathrm{new}} &= \max \left( 0, \phi^{\mathrm{old}} - \left( \frac{\partial^2 F}{\partial \phi^2} \right)^{-1} \frac{\partial F}{\partial \phi} \right) \\ &= \max \left( 0, \phi^{\mathrm{old}} - \frac{K(\Psi(\phi) - \Psi(K\phi)) - \sum_{k=1}^{K} \left( \Psi(\widehat{\phi}_k) - \Psi\left( \sum_{k'=1}^{K} \widehat{\phi}_{k'} \right) \right)}{K\left( \Psi^{(1)}(\phi) - K \Psi^{(1)}(K\phi) \right)} \right), \end{aligned} \tag{4.29}$$

where $\Psi_m(z) \equiv \frac{d^m}{dz^m} \Psi(z)$ is the *polygamma function* of order $m$.

The EVB learning for the GMM is summarized in Algorithm 7. If the prior hyperparameters are fixed and Step 3 in the algorithm is omitted, the algorithm reduces to the (nonempirical) VB learning algorithm.

### 4.1.2 Mixture of Exponential Families

It is well known that the Gaussian distribution is an example of the *exponential family distribution*:

$$p(\boldsymbol{x}|\boldsymbol{\tau}) = p(\boldsymbol{t}|\boldsymbol{\eta}) = \exp\left( \boldsymbol{\eta}^\top \boldsymbol{t} - A(\boldsymbol{\eta}) + B(\boldsymbol{t}) \right), \tag{4.30}$$

where $\boldsymbol{\eta} \in \boldsymbol{H}$ is the *natural parameter*, $\boldsymbol{\eta}^\top \boldsymbol{t}$ is its inner product with the vector $\boldsymbol{t} = \boldsymbol{t}(\boldsymbol{x}) = (t_1(\boldsymbol{x}), \ldots, t_M(\boldsymbol{x}))^\top$, and $A(\boldsymbol{\eta})$ and $B(\boldsymbol{t})$ are real-valued functions of the parameter $\boldsymbol{\eta}$ and the sufficient statistics $\boldsymbol{t}$, respectively (Brown, 1986) (see Eq. (1.27) in Section 1.2.3). Suppose functions $t_1, \ldots, t_M$ and the constant function, 1, are linearly independent and the number of parameters in a single component distribution, $p(\boldsymbol{t}|\boldsymbol{\eta})$, is $M$.

The VB framework for GMMs in Section 4.1.1 is generalized to a mixture of exponential family distributions as follows. The conditional conjugate prior distributions of $\boldsymbol{\alpha} \in \Delta^{K-1}$ and $\{\boldsymbol{\eta}_k\}_{k=1}^K$ are given by

$$p(\boldsymbol{\alpha}|\phi) = \mathrm{Dirichlet}_K(\boldsymbol{\alpha}; (\phi, \ldots, \phi)^\top), \tag{4.31}$$

$$p(\{\boldsymbol{\eta}_k\}_{k=1}^K|\boldsymbol{\nu}_0, \xi) = \prod_{k=1}^K \frac{1}{C(\xi, \boldsymbol{\nu}_0)} \exp\left(\xi(\boldsymbol{\nu}_0^\top \boldsymbol{\eta}_k - A(\boldsymbol{\eta}_k))\right), \tag{4.32}$$

where the function $C(\xi, \boldsymbol{\nu})$ of $\xi \in \mathbb{R}$ and $\boldsymbol{\nu} \in \mathbb{R}^M$ is defined by

$$C(\xi, \boldsymbol{\nu}) = \int_H \exp\left(\xi(\boldsymbol{\nu}^\top \boldsymbol{\eta} - A(\boldsymbol{\eta}))\right) d\boldsymbol{\eta}. \tag{4.33}$$

Constants $\xi > 0$, $\boldsymbol{\nu}_0 \in \mathbb{R}^M$, and $\phi > 0$ are the hyperparameters.

**VB Posterior for Mixture-of-Exponential-Family Models**

Here, we derive the VB posterior $r_w(\boldsymbol{w})$ for the mixture-of-exponential-family model using Eq. (4.2).

Using the complete data $\{\boldsymbol{x}^{(n)}, \boldsymbol{z}^{(n)}\}_{n=1}^N$, we put

$$\overline{N}_k = \sum_{n=1}^N \left\langle z_k^{(n)} \right\rangle_{r_{\mathcal{H}}(\mathcal{H})}, \tag{4.34}$$

$$\overline{\boldsymbol{t}}_k = \frac{1}{\overline{N}_k} \sum_{n=1}^N \left\langle z_k^{(n)} \right\rangle_{r_{\mathcal{H}}(\mathcal{H})} \boldsymbol{t}^{(n)}, \tag{4.35}$$

where $\boldsymbol{t}^{(n)} = \boldsymbol{t}(\boldsymbol{x}^{(n)})$. Note that the variables $\overline{N}_k$ and $\overline{\boldsymbol{t}}_k$ satisfy the constraints $\sum_{k=1}^K \overline{N}_k = N$ and $\sum_{k=1}^K \overline{N}_k \overline{\boldsymbol{t}}_k = \sum_{n=1}^N \boldsymbol{t}^{(n)}$. From Eq. (4.2) and the respective prior distributions, Eqs. (4.10) and (4.32), the VB posterior $r_w(\boldsymbol{w}) = r_\alpha(\boldsymbol{\alpha}) r_\eta(\{\boldsymbol{\eta}_k\}_{k=1}^K)$ is obtained as the product of the following two distributions:

$$r_\alpha(\boldsymbol{\alpha}) = \mathrm{Dirichlet}_K\left(\boldsymbol{\alpha}; (\widehat{\phi}_1, \ldots, \widehat{\phi}_K)^\top\right),$$

$$r_\eta(\{\boldsymbol{\eta}_k\}_{k=1}^K) = \prod_{k=1}^K \frac{1}{C(\widehat{\xi}_k, \widehat{\boldsymbol{\nu}}_k)} \exp\left(\widehat{\xi}_k(\widehat{\boldsymbol{\nu}}_k^\top \boldsymbol{\eta}_k - A(\boldsymbol{\eta}_k))\right), \tag{4.36}$$

where

$$\widehat{\phi}_k = \overline{N}_k + \phi, \tag{4.37}$$

$$\widehat{\boldsymbol{\nu}}_k = \frac{\overline{N}_k \overline{\boldsymbol{t}}_k + \xi \boldsymbol{\nu}_0}{\overline{N}_k + \xi}, \tag{4.38}$$

$$\widehat{\xi}_k = \overline{N}_k + \xi. \tag{4.39}$$

Let

$$\widehat{\boldsymbol{\eta}}_k = \langle \boldsymbol{\eta}_k \rangle_{r_\eta(\boldsymbol{\eta}_k)} = \frac{1}{\widehat{\xi}_k} \frac{\partial \log C(\widehat{\xi}_k, \widehat{\boldsymbol{\nu}}_k)}{\partial \boldsymbol{\nu}_k}. \tag{4.40}$$

It follows that

$$\langle A(\boldsymbol{\eta}_k) \rangle_{r_\eta(\boldsymbol{\eta}_k)} = \widehat{\boldsymbol{\eta}}_k^\top \widehat{\boldsymbol{\nu}}_k - \frac{\partial \log C(\widehat{\xi}_k, \widehat{\boldsymbol{\nu}}_k)}{\partial \xi_k}. \tag{4.41}$$

From Eq. (4.3), the VB posterior $r_{\mathcal{H}}(\mathcal{H})$ is given by

$$\begin{aligned} r_{\mathcal{H}}(\mathcal{H}) &= \prod_{n=1}^{N} r_z(\boldsymbol{z}^{(n)}) \\ &= \prod_{n=1}^{N} \mathrm{Multinomial}_{K,1}\left(\boldsymbol{z}^{(n)}; \widehat{\boldsymbol{z}}^{(n)}\right), \end{aligned}$$

where $\widehat{\boldsymbol{z}}^{(n)} \in \Delta^{K-1}$ is

$$\widehat{z}_k^{(n)} = \left\langle z_k^{(n)} \right\rangle_{r_{\mathcal{H}}(\mathcal{H})} = \frac{\overline{z}_k^{(n)}}{\sum_{k'=1}^{K} \overline{z}_{k'}^{(n)}}, \tag{4.42}$$

for

$$\overline{z}_k^{(n)} = \exp\left( \Psi(\widehat{\phi}_k) - \Psi\left( \sum_{k'=1}^{K} \widehat{\phi}_{k'} \right) + \widehat{\boldsymbol{\eta}}_k^\top \boldsymbol{t}^{(n)} - \langle A(\boldsymbol{\eta}_k) \rangle_{r_\eta(\boldsymbol{\eta}_k)} + B(\boldsymbol{t}^{(n)}) \right). \tag{4.43}$$

To obtain the preceding expression of $\overline{z}_k^{(n)}$, we used Eq. (4.19).

The free energy as a function of variational parameters is expressed as

$$\begin{aligned} F = {} & \log\left( \frac{\Gamma(\sum_{k=1}^{K} \widehat{\phi}_k)}{\prod_{k=1}^{K} \Gamma(\widehat{\phi}_k)} \right) - \log\left( \frac{\Gamma(K\phi)}{(\Gamma(\phi))^K} \right) - \sum_{k=1}^{K} \log C(\widehat{\xi}_k, \widehat{\boldsymbol{\nu}}_k) + K \log C(\xi, \boldsymbol{\nu}_0) \\ & + \sum_{n=1}^{N} \sum_{k=1}^{K} \widehat{z}_k^{(n)} \log \widehat{z}_k^{(n)} + \sum_{k=1}^{K} \left( \widehat{\phi}_k - \phi - \overline{N}_k \right) \left( \Psi(\widehat{\phi}_k) - \Psi(\textstyle\sum_{k'=1}^{K} \widehat{\phi}_{k'}) \right) \\ & + \sum_{k=1}^{K} \left[ \widehat{\boldsymbol{\eta}}_k^\top \left\{ \xi \left( \widehat{\boldsymbol{\nu}}_k - \boldsymbol{\nu}_0 \right) + \overline{N}_k \left( \widehat{\boldsymbol{\nu}}_k - \overline{\boldsymbol{t}}_k \right) \right\} + \left( \widehat{\xi}_k - \xi - \overline{N}_k \right) \frac{\partial \log C(\widehat{\xi}_k, \widehat{\boldsymbol{\nu}}_k)}{\partial \xi_k} \right] \\ & - \sum_{n=1}^{N} B(\boldsymbol{t}^{(n)}). \end{aligned} \tag{4.44}$$

The update rule of $\phi$ for the EVB learning is obtained by Eq. (4.29) as in the GMM. The partial derivatives of $F$ with respect to the hyperparameters $(\nu_0, \xi)$ are

$$\frac{\partial F}{\partial \nu_0} = K\frac{\partial \log C(\xi, \nu_0)}{\partial \nu_0} - \xi \sum_{k=1}^{K} \widehat{\boldsymbol{\eta}}_k, \tag{4.45}$$

$$\frac{\partial F}{\partial \xi} = \sum_{k=1}^{K} \left\{ \widehat{\boldsymbol{\eta}}_k^\top (\widehat{\nu}_k - \nu_0) - \frac{\partial \log C(\widehat{\xi}_k, \widehat{\nu}_k)}{\partial \xi_k} \right\} + K\frac{\partial \log C(\xi, \nu_0)}{\partial \xi}. \tag{4.46}$$

Equating these derivatives to zeros, we have the following stationary conditions:

$$\frac{1}{\xi}\frac{\partial \log C(\xi, \nu_0)}{\partial \nu_0} = \frac{1}{K}\sum_{k=1}^{K} \widehat{\boldsymbol{\eta}}_k, \tag{4.47}$$

$$\frac{\partial \log C(\xi, \nu_0)}{\partial \xi} = \left( \frac{1}{K}\sum_{k=1}^{K} \widehat{\boldsymbol{\eta}}_k \right)^\top \nu_0 - \frac{1}{K}\sum_{k=1}^{K} \langle A(\boldsymbol{\eta}_k) \rangle_{r_\eta(\boldsymbol{\eta}_k)}, \tag{4.48}$$

where we have used Eq. (4.41). If these equations are solved for $\nu_0$ and $\xi$, respectively, we obtain their update rules as in the case of the GMM. Otherwise, we need the Newton–Raphson steps to update them.

The EVB learning for the mixture of exponential families is summarized in Algorithm 8. If the prior hyperparameters are fixed and Step 3 in the algorithm is omitted, the algorithm reduces to the (nonempirical) VB learning algorithm.

**Algorithm 8** EVB learning for the mixture-of-exponential-family model.

1: Initialize the variational parameters $(\{\widehat{z}^{(n)}\}_{n=1}^N, \{\widehat{\phi}_k\}_{k=1}^K, \{\widehat{\nu}_k, \widehat{\xi}_k\}_{k=1}^K)$, and the hyperparameters $(\phi, \nu_0, \xi)$.

2: Apply (substitute the right-hand side into the left-hand side) Eqs. (4.43), (4.42), (4.34), (4.35), (4.37), (4.38), and (4.39) to update $\{\widehat{z}^{(n)}\}_{n=1}^N$, $\{\widehat{\phi}_k\}_{k=1}^K$, and $\{\widehat{\nu}_k, \widehat{\xi}_k\}_{k=1}^K$. Transform $\{\widehat{\nu}_k, \widehat{\xi}_k\}_{k=1}^K$ to $\{\widehat{\boldsymbol{\eta}}_k, \langle A(\boldsymbol{\eta}_k) \rangle_{r_\eta(\boldsymbol{\eta}_k)}\}_{k=1}^K$ by Eqs. (4.40) and (4.41).

3: Apply Eqs. (4.29), (4.47), and (4.48) to update $\phi$, $\nu_0$ and $\xi$, respectively.

4: Evaluate the free energy (4.44).

5: Iterate Steps 2 through 4 until convergence (until the energy decrease becomes smaller than a threshold).

### 4.1.3 Infinite Mixture Models

In 2000s, there was a revival of Bayesian nonparametric models to estimate the model complexity, e.g., the number of components in mixture models, by using a prior distribution for probability measures such as the *Dirichlet process (DP) prior*. The Bayesian nonparametric approach fits a single model adapting its complexity to the data. The VB framework plays an important role in achieving tractable inference for Bayesian nonparametric models. Here, we introduce the VB learning for the *stick-breaking* construction of the DP prior by instantiating the estimation of the number of components of the mixture model.

For the finite mixture model with $K$ components, we had the discrete latent variable,

$$z \in \{\boldsymbol{e}_1, \boldsymbol{e}_2, \ldots, \boldsymbol{e}_K\},$$

indicating the label of the component. We also assumed the multinomial distribution,

$$p(z|\boldsymbol{\alpha}) = \mathrm{Multinomial}_{K,1}(z; \boldsymbol{\alpha}) = \prod_{k=1}^{K} \alpha_k^{z_k}.$$

In the nonparametric Bayesian approach, we consider possibly an infinite number of components,

$$p(z|\boldsymbol{\alpha}) = \lim_{K \to \infty} \mathrm{Multinomial}_{K,1}(z; \boldsymbol{\alpha}),$$

and the following generation process of $\alpha_k$, called the stick-breaking process (Blei and Jordan, 2005; Gershman and Blei, 2012):

$$\alpha_k = v_k \prod_{l=1}^{k-1} (1 - v_l),$$

$$v_k \sim \mathrm{Beta}(1, \gamma),$$

where Beta($\alpha$,$\beta$) denotes the beta distribution with parameters $\alpha$ and $\beta$, and $\gamma$ is the scaling parameter.

To derive a tractable VB learning algorithm, the truncation level $T$ is usually introduced to the preceding process, which enforces $v_T = 1$. If the truncation level $T$ is sufficiently large, some components are left unused, and hence $T$ does not directly specify the number of components.

Then, the VB posterior $r(\mathcal{H}, \boldsymbol{v})$ for the latent variables and $\boldsymbol{v} = \{v_k\}_{k=1}^{T-1}$ is assumed to be factorized, $r_{\mathcal{H}}(\mathcal{H}) r_{\nu}(\boldsymbol{v})$, for which the free energy minimization implies further factorization:

$$r(\mathcal{H}, \boldsymbol{v}) = \prod_{n=1}^{N} r_z(z^{(n)}) \prod_{k=1}^{T-1} r_v(v_k),$$

where $r_z(z^{(n)})$ is the multinomial distribution as in the case of the finite mixture model, and $r_v(v_k)$ is the beta distribution because of the conditional conjugacy. To see this and how the VB learning algorithm is derived, we instantiate the GMM discussed in Section 4.1.1.

The free energy is decomposed as

$$F = \left\langle \log \frac{r_z(\{z^{(n)}\}_{n=1}^N) r_v(\boldsymbol{v}) r_\mu(\{\boldsymbol{\mu}_k\}_{k=1}^K)}{p(\boldsymbol{v}) p(\{\boldsymbol{\mu}_k\}_{k=1}^T)} \right\rangle_{r_z(\{z^{(n)}\}_{n=1}^N) r_v(\boldsymbol{v}) r_\mu(\{\boldsymbol{\mu}_k\}_{k=1}^K)}$$
$$- \left\langle \log p(\mathcal{D} | \{z^{(n)}\}_{n=1}^N, \boldsymbol{w}) \right\rangle_{r_z(\{z^{(n)}\}_{n=1}^N) r_v(\boldsymbol{v}) r_\mu(\{\boldsymbol{\mu}_k\}_{k=1}^K)}$$
$$- \left\langle \log p(\{z^{(n)}\}_{n=1}^N | \boldsymbol{v}) \right\rangle_{r_z(\{z^{(n)}\}_{n=1}^N) r_v(\boldsymbol{v})}.$$

These terms are computed in the same way as in Section 4.1.1 except for the last term, $\left\langle \log p(\{z^{(n)}\}_{n=1}^N | \boldsymbol{v}) \right\rangle_{r_z(\{z^{(n)}\}_{n=1}^N) r_v(\boldsymbol{v})} = \sum_{n=1}^N \left\langle \log p(z^{(n)} | \boldsymbol{v}) \right\rangle_{r_z(z^{(n)}) r_v(\boldsymbol{v})}$.

Let $c^{(n)}$ be the index $k$ such that $z_k^{(n)} = 1$ and $\theta$ be the indicator function. Then, we have

$$\left\langle \log p(z^{(n)} | \boldsymbol{v}) \right\rangle_{r_z(z^{(n)}) r_v(\boldsymbol{v})}$$
$$= \left\langle \log \prod_{k=1}^{\infty} (1 - v_k)^{\theta(c^{(n)} > k)} v_k^{\theta(c^{(n)} = k)} \right\rangle_{r_z(z^{(n)}) r_v(\boldsymbol{v})}$$
$$= \sum_{k=1}^{\infty} \left\{ r_z(c^{(n)} > k) \left\langle \log(1 - v_k) \right\rangle_{r_v(\boldsymbol{v})} + r_z(c^{(n)} = k) \left\langle \log v_k \right\rangle_{r_v(\boldsymbol{v})} \right\}$$
$$= \sum_{k=1}^{T-1} \left\{ r_z(c^{(n)} > k) \left\langle \log(1 - v_k) \right\rangle_{r_v(\boldsymbol{v})} + r_z(c^{(n)} = k) \left\langle \log v_k \right\rangle_{r_v(\boldsymbol{v})} \right\},$$

where we have used $\log v_T = 0$ and $r_z(c^{(n)} > T) = 0$.

Since the probabilities $r_z(c^{(n)} = k)$ and $r_z(c^{(n)} > k)$ are given by

$$r_z(c^{(n)} = k) = \widehat{z}_k^{(n)},$$
$$r_z(c^{(n)} > k) = \sum_{l=k+1}^{T} \widehat{z}_l^{(n)},$$

it follows from Eq. (4.12) that

$$\sum_{n=1}^{N} r_z(c^{(n)} = k) = \overline{N}_k,$$
$$\sum_{n=1}^{N} r_z(c^{(n)} > k) = \sum_{l=k+1}^{T} \overline{N}_l = N - \sum_{l=1}^{k} \overline{N}_l.$$

Now Eq. (4.3) in this case yields that

$$r_v(\boldsymbol{v}) \propto \sum_{n=1}^{N} \left\langle \log p(\boldsymbol{z}^{(n)}|\boldsymbol{v}) \right\rangle_{r_z(\boldsymbol{z}^{(n)})} p(\boldsymbol{v}).$$

It follows from similar manipulations to those just mentioned and the conditional conjugacy that

$$r_v(\boldsymbol{v}) = \prod_{k=1}^{T-1} \mathrm{Beta}(v_k; \widehat{\kappa}_k, \widehat{\lambda}_k), \tag{4.49}$$

i.e., for a fixed $r_{\mathcal{H}}(\mathcal{H})$, the optimal $r_v(\boldsymbol{v})$ is the beta distribution with the parameters,

$$\widehat{\kappa}_k = 1 + \overline{N}_k, \tag{4.50}$$

$$\widehat{\lambda}_k = \gamma + N - \sum_{l=1}^{k} \overline{N}_l. \tag{4.51}$$

The VB posterior $r_{\mathcal{H}}(\mathcal{H})$ is computed similarly except that the expectation

$$\langle \log \alpha_k \rangle_{r_\alpha(\boldsymbol{\alpha})} = \Psi(\overline{N}_k + \phi) - \Psi(N + K\phi)$$

in Eq. (4.19) for the finite mixture model is replaced by

$$\langle \log v_k \rangle_{r_v(v_k)} + \sum_{l=1}^{k-1} \langle \log(1 - v_l) \rangle_{r_v(v_l)} = \Psi(\widehat{\kappa}_k) - \Psi(\widehat{\kappa}_k + \widehat{\lambda}_k) + \sum_{l=1}^{k-1} \{\Psi(\widehat{\lambda}_l) - \Psi(\widehat{\kappa}_l + \widehat{\lambda}_l)\},$$

since

$$\langle \log v_k \rangle_{r_v(v_k)} = \Psi(\widehat{\kappa}_k) - \Psi(\widehat{\kappa}_k + \widehat{\lambda}_k),$$
$$\langle \log(1 - v_k) \rangle_{r_v(v_k)} = \Psi(\widehat{\lambda}_k) - \Psi(\widehat{\kappa}_k + \widehat{\lambda}_k),$$

for $r_v(v_k) = \mathrm{Beta}(v_k; \widehat{\kappa}_k, \widehat{\lambda}_k)$. In the case of the GMM, $\overline{z}_k^{(n)}$ in Eq. (4.21) is replaced with

$$\overline{z}_k^{(n)} = \exp\Bigg( \Psi(\widehat{\kappa}_k) - \Psi(\widehat{\kappa}_k + \widehat{\lambda}_k) + \sum_{l=1}^{k-1} \{\Psi(\widehat{\lambda}_l) - \Psi(\widehat{\kappa}_l + \widehat{\lambda}_l)\}$$
$$-\frac{1}{2} \left\| \boldsymbol{x}^{(n)} - \widehat{\boldsymbol{\mu}}_k \right\|^2 + M\widehat{\sigma}_k^2 \Bigg). \tag{4.52}$$

The free energy is given by

$$F = \sum_{k=1}^{T-1} \log\left( \frac{\Gamma(\widehat{\kappa}_k + \widehat{\lambda}_k)}{\Gamma(\widehat{\kappa}_k)\Gamma(\widehat{\lambda}_k)} \right) - (T-1)\log\gamma - \frac{M}{2} \sum_{k=1}^{T} \log\left(\xi \widehat{\sigma}_k^2\right) - \frac{TM}{2}$$
$$+ \sum_{n=1}^{N} \sum_{k=1}^{T} \widehat{z}_k^{(n)} \log \widehat{z}_k^{(n)} + \sum_{k=1}^{T-1} \left(\widehat{\kappa}_k - 1 - \overline{N}_k\right) \left\{ \Psi(\widehat{\kappa}_k) - \Psi(\widehat{\kappa}_k + \widehat{\lambda}_k) \right\}$$

$$+\sum_{k=1}^{T-1}\left\{\widehat{\lambda}_k-\gamma-\left(N-\sum_{l=1}^{k}\overline{N}_l\right)\right\}\left\{\Psi(\widehat{\lambda}_k)-\Psi(\widehat{\kappa}_k+\widehat{\lambda}_k)\right\}$$
$$+\sum_{k=1}^{T}\frac{\xi\left(\|\widehat{\boldsymbol{\mu}}_k-\boldsymbol{\mu}_0\|^2+M\widehat{\sigma}_k^2\right)}{2}+\sum_{k=1}^{T}\frac{\overline{N}_k\left(M\log(2\pi)+M\widehat{\sigma}_k^2\right)}{2}$$
$$+\sum_{k=1}^{T}\frac{\overline{N}_k\|\overline{\boldsymbol{x}}_k-\widehat{\boldsymbol{\mu}}_k\|^2+\sum_{n=1}^{N}\widehat{z}_k^{(n)}\|\boldsymbol{x}^{(n)}-\overline{\boldsymbol{x}}_k\|^2}{2}. \tag{4.53}$$

The VB learning algorithm is similar to Algorithm 7 for the finite GMM while the number of components $K$ is replaced with the truncation level $T$ throughout, the update rule (4.21) is replaced with Eq. (4.52), and $\{\widehat{\kappa}_k,\widehat{\lambda}_k\}_{k=1}^{T-1}$ are updated by Eqs. (4.50) and (4.51) instead of $\{\widehat{\phi}_k\}_{k=1}^{K}$.

By computing $\frac{\partial F}{\partial \gamma}$ and equating it to zero, the EVB learning for the hyperparameter $\gamma$ updates it as follows:

$$\gamma=\left[\frac{-1}{T-1}\sum_{k=1}^{T-1}\left\{\Psi(\widehat{\lambda}_k)-\Psi(\widehat{\kappa}_k+\widehat{\lambda}_k)\right\}\right]^{-1}, \tag{4.54}$$

which can replace the update rule of $\phi$ in Step 3 of Algorithm 7.

## 4.2 Other Latent Variable Models

In this section, we discuss more complex latent variable models than mixture models and derive VB learning algorithms for them. Although we focus on the models where the multinomial distribution is assumed on the observed data given latent variables, it is straightforward to replace it with other members of the exponential family.

### 4.2.1 Bayesian Networks

A *Bayesian network* is a probabilistic model defined by a graphical model expressing the relations among random variables by a graph and the conditional probabilities associated with them (Jensen, 2001). In this subsection, we focus on a Bayesian network whose states of all hidden nodes influence those of all observation nodes, and assume that it has $M$ observation nodes and $K$ hidden nodes. The graphical structure of this Bayesian network is called bipartite and presented in Figure 4.1.

The observation nodes are denoted by $\boldsymbol{x}=(\boldsymbol{x}_1,\ldots,\boldsymbol{x}_M)$, and the set of states of observation node $\boldsymbol{x}_j=(x_{j,1},\ldots,x_{j,Y_j})^\top\in\{\boldsymbol{e}_l\}_{l=1}^{Y_j}$ is $\{1,\ldots,Y_j\}$. The hidden

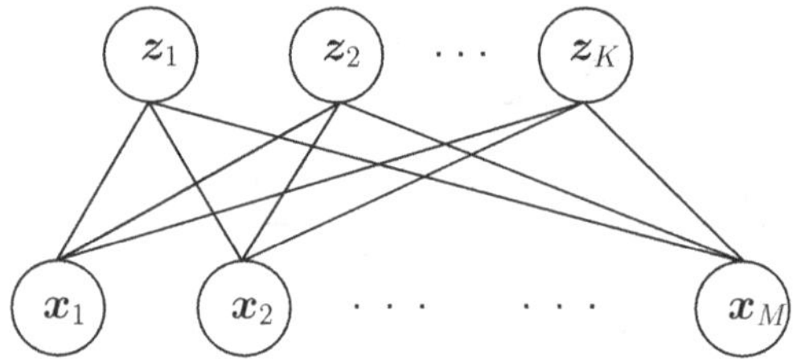


Figure 4.1 Graphical structure of the Bayesian network.

nodes are denoted by $z = (z_1, \ldots, z_K)$, and the set of states of hidden node $z_k = (z_{k,1}, \ldots, z_{k,T_k})^\top \in \{e_i\}_{i=1}^{T_k}$ is $\{1, \ldots, T_k\}$.

The probability that the state of the hidden node $z_k$ is $i$ $(1 \leq i \leq T_k)$ is expressed as

$$a_{(k,i)} = \mathrm{Prob}(z_k = e_i).$$

Then, $\boldsymbol{a}_k = (a_{(k,1)}, \ldots, a_{(k,T_k)})^\top \in \Delta^{T_k-1}$ for $k = 1, \ldots, K$.

The conditional probability that the $j$th observation node $\boldsymbol{x}_j$ is $l$ $(1 \leq l \leq Y_j)$, given the condition that the states of hidden nodes are $z = (z_1, \ldots, z_K)$, is denoted by

$$b_{(j,l|z)} = \mathrm{Prob}(\boldsymbol{x}_j = \boldsymbol{e}_l | z).$$

Then, $\boldsymbol{b}_{j|z} = (b_{(j,1|z)}, \ldots, b_{(j,Y_j|z)})^\top \in \Delta^{Y_j-1}$ for $j = 1, \ldots, M$. Define $\boldsymbol{b}_z = \{\boldsymbol{b}_{j|z}\}_{j=1}^M$ for $z \in \mathcal{Z} = \{z; z_k \in \{\boldsymbol{e}_i\}_{i=1}^{T_k}, k = 1, \ldots, K\}$. Let $\boldsymbol{w} = \{\{\boldsymbol{a}_k\}_{k=1}^K, \{\boldsymbol{b}_z\}_{z \in \mathcal{Z}}\}$ be the set of all parameters. Then, the joint probability that the states of observation nodes are $\boldsymbol{x} = (\boldsymbol{x}_1, \ldots, \boldsymbol{x}_M)$ and the states of hidden nodes are $z = (z_1, \ldots, z_K)$ is

$$p(\boldsymbol{x}, z|\boldsymbol{w}) = p(\boldsymbol{x}|\boldsymbol{b}_z) \prod_{k=1}^K \prod_{i=1}^{T_k} a_{(k,i)}^{z_{k,i}},$$

where

$$p(\boldsymbol{x}|\boldsymbol{b}_z) = \prod_{j=1}^M \prod_{l=1}^{Y_j} b_{(j,l|z)}^{x_{j,l}}.$$

Therefore, the marginal probability that the states of observation nodes are $\boldsymbol{x}$ is

$$\begin{aligned} p(\boldsymbol{x}|\boldsymbol{w}) &= \sum_{z \in \mathcal{Z}} p(\boldsymbol{x}, z|\boldsymbol{w}) \\ &= \sum_{z \in \mathcal{Z}} p(\boldsymbol{x}|\boldsymbol{b}_z) \prod_{k=1}^K \prod_{i=1}^{T_k} a_{(k,i)}^{z_{k,i}}, \end{aligned} \tag{4.55}$$

where we used the notation $\sum_{z \in \mathcal{Z}}$ for the summation over all states of hidden nodes. Let

$$M_{\text{obs}} = \sum_{j=1}^{M} (Y_j - 1),$$

which is the number of parameters to specify the conditional probability $p(\boldsymbol{x}|\boldsymbol{b}_z)$ of the states of all the observation nodes given the states of the hidden nodes. Then, the number of the parameters of the model, $D$, is

$$D = M_{\text{obs}} \prod_{k=1}^{K} T_k + \sum_{k=1}^{K} (T_k - 1). \tag{4.56}$$

We assume that the prior distribution $p(\boldsymbol{w})$ of the parameters $\boldsymbol{w} = \{\{\boldsymbol{a}_k\}_{k=1}^{K}, \{\boldsymbol{b}_z\}_{z \in \mathcal{Z}}\}$ is the conditional conjugate prior distribution. Then, $p(\boldsymbol{w})$ is given by $\left\{\prod_{k=1}^{K} p(\boldsymbol{a}_k|\phi)\right\}\left\{\prod_{z \in \mathcal{Z}} \prod_{j=1}^{M} p(\boldsymbol{b}_{j|z}|\xi)\right\}$, where

$$p(\boldsymbol{a}_k|\phi) = \text{Dirichlet}_{T_k}\left(\boldsymbol{a}_k; (\phi, \ldots, \phi)^\top\right), \tag{4.57}$$

$$p(\boldsymbol{b}_{j|z}|\xi) = \text{Dirichlet}_{Y_j}\left(\boldsymbol{b}_{j|z}; (\xi, \ldots, \xi)^\top\right), \tag{4.58}$$

are the Dirichlet distributions with hyperparameters $\phi > 0$ and $\xi > 0$.

### VB Posterior for Bayesian Networks

Let $\{\mathcal{D}, \mathcal{H}\}$ be the complete data with the observed data set $\mathcal{D} = \{\boldsymbol{x}^{(n)}\}_{n=1}^{N}$ and the corresponding hidden variables $\mathcal{H} = \{z^{(n)}\}_{n=1}^{N}$. Define the expected sufficient statistics:

$$\overline{N}^{z}_{(k,i_k)} = \sum_{n=1}^{N} \left\langle z^{(n)}_{k,i_k} \right\rangle_{r_{\mathcal{H}}(\mathcal{H})},$$

$$\overline{N}^{x}_{(j,l_j|z)} = \sum_{n=1}^{N} x^{(n)}_{j,l_j} r_z(z^{(n)} = z), \tag{4.59}$$

where

$$r_z(z^{(n)} = z) = \left\langle \prod_{k=1}^{K} z^{(n)}_{k,i_k} \right\rangle_{r_{\mathcal{H}}(\mathcal{H})} \tag{4.60}$$

is the estimated probability that $z^{(n)} = z = (\boldsymbol{e}_{i_1}, \ldots, \boldsymbol{e}_{i_K})$. Here $\boldsymbol{x}^{(n)}_j$ indicates the state of the $j$th observation node and $z^{(n)}_k$ indicates the state of the $k$th hidden node when the $n$th training sample is observed. From Eq. (4.2), the VB posterior distribution of parameters $\boldsymbol{w} = \{\{\boldsymbol{a}_k\}_{k=1}^{K}, \{\boldsymbol{b}_z\}_{z \in \mathcal{Z}}\}$ is given by

$$r_w(\boldsymbol{w}) = \left\{\prod_{k=1}^{K} r_a(\boldsymbol{a}_k)\right\}\left\{\prod_{z \in \mathcal{Z}} \prod_{j=1}^{M} r_b(\boldsymbol{b}_{j|z})\right\},$$

$$r_a(\boldsymbol{a}_k) = \mathrm{Dirichlet}_{T_k}\left(\boldsymbol{a}_k; \widehat{\boldsymbol{\phi}}_k\right), \tag{4.61}$$

$$r_b(\boldsymbol{b}_{j|z}) = \mathrm{Dirichlet}_{Y_j}\left(\boldsymbol{b}_{j|z}; \widehat{\boldsymbol{\xi}}_{j|z}\right), \tag{4.62}$$

where

$$\widehat{\boldsymbol{\phi}}_k = (\widehat{\phi}_{(k,1)}, \ldots, \widehat{\phi}_{(k,T_k)})^\top \quad (k = 1, \ldots, K),$$

$$\widehat{\phi}_{(k,i)} = \overline{N}^z_{(k,i)} + \phi \quad (i = 1, \ldots, T_k), \tag{4.63}$$

$$\widehat{\boldsymbol{\xi}}_{j|z} = (\widehat{\xi}_{(j,1|z)}, \ldots, \widehat{\xi}_{(j,Y_j|z)})^\top \quad (j = 1, \ldots, M, z \in \mathcal{Z}),$$

$$\widehat{\xi}_{(j,l|z)} = \overline{N}^x_{(j,l|z)} + \xi \quad (l = 1, \ldots, Y_j). \tag{4.64}$$

Note that if we define

$$\overline{N}^x_z = \sum_{n=1}^{N} r_z(z^{(n)} = z) = \sum_{n=1}^{N} \left\langle \prod_{k=1}^{K} z^{(n)}_{k,i_k} \right\rangle_{r_{\mathcal{H}}(\mathcal{H})},$$

for $z = (\boldsymbol{e}_{i_1}, \ldots, \boldsymbol{e}_{i_K}) \in \mathcal{Z}$, we have

$$\overline{N}^x_z = \sum_{l=1}^{Y_j} \overline{N}^x_{(j,l|z)}, \tag{4.65}$$

for $j = 1, \ldots, M$, and

$$\overline{N}^z_{(k,i)} = \sum_{z_{-k}} \overline{N}^x_z, \tag{4.66}$$

where $\sum_{z_{-k}}$ denotes the summation over $z_{k'}$ $(k' \neq k)$ other than $z_k = \boldsymbol{e}_i$.

It follows from Eqs. (4.61) and (4.62) that

$$\langle \log a_{(k,i)} \rangle_{r_a(\boldsymbol{a}_k)} = \Psi(\widehat{\phi}_{(k,i)}) - \Psi\left(\textstyle\sum_{i'=1}^{T_k} \widehat{\phi}_{(k,i')}\right) \quad (i = 1, \ldots, T_k),$$

for $k = 1, \ldots, K$ and

$$\left\langle \log b_{(j,l|z)} \right\rangle_{r_b(\boldsymbol{b}_{j|z})} = \Psi(\widehat{\xi}_{(j,l|z)}) - \Psi\left(\textstyle\sum_{l'=1}^{Y_j} \widehat{\xi}_{(j,l'|z)}\right) \quad (l = 1, \ldots, Y_j),$$

for $j = 1, \ldots, M$. From Eq. (4.3), the VB posterior distribution of the hidden variables is given by $r_{\mathcal{H}}(\mathcal{H}) = \prod_{n=1}^{N} r_z(z^{(n)})$, where for $z = (\boldsymbol{e}_{i_1}, \ldots, \boldsymbol{e}_{i_K})$,

$$\begin{aligned} r_z(z^{(n)} = z) &= \sum_{z^{(n)} \in \mathcal{Z}} r_z(z^{(n)}) \prod_{k=1}^{K} z^{(n)}_{k,i_k} \\ &\propto \exp\Bigg( \sum_{k=1}^{K} \left\{ \Psi(\widehat{\phi}_{(k,i_k)}) - \Psi\left(\textstyle\sum_{i'_k=1}^{T_k} \widehat{\phi}_{(k,i'_k)}\right) \right\} \\ &\qquad + \sum_{j=1}^{M} \left\{ \Psi(\widehat{\xi}_{(j,l^{(n)}_j|z)}) - \Psi\left(\textstyle\sum_{l'=1}^{Y_j} \widehat{\xi}_{(j,l'|z)}\right) \right\} \Bigg), \end{aligned} \tag{4.67}$$

if $\boldsymbol{x}^{(n)}_j = \boldsymbol{e}_{l^{(n)}_j}$.

The VB algorithm updates $\{\overline{N}^{x}_{(j,l_j|z)}\}$ using Eqs. (4.59) and (4.67) iteratively. The other expected sufficient statistics and variational parameters are computed by Eqs. (4.65), (4.66) and Eqs. (4.63), (4.64), respectively. The free energy as a function of the variational parameters is given by

$$\begin{aligned}
F = \sum_{k=1}^{K} \Bigg\{ &\log\left(\frac{\Gamma(\sum_{i=1}^{T_k} \widehat{\phi}_{(k,i)})}{\prod_{i=1}^{T_k} \Gamma(\widehat{\phi}_{(k,i)})}\right) - \log\left(\frac{\Gamma(T_k\phi)}{(\Gamma(\phi))^{T_k}}\right) \\
&+ \sum_{i=1}^{T_k} \left(\widehat{\phi}_{(k,i)} - \phi - \overline{N}^{z}_{(k,i)}\right)\left(\Psi(\widehat{\phi}_{(k,i)}) - \Psi\left(\textstyle\sum_{i'=1}^{T_k} \widehat{\phi}_{(k,i')}\right)\right)\Bigg\} \\
&+ \sum_{z\in\mathcal{Z}} \sum_{j=1}^{M} \Bigg\{ \log\left(\frac{\Gamma(\sum_{l=1}^{Y_j} \widehat{\xi}_{(j,l|z)})}{\prod_{l=1}^{Y_j} \Gamma(\widehat{\xi}_{(j,l|z)})}\right) - \log\left(\frac{\Gamma(Y_j\xi)}{(\Gamma(\xi))^{Y_j}}\right) \\
&+ \sum_{l=1}^{Y_j} \left(\widehat{\xi}_{(j,l|z)} - \xi - \overline{N}^{x}_{(j,l|z)}\right)\left(\Psi(\widehat{\xi}_{(j,l|z)}) - \Psi\left(\textstyle\sum_{l'=1}^{Y_j} \widehat{\xi}_{(j,l'|z)}\right)\right)\Bigg\} \\
&+ \sum_{n=1}^{N} \sum_{z\in\mathcal{Z}} r_z(z^{(n)} = z) \log r_z(z^{(n)} = z). \qquad (4.68)
\end{aligned}$$

The following update rule for the EVB learning of the hyperparameter $\phi$ is obtained in the same way as the update rule (4.29) for the GMM:

$$\phi^{\text{new}} = \max\left(0, \phi^{\text{old}} - \frac{\sum_{k=1}^{K}\left\{T_k(\Psi(\phi) - \Psi(T_k\phi)) - \sum_{i=1}^{T_k}\left(\Psi(\widehat{\phi}_{(k,i)}) - \Psi\left(\sum_{i'=1}^{T_k}\widehat{\phi}_{(k,i')}\right)\right)\right\}}{\sum_{k=1}^{K} T_k\left(\Psi^{(1)}(\phi) - T_k\Psi^{(1)}(T_k\phi)\right)}\right). \qquad (4.69)$$

Similarly, we obtain the following update rule of the hyperparameter $\xi$:

$$\xi^{\text{new}} = \max\left(0, \xi^{\text{old}} - \frac{\sum_{z\in\mathcal{Z}}\sum_{j=1}^{M}\left\{Y_j\left(\Psi(\xi) - \Psi(Y_j\xi)\right) - \sum_{l=1}^{Y_j}\left(\Psi(\widehat{\xi}_{(j,l|z)}) - \Psi\left(\sum_{l'=1}^{Y_k}\widehat{\xi}_{(j,l'|z)}\right)\right)\right\}}{\left(\prod_{k=1}^{K} T_k\right)\sum_{j=1}^{M} Y_j\left(\Psi^{(1)}(\xi) - Y_j\Psi^{(1)}(Y_j\xi)\right)}\right). \qquad (4.70)$$

Let $\widehat{\mathcal{S}} = \{r_z(z^{(n)} = z)\}_{n=1,z\in\mathcal{Z}}^{N} = \left\{\left\langle \prod_{k=1}^{K} z_{k,i_k}^{(n)} \right\rangle_{r_{\mathcal{H}}(\mathcal{H})}\right\}_{n=1,z\in\mathcal{Z}}^{N}$, $\widehat{\Phi} = \{\widehat{\boldsymbol{\phi}}_k\}_{k=1}^{K}$, and $\widehat{\Xi} = \{\widehat{\boldsymbol{\xi}}_{j|z}\}_{j=1,z\in\mathcal{Z}}^{M}$ be the sets of variational parameters. The EVB learning for the Bayesian network is summarized in Algorithm 9. If the prior hyperparameters are fixed and Step 3 in the algorithm is omitted, the algorithm reduces to the (nonempirical) VB learning algorithm.

### 4.2.2 Hidden Markov Models

*Hidden Markov models (HMMs)* have been widely used for sequence modeling in speech recognition, natural language processing, and so on (Rabiner,

**Algorithm 9** EVB learning for the Bayesian network.

1: Initialize the variational parameters $(\widehat{S}, \widehat{\Phi}, \widehat{\Xi})$ and the hyperparameters $(\phi, \xi)$.
2: Apply (substitute the right-hand side into the left-hand side) Eqs. (4.67), (4.59), (4.65), (4.66), (4.63), and (4.64) to update $\widehat{S}$, $\widehat{\Phi}$, and $\widehat{\Xi}$.
3: Apply Eqs. (4.69) and (4.70) to update $\phi$ and $\xi$, respectively.
4: Evaluate the free energy (4.68).
5: Iterate Steps 2 through 4 until convergence (until the energy decrease becomes smaller than a threshold).

1989). In this subsection, we consider discrete HMMs. Suppose a sequence $\mathcal{D} = (\boldsymbol{x}^{(1)}, \ldots, \boldsymbol{x}^{(T)})$ was observed. Each $\boldsymbol{x}^{(t)}$ is an $M$-dimensional binary vector ($M$-valued finite alphabet):

$$\boldsymbol{x}^{(t)} = (x_1^{(t)}, \ldots, x_M^{(t)}) \in \{\boldsymbol{e}_1, \ldots, \boldsymbol{e}_M\},$$

where if the output symbol at time $t$ is $m$, then $x_m^{(t)} = 1$, and otherwise 0. Moreover, $\boldsymbol{x}^{(t)}$ is produced in $K$-valued discrete hidden state $\boldsymbol{z}^{(t)}$. The sequence of hidden states $\mathcal{H} = (\boldsymbol{z}^{(1)}, \ldots, \boldsymbol{z}^{(T)})$ is generated by a first-order Markov process. Similarly, $\boldsymbol{z}^{(t)}$ is represented by a $K$-dimensional binary vector

$$\boldsymbol{z}^{(t)} = (z_1^{(t)}, \ldots, z_K^{(t)}) \in \{\boldsymbol{e}_1, \ldots, \boldsymbol{e}_K\},$$

where if the hidden state at time $t$ is $k$, then $z_k^{(t)} = 1$, and otherwise 0.

Without loss of generality, we assume that the initial state ($t = 1$) is the first one, namely $z_1^{(1)} = 1$. Then, the probability of a sequence is given by

$$p(\mathcal{D}|\boldsymbol{w}) = \sum_{\mathcal{H}} \prod_{m=1}^{M} b_{1,m}^{x_m^{(1)}} \prod_{t=2}^{T} \prod_{k=1}^{K} \prod_{l=1}^{K} a_{k,l}^{z_l^{(t)} z_k^{(t-1)}} \prod_{m=1}^{M} b_{k,m}^{z_k^{(t)} x_m^{(t)}}, \tag{4.71}$$

where $\sum_{\mathcal{H}}$ is taken all over possible values of hidden variables, and the model parameters, $\boldsymbol{w} = (\boldsymbol{A}, \boldsymbol{B})$, consist of the state transition probability matrix $\boldsymbol{A} = (\widetilde{\boldsymbol{a}}_1, \ldots, \widetilde{\boldsymbol{a}}_K)^\top$ and the emission probability matrix $\boldsymbol{B} = (\widetilde{\boldsymbol{b}}_1, \ldots, \widetilde{\boldsymbol{b}}_K)^T$ satisfying $\widetilde{\boldsymbol{a}}_k = (a_{k,1}, \ldots, a_{k,K})^\top \in \Delta^{K-1}$ and $\widetilde{\boldsymbol{b}}_k = (b_{k,1}, \ldots, b_{k,K})^T \in \Delta^{M-1}$ for $1 \leq k \leq K$, respectively. $a_{k,l}$ represents the transition probability from the $k$th hidden state to the $l$th hidden state and $b_{k,m}$ is the emission probability that alphabet $m$ is produced in the $k$th hidden state. Figure 4.2 illustrates an example of the state transition diagram of an HMM.

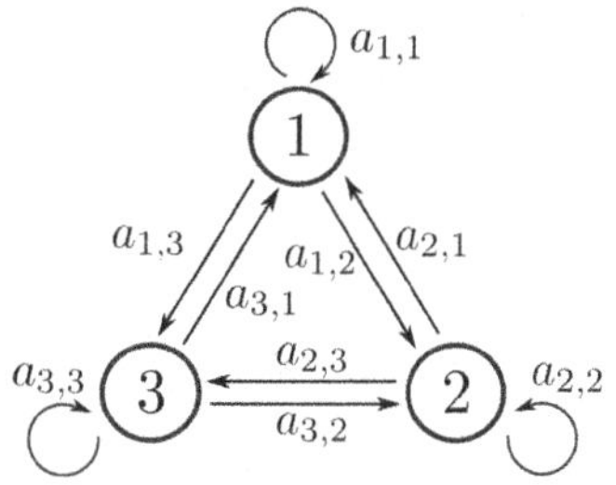


Figure 4.2 State transition diagram of an HMM.

The log-likelihood of the HMM for a sequence of complete data $\{\mathcal{D}, \mathcal{H}\}$ is defined by

$$\log p(\mathcal{D}, \mathcal{H}|\boldsymbol{w}) = \sum_{t=2}^{T} \sum_{k=1}^{K} \sum_{l=1}^{K} z_k^{(t)} z_l^{(t-1)} \log a_{k,l} + \sum_{t=1}^{T} \sum_{k=1}^{K} \sum_{m=1}^{M} z_k^{(t)} x_m^{(t)} \log b_{k,m}.$$

We assume that the prior distributions of the transition probability matrix $\boldsymbol{A}$ and the emission probability matrix $\boldsymbol{B}$ are the Dirichlet distributions with hyperparameters $\phi > 0$ and $\xi > 0$:

$$p(\boldsymbol{A}|\phi) = \prod_{k=1}^{K} \text{Dirichlet}_K \left( \widetilde{\boldsymbol{a}}_k; (\phi, \ldots, \phi)^\top \right), \tag{4.72}$$

$$p(\boldsymbol{B}|\xi) = \prod_{k=1}^{K} \text{Dirichlet}_M \left( \widetilde{\boldsymbol{b}}_k; (\xi, \ldots, \xi)^\top \right). \tag{4.73}$$

**VB Posterior for HMMs**

We define the expected sufficient statistics by

$$\overline{N}_k = \sum_{t=1}^{T} \left\langle z_k^{(t)} \right\rangle_{r_{\mathcal{H}}(\mathcal{H})}, \tag{4.74}$$

$$\overline{N}_{k,l}^{[z]} = \sum_{t=2}^{T} \left\langle z_l^{(t)} z_k^{(t-1)} \right\rangle_{r_{\mathcal{H}}(\mathcal{H})}, \tag{4.75}$$

$$\overline{N}_{k,m}^{[x]} = \sum_{t=1}^{T} \left\langle z_k^{(t)} \right\rangle_{r_{\mathcal{H}}(\mathcal{H})} x_m^{(t)}, \tag{4.76}$$

where the expected count $\overline{N}_k$ is constrained by $\overline{N}_k = \sum_l \overline{N}_{k,l}^{[z]}$. Then, the VB posterior distribution of parameters $r_w(\boldsymbol{w})$ is given by

$$r_A(\boldsymbol{A}) = \prod_{k=1}^{K} \mathrm{Dirichlet}_K\left(\widetilde{\boldsymbol{a}}_k; (\widehat{\phi}_{k,1}, \ldots, \widehat{\phi}_{k,K})^\top\right),$$

$$r_B(\boldsymbol{B}) = \prod_{k=1}^{K} \mathrm{Dirichlet}_M\left(\widetilde{\boldsymbol{b}}_k; (\widehat{\xi}_{k,1}, \ldots, \widehat{\xi}_{k,M})^\top\right),$$

where

$$\widehat{\phi}_{k,l} = \overline{N}^{[z]}_{k,l} + \phi, \tag{4.77}$$

$$\widehat{\xi}_{k,m} = \overline{N}^{[x]}_{k,m} + \xi. \tag{4.78}$$

The posterior distribution of hidden variables $r_{\mathcal{H}}(\mathcal{H})$ is given by

$$r_{\mathcal{H}}(\mathcal{H}) = \frac{1}{C_{\mathcal{H}}} \exp\left( \sum_{t=2}^{T} \sum_{k=1}^{K} \sum_{l=1}^{K} z_k^{(t)} z_l^{(t-1)} \langle \log a_{k,l} \rangle_{r_A(\boldsymbol{A})} + \sum_{t=1}^{T} \sum_{k=1}^{K} \sum_{m=1}^{M} z_k^{(t)} x_m^{(t)} \langle \log b_{k,m} \rangle_{r_B(\boldsymbol{B})} \right), \tag{4.79}$$

where $C_{\mathcal{H}}$ is the normalizing constant and

$$\langle \log a_{k,l} \rangle_{r_A(\boldsymbol{A})} = \Psi(\widehat{\phi}_{k,l}) - \Psi\left( \sum_{l'=1}^{K} \widehat{\phi}_{k,l'} \right),$$

$$\langle \log b_{k,m} \rangle_{r_B(\boldsymbol{B})} = \Psi(\widehat{\xi}_{k,m}) - \Psi\left( \sum_{m'=1}^{M} \widehat{\xi}_{k,m'} \right).$$

The expected sufficient statistics $\left\langle z_k^{(t)} \right\rangle_{r_{\mathcal{H}}(\mathcal{H})}$ and $\left\langle z_l^{(t)} z_k^{(t-1)} \right\rangle_{r_{\mathcal{H}}(\mathcal{H})}$ in Eqs. (4.74) through (4.76) can be efficiently computed in the order of $O(T)$ by the *forward–backward algorithm* (Beal, 2003). This algorithm can also compute $C_{\mathcal{H}}$. Thus, after the substitution of Eq. (4.3), the free energy is given by

$$\begin{aligned} F = \sum_{k=1}^{K} \Bigg\{ & \log\left( \frac{\Gamma(\sum_{l=1}^{K} \widehat{\phi}_{k,l})}{\prod_{l=1}^{K} \Gamma(\widehat{\phi}_{k,l})} \right) + \sum_{l=1}^{K} \left( \widehat{\phi}_{k,l} - \phi \right) \left( \Psi(\widehat{\phi}_{k,l}) - \Psi(\textstyle\sum_{l'=1}^{K} \widehat{\phi}_{k,l'}) \right) \\ & + \log\left( \frac{\Gamma(\sum_{m=1}^{M} \widehat{\xi}_{k,m})}{\prod_{m=1}^{M} \Gamma(\widehat{\xi}_{k,m})} \right) + \sum_{m=1}^{M} \left( \widehat{\xi}_{k,m} - \xi \right) \left( \Psi(\widehat{\xi}_{k,m}) - \Psi(\textstyle\sum_{m'=1}^{M} \widehat{\xi}_{k,m'}) \right) \Bigg\} \\ & - K \log\left( \frac{\Gamma(K\phi)}{(\Gamma(\phi))^K} \right) - K \log\left( \frac{\Gamma(M\xi)}{(\Gamma(\xi))^M} \right) - \log C_{\mathcal{H}}. \end{aligned} \tag{4.80}$$

The following update rule for the EVB learning of the hyperparameter $\phi$ is obtained in the same way as the update rule (4.29) for the GMM:

$$\phi^{\mathrm{new}} = \max\left(0, \phi^{\mathrm{old}} - \frac{K^2(\Psi(\phi) - \Psi(K\phi)) - \sum_{k=1}^{K} \sum_{l=1}^{K} \left( \Psi(\widehat{\phi}_{k,l}) - \Psi\left( \sum_{l'=1}^{K} \widehat{\phi}_{k,l'} \right) \right)}{K^2 \left( \Psi^{(1)}(\phi) - K\Psi^{(1)}(K\phi) \right)} \right). \tag{4.81}$$

**Algorithm 10** EVB learning for the hidden Markov model.

1: Initialize the variational parameters $(\widehat{S}, \widehat{\Phi}, \widehat{\Xi})$, and the hyperparameters $(\phi, \xi)$.
2: Apply the forward–backward algorithm to $r_{\mathcal{H}}(\mathcal{H})$ in Eq. (4.79) and compute $C_{\mathcal{H}}$.
3: Apply (substitute the right-hand side into the left-hand side) Eqs. (4.74), (4.75), (4.76), (4.77), and (4.78) to update $\widehat{\Phi}$, and $\widehat{\Xi}$.
4: Apply Eqs.(4.81) and (4.82) to update $\phi$ and $\xi$, respectively.
5: Evaluate the free energy (4.80).
6: Iterate Steps 2 through 5 until convergence (until the energy decrease becomes smaller than a threshold).

Similarly, we obtain the following update rule of the hyperparameter $\xi$:

$$\xi^{\text{new}} = \max\left(0, \xi^{\text{old}} - \frac{KM(\Psi(\xi)-\Psi(M\xi))-\sum_{k=1}^{K}\sum_{m=1}^{M}\left(\Psi(\widehat{\xi}_{k,m})-\Psi\left(\sum_{m'=1}^{M}\widehat{\xi}_{k,m'}\right)\right)}{KM\left(\Psi^{(1)}(\xi)-M\Psi^{(1)}(M\xi)\right)}\right). \tag{4.82}$$

Let

$$\widehat{S} = \left\{\left\{\left\langle z_k^{(t)}\right\rangle_{r_{\mathcal{H}}(\mathcal{H})}\right\}_{k=1}^{K}, \left\{\left\langle z_l^{(t)} z_k^{(t-1)}\right\rangle_{r_{\mathcal{H}}(\mathcal{H})}\right\}_{k,l=1}^{K}\right\}_{t=1}^{T},$$

$\widehat{\Phi} = \{\widehat{\phi}_{k,l}\}_{k,l=1}^{K}$, and $\widehat{\Xi} = \{\widehat{\xi}_{k,m}\}_{k,m=1}^{K,M}$ be the sets of variational parameters. The EVB learning for the HMM is summarized in Algorithm 10. If the prior hyperparameters are fixed, and Step 4 in the algorithm is omitted, the algorithm reduces to the (nonempirical) VB learning algorithm.

### 4.2.3 Probabilistic Context-Free Grammars

In this subsection, we discuss *probabilistic context-free grammars (PCFGs)*, which have been used for more complex sequence modeling applications than those with the Markov assumption in natural language processing, bioinformatics, and so on (Durbin et al., 1998). Without loss of generality, we can assume that the grammar is written by the Chomsky normal form. Let the model have $K$ nonterminal symbols and $M$ terminal symbols. The observation sequence of length $L$ is written by $\boldsymbol{X} = (\boldsymbol{x}^{(1)}, \ldots, \boldsymbol{x}^{(L)}) \in \{\boldsymbol{e}_1, \ldots, \boldsymbol{e}_M\}^L$. Then, the statistical model is defined by

$$p(\boldsymbol{X}|\boldsymbol{w}) = \sum_{\boldsymbol{Z} \in T(\boldsymbol{X})} p(\boldsymbol{X}, \boldsymbol{Z}|\boldsymbol{w}), \tag{4.83}$$

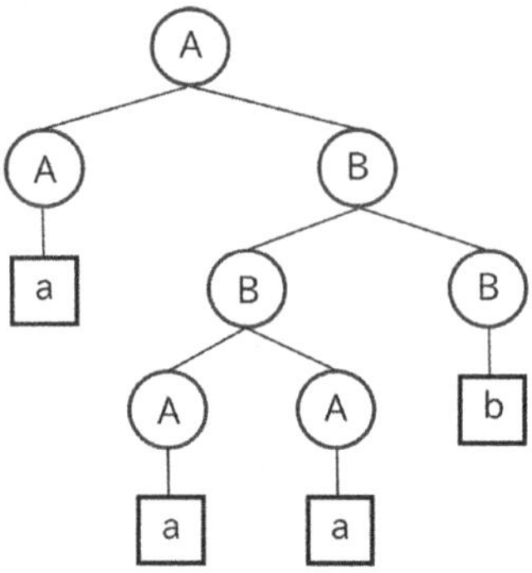


Figure 4.3 Derivation tree of PCFG. A and B are nonterminal symbols and a and b are terminal symbols.

$$p(\boldsymbol{X}, \boldsymbol{Z}|\boldsymbol{w}) = \prod_{i,j,k=1}^{K} \left(a_{i\to jk}\right)^{c^{\boldsymbol{Z}}_{i\to jk}} \prod_{l=1}^{L} \prod_{i=1}^{K} \prod_{m=1}^{M} (b_{i\to m})^{\widetilde{z}_i^{(l)} x_m^{(l)}},$$

$$\boldsymbol{w} = \{\{\boldsymbol{a}_i\}_{i=1}^{K}, \{\boldsymbol{b}_i\}_{i=1}^{K}\},$$

$$\boldsymbol{a}_i = \{a_{i\to jk}\}_{j,k=1}^{K} \quad (1 \le i \le K),$$

$$\boldsymbol{b}_i = \{b_{i\to m}\}_{m=1}^{M} \quad (1 \le i \le K),$$

where $T(\boldsymbol{X})$ is the set of derivation sequences that generate $\boldsymbol{X}$, and $\boldsymbol{Z}$ corresponds to a tree structure representing a derivation sequence. Figure 4.3 illustrates an example of the derivation tree of a PCFG model. The derivation sequence is summarized by $\{c^{\boldsymbol{Z}}_{i\to jk}\}_{i,j,k=1}^{K}$ and $\{\widetilde{z}_i^{(l)}\}_{l=1}^{L}$, where $c^{\boldsymbol{Z}}_{i\to jk}$ denotes the count of the transition rule from the nonterminal symbol $i$ to the pair of nonterminal symbols $(j,k)$ appearing in the derivation sequence $\boldsymbol{Z}$ and $\widetilde{\boldsymbol{z}}^{(l)} = (\widetilde{z}_1^{(l)}, \ldots, \widetilde{z}_K^{(l)})$ is the indicator of the (nonterminal) symbol generating the $l$th output symbol of $\boldsymbol{X}$. Moreover the parameter $a_{i\to jk}$ represents the probability that the nonterminal symbol $i$ emits the pair of nonterminal symbols $(j,k)$ and $b_{i\to m}$ represents the probability that the nonterminal symbol $i$ emits the terminal symbol $m$. The parameters, $\{\{\boldsymbol{a}_i\}_{i=1}^{K}, \{\boldsymbol{b}_i\}_{i=1}^{K}\}$, have constraints

$$a_{i\to ii} = 1 - \sum_{(j,k)\neq(i,i)} a_{i\to jk}, \quad b_{i\to M} = 1 - \sum_{m=1}^{M-1} b_{i\to m},$$

i.e., $\boldsymbol{a}_i \in \Delta^{K^2-1}$ and $\boldsymbol{b}_i \in \Delta^{M-1}$, respectively.

Let $\mathcal{D} = \{\boldsymbol{X}^{(1)}, \ldots, \boldsymbol{X}^{(N)}\}$ be a given training corpus and $\mathcal{H} = \{\boldsymbol{Z}^{(1)}, \ldots, \boldsymbol{Z}^{(N)}\}$ be the corresponding hidden derivation sequences. The log-likelihood for the complete sample $\{\mathcal{D}, \mathcal{H}\}$ is given by

$$\log p(\mathcal{D}, \mathcal{H}|\boldsymbol{w}) = \sum_{n=1}^{N} \left[ \sum_{i,j,k=1}^{K} c_{i\to jk}^{\boldsymbol{Z}^{(n)}} \log a_{i\to jk} + \sum_{l=1}^{L} \sum_{i=1}^{K} \sum_{m=1}^{M} \widetilde{z}_i^{(n,l)} x_m^{(n,l)} \log b_{i\to m} \right],$$

where $\boldsymbol{x}^{(n,l)}$ and $\widetilde{\boldsymbol{z}}^{(n,l)}$ are the indicators of the $l$th output symbol and the nonterminal symbol generating the $l$th output in the $n$th sequences, $\boldsymbol{X}^{(n)}$ and $\boldsymbol{Z}^{(n)}$, respectively.

We now turn to the VB learning for PCFGs (Kurihara and Sato, 2004). We assume that the prior distributions of parameters $\{\boldsymbol{a}_i\}_{i=1}^K$ and $\{\boldsymbol{b}_i\}_{i=1}^K$ are the Dirichlet distributions with hyperparameters $\phi$ and $\xi$:

$$p(\{\boldsymbol{a}_i\}_{i=1}^K|\phi) = \prod_{i=1}^{K} \mathrm{Dirichlet}_{K^2}\left(\boldsymbol{a}_i; (\phi, \ldots, \phi)^\top\right), \tag{4.84}$$

$$p(\{\boldsymbol{b}_i\}_{i=1}^K|\xi) = \prod_{i=1}^{K} \mathrm{Dirichlet}_{M}\left(\boldsymbol{b}_i; (\xi, \ldots, \xi)^\top\right). \tag{4.85}$$

**VB Posterior for PCFGs**

We define the expected sufficient statistics as follows:

$$\overline{N}_{i\to jk}^{z} = \sum_{n=1}^{N} \sum_{l=1}^{L} \left\langle c_{i\to jk}^{\boldsymbol{Z}^{(n)}} \right\rangle_{r_z(\boldsymbol{Z}^{(n)})}, \tag{4.86}$$

$$\overline{N}_{i}^{z} = \sum_{j,k=1}^{K} \overline{N}_{i\to jk}^{z},$$

$$\overline{N}_{i\to m}^{x} = \sum_{n=1}^{N} \sum_{l=1}^{L} \left\langle \widetilde{z}_i^{(n,l)} \right\rangle_{r_z(\boldsymbol{Z}^{(n)})} x_m^{(n,l)}, \tag{4.87}$$

$$\overline{N}_{i}^{x} = \sum_{m=1}^{M} \overline{N}_{i\to m}^{x}.$$

Then the VB posteriors of the parameters are given by

$$r_w(\boldsymbol{w}) = r_a(\{\boldsymbol{a}_i\}_{i=1}^K) r_b(\{\boldsymbol{b}_i\}_{i=1}^K),$$

$$r_a(\{\boldsymbol{a}_i\}_{i=1}^K) = \prod_{i=1}^{K} \mathrm{Dirichlet}_{K^2}\left(\boldsymbol{a}_i; (\widehat{\phi}_{i\to 11}, \ldots, \widehat{\phi}_{i\to KK})^\top\right), \tag{4.88}$$

$$r_b(\{\boldsymbol{b}_i\}_{i=1}^K) = \prod_{i=1}^{K} \mathrm{Dirichlet}_{M}\left(\boldsymbol{b}_i; (\widehat{\xi}_{i\to 1}, \ldots, \widehat{\xi}_{i\to M})^\top\right), \tag{4.89}$$

where

$$\widehat{\phi}_{i\to jk} = \overline{N}_{i\to jk}^{z} + \phi, \tag{4.90}$$

$$\widehat{\xi}_{i\to m} = \overline{N}_{i\to m}^{x} + \xi. \tag{4.91}$$

The VB posteriors of the hidden variables are given by

$$
\begin{aligned}
r_{\mathcal{H}}(\mathcal{H}) &= \prod_{n=1}^{N} r_z(\mathbf{Z}^{(n)}), \\
r_z(\mathbf{Z}^{(n)}) &= \frac{1}{C_{\mathbf{Z}^{(n)}}} \exp\left(\gamma_{\mathbf{Z}^{(n)}}\right), \\
\gamma_{\mathbf{Z}^{(n)}} &= \sum_{i,j,k=1}^{K} c_{i\to jk}^{\mathbf{Z}^{(n)}} \left\langle \log a_{i\to jk} \right\rangle_{r_a(\{\boldsymbol{a}_i\}_{i=1}^{K})} \\
&\quad + \sum_{l=1}^{L} \sum_{i=1}^{K} \sum_{m=1}^{M} \widehat{z}_i^{(n,l)} x_m^{(n,l)} \left\langle \log b_{i\to m} \right\rangle_{r_b(\{\boldsymbol{b}_i\}_{i=1}^{K})},
\end{aligned}
\tag{4.92}
$$

where $C_{\mathbf{Z}^{(n)}} = \sum_{\mathbf{Z}\in T(\boldsymbol{X}^{(n)})} \exp(\gamma_{\mathbf{Z}})$ is the normalizing constant and

$$
\begin{aligned}
\left\langle \log a_{i\to jk} \right\rangle_{r_a(\{\boldsymbol{a}_i\}_{i=1}^{K})} &= \Psi\left(\widehat{\phi}_{i\to jk}\right) - \Psi\left(\textstyle\sum_{j'=1}^{K}\sum_{k'=1}^{K} \widehat{\phi}_{i\to j'k'}\right), \\
\left\langle \log b_{i\to m} \right\rangle_{r_b(\{\boldsymbol{b}_i\}_{i=1}^{K})} &= \Psi\left(\widehat{\xi}_{i\to m}\right) - \Psi\left(\textstyle\sum_{m'=1}^{M} \widehat{\xi}_{i\to m'}\right).
\end{aligned}
$$

All the expected sufficient statistics and $C_{\mathbf{Z}^{(n)}}$ can be efficiently computed by the *inside–outside algorithm* (Kurihara and Sato, 2004). The free energy, after the substitution of Eq. (4.3), is given by

$$
\begin{aligned}
F = \sum_{i=1}^{K} \Bigg\{ &\log\left(\frac{\Gamma(\sum_{j,k=1}^{K} \widehat{\phi}_{i\to jk})}{\prod_{j,k=1}^{K} \Gamma(\widehat{\phi}_{i\to jk})}\right) \\
&+ \sum_{j,k=1}^{K} \left(\widehat{\phi}_{i\to jk} - \phi\right)\left(\Psi\left(\widehat{\phi}_{i\to jk}\right) - \Psi\left(\textstyle\sum_{j',k'=1}^{K} \widehat{\phi}_{i\to j'k'}\right)\right) \\
&+ \log\left(\frac{\Gamma\left(\sum_{m=1}^{M} \widehat{\xi}_{i\to m}\right)}{\prod_{m=1}^{M} \Gamma\left(\widehat{\xi}_{i\to m}\right)}\right) + \sum_{m=1}^{M} \left(\widehat{\xi}_{i\to m} - \xi\right)\left(\Psi\left(\widehat{\xi}_{i\to m}\right) - \Psi\left(\textstyle\sum_{m'=1}^{M} \widehat{\xi}_{i\to m'}\right)\right) \Bigg\} \\
&- K\log\left(\frac{\Gamma(K^2\phi)}{(\Gamma(\phi))^{K^2}}\right) - K\log\left(\frac{\Gamma(M\xi)}{(\Gamma(\xi))^{M}}\right) - \sum_{n=1}^{N} \log C_{\mathbf{Z}^{(n)}}.
\end{aligned}
\tag{4.93}
$$

The following update rules for the EVB learning of the hyperparameters $\phi$ and $\xi$ are obtained similarly to the HMM in Eqs. (4.81) and (4.82):

$$
\phi^{\mathrm{new}} = \max\left(0, \phi^{\mathrm{old}} - \frac{K^3(\Psi(\phi)-\Psi(K^2\phi)) - \sum_{i=1}^{K}\sum_{j,k=1}^{K}\left(\Psi\left(\widehat{\phi}_{i\to jk}\right) - \Psi\left(\sum_{j',k'=1}^{K} \widehat{\phi}_{i\to j'k'}\right)\right)}{K^3\left(\Psi^{(1)}(\phi) - K^2\Psi^{(1)}(K^2\phi)\right)}\right),
\tag{4.94}
$$

$$
\xi^{\mathrm{new}} = \max\left(0, \xi^{\mathrm{old}} - \frac{KM(\Psi(\xi)-\Psi(M\xi)) - \sum_{i=1}^{K}\sum_{m=1}^{M}\left(\Psi\left(\widehat{\xi}_{i\to m}\right) - \Psi\left(\sum_{m'=1}^{M} \widehat{\xi}_{i\to m'}\right)\right)}{KM\left(\Psi^{(1)}(\xi) - M\Psi^{(1)}(M\xi)\right)}\right).
\tag{4.95}
$$

**Algorithm 11** EVB learning for probabilistic context-free grammar.

1: Initialize the variational parameters $(\widehat{S}, \widehat{\Phi}, \widehat{\Xi})$ and the hyperparameters $(\phi, \xi)$.
2: Apply the inside–outside algorithm to $r_z(\mathbf{Z}^{(n)})$ in Eq. (4.92) and compute $C_{\mathbf{Z}^{(n)}}$ for $n = 1, \ldots, N$.
3: Apply (substitute the right-hand side into the left-hand side) Eqs. (4.86), (4.87), (4.90), and (4.91) to update $\widehat{\Phi}$, and $\widehat{\Xi}$.
4: Apply Eqs. (4.94) and (4.95) to update $\phi$ and $\xi$, respectively.
5: Evaluate the free energy (4.93).
6: Iterate Steps 2 through 5 until convergence (until the energy decrease becomes smaller than a threshold).

Let

$$\widehat{S} = \left\{ \left\{ \left\langle c^{\mathbf{Z}^{(n)}}_{i\to jk} \right\rangle_{r_z(\mathbf{Z}^{(n)})} \right\}_{i,j,k=1}^{K}, \left\{ \left\langle \overline{z}_i^{(n,l)} \right\rangle_{r_z(\mathbf{Z}^{(n)})} \right\}_{l=1}^{L} \right\}_{n=1}^{N},$$

$\widehat{\Phi} = \{\widehat{\phi}_{i\to jk}\}_{i,j,k=1}^{K}$, and $\widehat{\Xi} = \{\widehat{\xi}_{i\to m}\}_{i,m=1}^{K,M}$ be the sets of variational parameters. The EVB learning for the PCFG model is summarized in Algorithm 11. If the prior hyperparameters are fixed and Step 4 in the algorithm is omitted, the algorithm reduces to the (nonempirical) VB learning algorithm.

### 4.2.4 Latent Dirichlet Allocation

*Latent Dirichlet allocation (LDA)* (Blei et al., 2003) is a generative model successfully used in various applications such as text analysis (Blei et al., 2003), image analysis (Li and Perona, 2005), genomics (Bicego et al., 2010; Chen et al., 2010), human activity analysis (Huynh et al., 2008), and collaborative filtering (Krestel et al., 2009; Purushotham et al., 2012). Given word occurrences of documents in a corpora, LDA expresses each document as a mixture of multinomial distributions, each of which is expected to capture a *topic*. The extracted topics provide bases in a low-dimensional feature space, in which each document is compactly represented. This topic expression was shown to be useful for solving various tasks, including classification (Li and Perona, 2005), retrieval (Wei and Croft, 2006), and recommendation (Krestel et al., 2009).

In this subsection, we introduce the VB learning for tractable inference in the LDA model. Suppose that we observe $M$ documents, each of which consists of $N^{(m)}$ words. Each word is included in a vocabulary with size $L$. We assume that each word is associated with one of the $H$ topics, which is not observed.

We express the word occurrence by an $L$-dimensional indicator vector $\boldsymbol{w}$, where one of the entries is equal to one and the others are equal to zero. Similarly, we express the topic occurrence as an $H$-dimensional indicator vector $\boldsymbol{z}$. We define the following functions that give the item numbers chosen by $\boldsymbol{w}$ and $\boldsymbol{z}$, respectively:

$$\acute{l}(\boldsymbol{w}) = l \text{ if } w_l = 1 \text{ and } w_{l'} = 0 \text{ for } l' \neq l,$$
$$\acute{h}(\boldsymbol{z}) = h \text{ if } z_h = 1 \text{ and } z_{h'} = 0 \text{ for } h' \neq h.$$

In the LDA model (Blei et al., 2003), the word occurrence $\boldsymbol{w}^{(n,m)}$ of the $n$th position in the $m$th document is assumed to follow the multinomial distribution:

$$p(\boldsymbol{w}^{(n,m)}|\boldsymbol{\Theta}, \boldsymbol{B}) = \prod_{l=1}^{L} \left( (\boldsymbol{B}\boldsymbol{\Theta}^\top)_{l,m} \right)^{w_l^{(n,m)}} = (\boldsymbol{B}\boldsymbol{\Theta}^\top)_{\acute{l}(\boldsymbol{w}^{(n,m)}),m}, \tag{4.96}$$

where $\boldsymbol{\Theta} \in [0,1]^{M \times H}$ and $\boldsymbol{B} \in [0,1]^{L \times H}$ are parameter matrices to be estimated. The rows of $\boldsymbol{\Theta} = (\widetilde{\boldsymbol{\theta}}_1, \ldots, \widetilde{\boldsymbol{\theta}}_M)^\top$ and the columns of $\boldsymbol{B} = (\boldsymbol{\beta}_1, \ldots, \boldsymbol{\beta}_H)$ are probability mass vectors that sum up to one. That is, $\widetilde{\boldsymbol{\theta}}_m \in \Delta^{H-1}$ is the topic distribution of the $m$th document, and $\boldsymbol{\beta}_h \in \Delta^{L-1}$ is the word distribution of the $h$th topic.

Suppose that we observe the data $\mathcal{D} = \{\{\boldsymbol{w}^{(n,m)}\}_{n=1}^{N^{(m)}}\}_{m=1}^{M}$. Given the topic occurrence latent variable $\boldsymbol{z}^{(n,m)}$, the complete likelihood for each word is written as

$$p(\boldsymbol{w}^{(n,m)}, \boldsymbol{z}^{(n,m)}|\boldsymbol{\Theta}, \boldsymbol{B}) = p(\boldsymbol{w}^{(n,m)}|\boldsymbol{z}^{(n,m)}, \boldsymbol{B})p(\boldsymbol{z}^{(n,m)}|\boldsymbol{\Theta}), \tag{4.97}$$

where $p(\boldsymbol{w}^{(n,m)}|\boldsymbol{z}^{(n,m)}, \boldsymbol{B}) = \prod_{l=1}^{L}\prod_{h=1}^{H} (B_{l,h})^{w_l^{(n,m)} z_h^{(n,m)}}$, $p(\boldsymbol{z}^{(n,m)}|\boldsymbol{\Theta}) = \prod_{h=1}^{H} (\theta_{m,h})^{z_h^{(n,m)}}$.

We assume the Dirichlet priors on $\boldsymbol{\Theta}$ and $\boldsymbol{B}$:

$$p(\boldsymbol{\Theta}|\boldsymbol{\alpha}) = \prod_{m=1}^{M} \text{Dirichlet}_H(\widetilde{\boldsymbol{\theta}}_m; (\alpha_1, \ldots, \alpha_H)^\top), \tag{4.98}$$

$$p(\boldsymbol{B}|\boldsymbol{\eta}) = \prod_{h=1}^{H} \text{Dirichlet}_L(\boldsymbol{\beta}_h; (\eta_1, \ldots, \eta_L)^\top), \tag{4.99}$$

where $\boldsymbol{\alpha}$ and $\boldsymbol{\eta}$ are hyperparameters that control the prior sparsity.

### VB Posterior for LDA

For the set of all hidden variables $\mathcal{H} = \{\{\boldsymbol{z}^{(n,m)}\}_{n=1}^{N^{(m)}}\}_{m=1}^{M}$ and the parameter $\boldsymbol{w} = (\boldsymbol{\Theta}, \boldsymbol{B})$, we assume that our approximate posterior is factorized as Eq. (4.1). Thus, the update rule (4.2) yields the further factorization $r_{\Theta,B}(\boldsymbol{\Theta}, \boldsymbol{B}) = r_\Theta(\boldsymbol{\Theta})r_B(\boldsymbol{B})$ and the following update rules:

$$r_\Theta(\boldsymbol{\Theta}) \propto p(\boldsymbol{\Theta}|\alpha) \langle \log p(\mathcal{D}, \mathcal{H}|\boldsymbol{\Theta}, \boldsymbol{B}) \rangle_{r_B(\boldsymbol{B}) r_{\mathcal{H}}(\mathcal{H})}, \tag{4.100}$$

$$r_B(\boldsymbol{B}) \propto p(\boldsymbol{B}|\boldsymbol{\eta}) \langle \log p(\mathcal{D}, \mathcal{H}|\boldsymbol{\Theta}, \boldsymbol{B}) \rangle_{r(\boldsymbol{\Theta}) r_{\mathcal{H}}(\mathcal{H})}. \tag{4.101}$$

Define the expected sufficient statistics as

$$\overline{N}_h^{(m)} = \sum_{n=1}^{N^{(m)}} \left\langle z_h^{(n,m)} \right\rangle_{r_z\left(\{\{z^{(n,m)}\}_{n=1}^{N^{(m)}}\}_{m=1}^M\right)}, \tag{4.102}$$

$$\overline{W}_{l,h} = \sum_{m=1}^{M} \sum_{n=1}^{N^{(m)}} w_l^{(n,m)} \left\langle z_h^{(n,m)} \right\rangle_{r_z\left(\{\{z^{(n,m)}\}_{n=1}^{N^{(m)}}\}_{m=1}^M\right)}. \tag{4.103}$$

Then, the VB posterior distribution is given by the Dirichlet distributions:

$$r_\Theta(\boldsymbol{\Theta}) = \prod_{m=1}^{M} \mathrm{Dirichlet}\left(\widetilde{\boldsymbol{\theta}}_m; \widehat{\boldsymbol{\alpha}}_m\right), \tag{4.104}$$

$$r_B(\boldsymbol{B}) = \prod_{h=1}^{H} \mathrm{Dirichlet}\left(\boldsymbol{\beta}_h; \widehat{\boldsymbol{\eta}}_h\right), \tag{4.105}$$

where the variational parameters satisfy

$$\widehat{\alpha}_{m,h} = (\widehat{\boldsymbol{\alpha}}_m)_h = \overline{N}_h^{(m)} + \alpha_h, \tag{4.106}$$

$$\widehat{\eta}_{l,h} = (\widehat{\boldsymbol{\eta}}_h)_l = \overline{W}_{l,h} + \eta_l. \tag{4.107}$$

From the update rule (4.3), the VB posterior distribution of latent variables is given by the multinomial distribution:

$$r_z\left(\{\{z^{(n,m)}\}_{n=1}^{N^{(m)}}\}_{m=1}^M\right) = \prod_{m=1}^{M} \prod_{n=1}^{N^{(m)}} \mathrm{Multinomial}_{H,1}\left(z^{(n,m)}; \widehat{z}^{(n,m)}\right), \tag{4.108}$$

where the variational parameter $\widehat{z}^{(n,m)} \in \Delta^{H-1}$ is

$$\widehat{z}_h^{(n,m)} = \frac{\overline{z}_h^{(n,m)}}{\sum_{h'=1}^{H} \overline{z}_{h'}^{(n,m)}} \tag{4.109}$$

for

$$\overline{z}_h^{(n,m)} = \exp\left( \langle \log \Theta_{m,h} \rangle_{r_\Theta(\boldsymbol{\Theta})} + \sum_{l=1}^{L} w_l^{(n,m)} \langle \log B_{l,h} \rangle_{r_B(\boldsymbol{B})} \right). \tag{4.110}$$

We also have

$$\begin{aligned}
\left\langle z_h^{(n,m)} \right\rangle_{r_z\left(\{\{z^{(n,m)}\}_{n=1}^{N^{(m)}}\}_{m=1}^M\right)} &= \widehat{z}_h^{(n,m)}, \\
\langle \log \Theta_{m,h} \rangle_{r_\Theta(\boldsymbol{\Theta})} &= \Psi(\widehat{\alpha}_{m,h}) - \Psi\left(\textstyle\sum_{h'=1}^{H} \widehat{\alpha}_{m,h'}\right), \\
\langle \log B_{l,h} \rangle_{r_B(\boldsymbol{B})} &= \Psi(\widehat{\eta}_{l,h}) - \Psi\left(\textstyle\sum_{l'=1}^{L} \widehat{\eta}_{l',h}\right).
\end{aligned}$$

Iterating Eqs. (4.106), (4.107), and (4.110) provides a local minimum of the free energy, which is given as a function of the variational parameters by

$$F = \sum_{m=1}^{M}\left(\log\left(\frac{\Gamma(\sum_{h=1}^{H}\widehat{\alpha}_{m,h})}{\prod_{h=1}^{H}\Gamma(\widehat{\alpha}_{m,h})}\right) - \log\left(\frac{\Gamma(\sum_{h=1}^{H}\alpha_h)}{\prod_{h=1}^{H}\Gamma(\alpha_h)}\right)\right)$$
$$+ \sum_{h=1}^{H}\left(\log\left(\frac{\Gamma(\sum_{l=1}^{L}\widehat{\eta}_{l,h})}{\prod_{l=1}^{L}\Gamma(\widehat{\eta}_{l,h})}\right) - \log\left(\frac{\Gamma(\sum_{l=1}^{L}\eta_l)}{\prod_{l=1}^{L}\Gamma(\eta_l)}\right)\right)$$
$$+ \sum_{m=1}^{M}\sum_{h=1}^{H}\left(\widehat{\alpha}_{m,h} - (\overline{N}_h^{(m)} + \alpha_h)\right)\left(\Psi(\widehat{\alpha}_{m,h}) - \Psi(\textstyle\sum_{h'=1}^{H}\widehat{\alpha}_{m,h'})\right)$$
$$+ \sum_{h=1}^{H}\sum_{l=1}^{L}\left(\widehat{\eta}_{l,h} - (\overline{W}_{l,h} + \eta_l)\right)\left(\Psi(\widehat{\eta}_{l,h}) - \Psi(\textstyle\sum_{l'=1}^{L}\widehat{\eta}_{l',h})\right)$$
$$+ \sum_{m=1}^{M}\sum_{n=1}^{N^{(m)}}\sum_{h=1}^{H}\widehat{z}_h^{(n,m)}\log\widehat{z}_h^{(n,m)}. \tag{4.111}$$

The partial derivatives of the free energy with respect to $(\boldsymbol{\alpha}, \boldsymbol{\eta})$ are computed as follows:

$$\frac{\partial F}{\partial \alpha_h} = M\left(\Psi(\alpha_h) - \Psi(\textstyle\sum_{h'=1}^{H}\alpha_{h'})\right)$$
$$- \sum_{m=1}^{M}\left(\Psi(\widehat{\alpha}_{m,h}) - \Psi(\textstyle\sum_{h'=1}^{H}\widehat{\alpha}_{m,h'})\right), \tag{4.112}$$

$$\frac{\partial^2 F}{\partial \alpha_h \partial \alpha_{h'}} = M\left(\delta_{h,h'}\Psi^{(1)}(\alpha_h) - \Psi^{(1)}(\textstyle\sum_{h''=1}^{H}\alpha_{h''})\right), \tag{4.113}$$

$$\frac{\partial F}{\partial \eta_l} = H\left(\Psi(\eta_l) - \Psi(\textstyle\sum_{l'=1}^{L}\eta_{l'})\right)$$
$$- \sum_{h=1}^{H}\left(\Psi(\widehat{\eta}_{l,h}) - \Psi(\textstyle\sum_{l'=1}^{L}\widehat{\eta}_{l',h})\right), \tag{4.114}$$

$$\frac{\partial^2 F}{\partial \eta_l \partial \eta_{l'}} = H\left(\delta_{l,l'}\Psi^{(1)}(\eta_l) - \Psi^{(1)}(\textstyle\sum_{l''=1}^{L}\eta_{l''})\right), \tag{4.115}$$

where $\delta_{n,n'}$ is the *Kronecker delta*. Thus, we have the following Newton–Raphson steps to update the hyperparameters:

$$\boldsymbol{\alpha}^{\text{new}} = \mathbf{max}\left(\mathbf{0}, \boldsymbol{\alpha}^{\text{old}} - \left(\frac{\partial^2 F}{\partial\boldsymbol{\alpha}\partial\boldsymbol{\alpha}^\top}\right)^{-1}\frac{\partial F}{\partial\boldsymbol{\alpha}}\right), \tag{4.116}$$

$$\boldsymbol{\eta}^{\text{new}} = \mathbf{max}\left(\mathbf{0}, \boldsymbol{\eta}^{\text{old}} - \left(\frac{\partial^2 F}{\partial\boldsymbol{\eta}\partial\boldsymbol{\eta}^\top}\right)^{-1}\frac{\partial F}{\partial\boldsymbol{\eta}}\right), \tag{4.117}$$

where $\mathbf{max}(\cdot)$ is the max operator applied elementwise.

**Algorithm 12** EVB learning for latent Dirichlet allocation.

1: Initialize the variational parameters $(\{\{\widehat{z}^{(n,m)}\}_{n=1}^{N^{(m)}}\}_{m=1}^{M}, \{\widehat{\alpha}_m\}_{m=1}^{M}, \{\widehat{\eta}_h\}_{h=1}^{H})$, and the hyperparameters $(\alpha, \eta)$.
2: Apply (substitute the right-hand side into the left-hand side) Eqs. (4.110), (4.109), (4.102), (4.103), (4.106), and (4.107) to update $\{\{\widehat{z}^{(n,m)}\}_{n=1}^{N^{(m)}}\}_{m=1}^{M}$, $\{\widehat{\alpha}_m\}_{m=1}^{M}$, and $\{\widehat{\eta}_h\}_{h=1}^{H}$.
3: Apply Eqs. (4.116) and (4.117) to update $\alpha$ and $\eta$, respectively.
4: Evaluate the free energy (4.111).
5: Iterate Steps 2 through 4 until convergence (until the energy decrease becomes smaller than a threshold).

The EVB learning for LDA is summarized in Algorithm 12. If the prior hyperparameters are fixed and Step 3 in the algorithm is omitted, the algorithm reduces to the (nonempirical) VB learning algorithm.

We can also apply partially Bayesian (PB) learning by approximating the posterior of $\boldsymbol{\Theta}$ or $\boldsymbol{B}$ by the delta function (see Section 2.2.2). We call it PB-A learning if $\boldsymbol{\Theta}$ is marginalized and $\boldsymbol{B}$ is point-estimated, and PB-B learning if $\boldsymbol{B}$ is marginalized and $\boldsymbol{\Theta}$ is point-estimated. Note that the original VB algorithm for LDA proposed by Blei et al. (2003) corresponds to PB-A learing in our terminology. MAP learning, where both of $\boldsymbol{\Theta}$ and $\boldsymbol{B}$ are point-estimated, corresponds to the *probabilistic latent semantic analysis (pLSA)* (Hofmann, 2001), if we assume the flat prior $\alpha_h = \eta_l = 1$ (Girolami and Kaban, 2003).

# 5
# VB Algorithm under No Conjugacy

As discussed in Section 2.1.7, there are practical combinations of a model and a prior where conjugacy is no longer available. In this chapter, as a method for addressing nonconjugacy, we demonstrate *local variational approximation (LVA)*, also known as *direct site bounding*, for logistic regression and a sparsity-inducing prior (Jaakkola and Jordan, 2000; Girolami, 2001; Bishop, 2006; Seeger, 2008, 2009). Then we describe a general framework of LVA based on convex functions by using the associated Bregman divergence (Watanabe et al., 2011).

## 5.1 Logistic Regression

Let $\mathcal{D} = \{(\boldsymbol{x}^{(1)}, y^{(1)}), (\boldsymbol{x}^{(2)}, y^{(2)}), \ldots, (\boldsymbol{x}^{(N)}, y^{(N)})\}$ be the $N$ observations of the binary response variable $y^{(n)} \in \{0, 1\}$ and the input vector $\boldsymbol{x}^{(n)} \in \mathbb{R}^M$. The *logistic regression* model assumes the following Bernoulli model over $\boldsymbol{y} = (y^{(1)}, y^{(2)}, \ldots, y^{(N)})^\top$ given $\boldsymbol{X} = \{\boldsymbol{x}^{(1)}, \boldsymbol{x}^{(2)}, \ldots, \boldsymbol{x}^{(N)}\}$:

$$p(\boldsymbol{y}|\boldsymbol{X}, \boldsymbol{w}) = \prod_{n=1}^{N} \exp\left(y^{(n)}(\boldsymbol{w}^\top \boldsymbol{x}^{(n)}) - \log\left(1 + e^{\boldsymbol{w}^\top \boldsymbol{x}^{(n)}}\right)\right). \tag{5.1}$$

Let us consider the Bayesian learning of the parameter $\boldsymbol{w}$ assuming the Gaussian prior distribution:

$$p(\boldsymbol{w}) = \mathrm{Gauss}_M(\boldsymbol{w}; \boldsymbol{w}_0, \boldsymbol{S}_0^{-1}),$$

where $\boldsymbol{S}_0$ and $\boldsymbol{w}_0$ are the hyperparameters.

Gaussian approximations for the posterior distribution $p(\boldsymbol{w}|\mathcal{D}) \propto p(\boldsymbol{w}, \boldsymbol{y}|\boldsymbol{X})$ are obtained by LVA based on the facts that $-\log(e^{\sqrt{h}/2} + e^{-\sqrt{h}/2})$ is a convex function of $h$ and that $\log(1 + e^g)$ is a convex function of $g$. More specifically, because $\phi(h(w)) = -\log(e^{\sqrt{w^2}/2} + e^{-\sqrt{w^2}/2})$ is a convex function of $h(w) = w^2$ and $\psi(g(w)) = \log(1 + e^w)$ is a convex function of $g(w) = w$, they are bounded from below by their tangents at $h(\xi) = \xi^2$ and $g(\eta) = \eta$, respectively:

$$-\log\left(e^{\sqrt{w^2}/2} + e^{-\sqrt{w^2}/2}\right) \geq -\log\left(e^{\sqrt{\xi^2}/2} + e^{-\sqrt{\xi^2}/2}\right) - (w^2 - \xi^2)\frac{\tanh(\xi/2)}{4\xi},$$

$$\log(1 + e^w) \geq \log(1 + e^\eta) + (w - \eta)\frac{e^\eta}{1 + e^\eta}.$$

By substituting these bounds into the likelihood (5.1), we obtain the following bounds on $p(\boldsymbol{w}, \boldsymbol{y}|\boldsymbol{X})$:

$$\underline{p}(\boldsymbol{w}; \boldsymbol{\xi}) \leq p(\boldsymbol{w}, \boldsymbol{y}|\boldsymbol{X}) \leq \overline{p}(\boldsymbol{w}; \boldsymbol{\eta}),$$

where

$$\underline{p}(\boldsymbol{w}; \boldsymbol{\xi}) \equiv p(\boldsymbol{w}) \prod_{n=1}^{N} \exp\left(\left(y^{(n)} - \frac{1}{2}\right)\boldsymbol{w}^\top \boldsymbol{x}^{(n)} - \theta_n \left\{(\boldsymbol{w}^\top \boldsymbol{x}^{(n)})^2 - h_n\right\} - \log\left(e^{\frac{\sqrt{h_n}}{2}} + e^{-\frac{\sqrt{h_n}}{2}}\right)\right),$$

$$\overline{p}(\boldsymbol{w}; \boldsymbol{\eta}) \equiv p(\boldsymbol{w}) \prod_{n=1}^{N} \exp\left((y^{(n)} - \kappa_n)\boldsymbol{w}^\top \boldsymbol{x}^{(n)} - b(\kappa_n)\right).$$

Here we have put

$$\theta_n = \frac{\tanh(\sqrt{h_n}/2)}{4\sqrt{h_n}}, \tag{5.2}$$

$$\kappa_n = \frac{e^{g_n}}{1 + e^{g_n}}, \tag{5.3}$$

and $\{h_n\}_{n=1}^N$ and $\{g_n\}_{n=1}^N$ are the sets of variational parameters defined from $\boldsymbol{\xi} = (\xi_1, \xi_2, \ldots, \xi_M)^\top$ and $\boldsymbol{\eta} = (\eta_1, \eta_2, \ldots, \eta_M)^\top$ by the transformations $h_n = (\boldsymbol{\xi}^\top \boldsymbol{x}^{(n)})^2$ and $g_n = \boldsymbol{\eta}^\top \boldsymbol{x}^{(n)}$, respectively. We also defined the binary entropy function by $b(\kappa) = -\kappa \log \kappa - (1 - \kappa)\log(1 - \kappa)$ for $\kappa \in [0,\ 1]$.

Normalizing these bounds with respect to $\boldsymbol{w}$, we approximate the posterior by the Gaussian distributions as

$$q_\xi(\boldsymbol{w}; \boldsymbol{\xi}) = \mathrm{Gauss}_M(\underline{\boldsymbol{m}}, \underline{\boldsymbol{S}}^{-1}),$$

$$q_\eta(\boldsymbol{w}; \boldsymbol{\eta}) = \mathrm{Gauss}_M(\overline{\boldsymbol{m}}, \overline{\boldsymbol{S}}^{-1}),$$

whose mean and precision (inverse-covariance) matrix are respectively given by

$$\begin{aligned} \underline{\boldsymbol{m}} &= \underline{\boldsymbol{S}}^{-1}\left\{\boldsymbol{S}_0 \boldsymbol{w}_0 + \textstyle\sum_{n=1}^N (y^{(n)} - 1/2)\boldsymbol{x}^{(n)}\right\}, \\ \underline{\boldsymbol{S}} &= \boldsymbol{S}_0 + 2\textstyle\sum_{n=1}^N \theta_n \boldsymbol{x}^{(n)}\boldsymbol{x}^{(n)\top}, \end{aligned} \tag{5.4}$$

and

$$\begin{aligned}\overline{\boldsymbol{m}} &= \boldsymbol{w}_0 + \boldsymbol{S}_0^{-1}\textstyle\sum_{n=1}^{N}(y^{(n)} - \kappa_n)\boldsymbol{x}^{(n)},\\ \overline{\boldsymbol{S}} &= \boldsymbol{S}_0.\end{aligned} \tag{5.5}$$

We also obtain the bounds for the marginal likelihood, $\underline{Z}(\boldsymbol{\xi}) \equiv \int \underline{p}(\boldsymbol{w};\boldsymbol{\xi})d\boldsymbol{w}$ and $\overline{Z}(\boldsymbol{\eta}) \equiv \int \overline{p}(\boldsymbol{w};\boldsymbol{\eta})d\boldsymbol{w}$. These are respectively given in the forms of free energy bounds as follows:

$$\begin{aligned}\overline{F}(\boldsymbol{\xi}) &\equiv -\log \underline{Z}(\boldsymbol{\xi})\\ &= \frac{1}{2}\log|\underline{\boldsymbol{S}}| - \frac{1}{2}\log|\boldsymbol{S}_0| + \frac{\boldsymbol{w}_0^\top \boldsymbol{S}_0 \boldsymbol{w}_0}{2} - \frac{\underline{\boldsymbol{m}}^\top(\underline{\boldsymbol{S}})\underline{\boldsymbol{m}}}{2}\\ &\quad - \sum_{n=1}^{N}\left\{h_n\theta_n - \log\left(2\cosh\left(\frac{\sqrt{h_n}}{2}\right)\right)\right\},\end{aligned} \tag{5.6}$$

and

$$\begin{aligned}\underline{F}(\boldsymbol{\eta}) &\equiv -\log \overline{Z}(\boldsymbol{\eta})\\ &= \frac{\boldsymbol{w}_0^\top \boldsymbol{S}_0 \boldsymbol{w}_0}{2} - \frac{\overline{\boldsymbol{m}}^\top \boldsymbol{S}_0 \overline{\boldsymbol{m}}}{2} + \sum_{n=1}^{N} b(\kappa_n).\end{aligned}$$

We optimize the free energy bounds to determine the variational parameters. As will be discussed generally in Section 5.3.2, to decrease the upper-bound $\overline{F}(\boldsymbol{\xi})$, the EM algorithm is available, which instead maximizes

$$\left\langle \log \underline{p}(\boldsymbol{w};\boldsymbol{\xi})\right\rangle_{q_\xi(\boldsymbol{w};\boldsymbol{\xi}_o)},$$

where the expectation is taken with respect to the approximate posterior before updating with the variational parameters given by $\boldsymbol{\xi}_o$. The update rule of the variational parameters is specifically given by

$$\begin{aligned}h_n &= \left\langle(\boldsymbol{w}^\top \boldsymbol{x}^{(n)})^2\right\rangle_{q_\xi(\boldsymbol{w};\boldsymbol{\xi}_o)}\\ &= \boldsymbol{x}^{(n)\top}(\underline{\boldsymbol{S}}^{-1} + \underline{\boldsymbol{m}\boldsymbol{m}^\top})\boldsymbol{x}^{(n)},\end{aligned} \tag{5.7}$$

where $\underline{\boldsymbol{m}}$ and $\underline{\boldsymbol{S}}^{-1}$ are the mean and covariance matrix of $q_\xi(\boldsymbol{w};\boldsymbol{\xi}_o)$.

We can use the following gradient for the maximization of the lower-bound $\underline{F}(\boldsymbol{\eta})$:

$$\begin{aligned}\frac{\partial \underline{F}(\boldsymbol{\eta})}{\partial \kappa_n} &= \left\langle \boldsymbol{w}^\top \boldsymbol{x}^{(n)}\right\rangle_{q_\eta(\boldsymbol{w};\boldsymbol{\eta})} - \boldsymbol{\eta}^\top \boldsymbol{x}^{(n)}\\ &= \overline{\boldsymbol{m}}^\top \boldsymbol{x}^{(n)} - g_n.\end{aligned} \tag{5.8}$$

The Newton–Raphson step to update $\boldsymbol{\kappa} = (\kappa_1, \ldots, \kappa_N)^\top$ is given by

$$\boldsymbol{\kappa}^{\text{new}} = \boldsymbol{\kappa}^{\text{old}} - \left(\frac{\partial^2 \underline{F}}{\partial \boldsymbol{\kappa}\partial \boldsymbol{\kappa}^\top}\right)^{-1}\frac{\partial \underline{F}}{\partial \boldsymbol{\kappa}}, \tag{5.9}$$

**Algorithm 13** LVA algorithm for logistic regression.

1: Initialize the variational parameters $\{h_n\}_{n=1}^N$ and transform them to $\{\theta_n\}_{n=1}^N$ by Eq. (5.2).
2: Compute the approximate posterior mean and covariance matrix $(\underline{\boldsymbol{m}}, \underline{\boldsymbol{S}}^{-1})$ by Eq. (5.4).
3: Apply Eq. (5.7) to update $\{h_n\}_{n=1}^N$ and transform them to $\{\theta_n\}_{n=1}^N$ by Eq. (5.2).
4: Evaluate the free energy bound (5.6).
5: Iterate Steps 2 through 4 until convergence (until the decrease of the bound becomes smaller than a threshold).

where the $(n, n')$th entry of the Hessian matrix is given as follows:

$$\frac{\partial^2 \underline{F}(\boldsymbol{\eta})}{\partial \kappa_n \partial \kappa_{n'}} = -\boldsymbol{x}^{(n)\top} \boldsymbol{S}_0^{-1} \boldsymbol{x}^{(n')} - \delta_{n,n'} \left( \frac{1}{\kappa_n} + \frac{1}{1 - \kappa_n} \right).$$

The learning algorithm for logistic regression with LVA is summarized in Algorithm 13 in the case of $\overline{F}(\boldsymbol{\xi})$ minimization. To obtain the algorithm for $\underline{F}(\boldsymbol{\eta})$ maximization, the updated variables are replaced with $\{g_n\}_{n=1}^N$, $\{\kappa_n\}_{n=1}^N$, and $(\overline{\boldsymbol{m}}, \overline{\boldsymbol{S}}^{-1})$, and the update rule (5.7) in Step 3 is replaced with the Newton–Raphson step (5.9).

Recall the arguments in Section 2.1.7 that the VB posterior $r(\boldsymbol{w}; \widehat{\boldsymbol{\lambda}}) = q(\boldsymbol{w}; \boldsymbol{\xi})$ in Eq. (2.32) minimizes the upper-bound of the free energy (2.25) and the approximate posterior $r(\boldsymbol{w}; \widehat{\boldsymbol{\nu}}) = q(\boldsymbol{w}; \boldsymbol{\eta})$ in Eq. (2.58) maximizes the lower-bound of the objective function of EP (2.50). This means that the variational parameters are given by $\widehat{\boldsymbol{\lambda}} = (\underline{\boldsymbol{m}}, \underline{\boldsymbol{S}}^{-1})$ and $\widehat{\boldsymbol{\nu}} = (\overline{\boldsymbol{m}}, \overline{\boldsymbol{S}}^{-1})$ in the LVAs for VB and EP, respectively.

## 5.2 Sparsity-Inducing Prior

Another representative example where a nonconjugate prior is used is the linear regression model with a *sparsity-inducing prior* distribution (Girolami, 2001; Seeger, 2008, 2009). We discuss the linear regression model for i.i.d. data $\mathcal{D} = \{(\boldsymbol{x}^{(1)}, y^{(1)}), (\boldsymbol{x}^{(2)}, y^{(2)}), \ldots, (\boldsymbol{x}^{(N)}, y^{(N)})\}$, where for each observation, $\boldsymbol{x}^{(n)} \in \mathbb{R}^M$ is the input vector and $y^{(n)} \in \mathbb{R}$ is the response. Denoting $\boldsymbol{y} = (y^{(1)}, \ldots, y^{(N)})^\top$ and $\boldsymbol{X} = (\boldsymbol{x}^{(1)}, \ldots, \boldsymbol{x}^{(N)})^\top$, we assume the model,

$$p(\boldsymbol{y}|\boldsymbol{X}, \boldsymbol{w}) = \mathrm{Gauss}_N(\boldsymbol{y}; \boldsymbol{X}\boldsymbol{w}, \sigma^2 \boldsymbol{I}_N),$$

and the following sparsity-inducing prior with the $L_\beta$-norm:

$$p(\boldsymbol{w}) = \prod_{m=1}^{M} \frac{1}{C_{\beta,\gamma}} \exp\left(-\gamma |w_m|^\beta\right), \tag{5.10}$$

where $\gamma > 0$ and $0 < \beta \leq 2$ are the hyperparameters, and $C_{\beta,\gamma} = \frac{2}{\beta}\gamma^{1-1/\beta}\Gamma(1/\beta)$ is the normalizing constant. For $0 < \beta < 2$, the prior has a heavier tail than the Gaussian distribution ($\beta = 2$) and induces sparsity of the coefficients $\boldsymbol{w}$.

We apply the following inequality for $w, \xi \in \mathbb{R}$:

$$\left(w^2\right)^{\beta/2} \leq \frac{\beta}{2}\left(\xi^2\right)^{\frac{\beta}{2}-1}\left(w^2 - \xi^2\right),$$

which is obtained from the concavity of the function $f(x) = x^{\beta/2}$ for $x > 0$ and $0 < \beta < 2$. Introducing the variational parameter to each dimension, $\boldsymbol{\xi} = (\xi_1, \ldots, \xi_M)^\top$ and bounding the nonconjugate prior (5.10) by this inequality, we have

$$\begin{aligned} p(\boldsymbol{y}, \boldsymbol{w}|\boldsymbol{X}) &= p(\boldsymbol{y}|\boldsymbol{X}, \boldsymbol{w})p(\boldsymbol{w}) \\ &\geq \frac{1}{(2\pi\sigma^2)^{N/2} C_{\beta,\gamma}^M} \\ &\quad \cdot \exp\left(-\frac{1}{2\sigma^2}\sum_{n=1}^{N}(y^{(n)} - \boldsymbol{w}^\top \boldsymbol{x}^{(n)})^2 - \frac{\beta\gamma}{2}\sum_{m=1}^{M}\left(\xi_m^2\right)^{\frac{\beta}{2}-1}\left(w_m^2 - \xi_m^2\right)\right) \\ &\equiv \underline{p}(\boldsymbol{w}; \boldsymbol{\xi}). \end{aligned}$$

Normalizing the lower-bound $\underline{p}(\boldsymbol{w}; \boldsymbol{\xi})$, we obtain a Gaussian approximation to the posterior. This is in effect equivalent to assuming the Gaussian prior for $\boldsymbol{w}$:

$$\mathrm{Gauss}_M(\boldsymbol{w}; \boldsymbol{0}, \boldsymbol{S}_\xi^{-1}),$$

where $\boldsymbol{S}_\xi = \gamma\beta\mathbf{Diag}(\boldsymbol{\xi}^{\beta-1/2})$ for $\boldsymbol{\xi}^{\beta-1/2} \equiv \left(\left(\xi_1^2\right)^{\frac{\beta}{2}-1}, \ldots, \left(\xi_M^2\right)^{\frac{\beta}{2}-1}\right)^\top$.

The resulting Gaussian approximation to the posterior is

$$q_\xi(\boldsymbol{w}; \boldsymbol{\xi}) = \mathrm{Gauss}_M(\boldsymbol{w}; \underline{\boldsymbol{m}}, \underline{\boldsymbol{S}}^{-1}),$$

where

$$\underline{\boldsymbol{S}} = \boldsymbol{S}_\xi + \frac{1}{\sigma^2}\boldsymbol{X}^\top \boldsymbol{X}, \tag{5.11}$$

$$\underline{\boldsymbol{m}} = \frac{1}{\sigma^2}\underline{\boldsymbol{S}}^{-1}\boldsymbol{X}^\top \boldsymbol{y}. \tag{5.12}$$

**Algorithm 14** LVA algorithm for sparse linear regression.

1: Initialize the variational parameters $\{\xi_m^2\}_{m=1}^M$.
2: Compute the approximate posterior mean and covariance matrix $(\underline{m}, \underline{S}^{-1})$ by Eqs. (5.11) and (5.12).
3: Apply Eq. (5.14) to update $\{\xi_m^2\}_{m=1}^M$.
4: Evaluate the free energy bound (5.13).
5: Iterate Steps 2 through 4 until convergence (until the decrease of the bound becomes smaller than a threshold).

We also obtain the upper bound for the free energy:

$$\overline{F}(\xi) = -\log \int \underline{p}(w; \xi) dw$$
$$= \frac{N-M}{2}\log(2\pi) + \frac{\log|\underline{S}|}{2} + M \log C_{\beta,\gamma} - \sum_{n=1}^{N} \frac{(y^{(n)})^2}{2\sigma^2} + \frac{\gamma\beta}{2}\sum_{m=1}^{M}\left(\xi_m^2\right)^{\beta/2}, \tag{5.13}$$

which is optimized with respect to the variational parameter. The general framework in Section 5.3.2, which corresponds to the EM algorithm, provides the following update rule:

$$\xi_m^2 = \left\langle w_m^2 \right\rangle_{q_\xi(w;\xi_o)}$$
$$= (\underline{S}^{-1})_{mm} + \underline{m}_m^2, \tag{5.14}$$

where $\underline{m}$ and $\underline{S}^{-1}$ are the mean and covariance matrix of $q_\xi(w; \xi_o)$.

The learning algorithm for sparse linear regression with this approximation is summarized in Algorithm 14.

This approximation has been applied to the Laplace prior ($\beta = 1$) in Seeger (2008, 2009). LVA for another heavy-tailed distribution, $p(w) \propto \cosh^{-1/\beta}(\beta w)$, is discussed in Girolami (2001), which also bridges the Gaussian distribution ($\beta \to 0$) and the Laplace distribution ($\beta \to \infty$).

## 5.3 Unified Approach by Local VB Bounds

As discussed in Section 2.1.7, LVA for VB and LVA for EP form lower- and upper-bounds of the joint distribution $p(w, \mathcal{D})$, denoted by $\underline{p}(w; \xi)$ and $\overline{p}(w; \eta)$, respectively. If the bounds satisfying

$$\underline{p}(\boldsymbol{w};\boldsymbol{\xi}) \le p(\boldsymbol{w},\mathcal{D}), \tag{5.15}$$

$$\overline{p}(\boldsymbol{w};\boldsymbol{\eta}) \ge p(\boldsymbol{w},\mathcal{D}), \tag{5.16}$$

for all $\boldsymbol{w}$ and $\mathcal{D}$ are analytically integrable, then by normalizing the bounds instead of $p(\boldsymbol{w},\mathcal{D})$, LVAs approximate the posterior distribution by

$$q_{\xi}(\boldsymbol{w};\boldsymbol{\xi}) = \frac{\underline{p}(\boldsymbol{w};\boldsymbol{\xi})}{\underline{Z}(\boldsymbol{\xi})}, \tag{5.17}$$

$$q_{\eta}(\boldsymbol{w};\boldsymbol{\eta}) = \frac{\overline{p}(\boldsymbol{w};\boldsymbol{\eta})}{\overline{Z}(\boldsymbol{\eta})}, \tag{5.18}$$

respectively, where $\underline{Z}(\boldsymbol{\xi})$ and $\overline{Z}(\boldsymbol{\eta})$ are the normalization constants defined by

$$\underline{Z}(\boldsymbol{\xi}) = \int \underline{p}(\boldsymbol{w};\boldsymbol{\xi})d\boldsymbol{w},$$

$$\overline{Z}(\boldsymbol{\eta}) = \int \overline{p}(\boldsymbol{w};\boldsymbol{\eta})d\boldsymbol{w}.$$

Here $\boldsymbol{\xi}$ and $\boldsymbol{\eta}$ are called the variational parameters.

The respective approximations are optimized by estimating the variational parameters, $\boldsymbol{\xi}$ and $\boldsymbol{\eta}$ so that $\underline{Z}(\boldsymbol{\xi})$ is maximized and $\overline{Z}(\boldsymbol{\eta})$ is minimized since the inequalities

$$\underline{Z}(\boldsymbol{\xi}) \le Z \le \overline{Z}(\boldsymbol{\eta}) \tag{5.19}$$

hold by definition, where $Z = p(\mathcal{D})$ is the marginal likelihood.

To consider the respective LVAs in terms of information divergences in later sections, let us introduce the Bayes free energy,

$$F^{\mathrm{Bayes}} \equiv -\log Z,$$

and its lower- and upper-bounds, $\underline{F}(\boldsymbol{\eta}) = -\log \overline{Z}(\boldsymbol{\eta})$ and $\overline{F}(\boldsymbol{\xi}) = -\log \underline{Z}(\boldsymbol{\xi})$. By taking the negative logarithms on both sides of Eq. (5.19), we have

$$\underline{F}(\boldsymbol{\eta}) \le F^{\mathrm{Bayes}} \le \overline{F}(\boldsymbol{\xi}). \tag{5.20}$$

Hereafter, we follow the measure of the free energy and adopt the following terminology to refer to respective LVAs (5.18) and (5.17): the lower-bound maximization ($\underline{F}(\boldsymbol{\eta})$ maximization) and the upper-bound minimization ($\overline{F}(\boldsymbol{\xi})$ minimization).

### 5.3.1 Divergence Measures in LVA

Most of the existing LVA techniques are based on the convexity of the log-likelihood function or the log-prior (Jaakkola and Jordan, 2000; Bishop,

2006; Seeger, 2008, 2009). We describe these cases by using general convex functions, $\phi$ and $\psi$, and show that the objective functions,

$$\overline{F}(\xi) - F^{\text{Bayes}} = \log \frac{Z}{\underline{Z}(\xi)} \geq 0,$$

$$F^{\text{Bayes}} - \underline{F}(\eta) = \log \frac{\overline{Z}(\eta)}{Z} \geq 0,$$

to be minimized in the approximations (5.17) and (5.18), are decomposable into the sum of the KL divergence and the expected Bregman divergence.

Let $\phi$ and $\psi$ be twice differentiable real-valued strictly convex functions and denote by $d_\phi$ the Bregman divergence associated with the function $\phi$ (Banerjee et al., 2005):

$$d_\phi(\mathbf{v}_1, \mathbf{v}_2) = \phi(\mathbf{v}_1) - \phi(\mathbf{v}_2) - (\mathbf{v}_1 - \mathbf{v}_2)^\top \nabla\phi(\mathbf{v}_2) \geq 0, \tag{5.21}$$

where $\nabla\phi(\mathbf{v}_2)$ denotes the gradient vector of $\phi$ at $\mathbf{v}_2$.

Let us consider the case where $\phi$ and $\psi$ are respectively used to form the following bounds of the joint distribution $p(\mathbf{w}, \mathcal{D})$:

$$\underline{p}(\mathbf{w}; \boldsymbol{\xi}) = p(\mathbf{w}, \mathcal{D}) \exp\{-d_\phi(\boldsymbol{h}(\mathbf{w}), \boldsymbol{h}(\boldsymbol{\xi}))\}, \tag{5.22}$$

$$\overline{p}(\mathbf{w}; \boldsymbol{\eta}) = p(\mathbf{w}, \mathcal{D}) \exp\{d_\psi(\boldsymbol{g}(\mathbf{w}), \boldsymbol{g}(\boldsymbol{\eta}))\}, \tag{5.23}$$

where $\boldsymbol{h}$ and $\boldsymbol{g}$ are vector-valued functions of $\mathbf{w}$.[1]

Eq. (5.22) is interpreted as follows. $\log p(\mathbf{w}, \mathcal{D})$ includes a term that prevents analytic integration of $p(\mathbf{w}, \mathcal{D})$ with respect to $\mathbf{w}$. If such a term is expressed by the convex function $\phi$ of some function $\boldsymbol{h}$ transforming $\mathbf{w}$, it is replaced by the tangent hyperplane, $\phi(\boldsymbol{h}(\boldsymbol{\xi})) + (\boldsymbol{h}(\mathbf{w}) - \boldsymbol{h}(\boldsymbol{\xi}))^\top \nabla\phi(\boldsymbol{h}(\boldsymbol{\xi}))$, so that $\log \underline{p}(\mathbf{w}; \boldsymbol{\xi})$ makes a simpler function of $\mathbf{w}$, such as a quadratic function. Remember that if $\log \underline{p}(\mathbf{w}; \boldsymbol{\xi})$ is quadratic with respect to $\mathbf{w}$, $\underline{p}(\mathbf{w}; \boldsymbol{\xi})$ is analytically integrable by the Gaussian integral.

Rephrased in terms of the convex duality theory (Jordan et al., 1999; Bishop, 2006), $\phi(\boldsymbol{h}(\mathbf{w}))$ is replaced by its lower-bound,

$$\phi(\boldsymbol{h}(\mathbf{w})) \geq \phi(\boldsymbol{h}(\boldsymbol{\xi})) + (\boldsymbol{h}(\mathbf{w}) - \boldsymbol{h}(\boldsymbol{\xi}))^\top \nabla\phi(\boldsymbol{h}(\boldsymbol{\xi})) \tag{5.24}$$

$$= \boldsymbol{\theta}(\boldsymbol{\xi})^\top \boldsymbol{h}(\mathbf{w}) - \widetilde{\phi}(\boldsymbol{\theta}(\boldsymbol{\xi})), \tag{5.25}$$

where we have put $\boldsymbol{\theta}(\boldsymbol{\xi}) = \nabla\phi(\boldsymbol{h}(\boldsymbol{\xi}))$ and

$$\widetilde{\phi}(\boldsymbol{\theta}(\boldsymbol{\xi})) = \boldsymbol{\theta}(\boldsymbol{\xi})^\top \boldsymbol{h}(\boldsymbol{\xi}) - \phi(\boldsymbol{h}(\boldsymbol{\xi}))$$

$$= \max_{\boldsymbol{h}} \{\boldsymbol{\theta}(\boldsymbol{\xi})^\top \boldsymbol{h} - \phi(\boldsymbol{h})\}$$

[1] The functions $\boldsymbol{g}$ and $\boldsymbol{h}$ (also $\psi$ and $\phi$) can be dependent on $\mathcal{D}$ in this discussion. However, we denote them as if they were independent of $\mathcal{D}$ for simplicity. They are actually independent of $\mathcal{D}$ in the examples in Sections 5.1 and 5.2 and in most applications (Bishop, 2006; Seeger, 2008, 2009).

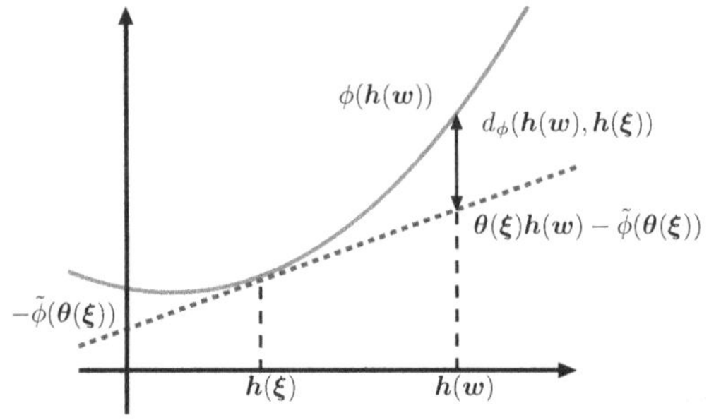


Figure 5.1 Convex function $\phi$ (solid curve), its tangent (dashed line), and the Bregman divergence (arrow).

is the conjugate function of $\phi$. The inequality (5.24) indicates the fact that the convex function $\phi$ is bounded globally by its tangent at $\boldsymbol{h}(\boldsymbol{\xi})$, which is equivalent to the nonnegativity of the Bregman divergence. In Eq. (5.25), the tangent is reparameterized by $\boldsymbol{\theta}(\boldsymbol{\xi})$, its gradient, instead of the contact point $\boldsymbol{h}(\boldsymbol{\xi})$, and its offset is given by $-\widetilde{\phi}(\boldsymbol{\theta}(\boldsymbol{\xi}))$. Figure 5.1 illustrates the relationship among the convex function $\phi$, its lower-bound, and the Bregman divergence.

We now describe the free energy bounds $\overline{F}(\boldsymbol{\xi})$ and $\underline{F}(\boldsymbol{\eta})$ in terms of information divergences. It follows from the definition (5.17) of the approximate posterior distribution that

$$\begin{aligned}\mathrm{KL}(q_\xi(\boldsymbol{w};\boldsymbol{\xi})\|p(\boldsymbol{w}|\mathcal{D})) &= \int q_\xi(\boldsymbol{w};\boldsymbol{\xi}) \log \frac{Z\underline{p}(\boldsymbol{w};\boldsymbol{\xi})}{\underline{Z}(\boldsymbol{\xi})p(\boldsymbol{w},\mathcal{D})} d\boldsymbol{w} \\ &= \log \frac{Z}{\underline{Z}(\boldsymbol{\xi})} - \left\langle d_\phi(\boldsymbol{h}(\boldsymbol{w}), \boldsymbol{h}(\boldsymbol{\xi}))\right\rangle_{q_\xi(\boldsymbol{w};\boldsymbol{\xi})}.\end{aligned}$$

We have a similar decomposition for $\mathrm{KL}(p(\boldsymbol{w}|\mathcal{D})\|q_\eta(\boldsymbol{w};\boldsymbol{\eta}))$ as well. Finally, we obtain the following expressions:[2]

$$\overline{F}(\boldsymbol{\xi}) - F^{\mathrm{Bayes}} = \left\langle d_\phi(\boldsymbol{h}(\boldsymbol{w}), \boldsymbol{h}(\boldsymbol{\xi}))\right\rangle_{q_\xi} + \mathrm{KL}(q_\xi\|p), \tag{5.26}$$

$$F^{\mathrm{Bayes}} - \underline{F}(\boldsymbol{\eta}) = \left\langle d_\psi(\boldsymbol{g}(\boldsymbol{w}), \boldsymbol{g}(\boldsymbol{\eta}))\right\rangle_{p} + \mathrm{KL}(p\|q_\eta). \tag{5.27}$$

Recall that $F^{\mathrm{Bayes}} + \mathrm{KL}(q_\xi\|p) = F$ is the free energy, which is further bounded by $\overline{F}(\widehat{\boldsymbol{\lambda}}, \boldsymbol{\xi})$ in Eq. (2.25). The expression (5.26) shows that the gap between $\min_{\widehat{\boldsymbol{\lambda}}} \overline{F}(\widehat{\boldsymbol{\lambda}}, \boldsymbol{\xi})$ and $F$ is the expected Bregman divergence $\left\langle d_\phi(\boldsymbol{h}(\boldsymbol{w}), \boldsymbol{h}(\boldsymbol{\xi}))\right\rangle_{q_\xi}$. Recall also that $F^{\mathrm{Bayes}} - \mathrm{KL}(p\|q_\eta) = E$ is the objective function of the EP problem (2.46) and that $\underline{F}(\boldsymbol{\eta}) = -\log \overline{Z}(\boldsymbol{\eta})$ is obtained as the maximum of its lower-bound, $\max_{\widehat{\boldsymbol{\nu}}} \underline{E}(\widehat{\boldsymbol{\nu}}, \boldsymbol{\eta})$ in Eq. (2.59), under a monotonic transformation.

[2] Hereafter in this section, we omit the notation "$(\boldsymbol{w}|\mathcal{D})$" if no confusion is likely.

The expression (5.27) shows that the gap between $E$ and $\max_{\widehat{\boldsymbol{v}}} \underline{E}(\widehat{\boldsymbol{v}}, \boldsymbol{\eta})$ is expressed by the expected Bregman divergence $\left\langle d_{\psi}(\boldsymbol{g}(\boldsymbol{w}), \boldsymbol{g}(\boldsymbol{\eta}))\right\rangle_p$ while the expectation is taken with respect to the true posterior.

Similarly, we also have the following decompositions:

$$\overline{F}(\boldsymbol{\xi}) - F^{\text{Bayes}} = \left\langle d_{\phi}(\boldsymbol{h}(\boldsymbol{w}), \boldsymbol{h}(\boldsymbol{\xi}))\right\rangle_p - \text{KL}(p\|q_{\xi}), \tag{5.28}$$

and

$$F^{\text{Bayes}} - \underline{F}(\boldsymbol{\eta}) = \left\langle d_{\psi}(\boldsymbol{g}(\boldsymbol{w}), \boldsymbol{g}(\boldsymbol{\eta}))\right\rangle_{q_\eta} - \text{KL}(q_\eta\|p).$$

Unlike Eqs. (5.26) and (5.27), the KL divergence is subtracted in these expressions. This again implies the affinities of LVAs by $\overline{F}$ minimization and $\underline{F}$ maximization to VB and EP, respectively.

### 5.3.2 Optimization of Approximations

In this subsection, we show that the conditions for the optimal variational parameters are generally given by the moment matching with respect to $\boldsymbol{h}(\boldsymbol{\xi})$ and $\boldsymbol{g}(\boldsymbol{\eta})$.

#### Optimal Variational Parameters

From Eqs. (5.22) and (5.23), we can see that the approximate posteriors,

$$\begin{aligned} q_{\xi}(\boldsymbol{w}; \boldsymbol{\xi}) &\propto p(\boldsymbol{w}, \mathcal{D}) \exp\{\boldsymbol{h}(\boldsymbol{w})^{\top} \nabla\phi(\boldsymbol{h}(\boldsymbol{\xi})) - \phi(\boldsymbol{h}(\boldsymbol{w}))\} \\ q_{\eta}(\boldsymbol{w}; \boldsymbol{\eta}) &\propto p(\boldsymbol{w}, \mathcal{D}) \exp\{-\boldsymbol{g}(\boldsymbol{w})^{\top} \nabla\psi(\boldsymbol{g}(\boldsymbol{\eta})) + \psi(\boldsymbol{g}(\boldsymbol{w}))\}, \end{aligned} \tag{5.29}$$

are members of the exponential family with natural parameters $\nabla\phi(\boldsymbol{h}(\boldsymbol{\xi}))$ and $\nabla\psi(\boldsymbol{g}(\boldsymbol{\eta}))$ (Section 1.2.3). Let

$$\boldsymbol{\theta}(\boldsymbol{\xi}) = \nabla\phi(\boldsymbol{h}(\boldsymbol{\xi})) \quad \text{and} \quad \boldsymbol{\kappa}(\boldsymbol{\eta}) = \nabla\psi(\boldsymbol{g}(\boldsymbol{\eta})).$$

The variational parameters are optimized so that $\overline{F}(\boldsymbol{\xi})$ is minimized and $\underline{F}(\boldsymbol{\eta})$ is maximized, respectively. In practice, however, they can be optimized directly with respect to $\boldsymbol{h}(\boldsymbol{\xi})$ and $\boldsymbol{g}(\boldsymbol{\eta})$ instead of $\boldsymbol{\xi}$ and $\boldsymbol{\eta}$. Applications of LVA, storing $\boldsymbol{h}(\boldsymbol{\xi})$ and $\boldsymbol{g}(\boldsymbol{\eta})$ as parameters, do not require $\boldsymbol{\xi}$ and $\boldsymbol{\eta}$ explicitly. Furthermore, we consider the gradient vectors of the free energy bounds with respect to $\boldsymbol{\theta}(\boldsymbol{\xi})$ and $\boldsymbol{\kappa}(\boldsymbol{\eta})$, which have one-to-one correspondence with $\boldsymbol{h}(\boldsymbol{\xi})$ and $\boldsymbol{g}(\boldsymbol{\eta})$, because they provide simple expressions of the gradient vectors. For notational simplicity, we drop the dependencies on $\boldsymbol{\xi}$ and $\boldsymbol{\eta}$ and denote as $\boldsymbol{\theta}$ and $\boldsymbol{\kappa}$.

The gradient of the upper bound with respect to $\boldsymbol{\theta}$ is given by[3]

$$\begin{aligned}
\nabla_{\boldsymbol{\theta}} \overline{F}(\boldsymbol{\xi}) &= \nabla_{\boldsymbol{\theta}} \left\{ -\log \int \underline{p}(\boldsymbol{w}; \boldsymbol{\xi}) d\boldsymbol{w} \right\} \\
&= -\int \frac{1}{\underline{Z}(\boldsymbol{\xi})} \frac{\partial \underline{p}(\boldsymbol{w}; \boldsymbol{\xi})}{\partial \boldsymbol{\theta}} d\boldsymbol{w} \\
&= -\frac{\partial \boldsymbol{h}(\boldsymbol{\xi})}{\partial \boldsymbol{\theta}} \int \frac{1}{\underline{Z}(\boldsymbol{\xi})} \frac{\partial \underline{p}(\boldsymbol{w}; \boldsymbol{\xi})}{\partial \boldsymbol{h}(\boldsymbol{\xi})} d\boldsymbol{w} \\
&= -\frac{\partial \boldsymbol{h}(\boldsymbol{\xi})}{\partial \boldsymbol{\theta}} \int \frac{\partial^2 \phi(\boldsymbol{h}(\boldsymbol{\xi}))}{\partial \boldsymbol{h} \partial \boldsymbol{h}^\top} (\boldsymbol{h}(\boldsymbol{w}) - \boldsymbol{h}(\boldsymbol{\xi})) q_{\xi}(\boldsymbol{w}; \boldsymbol{\xi}) d\boldsymbol{w} \\
&= -(\langle \boldsymbol{h}(\boldsymbol{w}) \rangle_{q_{\xi}} - \boldsymbol{h}(\boldsymbol{\xi})),
\end{aligned} \tag{5.30}$$

where we have used Eq. (5.22) and the fact that the matrix $\frac{\partial \boldsymbol{h}(\boldsymbol{\xi})}{\partial \boldsymbol{\theta}}$, whose $(i, j)$th entry is $\frac{\partial h_i(\boldsymbol{\xi})}{\partial \theta_j}$, is the inverse of the Hessian matrix $\frac{\partial^2 \phi(\boldsymbol{h}(\boldsymbol{\xi}))}{\partial \boldsymbol{h} \partial \boldsymbol{h}^\top}$. Similarly, we obtain

$$\nabla_{\boldsymbol{\kappa}} \underline{F}(\boldsymbol{\eta}) = \langle \boldsymbol{g}(\boldsymbol{w}) \rangle_{q_{\eta}} - \boldsymbol{g}(\boldsymbol{\eta}). \tag{5.31}$$

Hence, we can utilize gradient methods to minimize $\overline{F}(\boldsymbol{\xi})$ and maximize $\underline{F}(\boldsymbol{\eta})$. We can see that when $\boldsymbol{\xi}$ and $\boldsymbol{\eta}$ are optimized,

$$\boldsymbol{h}(\boldsymbol{\xi}) = \langle \boldsymbol{h}(\boldsymbol{w}) \rangle_{q_{\xi}} \quad \text{and} \quad \boldsymbol{g}(\boldsymbol{\eta}) = \langle \boldsymbol{g}(\boldsymbol{w}) \rangle_{q_{\eta}}$$

hold.

In practice, the variational parameter $\boldsymbol{h}(\boldsymbol{\xi})$ is iteratively updated so that $\overline{F}(\boldsymbol{\xi})$ is monotonically decreased. Recall the argument in Section 2.1.7 where LVA for VB was formulated as the joint minimization of $\overline{F}(\widehat{\boldsymbol{\lambda}}, \boldsymbol{\xi})$ over the approximate posterior $r(\boldsymbol{w}; \widehat{\boldsymbol{\lambda}})$ and $\boldsymbol{\xi}$. The free energy bound $\overline{F}(\boldsymbol{\xi}) = -\log \underline{Z}(\boldsymbol{\xi})$ was obtained as the minimum of $\overline{F}(\widehat{\boldsymbol{\lambda}}, \boldsymbol{\xi})$, which is attained by (see Eq. (2.32))

$$r(\boldsymbol{w}; \widehat{\boldsymbol{\lambda}}) = q(\boldsymbol{w}; \boldsymbol{\xi}).$$

Let $\boldsymbol{h}(\boldsymbol{\xi}_{\mathrm{o}})$ be a current estimate of $\boldsymbol{h}(\boldsymbol{\xi})$ and $\widehat{\boldsymbol{\lambda}}_{\mathrm{o}}$ be the variational parameter such that $r(\boldsymbol{w}; \widehat{\boldsymbol{\lambda}}_{\mathrm{o}}) = q(\boldsymbol{w}; \boldsymbol{\xi}_{\mathrm{o}})$. Then, updating $\boldsymbol{h}(\boldsymbol{\xi})$ to $\mathrm{argmin}_{\boldsymbol{h}(\boldsymbol{\xi})} \overline{F}(\widehat{\boldsymbol{\lambda}}_{\mathrm{o}}, \boldsymbol{\xi})$ decreases $\overline{F}(\boldsymbol{\xi})$ because

$$\overline{F}(\widehat{\boldsymbol{\lambda}}_{\mathrm{o}}, \boldsymbol{\xi}) \geq \overline{F}(\boldsymbol{\xi})$$

for all $\boldsymbol{\xi}$ and the equality holds when $\boldsymbol{\xi} = \boldsymbol{\xi}_{\mathrm{o}}$. More specifically, it follows for $\boldsymbol{h}(\widehat{\boldsymbol{\xi}}) = \mathrm{argmin}_{\boldsymbol{h}(\boldsymbol{\xi})} \overline{F}(\widehat{\boldsymbol{\lambda}}_{\mathrm{o}}, \boldsymbol{\xi})$ that

$$\overline{F}(\widehat{\boldsymbol{\xi}}) \leq \overline{F}(\widehat{\boldsymbol{\lambda}}_{\mathrm{o}}, \widehat{\boldsymbol{\xi}}) \leq \overline{F}(\widehat{\boldsymbol{\lambda}}_{\mathrm{o}}, \boldsymbol{\xi}_{\mathrm{o}}) = \overline{F}(\boldsymbol{\xi}_{\mathrm{o}}), \tag{5.32}$$

[3] We henceforth use the operator $\nabla$ with the subscript expressing for which variable the gradient is taken. That is, for a function $f(\boldsymbol{\theta})$, $\nabla_{\boldsymbol{\theta}} f(\boldsymbol{\theta}) = \frac{\partial f(\boldsymbol{\theta})}{\partial \boldsymbol{\theta}}$ denotes the vector whose $i$th element is $\frac{\partial f(\boldsymbol{\theta})}{\partial \theta_i}$.

which means that the bound is improved. This corresponds to the EM algorithm to decrease $\overline{F}(\boldsymbol{\xi})$ and yields the following specific update rule of $\boldsymbol{h}(\boldsymbol{\xi})$:

$$\begin{aligned}\boldsymbol{h}(\widehat{\boldsymbol{\xi}}) &= \underset{\boldsymbol{h}(\boldsymbol{\xi})}{\operatorname{argmin}}\, \overline{F}(\widehat{\boldsymbol{\lambda}}_{\text{o}}, \widehat{\boldsymbol{\xi}}) \\ &= \underset{\boldsymbol{h}(\boldsymbol{\xi})}{\operatorname{argmin}} \left\langle - \log \underline{p}(\boldsymbol{w}; \boldsymbol{\xi}) \right\rangle_{q_{\xi}(\boldsymbol{w}; \boldsymbol{\xi}_{\text{o}})} && (5.33) \\ &= \underset{\boldsymbol{h}(\boldsymbol{\xi})}{\operatorname{argmin}} \left\langle d_{\phi}(\boldsymbol{h}(\boldsymbol{w}), \boldsymbol{h}(\boldsymbol{\xi})) \right\rangle_{q_{\xi}(\boldsymbol{w}; \boldsymbol{\xi}_{\text{o}})} && (5.34) \\ &= \underset{\boldsymbol{h}(\boldsymbol{\xi})}{\operatorname{argmin}}\, d_{\phi}(\langle \boldsymbol{h}(\boldsymbol{w}) \rangle_{q_{q_{\xi}(\boldsymbol{w}; \boldsymbol{\xi}_{\text{o}})}}, \boldsymbol{h}(\boldsymbol{\xi})) && (5.35) \\ &= \langle \boldsymbol{h}(\boldsymbol{w}) \rangle_{q_{\xi}(\boldsymbol{w}; \boldsymbol{\xi}_{\text{o}})}, && (5.36)\end{aligned}$$

which is summarized as

$$\boldsymbol{h}(\widehat{\boldsymbol{\xi}}) = \langle \boldsymbol{h}(\boldsymbol{w}) \rangle_{q_{\xi}(\boldsymbol{w}; \boldsymbol{\xi}_{\text{o}})}. \tag{5.37}$$

The preceding lines of equations are basically derived by focusing on the terms depending on $\boldsymbol{h}(\boldsymbol{\xi})$. Eq. (5.33) follows from the definition of $\overline{F}(\widehat{\boldsymbol{\lambda}}_{\text{o}}, \widehat{\boldsymbol{\xi}})$ by Eq. (2.25). Eq. (5.34) follows from the definition of $\underline{p}(\boldsymbol{w}; \boldsymbol{\xi})$ by Eq. (5.15). Eq. (5.35) follows from the definition of the Bregman divergence (5.21) and the linearity of expectation. Eq. (5.36) follows from the nonnegativity of the Bregman divergence. Eqs. (5.34) through (5.36) are equivalent to the fact that the expected Bregman divergence is minimized by the mean (Banerjee et al., 2005).

The update rule (5.37) means that $\boldsymbol{h}(\boldsymbol{\xi})$ is updated to the expectation of $\boldsymbol{h}(\boldsymbol{w})$ with respect to the approximate posterior. Note here again that if we store $\boldsymbol{h}(\boldsymbol{\xi})$, $\boldsymbol{\xi}$ is not explicitly required.

The update rule (5.37) is an iterative substitution of $\boldsymbol{h}(\boldsymbol{\xi})$. To maximize the lower-bound $\underline{F}(\boldsymbol{\eta})$ in LVA for EP, such a simple update rule is not applicable in general. Thus, gradient-based optimization methods with the gradient (5.31) are usually used. The Newton–Raphson step to update $\boldsymbol{\kappa}$ is given by

$$\boldsymbol{\kappa}^{\text{new}} = \boldsymbol{\kappa}^{\text{old}} - \left(\nabla^2_{\boldsymbol{\kappa}} \underline{F}(\boldsymbol{\eta}^{\text{old}})\right)^{-1} \nabla_{\boldsymbol{\kappa}} \underline{F}(\boldsymbol{\eta}^{\text{old}}), \tag{5.38}$$

where the Hessian matrix is given as follows:

$$\begin{aligned}\nabla^2_{\boldsymbol{\kappa}} \underline{F}(\boldsymbol{\eta}) &= \frac{\partial^2 \underline{F}(\boldsymbol{\eta})}{\partial \boldsymbol{\kappa} \partial \boldsymbol{\kappa}^{\top}} \\ &= -\mathbf{Cov}(\boldsymbol{g}(\boldsymbol{w})) - \frac{\partial \boldsymbol{g}(\boldsymbol{\eta})}{\partial \boldsymbol{\kappa}} && (5.39)\end{aligned}$$

for the covariance matrix of $\boldsymbol{g}(\boldsymbol{w})$,

$$\mathbf{Cov}(\boldsymbol{g}(\boldsymbol{w})) = \left\langle \boldsymbol{g}(\boldsymbol{w}) \boldsymbol{g}(\boldsymbol{w})^{\top} \right\rangle_{q_{\eta}} - \langle \boldsymbol{g}(\boldsymbol{w}) \rangle_{q_{\eta}} \langle \boldsymbol{g}(\boldsymbol{w}) \rangle_{q_{\eta}}^{\top},$$

and $\frac{\partial \boldsymbol{g}(\boldsymbol{\eta})}{\partial \boldsymbol{\kappa}} = \left( \frac{\partial^2 \psi(\boldsymbol{g}(\boldsymbol{\eta}))}{\partial \boldsymbol{g} \partial \boldsymbol{g}^{\top}} \right)^{-1}$ holds in Eq. (5.39).

### 5.3.3 An Alternative View of VB for Latent Variable Models

In this subsection, we show that the VB learning for latent variable models can be viewed as a special case of LVA, where the log-sum-exp function is used to form the lower-bound of the log-likelihood (Jordan et al., 1999).

Let $\mathcal{H}$ be a vector of latent (unobserved) variables and consider the latent variable model,

$$p(\mathcal{D}, \boldsymbol{w}) = \sum_{\mathcal{H}} p(\mathcal{D}, \mathcal{H}, \boldsymbol{w}),$$

where $\sum_{\mathcal{H}}$ denotes the summation over all possible realizations of the latent variables. We have used the notation as if $\mathcal{H}$ were discrete in order to include examples such as GMMs and HMMs, where the likelihood function is given by $p(\mathcal{D}|\boldsymbol{w}) = \sum_{\mathcal{H}} p(\mathcal{D}, \mathcal{H}|\boldsymbol{w})$. In the case of a model with continuous latent variables, the summation $\sum_{\mathcal{H}}$ is simply replaced by the integration $\int d\mathcal{H}$. This includes, for example, the hierarchical prior distribution presented in Tipping (2001), where the prior distribution is defined by $p(\boldsymbol{w}) = \int p(\boldsymbol{w}|\mathcal{H})p(\mathcal{H})d\mathcal{H}$ with the hyperprior $p(\mathcal{H})$.

The Bayesian posterior distribution of the latent variables and the parameter $\boldsymbol{w}$ is

$$p(\mathcal{H}, \boldsymbol{w}|\mathcal{D}) = \frac{p(\mathcal{D}, \mathcal{H}, \boldsymbol{w})}{\sum_{\mathcal{H}} \int p(\mathcal{D}, \mathcal{H}, \boldsymbol{w}) d\boldsymbol{w}},$$

which is intractable when $Z = \sum_{\mathcal{H}} \int p(\mathcal{D}, \mathcal{H}, \boldsymbol{w}) d\boldsymbol{w}$ requires summation over exponentially many terms as in GMMs and HMMs or the analytically intractable integration. So is the posterior of the parameter $p(\boldsymbol{w}|\mathcal{D})$.

Let us consider an application of the local variational method for approximating $p(\boldsymbol{w}|\mathcal{D})$. By the convexity of the function $\log \sum_{\mathcal{H}} \exp(\cdot)$, the log-joint distribution is lower-bounded as follows:

$$\begin{aligned}
\log p(\mathcal{D}, \boldsymbol{w}) &= \log \sum_{\mathcal{H}} \exp\{\log p(\mathcal{D}, \mathcal{H}, \boldsymbol{w})\} \\
&\geq \log p(\mathcal{D}, \boldsymbol{\xi}) + \sum_{\mathcal{H}} \left( \log \frac{p(\mathcal{D}, \mathcal{H}, \boldsymbol{w})}{p(\mathcal{D}, \mathcal{H}, \boldsymbol{\xi})} \right) p(\mathcal{H}|\mathcal{D}, \boldsymbol{\xi}) \\
&= \log p(\mathcal{D}, \boldsymbol{w}) - \sum_{\mathcal{H}} p(\mathcal{H}|\mathcal{D}, \boldsymbol{\xi}) \log \frac{p(\mathcal{H}|\mathcal{D}, \boldsymbol{\xi})}{p(\mathcal{H}|\mathcal{D}, \boldsymbol{w})}, \qquad (5.40)
\end{aligned}$$

where $p(\mathcal{H}|\mathcal{D}, \boldsymbol{\xi}) = \frac{p(\mathcal{D}, \mathcal{H}, \boldsymbol{\xi})}{\sum_{\mathcal{H}} p(\mathcal{D}, \mathcal{H}, \boldsymbol{\xi})}$. This corresponds to the case where $\phi(\boldsymbol{h}) = \log \sum_n \exp(h_n)$ and $\boldsymbol{h}(\boldsymbol{w})$ is the vector-valued function that consists of the elements $\log p(\mathcal{D}, \mathcal{H}, \boldsymbol{w})$ for all possible $\mathcal{H}$. The vector $\boldsymbol{h}$ is infinite dimensional when $\mathcal{H}$ is continuous. Taking exponentials of the most right-hand side and left-hand side of Inequality (5.40) leads to Eqs. (5.22) and (5.15) with the Bregman divergence,

$$\begin{aligned} d_\phi(\boldsymbol{h}(\boldsymbol{w}), \boldsymbol{h}(\boldsymbol{\xi})) &= \sum_{\mathcal{H}} p(\mathcal{H}|\mathcal{D}, \boldsymbol{\xi}) \log \frac{p(\mathcal{H}|\mathcal{D}, \boldsymbol{\xi})}{p(\mathcal{H}|\mathcal{D}, \boldsymbol{w})} \\ &= \mathrm{KL}(p(\mathcal{H}|\mathcal{D}, \boldsymbol{\xi}) \| p(\mathcal{H}|\mathcal{D}, \boldsymbol{w})). \end{aligned}$$

From Eq. (5.26), we have

$$\begin{aligned} \overline{F}(\boldsymbol{\xi}) &= F^{\mathrm{Bayes}} + \mathrm{KL}(q_\xi(\boldsymbol{w}; \boldsymbol{\xi}) \| p(\boldsymbol{w}|\mathcal{D})) + \langle \mathrm{KL}(p(\mathcal{H}|\mathcal{D}, \boldsymbol{\xi}) \| p(\mathcal{H}|\mathcal{D}, \boldsymbol{w})) \rangle_{q_\xi(\boldsymbol{w};\boldsymbol{\xi})} \\ &= F^{\mathrm{Bayes}} + \mathrm{KL}(q_\xi(\boldsymbol{w}; \boldsymbol{\xi}) p(\mathcal{H}|\mathcal{D}, \boldsymbol{\xi}) \| p(\boldsymbol{w}, \mathcal{H}|\mathcal{D})), \end{aligned}$$

which is exactly the free energy of the factorized distribution $q_\xi(\boldsymbol{w}; \boldsymbol{\xi}) p(\mathcal{H}|\mathcal{D}, \boldsymbol{\xi})$. In fact, from Eqs. (5.29) and (5.40), the approximating posterior is given by

$$\begin{aligned} q_\xi(\boldsymbol{w}; \boldsymbol{\xi}) &\propto \exp\left\{ \sum_{\mathcal{H}} \log p(\mathcal{D}, \mathcal{H}, \boldsymbol{w}) p(\mathcal{H}|\mathcal{D}, \boldsymbol{\xi}) \right\} \\ &= \exp \left\langle \log p(\mathcal{D}, \mathcal{H}, \boldsymbol{w}) \right\rangle_{p(\mathcal{H}|\mathcal{D}, \boldsymbol{\xi})}. \end{aligned} \tag{5.41}$$

From Eq. (5.37), the EM update for the variational parameters $\boldsymbol{\xi}$ yields

$$\begin{aligned} &\log p(\mathcal{D}, \mathcal{H}, \boldsymbol{\xi}) = \langle \log p(\mathcal{D}, \mathcal{H}, \boldsymbol{w}) \rangle_{q_\xi(\boldsymbol{w}; \boldsymbol{\xi}_o)} \\ &\Rightarrow p(\mathcal{H}|\mathcal{D}, \boldsymbol{\xi}) \propto \exp \left\langle \log p(\mathcal{D}, \mathcal{H}, \boldsymbol{w}) \right\rangle_{q_\xi(\boldsymbol{w}; \boldsymbol{\xi}_o)}. \end{aligned} \tag{5.42}$$

Eqs. (5.41) and (5.42) are exactly the same as the VB algorithm for minimizing the free energy over the factorized distributions, Eqs. (4.2) and (4.3). In this example, we no longer have $\boldsymbol{\xi}$ satisfying Eq. (5.42) in general. However, if the model $p(\mathcal{H}, \mathcal{D}|\boldsymbol{w})$ and the prior $p(\boldsymbol{w})$ are included in the exponential family, $\boldsymbol{h}(\boldsymbol{\xi})$ as well as $p(\mathcal{H}|\mathcal{D}, \boldsymbol{\xi})$ and $q_\xi(\boldsymbol{w}; \boldsymbol{\xi})$ are expressed by expected sufficient statistics, the number of which is equal to the dimensionality of $\boldsymbol{w}$ (Beal, 2003). In that case, it is not necessary to obtain $\boldsymbol{\xi}$ explicitly but only to store and update the expected sufficient statistics instead.

# Part III

# Nonasymptotic Theory

# 6
# Global VB Solution of Fully Observed Matrix Factorization

Variational Bayesian (VB) learning has shown good performance in many applications. However, VB learning sometimes gives a seemingly different posterior and exhibits different sparsity behavior from full Bayesian learning. For example, Figure 6.1 compares the Bayes posterior (left) and the VB posterior (right) of $1 \times 1$ matrix factorization. VB posterior tries to approximate a *two-mode* Bayes posterior with a single-mode Gaussian, which results in the zero-mean Gaussian posterior with the VB estimator $\widehat{BA} = 0$. This behavior makes the VB estimator *exactly* sparse as shown in Figure 6.2: thresholding is observed for the VB estimator, while no thresholding is observed for the full Bayesian estimator. Mackay (2001) discussed the sparsity of VB learning as an artifact by showing *inappropriate* model pruning in mixture models. These facts might deprive the justification of VB learning based solely on the fact that it is a tractable approximation to Bayesian learning. Can we clarify the behavior of VB learning, and directly justify its use as an inference method? The nonasymptotic theory, introduced in Part III, gives some answer to this question.

In this chapter, we derive an analytic-form of the *global* VB solution of fully observed matrix factorization (MF). The analytic-form solution allows us to make intuitive discussion on the behavior of VB learning (Chapter 7), and further analysis gives theoretical guarantees of the performance of VB learning (Chapter 8). The analytic-form global solution naturally leads to efficient and reliable algorithms (Chapter 9), which are extended to other similar models (Chapters 10 and 11). Relation to MAP learning and partially Bayesian learning is also theoretically investigated (Chapter 12).

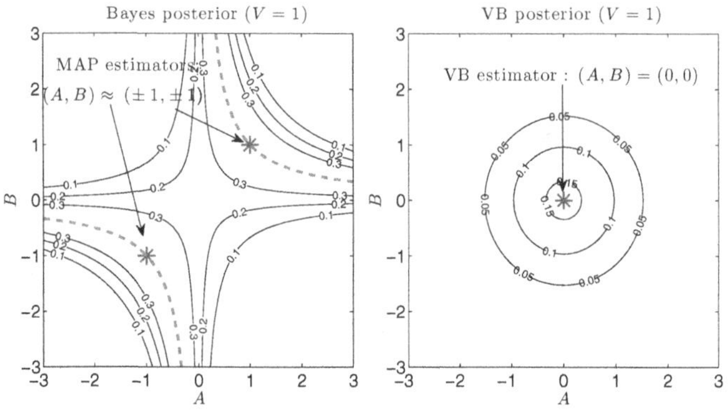


Figure 6.1 The Bayes posterior (left) and the VB posterior (right) of the $1 \times 1$ MF model $V = BA + \mathcal{E}$ with almost flat prior, when $V = 1$ is observed ($\mathcal{E}$ is the standard Gaussian noise). VB approximates the Bayes posterior having two modes by an origin-centered Gaussian, which induces sparsity.

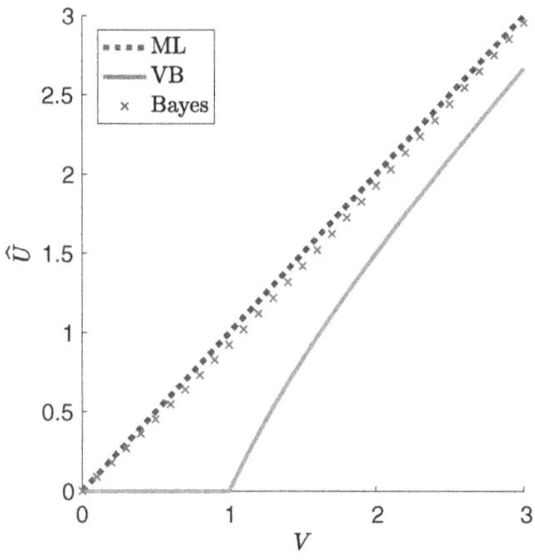


Figure 6.2 Behavior of the estimators of $\widehat{U} = \widehat{BA}$ as a function of the observed value $V$. The VB estimator is zero when $V \leq 1$, which indicates *exact* sparsity. On the other hand, the Bayesian estimator shows no sign of sparsity. The maximum likelihood estimator, i.e., $\widehat{U} = V$, is shown as a reference.

## 6.1 Problem Description

We first summarize the MF model and its VB learning algorithm, which was derived in Section 3.1. The likelihood and priors are given as

$$p(\boldsymbol{V}|\boldsymbol{A},\boldsymbol{B}) \propto \exp\left(-\frac{1}{2\sigma^2}\left\|\boldsymbol{V}-\boldsymbol{B}\boldsymbol{A}^\top\right\|_{\mathrm{Fro}}^2\right), \tag{6.1}$$

$$p(\boldsymbol{A}) \propto \exp\left(-\frac{1}{2}\mathrm{tr}\left(\boldsymbol{A}\boldsymbol{C}_A^{-1}\boldsymbol{A}^\top\right)\right), \tag{6.2}$$

$$p(\boldsymbol{B}) \propto \exp\left(-\frac{1}{2}\mathrm{tr}\left(\boldsymbol{B}\boldsymbol{C}_B^{-1}\boldsymbol{B}^\top\right)\right), \tag{6.3}$$

where the prior covariances are restricted to be diagonal:

$$\boldsymbol{C}_A = \mathbf{Diag}(c_{a_1}^2, \ldots, c_{a_H}^2),$$
$$\boldsymbol{C}_B = \mathbf{Diag}(c_{b_1}^2, \ldots, c_{b_H}^2),$$

for $c_{a_h}, c_{b_h} > 0, h = 1, \ldots, H$. Without loss of generality, we assume that the diagonal entries of the product $\boldsymbol{C}_A\boldsymbol{C}_B$ are arranged in nonincreasing order, i.e., $c_{a_h}c_{b_h} \geq c_{a_{h'}}c_{b_{h'}}$ for any pair $h < h'$. We assume that

$$L \leq M. \tag{6.4}$$

If $L > M$, we may simply redefine the transpose $\boldsymbol{V}^\top$ as $\boldsymbol{V}$ so that $L \leq M$ holds. Therefore, the assumption (6.4) does not impose any restriction.

We solve the following VB learning problem:

$$\widehat{r} = \underset{r}{\mathrm{argmin}}\, F(r) \quad \text{s.t.} \quad r(\boldsymbol{A}, \boldsymbol{B}) = r_A(\boldsymbol{A})r_B(\boldsymbol{B}), \tag{6.5}$$

where the objective function to be minimized is the free energy:

$$F = \left\langle \log \frac{r_A(\boldsymbol{A})r_B(\boldsymbol{B})}{p(\boldsymbol{V}|\boldsymbol{A}, \boldsymbol{B})p(\boldsymbol{A})p(\boldsymbol{B})} \right\rangle_{r_A(\boldsymbol{A})r_B(\boldsymbol{B})}.$$

The solution to the problem (6.5) is in the following form:

$$r_A(\boldsymbol{A}) = \mathrm{MGauss}_{M,H}(\boldsymbol{A}; \widehat{\boldsymbol{A}}, \boldsymbol{I}_M \otimes \widehat{\boldsymbol{\Sigma}}_A) \propto \exp\left(-\frac{\mathrm{tr}\left((\boldsymbol{A} - \widehat{\boldsymbol{A}})\widehat{\boldsymbol{\Sigma}}_A^{-1}(\boldsymbol{A} - \widehat{\boldsymbol{A}})^\top\right)}{2}\right), \tag{6.6}$$

$$r_B(\boldsymbol{B}) = \mathrm{MGauss}_{L,H}(\boldsymbol{B}; \widehat{\boldsymbol{B}}, \boldsymbol{I}_L \otimes \widehat{\boldsymbol{\Sigma}}_B) \propto \exp\left(-\frac{\mathrm{tr}\left((\boldsymbol{B} - \widehat{\boldsymbol{B}})\widehat{\boldsymbol{\Sigma}}_B^{-1}(\boldsymbol{B} - \widehat{\boldsymbol{B}})^\top\right)}{2}\right). \tag{6.7}$$

With the variational parameters $\widehat{\boldsymbol{A}}, \widehat{\boldsymbol{\Sigma}}_A, \widehat{\boldsymbol{B}}, \widehat{\boldsymbol{\Sigma}}_B$, the free energy can be explicitly written as

$$\begin{aligned}
2F = {} & LM\log(2\pi\sigma^2) + \frac{\left\|\boldsymbol{V} - \widehat{\boldsymbol{B}}\widehat{\boldsymbol{A}}^\top\right\|_{\mathrm{Fro}}^2}{\sigma^2} + M\log\frac{\det(\boldsymbol{C}_A)}{\det\left(\widehat{\boldsymbol{\Sigma}}_A\right)} + L\log\frac{\det(\boldsymbol{C}_B)}{\det\left(\widehat{\boldsymbol{\Sigma}}_B\right)} \\
& - (L+M)H + \mathrm{tr}\left\{\boldsymbol{C}_A^{-1}\left(\widehat{\boldsymbol{A}}^\top\widehat{\boldsymbol{A}} + M\widehat{\boldsymbol{\Sigma}}_A\right)\right\} + \mathrm{tr}\left\{\boldsymbol{C}_B^{-1}\left(\widehat{\boldsymbol{B}}^\top\widehat{\boldsymbol{B}} + L\widehat{\boldsymbol{\Sigma}}_B\right)\right\} \\
& + \sigma^{-2}\mathrm{tr}\left\{-\widehat{\boldsymbol{A}}^\top\widehat{\boldsymbol{A}}\widehat{\boldsymbol{B}}^\top\widehat{\boldsymbol{B}} + \left(\widehat{\boldsymbol{A}}^\top\widehat{\boldsymbol{A}} + M\widehat{\boldsymbol{\Sigma}}_A\right)\left(\widehat{\boldsymbol{B}}^\top\widehat{\boldsymbol{B}} + L\widehat{\boldsymbol{\Sigma}}_B\right)\right\}.
\end{aligned} \tag{6.8}$$

The stationary conditions for the variational parameters are given by

$$\widehat{A} = \sigma^{-2} V^\top \widehat{B}\widehat{\Sigma}_A, \tag{6.9}$$

$$\widehat{\Sigma}_A = \sigma^2 \left( \widehat{B}^\top \widehat{B} + L\widehat{\Sigma}_B + \sigma^2 C_A^{-1} \right)^{-1}, \tag{6.10}$$

$$\widehat{B} = \sigma^{-2} V \widehat{A}\widehat{\Sigma}_B, \tag{6.11}$$

$$\widehat{\Sigma}_B = \sigma^2 \left( \widehat{A}^\top \widehat{A} + M\widehat{\Sigma}_A + \sigma^2 C_B^{-1} \right)^{-1}. \tag{6.12}$$

In the subsequent sections, we derive the global solution to the problem (6.5) in an analytic form, which was obtained in Nakajima et al. (2013a, 2015).

## 6.2 Conditions for VB Solutions

With the explicit form (6.8) of the free energy, the VB learning problem (6.5) can be written as a minimization problem with respect to a *finite* number of variables:

$$\text{Given} \quad C_A, C_A \in \mathbb{D}_{++}^H, \quad \sigma^2 \in \mathbb{R}_{++},$$

$$\min_{\widehat{A},\widehat{B},\widehat{\Sigma}_A,\widehat{\Sigma}_B} \quad F \tag{6.13}$$

$$\text{s.t.} \quad \widehat{A} \in \mathbb{R}^{M\times H}, \widehat{B} \in \mathbb{R}^{L\times H}, \quad \widehat{\Sigma}_A, \widehat{\Sigma}_B \in \mathbb{S}_{++}^H. \tag{6.14}$$

We can easily show that the solution is a stationary point of the free energy.

**Lemma 6.1** *Any local solution of the problem* (6.13) *is a stationary point of the free energy* (6.8).

*Proof* Since

$$\left\| V - \widehat{B}\widehat{A}^\top \right\|_{\mathrm{Fro}}^2 \geq 0,$$

and

$$\mathrm{tr}\left\{ -\widehat{A}^\top \widehat{A}\widehat{B}^\top \widehat{B} + \left( \widehat{A}^\top \widehat{A} + M\widehat{\Sigma}_A \right)\left( \widehat{B}^\top \widehat{B} + L\widehat{\Sigma}_B \right) \right\}$$
$$= \mathrm{tr}\left\{ L\widehat{A}^\top \widehat{A}\widehat{\Sigma}_B + M\widehat{B}^\top \widehat{B}\widehat{\Sigma}_A + LM\widehat{\Sigma}_A\widehat{\Sigma}_B \right\} \geq 0,$$

the free energy (6.8) is lower-bounded as

$$2F \geq -M \log\det\left(\widehat{\Sigma}_A\right) - L\log\det\left(\widehat{\Sigma}_B\right)$$
$$+ \mathrm{tr}\left\{ C_A^{-1}\left( \widehat{A}^\top \widehat{A} + M\widehat{\Sigma}_A \right) \right\} + \mathrm{tr}\left\{ C_B^{-1}\left( \widehat{B}^\top \widehat{B} + L\widehat{\Sigma}_B \right) \right\} + \tau, \tag{6.15}$$

where $\tau$ is a finite constant. The right-hand side of Eq. (6.15) diverges to $+\infty$ if any entry of $\widehat{A}$ or $\widehat{B}$ goes to $+\infty$ or $-\infty$. Also it diverges if any eigenvalue of $\widehat{\Sigma}_A$ or $\widehat{\Sigma}_B$ goes to $+0$ or $\infty$. This implies that that no local solution exists on the boundary of (the closure of) the domain (6.14). Since the free energy is differentiable in the domain (6.14), any local minimizer is a stationary point.

For any observed matrix $V$, the free energy (6.8) can be finite, for example, at $\widehat{A} = \mathbf{0}_{M,H}$, $\widehat{B} = \mathbf{0}_{L,H}$, and $\widehat{\Sigma}_A = \widehat{\Sigma}_B = I_H$, where $\mathbf{0}_{D_1,D_2}$ denotes the $D_1 \times D_2$ matrix with all the entries equal to zero. Therefore, at least one minimizer always exists, which completes the proof of Lemma 6.1. □

Lemma 6.1 implies that the stationary conditions (6.9) through (6.12) are satisfied for any solution. Accordingly, we can obtain the global solution by finding all points that satisfy the stationary conditions. However, the condition involves $O(MH)$ unknowns, and therefore finding all such candidate points seems hard. The first step to tackle this problem is to find hidden separability, which enables us to decompose the problem so that each problem involves only $O(1)$ unknowns.

## 6.3 Irrelevant Degrees of Freedom

The most of the terms in the free energy (6.8) have symmetry, i.e., they are invariant with respect to the coordinate change shown in Eqs. (6.16) and (6.17). Assume that $(A^*, B^*, \Sigma_A^*, \Sigma_B^*)$ is a global solution of the VB problem (6.13), and let $F^* = F(A^*, B^*, \Sigma_A^*, \Sigma_B^*)$ be the minimum free energy. Consider the following rotation of the coordinate system for an arbitrary orthogonal matrix $\Omega \in \mathbb{R}^{H \times H}$:

$$\widehat{A} = A^* \Omega^\top, \qquad \widehat{\Sigma}_A = \Omega \Sigma_A^* \Omega^\top, \tag{6.16}$$

$$\widehat{B} = B^* \Omega^\top, \qquad \widehat{\Sigma}_B = \Omega \Sigma_B^* \Omega^\top. \tag{6.17}$$

We can easily confirm that the terms in Eq. (6.8) except the sixth and the seventh terms are invariant with respect to the rotation, and the free energy can be written as a function of $\Omega$ as follows:

$$2F(\Omega) = \mathrm{tr}\left\{ C_A^{-1} \Omega \left( \widehat{A}^\top \widehat{A} + M \widehat{\Sigma}_A \right) \Omega^\top \right\} + \mathrm{tr}\left\{ C_B^{-1} \Omega \left( \widehat{B}^\top \widehat{B} + L \widehat{\Sigma}_B \right) \Omega^\top \right\} + \mathrm{const.}$$

To find the irrelevant degrees of freedom, we consider *skewed* rotations that only affect a single term in Eq. (6.8).

Consider the following transform:

$$\widehat{A} = A^* C_A^{-1/2} \Omega^\top C_A^{1/2}, \qquad \Sigma_A = C_A^{1/2} \Omega C_A^{-1/2} \Sigma_A^* C_A^{-1/2} \Omega^\top C_A^{1/2}, \tag{6.18}$$

$$\widehat{\boldsymbol{B}} = \boldsymbol{B}^* \boldsymbol{C}_A^{1/2} \boldsymbol{\Omega}^\top \boldsymbol{C}_A^{-1/2}, \qquad \boldsymbol{\Sigma}_B = \boldsymbol{C}_A^{-1/2} \boldsymbol{\Omega} \boldsymbol{C}_A^{1/2} \boldsymbol{\Sigma}_B^* \boldsymbol{C}_A^{1/2} \boldsymbol{\Omega}^\top \boldsymbol{C}_A^{-1/2}. \tag{6.19}$$

Then, the free energy can be written as

$$2F(\boldsymbol{\Omega}) = \mathrm{tr}\left\{\boldsymbol{\Gamma}\boldsymbol{\Omega}\boldsymbol{\Phi}\boldsymbol{\Omega}^\top\right\} + \text{const.}, \tag{6.20}$$

where

$$\boldsymbol{\Gamma} = \boldsymbol{C}_A^{-1}\boldsymbol{C}_B^{-1},$$
$$\boldsymbol{\Phi} = \boldsymbol{C}_A^{1/2}\left(\boldsymbol{B}^{*\top}\boldsymbol{B}^* + L\boldsymbol{\Sigma}_B^*\right)\boldsymbol{C}_A^{1/2}.$$

By assumption, $\boldsymbol{\Omega} = \boldsymbol{I}_H$ is a minimizer of Eq. (6.20), i.e., $F(\boldsymbol{I}_H) = F^*$. Now we can use the following lemma:

**Lemma 6.2** *Let $\boldsymbol{\Gamma}, \boldsymbol{\Omega}, \boldsymbol{\Phi} \in \mathbb{R}^{H\times H}$ be a nondegenerate diagonal matrix, an orthogonal matrix, and a symmetric matrix, respectively. Let $\{\boldsymbol{\Lambda}^{(k)}, \boldsymbol{\Lambda}'^{(k)} \in \mathbb{R}^{H\times H}; k = 1, \ldots, K\}$ be arbitrary diagonal matrices. If a function*

$$G(\boldsymbol{\Omega}) = \mathrm{tr}\left\{\boldsymbol{\Gamma}\boldsymbol{\Omega}\boldsymbol{\Phi}\boldsymbol{\Omega}^\top + \sum_{k=1}^{K} \boldsymbol{\Lambda}^{(k)}\boldsymbol{\Omega}\boldsymbol{\Lambda}'^{(k)}\boldsymbol{\Omega}^\top\right\} \tag{6.21}$$

*is minimized (as a function of $\boldsymbol{\Omega}$, given $\boldsymbol{\Gamma}, \boldsymbol{\Phi}, \{\boldsymbol{\Lambda}^{(k)}, \boldsymbol{\Lambda}'^{(k)}\}$) at $\boldsymbol{\Omega} = \boldsymbol{I}_H$, then $\boldsymbol{\Phi}$ is diagonal. Here, $K$ can be any natural number including $K = 0$ (when only the first term exists).*

*Proof* Let

$$\boldsymbol{\Phi} = \boldsymbol{\Omega}'\boldsymbol{\Gamma}'\boldsymbol{\Omega}'^\top \tag{6.22}$$

be the eigenvalue decomposition of $\boldsymbol{\Phi}$. Let $\boldsymbol{\gamma}, \boldsymbol{\gamma}', \{\boldsymbol{\lambda}^{(k)}\}, \{\boldsymbol{\lambda}'^{(k)}\}$ be the vectors consist of the diagonal entries of $\boldsymbol{\Gamma}, \boldsymbol{\Gamma}', \{\boldsymbol{\Lambda}^{(k)}\}, \{\boldsymbol{\Lambda}'^{(k)}\}$, respectively, i.e.,

$$\boldsymbol{\Gamma} = \mathbf{Diag}(\boldsymbol{\gamma}), \quad \boldsymbol{\Gamma}' = \mathbf{Diag}(\boldsymbol{\gamma}'), \quad \boldsymbol{\Lambda}^{(k)} = \mathbf{Diag}(\boldsymbol{\lambda}^{(k)}), \quad \boldsymbol{\Lambda}'^{(k)} = \mathbf{Diag}(\boldsymbol{\lambda}'^{(k)}).$$

Then, Eq. (6.21) can be written as

$$G(\boldsymbol{\Omega}) = \mathrm{tr}\left\{\boldsymbol{\Gamma}\boldsymbol{\Omega}\boldsymbol{\Phi}\boldsymbol{\Omega}^\top + \sum_{k=1}^{K} \boldsymbol{\Lambda}^{(k)}\boldsymbol{\Omega}\boldsymbol{\Lambda}'^{(k)}\boldsymbol{\Omega}^\top\right\} = \boldsymbol{\gamma}^\top\boldsymbol{Q}\boldsymbol{\gamma}' + \sum_{k=1)}^{K} \boldsymbol{\lambda}^{(k)\top}\boldsymbol{R}\boldsymbol{\lambda}'^{(k)}, \tag{6.23}$$

where

$$\boldsymbol{Q} = (\boldsymbol{\Omega}\boldsymbol{\Omega}') \odot (\boldsymbol{\Omega}\boldsymbol{\Omega}'), \qquad \boldsymbol{R} = \boldsymbol{\Omega} \odot \boldsymbol{\Omega}.$$

Here, $\odot$ denotes the *Hadamard product*. Since $\boldsymbol{Q}$ as well as $\boldsymbol{R}$ is the Hadamard square of an orthogonal matrix, it is *doubly stochastic* (i.e., any of the columns and the rows sums up to one) (Marshall et al., 2009). Therefore, it can be seen that $\boldsymbol{Q}$ reassigns the components of $\boldsymbol{\gamma}$ to those of $\boldsymbol{\gamma}'$ when calculating the elementwise product in the first term of Eq. (6.23). The same applies to $\boldsymbol{R}$ and $\{\boldsymbol{\lambda}^{(k)}, \boldsymbol{\lambda}'^{(k)}\}$ in the second term. Naturally, rearranging the components of $\boldsymbol{\gamma}$ in nondecreasing order and the components of $\boldsymbol{\gamma}'$ in nonincreasing order minimizes $\boldsymbol{\gamma}^\top \boldsymbol{Q} \boldsymbol{\gamma}'$ (Ruhe, 1970; Marshall et al., 2009).

Using the expression (6.23) with $\boldsymbol{Q}$ and $\boldsymbol{R}$, we will prove that $\boldsymbol{\Phi}$ is diagonal if $\boldsymbol{\Omega} = \boldsymbol{I}_H$ minimizes Eq. (6.23). Let us consider a bilateral perturbation $\boldsymbol{\Omega} = \boldsymbol{\Delta}$ such that the $2 \times 2$ matrix $\boldsymbol{\Delta}_{(h,h')}$ consisting of the $h$th and the $h'$th columns and rows form an $2 \times 2$ orthogonal matrix

$$\boldsymbol{\Delta}_{(h,h')} = \begin{pmatrix} \cos\theta & -\sin\theta \\ \sin\theta & \cos\theta \end{pmatrix},$$

and the remaining entries coincide with those of the identity matrix. Then, the elements of $\boldsymbol{Q}$ become

$$Q_{i,j} = \begin{cases} (\Omega'_{h,j}\cos\theta - \Omega'_{h',j}\sin\theta)^2 & \text{if } i = h, \\ (\Omega'_{h,j}\sin\theta + \Omega'_{h',j}\cos\theta)^2 & \text{if } i = h', \\ \Omega'^2_{i,j} & \text{otherwise,} \end{cases}$$

and Eq. (6.23) can be written as a function of $\theta$:

$$G(\theta) = \sum_{j=1}^{H} \left\{ \gamma_h (\Omega'_{h,j}\cos\theta - \Omega'_{h',j}\sin\theta)^2 + \gamma_{h'} (\Omega'_{h,j}\sin\theta + \Omega'_{h',j}\cos\theta)^2 \right\} \gamma'_j$$
$$+ \sum_{k=1}^{K} \begin{pmatrix} \lambda_h^{(k)} & \lambda_{h'}^{(k)} \end{pmatrix} \begin{pmatrix} \cos^2\theta & \sin^2\theta \\ \sin^2\theta & \cos^2\theta \end{pmatrix} \begin{pmatrix} \lambda_h'^{(k)} \\ \lambda_{h'}'^{(k)} \end{pmatrix} + \text{const.} \tag{6.24}$$

Since Eq. (6.24) is differentiable at $\theta = 0$, our assumption that Eq. (6.23) is minimized when $\boldsymbol{\Omega} = \boldsymbol{I}_H$ requires that $\theta = 0$ is a stationary point of Eq. (6.24) for any $h \neq h'$. Therefore, it holds that

$$0 = \left.\frac{\partial G}{\partial \theta}\right|_{\theta=0} = 2 \sum_j \left\{ \gamma_h (\Omega'_{h,j}\cos\theta - \Omega'_{h',j}\sin\theta)(-\Omega'_{h,j}\sin\theta - \Omega'_{h',j}\cos\theta) \right.$$
$$\left. + \gamma_{h'} (\Omega'_{h,j}\sin\theta + \Omega'_{h',j}\cos\theta)(\Omega'_{h,j}\cos\theta - \Omega'_{h',j}\sin\theta) \right\} \gamma'_j$$
$$= 2(\gamma_{h'} - \gamma_h) \sum_j \Omega'_{h,j} \gamma'_j \Omega'_{h',j} = 2(\gamma_{h'} - \gamma_h) \Phi_{h,h'}. \tag{6.25}$$

In the last equality, we used Eq. (6.22). Since we assumed that $\boldsymbol{\Gamma}$ is nondegenerate ($\gamma_h \neq \gamma_{h'}$ for $h \neq h'$), Eq. (6.25) implies that $\boldsymbol{\Phi}$ is diagonal, which completes the proof of Lemma 6.2. □

Assume for simplicity that $\boldsymbol{\Gamma} = \boldsymbol{C}_A^{-1} \boldsymbol{C}_B^{-1}$ is nondegenerate, i.e., no pair of diagonal entries coincide, in Eq. (6.20). Then, since Eq. (6.20) is minimized at $\boldsymbol{\Omega} = \boldsymbol{I}_H$, Lemma 6.2 implies that $\boldsymbol{\Phi} = \boldsymbol{C}_A^{1/2} \left( \boldsymbol{B}^{*\top} \boldsymbol{B}^* + L\boldsymbol{\Sigma}_B^* \right) \boldsymbol{C}_A^{1/2}$ is diagonal. This means that $\boldsymbol{B}^{*\top} \boldsymbol{B}^* + L\boldsymbol{\Sigma}_B^*$ is diagonal. Thus, the stationary condition (6.10) implies that $\boldsymbol{\Sigma}_A^*$ is diagonal. Similarly, we can find that $\boldsymbol{\Sigma}_B^*$ is diaognal, if $\boldsymbol{\Gamma} = \boldsymbol{C}_A^{-1} \boldsymbol{C}_B^{-1}$ is nondegenerate.

To generalize the preceding discussion to degenerate cases, we need to consider an *equivalent solution*, defined as follows:

**Definition 6.3** (Equivalent solutions) We say that two points $(\widehat{\boldsymbol{A}}, \widehat{\boldsymbol{B}}, \widehat{\boldsymbol{\Sigma}}_A, \widehat{\boldsymbol{\Sigma}}_B)$ and $(\widehat{\boldsymbol{A}}', \widehat{\boldsymbol{B}}', \widehat{\boldsymbol{\Sigma}}_A', \widehat{\boldsymbol{\Sigma}}_B')$ are *equivalent* if both give the same free energy and the same mean prediction, i.e.,

$$F(\widehat{\boldsymbol{A}}, \widehat{\boldsymbol{B}}, \widehat{\boldsymbol{\Sigma}}_A, \widehat{\boldsymbol{\Sigma}}_B) = F(\widehat{\boldsymbol{A}}', \widehat{\boldsymbol{B}}', \widehat{\boldsymbol{\Sigma}}_A', \widehat{\boldsymbol{\Sigma}}_B') \quad \text{and} \quad \widehat{\boldsymbol{B}}\widehat{\boldsymbol{A}}^\top = \widehat{\boldsymbol{B}}'\widehat{\boldsymbol{A}}'^\top.$$

With this definition, we can obtain the following theorem (its proof is given in the next section):

**Theorem 6.4** *When $\boldsymbol{C}_A \boldsymbol{C}_B$ is nondegenerate (i.e., $c_{a_h} c_{b_h} > c_{a_{h'}} c_{b_{h'}}$ for any pair $h < h'$), any solution of the problem* (6.13) *has diagonal $\widehat{\boldsymbol{\Sigma}}_A$ and $\widehat{\boldsymbol{\Sigma}}_B$. When $\boldsymbol{C}_A \boldsymbol{C}_B$ is degenerate, any solution has an* equivalent *solution with diagonal $\widehat{\boldsymbol{\Sigma}}_A$ and $\widehat{\boldsymbol{\Sigma}}_B$.*

The result that the solution has diagonal $\widehat{\boldsymbol{\Sigma}}_A$ and $\widehat{\boldsymbol{\Sigma}}_B$ would be natural because we assumed the independent Gaussian priors on $\boldsymbol{A}$ and $\boldsymbol{B}$: the fact that any $\boldsymbol{V}$ can be decomposed into orthogonal singular components may imply that the observation $\boldsymbol{V}$ cannot convey any preference for singular-componentwise correlation. Note, however, that Theorem 6.4 does not necessarily hold when the observed matrix has missing entries.

Obviously, any VB solution (a solution of the problem (6.13)) with diagonal covariances can be found by solving the following problem:

$$\begin{aligned} &\text{Given} \quad \boldsymbol{C}_A, \boldsymbol{C}_A \in \mathbb{D}_{++}^H, \quad \sigma^2 \in \mathbb{R}_{++}, \\ &\min_{\widehat{\boldsymbol{A}}, \widehat{\boldsymbol{B}}, \widehat{\boldsymbol{\Sigma}}_A, \widehat{\boldsymbol{\Sigma}}_B} \quad F \qquad (6.26) \\ &\text{s.t.} \quad \widehat{\boldsymbol{A}} \in \mathbb{R}^{M \times H}, \widehat{\boldsymbol{B}} \in \mathbb{R}^{L \times H}, \quad \widehat{\boldsymbol{\Sigma}}_A, \widehat{\boldsymbol{\Sigma}}_B \in \mathbb{D}_{++}^H, \qquad (6.27) \end{aligned}$$

which is equivalent to solving the SimpleVB learning problem (3.30) with columnwise independence, introduced in Section 3.1.1. Theorem 6.4 states that, if $\boldsymbol{C}_A \boldsymbol{C}_B$ is nondegenerate, the set of VB solutions and the set of SimpleVB solutions are identical. When $\boldsymbol{C}_A \boldsymbol{C}_B$ is degenerate, the set of VB solutions is the union of the set of SimpleVB solutions and the set of their *equivalent* solutions with nondiagonal covariances. Actually, any VB solution can be obtained by rotating its *equivalent* SimpleVB solution (VB solution with diagonal covariances) (see Section 6.4.4). In practice, it is however sufficient to focus on the SimpleVB solutions, since *equivalent* solutions share the same free energy $F$ and the same mean prediction $\widehat{\boldsymbol{B}\boldsymbol{A}^\top}$. In this sense, we can conclude that the stronger *columnwise* independence constraint (3.29) does not degrade approximation accuracy, and the VB solution under the *matrixwise* independence (3.4) *essentially* agrees with the SimpleVB solution.

## 6.4 Proof of Theorem 6.4

In this section, we prove Theorem 6.4 by considering the following three cases separately:

**Case 1** When no pair of diagonal entries of $\boldsymbol{C}_A\boldsymbol{C}_B$ coincide.
**Case 2** When all diagonal entries of $\boldsymbol{C}_A\boldsymbol{C}_B$ coincide.
**Case 3** When (not all but) some pairs of diagonal entries of $\boldsymbol{C}_A\boldsymbol{C}_B$ coincide.

We will prove that, in Case 1, $\widehat{\boldsymbol{\Sigma}}_A$ and $\widehat{\boldsymbol{\Sigma}}_B$ are diagonal for any solution $(\widehat{\boldsymbol{A}}, \widehat{\boldsymbol{B}}, \widehat{\boldsymbol{\Sigma}}_A, \widehat{\boldsymbol{\Sigma}}_B)$, and that, in other cases, any solution has its *equivalent* solution with diagonal $\widehat{\boldsymbol{\Sigma}}_A$ and $\widehat{\boldsymbol{\Sigma}}_B$.

Remember our assumption that the diagonal entries $\{c_{a_h}c_{b_h}\}$ of $\boldsymbol{C}_A\boldsymbol{C}_B$ are arranged in nonincreasing order.

### 6.4.1 Proof for Case 1

Here, we consider the case where $c_{a_h}c_{b_h} > c_{a_{h'}}c_{b_{h'}}$ for any pair $h < h'$.

Assume that $(\boldsymbol{A}^*, \boldsymbol{B}^*, \boldsymbol{\Sigma}_A^*, \boldsymbol{\Sigma}_B^*)$ is a minimizer of the free energy (6.8), and consider the following variation defined with an arbitrary $H \times H$ orthogonal matrix $\boldsymbol{\Omega}$:

$$\widehat{\boldsymbol{A}} = \boldsymbol{A}^*\boldsymbol{C}_A^{-1/2}\boldsymbol{\Omega}^\top\boldsymbol{C}_A^{1/2}, \tag{6.28}$$

$$\widehat{\boldsymbol{B}} = \boldsymbol{B}^*\boldsymbol{C}_A^{1/2}\boldsymbol{\Omega}^\top\boldsymbol{C}_A^{-1/2}, \tag{6.29}$$

$$\widehat{\boldsymbol{\Sigma}}_A = \boldsymbol{C}_A^{1/2}\boldsymbol{\Omega}\boldsymbol{C}_A^{-1/2}\boldsymbol{\Sigma}_A^*\boldsymbol{C}_A^{-1/2}\boldsymbol{\Omega}^\top\boldsymbol{C}_A^{1/2}, \tag{6.30}$$

$$\widehat{\boldsymbol{\Sigma}}_B = \boldsymbol{C}_A^{-1/2}\boldsymbol{\Omega}\boldsymbol{C}_A^{1/2}\boldsymbol{\Sigma}_B^*\boldsymbol{C}_A^{1/2}\boldsymbol{\Omega}^\top\boldsymbol{C}_A^{-1/2}. \tag{6.31}$$

Note that this variation does not change $\widehat{\boldsymbol{B}\boldsymbol{A}^\top}$, and it holds that $(\widehat{\boldsymbol{A}}, \widehat{\boldsymbol{B}}, \widehat{\boldsymbol{\Sigma}}_A, \widehat{\boldsymbol{\Sigma}}_B) = (\boldsymbol{A}^*, \boldsymbol{B}^*, \boldsymbol{\Sigma}_A^*, \boldsymbol{\Sigma}_B^*)$ for $\boldsymbol{\Omega} = \boldsymbol{I}_H$. Then, the free energy (6.8) can be written as a function of $\boldsymbol{\Omega}$:

$$F(\boldsymbol{\Omega}) = \frac{1}{2}\mathrm{tr}\left\{\boldsymbol{C}_A^{-1}\boldsymbol{C}_B^{-1}\boldsymbol{\Omega}\boldsymbol{C}_A^{1/2}\left(\boldsymbol{B}^{*\top}\boldsymbol{B}^{*} + L\boldsymbol{\Sigma}_B^{*}\right)\boldsymbol{C}_A^{1/2}\boldsymbol{\Omega}^{\top}\right\} + \text{const.} \tag{6.32}$$

We define

$$\boldsymbol{\Phi} = \boldsymbol{C}_A^{1/2}\left(\boldsymbol{B}^{*\top}\boldsymbol{B}^{*} + L\boldsymbol{\Sigma}_B^{*}\right)\boldsymbol{C}_A^{1/2},$$

and rewrite Eq. (6.32) as

$$F(\boldsymbol{\Omega}) = \frac{1}{2}\mathrm{tr}\left\{\boldsymbol{C}_A^{-1}\boldsymbol{C}_B^{-1}\boldsymbol{\Omega}\boldsymbol{\Phi}\boldsymbol{\Omega}^{\top}\right\} + \text{const.} \tag{6.33}$$

The assumption that $(\boldsymbol{A}^{*}, \boldsymbol{B}^{*}, \boldsymbol{\Sigma}_A^{*}, \boldsymbol{\Sigma}_B^{*})$ is a minimizer requires that Eq. (6.33) is minimized when $\boldsymbol{\Omega} = \boldsymbol{I}_H$. Then, Lemma 6.2 (for $K = 0$) implies that $\boldsymbol{\Phi}$ is diagonal. Therefore,

$$\boldsymbol{C}_A^{-1/2}\boldsymbol{\Phi}\boldsymbol{C}_A^{-1/2}(= \boldsymbol{\Phi}\boldsymbol{C}_A^{-1}) = \boldsymbol{B}^{*\top}\boldsymbol{B}^{*} + L\boldsymbol{\Sigma}_B^{*}$$

is also diagonal. Consequently, Eq. (6.10) implies that $\boldsymbol{\Sigma}_A^{*}$ is diagonal.

Next, consider the following variation defined with an arbitrary $H \times H$ orthogonal matrix $\boldsymbol{\Omega}'$:

$$\begin{aligned}
\widehat{\boldsymbol{A}} &= \boldsymbol{A}^{*}\boldsymbol{C}_B^{1/2}\boldsymbol{\Omega}'^{\top}\boldsymbol{C}_B^{-1/2},\\
\widehat{\boldsymbol{B}} &= \boldsymbol{B}^{*}\boldsymbol{C}_B^{-1/2}\boldsymbol{\Omega}'^{\top}\boldsymbol{C}_B^{1/2},\\
\widehat{\boldsymbol{\Sigma}}_A &= \boldsymbol{C}_B^{-1/2}\boldsymbol{\Omega}'\boldsymbol{C}_B^{1/2}\boldsymbol{\Sigma}_A^{*}\boldsymbol{C}_B^{1/2}\boldsymbol{\Omega}'^{\top}\boldsymbol{C}_B^{-1/2},\\
\widehat{\boldsymbol{\Sigma}}_B &= \boldsymbol{C}_B^{1/2}\boldsymbol{\Omega}'\boldsymbol{C}_B^{-1/2}\boldsymbol{\Sigma}_B^{*}\boldsymbol{C}_B^{-1/2}\boldsymbol{\Omega}'^{\top}\boldsymbol{C}_B^{1/2}.
\end{aligned}$$

Then, the free energy as a function of $\boldsymbol{\Omega}'$ is given by

$$F(\boldsymbol{\Omega}') = \frac{1}{2}\mathrm{tr}\left\{\boldsymbol{C}_A^{-1}\boldsymbol{C}_B^{-1}\boldsymbol{\Omega}'\boldsymbol{C}_B^{1/2}\left(\boldsymbol{A}^{*\top}\boldsymbol{A}^{*} + M\boldsymbol{\Sigma}_A^{*}\right)\boldsymbol{C}_B^{1/2}\boldsymbol{\Omega}'^{\top}\right\} + \text{const.}$$

From this, we can similarly prove that $\boldsymbol{\Sigma}_B^{*}$ is also diagonal, which completes the proof for Case 1. □

### 6.4.2 Proof for Case 2

Here, we consider the case where $\boldsymbol{C}_A\boldsymbol{C}_B = c^2\boldsymbol{I}_H$ for some $c^2 \in \mathbb{R}_{++}$. In this case, there exist solutions with nondiagonal covariances. However, for any (or each) of those nondiagonal solutions, the equivalent class to which the (nondiagonal) solution belongs contains a solution with diagonal covariances.

We can easily show that the free energy (6.8) is invariant with respect to $\boldsymbol{\Omega}$ under the transformation (6.28) through (6.31). This arbitrariness forms an *equivalent* class of solutions. Since there exists $\boldsymbol{\Omega}$ that diagonalizes any given $\boldsymbol{\Sigma}_A^{*}$ through Eq. (6.30), each *equivalent* class involves a solution with diagonal

$\widehat{\boldsymbol{\Sigma}}_A$. In the following, we will prove that any solution with diagonal $\widehat{\boldsymbol{\Sigma}}_A$ has diagonal $\widehat{\boldsymbol{\Sigma}}_B$.

Assume that $(\boldsymbol{A}^*, \boldsymbol{B}^*, \boldsymbol{\Sigma}_A^*, \boldsymbol{\Sigma}_B^*)$ is a solution with diagonal $\boldsymbol{\Sigma}_A^*$, and consider the following variation defined with an arbitrary $H \times H$ orthogonal matrix $\boldsymbol{\Omega}$:

$$\widehat{\boldsymbol{A}} = \boldsymbol{A}^* \boldsymbol{C}_A^{-1/2} \boldsymbol{\Gamma}^{-1/2} \boldsymbol{\Omega}^\top \boldsymbol{\Gamma}^{1/2} \boldsymbol{C}_A^{1/2},$$
$$\widehat{\boldsymbol{B}} = \boldsymbol{B}^* \boldsymbol{C}_A^{1/2} \boldsymbol{\Gamma}^{1/2} \boldsymbol{\Omega}^\top \boldsymbol{\Gamma}^{-1/2} \boldsymbol{C}_A^{-1/2},$$
$$\widehat{\boldsymbol{\Sigma}}_A = \boldsymbol{C}_A^{1/2} \boldsymbol{\Gamma}^{1/2} \boldsymbol{\Omega} \boldsymbol{\Gamma}^{-1/2} \boldsymbol{C}_A^{-1/2} \boldsymbol{\Sigma}_A^* \boldsymbol{C}_A^{-1/2} \boldsymbol{\Gamma}^{-1/2} \boldsymbol{\Omega}^\top \boldsymbol{\Gamma}^{1/2} \boldsymbol{C}_A^{1/2},$$
$$\widehat{\boldsymbol{\Sigma}}_B = \boldsymbol{C}_A^{-1/2} \boldsymbol{\Gamma}^{-1/2} \boldsymbol{\Omega} \boldsymbol{\Gamma}^{1/2} \boldsymbol{C}_A^{1/2} \boldsymbol{\Sigma}_B^* \boldsymbol{C}_A^{1/2} \boldsymbol{\Gamma}^{1/2} \boldsymbol{\Omega}^\top \boldsymbol{\Gamma}^{-1/2} \boldsymbol{C}_A^{-1/2}.$$

Here, $\boldsymbol{\Gamma} = \mathbf{Diag}(\gamma_1, \ldots, \gamma_H)$ is an arbitrary nondegenerate ($\gamma_h \neq \gamma_{h'}$ for $h \neq h'$) positive-definite diagonal matrix. Then, the free energy can be written as a function of $\boldsymbol{\Omega}$:

$$F(\boldsymbol{\Omega}) = \frac{1}{2} \mathrm{tr} \Big\{ \boldsymbol{\Gamma} \boldsymbol{\Omega} \boldsymbol{\Gamma}^{-1/2} \boldsymbol{C}_A^{-1/2} \left( \boldsymbol{A}^{*\top} \boldsymbol{A}^* + M \boldsymbol{\Sigma}_A^* \right) \boldsymbol{C}_A^{-1/2} \boldsymbol{\Gamma}^{-1/2} \boldsymbol{\Omega}^\top$$
$$+ c^{-2} \boldsymbol{\Gamma}^{-1} \boldsymbol{\Omega} \boldsymbol{\Gamma}^{1/2} \boldsymbol{C}_A^{1/2} \left( \boldsymbol{B}^{*\top} \boldsymbol{B}^* + L \boldsymbol{\Sigma}_B^* \right) \boldsymbol{C}_A^{1/2} \boldsymbol{\Gamma}^{1/2} \boldsymbol{\Omega}^\top \Big\}. \tag{6.34}$$

We define

$$\boldsymbol{\Phi}_A = \boldsymbol{\Gamma}^{-1/2} \boldsymbol{C}_A^{-1/2} \left( \boldsymbol{A}^{*\top} \boldsymbol{A}^* + M \boldsymbol{\Sigma}_A^* \right) \boldsymbol{C}_A^{-1/2} \boldsymbol{\Gamma}^{-1/2},$$
$$\boldsymbol{\Phi}_B = c^{-2} \boldsymbol{\Gamma}^{1/2} \boldsymbol{C}_A^{1/2} \left( \boldsymbol{B}^{*\top} \boldsymbol{B}^* + L \boldsymbol{\Sigma}_B^* \right) \boldsymbol{C}_A^{1/2} \boldsymbol{\Gamma}^{1/2},$$

and rewrite Eq. (6.34) as

$$F(\boldsymbol{\Omega}) = \frac{1}{2} \mathrm{tr} \left\{ \boldsymbol{\Gamma} \boldsymbol{\Omega} \boldsymbol{\Phi}_A \boldsymbol{\Omega}^\top + \boldsymbol{\Gamma}^{-1} \boldsymbol{\Omega} \boldsymbol{\Phi}_B \boldsymbol{\Omega}^\top \right\}. \tag{6.35}$$

Since $\boldsymbol{\Sigma}_A^*$ is diagonal, Eq. (6.10) implies that $\boldsymbol{\Phi}_B$ is diagonal. The assumption that $(\boldsymbol{A}^*, \boldsymbol{B}^*, \boldsymbol{\Sigma}_A^*, \boldsymbol{\Sigma}_B^*)$ is a solution requires that Eq. (6.35) is minimized when $\boldsymbol{\Omega} = \boldsymbol{I}_H$. Accordingly, Lemma 6.2 implies that $\boldsymbol{\Phi}_A$ is diagonal. Consequently, Eq. (6.12) implies that $\boldsymbol{\Sigma}_B^*$ is diagonal.

Thus, we have proved that any solution has its *equivalent* solution with diagonal covariances, which completes the proof for Case 2. □

### 6.4.3 Proof for Case 3

Finally, we consider the case where $c_{a_h} c_{b_h} = c_{a_{h'}} c_{b_{h'}}$ for (not all but) some pairs $h \neq h'$. First, in the same way as Case 1, we can prove that $\widehat{\boldsymbol{\Sigma}}_A$ and $\widehat{\boldsymbol{\Sigma}}_B$ are block diagonal where the blocks correspond to the groups sharing the same $c_{a_h} c_{b_h}$. Next, we can apply the proof for Case 2 to each block, and show that any solution has its *equivalent* solution with diagonal $\widehat{\boldsymbol{\Sigma}}_A$ and $\widehat{\boldsymbol{\Sigma}}_B$. Combining these results completes the proof of Theorem 6.4. □

### 6.4.4 General Expression

In summary, for any minimizer of Eq. (6.8), the covariances can be written in the following form:

$$\widehat{\boldsymbol{\Sigma}}_A = \boldsymbol{C}_A^{1/2}\boldsymbol{\Theta}\boldsymbol{C}_A^{-1/2}\boldsymbol{\Gamma}_{\widehat{\Sigma}_A}\boldsymbol{C}_A^{-1/2}\boldsymbol{\Theta}^\top\boldsymbol{C}_A^{1/2}(= \boldsymbol{C}_B^{-1/2}\boldsymbol{\Theta}\boldsymbol{C}_B^{1/2}\boldsymbol{\Gamma}_{\widehat{\Sigma}_A}\boldsymbol{C}_B^{1/2}\boldsymbol{\Theta}^\top\boldsymbol{C}_B^{-1/2}), \quad (6.36)$$

$$\widehat{\boldsymbol{\Sigma}}_B = \boldsymbol{C}_A^{-1/2}\boldsymbol{\Theta}\boldsymbol{C}_A^{1/2}\boldsymbol{\Gamma}_{\widehat{\Sigma}_B}\boldsymbol{C}_A^{1/2}\boldsymbol{\Theta}^\top\boldsymbol{C}_A^{-1/2}(= \boldsymbol{C}_B^{1/2}\boldsymbol{\Theta}\boldsymbol{C}_B^{-1/2}\boldsymbol{\Gamma}_{\widehat{\Sigma}_B}\boldsymbol{C}_B^{-1/2}\boldsymbol{\Theta}^\top\boldsymbol{C}_B^{1/2}). \quad (6.37)$$

Here, $\boldsymbol{\Gamma}_{\widehat{\Sigma}_A}$ and $\boldsymbol{\Gamma}_{\widehat{\Sigma}_B}$ are positive-definite diagonal matrices, and $\boldsymbol{\Theta}$ is a block diagonal matrix such that the blocks correspond to the groups sharing the same $c_{a_h}c_{b_h}$, and each block consists of an orthogonal matrix. Furthermore, if there exists a solution with $(\widehat{\boldsymbol{\Sigma}}_A, \widehat{\boldsymbol{\Sigma}}_B)$ written in the form of Eqs. (6.36) and (6.37) with a certain set of $(\boldsymbol{\Gamma}_{\widehat{\Sigma}_A}, \boldsymbol{\Gamma}_{\widehat{\Sigma}_B}, \boldsymbol{\Theta})$, then there also exist its *equivalent* solutions with the same $(\boldsymbol{\Gamma}_{\widehat{\Sigma}_A}, \boldsymbol{\Gamma}_{\widehat{\Sigma}_B})$ for *any* $\boldsymbol{\Theta}$. Focusing on the solution with $\boldsymbol{\Theta} = \boldsymbol{I}_H$ as the representative of each *equivalent* class, we can assume that $\widehat{\boldsymbol{\Sigma}}_A$ and $\widehat{\boldsymbol{\Sigma}}_B$ are diagonal without loss of generality.

## 6.5 Problem Decomposition

As discussed in Section 6.3, Theorem 6.4 allows us to focus on the solutions that have diagonal posterior covariances, i.e., $\widehat{\boldsymbol{\Sigma}}_A, \widehat{\boldsymbol{\Sigma}}_B \in \mathbb{D}_{++}^H$. For any solution with diagonal covariances, the stationary conditions (6.10) and (6.12) (with Lemma 6.1) imply that $\widehat{\boldsymbol{A}}^\top\widehat{\boldsymbol{A}}$ and $\widehat{\boldsymbol{B}}^\top\widehat{\boldsymbol{B}}$ are also diagonal, which means that the column vectors of $\widehat{\boldsymbol{A}}$, as well as $\widehat{\boldsymbol{B}}$, are orthogonal to each other. In such a case, the free energy (6.8) depends on the column vectors of $\widehat{\boldsymbol{A}}$ and $\widehat{\boldsymbol{B}}$ only through the second term

$$\sigma^{-2}\left\|\boldsymbol{V} - \widehat{\boldsymbol{B}}\widehat{\boldsymbol{A}}^\top\right\|_{\mathrm{Fro}}^2,$$

which coincides with the objective function for the singular value decomposition (SVD). This leads to the following lemma:

**Lemma 6.5** *Let*

$$\boldsymbol{V} = \sum_{h=1}^{L} \gamma_h \boldsymbol{\omega}_{b_h}\boldsymbol{\omega}_{a_h}^\top \quad (6.38)$$

*be the SVD of* $\boldsymbol{V}$, *where* $\gamma_h$ $(\geq 0)$ *is the hth largest singular value, and* $\boldsymbol{\omega}_{a_h}$ *and* $\boldsymbol{\omega}_{b_h}$ *are the associated right and left singular vectors. Then, any VB solution (with diagonal posterior covariances) can be written as*

$$\widehat{\boldsymbol{B}\boldsymbol{A}^\top} = \sum_{h=1}^{H} \widehat{\gamma}_h^{\mathrm{VB}} \boldsymbol{\omega}_{b_h} \boldsymbol{\omega}_{a_h}^\top \tag{6.39}$$

*for some* $\{\widehat{\gamma}_h^{\mathrm{VB}} \geq 0\}$.

Thanks to Theorem 6.4 and Lemma 6.5, the variational parameters $\widehat{\boldsymbol{A}} = (\widehat{\boldsymbol{a}}_1, \ldots, \widehat{\boldsymbol{a}}_H), \widehat{\boldsymbol{B}} = (\widehat{\boldsymbol{b}}_1, \ldots, \widehat{\boldsymbol{b}}_H), \widehat{\boldsymbol{\Sigma}}_A, \widehat{\boldsymbol{\Sigma}}_B$ can be expressed as

$$\begin{aligned}
\widehat{\boldsymbol{a}}_h &= \widehat{a}_h \boldsymbol{\omega}_{a_h}, \\
\widehat{\boldsymbol{b}}_h &= \widehat{b}_h \boldsymbol{\omega}_{b_h}, \\
\widehat{\boldsymbol{\Sigma}}_A &= \mathbf{Diag}\left(\widehat{\sigma}_{a_1}^2, \ldots, \widehat{\sigma}_{a_H}^2\right), \\
\widehat{\boldsymbol{\Sigma}}_B &= \mathbf{Diag}\left(\widehat{\sigma}_{b_1}^2, \ldots, \widehat{\sigma}_{b_H}^2\right),
\end{aligned}$$

with a new set of unknowns $\{\widehat{a}_h, \widehat{b}_h \in \mathbb{R}, \widehat{\sigma}_{a_h}^2, \widehat{\sigma}_{b_h}^2 \in \mathbb{R}_{++}\}_{h=1}^H$. Thus, the following holds:

**Corollary 6.6** *The VB posterior can be written as*

$$r(\boldsymbol{A}, \boldsymbol{B}) = \prod_{h=1}^{H} \mathrm{Gauss}_M(\boldsymbol{a}_h; \widehat{a}_h \boldsymbol{\omega}_{a_h}, \widehat{\sigma}_{a_h}^2 \boldsymbol{I}_M) \prod_{h=1}^{H} \mathrm{Gauss}_L(\boldsymbol{b}_h; \widehat{b}_h \boldsymbol{\omega}_{b_h}, \widehat{\sigma}_{b_h}^2 \boldsymbol{I}_L), \tag{6.40}$$

*where* $\{\widehat{a}_h, \widehat{b}_h, \widehat{\sigma}_{a_h}^2, \widehat{\sigma}_{b_h}^2\}_{h=1}^H$ *are the solution of the following minimization problem:*

$$\begin{aligned}
&\textit{Given} \quad \sigma^2 \in \mathbb{R}_{++}, \quad \{c_{a_h}^2, c_{b_h}^2 \in \mathbb{R}_{++}\}_{h=1}^H, \\
&\min_{\{\widehat{a}_h, \widehat{b}_h, \widehat{\sigma}_{a_h}^2, \widehat{\sigma}_{b_h}^2\}_{h=1}^H} \quad F \\
&\textit{s.t.} \quad \{\widehat{a}_h, \widehat{b}_h \in \mathbb{R}, \quad \widehat{\sigma}_{a_h}^2, \widehat{\sigma}_{b_h}^2 \in \mathbb{R}_{++}\}_{h=1}^H.
\end{aligned} \tag{6.41}$$

*Here, F is the free energy* (6.8)*, which can be written as*

$$2F = LM \log(2\pi\sigma^2) + \frac{\sum_{h=1}^{L} \gamma_h^2}{\sigma^2} + \sum_{h=1}^{H} 2F_h, \tag{6.42}$$

$$\begin{aligned}
\textit{where} \quad 2F_h = {} & M \log \frac{c_{a_h}^2}{\widehat{\sigma}_{a_h}^2} + L \log \frac{c_{b_h}^2}{\widehat{\sigma}_{b_h}^2} + \frac{\widehat{a}_h^2 + M\widehat{\sigma}_{a_h}^2}{c_{a_h}^2} + \frac{\widehat{b}_h^2 + L\widehat{\sigma}_{b_h}^2}{c_{b_h}^2} \\
& - (L + M) + \frac{-2\widehat{a}_h \widehat{b}_h \gamma_h + \left(\widehat{a}_h^2 + M\widehat{\sigma}_{a_h}^2\right)\left(\widehat{b}_h^2 + L\widehat{\sigma}_{b_h}^2\right)}{\sigma^2}.
\end{aligned} \tag{6.43}$$

Importantly, the free energy (6.42) depends on the variational parameters $\{\widehat{a}_h, \widehat{b}_h, \widehat{\sigma}^2_{a_h}, \widehat{\sigma}^2_{b_h}\}_{h=1}^H$ only through the third term, and the third term is decomposed into $H$ terms, each of which only depends on the variational parameters $(\widehat{a}_h, \widehat{b}_h, \widehat{\sigma}^2_{a_h}, \widehat{\sigma}^2_{b_h})$ for the $h$th singular component. Accordingly, given the noise variance $\sigma^2$, we can separately minimize the free energy (6.43), which involves only four unknowns, for each singular component.

## 6.6 Analytic Form of Global VB Solution

The stationary conditions of Eq. (6.43) are given by

$$\widehat{a}_h = \frac{\widehat{\sigma}^2_{a_h}}{\sigma^2} \gamma_h \widehat{b}_h, \tag{6.44}$$

$$\widehat{\sigma}^2_{a_h} = \sigma^2 \left( \widehat{b}_h^2 + L\widehat{\sigma}^2_{b_h} + \frac{\sigma^2}{c^2_{a_h}} \right)^{-1}, \tag{6.45}$$

$$\widehat{b}_h = \frac{\widehat{\sigma}^2_{b_h}}{\sigma^2} \gamma_h \widehat{a}_h, \tag{6.46}$$

$$\widehat{\sigma}^2_{b_h} = \sigma^2 \left( \widehat{a}_h^2 + M\widehat{\sigma}^2_{a_h} + \frac{\sigma^2}{c^2_{b_h}} \right)^{-1}, \tag{6.47}$$

which form is a *polynomial system*, a set of polynomial equations, on the four unknowns $(\widehat{a}_h, \widehat{b}_h, \widehat{\sigma}^2_{a_h}, \widehat{\sigma}^2_{b_h})$. Since Lemma 6.1 guarantees that any minimizer is a stationary point, we can obtain the global solution by finding all points that satisfy the stationary conditions (6.44) through (6.47) and comparing the free energy (6.43) at those points.

This leads to the following theorem and corollary:

**Theorem 6.7** *The VB solution is given by*

$$\widehat{U}^{\mathrm{VB}} = \sum_{h=1}^{H} \widehat{\gamma}_h^{\mathrm{VB}} \omega_{b_h} \omega_{a_h}^\top, \qquad \textit{where} \qquad \widehat{\gamma}_h^{\mathrm{VB}} = \begin{cases} \breve{\gamma}_h^{\mathrm{VB}} & \textit{if } \gamma_h \geq \underline{\gamma}_h^{\mathrm{VB}}, \\ 0 & \textit{otherwise}, \end{cases} \tag{6.48}$$

*for*

$$\underline{\gamma}_h^{\mathrm{VB}} = \sigma \sqrt{ \frac{(L+M)}{2} + \frac{\sigma^2}{2c^2_{a_h} c^2_{b_h}} + \sqrt{ \left( \frac{(L+M)}{2} + \frac{\sigma^2}{2c^2_{a_h} c^2_{b_h}} \right)^2 - LM } }, \tag{6.49}$$

$$\breve{\gamma}_h^{\mathrm{VB}} = \gamma_h \left( 1 - \frac{\sigma^2}{2\gamma_h^2} \left( M + L + \sqrt{ (M-L)^2 + \frac{4\gamma_h^2}{c^2_{a_h} c^2_{b_h}} } \right) \right). \tag{6.50}$$

**Corollary 6.8** *The VB posterior is given by Eq.* (6.40) *with the following variational parameters: if* $\gamma_h > \underline{\gamma}_h^{\mathrm{VB}}$,

$$\widehat{a}_h = \pm\sqrt{\breve{\gamma}_h^{\mathrm{VB}}\widehat{\delta}_h^{\mathrm{VB}}}, \quad \widehat{b}_h = \pm\sqrt{\frac{\breve{\gamma}_h^{\mathrm{VB}}}{\widehat{\delta}_h^{\mathrm{VB}}}}, \quad \widehat{\sigma}_{a_h}^2 = \frac{\sigma^2\widehat{\delta}_h^{\mathrm{VB}}}{\gamma_h}, \quad \widehat{\sigma}_{b_h}^2 = \frac{\sigma^2}{\gamma_h\widehat{\delta}_h^{\mathrm{VB}}}, \tag{6.51}$$

*where*

$$\widehat{\delta}_h^{\mathrm{VB}}\left(\equiv\frac{\widehat{a}_h}{\widehat{b}_h}\right) = \frac{c_{a_h}}{\sigma^2}\left(\gamma_h - \breve{\gamma}_h^{\mathrm{VB}} - \frac{L\sigma^2}{\gamma_h}\right), \tag{6.52}$$

*and otherwise,*

$$\widehat{a}_h = 0, \quad \widehat{b}_h = 0, \quad \widehat{\sigma}_{a_h}^2 = c_{a_h}^2\left(1 - \frac{L\widehat{\zeta}_h^{\mathrm{VB}}}{\sigma^2}\right), \quad \widehat{\sigma}_{b_h}^2 = c_{b_h}^2\left(1 - \frac{M\widehat{\zeta}_h^{\mathrm{VB}}}{\sigma^2}\right), \tag{6.53}$$

*where*

$$\widehat{\zeta}_h^{\mathrm{VB}}\left(\equiv\widehat{\sigma}_{a_h}^2\widehat{\sigma}_{b_h}^2\right) = \frac{\sigma^2}{2LM}\left(L + M + \frac{\sigma^2}{c_{a_h}^2c_{b_h}^2} - \sqrt{\left(L + M + \frac{\sigma^2}{c_{a_h}^2c_{b_h}^2}\right)^2 - 4LM}\right). \tag{6.54}$$

Theorem 6.7 states that the VB solution for fully observed MF is a truncated shrinkage SVD with the truncation threshold and the shrinkage estimator given by Eqs. (6.49) and (6.50), respectively. Corollary 6.8 completely specifies the VB posterior.[1]

These results give insights into the behavior of VB learning; for example, they explain why a sparse solution is obtained, and what are similarities and differences between the Bayes posterior and the VB posterior, which will be discussed in Chapter 7. The results also form the basis of further analysis on the global empirical VB solution (Section 6.8), which will be used for performance guarantee (Chapter 8), and global (or efficient local) solvers for multilinear models (Chapters 9, 10, and 11). Before moving on, we give the proofs of the theorem and the corollary in the next section.

## 6.7 Proofs of Theorem 6.7 and Corollary 6.8

We will find all stationary points that satisfy Eqs. (6.44) through (6.47), and compare the free energy (6.43).

[1] The similarity between $(\underline{\gamma}_h^{\mathrm{VB}})^2$ and $LM\widehat{\zeta}_h^{\mathrm{VB}}$ comes from the fact that they are the two different solutions of the same quadratic equations, i.e., Eq. (6.79) with respect to $(\underline{\gamma}_h^{\mathrm{VB}})^2$ and (6.77) with respect to $LM\widehat{\zeta}_h^{\mathrm{VB}}$.

By using Eqs. (6.45) and (6.47), the free energy (6.43) can be simplified as

$$F_h = M \log \frac{c_{a_h}^2}{\widehat{\sigma}_{a_h}^2} + L \log \frac{c_{b_h}^2}{\widehat{\sigma}_{b_h}^2} + \frac{1}{\sigma^2}\left(\widehat{a}_h^2 + M\widehat{\sigma}_{a_h}^2 + \frac{\sigma^2}{c_{b_h}^2}\right)\left(\widehat{b}_h^2 + L\widehat{\sigma}_{b_h}^2 + \frac{\sigma^2}{c_{a_h}^2}\right)$$
$$- (L + M) + \frac{-2\widehat{a}_h \widehat{b}_h \gamma_h}{\sigma^2} - \frac{\sigma^2}{c_{a_h}^2 c_{b_h}^2}$$
$$= M \log \frac{c_{a_h}^2}{\widehat{\sigma}_{a_h}^2} + L \log \frac{c_{b_h}^2}{\widehat{\sigma}_{b_h}^2} + \frac{\sigma^2}{\widehat{\sigma}_{a_h}^2 \widehat{\sigma}_{b_h}^2} - \frac{2\widehat{a}_h \widehat{b}_h \gamma_h}{\sigma^2} - \left(L + M + \frac{\sigma^2}{c_{a_h}^2 c_{b_h}^2}\right). \tag{6.55}$$

The stationary conditions (6.44) through (6.47) imply two possibilities of stationary points.

### 6.7.1 Null Stationary Point

If $\widehat{a}_h = 0$ or $\widehat{b}_h = 0$, Eqs. (6.44) and (6.46) require that $\widehat{a}_h = 0$ and $\widehat{b}_h = 0$. In this case, Eqs. (6.45) and (6.47) lead to

$$\widehat{\sigma}_{a_h}^2 = c_{a_h}^2 \left(1 - \frac{L\widehat{\sigma}_{a_h}^2 \widehat{\sigma}_{b_h}^2}{\sigma^2}\right), \tag{6.56}$$

$$\widehat{\sigma}_{b_h}^2 = c_{b_h}^2 \left(1 - \frac{M\widehat{\sigma}_{a_h}^2 \widehat{\sigma}_{b_h}^2}{\sigma^2}\right). \tag{6.57}$$

Multiplying Eqs. (6.56) and (6.57), we have

$$\left(1 - \frac{L\widehat{\sigma}_{a_h}^2 \widehat{\sigma}_{b_h}^2}{\sigma^2}\right)\left(1 - \frac{M\widehat{\sigma}_{a_h}^2 \widehat{\sigma}_{b_h}^2}{\sigma^2}\right) = \frac{\widehat{\sigma}_{a_h}^2 \widehat{\sigma}_{b_h}^2}{c_{a_h}^2 c_{b_h}^2}, \tag{6.58}$$

and therefore

$$\frac{LM}{\sigma^2}\widehat{\sigma}_{a_h}^4 \widehat{\sigma}_{b_h}^4 - \left(L + M + \frac{\sigma^2}{c_{a_h}^2 c_{b_h}^2}\right)\widehat{\sigma}_{a_h}^2 \widehat{\sigma}_{b_h}^2 + \sigma^2 = 0. \tag{6.59}$$

Solving the quadratic equation (6.59) with respect to $\widehat{\sigma}_{a_h}^2 \widehat{\sigma}_{b_h}^2$ and checking the signs of $\widehat{\sigma}_{a_h}^2$ and $\widehat{\sigma}_{b_h}^2$, we have the following lemma:

**Lemma 6.9** *For any $\gamma_h \geq 0$ and $c_{a_h}^2, c_{b_h}^2, \sigma^2 \in \mathbb{R}_{++}$, the null stationary point given by Eq. (6.53) exists with the following free energy:*

$$F_h^{\text{VB-Null}} = -M \log\left(1 - \frac{L}{\sigma^2}\widehat{\zeta}_h^{\text{VB}}\right) - L \log\left(1 - \frac{M}{\sigma^2}\widehat{\zeta}_h^{\text{VB}}\right) - \frac{LM}{\sigma^2}\widehat{\zeta}_h^{\text{VB}}, \tag{6.60}$$

*where $\widehat{\zeta}_h^{\text{VB}}$ is defined by Eq. (6.54).*

*Proof* Eq. (6.59) has two positive real solutions:

$$\widehat{\sigma}^2_{a_h}\widehat{\sigma}^2_{b_h} = \frac{\sigma^2}{2LM}\left(L + M + \frac{\sigma^2}{c^2_{a_h}c^2_{b_h}} \pm \sqrt{\left(L + M + \frac{\sigma^2}{c^2_{a_h}c^2_{b_h}}\right)^2 - 4LM}\right).$$

The larger solution (with the plus sign) is decreasing with respect to $c^2_{a_h}c^2_{b_h}$, and lower-bounded as $\widehat{\sigma}^2_{a_h}\widehat{\sigma}^2_{b_h} > \sigma^2/L$. The smaller solution (with the minus sign) is increasing with respect to $c^2_{a_h}c^2_{b_h}$, and upper-bounded as $\widehat{\sigma}^2_{a_h}\widehat{\sigma}^2_{b_h} < \sigma^2/M$.

For $\widehat{\sigma}^2_{a_h}$ and $\widehat{\sigma}^2_{b_h}$ to be positive, Eqs. (6.56) and (6.57) require that

$$\widehat{\sigma}^2_{a_h}\widehat{\sigma}^2_{b_h} < \frac{\sigma^2}{M}, \tag{6.61}$$

which is violated by the larger solution, while satisfied by the smaller solution. With the smaller solution (6.54), Eqs. (6.56) and (6.57) give the stationary point given by Eq. (6.53).

Using Eq. (6.59), we can easily derive Eq. (6.60) from Eq. (6.55), which completes the proof of Lemma 6.9.

□

### 6.7.2 Positive Stationary Point

Assume that $\widehat{a}_h, \widehat{b}_h \neq 0$. In this case, Eqs. (6.44) and (6.46) imply that $\widehat{a}_h$ and $\widehat{b}_h$ have the same sign. Define

$$\widehat{\gamma}_h = \widehat{a}_h\widehat{b}_h > 0, \tag{6.62}$$

$$\widehat{\delta}_h = \frac{\widehat{a}_h}{\widehat{b}_h} > 0. \tag{6.63}$$

From Eqs. (6.44) and (6.46), we have

$$\sigma^2_{a_h} = \frac{\sigma^2}{\gamma_h}\widehat{\delta}_h, \tag{6.64}$$

$$\sigma^2_{b_h} = \frac{\sigma^2}{\gamma_h}\widehat{\delta}_h^{-1}. \tag{6.65}$$

Substituting Eqs. (6.64) and (6.65) into Eqs. (6.45) and (6.47) gives

$$\gamma_h\widehat{\delta}_h^{-1} = \left(\widehat{\gamma}_h\widehat{\delta}_h^{-1} + L\frac{\sigma^2}{\gamma_h}\widehat{\delta}_h^{-1} + \frac{\sigma^2}{c^2_{a_h}}\right), \tag{6.66}$$

$$\gamma_h\widehat{\delta}_h = \left(\widehat{\gamma}_h\widehat{\delta}_h + M\frac{\sigma^2}{\gamma_h}\widehat{\delta}_h + \frac{\sigma^2}{c^2_{b_h}}\right), \tag{6.67}$$

and therefore,

$$\widehat{\delta}_h = \frac{c_{a_h}}{\sigma^2}\left(\gamma_h - \widehat{\gamma}_h - \frac{L\sigma^2}{\gamma_h}\right), \tag{6.68}$$

$$\widehat{\delta}_h^{-1} = \frac{c_{b_h}}{\sigma^2}\left(\gamma_h - \widehat{\gamma}_h - \frac{M\sigma^2}{\gamma_h}\right). \tag{6.69}$$

Multiplying Eqs. (6.68) and (6.69), we have

$$\left(\gamma_h - \widehat{\gamma}_h - \frac{L\sigma^2}{\gamma_h}\right)\left(\gamma_h - \widehat{\gamma}_h - \frac{M\sigma^2}{\gamma_h}\right) = \frac{\sigma^4}{c_{a_h}c_{b_h}}, \tag{6.70}$$

and therefore

$$\widehat{\gamma}_h^2 - \left(2\gamma_h - \frac{(L+M)\sigma^2}{\gamma_h}\right)\widehat{\gamma}_h + \left(\gamma_h - \frac{L\sigma^2}{\gamma_h}\right)\left(\gamma_h - \frac{M\sigma^2}{\gamma_h}\right) - \frac{\sigma^4}{c_{a_h}c_{b_h}} = 0. \tag{6.71}$$

By solving the quadratic equation (6.71) with respect to $\widehat{\gamma}_h$, and checking the signs of $\widehat{\gamma}_h, \widehat{\delta}_h, \widehat{\sigma}^2_{a_h}$, and $\widehat{\sigma}^2_{b_h}$, we have the following lemma:

**Lemma 6.10** *If and only if $\gamma_h > \underline{\gamma}_h^{\mathrm{VB}}$, where $\underline{\gamma}_h^{\mathrm{VB}}$ is defined by Eq.* (6.49), *the positive stationary point given by Eq.* (6.51) *exists with the following free energy:*

$$\begin{aligned} F_h^{\mathrm{VB-Posi}} = & -M\log\left(1-\left(\frac{\breve{\gamma}_h^{\mathrm{VB}}}{\gamma_h}+\frac{L\sigma^2}{\gamma_h^2}\right)\right) - L\log\left(1-\left(\frac{\breve{\gamma}_h^{\mathrm{VB}}}{\gamma_h}+\frac{M\sigma^2}{\gamma_h^2}\right)\right) \\ & - \frac{\gamma_h^2}{\sigma^2}\left(\frac{\breve{\gamma}_h^{\mathrm{VB}}}{\gamma_h}+\frac{L\sigma^2}{\gamma_h^2}\right)\left(\frac{\breve{\gamma}_h^{\mathrm{VB}}}{\gamma_h}+\frac{M\sigma^2}{\gamma_h^2}\right), \end{aligned} \tag{6.72}$$

*where $\breve{\gamma}_h^{\mathrm{VB}}$ is defined by Eq. (6.50).*

*Proof* Since $\widehat{\delta}_h > 0$, Eqs. (6.68) and (6.69) require that

$$\widehat{\gamma}_h < \gamma_h - \frac{L\sigma^2}{\gamma_h}, \tag{6.73}$$

and therefore, the positive stationary point exists only when

$$\gamma_h > \sqrt{M}\sigma. \tag{6.74}$$

Let us assume that Eq. (6.74) holds.

Eq. (6.71) has two solutions:

$$\widehat{\gamma}_h = \frac{1}{2}\left(2\gamma_h - \frac{(L+M)\sigma^2}{\gamma_h} \pm \sqrt{\left(\frac{(M-L)\sigma^2}{\gamma_h}\right)^2 + \frac{4\sigma^4}{c_{a_h}^2 c_{b_h}^2}}\right).$$

The larger solution with the plus sign is positive, decreasing with respect to $c_{a_h}^2 c_{b_h}^2$, and lower-bounded as $\widehat{\gamma}_h > \gamma_h - L\sigma^2/\gamma_h$, which violates the condition (6.73).

The smaller solution, Eq. (6.50), with the minus sign is positive if the intercept of the left-hand side in Eq. (6.71) is positive, i.e.,

$$\left(\gamma_h - \frac{L\sigma^2}{\gamma_h}\right)\left(\gamma_h - \frac{M\sigma^2}{\gamma_h}\right) - \frac{\sigma^4}{c_{a_h}^2 c_{b_h}^2} > 0. \tag{6.75}$$

From the condtion (6.75), we obtain the threshold (6.49) for the existence of the positive stationary point. Note that $\underline{\gamma}_h^{\mathrm{VB}} > \sqrt{M}\sigma$, and therefore, Eq. (6.74) holds whenever $\gamma_h > \underline{\gamma}_h^{\mathrm{VB}}$.

Assume that $\gamma_h > \underline{\gamma}_h^{\mathrm{VB}}$. Then, with the solution (6.50), $\widehat{\delta}_h$, given by Eq. (6.68), and $\widehat{\sigma}_{a_h}^2$ and $\widehat{\sigma}_{b_h}^2$, given by Eqs. (6.64) and (6.65), are all positive. Thus, we obtain the positive stationary point (6.51).

Substituting Eqs. (6.64) and (6.65), and then Eqs. (6.68) and (6.69), into the free energy (6.55), we have

$$\begin{aligned} F_h^{\mathrm{VB\text{-}Posi}} = & -M\log\left(1 - \frac{\breve{\gamma}_h^{\mathrm{VB}}}{\gamma_h} - \frac{L\sigma^2}{\gamma_h^2}\right) - L\log\left(1 - \frac{\breve{\gamma}_h^{\mathrm{VB}}}{\gamma_h} - \frac{M\sigma^2}{\gamma_h^2}\right) \\ & + \frac{-2\gamma_h\breve{\gamma}_h^{\mathrm{VB}}}{\sigma^2} + \frac{\gamma_h^2}{\sigma^2} - \left(L + M + \frac{\sigma^2}{c_{a_h}^2 c_{b_h}^2}\right). \end{aligned} \tag{6.76}$$

Using Eq. (6.70), we can eliminate the direct dependency on $c_{a_h}^2 c_{b_h}^2$, and express the free energy (6.76) as a function of $\breve{\gamma}_h^{\mathrm{VB}}$. This results in Eq. (6.72), and completes the proof of Lemma 6.10. □

### 6.7.3 Useful Relations

Let us summarize some useful relations between variables, which are used in the subsequent sections. $\widehat{\zeta}_h^{\mathrm{VB}}$, $\breve{\gamma}_h^{\mathrm{VB}}$, and $\underline{\gamma}_h^{\mathrm{VB}}$, derived from Eqs. (6.58), (6.70), and the constant part of Eq. (6.71), respectively, satisfy the following:

$$\left(1 - \frac{L\widehat{\zeta}_h^{\mathrm{VB}}}{\sigma^2}\right)\left(1 - \frac{M\widehat{\zeta}_h^{\mathrm{VB}}}{\sigma^2}\right) - \frac{\widehat{\zeta}_h^{\mathrm{VB}}}{c_{a_h}^2 c_{b_h}^2} = 0, \tag{6.77}$$

$$\left(\gamma_h - \breve{\gamma}_h^{\mathrm{VB}} - \frac{L\sigma^2}{\gamma_h}\right)\left(\gamma_h - \breve{\gamma}_h^{\mathrm{VB}} - \frac{M\sigma^2}{\gamma_h}\right) - \frac{\sigma^4}{c_{a_h} c_{b_h}} = 0, \tag{6.78}$$

$$\left(\underline{\gamma}_h^{\mathrm{VB}} - \frac{L\sigma^2}{\underline{\gamma}_h^{\mathrm{VB}}}\right)\left(\underline{\gamma}_h^{\mathrm{VB}} - \frac{M\sigma^2}{\underline{\gamma}_h^{\mathrm{VB}}}\right) - \frac{\sigma^4}{c_{a_h} c_{b_h}} = 0. \tag{6.79}$$

From Eqs. (6.54) and (6.49), we find that

$$\underline{\gamma}_h^{\mathrm{VB}} = \sqrt{\left((L+M)\sigma^2 + \frac{\sigma^4}{c_{a_h}^2 c_{b_h}^2}\right) - LM\widehat{\zeta}_h^{\mathrm{VB}}}, \tag{6.80}$$

which is useful when comparing the free energies of the null and the positive stationary points.

### 6.7.4 Free Energy Comparison

Lemmas 6.9 and 6.10 imply that, when $\gamma_h \leq \underline{\gamma}_h^{\mathrm{VB}}$, the null stationary point is only the stationary point, and therefore the global solution. When $\gamma_h > \underline{\gamma}_h^{\mathrm{VB}}$, both of the null and the positive stationary points exist, and therefore identifying the global solution requires us to compare their free energies, given by Eqs. (6.60) and (6.72).

Given the observed singular value $\gamma_h \geq 0$, we view the free energy as a function of $c_{a_h}^2 c_{b_h}^2$. We also view the threshold $\underline{\gamma}_h^{\mathrm{VB}}$ as a function of $c_{a_h}^2 c_{b_h}^2$. We find from Eq. (6.49) that $\underline{\gamma}_h^{\mathrm{VB}}$ is decreasing and lower-bounded by $\underline{\gamma}_h^{\mathrm{VB}} > \sqrt{M}\sigma$. Therefore, when $\gamma_h \leq \sqrt{M}\sigma$, $\underline{\gamma}_h^{\mathrm{VB}}$ never gets smaller than $\gamma_h$ for any $c_{a_h}^2 c_{b_h}^2 > 0$. When $\gamma_h > \sqrt{M}\sigma$, on the other hand, there is a threshold $\underline{c}_{a_h}\underline{c}_{b_h}$ such that $\gamma_h > \underline{\gamma}_h^{\mathrm{VB}}$ if $c_{a_h}^2 c_{b_h}^2 > \underline{c}_{a_h}\underline{c}_{b_h}$. Eq. (6.79) implies that the threshold is given by

$$\underline{c}_{a_h}\underline{c}_{b_h} = \frac{\sigma^4}{\gamma_h^2\left(1 - \frac{L\sigma^2}{\gamma_h^2}\right)\left(1 - \frac{M\sigma^2}{\gamma_h^2}\right)}. \tag{6.81}$$

We have the following lemma:

**Lemma 6.11** *For any $\gamma_h \geq 0$ and $c_{a_h}^2 c_{b_h}^2 > 0$, the derivative of the free energy* (6.60) *at the null stationary point with respect to $c_{a_h}^2 c_{b_h}^2$ is given by*

$$\frac{\partial F_h^{\mathrm{VB-Null}}}{\partial c_{a_h}^2 c_{b_h}^2} = \frac{LM\widehat{\zeta}_h^{\mathrm{VB}}}{\sigma^2 c_{a_h}^2 c_{b_h}^2}. \tag{6.82}$$

*For $\gamma_h > M/\sigma^2$ and $c_{a_h}^2 c_{b_h}^2 > \underline{c}_{a_h}\underline{c}_{b_h}$, the derivative of the free energy* (6.72) *at the positive stationary point with respect to $c_{a_h}^2 c_{b_h}^2$ is given by*

$$\frac{\partial F_h^{\mathrm{VB-Posi}}}{\partial c_{a_h}^2 c_{b_h}^2} = \frac{\gamma_h^2}{\sigma^2 c_{a_h}^2 c_{b_h}^2}\left(\frac{(\breve{\gamma}_h^{\mathrm{VB}})^2}{\gamma_h^2} - \left(1 - \frac{(L+M)\sigma^2}{\gamma_h^2}\right)\frac{\breve{\gamma}_h^{\mathrm{VB}}}{\gamma_h} + \frac{LM\sigma^4}{\gamma_h^4}\right). \quad (6.83)$$

*The derivative of the difference is negative, i.e.,*

$$\frac{\partial(F_h^{\mathrm{VB-Posi}} - F_h^{\mathrm{VB-Null}})}{\partial c_{a_h}^2 c_{b_h}^2} = -\frac{1}{\sigma^2 c_{a_h}^2 c_{b_h}^2}\left(\gamma_h\left(\gamma_h - \breve{\gamma}_h^{\mathrm{VB}}\right) - (\underline{\gamma}_h^{\mathrm{VB}})^2\right) < 0. \quad (6.84)$$

*Proof* By differentiating Eqs. (6.60), (6.54), (6.72), and (6.50), we have

$$\begin{aligned}\frac{\partial F_h^{\mathrm{VB-Null}}}{\partial \widehat{\zeta}_h^{\mathrm{VB}}} &= \frac{LM}{\sigma^2\left(1 - \frac{L}{\sigma^2}\widehat{\zeta}_h^{\mathrm{VB}}\right)} + \frac{LM}{\sigma^2\left(1 - \frac{M}{\sigma^2}\widehat{\zeta}_h^{\mathrm{VB}}\right)} - \frac{LM}{\sigma^2} \\ &= \frac{LM c_{a_h}^2 c_{b_h}^2 \left(1 + \frac{\sqrt{LM}}{\sigma^2}\widehat{\zeta}_h^{\mathrm{VB}}\right)\left(1 - \frac{\sqrt{LM}}{\sigma^2}\widehat{\zeta}_h^{\mathrm{VB}}\right)}{\sigma^2 \widehat{\zeta}_h^{\mathrm{VB}}}, \end{aligned} \quad (6.85)$$

$$\begin{aligned}\frac{\partial \widehat{\zeta}_h^{\mathrm{VB}}}{\partial c_{a_h}^2 c_{b_h}^2} &= \frac{\sigma^2}{2LM}\left(-\frac{\sigma^2}{c_{a_h}^4 c_{b_h}^4} + \frac{2\sigma^2\left(L + M + \frac{\sigma^2}{c_{a_h}^2 c_{b_h}^2}\right)}{2c_{a_h}^4 c_{b_h}^4 \sqrt{\left(L + M + \frac{\sigma^2}{c_{a_h}^2 c_{b_h}^2}\right)^2 - 4LM}}\right) \\ &= \frac{1}{c_{a_h}^4 c_{b_h}^4}\left(\frac{(\widehat{\zeta}_h^{\mathrm{VB}})^2}{\left(1 - \frac{\sqrt{LM}\widehat{\zeta}_h^{\mathrm{VB}}}{\sigma^2}\right)\left(1 + \frac{\sqrt{LM}\widehat{\zeta}_h^{\mathrm{VB}}}{\sigma^2}\right)}\right), \end{aligned} \quad (6.86)$$

$$\begin{aligned}\frac{\partial F_h^{\mathrm{VB-Posi}}}{\partial \breve{\gamma}_h^{\mathrm{VB}}} &= \frac{M}{\gamma_h\left(1 - \left(\frac{\breve{\gamma}_h^{\mathrm{VB}}}{\gamma_h} + \frac{L\sigma^2}{\gamma_h^2}\right)\right)} + \frac{L}{\gamma_h\left(1 - \left(\frac{\breve{\gamma}_h^{\mathrm{VB}}}{\gamma_h} + \frac{M\sigma^2}{\gamma_h^2}\right)\right)} \\ &\qquad - \frac{\gamma_h}{\sigma^2}\left(\frac{2\breve{\gamma}_h^{\mathrm{VB}}}{\gamma_h} + \frac{(L+M)\sigma^2}{\gamma_h^2}\right) \\ &= \frac{2c_{a_h}^2 c_{b_h}^2 \gamma_h^3 \left(1 - \left(\frac{\breve{\gamma}_h^{\mathrm{VB}}}{\gamma_h} + \frac{(L+M)\sigma^2}{2\gamma_h^2}\right)\right)\left(\frac{(\breve{\gamma}_h^{\mathrm{VB}})^2}{\gamma_h^2} - \left(1 - \frac{(L+M)\sigma^2}{\gamma_h^2}\right)\frac{\breve{\gamma}_h^{\mathrm{VB}}}{\gamma_h} + \frac{LM\sigma^4}{\gamma_h^4}\right)}{\sigma^6}, \end{aligned} \quad (6.87)$$

$$\begin{aligned}\frac{\partial \widehat{\gamma}_h}{\partial c_{a_h}^2 c_{b_h}^2} &= \frac{4\gamma_h^2\sigma^2}{4\gamma_h c_{a_h}^4 c_{b_h}^4 \sqrt{(M-L)^2 + \frac{4\gamma_h^2}{c_{a_h}^2 c_{b_h}^2}}} \\ &= \frac{\sigma^4}{2\gamma_h c_{a_h}^4 c_{b_h}^4 \left(1 - \left(\frac{\breve{\gamma}_h^{\mathrm{VB}}}{\gamma_h} + \frac{(L+M)\sigma^2}{2\gamma_h^2}\right)\right)}. \end{aligned} \quad (6.88)$$

Here, we used Eqs. (6.54) and (6.77) to obtain Eqs. (6.85) and (6.86), and Eqs. (6.50) and (6.78) to obtain Eqs. (6.87) and (6.88), respectively. Eq. (6.82) is obtained by multiplying Eqs. (6.85) and (6.86), while Eq. (6.83) is obtained by multiplying Eqs. (6.87) and (6.88).

Taking the difference between the derivatives (6.82) and (6.83), and then using Eqs. (6.78) and (6.80), we have

$$\frac{\partial(F_h^{\text{VB-Posi}} - F_h^{\text{VB-Null}})}{\partial c_{a_h}^2 c_{b_h}^2} = \frac{\partial F_h^{\text{VB-Posi}}}{\partial c_{a_h}^2 c_{b_h}^2} - \frac{\partial F_h^{\text{VB-Null}}}{\partial c_{a_h}^2 c_{b_h}^2}$$
$$= -\frac{1}{\sigma^2 c_{a_h}^2 c_{b_h}^2}\left(\gamma_h(\gamma_h - \widehat{\gamma}_h) - (\underline{\gamma}_h^{\text{VB}})^2\right). \tag{6.89}$$

The following can be obtained from Eqs. (6.78) and (6.79), respectively:

$$\left(\gamma_h(\gamma_h - \breve{\gamma}_h^{\text{VB}}) - \frac{(L+M)\sigma^2}{2}\right)^2 = \frac{(L+M)^2\sigma^4}{4} - LM\sigma^4 + \frac{\sigma^4}{c_{a_h}^2 c_{b_h}^2}\gamma_h^2, \tag{6.90}$$

$$\left((\underline{\gamma}_h^{\text{VB}})^2 - \frac{(L+M)\sigma^2}{2}\right)^2 = \frac{(L+M)^2\sigma^4}{4} - LM\sigma^4 + \frac{\sigma^4}{c_{a_h}^2 c_{b_h}^2}(\underline{\gamma}_h^{\text{VB}})^2. \tag{6.91}$$

Eqs. (6.90) and (6.91) imply that

$$\gamma_h(\gamma_h - \breve{\gamma}_h^{\text{VB}}) > (\underline{\gamma}_h^{\text{VB}})^2 \qquad \text{when} \qquad \gamma_h > \underline{\gamma}_h^{\text{VB}}.$$

Therefore, Eq. (6.89) is negative, which completes the proof of Lemma 6.11. □

It is easy to show that the null stationary point (6.53) and the positive stationary point (6.51) coincide with each other at $c_{a_h}^2 c_{b_h}^2 \to \underline{c}_{a_h}\underline{c}_{b_h} + 0$. Here, $+0$ means that it approaches to zero from the positive side. Therefore,

$$\lim_{c_{a_h}^2 c_{b_h}^2 \to \underline{c}_{a_h}\underline{c}_{b_h}+0} \left(F_h^{\text{VB-Posi}} - F_h^{\text{VB-Null}}\right) = 0. \tag{6.92}$$

Eqs. (6.84) and (6.92) together imply that

$$F_h^{\text{VB-Posi}} - F_h^{\text{VB-Null}} < 0 \quad \text{for} \quad c_{a_h}^2 c_{b_h}^2 > \underline{c}_{a_h}\underline{c}_{b_h}, \tag{6.93}$$

which results in the following lemma:

**Lemma 6.12** *The positive stationary point is the global solution (the global minimizer of the free energy* (6.43) *for fixed $c_{a_h}$ and $c_{b_h}$) whenever it exists.*

Combining Lemmas 6.9, 6.10, and 6.12 completes the proof of Theorem 6.7 and Corollary 6.8. □

Figure 6.3 illustrates the behavior of the free energies.

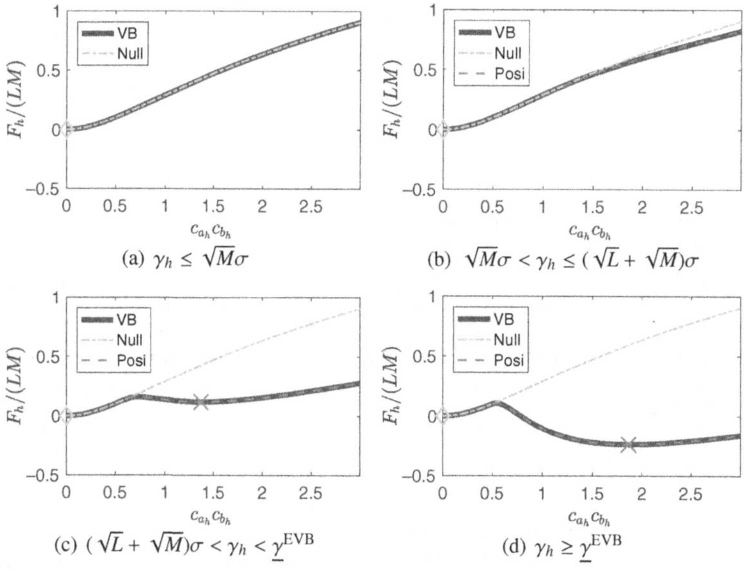


Figure 6.3 Behavior of the free energies (6.60) and (6.72) at the null and the positive stationary points as functions of $c_{a_h}c_{b_h}$, when $L = M = H = 1$ and $\sigma^2 = 1$. The curve labeled as "VB" shows the VB free energy $F_h = \min(F_h^{\text{VB-Null}}, F_h^{\text{VB-Posi}})$ at the global solution, given $c_{a_h}c_{b_h}$. If $\gamma_h \le \sqrt{M}\sigma$, only the null stationary point exists for any $c_{a_h}c_{b_h} > 0$. Otherwise, the positive stationary point exists for $c_{a_h}c_{b_h} > \underline{c}_{a_h}\underline{c}_{b_h}$, and it is the global minimum whenever it exists. In empirical VB learning where $c_{a_h}c_{b_h}$ is also optimized, $c_{a_h}c_{b_h} \to 0$ (indicated by a diamond) is the unique local minimum if $\gamma_h \le (\sqrt{L} + \sqrt{M})\sigma$. Otherwise, a positive local minimum also exists (indicated by a cross), and it is the global minimum if and only if $\gamma_h \ge \underline{\gamma}^{\mathrm{EVB}}$ (see Section 6.9 for detailed discussion).

## 6.8 Analytic Form of Global Empirical VB Solution

In this section, we will solve the empirical VB (EVB) problem where the prior covariances are also estimated from observation, i.e.,

$$\widehat{r} = \underset{r, \boldsymbol{C}_A, \boldsymbol{C}_B}{\operatorname{argmin}} F(r) \quad \text{s.t.} \quad r(\boldsymbol{A}, \boldsymbol{B}) = r_A(\boldsymbol{A}) r_B(\boldsymbol{B}). \tag{6.94}$$

Since the solution of the EVB problem is a VB solution with some values for the prior covariances $\boldsymbol{C}_A, \boldsymbol{C}_B$, the empirical VB posterior is in the same form as the VB posterior (6.40). Accordingly, the problem (6.94) can be written with the variational parameters $\{\widehat{a}_h, \widehat{b}_h, \widehat{\sigma}^2_{a_h}, \widehat{\sigma}^2_{b_h}\}_{h=1}^H$ as follows:

$$\text{Given} \quad \sigma^2 \in \mathbb{R}_{++},$$

$$\min_{\{\widehat{a}_h, \widehat{b}_h, \widehat{\sigma}^2_{a_h}, \widehat{\sigma}^2_{b_h}, c^2_{a_h}, c^2_{b_h}\}_{h=1}^H} F \tag{6.95}$$

$$\text{s.t.} \quad \{\widehat{a}_h, \widehat{b}_h \in \mathbb{R}, \quad \widehat{\sigma}^2_{a_h}, \widehat{\sigma}^2_{b_h}, c^2_{a_h}, c^2_{b_h} \in \mathbb{R}_{++}\}_{h=1}^H, \tag{6.96}$$

where the free energy $F$ is given by Eq. (6.42).

Solving the empirical VB problem (6.95) is not much harder than the VB problem (6.41) because the objective is still separable into $H$ singular components when the prior variances $\{c^2_{a_h}, c^2_{b_h}\}$ are also optimized. More specifically, we can obtain the empirical VB solution by minimizing the *componentwise free energy* (6.43) with respect to the only six unknowns $(\widehat{a}_h, \widehat{b}_h, \widehat{\sigma}^2_{a_h}, \widehat{\sigma}^2_{b_h}, c^2_{a_h}, c^2_{b_h})$ for each $h$th component. On the other hand, analyzing the VB estimator for the noise variance $\sigma^2$ is hard, since $F_h$ for all $h = 1, \ldots, H$ depend on $\sigma^2$ and therefore the free energy (6.42) is not separable. We postpone the analysis of this full empirical VB learning to Chapter 8, where the theoretical performance guarantee is derived.

For the problem (6.95), the stationary points of the free energy (6.43) satisfy Eqs. (6.44) through (6.47) along with Eqs. (3.26) and (3.27), which are written with the new set of variational parameters as

$$c^2_{a_h} = \widehat{a}^2_h/M + \widehat{\sigma}^2_{a_h}, \tag{6.97}$$

$$c^2_{b_h} = \widehat{b}^2_h/L + \widehat{\sigma}^2_{b_h}. \tag{6.98}$$

However, unlike the VB solution, for which Lemma 6.1 holds, we cannot assume that the EVB solution is a stationary point, since the free energy $F_h$ given by Eq. (6.43) does not necessarily diverge to $+\infty$ when approaching the domain boundary (6.96). More specifically, $F_h$ can converge to a finite value, for example, for $\widehat{a}_h = \widehat{b}_h = 0, \widehat{\sigma}^2_{a_h}, \widehat{\sigma}^2_{b_h}, c^2_{a_h}, c^2_{b_h} \to +0$. Taking this into account, we can obtain the following theorem:

**Theorem 6.13** *Let*

$$\alpha = \frac{L}{M} \qquad (0 < \alpha \leq 1), \tag{6.99}$$

*and let $\underline{\tau} = \underline{\tau}(\alpha)$ be the unique zero-cross point of the following decreasing function:*

$$\Xi(\tau; \alpha) = \Phi(\tau) + \Phi\left(\frac{\tau}{\alpha}\right), \qquad \textit{where} \qquad \Phi(z) = \frac{\log(z+1)}{z} - \frac{1}{2}. \tag{6.100}$$

*Then, the EVB solution is given by*

$$\widehat{U}^{\mathrm{EVB}} = \sum_{h=1}^{H} \widehat{\gamma}_h^{\mathrm{EVB}} \omega_{b_h} \omega_{a_h}^{\top}, \qquad \textit{where} \qquad \widehat{\gamma}_h^{\mathrm{EVB}} = \begin{cases} \breve{\gamma}_h^{\mathrm{EVB}} & \textit{if } \gamma_h \geq \underline{\gamma}^{\mathrm{EVB}}, \\ 0 & \textit{otherwise}, \end{cases} \tag{6.101}$$

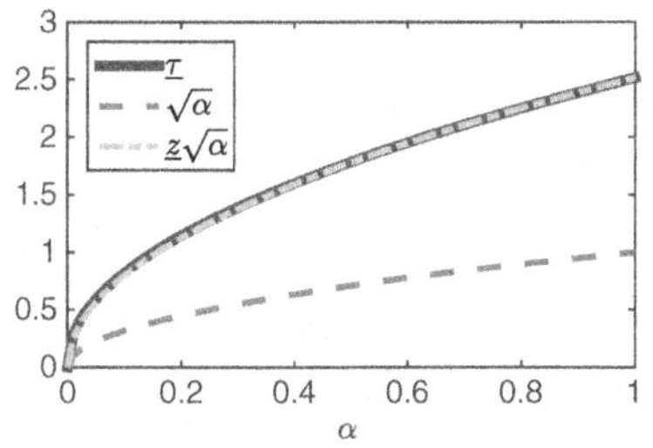


Figure 6.4 Values of $\underline{\tau}(\alpha)$, $\sqrt{\alpha}$, and $\underline{z}\sqrt{\alpha}$.

*for*

$$\underline{\gamma}^{\mathrm{EVB}} = \sigma\sqrt{M\left(1+\underline{\tau}\right)\left(1+\frac{\alpha}{\underline{\tau}}\right)}, \tag{6.102}$$

$$\breve{\gamma}_h^{\mathrm{EVB}} = \frac{\gamma_h}{2}\left(1-\frac{(M+L)\sigma^2}{\gamma_h^2}+\sqrt{\left(1-\frac{(M+L)\sigma^2}{\gamma_h^2}\right)^2-\frac{4LM\sigma^4}{\gamma_h^4}}\right). \tag{6.103}$$

The EVB threshold (6.102) involves $\underline{\tau}$, which needs to be numerically computed. However, we can easily prepare a table of the values for $0 < \alpha \leq 1$ beforehand, like the cumulative Gaussian probability used in statistical tests (Pearson, 1914). Alternatively, $\underline{\tau} \approx \underline{z}\sqrt{\alpha}$ is a good approximation, where $\underline{z} \approx 2.5129$ is the unique zero-cross point of $\Phi(z)$, as seen in Figure 6.4. We can show that $\underline{\tau}$ lies in the following range (Lemma 6.18 in Section 6.9):

$$\sqrt{\alpha} < \underline{\tau} < \underline{z}. \tag{6.104}$$

We will see in Chapter 8 that $\underline{\tau}$ is an important quantity in describing the behavior of the full empirical VB solution where the noise variance $\sigma^2$ is also estimated from observation.

In Section 6.9, we give the proof of Theorem 6.13. Then, in Section 6.10, some corollaries obtained and variables defined in the proof are summarized, which will be used in Chapter 8.

## 6.9 Proof of Theorem 6.13

In this section, we prove Theorem 6.13, which provides explicit forms, Eqs. (6.102) and (6.103), of the EVB threshold $\underline{\gamma}^{\mathrm{EVB}}$ and the EVB shrinkage estimator $\breve{\gamma}_h^{\mathrm{EVB}}$. In fact, we can easily obtain Eq. (6.103) in an intuitive way, by using some of the results obtained in Section 6.7. After that, by expressing the free energy $F_h$ with normalized versions of the observation and the estimator, we derive Eq. (6.102).

### 6.9.1 EVB Shrinkage Estimator

Eqs. (6.60) and (6.72) imply that the free energy does not depend on the ratio $c_{a_h}/c_{b_h}$ between the hyperparameters. Accordingly, we fix the ratio to $c_{a_h}/c_{b_h} = 1$ without loss of generality. Lemma 6.11 allows us to minimize the free energy with respect to $c_{a_h}c_{b_h}$ in a straightforward way.

Let us regard the free energies (6.60) and (6.72) at the null and the positive stationary points as functions of $c_{a_h}c_{b_h}$ (see Figure 6.3). Then, we find from Eq. (6.82) that

$$\frac{\partial F_h^{\mathrm{VB-Null}}}{\partial c_{a_h}^2 c_{b_h}^2} > 0,$$

which implies that the free energy (6.60) at the null stationary point is increasing. Using Lemma 6.9, we thus have the following lemma:

**Lemma 6.14** *For any given $\gamma_h \geq 0$ and $\sigma^2 > 0$, the null EVB local solution, given by*

$$\widehat{a}_h = 0, \quad \widehat{b}_h = 0, \quad \widehat{\sigma}_{a_h}^2 = \sqrt{\widehat{\zeta}^{\mathrm{EVB}}}, \quad \widehat{\sigma}_{b_h}^2 = \sqrt{\widehat{\zeta}^{\mathrm{EVB}}}, \quad c_{a_h}c_{b_h} = \sqrt{\widehat{\zeta}^{\mathrm{EVB}}},$$
$$\text{where} \qquad \widehat{\zeta}^{\mathrm{EVB}} \to +0,$$

*exists, and its free energy is given by*

$$F_h^{\mathrm{EVB-Null}} \to +0. \tag{6.105}$$

When $\gamma_h \geq (\sqrt{L} + \sqrt{M})\sigma$, the derivative (6.83) of the free energy (6.72) at the positive stationary point can be further factorized as

$$\frac{\partial F_h^{\mathrm{VB-Posi}}}{\partial c_{a_h}^2 c_{b_h}^2} = \frac{\gamma_h}{\sigma^2 c_{a_h}^2 c_{b_h}^2}\left(\breve{\gamma}_h^{\mathrm{VB}} - \acute{\gamma}_h\right)\left(\breve{\gamma}_h^{\mathrm{VB}} - \breve{\gamma}_h^{\mathrm{EVB}}\right), \tag{6.106}$$

$$\text{where} \quad \acute{\gamma}_h = \frac{\gamma_h}{2}\left(1 - \frac{(L+M)\sigma^2}{\gamma_h^2} - \sqrt{\left(1 - \frac{(L+M)\sigma^2}{\gamma_h^2}\right)^2 - \frac{4LM\sigma^4}{\gamma_h^4}}\right), \tag{6.107}$$

and $\breve{\gamma}_h^{\mathrm{EVB}}$ is given by Eq. (6.103). The VB shrinkage estimator (6.50) is an increasing function of $c_{a_h} c_{b_h}$ ranging over

$$0 < \breve{\gamma}_h^{\mathrm{VB}} < \gamma_h - \frac{M\sigma^2}{\gamma_h},$$

and both of Eqs. (6.107) and (6.103) are in this range, i.e.,

$$0 < \acute{\gamma}_h \leq \breve{\gamma}_h^{\mathrm{EVB}} < \gamma_h - \frac{M\sigma^2}{\gamma_h}.$$

Therefore Eq. (6.106) leads to the following lemma:

**Lemma 6.15** *If $\gamma_h \leq (\sqrt{L} + \sqrt{M})\sigma$, the free energy $F_h^{\mathrm{VB-Posi}}$ at the positive stationary point is monotonically increasing. Otherwise,*

$$F_h^{\mathrm{VB-Posi}} \textit{ is } \begin{cases} \textit{increasing} & \textit{for} \quad \breve{\gamma}_h^{\mathrm{VB}} < \acute{\gamma}_h, \\ \textit{decreasing} & \textit{for} \quad \acute{\gamma}_h < \breve{\gamma}_h^{\mathrm{VB}} < \breve{\gamma}_h^{\mathrm{EVB}}, \\ \textit{increasing} & \textit{for} \quad \breve{\gamma}_h^{\mathrm{VB}} > \breve{\gamma}_h^{\mathrm{EVB}}, \end{cases}$$

*and therefore minimized at $\breve{\gamma}_h^{\mathrm{VB}} = \breve{\gamma}_h^{\mathrm{EVB}}$.*

We can see this behavior of the free energy in Figure 6.3.

The derivative (6.83) is zero when $\breve{\gamma}_h^{\mathrm{VB}} = \breve{\gamma}_h^{\mathrm{EVB}}$, which leads to

$$\left(\breve{\gamma}_h^{\mathrm{EVB}} + \frac{L\sigma^2}{\gamma_h}\right)\left(\breve{\gamma}_h^{\mathrm{EVB}} + \frac{M\sigma^2}{\gamma_h}\right) = \gamma_h \breve{\gamma}_h^{\mathrm{EVB}}. \tag{6.108}$$

Using Eq. (6.108), we obtain the following lemma:

**Lemma 6.16** *If and only if*

$$\gamma_h \geq \underline{\gamma}^{\mathrm{local-EVB}} \equiv (\sqrt{L} + \sqrt{M})\sigma, \tag{6.109}$$

*the positive EVB local solution given by*

$$\widehat{a}_h = \pm\sqrt{\breve{\gamma}_h^{\mathrm{EVB}} \widehat{\delta}_h^{\mathrm{EVB}}}, \quad \widehat{b}_h = \pm\sqrt{\frac{\breve{\gamma}_h^{\mathrm{EVB}}}{\widehat{\delta}_h^{\mathrm{EVB}}}}, \tag{6.110}$$

$$\widehat{\sigma}_{a_h}^2 = \frac{\sigma^2 \widehat{\delta}_h^{\mathrm{EVB}}}{\gamma_h}, \quad \widehat{\sigma}_{b_h}^2 = \frac{\sigma^2}{\gamma_h \widehat{\delta}_h^{\mathrm{EVB}}}, \quad c_{a_h} c_{b_h} = \sqrt{\frac{\gamma_h \breve{\gamma}_h^{\mathrm{EVB}}}{LM}}, \tag{6.111}$$

*exists with the following free energy:*

$$F_h^{\mathrm{EVB-Posi}} = M \log\left(\frac{\gamma_h \breve{\gamma}_h^{\mathrm{EVB}}}{M\sigma^2} + 1\right) + L \log\left(\frac{\gamma_h \breve{\gamma}_h^{\mathrm{EVB}}}{L\sigma^2} + 1\right) - \frac{\gamma_h \breve{\gamma}_h^{\mathrm{EVB}}}{\sigma^2}. \tag{6.112}$$

*Here,*

$$\widehat{\delta}_h^{\mathrm{EVB}} = \sqrt{\frac{M\breve{\gamma}_h^{\mathrm{EVB}}}{L\gamma_h}} \left(1 + \frac{L\sigma^2}{\gamma_h \breve{\gamma}_h^{\mathrm{EVB}}}\right), \tag{6.113}$$

*and* $\breve{\gamma}_h^{\mathrm{EVB}}$ *is given by Eq.* (6.103).

*Proof* Lemma 6.15 immediately leads to the EVB shrinkage estimator (6.103). We can find the value of $c_{a_h} c_{b_h}$ at the positive EVB local solution by combining the condition (6.78) for the VB estimator and the condition (6.108) for the EVB estimator. Specifically, by using the condition (6.108), the condition (6.78) for $\breve{\gamma}_h^{\mathrm{VB}}$ replaced with $\breve{\gamma}_h^{\mathrm{EVB}}$ can be written as

$$\left(\gamma_h - \frac{\gamma_h \breve{\gamma}_h^{\mathrm{EVB}}}{\breve{\gamma}_h^{\mathrm{EVB}} + \frac{M\sigma^2}{\gamma_h}}\right)\left(\gamma_h - \frac{\gamma_h \breve{\gamma}_h^{\mathrm{EVB}}}{\breve{\gamma}_h^{\mathrm{EVB}} + \frac{L\sigma^2}{\gamma_h}}\right) = \frac{\sigma^4}{c_{a_h}^2 c_{b_h}^2},$$

and therefore

$$\left(\frac{M\sigma^2}{\breve{\gamma}_h^{\mathrm{EVB}} + \frac{M\sigma^2}{\gamma_h}}\right)\left(\frac{L\sigma^2}{\breve{\gamma}_h^{\mathrm{EVB}} + \frac{L\sigma^2}{\gamma_h}}\right) = \frac{\sigma^4}{c_{a_h}^2 c_{b_h}^2}.$$

Applying the condition (6.108) again gives

$$\frac{LM\sigma^4}{\gamma_h \breve{\gamma}_h^{\mathrm{EVB}}} = \frac{\sigma^4}{c_{a_h}^2 c_{b_h}^2},$$

which leads to the last equation in Eq. (6.111).

Similarly, using the condition (6.108), Eq. (6.52) for $\breve{\gamma}_h^{\mathrm{VB}}$ replaced with $\breve{\gamma}_h^{\mathrm{EVB}}$ is written as

$$\begin{aligned}
\widehat{\delta}_h &= \frac{c_{a_h}^2}{\sigma^2}\left(\gamma_h - \frac{\gamma_h \breve{\gamma}_h^{\mathrm{EVB}}}{\breve{\gamma}_h^{\mathrm{EVB}} + \frac{M\sigma^2}{\gamma_h}}\right) \\
&= \frac{c_{a_h}^2}{\sigma^2}\left(\frac{M\sigma^2}{\breve{\gamma}_h^{\mathrm{EVB}} + \frac{M\sigma^2}{\gamma_h}}\right) \\
&= \frac{c_{a_h}^2 M}{\gamma_h}\left(\frac{\gamma_h \breve{\gamma}_h^{\mathrm{EVB}} + L\sigma^2}{\gamma_h \breve{\gamma}_h^{\mathrm{EVB}}}\right) \\
&= \frac{c_{a_h}^2 M}{\gamma_h}\left(1 + \frac{L\sigma^2}{\gamma_h \breve{\gamma}_h^{\mathrm{EVB}}}\right).
\end{aligned}$$

Using the assumption that $c_{a_h} = c_{b_h}$ and therefore $c_{a_h}^2 = c_{a_h} c_{b_h}$, we obtain Eq. (6.113). Eq. (6.110) and the first two equations in Eq. (6.111) are simply obtained from Lemma 6.10.

Finally, applying Eq. (6.108) to the free energy (6.72), we have

$$F_h^{\mathrm{EVB-Posi}} = -M\log\left(1 - \frac{\gamma_h \breve{\gamma}_h^{\mathrm{EVB}}}{\gamma_h \breve{\gamma}_h^{\mathrm{EVB}} + M\sigma^2}\right) - L\log\left(1 - \frac{\gamma_h \breve{\gamma}_h^{\mathrm{EVB}}}{\gamma_h \breve{\gamma}_h^{\mathrm{EVB}} + L\sigma^2}\right)$$
$$- \frac{\gamma_h \breve{\gamma}_h^{\mathrm{EVB}}}{\sigma^2},$$

which leads to Eq. (6.112). This completes the proof of Lemma 6.16. □

In Figure 6.3, the positive EVB local solution at $c_{a_h} c_{b_h} = \sqrt{\gamma_h \breve{\gamma}_h^{\mathrm{EVB}}/(LM)}$ is indicated by a cross if it exists.

### 6.9.2 EVB Threshold

Lemmas 6.14 and 6.16 state that, if $\gamma_h \leq \underline{\gamma}^{\mathrm{local-EVB}}$, only the null EVB local solution exists, and therefore it is the global EVB solution. In this section, assuming that $\gamma_h \geq \underline{\gamma}^{\mathrm{local-EVB}}$, we compare the free energy (6.105) at the null EVB local solution and the free energy (6.112) at the positive EVB local solution. Since $F_h^{\mathrm{EVB-Null}} \to +0$, we simply consider the situation where $F_h^{\mathrm{EVB-Posi}} \leq 0$. Eq. (6.108) gives

$$\left(\gamma_h \breve{\gamma}_h^{\mathrm{EVB}} + L\sigma^2\right)\left(1 + \frac{M\sigma^2}{\gamma_h \breve{\gamma}_h^{\mathrm{EVB}}}\right) = \gamma_h^2. \tag{6.114}$$

By using Eqs. (6.103) and (6.109), we have

$$\begin{aligned} \gamma_h \breve{\gamma}_h^{\mathrm{EVB}} &= \frac{1}{2}\Bigg(\gamma_h^2 - \left(\underline{\gamma}^{\mathrm{local-EVB}}\right)^2 + 2\sqrt{LM}\sigma^2 \\ &\quad + \sqrt{\left(\gamma_h^2 - \left(\underline{\gamma}^{\mathrm{local-EVB}}\right)^2\right)\left(\gamma_h^2 - \left(\underline{\gamma}^{\mathrm{local-EVB}}\right)^2 + 4\sqrt{LM}\sigma^2\right)}\Bigg) \\ &\geq \sqrt{LM}\sigma^2. \end{aligned} \tag{6.115}$$

Remember the definition of $\alpha$ (Eq. (6.99))

$$\alpha = \frac{L}{M} \qquad (0 < \alpha \leq 1),$$

and let

$$x_h = \frac{\gamma_h^2}{M\sigma^2}, \tag{6.116}$$

$$\tau_h = \frac{\gamma_h \breve{\gamma}_h^{\mathrm{EVB}}}{M\sigma^2}. \tag{6.117}$$

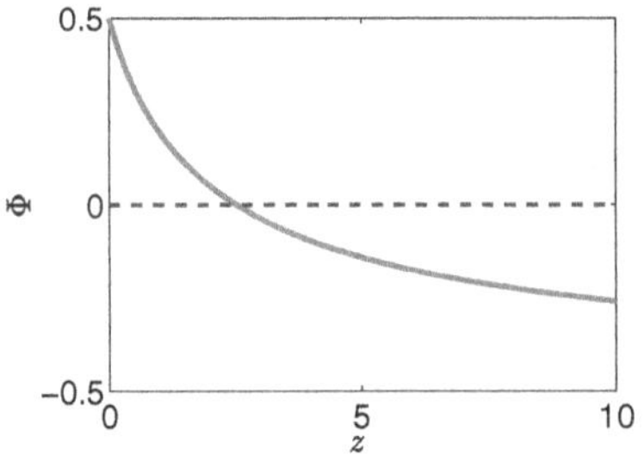


Figure 6.5 $\Phi(z) = \frac{\log(z+1)}{z} - \frac{1}{2}$.

Eqs. (6.114) and (6.103) imply the following mutual relations between $x_h$ and $\tau_h$:

$$x_h \equiv x(\tau_h; \alpha) = (1 + \tau_h)\left(1 + \frac{\alpha}{\tau_h}\right), \tag{6.118}$$

$$\tau_h \equiv \tau(x_h; \alpha) = \frac{1}{2}\left(x_h - (1 + \alpha) + \sqrt{(x_h - (1 + \alpha))^2 - 4\alpha}\right). \tag{6.119}$$

Eqs. (6.109) and (6.115) lead to

$$x_h \geq \underline{x}^{\text{local}} = \frac{(\underline{\gamma}^{\text{local-EVB}})^2}{M\sigma^2} = x(\sqrt{\alpha}; \alpha) = (1 + \sqrt{\alpha})^2, \tag{6.120}$$

$$\tau_h \geq \underline{\tau}^{\text{local}} = \sqrt{\alpha}. \tag{6.121}$$

Then, using $\Xi(\tau; \alpha)$ defined by Eq. (6.100), we can rewrite Eq. (6.112) as

$$\begin{aligned} F_h^{\text{EVB-Posi}} &= M \log(\tau_h + 1) + L \log\left(\frac{\tau_h}{\alpha} + 1\right) - M\tau_h \\ &= M\tau_h \Xi(\tau; \alpha). \end{aligned} \tag{6.122}$$

The following holds for $\Phi(z)$ (which is also defined in Eq. (6.100)):

**Lemma 6.17** *$\Phi(z)$ is decreasing for $z > 0$.*

*Proof* The derivative is

$$\frac{\partial \Phi}{\partial z} = \frac{1 - \frac{1}{z+1} - \log(z+1)}{z^2},$$

which is negative for $z > 0$ because

$$\frac{1}{z+1} + \log(z+1) > 1.$$

This completes the proof of Lemma 6.17. □

Figure 6.5 shows the profile of $\Phi(z)$. Since $\Phi(z)$ is decreasing, $\Xi(\tau;\alpha)$ is also decreasing with respect to $\tau$. It holds that, for any $0 < \alpha \leq 1$,

$$\lim_{\tau\to 0} \Xi(\tau;\alpha) = 1,$$
$$\lim_{\tau\to\infty} \Xi(\tau;\alpha) = -1.$$

Therefore, $\Xi(\tau;\alpha)$ has a unique zero-cross point $\underline{\tau}$, such that

$$\Xi(\tau;\alpha) \leq 0 \qquad \text{if and only if} \qquad \tau \geq \underline{\tau}. \tag{6.123}$$

Then, we can prove the following lemma:

**Lemma 6.18** *The unique zero-cross point $\underline{\tau}$ of $\Xi(\tau;\alpha)$ lies in the following range:*

$$\sqrt{\alpha} < \underline{\tau} < \underline{z},$$

*where $\underline{z} \approx 2.5129$ is the unique zero-cross point of $\Phi(z)$.*

*Proof* Since $\Phi(z)$ is decreasing, $\Xi(\tau;\alpha)$ is upper-bounded by

$$\Xi(\tau;\alpha) = \Phi(\tau) + \Phi\left(\frac{\tau}{\alpha}\right) \leq 2\Phi(\tau) = \Xi(\tau;1).$$

Therefore, the unique zero-cross point $\underline{\tau}$ of $\Xi(\tau;\alpha)$ is no greater than the unique zero-cross point $\underline{z}$ of $\Phi(z)$, i.e.,

$$\underline{\tau} \leq \underline{z}.$$

For obtaining the lower-bound $\underline{\tau} > \sqrt{\alpha}$, it suffices to show that $\Xi(\sqrt{\alpha};\alpha) > 0$. Let us prove that the following function is decreasing and positive for $0 < \alpha \leq 1$:

$$g(\alpha) \equiv \frac{\Xi\left(\sqrt{\alpha};\alpha\right)}{\sqrt{\alpha}}.$$

From the definition (6.100) of $\Xi(\tau;\alpha)$, we have

$$g(\alpha) = \left(1 + \frac{1}{\alpha}\right)\log(\sqrt{\alpha}+1) - \log\sqrt{\alpha} - \frac{1}{\sqrt{\alpha}}.$$

The derivative is given by

$$\begin{aligned}
\frac{\partial g}{\partial\sqrt{\alpha}} &= \frac{\left(1+\frac{1}{\alpha}\right)}{\sqrt{\alpha}+1} - \frac{2}{\alpha^{3/2}}\log(\sqrt{\alpha}+1) - \frac{1}{\sqrt{\alpha}} + \frac{1}{\alpha} \\
&= -\frac{2}{\alpha^{3/2}}\left(\log(\sqrt{\alpha}+1) + \frac{1}{\sqrt{\alpha}+1} - 1\right) \\
&< 0,
\end{aligned}$$

which implies that $g(\alpha)$ is decreasing. Since

$$g(1) = 2\log 2 - 1 \approx 0.3863 > 0,$$

$g(\alpha)$ is positive for $0 < \alpha \leq 1$, which completes the proof of Lemma 6.18. □

Since Eq. (6.118) is increasing with respect to $\tau_h$ ($> \sqrt{\alpha}$), the thresholding condition $\tau \geq \underline{\tau}$ in Eq. (6.123) can be expressed in terms of $x$:

$$\Xi(\tau(x); \alpha) \leq 0 \qquad \text{if and only if} \qquad x \geq \underline{x}, \tag{6.124}$$

$$\text{where} \qquad \underline{x} \equiv x(\underline{\tau}; \alpha) = \left(1 + \underline{\tau}\right)\left(1 + \frac{\alpha}{\underline{\tau}}\right). \tag{6.125}$$

Using Eqs. (6.116) and (6.122), we have

$$F_h^{\mathrm{EVB-Posi}} \leq 0 \qquad \text{if and only if} \qquad \gamma_h \geq \underline{\gamma}^{\mathrm{EVB}}, \tag{6.126}$$

where $\underline{\gamma}^{\mathrm{EVB}}$ is defined by Eq. (6.102). Thus, we have the following lemma:

**Lemma 6.19** *The positive EVB local solution is the global EVB solution if and only if* $\gamma_h \geq \underline{\gamma}^{\mathrm{EVB}}$.

Combining Lemmas 6.14, 6.16, and 6.19 completes the proof of Theorem 6.13. □

Figure 6.6 shows estimators and thresholds for $L = M = H = 1$ and $\sigma^2 = 1$. The curves indicate the VB solution $\widehat{\gamma}_h^{\mathrm{VB}}$ given by Eq. (6.48), the EVB solution $\widehat{\gamma}_h^{\mathrm{EVB}}$ given by Eq. (6.101), the EVB positive local minimizer $\breve{\gamma}_h^{\mathrm{EVB}}$ given by Eq. (6.103), and the EVB positive local maximizer $\acute{\gamma}_h$ given by Eq. (6.107), respectively. The arrows indicate the VB threshold $\underline{\gamma}_h^{\mathrm{VB}}$ given by Eq. (6.49), the local EVB threshold $\underline{\gamma}^{\mathrm{local-EVB}}$ given by Eq. (6.109), and the EVB threshold $\underline{\gamma}^{\mathrm{EVB}}$ given by Eq. (6.102), respectively.

## 6.10 Summary of Intermediate Results

In the rest of this section, we summarize some intermediate results obtained in Section 6.9, which are useful for further analysis (mainly in Chapter 8).

Summarizing Eqs. (6.109), (6.114), and (6.115) leads to the following corollary:

**Corollary 6.20** *The EVB shrinkage estimator* (6.103) *is a stationary point of the free energy* (6.43), *which exists if and only if*

$$\gamma_h \geq \underline{\gamma}^{\mathrm{local-EVB}} \equiv (\sqrt{L} + \sqrt{M})\sigma, \tag{6.127}$$

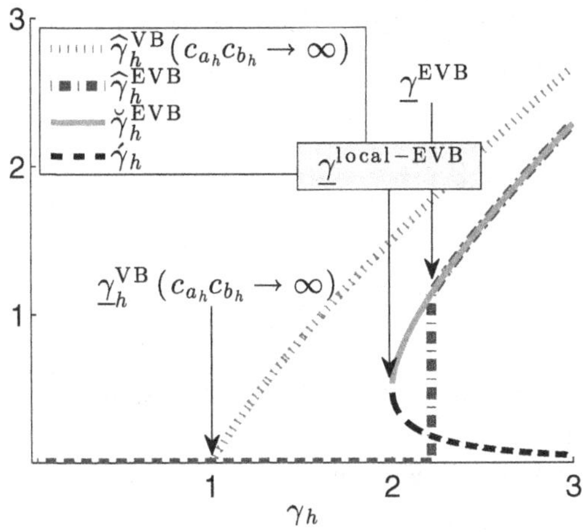


Figure 6.6 Estimators and thresholds for $L = M = H = 1$ and $\sigma^2 = 1$.

*and satisfies the following equation:*

$$\left(\gamma_h \breve{\gamma}_h^{\mathrm{EVB}} + L\sigma^2\right)\left(1 + \frac{M\sigma^2}{\gamma_h \breve{\gamma}_h^{\mathrm{EVB}}}\right) = \gamma_h^2. \tag{6.128}$$

*It holds that*

$$\gamma_h \breve{\gamma}_h^{\mathrm{EVB}} \geq \sqrt{LM}\sigma^2. \tag{6.129}$$

Combining Lemmas 6.14, 6.16, and 6.19 leads to the following corollary:

**Corollary 6.21** *The minimum free energy achieved under EVB learning is given by Eq. (6.42) with*

$$2F_h = \begin{cases} M \log\left(\frac{\gamma_h \breve{\gamma}_h^{\mathrm{EVB}}}{M\sigma^2} + 1\right) + L \log\left(\frac{\gamma_h \breve{\gamma}_h^{\mathrm{EVB}}}{L\sigma^2} + 1\right) - \frac{\gamma_h \breve{\gamma}_h^{\mathrm{EVB}}}{\sigma^2} & \text{if } \gamma_h \geq \underline{\gamma}^{\mathrm{EVB}}, \\ +0 & \text{otherwise.} \end{cases} \tag{6.130}$$

Corollary 6.20 together with Theorem 6.13 implies that when

$$\underline{\gamma}^{\mathrm{local-EVB}} \leq \gamma_h < \underline{\gamma}^{\mathrm{EVB}},$$

a stationary point (called the *positive* EVB local solution and specified by Lemma 6.16) exists at Eq. (6.103), but it is not the global minimum. Actually, a local minimum (called the *null* EVB local solution and specified by Lemma 6.14) with $F_h = +0$ always exists. The stationary point at Eq. (6.103) is a *nonglobal* local minimum when $\underline{\gamma}^{\mathrm{local-EVB}} \leq \gamma_h < \underline{\gamma}^{\mathrm{EVB}}$ and the global minimum when $\gamma_h \geq \underline{\gamma}^{\mathrm{EVB}}$ (see Figure 6.3 with its caption). This phase transition induces the free energy thresholding observed in Corollary 6.21.

We define a *local-EVB estimator* by

$$\widehat{U}^{\mathrm{local-EVB}} = \sum_{h=1}^{H} \widehat{\gamma}_h^{\mathrm{local-EVB}} \omega_{b_h} \omega_{a_h}^{\top},$$

$$\text{where} \quad \widehat{\gamma}_h^{\mathrm{local-EVB}} = \begin{cases} \breve{\gamma}_h^{\mathrm{EVB}} & \text{if } \gamma_h \geq \underline{\gamma}^{\mathrm{local-EVB}}, \\ 0 & \text{otherwise,} \end{cases} \tag{6.131}$$

and call $\underline{\gamma}^{\mathrm{local-EVB}}$ a local-EVB threshold. This estimator gives the positive EVB local solution, whenever it exists, for each singular component. There is an interesting relation between the *local*-EVB solution and an alternative dimensionality selection method (Hoyle, 2008), which will be discussed in Section 8.6.

Rescaling the quantities related to the squared singular value by $M\sigma^2$—to which the contribution from noise (each eigenvalue of $\mathcal{E}^{\top}\mathcal{E}$) scales linearly—simplifies expressions. Assume that the condition (6.127) holds, and define

$$x_h = \frac{\gamma_h^2}{M\sigma^2}, \tag{6.132}$$

$$\tau_h = \frac{\gamma_h \breve{\gamma}_h^{\mathrm{EVB}}}{M\sigma^2}, \tag{6.133}$$

which are used as a rescaled observation and a rescaled EVB estimator, respectively. Eqs. (6.128) and (6.103) specify the mutual relations between them:

$$x_h \equiv x(\tau_h; \alpha) = (1 + \tau_h)\left(1 + \frac{\alpha}{\tau_h}\right), \tag{6.134}$$

$$\tau_h \equiv \tau(x_h; \alpha) = \frac{1}{2}\left(x_h - (1 + \alpha) + \sqrt{(x_h - (1 + \alpha))^2 - 4\alpha}\right). \tag{6.135}$$

With these rescaled variables, the condition (6.127), as well as (6.129), for the existence of the positive local-EVB solution $\breve{\gamma}_h^{\mathrm{EVB}}$ is expressed as

$$x_h \geq \underline{x}^{\mathrm{local}} = \frac{(\underline{\gamma}^{\mathrm{local-EVB}})^2}{M\sigma^2} = x(\sqrt{\alpha}; \alpha) = (1 + \sqrt{\alpha})^2, \tag{6.136}$$

$$\tau_h \geq \underline{\tau}^{\mathrm{local}} = \sqrt{\alpha}. \tag{6.137}$$

The EVB threshold (6.102) is expressed as

$$\underline{x} = \frac{(\underline{\gamma}^{\mathrm{EVB}})^2}{M\sigma^2} = x(\underline{\tau}; \alpha) = \left(1 + \underline{\tau}\right)\left(1 + \frac{\alpha}{\underline{\tau}}\right), \tag{6.138}$$

and the free energy (6.130) is expressed as

$$F_h = M\tau_h \cdot \min\left(0, \Xi\left(\tau_h; \alpha\right)\right), \tag{6.139}$$

where $\Xi(\tau; \alpha)$ is defined by Eq. (6.100).

The preceding rescaled expressions give an intuition of Theorem 6.13: the EVB solution $\widehat{\gamma}_h^{\mathrm{EVB}}$ is positive if and only if the positive local-EVB solution $\breve{\gamma}_h^{\mathrm{EVB}}$ exists (i.e., $x_h \geq \underline{x}^{\mathrm{local}}$), and the free energy $\Xi\left(\tau(x_h; \alpha); \alpha\right)$ at the local-EVB solution is nonpositive (i.e., $\tau(x_h; \alpha) \geq \underline{\tau}$ or equivalently $x_h \geq \underline{x}$ ).

# 7

# Model-Induced Regularization and Sparsity Inducing Mechanism

Variational Bayesian (VB) learning often shows the automatic relevance determination (ARD) property—the solution is sparse with unnecessary components eliminated automatically. In this chapter, we try to elucidate the sparsity inducing mechanism of VB learning, based on the global analytic solutions derived in Chapter 6. We argue that the ARD property is induced by the model-induced regularization (MIR), which all Bayesian learning methods possess when *unidentifiable* models are involved, and that MIR is enhanced by the independence constraint (imposed for computational tractability), which induces phase transitions making the solution (exactly) sparse (Nakajima and Sugiyama, 2011).

We first show the VB solution for special cases where the MIR effect is visible in the solution form. Then we illustrate the behavior of the posteriors and estimators in the one-dimensional case, comparing VB learning with maximum a posteriori (MAP) learning and Bayesian learning. After that, we explain MIR, and how it is enhanced in VB learning through phase transitions.

## 7.1 VB Solutions for Special Cases

Here we discuss two special cases of fully observed matrix factorization (MF), in which the VB solution is simple and intuitive.

### Almost Flat Prior

When $c_{a_h} c_{b_h} \to \infty$ (i.e., the prior is *almost* flat), the VB solution given by Theorem 6.7 in Chapter 6 has a simple form.

**Corollary 7.1** *The VB solution of the fully observed matrix factorization model* (6.1) *through* (6.3) *is given by*

$$\widehat{U}^{\mathrm{VB}} = \sum_{h=1}^{H} \widehat{\gamma}_h^{\mathrm{VB}} \omega_{b_h} \omega_{a_h}^{\top}, \tag{7.1}$$

*where the estimator* $\widehat{\gamma}_h^{\mathrm{VB}}$ *corresponding to the hth largest singular value is upper-bounded as*

$$\widehat{\gamma}_h^{\mathrm{VB}} < \max\left\{0, \left(1 - \frac{\max(L, M)\sigma^2}{\gamma_h^2}\right)\gamma_h\right\}. \tag{7.2}$$

*For the almost flat prior (i.e.,* $c_{a_h} c_{b_h} \to \infty$*), the equality holds, i.e.,*

$$\lim_{c_{a_h} c_{b_h} \to \infty} \widehat{\gamma}_h^{\mathrm{VB}} = \max\left\{0, \left(1 - \frac{\max(L, M)\sigma^2}{\gamma_h^2}\right)\gamma_h\right\}. \tag{7.3}$$

*Proof* It is clear that the threshold (6.49) is decreasing and the shrinkage factor (6.50) is increasing with respect to $c_{a_h} c_{b_h}$. Therefore, $\widehat{\gamma}_h^{\mathrm{VB}}$ is largest for $c_{a_h} c_{b_h} \to \infty$. In this limit, Eqs. (6.49) and (6.50) are reduced to

$$\begin{aligned}
\lim_{c_{a_h} c_{b_h} \to \infty} \underline{\gamma}_h^{\mathrm{VB}} &= \sigma \sqrt{\frac{(L+M)}{2} + \sqrt{\left(\frac{(L+M)}{2}\right)^2 - LM}} \\
&= \sigma \sqrt{\max(L, M)}, \\
\lim_{c_{a_h} c_{b_h} \to \infty} \breve{\gamma}_h^{\mathrm{VB}} &= \gamma_h \left(1 - \frac{\sigma^2}{2\gamma_h^2}\left(M + L + \sqrt{(M-L)^2}\right)\right) \\
&= \left(1 - \frac{\max(L, M)\sigma^2}{\gamma_h^2}\right)\gamma_h,
\end{aligned}$$

which prove the corollary. □

The form of the VB solution (7.3) in the limit is known as the *positive-part James–Stein (PJS) estimator* (James and Stein, 1961), operated on each singular component separately (see Appendix A for its interesting property and the relation to Bayesian learning). A counterintuitive fact—a shrinkage is observed even in the limit of the flat prior—will be explained in terms of MIR in Section 7.3.

### Square Matrix

When $L = M$ (i.e., the observed matrix $\boldsymbol{V}$ is square), the VB solution is intuitive, so that the shrinkage caused by MIR and the shrinkage caused by the prior are separately visible in its formula.

**Corollary 7.2** *When $L = M$, the VB solution is given by Eq.* (7.1) *with*

$$\widehat{\gamma}_h^{\mathrm{VB}} = \max\left\{0, \left(1 - \frac{M\sigma^2}{\gamma_h^2}\right)\gamma_h - \frac{\sigma^2}{c_{a_h}c_{b_h}}\right\}. \tag{7.4}$$

*Proof* When $L = M$, Eqs. (6.49) and (6.50) can be written as

$$\underline{\gamma}_h^{\mathrm{VB}} = \sigma\sqrt{M + \frac{\sigma^2}{2c_{a_h}^2c_{b_h}^2} + \sqrt{\left(M + \frac{\sigma^2}{2c_{a_h}^2c_{b_h}^2}\right)^2 - M^2}},$$

$$\begin{aligned}\breve{\gamma}_h^{\mathrm{VB}} &= \gamma_h\left(1 - \frac{\sigma^2}{2\gamma_h^2}\left(2M + \sqrt{\frac{4\gamma_h^2}{c_{a_h}^2c_{b_h}^2}}\right)\right)\\ &= \gamma_h\left(1 - \frac{\sigma^2}{\gamma_h^2}\left(M + \frac{\gamma_h}{c_{a_h}c_{b_h}}\right)\right).\end{aligned}$$

We can confirm that $\breve{\gamma}_h^{\mathrm{VB}} \leq 0$ when $\gamma_h \leq \underline{\gamma}_h^{\mathrm{VB}}$, which proves the corollary. Actually, we can confirm that $\breve{\gamma}_h^{\mathrm{VB}} = 0$ when $\gamma_h = \underline{\gamma}_h^{\mathrm{VB}}$, and $\breve{\gamma}_h^{\mathrm{VB}} < 0$ when $\gamma_h < \underline{\gamma}_h^{\mathrm{VB}}$ for any $L, M, c_{a_h}^2$, and $c_{b_h}^2$. □

In the VB solution (7.4), we can identify the PJS shrinkage and a constant shrinkage. The PJS shrinkage can be considered to be caused by MIR since it appears even with the flat prior, while the constant shrinkage $-\sigma^2/(c_{a_h}c_{b_h})$ is considered to be caused by the prior since it appears in MAP learning (see Theorem 12.1 in Chapter 12).

The empirical VB (EVB) solution is also simple for square matrices. The following corollary is obtained from Theorem 6.13 in Chapter 6:

**Corollary 7.3** *When $L = M$, the global EVB solution is given by*

$$\widehat{\gamma}_h^{\mathrm{EVB}} = \begin{cases}\breve{\gamma}_h^{\mathrm{EVB}} & \text{if } \gamma_h > \underline{\gamma}^{\mathrm{EVB}},\\ 0 & \text{otherwise,}\end{cases}$$

*where*

$$\underline{\gamma}^{\mathrm{EVB}} = \sigma\sqrt{M\left(2 + \underline{\tau}(1) + \frac{1}{\underline{\tau}(1)}\right)},$$

$$\breve{\gamma}_h^{\mathrm{EVB}} = \frac{\gamma_h}{2}\left(1 - \frac{2M\sigma^2}{\gamma_h^2} + \sqrt{1 - \frac{4M\sigma^2}{\gamma_h^2}}\right).$$

*Proof* When $L = M$, Eqs. (6.102) and (6.103) can be written as

$$\underline{\gamma}^{\mathrm{EVB}} = \sigma \sqrt{M\left(2 + \underline{\tau}(1) + \frac{1}{\underline{\tau}(1)}\right)},$$

$$\breve{\gamma}_h^{\mathrm{EVB}} = \frac{\gamma_h}{2}\left(1 - \frac{2M\sigma^2}{\gamma_h^2} + \sqrt{\left(1 - \frac{2M\sigma^2}{\gamma_h^2}\right)^2 - \frac{4M^2\sigma^4}{\gamma_h^4}}\right)$$
$$= \frac{\gamma_h}{2}\left(1 - \frac{2M\sigma^2}{\gamma_h^2} + \sqrt{1 - \frac{4M\sigma^2}{\gamma_h^2}}\right),$$

which completes the proof. □

## 7.2 Posteriors and Estimators in a One-Dimensional Case

In order to illustrate how strongly Bayesian learning and its approximation methods are regularized, we depict posteriors and estimators in the MF model for $L = M = H = 1$ (i.e., $U$, $V$, $A$, and $B$ are merely scalars):

$$p(V|A, B) = \frac{1}{\sqrt{2\pi\sigma^2}} \exp\left(-\frac{(V - BA)^2}{2\sigma^2}\right). \tag{7.5}$$

In this model, we can visualize the *unidentifiability* of the MF model as *equivalence classes*—a set of points $(A, B)$ on which the product is unchanged, i.e., $U = BA$, represents the same distribution (see Figure 7.1). When $U = 0$, the equivalence class has a "cross-shape" profile on the $A$- and $B$-axes; otherwise, it forms a pair of hyperbolic curves. This redundant structure in the

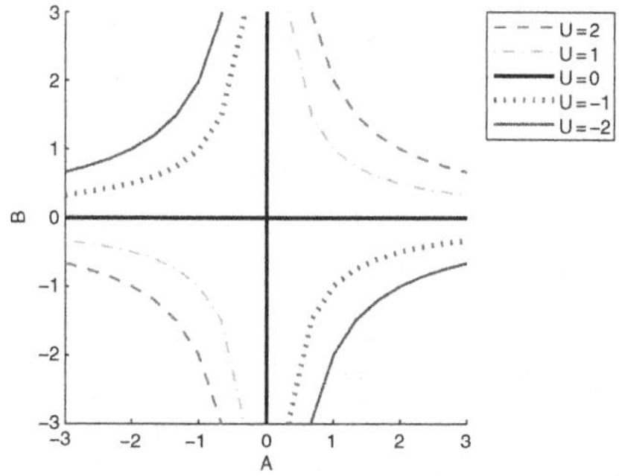


Figure 7.1 Equivalence class structure of the one-dimensional MF model. Any $A$ and $B$ such that their product is unchanged give the same $U$.

parameter space is the origin of MIR, and highly influences the phase transition phenomenon in VB learning, as we will see shortly.

With Gaussian priors,

$$p(A) = \frac{1}{\sqrt{2\pi c_a^2}} \exp\left(-\frac{A^2}{2c_a^2}\right), \tag{7.6}$$

$$p(B) = \frac{1}{\sqrt{2\pi c_b^2}} \exp\left(-\frac{B^2}{2c_b^2}\right), \tag{7.7}$$

the Bayes posterior is proportional to

$$\begin{aligned} p(A, B|V) &\propto p(V|A, B)p(A)p(B) \\ &\propto \exp\left(-\frac{1}{2\sigma^2}(V - BA)^2 - \frac{A^2}{2c_a^2} - \frac{B^2}{2c_b^2}\right). \end{aligned} \tag{7.8}$$

Figure 7.2 shows the contour of the *unnormalized* Bayes posterior (7.8) when $V = 0, 1, 2$ are observed, the noise variance is $\sigma^2 = 1$, and the prior covariances are set to $c_a = c_b = 100$ (i.e., almost flat priors). We can see that the equivalence class structure is reflected in the Bayes posterior: when $V = 0$, the surface of the Bayes posterior has a cross-shaped profile and its maximum is at the origin; when $V > 0$, the surface is divided into the positive orthant (i.e., $A, B > 0$) and the negative orthant (i.e., $A, B < 0$), and the two "modes" get farther as $V$ increases.

## MAP Solution

Let us first investigate the behavior of the *MAP estimator*, which coincides with the maximum likelihood (ML) estimator when the priors are flat. For finite $c_a$ and $c_b$, the MAP solution can be expressed as

$$\widehat{A}^{\mathrm{MAP}} = \pm\sqrt{\frac{c_a}{c_b} \max\left\{0, |V| - \frac{\sigma^2}{c_a c_b}\right\}},$$

$$\widehat{B}^{\mathrm{MAP}} = \pm\mathrm{sign}(V)\sqrt{\frac{c_b}{c_a} \max\left\{0, |V| - \frac{\sigma^2}{c_a c_b}\right\}},$$

where $\mathrm{sign}(\cdot)$ denotes the sign of a scalar (see Corollary 12.2 in Chapter 12 for derivation). In Figure 7.2, the asterisks indicate the MAP estimators, and the dashed curves indicate the ML estimators (the modes of the contour of Eq. (7.8) when $c_a = c_b \to \infty$). When $V = 0$, the Bayes posterior takes the maximum value on the $A$- and $B$-axes, which results in the MAP estimator equal to $\widehat{U}^{\mathrm{MAP}} (= \widehat{B}^{\mathrm{MAP}} \widehat{A}^{\mathrm{MAP}}) = 0$. When $V = 1$, the profile of the Bayes posterior is hyperbolic and the maximum value is achieved on the hyperbolic curves in the positive orthant (i.e., $A, B > 0$) and the negative orthant (i.e., $A, B < 0$);

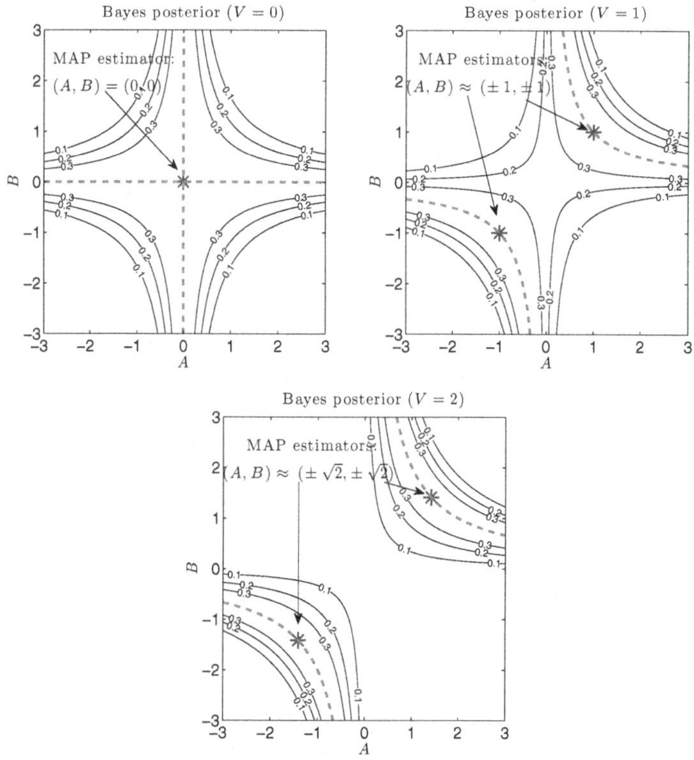


Figure 7.2 (Unnormalized) Bayes posteriors for $c_a = c_b = 100$ (i.e., almost flat priors). The asterisks are the MAP estimators, and the dashed curves indicate the ML estimators (the modes of the contour when $c_a = c_b = c \to \infty$).

in either case, $\widehat{U}^{\mathrm{MAP}} \approx 1$ ($\lim_{c_a, c_b \to \infty} \widehat{U}^{\mathrm{MAP}} = 1$). When $V = 2$, a similar multimodal structure is observed and the MAP estimator is $\widehat{U}^{\mathrm{MAP}} \approx 2$ ($\lim_{c_a, c_b \to \infty} \widehat{U}^{\mathrm{MAP}} = 2$). From these plots, we can visually confirm that the MAP estimator with almost flat priors ($c_a = c_b = 100$) approximately agrees with the ML estimator: $\widehat{U}^{\mathrm{MAP}} \approx \widehat{U}^{\mathrm{ML}} = V$ ($\lim_{c_a, c_b \to \infty} \widehat{U}^{\mathrm{MAP}} = \widehat{U}^{\mathrm{ML}}$). We will use the ML estimator as an *unregularized* reference in the following discussion.

Figure 7.3 shows the contour of the Bayes posterior when $c_a = c_b = 2$. The MAP estimators shift from the ML solutions (dashed curves) toward the origin, and they are more clearly contoured as peaks.

### VB Solution

Next we depict the VB posterior, given by Corollary 6.8 in Chapter 6. When $L = M = H = 1$, the VB solution is given by

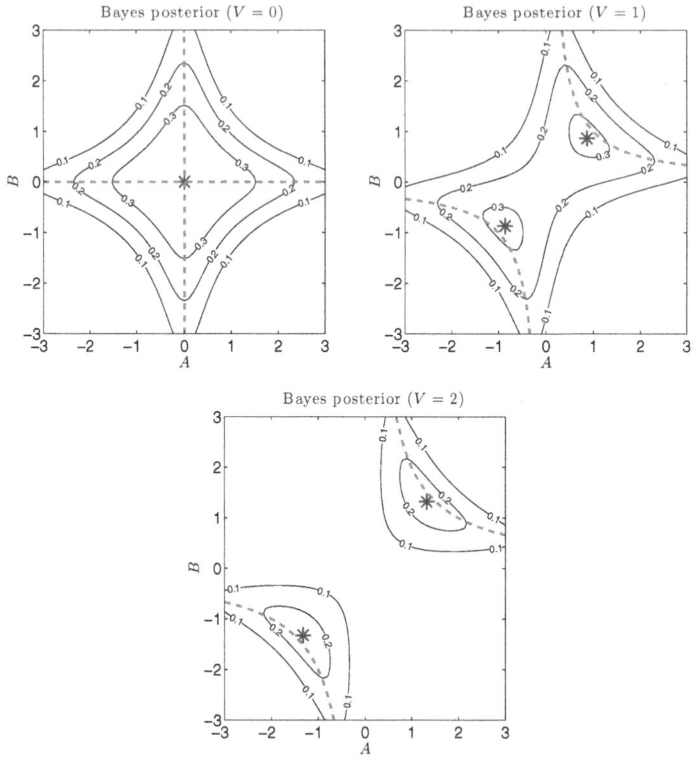


Figure 7.3 (Unnormalized) Bayes posteriors for $c_a = c_b = 2$. The dashed curves indicating the ML estimators are identical to those in Figure 7.2.

$$r(A, B) = \begin{cases} \mathrm{Gauss}_1\left(A; \pm\sqrt{\breve{\gamma}^{\mathrm{VB}} \frac{c_a}{c_b}}, \frac{\sigma^2 c_a}{|V| c_b}\right) \mathrm{Gauss}_1\left(B; \pm\mathrm{sign}(V)\sqrt{\breve{\gamma}^{\mathrm{VB}} \frac{c_b}{c_a}}, \frac{\sigma^2 c_b}{|V| c_a}\right) & \text{if } |V| \geq \underline{\gamma}^{\mathrm{VB}}, \\ \mathrm{Gauss}_1\left(A; 0, c_a^2 \widehat{\kappa}^{\mathrm{VB}}\right) \mathrm{Gauss}_1\left(B; 0, c_b^2 \widehat{\kappa}^{\mathrm{VB}}\right) & \text{otherwise}, \end{cases} \tag{7.9}$$

where

$$\underline{\gamma}^{\mathrm{VB}} = \sigma\sqrt{1 + \frac{\sigma^2}{2c_a^2 c_b^2} + \sqrt{\left(1 + \frac{\sigma^2}{2c_a^2 c_b^2}\right)^2 - 1}},$$

$$\breve{\gamma}^{\mathrm{VB}} = \left(1 - \frac{\sigma^2}{V^2}\right)|V| - \frac{\sigma^2}{c_a c_b},$$

$$\widehat{\kappa}^{\mathrm{VB}} = -\frac{\sigma^2}{2c_a^2 c_b^2} + \sqrt{\left(1 + \frac{\sigma^2}{2c_a^2 c_b^2}\right)^2 - 1}.$$

Figure 7.4 shows the contour of the VB posterior (7.9) when $V = 0, 1, 2$ are observed, the noise variance is $\sigma^2 = 1$, and the prior covariances are

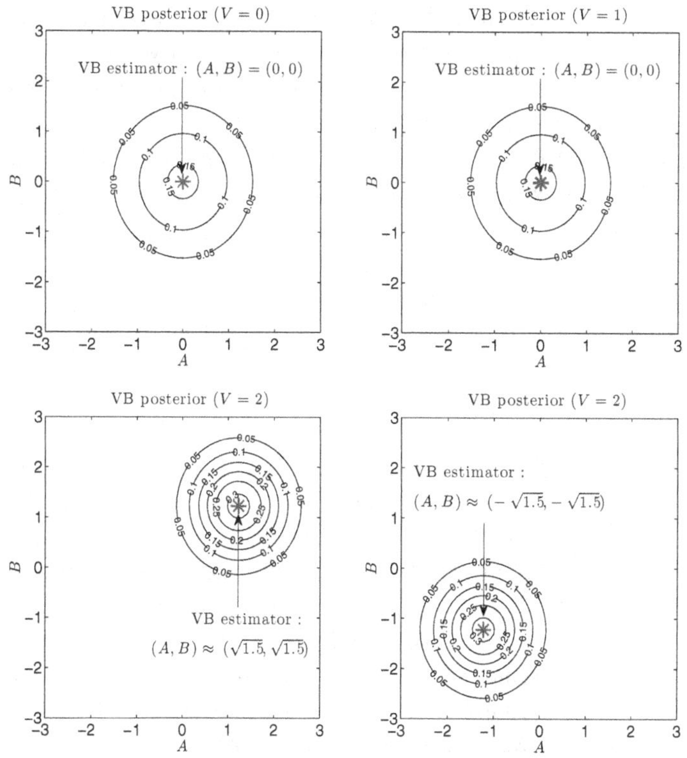


Figure 7.4 VB solutions for $c_a = c_b = 100$ (i.e., almost flat priors). When $V = 2$, VB learning gives either one of the two solutions shown in the bottom row.

set to $c_a = c_b = 100$ (i.e., almost flat priors). When $V = 0$, the cross-shaped contour of the Bayes posterior (see Figure 7.2) is approximated by a spherical Gaussian distribution located at the origin. Thus, the VB estimator is $\widehat{U}^{\mathrm{VB}} = 0$, which coincides with the MAP estimator. When $V = 1$, two hyperbolic "modes" of the Bayes posterior are approximated again by a spherical Gaussian distribution located at the origin. Thus, the VB estimator is still $\widehat{U}^{\mathrm{VB}} = 0$, which differs from the MAP estimator $\widehat{U}^{\mathrm{MAP}} \approx 1$.

$V = \underline{\gamma}^{\mathrm{VB}} \approx \sqrt{M\sigma^2} = 1$ ($\lim_{c_a, c_b \to \infty} \underline{\gamma}^{\mathrm{VB}} = \sqrt{M\sigma^2}$) is actually a transition point of the VB solution. When $V$ is not larger than the threshold $\underline{\gamma}^{\mathrm{VB}} \approx 1$, VB learning tries to approximate the two "modes" of the Bayes posterior by the origin-centered Gaussian distribution. When $V$ goes beyond the threshold $\underline{\gamma}^{\mathrm{VB}} \approx 1$, the "distance" between two hyperbolic modes of the Bayes posterior becomes so large that VB learning chooses to approximate one of those two modes in the positive and the negative orthants. As such, the symmetry is broken spontaneously and the VB estimator is detached from the origin. The bottom row of Figure 7.4 shows the contour of the two possible VB posteriors when $V = 2$. Note that the VB estimator, $\widehat{U}^{\mathrm{VB}} \approx 3/2$, is the same for both cases, and differs from the MAP estimator $\widehat{U}^{\mathrm{MAP}} \approx 2$.

In general, the VB estimator is closer to the origin than the MAP estimator, and the relative difference between them tends to shrink as $V$ increases.

### Bayesian Estimator

The full Bayesian estimator is defined as the mean of the Bayes posterior (see Eq. (1.7)). In the MF model with $L = M = H = 1$, the Bayesian estimator is expressed as

$$\widehat{U}^{\mathrm{Bayes}} = \langle BA \rangle_{p(V|A,B)p(A)p(B)/p(V)} \,. \tag{7.10}$$

If $V = 0, 1, 2, 3$ are observed, the Bayesian estimator with almost flat priors are $\widehat{U}^{\mathrm{Bayes}} = 0, 0.92, 1.93, 2.95$, respectively, which were numerically computed.[1] Compared with the MAP estimator (with almost flat priors), which gives $\widehat{U}^{\mathrm{MAP}} = 0, 1, 2, 3$, respectively, the Bayesian estimator is slightly shrunken.

### EVB Solution

Next we consider the empirical Bayesian solutions, where the hyperparameters $c_a, c_b$ are also estimated from observation (the noise variance $\sigma^2$ is still treated as a given constant). We fix the ratio between the prior variances to $c_a/c_b = 1$.

From Corollary 7.3 and Eq. (7.9), we obtain the EVB posterior for $L = M = H = 1$ as follows:

$$r(A,B) = \begin{cases} \mathrm{Gauss}_1\left(A; \pm\sqrt{\breve{\gamma}^{\mathrm{EVB}}}, \frac{\sigma^2}{|V|}\right) \mathrm{Gauss}_1\left(B; \pm\mathrm{sign}(V)\sqrt{\breve{\gamma}^{\mathrm{EVB}}}, \frac{\sigma^2}{|V|}\right) & \\ \qquad\qquad\qquad\qquad\qquad\qquad\qquad\qquad \text{if } |V| \geq \underline{\gamma}^{\mathrm{EVB}}, & \\ \mathrm{Gauss}_1\left(A; 0, +0\right) \mathrm{Gauss}_1\left(B; 0, +0\right) \qquad\qquad \text{otherwise}, & \end{cases} \tag{7.11}$$

where

$$\underline{\gamma}^{\mathrm{EVB}} = \sigma\sqrt{2 + \underline{\tau}(1) + \frac{1}{\underline{\tau}(1)}} \approx \sigma\sqrt{2 + 2.5129 + \frac{1}{2.5129}} \approx 2.216\sigma,$$

$$\breve{\gamma}^{\mathrm{EVB}} = \frac{|V|}{2}\left(1 - \frac{2\sigma^2}{V^2} + \sqrt{1 - \frac{4\sigma^2}{V^2}}\right).$$

[1] More precisely, we numerically calculated the Bayesian estimator (7.10) by sampling $A$ and $B$ from the almost flat priors $p(A)p(B)$ for $c_a = c_b = 100$ and computing the ratio between the sample averages of $BA \cdot p(V|A,B)$ and $p(V|A,B)$.

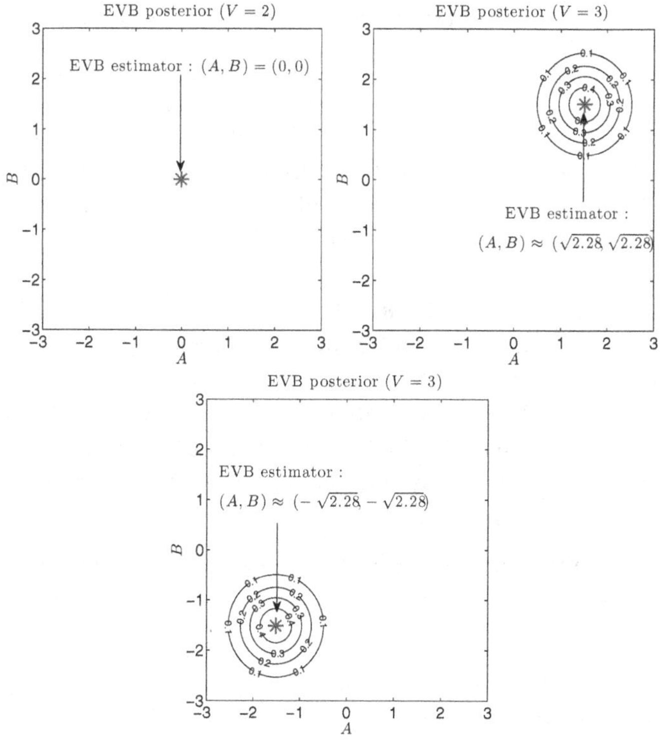


Figure 7.5 EVB solutions. Top-left: When $V = 2$, the EVB posterior is reduced to the Dirac delta function located at the origin. Top-right and bottom: When $V = 3$, the EVB posterior is detached from the origin, and located at $(A, B) \approx (\sqrt{2.28}, \sqrt{2.28})$ or $(A, B) \approx (-\sqrt{2.28}, -\sqrt{2.28})$, both of which yield the same EVB estimator $\widehat{U}^{\mathrm{EVB}} \approx 2.28$.

Figure 7.5 shows the EVB posterior when $V = 2, 3$ are observed, and the noise variance is $\sigma^2 = 1$. When $V = 2 < \underline{\gamma}^{\mathrm{EVB}}$, the EVB posterior is given by the Dirac delta function located at the origin, resulting in the EVB estimator equal to $\widehat{U}^{\mathrm{EVB}} = 0$ (top-left graph). On the other hand, when $V = 3 > \underline{\gamma}^{\mathrm{EVB}}$, the EVB posterior is a Gaussian located in the top-right region or bottom-left region, and the EVB estimator is $\widehat{U}^{\mathrm{EVB}} \approx 2.28$ for both solutions (top-right and bottom graphs).

## Empirical Bayesian Estimator

The *empirical Bayesian (EBayes) estimator* (introduced in Section 1.2.7) is the Bayesian estimator,

$$\widehat{U}^{\mathrm{EBayes}} = \langle BA \rangle_{p(V|A,B)p(A;\widehat{c}_a)p(B;\widehat{c}_b)/p(V;\widehat{c}_a,\widehat{c}_b)},$$

with the hyperparameters estimated by minimizing the *Bayes free energy* $F^{\mathrm{Bayes}}(V; c_a, c_b) \equiv -\log p(V; c_a, c_b)$, i.e.,

$$(\widehat{c}_a, \widehat{c}_b) = \underset{(c_a, c_b)}{\mathrm{argmin}}\, F^{\mathrm{Bayes}}(V; c_a, c_b).$$

When $V = 0, 1, 2, 3$ are observed, the EBayes estimators are $0.00, 0.00, 1.25, 2.58$ (with the prior variance estimators given by $\widehat{c}_a = \widehat{c}_b \approx 0.0, 0.0, 1.4, 2.1$), respectively, which were numerically computed.[2]

### Behavior of Estimators

Figure 7.6 shows the behavior of estimators, including the MAP estimator $\widehat{U}^{\mathrm{MAP}}$, the VB estimator $\widehat{U}^{\mathrm{VB}}$, the Bayesian estimator $\widehat{U}^{\mathrm{Bayes}}$, the EVB estimator $\widehat{U}^{\mathrm{EVB}}$, and the EBayes estimator $\widehat{U}^{\mathrm{EBayes}}$, when the noise variance is $\sigma^2 = 1$. For nonempirical Bayesian estimators, i.e., the MAP, the VB, and the Bayesian estimators, the hyperparameters are set to $c_a = c_b = 100$ (i.e., almost flat priors). Overall, the solutions satisfy

$$\widehat{U}^{\mathrm{EVB}} < \widehat{U}^{\mathrm{EBayes}} < \widehat{U}^{\mathrm{VB}} < \widehat{U}^{\mathrm{Bayes}} < \widehat{U}^{\mathrm{MAP}} (\approx \widehat{U}^{\mathrm{ML}}),$$

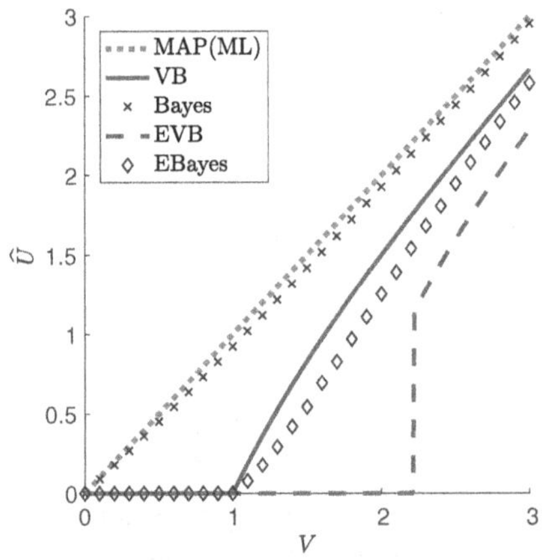


Figure 7.6 Behavior of the MAP estimator $\widehat{U}^{\mathrm{MAP}}$, the VB estimator $\widehat{U}^{\mathrm{VB}}$, the Bayesian estimator $\widehat{U}^{\mathrm{Bayes}}$, the EVB estimator $\widehat{U}^{\mathrm{EVB}}$, and the EBayes estimator $\widehat{U}^{\mathrm{EBayes}}$, when the noise variance is $\sigma^2 = 1$. For the MAP, the VB, and the Bayesian estimators, the hyperparameters are set to $c_a = c_b = 100$ (i.e., almost flat priors).

[2] For $c_a c_b = 10^{-2.00}, 10^{-1.99}, \ldots, 10^{1.00}$, we numerically computed the Bayes free energy, and chose its minimizer $\widehat{c}_a\widehat{c}_b$, with which the Bayesian estimator was computed.

which shows the strength of the regularization effect of each method. Naturally, the empirical Bayesian variants are more regularized than their nonempirical Bayesian counterparts with almost flat priors.

With almost flat priors, the MAP estimator is almost identical to the ML estimator, $\widehat{U}^{\mathrm{MAP}} \approx \widehat{U}^{\mathrm{ML}} = V$, meaning that it is unregularized. We see in Figure 7.6 that the Bayesian estimator $\widehat{U}^{\mathrm{Bayes}}$ is regularized even with almost flat priors. Furthermore, the VB estimator $\widehat{U}^{\mathrm{VB}}$ shows thresholding behavior, which leads to exact sparsity in multidimensional cases. Exact sparsity also appears in EVB learning and EBayes learning. In the subsequent sections, we explain those observations in terms of model-induced regularization and phase transitions.

## 7.3 Model-Induced Regularization

In this section, we explain the origin of the shrinkage of the Bayesian estimator, observed in Section 7.2. The shrinkage is caused by an implicit regularization effect, called model-induced regularization (MIR), which is strongly related to unidentifiability of statistical models.

### 7.3.1 Unidentifiable Models

*Identifiability* is formally defined as follows:

**Definition 7.4** (Identifiability of statistical models) A statistical model $p(\cdot|\boldsymbol{w})$ parameterized by $\boldsymbol{w} \in \mathcal{W}$ is said to be identifiable, if the mapping $\boldsymbol{w} \mapsto p(\cdot|\boldsymbol{w})$ is one-to-one, i.e.,

$$p(\cdot|\boldsymbol{w}_1) = p(\cdot|\boldsymbol{w}_2) \iff \boldsymbol{w}_1 = \boldsymbol{w}_2 \qquad \text{for any } \boldsymbol{w}_1, \boldsymbol{w}_2 \in \mathcal{W}.$$

Otherwise, it is said to be unidentifiable.[3]

Many popular statistical models are unidentifiable.

**Example 7.5** The MF model (introduced in Section 3.1) is unidentifiable, because the model distribution

$$p(\boldsymbol{V}|\boldsymbol{A}, \boldsymbol{B}) \propto \exp\left(-\frac{1}{2\sigma^2}\left\|\boldsymbol{V} - \boldsymbol{B}\boldsymbol{A}^\top\right\|_{\mathrm{Fro}}^2\right) \tag{7.12}$$

[3] Distributions are identified in weak topology in distribution, i.e., $p(\boldsymbol{x}|\boldsymbol{w}_1)$ is identified with $p(\boldsymbol{x}|\boldsymbol{w}_2)$ if $\int f(\boldsymbol{x})p(\boldsymbol{x}|\boldsymbol{w}_1)d\boldsymbol{x} = \int f(\boldsymbol{x})p(\boldsymbol{x}|\boldsymbol{w}_2)d\boldsymbol{x}$ for any bounded continuous function $f(\boldsymbol{x})$.

is invariant to the following transformation $(\boldsymbol{A}, \boldsymbol{B}) \mapsto (\boldsymbol{A}\boldsymbol{T}^\top, \boldsymbol{B}\boldsymbol{T}^{-1})$ for any nonsingular matrix $\boldsymbol{T} \in \mathbb{R}^{H \times H}$.

**Example 7.6** The *multilayer neural network* model is unidentifiable. Consider a three-layer neural network with $H$ hidden units:

$$p(\boldsymbol{y}|\boldsymbol{x}, \boldsymbol{A}, \boldsymbol{B}) \propto \exp\left(-\frac{1}{2\sigma^2}\|\boldsymbol{y} - \boldsymbol{f}(\boldsymbol{x}; \boldsymbol{A}, \boldsymbol{B})\|^2\right), \quad \boldsymbol{f}(\boldsymbol{x}; \boldsymbol{A}, \boldsymbol{B}) = \sum_{h=1}^{H} \boldsymbol{b}_h \cdot \psi\left(\boldsymbol{a}_h^\top \boldsymbol{x}\right), \tag{7.13}$$

where $\boldsymbol{x} \in \mathbb{R}^M$ is an input vector, $\boldsymbol{y} \in \mathbb{R}^L$ is an output vector, $\boldsymbol{A} = (\boldsymbol{a}_1, \ldots, \boldsymbol{a}_H) \in \mathbb{R}^{M \times H}$ and $\boldsymbol{B} = (\boldsymbol{b}_1, \ldots, \boldsymbol{b}_H) \in \mathbb{R}^{L \times H}$ are the weight parameters to be estimated, and $\psi(\cdot)$ is an antisymmetric nonlinear activation function such as $\tanh(\cdot)$. This model expresses the identical distribution on each of the following sets of points in the parameter space:

$$\{\boldsymbol{a}_h \in \mathbb{R}^M, \boldsymbol{b}_h = \boldsymbol{0}\} \cup \{\boldsymbol{a}_h = \boldsymbol{0}, \boldsymbol{b}_h \in \mathbb{R}^L\} \text{ for any } h,$$
$$\{\boldsymbol{a}_h = \boldsymbol{a}_{h'}, \boldsymbol{b}_h, \boldsymbol{b}_{h'} \in \mathbb{R}^L, \boldsymbol{b}_h + \boldsymbol{b}_{h'} = \text{const.}\} \text{ for any pair } h, h'.$$

In other words, the model is invariant for any $\boldsymbol{a}_h \in \mathbb{R}^M$ if $\boldsymbol{b}_h = \boldsymbol{0}$, for any $\boldsymbol{b}_h \in \mathbb{R}^L$ if $\boldsymbol{a}_h = \boldsymbol{0}$, and for any $\boldsymbol{b}_h, \boldsymbol{b}_{h'} \in \mathbb{R}^L$ as long as $\boldsymbol{b}_h + \boldsymbol{b}_{h'}$ is unchanged and $\boldsymbol{a}_h = \boldsymbol{a}_{h'}$.

**Example 7.7** (Mixture models) The *mixture model* (introduced as Example 1.3 in Section 1.1.4) is generally unidentifiable. The model distribution is given as

$$p(\boldsymbol{x}|\boldsymbol{\alpha}, \{\boldsymbol{\tau}_k\}) = \sum_{k=1}^{K} \alpha_k p(\boldsymbol{x}|\boldsymbol{\tau}_k), \tag{7.14}$$

where $\boldsymbol{x} \in \mathcal{X}$ is an observed random variable, and $\boldsymbol{\alpha} = (\alpha_1, \ldots, \alpha_K)^\top \in \Delta^{K-1}$ and $\{\boldsymbol{\tau}_k \in \mathcal{T}\}_{k=1}^K$ are the parameters to be estimated. This model expresses the identical distribution on each of the following sets of points in the parameter space:

$$\{\alpha_k = 0, \boldsymbol{\tau}_k \in \mathcal{T}\} \qquad \text{for any } k,$$
$$\{\alpha_k, \alpha_{k'} \in [0, 1], \alpha_k + \alpha_{k'} = \text{const.}, \boldsymbol{\tau}_k = \boldsymbol{\tau}_{k'}\} \text{ for any pair } k, k'.$$

Namely, if the mixing weight $\alpha_k$ is zero for the $k$th mixture component, the corresponding component parameter $\boldsymbol{\tau}_k$ does not affect the model distribution, and if there are two identical components $\boldsymbol{\tau}_k = \boldsymbol{\tau}_{k'}$, the balance between the corresponding mixture weights, $\alpha_k$ and $\alpha_{k'}$, are arbitrary.

Readers might have noticed that, in the multilayer neural network (Example 7.6) and the mixture model (Example 7.7), the model expressed by the unidentifiable sets of points corresponds to the model with fewer components or smaller degrees of freedom. For example, if $\boldsymbol{a}_h = \boldsymbol{0}$ or $\boldsymbol{b}_h = \boldsymbol{0}$ in the neural network with $H$ hidden units, the model is reduced to the neural network with $H - 1$ hidden units. If two hidden units receive the identical input, i.e., $\psi\left(\boldsymbol{a}_h^\top \boldsymbol{x}\right) = \psi\left(\boldsymbol{a}_{h'}^\top \boldsymbol{x}\right)$ for any $\boldsymbol{x} \in \mathbb{R}^M$, they can be combined into a single unit with its output weight equal to the sum of the original output weights, i.e., $\boldsymbol{b}_h + \boldsymbol{b}_{h'} \rightarrow \boldsymbol{b}_h$. Thus, the model is again reduced to the neural network with $H - 1$ hidden units. The same applies to the mixture models and many other popular statistical models, including Bayesian networks, hidden Markov models, and latent Dirichlet allocation, which were introduced in Chapter 4. As will be explained shortly, this nesting structure—simpler models correspond to unidentifiable sets of points in the parameter space of more complex models—is essential for MIR.

### 7.3.2 Singularities

Continuous points denoting the same distribution are called *singularities*, on which the *Fisher information*,

$$\mathbb{S}_+^D \ni \boldsymbol{F} = \int \frac{\partial \log p(\boldsymbol{x}|\boldsymbol{w})}{\partial \boldsymbol{w}} \left( \frac{\partial \log p(\boldsymbol{x}|\boldsymbol{w})}{\partial \boldsymbol{w}} \right)^\top p(\boldsymbol{x}|\boldsymbol{w}) d\boldsymbol{x}, \tag{7.15}$$

is *singular*, i.e., it has at least one zero eigenvalue. This is a natural consequence from the fact that the Fisher information corresponds to the *metric* when the distance between two points in the parameter space is measured by the *KL divergence* (Jeffreys, 1946), i.e., it holds that

$$\mathrm{KL}\left(p(\boldsymbol{x}|\boldsymbol{w}) \| p(\boldsymbol{x}|\boldsymbol{w} + \Delta \boldsymbol{w})\right) = \frac{1}{2} \Delta \boldsymbol{w}^\top \boldsymbol{F} \Delta \boldsymbol{w} + O(\|\Delta \boldsymbol{w}\|^3)$$

for a small change $\Delta \boldsymbol{w}$ of the parameter. On the singularities, there is at least one direction in which the small change $\Delta \boldsymbol{w}$ does not affect the distribution, implying that the Fisher metric $\boldsymbol{F}$ is singular. This means that the *volume element*, proportional to the determinant of the Fisher metric, is zero on the singularities, while it is positive on the regular points (see Appendix B for more details on the Fisher metric and the volume element in the parameter space).

This strong nonuniformity of (the density of) the volume element affects the behavior of Bayesian learning. For this reason, statistical models having singularities in their parameter space are called *singular models* and distinguished

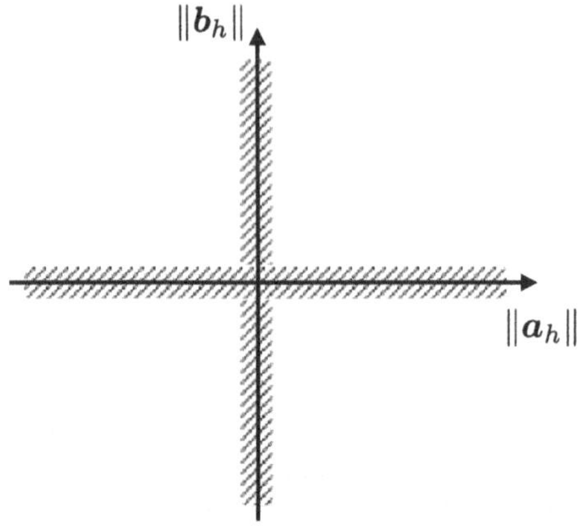


Figure 7.7 Singularities of a neural network model.

from the *regular models* in statistical learning theory (Watanabe, 2009). There are two aspects of how singularities affect the learning properties. In this chapter, we focus on one aspect that leads to MIR. The other aspect will be discussed in Chapter 13.

Figure 7.7 illustrates the singularities in the parameter space of the three-layer neural network (7.13) with $H = 1$ hidden unit (see Example 7.6). The horizontal axis corresponds to an arbitrary direction of $\boldsymbol{a}_h \in \mathbb{R}^M$, while the vertical axis corresponds to an arbitrary direction of $\boldsymbol{b}_h \in \mathbb{R}^L$. The shadowed locations correspond to the singularities. Importantly, all points on the singularities express the *identical* neural network model with no ($H = 0$) hidden unit, while each *regular* point expresses a *different* neural network model with $H = 1$ hidden unit. This illustration gives an intuition that the neighborhood of the smaller model ($H = 0$) is broader than the neighborhood of the larger model ($H = 1$) in the parameter space.

Consider the *Jeffreys prior*,

$$p^{\mathrm{Jef}}(\boldsymbol{w}) \propto \sqrt{\det(\boldsymbol{F})}, \tag{7.16}$$

which is the *uniform prior* in the space of distributions when the distance is measured by the KL divergence (see Appendix B). As discussed previously, the Fisher information is singular on the singularities, giving $p^{\mathrm{Jef}}(\boldsymbol{w}) = 0$ for the smaller model (with $H = 0$), while the Fisher information is regular on the other points, giving $p^{\mathrm{Jef}}(\boldsymbol{w}) > 0$ for the larger model (with $H = 1$). Also in the neighborhood of the singularities, the Fisher information has similar values and it holds that $p^{\mathrm{Jef}}(\boldsymbol{w}) \ll 1$. This means that, in comparison with the

Jeffreys prior, the flat priors on $\boldsymbol{a}_h$ and $\boldsymbol{b}_h$—the uniform prior *in the parameter space*—put much more mass to the smaller model and its neighborhood. A consequence is that, if we apply Bayesian learning with the flat prior, the overweighted singularities and their neighborhood pull the estimator to the smaller model *through the integral computation*, which induces implicit regularization—MIR. The same argument holds for mixture models (Example 7.6), and other popular models, including Bayesian networks, hidden Markov models, and latent Dirichlet allocation.

In summary, MIR occurs in general singular models for the following reasons:

- There is strong nonuniformity in (the density of) the volume element around the singularities.
- Singularities correspond to the model with fewer degrees of freedom than the regular points.

This structure in the parameter space makes the flat prior favor smaller models in Bayesian learning, which appears as MIR. Note that MIR does not occur in point-estimation methods, including ML estimation and MAP learning, since the nonuniformity of the volume element affects the estimator only through *integral computations*.

MIR also occurs in the MF model (Example 7.6), which will be investigated in the next subsection with a generalization of the Jeffreys prior.

### 7.3.3 MIR in one-Dimensional Matrix Factorization

In Section 7.2, we numerically observed MIR—the Bayesian estimator is shrunken even with the almost flat prior in the one-dimensional MF model. However, in the MF model, the original definition (7.16) of the Jeffreys prior is zero everywhere in the parameter space because of the equivalence class structure (Figure 7.1), and therefore, it provides no information on MIR. To evaluate the nonuniformity of the volume element, we redefine the (generalized) Jeffreys prior by ignoring the zero *common* eigenvalues, i.e.,

$$p^{\mathrm{Jef}}(\boldsymbol{w}) \propto \sqrt{\prod_{d=1}^{\overline{D}} \lambda_d}, \tag{7.17}$$

where $\lambda_d$ is the $d$th largest eigenvalue of the Fisher metric $\boldsymbol{F}$, and $\overline{D}$ is the maximum number of positive eigenvalues over the whole parameter space.

Let us consider the *nonfactorizing* model,

$$p(V|U) = \mathrm{Gauss}_1(V; U, \sigma^2) \propto \exp\left(-\frac{1}{2\sigma^2}(V - U)^2\right), \tag{7.18}$$

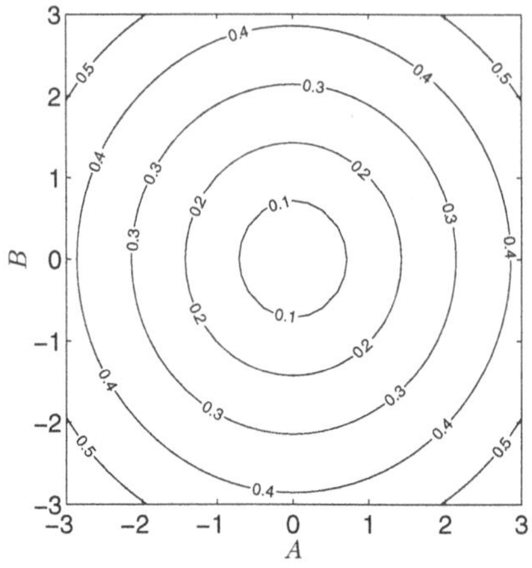


Figure 7.8 The (unnormalized) Jeffreys noninformative prior (7.20) of the one-dimensional MF model (7.5).

where $U$ itself is the parameter to be estimated. The Jeffreys prior for this model is uniform (see Example B.1 in Appendix B for derivation):

$$p^{\mathrm{Jef}}(U) \propto 1. \tag{7.19}$$

On the other hand, the Jeffreys prior for the MF model (7.5) is given as follows (see Example B.2 in Appendix B for derivation):

$$p^{\mathrm{Jef}}(A, B) \propto \sqrt{A^2 + B^2}, \tag{7.20}$$

which is illustrated in Figure 7.8. Note that the Jeffreys priors (7.19) and (7.20) for both cases are *improper*, meaning that they cannot be normalized since their integrals diverge.

Jeffreys (1946) stated that the both combinations, the *nonfactorizing* model (7.18) with its Jeffreys prior (7.19) and the MF model (7.5) with its Jeffreys prior (7.20) give the equivalent Bayesian estimator. We can easily show that the former combination, Eqs. (7.18) and (7.19), gives an unregularized solution. Thus, the Bayesian estimator in the MF model (7.5) with its Jeffreys prior (7.20) is also unregularized. Since the flat prior on $(A, B)$ has more probability mass around the origin than the Jeffreys prior (7.20) (see Figure 7.8), it favors smaller $|U|$ and regularizes the Bayesian estimator.

Although MIR appears also in regular models unless the Jeffreys prior is flat in the parameter space, its effect is prominent in singular models with unidentifiability, since the difference between the flat prior and the Jeffreys prior is large.

### 7.3.4 Evidence View of Unidentifiable Models

MIR works as Occam's razor in general. MacKay (1992) explained, with the illustration shown in the left panel of Figure 7.9, that evidence-based (i.e., free-energy-minimization-based) model selection is naturally equipped with Occam's razor. In the figure, the horizontal axis denotes the space of the observed data set $\mathcal{D}$. $\mathcal{H}_1$ and $\mathcal{H}_2$ denote a simple hypothesis and a more complex hypothesis, respectively. For example, in the MF model, the observed data set corresponds to the observed matrix, i.e., $\mathcal{D} = \boldsymbol{V}$, $\mathcal{H}_1$ corresponds to a lower-rank model, and $\mathcal{H}_2$ corresponds to a higher-rank model. The vertical axis indicates the evidence or marginal likelihood,

$$p(\mathcal{D}|\mathcal{H}_t) = \left\langle p(\mathcal{D}|\boldsymbol{w}_{\mathcal{H}_t}, \mathcal{H}_t) \right\rangle_{p(\boldsymbol{w}_{\mathcal{H}_t})} \qquad \text{for} \qquad t = 1, 2, \tag{7.21}$$

where $\boldsymbol{\theta}_{\mathcal{H}_t}$ denotes the unknown parameters that the hypothesis $\mathcal{H}_t$ has.

Since $\mathcal{H}_1$ is simple, it covers a limited area of the space of $\mathcal{D}$ (meaning that it can explain only a simple phenomenon), while $\mathcal{H}_2$ covers a broader area. The illustration implies that, because of the normalization, it holds that

$$p(\mathcal{D}|\mathcal{H}_1) > p(\mathcal{D}|\mathcal{H}_2) \qquad \text{for} \qquad \mathcal{D} \in C_1,$$

where $C_1$ denotes the observed data region where $\mathcal{H}_1$ can explain the data well. This view gives an intuition on why evidence-based model selection prefers simpler models when the observed data can be well explained by them.

However, this view does not explain MIR, which is observed even without any model selection procedure. In fact, the illustration in the left panel of Figure 7.9 is not accurate for unidentifiable models unless the Jeffreys prior is adopted (note that a hypothesis consists of a model and a prior). The right illustration of Figure 7.9 is a more accurate view for unidentifiable models. When $\mathcal{H}_2$ is a complex unidentifiable model nesting $\mathcal{H}_1$ as a simpler model in

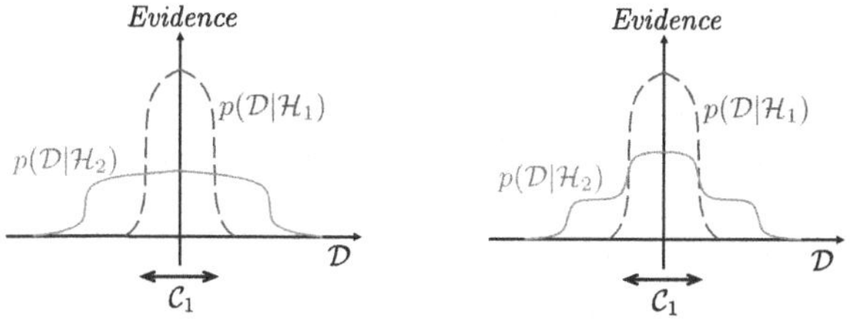


Figure 7.9 Left: The evidence view by MacKay (1992), which gives an intuition on why evidence-based model selection prefers simpler models. Right: A more accurate view for *unidentifiable* models. Simpler models are preferred *even without explicit model selection.*

its parameter space, its evidence $p(\mathcal{D}|\mathcal{H}_2)$ has a bump covering the region $C_1$ if the flat prior is adopted. This is because the flat prior typically places large weights on the singularities representing the simpler model $\mathcal{H}_1$.

## 7.4 Phase Transition in VB Learning

In Section 7.3, we explained MIR, which shrinks the Bayesian estimator. We can expect that VB learning, which involves integral computations, inherits this property. However, we observe in Figure 7.6 that VB learning behaves differently from Bayesian learning. Actually, the Bayesian estimator behaves more similarly to the ML estimator (the MAP estimator with almost flat priors), rather than the VB estimator. A remarkable difference is that the VB estimator, which is upper-bounded by the PJS estimator (7.2), shows exact sparsity, i.e., the estimator can be zero for nonzero observation $|V|$. In this section, we explain that this gap is caused by a phase transition phenomenon in VB learning.

The middle graph in Figure 7.2 shows the Bayes posterior when $V = 1$. The probability mass in the first and the third quadrants pulls the product $U = BA$ toward the positive direction, and the mass in the second and the fourth quadrants toward the negative direction. Since the Bayes posterior is skewed and more mass is placed in the first and the third quadrants, the Bayesian estimator $\widehat{U}^{\mathrm{Bayes}} = \langle BA \rangle_{p(A,B|V)}$ is positive. This is true even if $V > 0$ is very small, and therefore, no thresholding occurs in Bayesian learning—the Bayesian estimator is not sparse.

On the other hand, the VB posterior (the top-right graph of Figure 7.4) is prohibited to be skewed because of the independent constraint, which causes the following phase transition phenomenon. As seen in Figure 7.2, the Bayes posterior has two modes unless $V = 0$, and the distance between the two modes increases as $|V|$ increases. Since the VB posterior tries to approximate the Bayes posterior with a single uncorrelated distribution, it stays at the origin if the two modes are close to each other so that covering both modes minimizes the free energy. The VB posterior detaches from the origin if the two modes get far apart so that approximating either one of the modes minimizes the free energy. This phase transition mechanism makes the VB estimator exactly sparse. The profile of the Bayes posterior (the middle graph of Figure 7.2) implies that, if we restrict the posterior to be Gaussian, but allow it to have correlation between $A$ and $B$, exact sparsity will not appear. In this sense, we can say that MIR is enhanced by the independence constraint, which was imposed for computational tractability.

Mackay (2001) pointed out that there are cases where VB learning prunes model components *inappropriately*, by giving a toy example of a mixture of Gaussians. Note that *appropriateness* was measured in terms of the similarity to full Bayesian learning. He plotted the free energy of the mixture of Gaussians as a function of hidden responsibility variables—the probabilities that each sample belongs to each Gaussian component—and argued that VB learning sometimes favors simpler models too much. In this case, degrees of freedom are pruned when spontaneous symmetry breaking (a phase transition) occurs. Interestingly, in the MF model, degrees of freedom are pruned when spontaneous symmetry breaking does *not* occur, as explained earlier.

Eq. (7.3) implies that the symmetry breaking occurs when $V > \underline{\gamma}_h^{\mathrm{VB}} \approx \sqrt{M\sigma^2} = 1$, which coincides with the average contribution of noise to the observed singular values over all singular components—more accurately, $\sqrt{M\sigma^2}$ is the square root of the average eigenvalues of the Wishart matrix $\mathcal{E}\mathcal{E}^\top \sim \mathrm{Wishart}_L(\sigma^2 \boldsymbol{I}_L, M)$.[4] In this way, VB learning discards singular components dominated by noise.

Given that the full Bayesian estimator in MF is not sparse (see Figure 7.6), one might argue that the sparsity of VB learning is an *inappropriate* artifact. On the other hand, given that automatic model pruning by VB learning has been acknowledged as a practically useful property (Bishop, 1999b; Bishop and Tipping, 2000; Sato et al., 2004; Babacan et al., 2012b), one might also argue that *appropriateness* should be measured in terms of performance. Motivated by the latter idea, performance analysis has been carried out (Nakajima et al., 2015), which will be detailed in Chapter 8.

In the empirical Bayesian scenario, where the prior variances $c_a, c_b$ are also estimated from observation, Bayesian learning also gives a sparse solution, which is shown as diamonds (labeled as "EBayes") in Figure 7.6. This is somewhat natural since, in empirical Bayesian learning, the dependency between $A$ and $c_a^{-2}$ (as well as $B$ and $c_b^{-2}$) in the prior (7.6) (in the prior (7.7)) and hence in the Bayes posterior is broken—the point-estimation of $c_a^2$ (as well as $c_a^2$) forces it to be independent of all other parameters. This forced independence causes a similar phase transition phenomenon to the one caused by the independence constraint between $A$ and $B$ in the (nonempirical) VB learning, and results in exact sparsity of the EBayes estimator.

EVB learning has a different transition point, and tends to give a sparser solution than VB learning. A notable difference from the VB estimator is that the EVB estimator is no longer continuous as a function of the observation $V$. This comes from the fact that, when $|V| > \underline{\gamma}^{\text{local-EVB}}$, there exist two local

[4] It holds that $\mathcal{E}\mathcal{E}^\top \sim \mathrm{Wishart}_L(\sigma^2 \boldsymbol{I}_L, M)$ if $\mathcal{E} \sim \mathrm{Gauss}_L(\boldsymbol{0}, \sigma^2 \boldsymbol{I}_L)$.

solutions (see Figure 6.3), but the global solution is $\widehat{U}^{\mathrm{EVB}} = 0$ until the observed amplitude $|V|$ exceeds $\underline{\gamma}^{\mathrm{EVB}} (> \underline{\gamma}^{\mathrm{local-EVB}})$. When the positive local solution $\breve{\gamma}^{\mathrm{EVB}}$ becomes the global solution, it is already distant from the origin, which makes the estimator noncontinuous at the thresholding point (see the dashed curve labeled as "EVB" in Figure 7.6).

## 7.5 Factorization as ARD Model

As shown in Section 7.1, MIR in VB learning for the MF model appears as PJS shrinkage. We can see this as a natural consequence from the equivalence between the MF model and the ARD model (Neal, 1996).

Assume that $\boldsymbol{C}_A = \boldsymbol{I}_H$ in the MF model (6.1) through (6.3), and consider the following transformation: $\boldsymbol{B}\boldsymbol{A}^\top \mapsto \boldsymbol{U} \in \mathbb{R}^{L \times M}$. Then, the likelihood (6.1) and the prior (6.2) on $\boldsymbol{A}$ become

$$p(\boldsymbol{V}|\boldsymbol{U}) \propto \exp\left(-\frac{1}{2\sigma^2} \|\boldsymbol{V} - \boldsymbol{U}\|_{\mathrm{Fro}}^2\right), \tag{7.22}$$

$$p(\boldsymbol{U}|\boldsymbol{B}) \propto \exp\left(-\frac{1}{2}\mathrm{tr}\left(\boldsymbol{U}^\top (\boldsymbol{B}\boldsymbol{B}^\top)^\dagger \boldsymbol{U}\right)\right), \tag{7.23}$$

where $\dagger$ denotes the Moore–Penrose generalized inverse of a matrix. The prior (6.3) on $\boldsymbol{B}$ is kept unchanged. $p(\boldsymbol{U}|\boldsymbol{B})$ in Eq. (7.23) is so-called the ARD prior with the covariance hyperparameter $\boldsymbol{B}\boldsymbol{B}^\top \in \mathbb{R}^{L \times L}$. It is known that this prior induces the ARD property—empirical Bayesian learning, where the prior covariance hyperparameter $\boldsymbol{B}\boldsymbol{B}^\top$ is estimated from observation by maximizing the marginal likelihood (or minimizing the free energy), induces strong regularization and sparsity (Neal, 1996). Efron and Morris (1973) showed that this particular model gives the JS shrinkage estimator as an empirical Bayesian estimator (see Appendix A).

This equivalence can explain the sparsity-inducing terms (3.113) through (3.116), introduced for sparse additive matrix factorization (SAMF) in Section 3.5. The ARD prior (7.23) induces low-rankness on $\boldsymbol{U}$ if no restriction on $\boldsymbol{B}\boldsymbol{B}^\top$ is imposed. We can similarly show that, $(\gamma_l^e)^2$ in Eq. (3.114) corresponds to the prior variance shared by the entries in $\widetilde{\boldsymbol{u}}_l \equiv \gamma_l^e \widetilde{\boldsymbol{d}}_l \in \mathbb{R}^M$, that $(\gamma_m^d)^2$ in Eq. (3.115) corresponds to the prior variance shared by the entries in $\boldsymbol{u}_m \equiv \gamma_m^d \boldsymbol{e}_m \in \mathbb{R}^L$, and that $E_{l,m}^2$ in Eq. (3.116) corresponds to the prior variance on $U_{l,m} \equiv E_{l,m} D_{l,m} \in \mathbb{R}$, respectively. This explains why the factorization forms in Eqs. (3.113) through (3.116) induce low-rank, rowwise, columnwise, and elementwise sparsity, respectively. If we employ the sparse matrix factorization (SMF) term (3.117), ARD occurs in each partition, which induces partitionwise sparsity and low-rank sparsity within each partition.

# 8

# Performance Analysis of VB Matrix Factorization

In this chapter, we further analyze the behavior of VB learning in the fully observed MF model, introduced in Section 3.1. Then, we derive a theoretical guarantee for rank estimation (Nakajima et al., 2015), which corresponds to the hidden dimensionality selection in principal component analysis (PCA).

In Chapter 6, we derived an analytic-form solution (Theorem 6.13) of EVB learning, where the prior variances are also estimated from observation. When discussing the dimensionality selection performance in PCA, it is more practical to assume that the noise variance $\sigma^2$ is estimated, since it is unknown in many situations. To this end, we first analyze the behavior of the noise variance estimator. After that, based on the random matrix theory, we derive a theoretical guarantee of dimensionality selection performance, and show numerical results validating the theory. We also discuss the relation to an alternative dimensionality selection method (Hoyle, 2008) based on the Laplace approximation.

In the following analysis, we use some results in Chapter 6. Specifically, we mostly rely on Theorem 6.13 along with the corollaries and the equations summarized in Section 6.10.

## 8.1 Objective Function for Noise Variance Estimation

Let us consider the *complete* empirical VB problem, where all the variational parameters and the hyperparameters are estimated in the free energy minimization framework:

$$\min_{\{\widehat{a}_h, \widehat{b}_h, \widehat{\sigma}^2_{a_h}, \widehat{\sigma}^2_{b_h}, c^2_{a_h}, c^2_{b_h}\}_{h=1}^H, \sigma^2} F \tag{8.1}$$

$$\text{s.t.} \quad \{\widehat{a}_h, \widehat{b}_h \in \mathbb{R}, \quad \widehat{\sigma}^2_{a_h}, \widehat{\sigma}^2_{b_h}, c^2_{a_h}, c^2_{b_h} \in \mathbb{R}_{++}\}_{h=1}^H, \sigma^2 \in \mathbb{R}_{++}.$$

Here, the free energy is given by

$$2F = LM\log(2\pi\sigma^2) + \frac{\sum_{h=1}^{L}\gamma_h^2}{\sigma^2} + \sum_{h=1}^{H} 2F_h, \tag{8.2}$$

where
$$2F_h = M\log\frac{c_{a_h}^2}{\widehat{\sigma}_{a_h}^2} + L\log\frac{c_{b_h}^2}{\widehat{\sigma}_{b_h}^2} + \frac{\widehat{a}_h^2 + M\widehat{\sigma}_{a_h}^2}{c_{a_h}^2} + \frac{\widehat{b}_h^2 + L\widehat{\sigma}_{b_h}^2}{c_{b_h}^2}$$
$$- (L+M) + \frac{-2\widehat{a}_h\widehat{b}_h\gamma_h + \left(\widehat{a}_h^2 + M\widehat{\sigma}_{a_h}^2\right)\left(\widehat{b}_h^2 + L\widehat{\sigma}_{b_h}^2\right)}{\sigma^2}. \tag{8.3}$$

Note that we are focusing on the solution with diagonal posterior covariances without loss of generality (see Theorem 6.4).

We have already obtained the empirical VB estimator (Theorem 6.13) and the minimum free energy (Corollary 6.21) for given $\sigma^2$. By using those results, we can express the free energy (8.2) as a function of $\sigma^2$. With the rescaled expressions (6.132) through (6.138), the free energy can be written in a simple form, which leads to the following theorem:

**Theorem 8.1** *The noise variance estimator, denoted by* $\widehat{\sigma}^{2\,\mathrm{EVB}}$, *is the global minimizer of*

$$\Omega(\sigma^{-2})\left(\equiv \frac{2F(\sigma^{-2})}{LM} + \mathrm{const.}\right) = \frac{1}{L}\left(\sum_{h=1}^{H}\psi\left(\frac{\gamma_h^2}{M\sigma^2}\right) + \sum_{h=H+1}^{L}\psi_0\left(\frac{\gamma_h^2}{M\sigma^2}\right)\right), \tag{8.4}$$

*where*

$$\psi(x) = \psi_0(x) + \theta\left(x > \underline{x}\right)\psi_1(x), \tag{8.5}$$

$$\psi_0(x) = x - \log x, \tag{8.6}$$

$$\psi_1(x) = \log(\tau(x;\alpha) + 1) + \alpha\log\left(\frac{\tau(x;\alpha)}{\alpha} + 1\right) - \tau(x;\alpha). \tag{8.7}$$

*Here,* $\underline{x}$ *is given by*

$$\underline{x} = \left(1 + \underline{\tau}\right)\left(1 + \frac{\alpha}{\underline{\tau}}\right), \tag{8.8}$$

*where* $\underline{\tau}$ *is defined in Theorem 6.13,* $\tau(x;\alpha)$ *is a function of* $x$ $(> \underline{x})$ *defined by*

$$\tau(x;\alpha) = \frac{1}{2}\left(x - (1+\alpha) + \sqrt{(x-(1+\alpha))^2 - 4\alpha}\right), \tag{8.9}$$

*and* $\theta(\cdot)$ *denotes the indicator function such that* $\theta(\mathrm{condition}) = 1$ *if the condition is true and* $\theta(\mathrm{condition}) = 0$ *otherwise.*

*Proof* By using Lemma 6.14 and Lemma 6.16, the free energy (8.2) can be written as a function of $\sigma^2$ as follows:

$$2F = LM\log(2\pi\sigma^2) + \frac{\sum_{h=1}^{L}\gamma_h^2}{\sigma^2} + \sum_{h=1}^{H}\theta\left(\gamma_h > \underline{\gamma}^{\mathrm{EVB}}\right)F_h^{\mathrm{EVB-Posi}}, \tag{8.10}$$

where $F_h^{\mathrm{EVB-Posi}}$ is given by Eq. (6.112). By using Eqs. (6.133) and (6.135), Eq. (6.112) can be written as

$$\begin{aligned} F_h^{\mathrm{EVB-Posi}} &= M\log(\tau_h + 1) + L\log\left(\frac{\tau_h}{\alpha} + 1\right) - M\tau_h \\ &= M\psi_1(x_h). \end{aligned} \tag{8.11}$$

Therefore, Eq. (8.10) is written as

$$\begin{aligned} 2F &= M\left\{\sum_{h=1}^{L}\log\left(\frac{2\pi\gamma_h^2}{M}\right) + \sum_{h=1}^{L}\left(\log\left(\frac{M\sigma^2}{\gamma_h^2}\right) + \frac{\gamma_h^2}{M\sigma^2}\right)\right. \\ &\quad \left. + \sum_{h=1}^{H}\theta\left(\gamma_h > \underline{\gamma}^{\mathrm{EVB}}\right)\frac{F_h^{\mathrm{EVB-Posi}}}{M}\right\} \\ &= M\left\{\sum_{h=1}^{L}\log\left(\frac{2\pi\gamma_h^2}{M}\right) + \sum_{h=1}^{L}\psi_0(x_h) + \sum_{h=1}^{H}\theta\left(x_h > \underline{x}\right)\psi_1(x_h)\right\}. \end{aligned}$$

Note that the first term in the curly braces is constant with respect to $\sigma^2$. By defining

$$\Omega = \frac{2F}{LM} - \frac{1}{L}\sum_{h=1}^{L}\log\left(\frac{2\pi\gamma_h^2}{M}\right),$$

we obtain Eq. (8.4), which completes the proof of Theorem 8.1. □

The functions $\psi_0(x)$ and $\psi(x)$ are depicted in Figure 8.1. We can confirm the convexity of $\psi_0(x)$ and the quasiconvexity of $\psi(x)$ (Lemma 8.4 in Section 8.3), which are useful properties in the subsequent analysis.[1]

## 8.2 Bounds of Noise Variance Estimator

Let $\widehat{H}^{\mathrm{EVB}}$ be the estimated rank by EVB learning, i.e., the rank of the EVB estimator $\widehat{\boldsymbol{U}}^{\mathrm{EVB}}$, such that $\widehat{\gamma}_h^{\mathrm{EVB}} > 0$ for $h = 1, \ldots, \widehat{H}^{\mathrm{EVB}}$, and $\widehat{\gamma}_h^{\mathrm{EVB}} = 0$ for

[1] A function $f : \mathcal{X} \mapsto \mathbb{R}$ on the domain $\mathcal{X}$ being a convex subset of a real vector space is said to be *quasiconvex* if $f(\lambda\boldsymbol{x}_1 + (1-\lambda)\boldsymbol{x}_2) \leq \max(f(\boldsymbol{x}_1), f(\boldsymbol{x}_2))$ for all $\boldsymbol{x}_1, \boldsymbol{x}_2 \in \mathcal{X}$ and $\lambda \in [0, 1]$. It is furthermore said to be *strictly quasiconxex* if $f(\lambda\boldsymbol{x}_1 + (1-\lambda)\boldsymbol{x}_2) < \max(f(\boldsymbol{x}_1), f(\boldsymbol{x}_2))$ for all $\boldsymbol{x}_1 \neq \boldsymbol{x}_2$ and $\lambda \in (0, 1)$. Intuitively, a strictly quasiconvex function does not have more than one local minima.

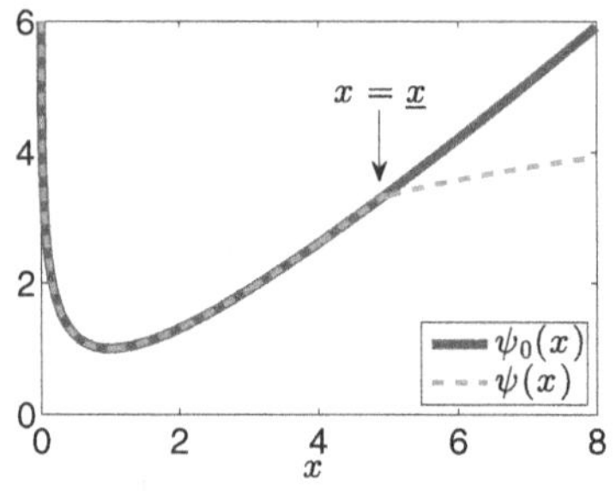


Figure 8.1 $\psi_0(x)$ and $\psi(x)$.

$h = \widehat{H}^{\mathrm{EVB}} + 1, \ldots, H$. By further analyzing the objective (8.4), we can derive bounds of the estimated rank and the noise variance estimator:

**Theorem 8.2** $\widehat{H}^{\mathrm{EVB}}$ *is upper-bounded as*

$$\widehat{H}^{\mathrm{EVB}} \leq \overline{H} = \min\left(\left\lceil \frac{L}{1+\alpha} \right\rceil - 1, H\right), \tag{8.12}$$

*and the noise variance estimator* $\widehat{\sigma}^{2\,\mathrm{EVB}}$ *is bounded as follows:*

$$\max\left(\underline{\sigma}^2_{\overline{H}+1}, \frac{\sum_{h=\overline{H}+1}^{L} \gamma_h^2}{M\left(L-\overline{H}\right)}\right) \leq \widehat{\sigma}^{2\,\mathrm{EVB}} \leq \frac{1}{LM}\sum_{h=1}^{L}\gamma_h^2, \tag{8.13}$$

$$\textit{where} \quad \underline{\sigma}^2_h = \begin{cases} \infty & \textit{for } h = 0, \\ \frac{\gamma_h^2}{M\underline{x}} & \textit{for } h = 1, \ldots, L, \\ 0 & \textit{for } h = L+1. \end{cases} \tag{8.14}$$

Theorem 8.2 states that EVB learning discards the $(L - \lceil L/(1+\alpha) \rceil + 1)$ smallest components, regardless of the observed singular values $\{\gamma_h\}_{h=1}^{L}$. For example, half of the components are always discarded when the matrix is square (i.e., $\alpha = L/M = 1$). The smallest singular value $\gamma_L$ is always discarded, and $\widehat{\sigma}^{2\,\mathrm{EVB}} \geq \gamma_L^2/M$ always holds.

Given the EVB estimators $\{\widehat{\gamma}_h^{\mathrm{EVB}}\}_{h=1}^{H}$ for the singular values, the noise variance estimator $\widehat{\sigma}^{2\,\mathrm{EVB}}$ is specified by the following corollary:

**Corollary 8.3** *The EVB estimator for the noise variance satisfies the following equality:*

$$\widehat{\sigma}^{2\,\mathrm{EVB}} = \frac{1}{LM}\left(\sum_{l=1}^{L}\gamma_l^2 - \sum_{h=1}^{H}\gamma_h\widehat{\gamma}_h^{\mathrm{EVB}}\right). \tag{8.15}$$

This corollary can be used for implementing a global EVB solver (see Chapter 9). In the next section we give the proofs of the theorem and the corollary.

## 8.3 Proofs of Theorem 8.2 and Corollary 8.3

First, we show nice properties of the functions, $\psi(x)$ and $\psi_0(x)$, which are defined by Eqs. (8.5) and (8.6), respectively, and depicted in Figure 8.1:

**Lemma 8.4** *The following hold for $x > 0$: $\psi_0(x)$ is differentiable and strictly convex; $\psi(x)$ is continuous and strictly quasiconvex; $\psi(x)$ is differentiable except $x = \underline{x}$, at which $\psi(x)$ has a discontinuously decreasing derivative, i.e., $\lim_{x \to \underline{x}-0} \partial\psi/\partial x > \lim_{x \to \underline{x}+0} \partial\psi/\partial x$; both of $\psi_0(x)$ and $\psi(x)$ are minimized at $x = 1$. For $x > \underline{x}$, $\psi_1(x)$ is negative and decreasing.*

*Proof* Since

$$\frac{\partial \psi_0}{\partial x} = 1 - \frac{1}{x}, \tag{8.16}$$
$$\frac{\partial^2 \psi_0}{\partial x^2} = \frac{1}{x^2} > 0,$$

$\psi_0(x)$ is differentiable and strictly convex for $x > 0$ with its minimizer at $x = 1$. $\psi_1(x)$ is continuous for $x \geq \underline{x}$, and Eq. (8.11) implies that $\psi_1(x_h) \propto F_h^{\mathrm{EVB-Posi}}$. Accordingly, $\psi_1(x) \leq 0$ for $x \geq \underline{x}$, where the equality holds when $x = \underline{x}$. This equality implies that $\psi(x)$ is continuous. Since $\underline{x} > 1$, $\psi(x)$ shares the same minimizer as $\psi_0(x)$ at $x = 1$ (see Figure 8.1).

Hereafter, we investigate $\psi_1(x)$ and $\psi(x)$ for $x \geq \underline{x}$. By differentiating Eqs. (8.7) and (6.135), respectively, we have

$$\frac{\partial \psi_1}{\partial \tau} = -\left( \frac{\frac{\tau^2}{\alpha} - 1}{(\tau + 1)\left(\frac{\tau}{\alpha} + 1\right)} \right) < 0, \tag{8.17}$$

$$\frac{\partial \tau}{\partial x} = \frac{1}{2}\left( 1 + \frac{x - (1 + \alpha)}{\sqrt{(x - (1 + \alpha))^2 - 4\alpha}} \right) > 0. \tag{8.18}$$

Substituting Eq. (6.134) into Eq. (8.18), we have

$$\frac{\partial \tau}{\partial x} = \frac{\tau^2}{\alpha\left(\frac{\tau^2}{\alpha} - 1\right)}. \tag{8.19}$$

Multiplying Eqs. (8.17) and (8.19) gives

$$\frac{\partial \psi_1}{\partial x} = \frac{\partial \psi_1}{\partial \tau}\frac{\partial \tau}{\partial x} = -\left(\frac{\tau^2}{\alpha(\tau+1)\left(\frac{\tau}{\alpha}+1\right)}\right) = -\frac{\tau}{x} < 0, \tag{8.20}$$

which implies that $\psi_1(x)$ is decreasing for $x > \underline{x}$.

Let us focus on the thresholding point of $\psi(x)$ at $x = \underline{x}$. Eq. (8.20) does not converge to zero for $x \to \underline{x} + 0$ but stay negative. On the other hand, $\psi_0(x)$ is differentiable at $x = \underline{x}$. Consequently, $\psi(x)$ has a discontinuously decreasing derivative, i.e., $\lim_{x\to\underline{x}-0} \partial\psi/\partial x > \lim_{x\to\underline{x}+0} \partial\psi/\partial x$, at $x = \underline{x}$.

Finally, we prove the strict quasiconvexity of $\psi(x)$. Taking the sum of Eqs. (8.16) and (8.20) gives

$$\frac{\partial \psi}{\partial x} = \frac{\partial \psi_0}{\partial x} + \frac{\partial \psi_1}{\partial x} = 1 - \frac{1+\tau}{x} = 1 - \frac{1+\tau}{1+\tau+\alpha+\alpha\tau^{-1}} > 0.$$

This means that $\psi(x)$ is increasing for $x > \underline{x}$. Since $\psi_0(x)$ is strictly convex and increasing at $x = \underline{x}$, and $\psi(x)$ is continuous, $\psi(x)$ is strictly quasiconvex. This completes the proof of Lemma 8.4. □

Lemma 8.4 implies that our objective (8.4) is a sum of quasiconvex functions with respect to $\sigma^{-2}$. Therefore, its minimizer can be bounded by the smallest one and the largest one among the set collecting the minimizer from each quasiconvex function:

**Lemma 8.5** *$\Omega(\sigma^{-2})$ has at least one global minimizer, and any of its local minimizers is bounded as*

$$\frac{M}{\gamma_1^2} \leq \widehat{\sigma}^{-2} \leq \frac{M}{\gamma_L^2}. \tag{8.21}$$

*Proof* The strict convexity of $\psi_0(x)$ and the strict quasiconvexity of $\psi(x)$ also hold for $\psi_0(\gamma_h^2\sigma^{-2}/M)$ and $\psi(\gamma_h^2\sigma^{-2}/M)$ as functions of $\sigma^{-2}$ (for $\gamma_h > 0$). Because of the different scale factor $\gamma_h^2/M$ for each $h = 1\ldots, L$, each of $\psi_0(\gamma_h^2\sigma^{-2}/M)$ and $\psi(\gamma_h^2\sigma^{-2}/M)$ has a minimizer at a different position:

$$\sigma^{-2} = \frac{M}{\gamma_h^2}.$$

The strict quasiconvexity of $\psi_0$ and $\psi$ guarantees that $\Omega(\sigma^{-2})$ is decreasing for

$$0 < \sigma^{-2} < \frac{M}{\gamma_1^2}, \tag{8.22}$$

and increasing for

$$\frac{M}{\gamma_L^2} < \sigma^{-2} < \infty. \tag{8.23}$$

This proves Lemma 8.5. □

$\Omega(\sigma^{-2})$ has at most $H$ nondifferentiable points, which come from the nondifferentiable point $x = \underline{x}$ of $\psi(x)$. The values

$$\underline{\sigma}_h^{-2} = \begin{cases} 0 & \text{for } h = 0, \\ \frac{M\underline{x}}{\gamma_h^2} & \text{for } h = 1, \ldots, L, \\ \infty & \text{for } h = L + 1, \end{cases} \tag{8.24}$$

defined in Eq. (8.14) for $h = 1, \ldots, H$ actually correspond to those points.

Lemma 8.4 states that, at $x = \underline{x}$, $\psi(x)$ has a discontinuously decreasing derivative and neither $\psi_0(x)$ nor $\psi(x)$ has a discontinuously increasing derivative at any point. Therefore, none of those nondifferentiable points can be a local minimum. Consequently, we have the following lemma:

**Lemma 8.6** *$\Omega(\sigma^{-2})$ has no local minimizer at $\sigma^{-2} = \underline{\sigma}_h^{-2}$ for $h = 1, \ldots, H$, and therefore any of its local minimizers is a stationary point.*

Then, Theorem 6.13 leads to the following lemma:

**Lemma 8.7** *The estimated rank is $\widehat{H} = h$ if and only if the inverse noise variance estimator lies in the range*

$$\widehat{\sigma}^{-2} \in \mathcal{B}_h \equiv \left\{\sigma^{-2}; \underline{\sigma}_h^{-2} < \sigma^{-2} < \underline{\sigma}_{h+1}^{-2}\right\}. \tag{8.25}$$

Figure 8.2 shows quasiconvex functions $\{\psi(\gamma_h^2\sigma^{-2}/M)\}_{h=1}^H$ and their average $\Omega(\sigma^{-2})$ in two exemplary cases for $H = L$. In the left case, the inverse noise variance estimator $\widehat{\sigma}^{-2}$ is smaller than the inverse threshold $\underline{\sigma}_1^{-2}$ for the largest singular value, and therefore no EVB estimator $\widehat{\gamma}_h$ is positive, i.e., $\widehat{H} = 0$. In the right case, it holds that $\underline{\sigma}_1^{-2} < \widehat{\sigma}^{-2} < \underline{\sigma}_2^{-2}$, and therefore $\widehat{\gamma}_1$ is positive and the others are zero, i.e., $\widehat{H} = 1$.

We have the following lemma:

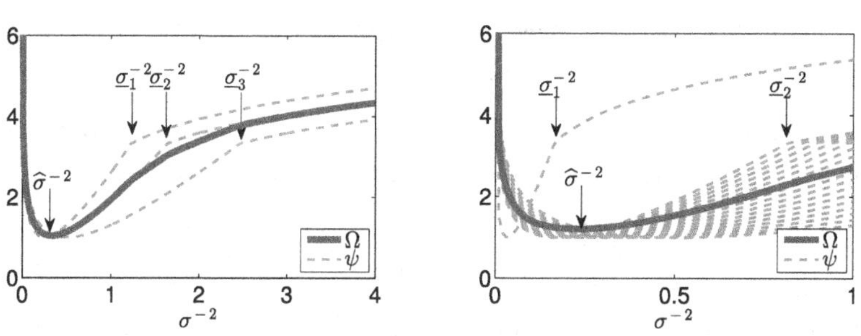


Figure 8.2 $\{\psi(\gamma_h^2\sigma^{-2}/M)\}_{h=1}^H$ and $\Omega(\sigma^{-2})$ in two examplary cases for $H = L$. Left: the case where $\gamma_h^2/M = 4, 3, 2$ for $h = 1, 2, 3$. Right: the case where $\gamma_1^2/M = 30$, $\gamma_h^2/M = 6, 5.75, 5.5, \ldots, 2.0$ for $h = 2, \ldots, 18$.

**Lemma 8.8** *The derivative of $\Omega(\sigma^{-2})$ is given by*

$$\Theta \equiv \frac{\partial \Omega}{\partial \sigma^{-2}} = -\sigma^2 + \frac{\sum_{h=1}^{\widehat{H}} \gamma_h \left(\gamma_h - \breve{\gamma}_h^{\mathrm{EVB}}\right) + \sum_{h=\widehat{H}+1}^{L} \gamma_h^2}{LM}, \tag{8.26}$$

*where $\widehat{H}$ is a function of $\sigma^{-2}$ defined by*

$$\widehat{H} = \widehat{H}(\sigma^{-2}) = h \qquad \text{if} \qquad \sigma^{-2} \in \mathcal{B}_h. \tag{8.27}$$

*Proof* The derivative of Eq. (8.4) with respect to $\sigma^{-2}$ is given by

$$\frac{\partial \Omega}{\partial \sigma^{-2}} = \frac{1}{L}\left( \sum_{h=1}^{H} \frac{\gamma_h^2}{M}\frac{\partial \psi}{\partial x} + \sum_{h=H+1}^{L} \frac{\gamma_h^2}{M}\frac{\partial \psi_0}{\partial x} \right). \tag{8.28}$$

By using Eqs. (8.16) and (8.20), Eq. (8.28) can be written as

$$\begin{aligned}
\frac{\partial \Omega}{\partial \sigma^{-2}} &= \frac{1}{L}\left( \sum_{h=1}^{L} \frac{\gamma_h^2}{M}\frac{\partial \psi_0}{\partial x} + \sum_{h=1}^{H} \theta\left(x_h \geq \underline{x}\right) \frac{\gamma_h^2}{M}\frac{\partial \psi_1}{\partial x} \right) \\
&= \frac{1}{L}\left( \sum_{h=1}^{L} \frac{\gamma_h^2}{M}\left(1 - \frac{1}{x_h}\right) - \sum_{h=1}^{H} \theta\left(x_h \geq \underline{x}\right) \frac{\gamma_h^2 \tau_h}{M x_h} \right) \\
&= \frac{\sum_{h=1}^{L} \gamma_h^2}{LM} - \sigma^2 - \frac{1}{L}\sum_{h=1}^{H} \theta\left(\tau_h \geq \underline{\tau}\right) \sigma^2 \tau_h.
\end{aligned} \tag{8.29}$$

Here we also used the definition (6.132) of $x_h$. Using Eq. (6.133), Eq. (8.29) can be written as

$$\begin{aligned}
\frac{\partial \Omega}{\partial \sigma^{-2}} &= \frac{\sum_{h=1}^{L} \gamma_h^2}{LM} - \sigma^2 - \sum_{h=1}^{H} \theta\left(\gamma_h \geq \underline{\gamma}^{\mathrm{EVB}}\right) \frac{\gamma_h \breve{\gamma}_h^{\mathrm{EVB}}}{LM} \\
&= -\sigma^2 + \frac{\sum_{h=1}^{H} \gamma_h \left(\gamma_h - \widehat{\gamma}_h^{\mathrm{EVB}}\right) + \sum_{h=H+1}^{L} \gamma_h^2}{LM}.
\end{aligned} \tag{8.30}$$

Here we also used the definition (6.101) of $\widehat{\gamma}_h^{\mathrm{EVB}}$. Using the definition (8.27) and Lemma 8.7, we can replace $\widehat{\gamma}_h^{\mathrm{EVB}}$ and $H$ with $\breve{\gamma}_h^{\mathrm{EVB}}$ and $\widehat{H}$, respectively, which completes the proof of Lemma 8.8. □

Note that Eq. (8.26) involves the shrinkage estimator $\breve{\gamma}_h^{\mathrm{EVB}}$, which is a function of $\sigma^{-2}$ (see Eq. (6.103)). For each hypothetical $\widehat{H}$, the solutions of the equation

$$\Theta = 0 \tag{8.31}$$

lying in $\sigma^{-2} \in \mathcal{B}_{\widehat{H}}$ are stationary points, and hence candidates for the global minimum. If we can solve Eq. (8.31) for all $\widehat{H} = 1, \ldots, H$, we can obtain the global solution by evaluating the objective (8.4) at each obtained stationary point. However, solving Eq. (8.31) is computationally hard unless $\widehat{H}$ is small.[2] Based on Lemma 8.8, we will obtain tighter bounds than Lemma 8.5.

Since

$$\gamma_h - \breve{\gamma}_h^{\mathrm{EVB}} > 0,$$

Eq. (8.26) is upper-bounded by

$$\Theta \leq -\sigma^2 + \sum_{h=1}^{L} \frac{\gamma_h^2}{LM},$$

which leads to the upper-bound given in Eq. (8.13). Actually, if

$$\left( \sum_{h=1}^{L} \frac{\gamma_h^2}{LM} \right)^{-1} \in \mathcal{B}_0,$$

then

$$\widehat{H} = 0,$$

$$\widehat{\sigma^2} = \sum_{h=1}^{L} \frac{\gamma_h^2}{LM},$$

is a local minimum.

The following lemma is easily obtained from Eq. (6.103) by using the inequalities $z_1 < \sqrt{z_1^2 - z_2^2} < z_1 - z_2$ for $z_1 > z_2 > 0$:

**Lemma 8.9** *For $\gamma_h \geq \underline{\gamma}^{\mathrm{EVB}}$, the EVB shrinkage estimator* (6.103) *can be bounded as follows:*

$$\gamma_h - \frac{(\sqrt{M} + \sqrt{L})^2 \sigma^2}{\gamma_h} < \breve{\gamma}_h^{\mathrm{EVB}} < \gamma_h - \frac{(M + L)\sigma^2}{\gamma_h}. \tag{8.32}$$

This lemma is important for our analysis, because it allows us to bound the most complicated part of Eq. (8.26) by quantities independent of $\gamma_h$, i.e.,

$$(M + L)\sigma^2 < \gamma_h \left( \gamma_h - \breve{\gamma}_h^{\mathrm{EVB}} \right) < (\sqrt{M} + \sqrt{L})^2 \sigma^2. \tag{8.33}$$

Using Eq. (8.33), we obtain the following lemma:

**Lemma 8.10** *Any local minimizer exists in $\sigma^{-2} \in \mathcal{B}_{\widehat{H}}$ such that*

$$\widehat{H} < \frac{L}{1 + \alpha},$$

[2] It is easy to derive a closed-form solution for $\widehat{H} = 0, 1$.

*and the following holds for any local minimizer lying in* $\sigma^{-2} \in \mathcal{B}_{\widehat{H}}$*:*

$$\widehat{\sigma^2} \geq \frac{\sum_{h=\widehat{H}+1}^{L} \gamma_h^2}{LM - \widehat{H}(M+L)}.$$

*Proof* By substituting the lower-bound in Eq. (8.33) into Eq. (8.26), we obtain

$$\Theta \geq -\sigma^2 + \frac{\widehat{H}(L+M)\sigma^2 + \sum_{h=\widehat{H}+1}^{L} \gamma_h^2}{LM}.$$

This implies that $\Theta > 0$ unless the following hold:

$$\widehat{H} < \frac{LM}{L+M} = \frac{L}{1+\alpha},$$

$$\sigma^2 \geq \frac{\sum_{h=\widehat{H}+1}^{L} \gamma_h^2}{LM - \widehat{H}(L+M)}.$$

Therefore, no local minimum exists if either of these conditions is violated. This completes the proof of Lemma 8.10. □

It holds that

$$\frac{\sum_{h=\widehat{H}+1}^{L} \gamma_h^2}{LM - \widehat{H}(M+L)} \geq \frac{\sum_{h=\widehat{H}+1}^{L} \gamma_h^2}{M(L-\widehat{H})}, \tag{8.34}$$

of which the right-hand side is decreasing with respect to $\widehat{H}$. Combining Lemmas 8.5, 8.6, 8.7, and 8.10 and Eq. (8.34) completes the proof of Theorem 8.2. Corollary 8.3 is easily obtained from Lemmas 8.6 and 8.8.

## 8.4 Performance Analysis

To analyze the behavior of the EVB solution in the fully observed MF model, we rely on the *random matrix theory* (Marčenko and Pastur, 1967; Wachter, 1978; Johnstone, 2001; Bouchaud and Potters, 2003; Hoyle and Rattray, 2004; Baik and Silverstein, 2006), which describes the distribution of the singular values of random matrices in the limit when the matrix size goes to infinity. We first introduce some results obtained in the random matrix theory and then apply them to our analysis.

### 8.4.1 Random Matrix Theory

Assume that the observed matrix $\boldsymbol{V}$ is generated from the *spiked covariance model* (Johnstone, 2001):

$$\boldsymbol{V} = \boldsymbol{U}^* + \mathcal{E}, \tag{8.35}$$

where $U^* \in \mathbb{R}^{L\times M}$ is a *true* signal matrix with rank $H^*$ and singular values $\{\gamma_h^*\}_{h=1}^{H^*}$, and $\mathcal{E} \in \mathbb{R}^{L\times M}$ is a random matrix such that each element is independently drawn from a distribution with mean zero and variance $\sigma^{*2}$ (not necessarily Gaussian). As the observed singular values $\{\gamma_h\}_{h=1}^{L}$ of $V$, the true singular values $\{\gamma_h^*\}_{h=1}^{H^*}$ are also assumed to be arranged in the nonincreasing order.

We define normalized versions of the observed and the true singular values:

$$y_h = \frac{\gamma_h^2}{M\sigma^{*2}} \qquad \text{for} \qquad h = 1,\dots,L, \tag{8.36}$$

$$\nu_h^* = \frac{\gamma_h^{*2}}{M\sigma^{*2}} \qquad \text{for} \qquad h = 1,\dots,H^*. \tag{8.37}$$

In other words, $\{y_h\}_{h=1}^{L}$ are the eigenvalues of $VV^\top/(M\sigma^{*2})$, and $\{\nu_h^*\}_{h=1}^{H^*}$ are the eigenvalues of $U^*U^{*\top}/(M\sigma^{*2})$. Note the difference between $x_h$, defined by Eq. (6.132), and $y_h$: $x_h$ is the squared observed singular value normalized with the model noise variance $\sigma^2$, which is to be estimated, while $y_h$ is the one normalized with the true noise variance $\sigma^{*2}$.

Define the empirical distribution of the observed eigenvalues $\{y_h\}_{h=1}^{L}$ by

$$p(y) = \frac{1}{L}\sum_{h=1}^{L}\delta(y - y_h), \tag{8.38}$$

where $\delta(y)$ denotes the Dirac delta function. When $H^* = 0$, the observed matrix $V = \mathcal{E}$ consists only of noise, and its singular value distribution in the *large-scale limit* is specified by the following proposition:

**Proposition 8.11** *(Marčenko and Pastur, 1967; Wachter, 1978) In the large-scale limit when L and M go to infinity with its ratio $\alpha = L/M$ fixed, the empirical distribution of the eigenvalue y of $\mathcal{E}\mathcal{E}^\top/(M\sigma^{*2})$ almost surely converges to*

$$p(y) \to p^{\mathrm{MP}}(y) \equiv \frac{\sqrt{(y-\underline{y})(\overline{y}-y)}}{2\pi\alpha y}\theta(\underline{y} < y < \overline{y}), \tag{8.39}$$

*where* $$\overline{y} = (1+\sqrt{\alpha})^2, \qquad \underline{y} = (1-\sqrt{\alpha})^2, \tag{8.40}$$

*and $\theta(\cdot)$ is the indicator function, defined in Theorem 8.1.*[3]

Figure 8.3 shows Eq. (8.39), which we call the *Marčenko–Pastur (MP) distribution*, for $\alpha = 0.1, 1$. The mean $\langle y\rangle_{p^{\mathrm{MP}}(y)} = 1$ (which is constant for

[3] Convergence is in weak topology in distribution, i.e., $p(y)$ almost surely converges to $p^{\mathrm{MP}}(y)$ so that $\int f(y)p(y)dy = \int f(y)p^{\mathrm{MP}}(y)dy$ for any bounded continuous function $f(y)$.

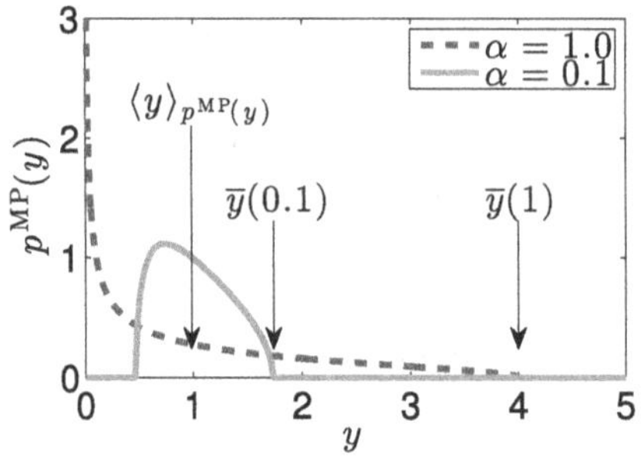


Figure 8.3 Marčenko–Pastur distirbution.

any $0 < \alpha \leq 1$) and the upper-limits $\overline{y} = \overline{y}(\alpha)$ of the support for $\alpha = 0.1, 1$ are indicated by arrows. Proposition 8.11 states that the probability mass is concentrated in the range between $\underline{y} \leq y \leq \overline{y}$. Note that the MP distribution appears for a *single* sample matrix; differently from standard "large-sample" theories, Proposition 8.11 does not require one to average over many sample matrices.[4] This single-sample property of the MP distribution is highly useful in our analysis because the MF model usually assumes a single observed matrix $\boldsymbol{V}$. We call the (unnormalized) singular value corresponding to the upper-limit $\overline{y}$, i.e.,

$$\overline{\gamma}^{\mathrm{MPUL}} = \sqrt{M\sigma^{*2} \cdot \overline{y}} = (\sqrt{L} + \sqrt{M})\sigma^{*}, \tag{8.41}$$

the *Marčenko–Pastur upper limit (MPUL)*.

When $H^{*} > 0$, the true signal matrix $\boldsymbol{U}^{*}$ affects the singular value distribution of $\boldsymbol{V}$. However, if $H^{*} \ll L$, the distribution can be approximated by a mixture of spikes (delta functions) and the MP distribution $p^{\mathrm{MP}}(y)$. Let $H^{**}$ $(\leq H^{*})$ be the number of singular values of $\boldsymbol{U}^{*}$ greater than $\gamma_h^{*} > \alpha^{1/4}\sqrt{M}\sigma^{*}$, i.e.,

$$\nu_{H^{**}}^{*} > \sqrt{\alpha} \qquad \text{and} \qquad \nu_{H^{**}+1}^{*} \leq \sqrt{\alpha}. \tag{8.42}$$

Then, the following proposition holds:

**Proposition 8.12** *(Baik and Silverstein, 2006) In the large-scale limit when $L$ and $M$ go to infinity with finite $\alpha$ and $H^{*}$, it almost surely holds that*

$$y_h = y_h^{\mathrm{Sig}} \equiv \left(1 + \nu_h^{*}\right)\left(1 + \frac{\alpha}{\nu_h^{*}}\right) \qquad \textit{for} \qquad h = 1, \ldots, H^{**}, \tag{8.43}$$

$$y_{H^{**}+1} = \overline{y}, \qquad \textit{and} \qquad y_L = \underline{y}.$$

[4] This property is called *self-averaging* (Bouchaud and Potters, 2003).

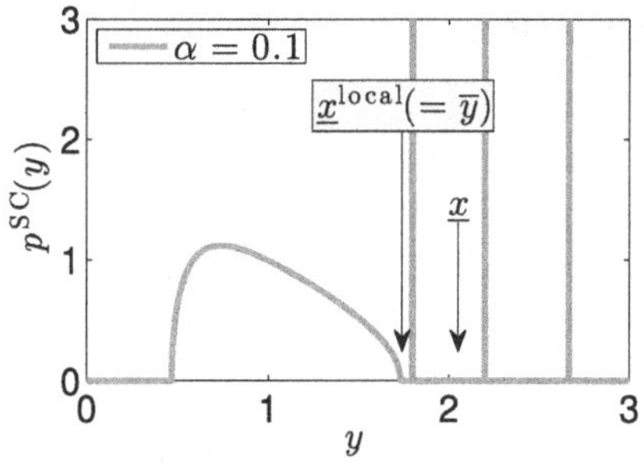


Figure 8.4 Spiked covariance distribution when $\{\nu_h^*\}_{h=1}^{H^{**}} = \{1.5, 1.0, 0.5\}$.

Furthermore, Hoyle and Rattray (2004) argued that, when $L$ and $M$ are large (but finite) and $H^* \ll L$, the empirical distribution of the eigenvalue $y$ of $\boldsymbol{V}\boldsymbol{V}^\top/(M\sigma^{*2})$, is accurately approximated by

$$p(y) \approx p^{\mathrm{SC}}(y) \equiv \frac{1}{L}\sum_{h=1}^{H^{**}} \delta\left(y - y_h^{\mathrm{Sig}}\right) + \frac{L - H^{**}}{L} p^{\mathrm{MP}}(y). \tag{8.44}$$

Figure 8.4 shows Eq. (8.44), which we call the *spiked covariance (SC) distribution*, for $\alpha = 0.1$, $H^{**} = 3$, and $\{\nu_h^*\}_{h=1}^{H^{**}} = \{1.5, 1.0, 0.5\}$. The SC distribution is irrespective of $\{\nu_h^*\}_{h=H^{**}+1}^{H^*}$, which satisfy $0 < \nu_h^* \leq \sqrt{\alpha}$ (see the definition (8.42) of $H^{**}$).

Proposition 8.12 states that in the large-scale limit, the large signal components such that $\nu_h^* > \sqrt{\alpha}$ appear outside the support of the MP distribution as spikes, while the other small signals are indistinguishable from the MP distribution (note that Eq. (8.43) implies that $y_h^{\mathrm{Sig}} > \overline{y}$ for $\nu_h^* > \sqrt{\alpha}$). This implies that any PCA method fails to recover the true dimensionality, unless

$$\nu_{H^*}^* > \sqrt{\alpha}. \tag{8.45}$$

The condition (8.45) requires that $\boldsymbol{U}^*$ has no small positive singular value such that $0 < \nu_h^* \leq \sqrt{\alpha}$, and therefore $H^{**} = H^*$.

The approximation (8.44) allows us to investigate more practical situations where the matrix size is finite. In Sections 8.4.2 and 8.4.4, respectively, we provide two theorems: one is based on Proposition 8.12 and guarantees perfect rank (PCA dimensionality) recovery of EVB learning in the large-scale limit, and the other one assumes that the approximation (8.44) exactly holds and provides a more realistic condition for perfect recovery.

### 8.4.2 Perfect Rank Recovery Condition in Large-Scale Limit

Now, we are almost ready for clarifying the behavior of the EVB solution. We assume that the model rank is set to be large enough, i.e., $H^* \leq H \leq L$, and all model parameters including the noise variance are estimated from observation (i.e., complete EVB learning). The last proposition on which our analysis relies is related to the property, called the *strong unimodality*, of the *log-concave distributions*:

**Proposition 8.13** *(Ibragimov, 1956; Dharmadhikari and Joag-Dev, 1988) The convolution*

$$g(s) = \langle f(s+t) \rangle_{p(t)} = \int f(s+t)p(t)dt$$

*is quasiconvex, if $p(t)$ is a log-concave distribution, and $f(t)$ is a quasiconvex function.*

In the large-scale limit, the summation over $h = 1, \ldots, L$ in the objective $\Omega(\sigma^{-2})$, given by Eq. (8.4), for noise variance estimation can be replaced with the expectation over the MP distribution $p^{\mathrm{MP}}(y)$. By scaling variables, the objective can be written as a convolution with a scaled version of the MP distribution, which turns out to be log-concave. Accordingly, we can use Proposition 8.13 to show that $\Omega(\sigma^{-2})$ is quasiconvex, which means that the noise variance estimation by EVB learning can be accurately performed by a local search algorithm. Combining this result with Proposition 8.12, we obtain the following theorem:

**Theorem 8.14** *In the large-scale limit when L and M go to infinity with finite $\alpha$ and $H^*$, EVB learning almost surely recovers the true rank, i.e., $\widehat{H}^{\mathrm{EVB}} = H^*$, if and only if*

$$\nu^*_{H^*} \geq \underline{\tau}, \tag{8.46}$$

*where $\underline{\tau}$ is defined in Theorem 6.13.*

Furthermore, the following corollary completely describes the behavior of the EVB solution in the large-scale limit:

**Corollary 8.15** *In the large-scale limit, the objective $\Omega(\sigma^{-2})$, defined by Eq. (8.4), for the noise variance estimation converges to a quasiconvex function, and it almost surely holds that*

$$\widehat{\tau}_h^{\mathrm{EVB}} \left( \equiv \frac{\gamma_h \widehat{\gamma}_h^{\mathrm{EVB}}}{M \widehat{\sigma}^{2\,\mathrm{EVB}}} \right) = \begin{cases} \nu^*_h & \text{if } \nu^*_h \geq \underline{\tau}, \\ 0 & \text{otherwise}, \end{cases} \tag{8.47}$$

$$\widehat{\sigma}^{2\,\mathrm{EVB}} = \sigma^{*2}.$$

One may get intuition of Eqs. (8.46) and (8.47) by comparing Eqs. (8.8) and (6.134) with Eq. (8.43): The estimator $\tau_h$ has the same relation to the observation $x_h$ as the true signal $v_h^*$, and hence is an unbiased estimator of the signal. However, Theorem 8.14 does not even approximately hold in practical situations with moderate-sized matrices (see the numerical validation in Section 8.5). After proving Theorem 8.14 and Corollary 8.15, we will derive a more practical condition for perfect recovery in Section 8.4.4.

### 8.4.3 Proofs of Theorem 8.14 and Corollary 8.15

In the large-scale limit, we can substitute the expectation $\langle f(y)\rangle_{p(y)}$ for the summation $L^{-1}\sum_{h=1}^{L} f(y_h)$. We can also substitute the MP distribution $p^{\mathrm{MP}}(y)$ for $p(y)$ for the expectation, since the contribution from the $H^*$ signal components converges to zero. Accordingly, our objective (8.4) converges to

$$\begin{aligned} \Omega(\sigma^{-2}) \to \Omega^{\mathrm{LSL}}(\sigma^{-2}) &\equiv \int_{\kappa}^{\overline{y}} \psi\left(\sigma^{*2}\sigma^{-2}y\right) p^{\mathrm{MP}}(y)dy + \int_{\underline{y}}^{\kappa} \psi_0\left(\sigma^{*2}\sigma^{-2}y\right) p^{\mathrm{MP}}(y)dy \\ &= \Omega^{\mathrm{LSL-Full}}(\sigma^{-2}) - \int_{\max(\underline{x}\sigma^2/\sigma^{*2},\underline{y})}^{\kappa} \psi_1\left(\sigma^{*2}\sigma^{-2}y\right) p^{\mathrm{MP}}(y)dy, \end{aligned} \tag{8.48}$$

where

$$\Omega^{\mathrm{LSL-Full}}(\sigma^{-2}) \equiv \int_{\underline{y}}^{\overline{y}} \psi\left(\sigma^{*2}\sigma^{-2}y\right) p^{\mathrm{MP}}(y)dy, \tag{8.49}$$

and $\kappa$ is a constant satisfying

$$\frac{H}{L} = \int_{\kappa}^{\overline{y}} p^{\mathrm{MP}}(y)dy \qquad (\underline{y} \le \kappa \le \overline{y}). \tag{8.50}$$

Note that $\underline{x}$, $\underline{y}$, and $\overline{y}$ are defined by Eqs. (8.8) and (8.40), and it holds that

$$\underline{x} > \overline{y}. \tag{8.51}$$

We first investigate Eq. (8.49), which corresponds to the objective for the full-rank model (i.e., $H = L$). Let

$$s = \log(\sigma^{-2}),$$
$$t = \log y \qquad \left(dt = \tfrac{1}{y}dy\right).$$

Then Eq. (8.49) is written as a convolution:

$$\begin{aligned} \widetilde{\Omega}^{\mathrm{LSL-Full}}(s) \equiv \Omega^{\mathrm{LSL-Full}}(e^s) &= \int \psi\left(\sigma^{*2}e^{s+t}\right) e^t p^{\mathrm{MP}}(e^t)dt \\ &= \int \widetilde{\psi}(s+t) p^{\mathrm{LSMP}}(t)dt, \end{aligned} \tag{8.52}$$

where

$$\widetilde{\psi}(s) = \psi(\sigma^{*2} e^s),$$
$$p^{\mathrm{LSMP}}(t) = e^t p^{\mathrm{MP}}(e^t)$$
$$= \frac{\sqrt{(e^t - \underline{y})(\overline{y} - e^t)}}{2\pi\alpha} \theta(\underline{y} < e^t < \overline{y}). \tag{8.53}$$

Since Lemma 8.4 states that $\psi(x)$ is quasiconvex, its composition $\widetilde{\psi}(s)$ with the nondecreasing function $\sigma^{*2} e^s$ is also quasiconvex.

The following holds for $p^{\mathrm{LSMP}}(t)$, which we call a log-scaled MP (LSMP) distribution:

**Lemma 8.16** *The LSMP distribution* (8.53) *is log-concave.*

*Proof* Focusing on the support,

$$\log \underline{y} < t < \log \overline{y},$$

of the LSMP distribution (8.53), we define

$$f(t) \equiv 2 \log p^{\mathrm{LSMP}}(t) = 2 \log \frac{\sqrt{(e^t - \underline{y})(\overline{y} - e^t)}}{2\pi\alpha}$$
$$= \log(-e^{2t} + (\underline{y} + \overline{y})e^t - \underline{y}\overline{y}) + \mathrm{const.}$$

Let

$$u(t) \equiv (e^t - \underline{y})(\overline{y} - e^t) = -e^{2t} + (\underline{y} + \overline{y})e^t - \underline{y}\overline{y} > 0, \tag{8.54}$$

and let

$$v(t) \equiv \frac{\partial u}{\partial t} = -2e^{2t} + (\underline{y} + \overline{y})e^t = u - e^{2t} + \underline{y}\overline{y},$$
$$w(t) \equiv \frac{\partial^2 u}{\partial t^2} = -4e^{2t} + (\underline{y} + \overline{y})e^t = v - 2e^{2t},$$

be the first and the second derivatives of $u$. Then, the first and the second derivatives of $f(t)$ are given by

$$\frac{\partial f}{\partial t} = \frac{v}{u},$$
$$\frac{\partial^2 f}{\partial t^2} = \frac{uw - v^2}{u^2}$$
$$= -\frac{e^t \left( (\underline{y} + \overline{y})e^{2t} - 4\underline{y}\overline{y}e^t + (\underline{y} + \overline{y})\underline{y}\overline{y} \right)}{u^2}$$

$$= -\frac{e^t(\underline{y}+\overline{y})}{u^2}\left(\left(e^t - \frac{2\underline{y}\overline{y}}{(\underline{y}+\overline{y})}\right)^2 + \frac{\underline{y}\overline{y}\left(\overline{y}-\underline{y}\right)^2}{(\underline{y}+\overline{y})^2}\right)$$

$$\leq 0.$$

This proves the log-concavity of the LSMP dsitribution $p^{\mathrm{LSMP}}(t)$, and completes the proof of Lemma 8.16. □

Lemma 8.16 and Proposition 8.13 imply that $\widetilde{\Omega}^{\mathrm{LSL-Full}}(s)$ is quasiconvex, and therefore its composition $\Omega^{\mathrm{LSL-Full}}(\sigma^{-2})$ with the nondecreasing function $\log(\sigma^{-2})$ is quasiconvex. The minimizer of $\Omega^{\mathrm{LSL-Full}}(\sigma^{-2})$ can be found by evaluating the derivative $\Theta$, given by Eq. (8.26), in the large-scale limit:

$$\Theta^{\mathrm{Full}} \to \Theta^{\mathrm{LSL-Full}} = -\sigma^2 + \sigma^{*2}\int_{\underline{y}}^{\overline{y}} y \cdot p^{\mathrm{MP}}(y)dy$$
$$- \int_{\underline{x}\sigma^2/\sigma^{*2}}^{\overline{y}} \tau(\sigma^{*2}\sigma^{-2}y;\alpha)p^{\mathrm{MP}}(y)dy. \tag{8.55}$$

Here, we used Eqs. (6.133) and (8.9). In the range

$$0 < \sigma^{-2} < \frac{\underline{x}\sigma^{*-2}}{\overline{y}} \qquad \left(\text{i.e.,} \quad \frac{\underline{x}\sigma^2}{\sigma^{*2}} > \overline{y}\right), \tag{8.56}$$

the third term in Eq. (8.55) is zero. Therefore, Eq. (8.55) is increasing with respect to $\sigma^{-2}$, and zero when

$$\sigma^2 = \sigma^{*2}\int_{\underline{y}}^{\overline{y}} y \cdot p^{\mathrm{MP}}(y)dy = \sigma^{*2}.$$

Accordingly, $\Omega^{\mathrm{LSL-Full}}(\sigma^{-2})$ is strictly convex in the range (8.56). Eq. (8.51) implies that the point $\sigma^{-2} = \sigma^{*-2}$ is contained in the region (8.56), and therefore it is a local minimum of $\Omega^{\mathrm{LSL-Full}}(\sigma^{-2})$. Combined with the quasiconvexity of $\Omega^{\mathrm{LSL-Full}}(\sigma^{-2})$, we have the following lemma:

**Lemma 8.17** *The objective $\Omega^{\mathrm{LSL-Full}}(\sigma^{-2})$ for the full-rank model $H = L$ in the large-scale limit is quasiconvex with its minimizer at $\sigma^{-2} = \sigma^{*-2}$. It is strictly convex in the range* (8.56).

For any $\kappa$ ($\underline{y} < \kappa < \overline{y}$), the second term in Eq. (8.48) is zero in the range (8.56), which includes its minimizer at $\sigma^{-2} = \sigma^{*-2}$. Since Lemma 8.4 states that $\psi_1(x)$ is decreasing for $x > \underline{x}$, the second term in Eq. (8.48) is nondecreasing in the region where

$$\left(\sigma^{*-2} <\right)\frac{\underline{x}\sigma^{*-2}}{\overline{y}} \leq \sigma^{-2} < \infty.$$

Therefore, the quasi-convexity of $\Omega^{\mathrm{LSL-Full}}$ is inherited to $\Omega^{\mathrm{LSL}}$:

**Lemma 8.18** *The objective $\Omega^{\mathrm{LSL}}(\sigma^{-2})$ for noise variance estimation in the large-scale limit is quasiconvex with its minimizer at $\sigma^{-2} = \sigma^{*-2}$. $\Omega^{\mathrm{LSL}}(\sigma^{-2})$ is strictly convex in the range* (8.56).

Thus we have proved that EVB learning accurately estimates the noise variance in the large-scale limit:

$$\widehat{\sigma}^{2\,\mathrm{EVB}} = \sigma^{*2}. \tag{8.57}$$

Assume that Eq. (8.45) holds. Then Proposition 8.12 guarantees that, in the large-scale limit, the following hold:

$$\frac{\gamma_{H^*}^2}{M\sigma^{*2}} \equiv y_{H^*} = (1 + \nu_{H^*}^*)\left(1 + \frac{\alpha}{\nu_{H^*}^*}\right), \tag{8.58}$$

$$\frac{\gamma_{H^*+1}^2}{M\sigma^{*2}} \equiv y_{H^*+1} = \overline{y} = (1 + \sqrt{\alpha})^2. \tag{8.59}$$

Remember that the EVB threshold is given by Eq. (8.8), i.e.,

$$\frac{(\underline{\gamma}^{\mathrm{EVB}})^2}{M\widehat{\sigma}^{2\,\mathrm{EVB}}} \equiv \underline{x} = \left(1 + \underline{\tau}\right)\left(1 + \frac{\alpha}{\underline{\tau}}\right). \tag{8.60}$$

Since Lemma 8.18 states that $\widehat{\sigma}^{2\,\mathrm{EVB}} = \sigma^{*2}$, comparing Eqs. (8.58) and (8.59) with Eq. (8.60) results in the following lemma:

**Lemma 8.19** *It almost surely holds that*

$$\gamma_{H^*} \geq \underline{\gamma}^{\mathrm{EVB}} \qquad \text{if and only if} \qquad \nu_{H^*}^* \geq \underline{\tau}, \tag{8.61}$$
$$\gamma_{H^*+1} < \underline{\gamma}^{\mathrm{EVB}} \qquad \text{for any} \qquad \{\nu_h^*\}.$$

This completes the proof of Theorem 8.14. Comparing Eqs. (6.134) and (8.43) under Lemmas 8.18 and 8.19 proves Corollary 8.15. □

### 8.4.4 Practical Condition for Perfect Rank Recovery

Theorem 8.14 rigorously holds in the large-scale limit. However, it does not describe the behavior of the EVB solution very accurately in practical finite matrix-size cases. We can obtain a more practical condition for perfect recovery by relying on the approximation (8.44). We can prove the following theorem:

**Theorem 8.20** *Let*

$$\xi = \frac{H^*}{L}$$

*be the relevant rank ratio, and assume that*

$$p(y) = p^{\mathrm{SC}}(y). \tag{8.62}$$

*Then, EVB learning recovers the true rank, i.e.,* $\widehat{H}^{\mathrm{EVB}} = H^*$, *if the following two inequalities hold:*

$$\xi < \frac{1}{\underline{x}}, \tag{8.63}$$

$$\nu^*_{H^*} > \frac{\left(\frac{\underline{x}-1}{1-\underline{x}\xi} - \alpha\right) + \sqrt{\left(\frac{\underline{x}-1}{1-\underline{x}\xi} - \alpha\right)^2 - 4\alpha}}{2}, \tag{8.64}$$

*where* $\underline{x}$ *is defined by Eq. (8.8).*

Note that, in the large-scale limit, $\xi$ converges to zero, and the sufficient condition, Eqs. (8.63) and (8.64), in Theorem 8.20 is equivalent to the necessary and sufficient condition (8.46) in Theorem 8.14.

Theorem 8.20 only requires that the SC distribution (8.44) well approximates the observed singular value distribution. Accordingly, it well describes the dependency of the EVB solution on $\xi$, which will be shown in numerical validation in Section 8.5. Theorem 8.20 states that, if the true rank $H^*$ is small enough compared with $L$ and the smallest signal $\nu^*_{H^*}$ is large enough, EVB learning perfectly recovers the true rank.

The following corollary also supports EVB learning:

**Corollary 8.21** *Under the assumption* (8.62) *and the conditions* (8.63) *and* (8.64), *the objective* $\Omega(\sigma^{-2})$ *for the noise variance estimation has no local minimum (no stationary point if* $\xi > 0$*) that results in a wrong estimated rank* $\widehat{H}^{\mathrm{EVB}} \neq H^*$.

This corollary states that, although the objective function (8.4) is nonconvex and possibly multimodal in general, any local minimum leads to the correct estimated rank. Therefore, perfect recovery does not require global search, but only local search, for noise variance estimation, if $L$ and $M$ are sufficiently large so that we can warrant Eq. (8.62).

In the next section, we give the proofs of Theorem 8.20 and Corollary 8.21, and then show numerical experiments that support the theory.

### 8.4.5 Proofs of Theorem 8.20 and Corollary 8.21

We regroup the terms in Eq. (8.4) as follows:

$$\Omega(\sigma^{-2}) = \Omega_1(\sigma^{-2}) + \Omega_0(\sigma^{-2}), \tag{8.65}$$

where

$$\Omega_1(\sigma^{-2}) = \frac{1}{H^*} \sum_{h=1}^{H^*} \psi\left(\frac{\gamma_h^2}{M}\sigma^{-2}\right), \tag{8.66}$$

$$\Omega_0(\sigma^{-2}) = \frac{1}{L - H^*} \left( \sum_{h=H^*+1}^{H} \psi\left(\frac{\gamma_h^2}{M}\sigma^{-2}\right) + \sum_{h=H+1}^{L} \psi_0\left(\frac{\gamma_h^2}{M}\sigma^{-2}\right) \right). \tag{8.67}$$

In the following, assuming that Eq. (8.62) holds and

$$y_{H^*} > \overline{y}, \tag{8.68}$$

we derive a sufficient condition for any local minimizer to lie only in $\sigma^{-2} \in \mathcal{B}_{H^*}$, with which Lemma 8.7 proves Theorem 8.20.

Under the assumption (8.62) and the condition (8.68), $\Omega_0(\sigma^{-2})$, defined by Eq. (8.67), is equivalent to the objective $\Omega^{\mathrm{LSL}}(\sigma^{-2})$ in the large-scale limit. Using Lemma 8.18, and noting that

$$\underline{\sigma}_{H^*+1}^{-2} = \frac{M\underline{x}}{\gamma_{H^*+1}^{2}} = \frac{\underline{x}\sigma^{*-2}}{\overline{y}} > \sigma^{*-2}, \tag{8.69}$$

we have the following lemma:

**Lemma 8.22** *$\Omega_0(\sigma^{-2})$ is quasiconvex with its minimizer at*

$$\sigma^{-2} = \sigma^{*-2}.$$

*$\Omega_0(\sigma^{-2})$ is strictly convex in the range*

$$0 < \sigma^{-2} < \underline{\sigma}_{H^*+1}^{-2}.$$

Using Lemma 8.22 and the strict quasiconvexity of $\psi(x)$, we can deduce the following lemma:

**Lemma 8.23** *$\Omega(\sigma^{-2})$ is nondecreasing (increasing if $\xi > 0$) in the range $\underline{\sigma}_{H^*+1}^{2} < \sigma^{-2} < \infty$.*

*Proof* Lemma 8.22 states that $\Omega_0(\sigma^{-2})$, defined by Eq. (8.67), is quasiconvex with its minimizer at

$$\sigma^{-2} = \left( \frac{\sum_{h=H^*+1}^{L} \gamma_h^2}{(L - H^*)M} \right)^{-1} = \sigma^{*-2}.$$

Since $\Omega_1(\sigma^{-2})$, defined by Eq. (8.66), is the sum of strictly quasiconvex functions with their minimizers at $\sigma^{-2} = M/\gamma_h^2 < \sigma^{*-2}$ for $h = 1, \ldots, H^*$, our objective $\Omega(\sigma^{-2})$, given by Eq. (8.65), is nondecreasing (increasing if $H^* > 0$) for

$$\sigma^{-2} \geq \sigma^{*-2}.$$

Since Eq. (8.69) implies that $\underline{\sigma}_{H^*+1}^{-2} > \sigma^{*-2}$, $\Omega(\sigma^{-2})$ is nondecreasing (increasing if $\xi > 0$) for $\sigma^{-2} > \underline{\sigma}_{H^*+1}^{-2}$, which completes the proof of Lemma 8.23. □

Using the bounds given by Eq. (8.33) and Lemma 8.22, we also obtain the following lemma:

**Lemma 8.24** *$\Omega(\sigma^{-2})$ is increasing at $\sigma^{-2} = \underline{\sigma}_{H^*+1}^{2} - 0$.*[5] *It is decreasing at $\sigma^{-2} = \underline{\sigma}_{H^*}^{2} + 0$ if the following hold:*

$$\xi < \frac{1}{(1+\sqrt{\alpha})^2}, \tag{8.70}$$

$$y_{H^*} > \frac{\underline{x}(1-\xi)}{1-\xi(1+\sqrt{\alpha})^2}. \tag{8.71}$$

*Proof* Lemma 8.22 states that $\Omega_0(\sigma^{-2})$ is strictly convex in the range $0 < \sigma^{-2} < \underline{\sigma}_{H^*+1}^{2}$, and minimized at $\sigma^{-2} = \sigma^{*-2}$. Since Eq. (8.69) implies that $\sigma^{*-2} < \underline{\sigma}_{H^*+1}^{2}$, $\Omega_0(\sigma^{-2})$ is increasing at $\sigma^{-2} = \underline{\sigma}_{H^*+1}^{2} - 0$. Since $\Omega_1(\sigma^{-2})$ is the sum of strictly quasiconvex functions with their minimizers at $\sigma^{-2} = M/\gamma_h^2 < \sigma^{*-2}$ for $h = 1, \ldots, H^*$, $\Omega(\sigma^{-2})$ is also increasing at $\sigma^{-2} = \underline{\sigma}_{H^*+1}^{2} - 0$.

Let us investigate the sign of the derivative $\Theta$ of $\Omega(\sigma^{-2})$ at $\sigma^{-2} = \underline{\sigma}_{H^*}^{2} + 0 \in \mathcal{B}_{H^*}$. Substituting the upper-bound in Eq. (8.33) into Eq. (8.26), we have

$$\begin{aligned} \Theta &< -\sigma^2 + \frac{H^*(\sqrt{L}+\sqrt{M})^2\sigma^2 + \sum_{h=H^*+1}^{L}\gamma_h^2}{LM} \\ &= -\sigma^2 + \frac{H^*(\sqrt{L}+\sqrt{M})^2\sigma^2 + (L-H^*)M\sigma^{*2}}{LM}. \end{aligned} \tag{8.72}$$

The right-hand side of Eq. (8.72) is negative if the following hold:

$$\xi = \frac{H^*}{L} < \frac{M}{(\sqrt{L}+\sqrt{M})^2} = \frac{1}{(1+\sqrt{\alpha})^2}, \tag{8.73}$$

$$\sigma^2 > \frac{(L-H^*)M\sigma^{*2}}{LM - H^*(\sqrt{L}+\sqrt{M})^2} = \frac{(1-\xi)\sigma^{*2}}{1-\xi(1+\sqrt{\alpha})^2}. \tag{8.74}$$

Assume that the first condition (8.73) holds. Then the second condition (8.74) holds at $\sigma^{-2} = \underline{\sigma}_{H^*}^{2} + 0$, if

$$\underline{\sigma}_{H^*}^{-2} < \frac{1-\xi(1+\sqrt{\alpha})^2}{(1-\xi)}\sigma^{*-2},$$

[5] By "−0" we denote an arbitrarily large negative value.

or equivalently,

$$y_{H^*} = \frac{\gamma_{H^*}^2}{M\sigma^{*2}} = \underline{x} \cdot \frac{\underline{\sigma}_{H^*}^2}{\sigma^{*2}} > \frac{\underline{x}(1-\xi)}{1-\xi(1+\sqrt{\alpha})^2},$$

which completes the proof of Lemma 8.24. □

Finally, we obtain the following lemma:

**Lemma 8.25** *$\Omega(\sigma^{-2})$ is decreasing in the range $0 < \sigma^{-2} < \underline{\sigma}_{H^*}^2$ if the following hold:*

$$\xi < \frac{1}{\underline{x}}, \tag{8.75}$$

$$y_{H^*} > \frac{\underline{x}(1-\xi)}{1-\underline{x}\xi}. \tag{8.76}$$

*Proof* In the range $0 < \sigma^{-2} < \underline{\sigma}_{H^*}^2$, the estimated rank (8.27) is bounded as

$$0 \le \widehat{H} \le H^* - 1. \tag{8.77}$$

Substituting the upper-bound in Eq. (8.33) into Eq. (8.26), we have

$$\begin{aligned}\Theta &< -\sigma^2 + \frac{\widehat{H}(\sqrt{L}+\sqrt{M})^2\sigma^2 + \sum_{h=\widehat{H}+1}^{H^*}\gamma_h^2 + \sum_{h=H^*+1}^{L}\gamma_h^2}{LM}\\ &= -\sigma^2 + \frac{\widehat{H}(\sqrt{L}+\sqrt{M})^2\sigma^2 + \sum_{h=\widehat{H}+1}^{H^*}\gamma_h^2 + (L-H^*)M\sigma^{*2}}{LM}.\end{aligned} \tag{8.78}$$

The right-hand side of Eq. (8.78) is negative, if the following hold:

$$\frac{\widehat{H}}{L} < \frac{M}{(\sqrt{L}+\sqrt{M})^2} = \frac{1}{(1+\sqrt{\alpha})^2}, \tag{8.79}$$

$$\sigma^2 > \frac{\sum_{h=\widehat{H}+1}^{H^*}\gamma_h^2 + (L-H^*)M\sigma^{*2}}{LM - \widehat{H}(\sqrt{L}+\sqrt{M})^2}. \tag{8.80}$$

Assume that

$$\xi = \frac{H^*}{L} < \frac{1}{(1+\sqrt{\alpha})^2}.$$

Then both of the conditions (8.79) and (8.80) hold for $\sigma^{-2} \in (0, \underline{\sigma}_{H^*}^2)$, if the following holds:

$$\underline{\sigma}_{\widehat{H}+1}^{-2} < \frac{LM - \widehat{H}(\sqrt{L}+\sqrt{M})^2}{\sum_{h=\widehat{H}+1}^{H^*}\gamma_h^2 + (L-H^*)M\sigma^{*2}} \quad \text{for} \quad \widehat{H} = 0, \ldots, H^*-1. \tag{8.81}$$

Since the sum $\sum_{h=\widehat{H}+1}^{H^*} \gamma_h^2$ in the right-hand side of Eq. (8.81) is upper-bounded as

$$\sum_{h=\widehat{H}+1}^{H^*} \gamma_h^2 \leq (H^* - \widehat{H})\gamma_{\widehat{H}+1}^2,$$

Eq. (8.81) holds if

$$\begin{aligned} \underline{\sigma}_{\widehat{H}+1}^{-2} &< \frac{LM - \widehat{H}(\sqrt{L}+\sqrt{M})^2}{(H^* - \widehat{H})\gamma_{\widehat{H}+1}^2 + (L - H^*)L\sigma^{*2}} \\ &= \frac{1 - \frac{\widehat{H}}{L}(1+\sqrt{\alpha})^2}{(\xi - \frac{\widehat{H}}{L})\frac{\gamma_{\widehat{H}+1}^2}{M} + (1-\xi)\sigma^{*2}} \quad \text{for} \quad \widehat{H} = 0, \ldots, H^* - 1. \end{aligned} \tag{8.82}$$

Using Eq. (8.24), the condition (8.82) is rewritten as

$$\frac{\gamma_{\widehat{H}+1}^2}{M\underline{x}} > \frac{(\xi - \frac{\widehat{H}}{L})\frac{\gamma_{\widehat{H}+1}^2}{M} + (1-\xi)\sigma^{*2}}{1 - \frac{\widehat{H}}{L}(1+\sqrt{\alpha})^2}$$

$$\left(1 - \frac{\widehat{H}}{L}(1+\sqrt{\alpha})^2\right)\frac{\gamma_{\widehat{H}+1}^2}{M\sigma^{*2}} > \left(\xi\underline{x} - \frac{\widehat{H}}{L}\underline{x}\right)\frac{\gamma_{\widehat{H}+1}^2}{M\sigma^{*2}} + (1-\xi)\underline{x},$$

or equivalently

$$y_{\widehat{H}+1} = \frac{\gamma_{\widehat{H}+1}^2}{M\sigma^{*2}} > \frac{(1-\xi)\,\underline{x}}{\left(1 - \xi\underline{x} + \frac{\widehat{H}}{L}\left(\underline{x} - (1+\sqrt{\alpha})^2\right)\right)} \quad \text{for} \quad \widehat{H} = 0, \ldots, H^* - 1. \tag{8.83}$$

Note that $\underline{x} > \overline{y} = (1+\sqrt{\alpha})^2$. Further bounding both sides, we have the following sufficient condition for Eq. (8.83) to hold:

$$y_{H^*} > \frac{(1-\xi)\underline{x}}{\max\left(0, 1 - \xi\underline{x}\right)}. \tag{8.84}$$

Thus we obtain the conditions (8.75) and (8.76) for $\Theta$ to be negative for $\sigma^{-2} \in (0, \underline{\sigma}_{H^*}^2)$, which completes the proof of Lemma 8.25. □

Lemmas 8.23, 8.24, and 8.25 together state that, if all the conditions (8.68) and (8.70) through (8.76) hold, at least one local minimum exists in the correct range $\sigma^{-2} \in \mathcal{B}_{H^*}$, and no local minimum (no stationary point if $\xi > 0$) exists outside the correct range. Therefore, we can estimate the correct rank $\widehat{H}^{\mathrm{EVB}} = H^*$ by using a local search algorithm for noise variance estimation. Choosing the tightest conditions, we have the following lemma:

**Lemma 8.26** *$\Omega(\sigma^{-2})$ has a global minimum in $\sigma^{-2} \in \mathcal{B}_{H^*}$, and no local minimum (no stationary point if $\xi > 0$) outside $\mathcal{B}_{H^*}$, if the following hold:*

$$\xi < \frac{1}{\underline{x}},$$

$$y_{H^*} = \frac{\gamma_{H^*}^2}{M\sigma^{*2}} > \frac{\underline{x}(1-\xi)}{1-\underline{x}\xi}. \tag{8.85}$$

Using Eq. (8.43), Eq. (8.85) can be written with the *true* signal amplitude as follows:

$$(1 + \nu_{H^*}^*)\left(1 + \frac{\alpha}{\nu_{H^*}^*}\right) - \frac{\underline{x}(1-\xi)}{1-\underline{x}\xi} > 0. \tag{8.86}$$

The left-hand side of Eq. (8.86) can be factorized as follows:

$$\frac{1}{\nu_{H^*}^*}\left(\nu_{H^*}^* - \frac{\left(\frac{\underline{x}(1-\xi)}{1-\underline{x}\xi} - (1+\alpha)\right) + \sqrt{\left(\frac{\underline{x}(1-\xi)}{1-\underline{x}\xi} - (1+\alpha)\right)^2 - 4\alpha}}{2}\right)$$

$$\cdot\left(\nu_{H^*}^* - \frac{\left(\frac{\underline{x}(1-\xi)}{1-\underline{x}\xi} - (1+\alpha)\right) - \sqrt{\left(\frac{\underline{x}(1-\xi)}{1-\underline{x}\xi} - (1+\alpha)\right)^2 - 4\alpha}}{2}\right) > 0. \tag{8.87}$$

When Eq. (8.45) holds, the last factor in the left-hand side in Eq. (8.87) is positive. Therefore, we have the following condition:

$$\nu_{H^*}^* > \frac{\left(\frac{\underline{x}(1-\xi)}{1-\underline{x}\xi} - (1+\alpha)\right) + \sqrt{\left(\frac{\underline{x}(1-\xi)}{1-\underline{x}\xi} - (1+\alpha)\right)^2 - 4\alpha}}{2}$$

$$= \frac{\left(\frac{\underline{x}-1}{1-\underline{x}\xi} - \alpha\right) + \sqrt{\left(\frac{\underline{x}-1}{1-\underline{x}\xi} - \alpha\right)^2 - 4\alpha}}{2}. \tag{8.88}$$

Lemma 8.26 with the condition (8.85) replaced with the condition (8.88) leads to Theorem 8.20 and Corollary 8.21.

## 8.5 Numerical Verification

Figure 8.5 shows numerical simulation results for $M = 200$ and $L = 20, 100, 200$. $\mathcal{E}$ was drawn from the independent Gaussian distribution with mean 0 and variance $\sigma^{*2} = 1$, and *true* signal singular values $\{\gamma_h^*\}_{h=1}^{H^*}$ were

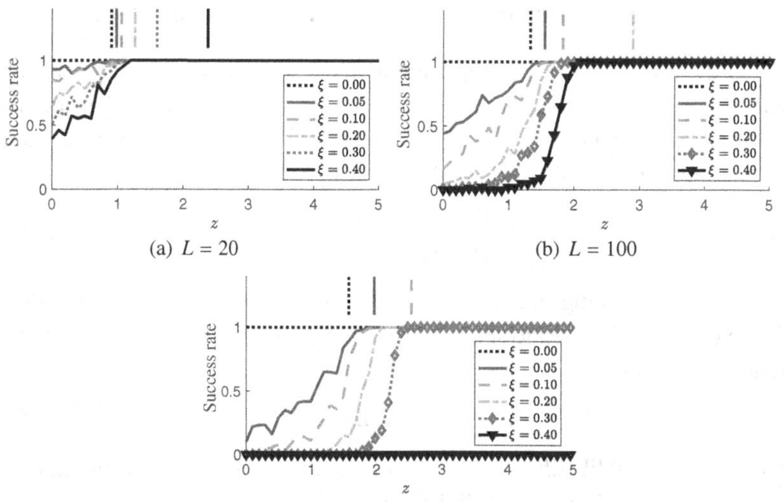


(a) $L = 20$

(b) $L = 100$

(c) $L = 200$

Figure 8.5 Success rate of rank recovery in numerical simulation for $M = 200$. The horizontal axis indicates the lower limit of the support of the simulated true signal distribution, i.e., $z \approx \sqrt{v^*_{H^*}}$. The recovery condition (8.64) for finite-sized matrices is indicated by a vertical bar with the same line style for each $\xi$. The leftmost vertical bar, which corresponds to the condition (8.64) for $\xi = 0$, coincides with the recovery condition (8.46) for infinite-sized matrices.

drawn from the uniform distribution on $[z\sqrt{M}\sigma^*, 10\sqrt{M}\sigma^*]$ for different $z$, which is indicated by the horizontal axis. We used Algorithm 16, which will be introduced in Chapter 9, to compute the global EVB solution.

The vertical axis indicates the success rate of rank recovery over 100 trials, i.e., the proportion of the trials giving $\widehat{H}^{\mathrm{EVB}} = H^*$. If the condition (8.63) on $\xi$ is violated, the corresponding curve is depicted with markers. Otherwise, the condition (8.64) on $v^*_{H^*}(= \gamma^{*2}_{H^*}/(M\sigma^{*2}))$ is indicated by a vertical bar with the same line style for each $\xi$. In other words, Theorem 8.20 states that the success rate should be equal to one if $z$ $(> \gamma^*_{H^*}/(\sqrt{M}\sigma^{*2}))$ is larger than the value indicated by the vertical bar. The leftmost vertical bar, which corresponds to the condition (8.64) for $\xi = 0$, coincides with the recovery condition (8.46), given by Theorem 8.14, for infinite-sized matrices.

We see that Theorem 8.20 with the condition (8.64) approximately holds for these moderate-sized matrices, while Theorem 8.14 with the condition (8.46), which does not depend on the relevant rank ratio $\xi$, immediately breaks for positive $\xi$.

## 8.6 Comparison with Laplace Approximation

Here, we compare EVB learning with an alternative dimensionality selection method (Hoyle, 2008) based on the *Laplace approximation (LA)*. Consider the PCA application, where $D$ denotes the dimensionality of the observation space, and $N$ denotes the number of samples, i.e., in our MF notation to keep $L \leq M$,

$$\begin{aligned} L = D, M = N \quad & \text{if} \quad D \leq N, \\ L = N, M = D \quad & \text{if} \quad D > N. \end{aligned} \tag{8.89}$$

Right after Tipping and Bishop (1999) proposed the *probabilistic PCA*, Bishop (1999a) proposed to select the PCA dimensionality by maximizing the marginal likelihood:

$$p(\boldsymbol{V}) = \langle p(\boldsymbol{V}|\boldsymbol{A}, \boldsymbol{B}) \rangle_{p(\boldsymbol{A})p(\boldsymbol{B})} \cdot \tag{8.90}$$

Since the marginal likelihood (8.90) is computationally intractable, he approximated it by LA, and suggested Gibbs sampling and VB learning as alternatives. The VB variant, of which the model is almost the same as the MF defined by Eqs. (6.1) through (6.3), was also proposed by himself (Bishop, 1999b) along with a standard local solver similar to Algorithm 1 in Chapter 3.

The LA-based approach was polished in Minka (2001a), by introducing a conjugate prior[6] on $\boldsymbol{B}$ to $p(\boldsymbol{V}|\boldsymbol{B}) = \langle p(\boldsymbol{V}|\boldsymbol{A}, \boldsymbol{B}) \rangle_{p(\boldsymbol{A})}$, and ignoring the non-leading terms that do not grow fast as the number $N$ of samples goes to infinity. Hoyle (2008) pointed out that Minka's method is inaccurate when $D \gg N$, and proposed the *overlap (OL) method*, a further polished variant of the LA-based approach. A notable difference of the OL method from most of the LA-based methods is that the OL method applies LA around a more accurate estimator than the MAP estimator.[7] Thanks to the use of the accurate estimator, the OL method behaves *optimally* in the large-scale limit when $D$ and $N$ go to infinity, while Minka's method does not. We will clarify the meaning of the optimality and discuss it in more detail in Section 8.7.

The OL method minimizes an approximation to the negative logarithm of the marginal likelihood (8.90), which depends on estimators for $\lambda_h = b_h^2 + \sigma^2$ and $\sigma^2$ computed by an iterative algorithm, over the hypothetical model rank $H = 0, \ldots, L$ (see Appendix C for the detailed computational procedure). Figure 8.6 shows numerical simulation results that compare EVB learning and the OL method: Figure 8.6(a) shows the success rate for the no-signal case $\xi = 0$ ($H^* = 0$), while Figures 8.6(b) through 8.6(f) show the success rate for $\xi = 0.05$ and $D = 20, 100, 200, 400$, and $1,000$, respectively. We also show

[6] This conjugate prior does not satisfy the implicit requirement, footnoted in Section 1.2.4, that the moments of the family member can be computed analytically.

[7] As explained in Section 2.2.1, LA is usually applied around the MAP estimator.

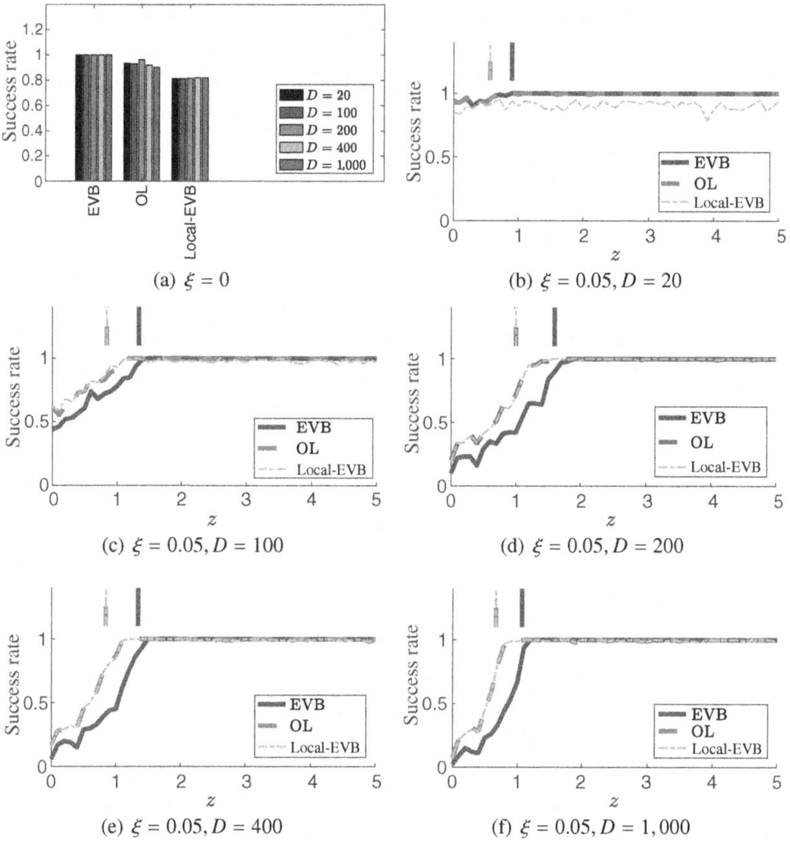


Figure 8.6 Success rate of PCA dimensionality recovery by (global) EVB learning, the OL method, and the local-EVB estimator for $N = 200$. Vertical bars indicate the recovery conditions, Eq. (8.46) for EVB learning, and Eq. (8.95) for the OL method and the local-EVB estimator, in the large-scale limit.

the performance of the *local-EVB estimator* (6.131), which was computed by a local solver (Algorithm 18 introduced in Chapter 9). For the OL method and the local-EVB estimator, we initialized the noise variance estimator to $10^{-4} \cdot \sum_{h=1}^{L} \gamma_h^2/(LM)$.

In comparison with the OL method, EVB learning shows its conservative nature: It exhibits almost zero false positive rate (Figure 8.6(a)) at the expense of low sensitivity (Figures 8.6(c) through 8.6(f)). Actually, because of its low sensitivity, EVB learning does not behave optimally in the large-scale limit. The local-EVB estimator, on the other hand, behaves similarly to the OL method, for which the reason will be elucidated in the next section.

## 8.7 Optimality in Large-Scale Limit

Consider the large-scale limit, and assume that the model rank $H$ is set to be large enough but finite so that $H \geq H^*$ and $H/L \to 0$. Then the rank estimation procedure, detailed in Appendix C, by the OL method is reduced to counting the number of components such that $\widehat{\lambda}_h^{\text{OL-LSL}} > \widehat{\sigma}^{2\,\text{OL-LSL}}$, i.e.,

$$\widehat{H}^{\text{OL-LSL}} = \sum_{h=1}^{L} \theta\left(\widehat{\lambda}_h^{\text{OL-LSL}} > \widehat{\sigma}^{2\,\text{OL-LSL}}\right), \tag{8.91}$$

where $\theta(\cdot)$ is the indicator function defined in Theorem 8.1. Here $\widehat{\lambda}_h^{\text{OL-LSL}}$ and $\widehat{\sigma}^{2\,\text{OL-LSL}}$ are computed by iterating the following updates until convergence:

$$\widehat{\lambda}_h^{\text{OL-LSL}} = \begin{cases} \breve{\lambda}_h^{\text{OL-LSL}} & \text{if } \gamma_h \geq \underline{\gamma}^{\text{local-EVB}}, \\ \widehat{\sigma}^{2\,\text{OL-LSL}} & \text{otherwise}, \end{cases} \tag{8.92}$$

$$\widehat{\sigma}^{2\,\text{OL-LSL}} = \frac{1}{(M-H)}\left(\sum_{l=1}^{L} \frac{\gamma_l^2}{L} - \sum_{h=1}^{H} \widehat{\lambda}_h^{\text{OL-LSL}}\right), \tag{8.93}$$

where

$$\breve{\lambda}_h^{\text{OL-LSL}} = \frac{\gamma_h^2}{2L}\left(1 - \frac{(M-L)\widehat{\sigma}^{2\,\text{OL-LSL}}}{\gamma_h^2} + \sqrt{\left(1 - \frac{(M-L)\widehat{\sigma}^{2\,\text{OL-LSL}}}{\gamma_h^2}\right)^2 - \frac{4L\widehat{\sigma}^{2\,\text{OL-LSL}}}{\gamma_h^2}}\right). \tag{8.94}$$

The OL method evaluates its objective, which approximates the negative logarithm of the marginal likelihood (8.90), after the updates (8.92) and (8.93) converge for each hypothetical $H$, and adopts the minimizer $\widehat{H}^{\text{OL-LSL}}$ as the rank estimator. However, Hoyle (2008) proved that, in the large-scale limit, the objective decreases as $H$ increases, as long as Eq. (8.94) is a real number (or equivalently $\gamma_h \geq \underline{\gamma}^{\text{local-EVB}}$ holds) for all $h = 1, \ldots, H$ at the convergence. Accordingly, Eq. (8.91) holds.

Interestingly, the threshold in Eq. (8.92) coincides with the local-EVB threshold (6.127). Moreover, the updates (8.92) and (8.93) for the OL method are equivalent to the updates (9.29) and (9.30) for the local-EVB estimator (Algorithm 18) with the following correspondence:

$$\widehat{\lambda}_h^{\text{OL-LSL}} = \frac{\gamma_h \widehat{\gamma}_h^{\text{local-EVB}}}{L} + \widehat{\sigma}^{2\,\text{local-EVB}},$$

$$\widehat{\sigma}^{2\,\text{OL-LSL}} = \widehat{\sigma}^{2\,\text{local-EVB}}.$$

Thus, the rank estimation procedure by the OL method and that by the local-EVB estimator are equivalent, and therefore $\widehat{H}^{\text{OL-LSL}} = \widehat{H}^{\text{local-EVB}}$ in the large-scale limit.

If the noise variance is accurately estimated, i.e., $\widehat{\sigma}^2 = \sigma^{*2}$, the threshold $\underline{\gamma}^{\mathrm{local-EVB}}$ both for the OL method and the local-EVB estimator coincides with the MPUL (8.41), which corresponds to the minimum detectable observed singular value. By using this fact, the optimality of the OL method in the large-scale limit was shown:

**Proposition 8.27** *(Hoyle, 2008) In the large-scale limit, when L and M go to infinity with finite $\alpha$, $H^*$, and $H$ ($\geq H^*$)[8], the OL method almost surely recovers the true rank, i.e., $\widehat{H}^{\mathrm{OL-LSL}} = H^*$, if and only if*

$$\nu^*_{H^*} > \sqrt{\alpha}. \tag{8.95}$$

*It almost surely holds that*

$$\frac{\widehat{\lambda}_h^{\mathrm{OL-LSL}}}{\widehat{\sigma}^{2\ \mathrm{OL-LSL}}} - 1 = \nu^*_h,$$
$$\widehat{\sigma}^{2\ \mathrm{OL-LSL}} = \sigma^{*2}.$$

The condition (8.95) coincides with the condition (8.45), which any PCA method requires for perfect dimensionality recovery. In this sense, the OL method, as well as the local-EVB estimator, is optimal in the large-scale limit.

On the other hand, Theorem 8.14 implies that (global) EVB learning is not optimal in the large-scale limit but more conservative (see the difference between $\underline{\tau}$ and $\sqrt{\alpha}$ in Figure 6.4). In Figure 8.6, the conditions for perfect dimensionality recovery in the large-scale limit are indicated by vertical bars:

$$z = \sqrt{\underline{\tau}} \text{ for EVB, and } z = \sqrt{\underline{\tau}^{\mathrm{local}}} = \alpha^{1/4} \text{ for OL and local-EVB.}$$

All methods accurately estimate the noise variance in the large-scale limit, i.e.,

$$\widehat{\sigma}^{2\ \mathrm{EVB}} = \widehat{\sigma}^{2\ \mathrm{OL-LSL}} = \widehat{\sigma}^{2\ \mathrm{local-EVB}} = \sigma^{*2}.$$

Taking this into account, we indicate the recovery conditions in Figure 8.4 by arrows at

$$y = \underline{x} \text{ for EVB, and } y = \underline{x}^{\mathrm{local}} (= \overline{y}) \text{ for OL and local-EVB,}$$

respectively. Figure 8.4 implies that, in this particular case, EVB learning discards the third spike coming from the third true signal $\nu^*_3 = 0.5$, while the OL method and the local-EVB estimator successfully capture it as a signal.

When the matrix size is finite, the conservative nature of EVB learning is not always bad, since it offers almost zero false positive rate, which makes

[8] Unlike our analysis in Section 8.4, Hoyle (2008) assumed $H/L \to 0$ to prove that the OL method accurately estimates the noise variance.

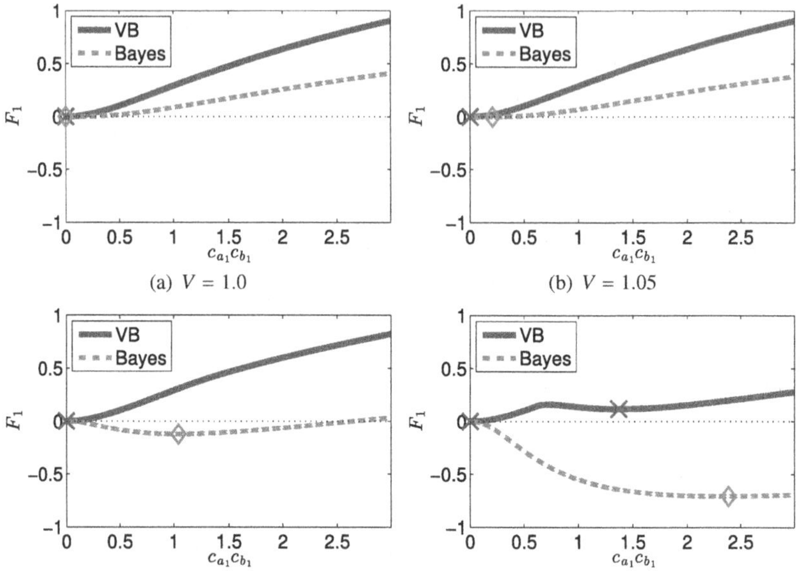


(a) $V = 1.0$ (b) $V = 1.05$ (c) $V = 1.5$ (d) $V = 2.1$

Figure 8.7 The VB free energy contribution (6.55) from the (first) component and its counterpart (8.96) of Bayesian learning for $L = M = H = 1$ and $\sigma^2 = 1$. Markers indicate the local minima.

Theorem 8.20 approximately hold for finite cases, as seen in Figures 8.5 and 8.6. However, the fact that not (global) EVB learning but the local-EVB estimator is optimal in the large-scale limit might come from inaccurate approximation to the Bayes posterior by the VB posterior. Having this in mind, we discuss the difference between VB learning and full Bayesian learning in the remainder of this section.

Figure 8.7 shows the VB free energy contribution (6.55) from the (first) component as a function of $c_a c_b$, and its counterpart of Bayesian learning:

$$2F_1^{\text{Bayes}} = -2\log\langle p(V|A,B)\rangle_{p(A)p(B)} - \left(\log(2\pi\sigma^2) + \frac{V^2}{\sigma^2}\right), \tag{8.96}$$

which was numerically computed. We see that the minimizer (shown as a diamond) of the Bayes free energy is at $c_a c_b \to +0$ until $V$ exceeds 1.

The difference in behavior between EVB learning and the local-EVB estimator appears in the nonempty range of the observed value $V$ where the positive local solution exists but gives positive free energy. Figure 8.7(d) shows this case, where a bump exists between two local minima (indicated by crosses). On the other hand, such multimodality is not observed in empirical full Bayesian learning (see the dashed curves in Figures 8.7(a) through 8.7(d)). We can say that this multimodality in EVB learning with a bump between two local minima is induced by the independence constraint for VB learning. We further guess that it is this bump that pushes the EVB threshold from the optimal point (at the local-EVB threshold) to a larger value. Further investigation is necessary to fully understand this phenomenon.

# 9
# Global Solver for Matrix Factorization

The analytic-form solutions, derived in Chapter 6, for VB learning and EVB learning in fully observed MF can naturally be used to develop efficient and reliable VB solvers. Some properties, shown in Chapter 8, can also be incorporated when the noise variance is unknown and to be estimated.

In this chapter, we introduce global solvers for VB learning and EVB learning (Nakajima et al., 2013a, 2015), and how to extend them to more general cases with missing entries and nonconjugate likelihoods (Seeger and Bouchard, 2012).

## 9.1 Global VB Solver for Fully Observed MF

We consider the MF model, introduced in Section 3.1:

$$p(\boldsymbol{V}|\boldsymbol{A},\boldsymbol{B}) \propto \exp\left(-\frac{1}{2\sigma^2}\left\|\boldsymbol{V}-\boldsymbol{B}\boldsymbol{A}^\top\right\|_{\mathrm{Fro}}^2\right), \tag{9.1}$$

$$p(\boldsymbol{A}) \propto \exp\left(-\frac{1}{2}\mathrm{tr}\left(\boldsymbol{A}\boldsymbol{C}_A^{-1}\boldsymbol{A}^\top\right)\right), \tag{9.2}$$

$$p(\boldsymbol{B}) \propto \exp\left(-\frac{1}{2}\mathrm{tr}\left(\boldsymbol{B}\boldsymbol{C}_B^{-1}\boldsymbol{B}^\top\right)\right), \tag{9.3}$$

where $\boldsymbol{V} \in \mathbb{R}^{L\times M}$ is an observed matrix;

$$\boldsymbol{A} = (\boldsymbol{a}_1,\ldots,\boldsymbol{a}_H) = (\widetilde{\boldsymbol{a}}_1,\ldots,\widetilde{\boldsymbol{a}}_M)^\top \in \mathbb{R}^{M\times H},$$

$$\boldsymbol{B} = (\boldsymbol{b}_1,\ldots,\boldsymbol{b}_H) = \left(\widetilde{\boldsymbol{b}}_1,\ldots,\widetilde{\boldsymbol{b}}_L\right)^\top \in \mathbb{R}^{L\times H},$$

are parameter matrices; and $C_A$, $C_B$, and $\sigma^2$ are hyperparameters. The prior covariance hyperparameters are restricted to be diagonal:

$$C_A = \mathbf{Diag}(c_{a_1}^2, \ldots, c_{a_H}^2),$$
$$C_B = \mathbf{Diag}(c_{b_1}^2, \ldots, c_{b_H}^2).$$

Our VB solver gives the global solution to the following minimization problem,

$$\widehat{r} = \underset{r}{\mathrm{argmin}} F(r) \quad \text{s.t.} \quad r(\boldsymbol{A}, \boldsymbol{B}) = r_A(\boldsymbol{A}) r_B(\boldsymbol{B}), \tag{9.4}$$

of the free energy

$$F(r) = \left\langle \log \frac{r_A(\boldsymbol{A}) r_B(\boldsymbol{B})}{p(\boldsymbol{V}|\boldsymbol{A}, \boldsymbol{B}) p(\boldsymbol{A}) p(\boldsymbol{B})} \right\rangle_{r_A(\boldsymbol{A}) r_B(\boldsymbol{B})}. \tag{9.5}$$

Assume that $L \leq M$ without loss of generality, and let

$$\boldsymbol{V} = \sum_{h=1}^{L} \gamma_h \boldsymbol{\omega}_{b_h} \boldsymbol{\omega}_{a_h}^\top \tag{9.6}$$

be the singular value decomposition (SVD) of the observed matrix $\boldsymbol{V} \in \mathbb{R}^{L \times M}$. According to Theorem 6.7, the VB solution is given by

$$\widehat{\boldsymbol{U}}^{\mathrm{VB}} = \widehat{\boldsymbol{B}\boldsymbol{A}}^\top = \sum_{h=1}^{H} \widehat{\gamma}_h^{\mathrm{VB}} \boldsymbol{\omega}_{b_h} \boldsymbol{\omega}_{a_h}^\top, \quad \text{where} \quad \widehat{\gamma}_h^{\mathrm{VB}} = \begin{cases} \breve{\gamma}_h^{\mathrm{VB}} & \text{if } \gamma_h \geq \underline{\gamma}_h^{\mathrm{VB}}, \\ 0 & \text{otherwise,} \end{cases} \tag{9.7}$$

for

$$\underline{\gamma}_h^{\mathrm{VB}} = \sigma \sqrt{\frac{(L+M)}{2} + \frac{\sigma^2}{2 c_{a_h}^2 c_{b_h}^2} + \sqrt{\left( \frac{(L+M)}{2} + \frac{\sigma^2}{2 c_{a_h}^2 c_{b_h}^2} \right)^2 - LM}}, \tag{9.8}$$

$$\breve{\gamma}_h^{\mathrm{VB}} = \gamma_h \left( 1 - \frac{\sigma^2}{2\gamma_h^2} \left( M + L + \sqrt{(M-L)^2 + \frac{4\gamma_h^2}{c_{a_h}^2 c_{b_h}^2}} \right) \right). \tag{9.9}$$

Corollary 6.8 completely specifies the VB posterior, which is written as

$$r(\boldsymbol{A}, \boldsymbol{B}) = \prod_{h=1}^{H} \mathrm{Gauss}_M(\boldsymbol{a}_h; \widehat{a}_h \boldsymbol{\omega}_{a_h}, \widehat{\sigma}_{a_h}^2 \boldsymbol{I}_M) \prod_{h=1}^{H} \mathrm{Gauss}_L(\boldsymbol{b}_h; \widehat{b}_h \boldsymbol{\omega}_{b_h}, \widehat{\sigma}_{b_h}^2 \boldsymbol{I}_L) \tag{9.10}$$

with the following variational parameters: if $\gamma_h > \underline{\gamma}_h^{\mathrm{VB}}$,

$$\widehat{a}_h = \pm \sqrt{\breve{\gamma}_h^{\mathrm{VB}} \widehat{\delta}_h^{\mathrm{VB}}}, \quad \widehat{b}_h = \pm \sqrt{\frac{\breve{\gamma}_h^{\mathrm{VB}}}{\widehat{\delta}_h^{\mathrm{VB}}}}, \quad \widehat{\sigma}_{a_h}^2 = \frac{\sigma^2 \widehat{\delta}_h^{\mathrm{VB}}}{\gamma_h}, \quad \widehat{\sigma}_{b_h}^2 = \frac{\sigma^2}{\gamma_h \widehat{\delta}_h^{\mathrm{VB}}}, \tag{9.11}$$

**Algorithm 15** Global VB solver for fully observed matrix factorization.

1: Transpose $V \to V^\top$ if $L > M$, and set $H (\leq L)$ to a sufficiently large value.
2: Compute the SVD (9.6) of $V$.
3: Apply Eqs. (9.7) through (9.9) to get the VB estimator.
4: If necessary, compute the variational parameters by using Eqs. (9.11) through (9.14), which specify the VB posterior (9.10). We can also evaluate the free energy by using Eqs. (9.15) and (9.16).

where
$$\widehat{\delta}_h^{\mathrm{VB}} \left( \equiv \frac{\widehat{a}_h}{\widehat{b}_h} \right) = \frac{c_{a_h}}{\sigma^2} \left( \gamma_h - \breve{\gamma}_h^{\mathrm{VB}} - \frac{L\sigma^2}{\gamma_h} \right), \tag{9.12}$$

and otherwise,

$$\widehat{a}_h = 0, \quad \widehat{b}_h = 0, \quad \widehat{\sigma}_{a_h}^2 = c_{a_h}^2 \left( 1 - \frac{L\widehat{\zeta}_h^{\mathrm{VB}}}{\sigma^2} \right), \quad \widehat{\sigma}_{b_h}^2 = c_{b_h}^2 \left( 1 - \frac{M\widehat{\zeta}_h^{\mathrm{VB}}}{\sigma^2} \right), \tag{9.13}$$

where

$$\widehat{\zeta}_h^{\mathrm{VB}} \left( \equiv \widehat{\sigma}_{a_h}^2 \widehat{\sigma}_{b_h}^2 \right) = \frac{\sigma^2}{2LM} \left( L + M + \frac{\sigma^2}{c_{a_h}^2 c_{b_h}^2} - \sqrt{\left( L + M + \frac{\sigma^2}{c_{a_h}^2 c_{b_h}^2} \right)^2 - 4LM} \right). \tag{9.14}$$

The free energy can be written as

$$2F = LM \log(2\pi\sigma^2) + \frac{\sum_{h=1}^L \gamma_h^2}{\sigma^2} + \sum_{h=1}^H 2F_h, \tag{9.15}$$

where
$$2F_h = M \log \frac{c_{a_h}^2}{\widehat{\sigma}_{a_h}^2} + L \log \frac{c_{b_h}^2}{\widehat{\sigma}_{b_h}^2} + \frac{\widehat{a}_h^2 + M\widehat{\sigma}_{a_h}^2}{c_{a_h}^2} + \frac{\widehat{b}_h^2 + L\widehat{\sigma}_{b_h}^2}{c_{b_h}^2} - (L + M) + \frac{-2\widehat{a}_h \widehat{b}_h \gamma_h + \left( \widehat{a}_h^2 + M\widehat{\sigma}_{a_h}^2 \right) \left( \widehat{b}_h^2 + L\widehat{\sigma}_{b_h}^2 \right)}{\sigma^2}. \tag{9.16}$$

Based on these results, we can straightforwardly construct a global solver for VB learning, which is given in Algorithm 15.

## 9.2 Global EVB Solver for Fully Observed MF

EVB learning, where the hyperparameters $C_A, C_A$, and $\sigma^2$ are also estimated from observation, solves the following minimization problem,

$$\widehat{r} = \underset{r, C_A, C_A, \sigma^2}{\mathrm{argmin}} \, F \quad \text{s.t.} \quad r(A, B) = r_A(A) r_B(B), \tag{9.17}$$

of the free energy (9.5).

According to Theorem 6.13, given the noise variance $\sigma^2$, the EVB solution can be written as

$$\widehat{U}^{\mathrm{EVB}} = \sum_{h=1}^{H} \widehat{\gamma}_h^{\mathrm{EVB}} \omega_{b_h} \omega_{a_h}^{\top}, \quad \text{where} \quad \widehat{\gamma}_h^{\mathrm{EVB}} = \begin{cases} \breve{\gamma}_h^{\mathrm{EVB}} & \text{if } \gamma_h \geq \underline{\gamma}^{\mathrm{EVB}}, \\ 0 & \text{otherwise,} \end{cases} \tag{9.18}$$

for

$$\underline{\gamma}^{\mathrm{EVB}} = \sigma \sqrt{M\left(1+\underline{\tau}\right)\left(1+\frac{\alpha}{\underline{\tau}}\right)},$$

$$\breve{\gamma}_h^{\mathrm{EVB}} = \frac{\gamma_h}{2}\left(1 - \frac{(M+L)\sigma^2}{\gamma_h^2} + \sqrt{\left(1 - \frac{(M+L)\sigma^2}{\gamma_h^2}\right)^2 - \frac{4LM\sigma^4}{\gamma_h^4}}\right).$$

Here

$$\alpha = \frac{L}{M} \qquad (0 < \alpha \leq 1),$$

is the "squaredness" of the observed matrix $\boldsymbol{V}$, and $\underline{\tau} = \underline{\tau}(\alpha)$ is the unique zero-cross point of the following function:

$$\Xi(\tau; \alpha) = \Phi(\tau) + \Phi\left(\frac{\tau}{\alpha}\right), \qquad \text{where} \qquad \Phi(z) = \frac{\log(z+1)}{z} - \frac{1}{2}. \tag{9.19}$$

Summarizing Lemmas 6.14, 6.16, and 6.19, the EVB posterior is completely specified by Eq. (9.10) with the variational parameters given as follows: If $\gamma_h \geq \underline{\gamma}^{\mathrm{EVB}}$,

$$\widehat{a}_h = \pm\sqrt{\breve{\gamma}_h^{\mathrm{EVB}} \widehat{\delta}_h^{\mathrm{EVB}}}, \quad \widehat{b}_h = \pm\sqrt{\frac{\breve{\gamma}_h^{\mathrm{EVB}}}{\widehat{\delta}_h^{\mathrm{EVB}}}}, \tag{9.20}$$

$$\widehat{\sigma}_{a_h}^2 = \frac{\sigma^2 \widehat{\delta}_h^{\mathrm{EVB}}}{\gamma_h}, \quad \widehat{\sigma}_{b_h}^2 = \frac{\sigma^2}{\gamma_h \widehat{\delta}_h^{\mathrm{EVB}}}, \quad c_{a_h} c_{b_h} = \sqrt{\frac{\gamma_h \breve{\gamma}_h^{\mathrm{EVB}}}{LM}}, \tag{9.21}$$

$$\text{where} \qquad \widehat{\delta}_h^{\mathrm{EVB}} = \sqrt{\frac{M \breve{\gamma}_h^{\mathrm{EVB}}}{L \gamma_h}}\left(1 + \frac{L\sigma^2}{\gamma_h \breve{\gamma}_h^{\mathrm{EVB}}}\right), \tag{9.22}$$

and otherwise

$$\widehat{a}_h = 0, \quad \widehat{b}_h = 0, \quad \widehat{\sigma}_{a_h}^2 = \sqrt{\widehat{\zeta}^{\mathrm{EVB}}}, \quad \widehat{\sigma}_{b_h}^2 = \sqrt{\widehat{\zeta}^{\mathrm{EVB}}}, \quad c_{a_h} c_{b_h} = \sqrt{\widehat{\zeta}^{\mathrm{EVB}}}, \tag{9.23}$$

$$\text{where} \qquad \widehat{\zeta}^{\mathrm{EVB}} \to +0. \tag{9.24}$$

To use the preceding result, we need to prepare a table of $\underline{\tau}$ by computing the zero-cross point of Eq. (9.19) as a function of $\alpha$. A simple approximation $\underline{\tau} \approx \underline{z}\sqrt{\alpha} \approx 2.5129\sqrt{\alpha}$ is a reasonable alternative (see Figure 6.4).

For noise variance estimation, we can use Theorems 8.1 and 8.2, derived in Chapter 8. Specifically, after performing the SVD (9.6), we first estimate the noise variance by solving the following problem:

$$\widehat{\sigma}^{2\ \mathrm{EVB}} = \underset{\sigma^2}{\mathrm{argmin}}\ \Omega(\sigma^{-2}), \tag{9.25}$$

$$\text{s.t.} \qquad \max\left(\underline{\sigma}^2_{\overline{H}+1}, \frac{\sum_{h=\overline{H}+1}^{L} \gamma_h^2}{M\left(L-\overline{H}\right)}\right) \leq \sigma^2 \leq \frac{1}{LM}\sum_{h=1}^{L}\gamma_h^2, \tag{9.26}$$

where

$$\Omega(\sigma^{-2}) = \frac{1}{L}\left(\sum_{h=1}^{H}\psi\left(\frac{\gamma_h^2}{M\sigma^2}\right) + \sum_{h=H+1}^{L}\psi_0\left(\frac{\gamma_h^2}{M\sigma^2}\right)\right), \tag{9.27}$$

$$\psi(x) = \psi_0(x) + \theta\left(x > \underline{x}\right)\psi_1(x),$$

$$\psi_0(x) = x - \log x,$$

$$\psi_1(x) = \log\left(\tau(x;\alpha) + 1\right) + \alpha \log\left(\frac{\tau(x;\alpha)}{\alpha} + 1\right) - \tau(x;\alpha),$$

$$\underline{x} = \left(1 + \underline{\tau}\right)\left(1 + \frac{\alpha}{\underline{\tau}}\right),$$

$$\tau(x;\alpha) = \frac{1}{2}\left(x - (1+\alpha) + \sqrt{(x-(1+\alpha))^2 - 4\alpha}\right),$$

$$\underline{\sigma}_h^2 = \begin{cases} \infty & \text{for } h = 0, \\ \frac{\gamma_h^2}{M\underline{x}} & \text{for } h = 1, \ldots, L, \\ 0 & \text{for } h = L+1, \end{cases}$$

$$\overline{H} = \min\left(\left\lceil \frac{L}{1+\alpha} \right\rceil - 1, H\right).$$

Problem (9.25) is simply a one-dimensional search for the minimizer of the function $\Omega(\sigma^{-2})$, which is typically smooth. Note also that, if the matrix size is large enough, Corollary 8.21 states that any local minimizer is accurate enough to estimate the correct rank. Given the estimated noise variance $\sigma^2 = \widehat{\sigma}^{2\ \mathrm{EVB}}$, Eq. (9.18) gives the EVB solution.

Algorithm 16 summarizes the procedure explained in the preceding discussion. This algorithm gives the global solution, provided that the global solution to the one-dimensional search problem (9.25) is attained. If the noise variance $\sigma^2$ is known, we should simply skip Step 4.

**Algorithm 16** Global EVB solver for fully observed matrix facrtorization.

1: Transpose $V \to V^\top$ if $L > M$, and set $H$ ($\leq L$) to a sufficiently large value.
2: Refer to the table of $\underline{\tau}(\alpha)$ at $\alpha = L/M$ (or use a simple approximation $\underline{\tau} \approx 2.5129\sqrt{\alpha}$).
3: Compute the SVD (9.6) of $V$.
4: Solve the one-dimensional search problem (9.25) to get $\widehat{\sigma}^{2\,\mathrm{EVB}}$.
5: Apply Eq. (9.18) to get the EVB estimator $\{\widehat{\gamma}_h^{\mathrm{EVB}}\}_{h=1}^H$ for $\sigma^2 = \widehat{\sigma}^{2\,\mathrm{EVB}}$.
6: If necessary, compute the variational parameters and the hyperparameters by using Eqs. (9.20) through (9.24), which specify the EVB posterior (9.10). We can also evaluate the free energy by using Eqs. (9.15) and (9.16), noting that $F_h \to +0$ for $h$ such that $\gamma_h < \underline{\gamma}^{\mathrm{EVB}}$.

**Algorithm 17** Iterative EVB solver for fully observed matrix factorization.

1: Transpose $V \to V^\top$ if $L > M$, and set $H$ ($\leq L$) to a sufficiently large value.
2: Refer to the table of $\underline{\tau}(\alpha)$ at $\alpha = L/M$ (or use a simple approximation $\underline{\tau} \approx 2.5129\sqrt{\alpha}$).
3: Compute the SVD (9.6) of $V$.
4: Initialize the noise variance $\widehat{\sigma}^{2\,\mathrm{EVB}}$ to the lower bound in Eq. (9.26).
5: Apply Eq. (9.18) to update the EVB estimator $\{\widehat{\gamma}_h^{\mathrm{EVB}}\}_{h=1}^H$.
6: Apply Eq. (9.28) to update the noise variance estimator $\widehat{\sigma}^{2\,\mathrm{EVB}}$.
7: Compute the variational parameters and the hyperparameters by using Eqs. (9.20) through (9.24).
8: Evaluate the free energy (9.15), noting that $F_h \to +0$ for $h$ such that $\gamma_h < \underline{\gamma}^{\mathrm{EVB}}$.
9: Iterate Steps 5 through 8 until convergence (until the energy decrease becomes smaller than a threshold).

Another implementation is to iterate Eq. (9.18) and

$$\widehat{\sigma}^{2\,\mathrm{EVB}} = \frac{1}{LM}\left(\sum_{l=1}^{L}\gamma_l^2 - \sum_{h=1}^{H}\gamma_h\widehat{\gamma}_h^{\mathrm{EVB}}\right) \tag{9.28}$$

in turn. Note that Eq. (9.28) was derived in Corollary 8.3 and can be used as an update rule for the noise variance estimator, given the current EVB estimators $\{\widehat{\gamma}_h^{\mathrm{EVB}}\}_{h=1}^H$. Although it is not guaranteed, this iterative algorithm (Algorithm 17) tends to converge to the global solution if we initialize the noise variance $\widehat{\sigma}^{2\,\mathrm{EVB}}$ to be sufficiently small (Nakajima et al., 2015). We recommend to initialize it to the lower-bound given in Eq. (9.26).

**Algorithm 18** Local-EVB solver for fully observed matrix factorization.

1: Transpose $\boldsymbol{V} \rightarrow \boldsymbol{V}^\top$ if $L > M$, and set $H\ (\leq L)$ to a sufficiently large value.
2: Refer to the table of $\underline{\tau}(\alpha)$ at $\alpha = L/M$ (or use a simple approximation $\underline{\tau} \approx 2.5129\sqrt{\alpha}$).
3: Compute the SVD (9.6) of $\boldsymbol{V}$.
4: Initialize the noise variance $\widehat{\sigma}^{2\ \mathrm{local-EVB}}$ to the lower-bound in Eq. (9.26).
5: Apply Eq. (9.29) to update the local-EVB estimator $\{\widehat{\gamma}_h^{\mathrm{local-EVB}}\}_{h=1}^H$.
6: Apply Eq. (9.30) to update the noise variance estimator $\widehat{\sigma}^{2\ \mathrm{local-EVB}}$.
7: Compute the variational parameters and the hyperparameters by using Eqs. (9.20) through (9.24).
8: Evaluate the free energy (9.15), noting that $F_h \rightarrow +0$ for $h$ such that $\gamma_h < \underline{\gamma}^{\mathrm{local-EVB}}$.
9: Iterate Steps 5 through 8 until convergence (until the energy decrease becomes smaller than a threshold).

Finally, we introduce an iterative solver, in Algorithm 18, for the *local-EVB estimator* (6.131), which iterates the following updates:

$$\widehat{\gamma}_h^{\mathrm{local-EVB}} = \begin{cases} \breve{\gamma}_h^{\mathrm{EVB}} & \text{if } \gamma_h \geq \underline{\gamma}^{\mathrm{local-EVB}}, \\ 0 & \text{otherwise}, \end{cases} \tag{9.29}$$

$$\widehat{\sigma}^{2\ \mathrm{local-EVB}} = \frac{1}{LM}\left(\sum_{l=1}^{L} \gamma_l^2 - \sum_{h=1}^{H} \gamma_h \widehat{\gamma}_h^{\mathrm{local-EVB}}\right), \tag{9.30}$$

where

$$\underline{\gamma}^{\mathrm{local-EVB}} \equiv \left(\sqrt{L} + \sqrt{M}\right)\sigma \tag{9.31}$$

is the local-EVB threshold, defined by Eq. (6.127). If we initialize the noise variance $\widehat{\sigma}^{2\ \mathrm{local-EVB}}$ to be sufficiently small, this algorithm tends to retain the positive local-EVB solution for each $h$ if it exists, and therefore does not necessarily converge to the global EVB solution. The interesting relation between the local-EVB estimator and the overlap (OL) method (Hoyle, 2008), an alternative dimensionality selection method based on the Laplace approximation, was discussed in Section 8.7.

## 9.3 Empirical Comparison with the Standard VB Algorithm

Here we see how efficient the global solver (Algorithm 16) is in comparison with the standard VB algorithm (Algorithm 1 in Section 3.1) on artificial and benchmark data.

### 9.3.1 Experiment on Artificial Data

We first created an artificial data set (*Artificial1*) with the data matrix size $L = 100$ and $M = 300$, and the true rank $H^* = 20$. We randomly drew *true* matrices $\boldsymbol{A}^* \in \mathbb{R}^{M \times H^*}$ and $\boldsymbol{B}^* \in \mathbb{R}^{L \times H^*}$ so that each entry of $\boldsymbol{A}^*$ and $\boldsymbol{B}^*$ follows $\mathrm{Gauss}_1(0, 1)$, where $\mathrm{Gauss}_1(\mu, \sigma^2)$ denotes the one-dimensional Gaussian distribution with mean $\mu$ and variance $\sigma^2$. An observed matrix $\boldsymbol{V}$ was created by adding noise subject to $\mathrm{Gauss}_1(0, 1)$ to each entry of $\boldsymbol{B}^* \boldsymbol{A}^{*\top}$.

We evaluated the performance under the complete empirical Bayesian scenario, where all variational parameters and hyperparameters are estimated from observation. We used the full-rank model (i.e., $H = \min(L, M)$), expecting that irrelevant $H - H^*$ components will be automatically trimmed out by the automatic relevance determination (ARD) effect (see Chapters 7 and 8).

We compare the global solver (Algorithm 16) and the standard VB algorithm (Algorithm 1 in Section 3.1), and show the free energy, the computation time, and the estimated rank over iterations in Figure 9.1. For the standard VB algorithm, initial values were set in the following way: $\widehat{\boldsymbol{A}}$ and $\widehat{\boldsymbol{B}}$ are randomly created so that each entry follows $\mathrm{Gauss}_1(0, 1)$. Other variables are set to $\widehat{\boldsymbol{\Sigma}}_A = \widehat{\boldsymbol{\Sigma}}_B = \boldsymbol{C}_A = \boldsymbol{C}_B = \boldsymbol{I}_H$ and $\sigma^2 = 1$. Note that we rescale $\boldsymbol{V}$ so that $\|\boldsymbol{V}\|^2_{\mathrm{Fro}} / (LM) = 1$, before starting iterations. We ran the standard algorithm 10

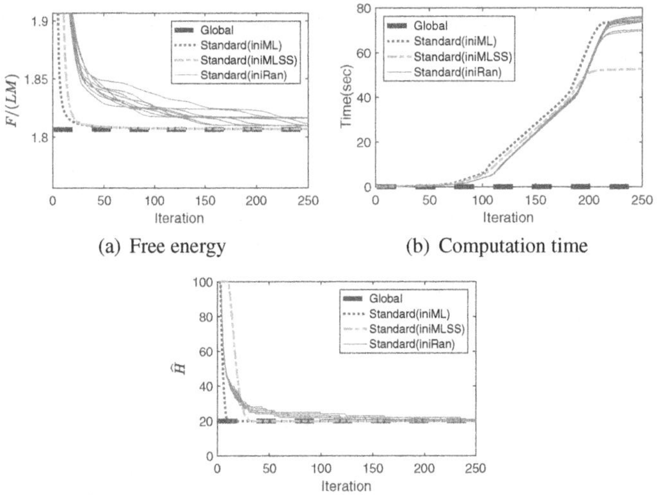


(a) Free energy

(b) Computation time

(c) Estimated rank

Figure 9.1 Experimental results on the *Artificial1* data, where the data matrix size is $L = 100$ and $M = 300$, and the true rank is $H^* = 20$.

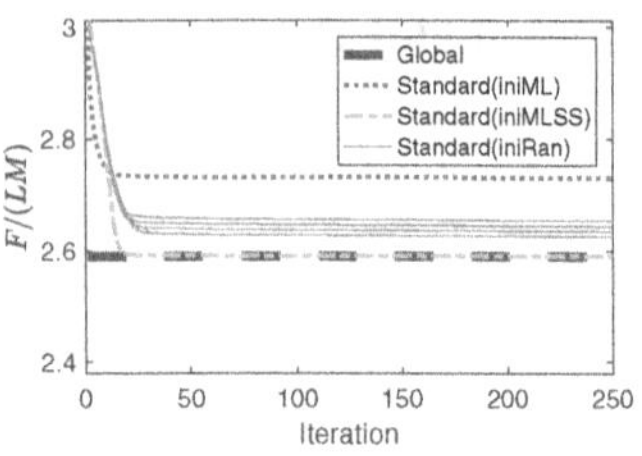


(a) Free energy

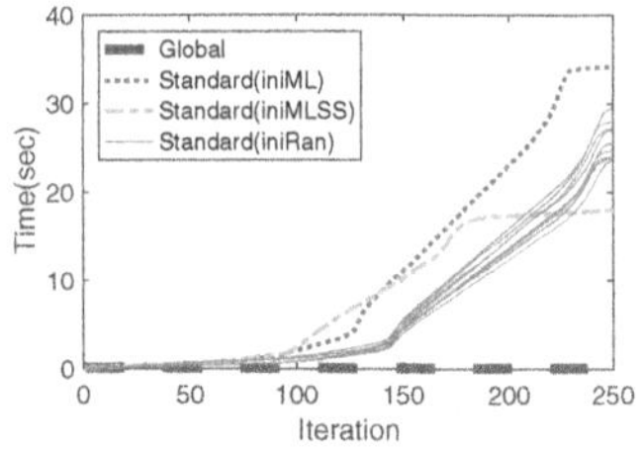


(b) Computation time

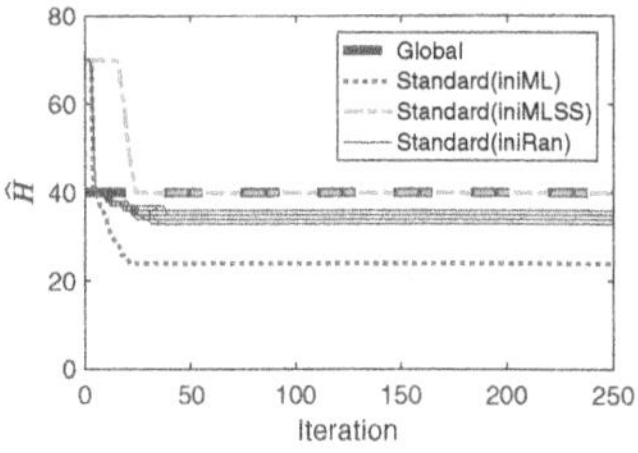


(c) Estimated rank

Figure 9.2 Experimental results on the *Artificial2* data set ($L$ = 70, $M$ = 300, and $H^*$ = 40).

times, starting from different initial points, and each trial is plotted by a solid curve labeled as "Standard(iniRan)" in Figure 9.1.

The global solver has no iteration loop, and therefore the corresponding dashed line labeled as "Global" is constant over iterations. We see that the global solver finds the true rank $\widehat{H} = H^* = 20$ immediately ($\sim$ 0.1 sec on average over 10 trials), while the standard iterative algorithm does not converge in 60 sec.

Figure 9.2 shows experimental results on another artificial data set (*Artificial2*) where $L = 70$, $M = 300$, and $H^* = 40$. In this case, all the 10 trials of the standard algorithm are trapped at local minima. We empirically observed that the local minimum problem tends to be more critical when $H^*$ is large (close to $H$).

We also evaluated the standard algorithm with different initialization schemes. The curve labeled as "Standard(iniML)" indicates the standard algorithm starting from the maximum likelihood (ML) solution: $(\widehat{\boldsymbol{a}}_h, \widehat{\boldsymbol{b}}_h) = (\sqrt{\gamma_h}\boldsymbol{\omega}_{a_h}, \sqrt{\gamma_h}\boldsymbol{\omega}_{b_h})$. The initial values for other variables are the same as the random initialization. Figures 9.1 and 9.2 show that the ML initialization generally makes convergence faster than the random initialization, but suffers from the local minimum problem more severely—it tends to converge to a worse local minimum.

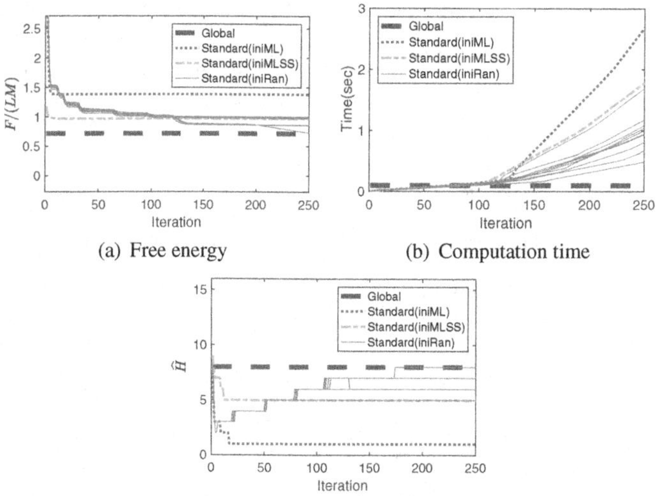


(a) Free energy

(b) Computation time

(c) Estimated rank

Figure 9.3 Experimental results on the *Glass* data set ($L = 9, M = 214$).

We observed that starting from a small noise variance tends to alleviate the local minimum problem at the expense of slightly slower convergence. The curve labeled as "Standard(iniMLSS)" indicates the standard algorithm starting from the ML solution with a small noise variance $\sigma^2 = 0.0001$. We see in Figures 9.1 and 9.2 that this initialization improves the quality of solutions, and successfully finds the true rank for these artificial data sets. However, we will show in Section 9.3.2 that this scheme still suffers from the local minimum problem on benchmark datasets.

### 9.3.2 Experiment on Benchmark Data

Figures 9.3 through 9.5 show the experimental results on the *Glass*, the *Satimage*, and the *Spectf* data sets available from the University of California, Irvine (UCI) repository (Asuncion and Newman, 2007). A similar tendency to the artificial data experiment (Figures 9.1 and 9.2) is observed: "Standard(iniRan)" converges slowly, and is often trapped at a local minimum with a *wrong* estimated rank;[1] "Standard(iniML)" converges slightly faster but to a worse local minimum; and "Standard(iniMLSS)" tends to give a better solution. Unlike the artificial data experiment, "Standard(iniMLSS)" fails to

[1] Since the *true* ranks of the benchmark data sets are unknown, we mean by a *wrong* rank a rank different from the one giving the lowest free energy.

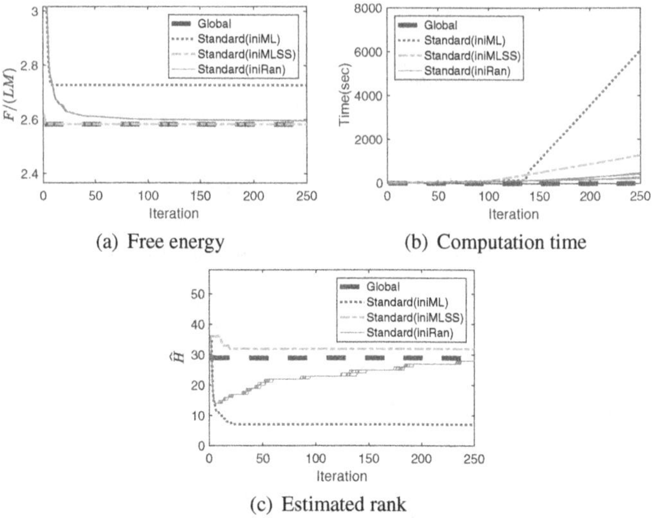


(a) Free energy (b) Computation time

(c) Estimated rank

Figure 9.4 Experimental results on the *Satimage* data set ($L = 36, M = 6435$).

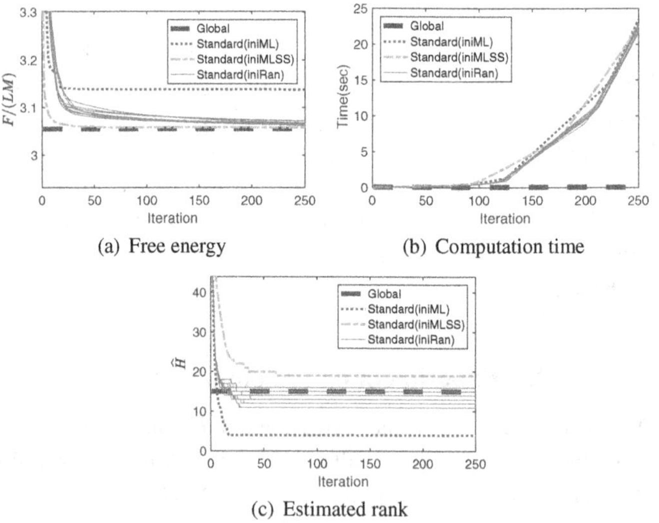


(a) Free energy (b) Computation time

(c) Estimated rank

Figure 9.5 Experimental results on the *Spectf* data set ($L = 44, M = 267$).

find the *correct* rank in these benchmark data sets. We also conducted experiments on other benchmark data sets and found that the standard VB algorithm generally converges slowly, and sometimes suffers from the local minimum problem, while the global solver gives the global solution immediately.

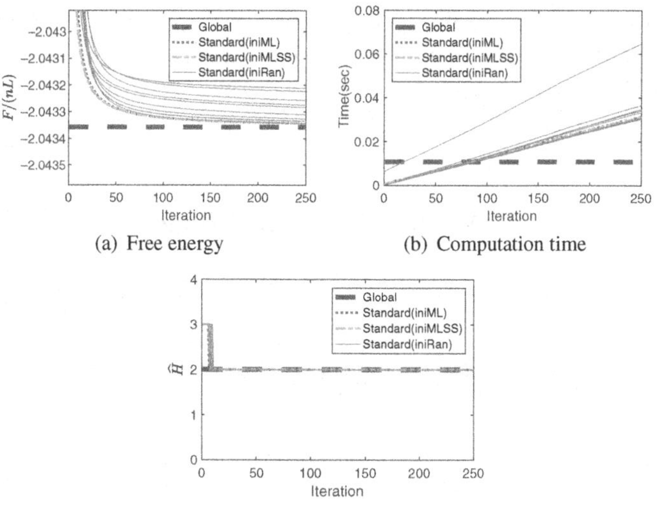


(a) Free energy

(b) Computation time

(c) Estimated rank

Figure 9.6 Experimental results on the *Concrete Slump Test* data set (an RRR task with $L = 3, M = 7$).

Finally, we applied EVB learning to the *reduced rank regression* (RRR) model (see Section 3.1.2), of which the model likelihood is given by Eq. (3.36). Figure 9.6 shows the results on the *Concrete Slump Test* data set, where we centered the $L = 3$-dimensional outputs and prewhitened the $M = 7$-dimensional inputs. We also standardized the outputs so that the variance of each element is equal to one. Note that we cannot directly apply Algorithm 16 for the RRR model. Instead, we use Algorithm 16 with a fixed noise variance (skipping Step 4) and apply one-dimensional search to minimize the free energy (3.42), in order to estimate the *rescaled* noise variance $\sigma^2$. For the standard VB algorithm, the rescaled noise variance should be updated by Eq. (3.43), instead of Eq. (3.28). The original noise variance $\sigma'^2$ is recovered by Eq. (3.40) for both cases.

Overall, the global solver showed excellent performance over the standard VB algorithm.

## 9.4 Extension to Nonconjugate MF with Missing Entries

The global solvers introduced in Section 9.1 can be directly applied only for the fully observed isotropic Gaussian likelihood (9.1). However, the global solver can be used as a subroutine to develop efficient algorithms for more general cases. In this section, we introduce the approach by Seeger and Bouchard (2012), where an *iterative singular value shrinkage* algorithm was proposed, based on the global VB solver (Algorithm 15) and local variational approximation (Section 2.1.7).

### 9.4.1 Nonconjugate MF Model

Consider the following model:

$$p(\boldsymbol{V}|\boldsymbol{A},\boldsymbol{B}) = \prod_{l=1}^{L}\prod_{m=1}^{M}\phi_{l,m}(V_{l,m}, \widetilde{\boldsymbol{b}}_l^{\top}\widetilde{\boldsymbol{a}}_m), \tag{9.32}$$

$$p(\boldsymbol{A}) \propto \exp\left(-\frac{1}{2}\mathrm{tr}\left(\boldsymbol{A}\boldsymbol{C}_A^{-1}\boldsymbol{A}^{\top}\right)\right), \tag{9.33}$$

$$p(\boldsymbol{B}) \propto \exp\left(-\frac{1}{2}\mathrm{tr}\left(\boldsymbol{B}\boldsymbol{C}_B^{-1}\boldsymbol{B}^{\top}\right)\right), \tag{9.34}$$

where $\phi_{l,m}(v|u)$ is a function of $v$ and $u$, and satisfy

$$-\frac{\partial^2 \log \phi_{l,m}(v,u)}{\partial u^2} \leq \frac{1}{\sigma^2} \tag{9.35}$$

for any $l, m, v$, and $u$.

The function $\phi_{l,m}(v,u)$ corresponds to the model distribution of the $(l,m)$th entry $v$ of $\boldsymbol{V}$ parameterized by $u$. If

$$\phi_{l,m}(v,u) = \mathrm{Gauss}_1(v; u, \sigma^2) \tag{9.36}$$

for all $l$ and $m$, the model (9.32) through (9.34) is reduced to the fully observed isotropic Gaussian MF model (9.1)–(9.3), and the Hessian,

$$-\frac{\partial^2 \log \phi_{l,m}(v,u)}{\partial u^2} = \frac{1}{\sigma^2},$$

of the negative log-likelihood is a constant with respect to $v$ and $u$.

The model (9.32) through (9.34) can cover the case with missing entries by setting the noise variance in Eq. (9.36) to $\sigma^2 \to \infty$ for the unobserved entries (the condition (9.35) is tight for the smallest $\sigma^2$). Other one-dimensional distributions, including the *Bernoulli distribution* with *sigmoid* parameterization and the *Poisson distribution*, satisfy the condition (9.35) for a certain $\sigma^2$, which will be introduced in Section 9.4.3.

### 9.4.2 Local Variational Approximation for Non-conjugate MF

The VB learning problem (9.4) minimizes the free energy, which can be written as

$$F(r) = \left\langle \log \frac{r_A(\boldsymbol{A}) r_B(\boldsymbol{B})}{\prod_{l=1}^{L} \prod_{m=1}^{M} \phi_{l,m}(V_{l,m}, \widetilde{\boldsymbol{b}}_l^\top \widetilde{\boldsymbol{a}}_m) p(\boldsymbol{A}) p(\boldsymbol{B})} \right\rangle_{r_A(\boldsymbol{A}) r_B(\boldsymbol{B})}. \tag{9.37}$$

In order to make the global VB solver applicable as a subroutine, we instead solve the following joint minimization problem,

$$\widehat{r} = \underset{r, \Xi}{\operatorname{argmin}}\ \overline{F}(r, \Xi) \quad \text{s.t.} \quad r(\boldsymbol{A}, \boldsymbol{B}) = r_A(\boldsymbol{A}) r_B(\boldsymbol{B}), \tag{9.38}$$

of an upper-bound of the free energy,

$$F \leq \overline{F}(r, \Xi) \equiv \left\langle \log \frac{r_A(\boldsymbol{A}) r_B(\boldsymbol{B})}{\prod_{l=1}^{L} \prod_{m=1}^{M} \underline{\phi}_{l,m}(V_{l,m}, \widetilde{\boldsymbol{b}}_l^\top \widetilde{\boldsymbol{a}}_m, \Xi_{l,m}) p(\boldsymbol{A}) p(\boldsymbol{B})} \right\rangle_{r_A(\boldsymbol{A}) r_B(\boldsymbol{B})}, \tag{9.39}$$

where

$$\underline{\phi}_{l,m}(v, u, \xi) \leq \phi_{l,m}(v, u) \tag{9.40}$$

is a lower-bound of the likelihood parameterized with variational parameters $\Xi \in \mathbb{R}^{L \times M}$.

The condition (9.35) allows us to form a parametric lower-bound in the (unnormalized) isotropic Gaussian form, which we derive as follows. Any function $f(x)$ with bounded curvature $\frac{\partial^2 f}{\partial x^2} \leq \kappa$ can be upper-bounded by the following quadratic function:

$$f(x) \leq \kappa (x - \xi)^2 + \left. \frac{\partial f}{\partial x} \right|_{x=\xi} (x - \xi) + f(\xi) \quad \text{for any } \xi \in \mathbb{R}. \tag{9.41}$$

Therefore, it holds that, for any $\xi \in \mathbb{R}$,

$$-\log \phi_{l,m}(v, u) \leq \frac{(u - \xi)^2}{2\sigma^2} + g(v, \xi)(u - \xi) - \log \phi_{l,m}(v, \xi), \tag{9.42}$$

where

$$g(v, \xi) = - \left. \frac{\partial \log \phi_{l,m}(v, u)}{\partial u} \right|_{u=\xi}.$$

The left graph in Figure 9.7 shows the parametric quadratic upper-bounds (9.42) for the Bernoulli likelihood with sigmoid parameterization.

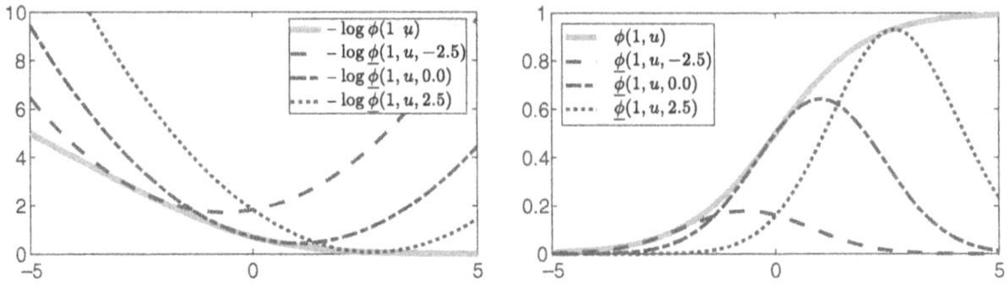


Figure 9.7 Parametric quadratic (Gaussian-form) bounds for the Bernoulli likelihood with sigmoid parameterization, $\phi(v,u) = e^{vu}/(1+e^{u})$, for $v = 1$. Left: the negative log-likelihood (the left-hand side of Eq. (9.42)) and its quadratic upper-bounds (the right-hand side of Eq. (9.42)) for $\xi = -2.5, 0.0, 2.5$. Right: the likelihood function $\phi(1,\xi)$ and its Gaussian-form lower-bounds (9.43) for $\xi = -2.5, 0.0, 2.5$.

Since $\log(\cdot)$ is a monotonic function, we can adopt the following parametric lower-bound of $\phi_{l,m}(v,u)$:

$$\begin{aligned}
\underline{\phi}_{l,m}(u,\xi) &= \exp\left(-\left(\frac{(u-\xi)^2}{2\sigma^2} + g(v,\xi)(u-\xi) - \log\phi_{l,m}(v,\xi)\right)\right) \\
&= \phi_{l,m}(v,\xi)\exp\left(-\tfrac{1}{2\sigma^2}\left((u-\xi)^2 + 2\sigma^2 g(v,\xi)(u-\xi)\right)\right) \\
&= \phi_{l,m}(v,\xi)\exp\left(-\tfrac{1}{2\sigma^2}\left(\left(u-\xi+\sigma^2 g(v,\xi)\right)^2 - \left(\sigma^2 g(v,\xi)\right)^2\right)\right) \\
&= \phi_{l,m}(v,\xi)\exp\left(\tfrac{\sigma^2}{2}g^2(v,\xi)\right)\exp\left(-\tfrac{1}{2\sigma^2}\left((\xi-\sigma^2 g(v,\xi))-u\right)^2\right) \\
&= \sqrt{2\pi\sigma^2}\,\phi_{l,m}(v,\xi)\exp\left(\tfrac{\sigma^2}{2}g^2(v,\xi)\right)\mathrm{Gauss}_1\left(\xi-\sigma^2 g(v,\xi); u, \sigma^2\right).
\end{aligned} \tag{9.43}$$

The right graph in Figure 9.7 shows the parametric Gaussian-form lower-bounds (9.43) for the Bernoulli likelihood with sigmoid parameterization.

Substituting Eq. (9.43) into Eq. (9.39) gives

$$\begin{aligned}
\overline{F}(r,\Xi) &= -\sum_{l=1}^{L}\sum_{m=1}^{M}\left(\frac{1}{2}\log(2\pi\sigma^2) + \log\phi_{l,m}(V_{l,m},\Xi_{l,m}) + \frac{\sigma^2}{2}g^2(V_{l,m},\Xi_{l,m})\right) \\
&\quad + \left\langle \log\frac{r_A(\boldsymbol{A})r_B(\boldsymbol{B})}{\prod_{l=1}^{L}\prod_{m=1}^{M}\mathrm{Gauss}_1(\Xi_{l,m}-\sigma^2 g(V_{l,m},\Xi_{l,m}); \widetilde{\boldsymbol{b}}_l^{\top}\widetilde{\boldsymbol{a}}_m, \sigma^2)p(\boldsymbol{A})p(\boldsymbol{B})}\right\rangle_{r_A(\boldsymbol{A})r_B(\boldsymbol{B})} \\
&= -\sum_{l=1}^{L}\sum_{m=1}^{M}\left(\frac{1}{2}\log(2\pi\sigma^2) + \log\phi_{l,m}(V_{l,m},\Xi_{l,m}) + \frac{\sigma^2}{2}g^2(V_{l,m},\Xi_{l,m})\right) \\
&\quad + \left\langle \log\frac{r_A(\boldsymbol{A})r_B(\boldsymbol{B})}{(2\pi\sigma^2)^{-LM/2}\exp\left(-\frac{1}{2\sigma^2}\|\check{\boldsymbol{V}}-\boldsymbol{B}\boldsymbol{A}^{\top}\|_{\mathrm{Fro}}^2\right)p(\boldsymbol{A})p(\boldsymbol{B})}\right\rangle_{r_A(\boldsymbol{A})r_B(\boldsymbol{B})},
\end{aligned} \tag{9.44}$$

where $\check{\boldsymbol{V}} \in \mathbb{R}^{L\times M}$ is a matrix such that

$$\check{V}_{l,m} = \Xi_{l,m} - \sigma^2 g(V_{l,m},\Xi_{l,m}). \tag{9.45}$$

The first term in Eq. (9.44) does not depend on $r$ and the second term is equal to the free energy of the fully observed isotropic Gaussian MF model with the observed matrix $\boldsymbol{V}$ replaced with $\check{\boldsymbol{V}}$. Therefore, given the variational parameter $\boldsymbol{\Xi}$, we can partially solve the minimization problem (9.38) with respect to $r$ by applying the global VB solver (Algorithm 15). The solution is Gaussian in the following form:

$$r(\boldsymbol{A}, \boldsymbol{B}) = r_A(\boldsymbol{A}) r_B(\boldsymbol{B}), \qquad \text{where} \tag{9.46}$$

$$r_A(\boldsymbol{A}) = \mathrm{MGauss}_{M,H}(\boldsymbol{A}; \widehat{\boldsymbol{A}}, \boldsymbol{I}_M \otimes \widehat{\boldsymbol{\Sigma}}_A) \propto \exp\left( -\frac{\mathrm{tr}\left( (\boldsymbol{A}-\widehat{\boldsymbol{A}}) \widehat{\boldsymbol{\Sigma}}_A^{-1} (\boldsymbol{A}-\widehat{\boldsymbol{A}})^\top \right)}{2} \right), \tag{9.47}$$

$$r_B(\boldsymbol{B}) = \mathrm{MGauss}_{L,H}(\boldsymbol{B}; \widehat{\boldsymbol{B}}, \boldsymbol{I}_L \otimes \widehat{\boldsymbol{\Sigma}}_B) \propto \exp\left( -\frac{\mathrm{tr}\left( (\boldsymbol{B}-\widehat{\boldsymbol{B}}) \widehat{\boldsymbol{\Sigma}}_B^{-1} (\boldsymbol{B}-\widehat{\boldsymbol{B}})^\top \right)}{2} \right). \tag{9.48}$$

Here the mean and the covariance parameters $\widehat{\boldsymbol{A}}, \widehat{\boldsymbol{\Sigma}}_A, \widehat{\boldsymbol{B}}, \widehat{\boldsymbol{\Sigma}}_B$ are another set of variational parameters.

Given the optimal $r$ specified by Eq. (9.46), the free energy bound (9.44) is written (as a function of $\boldsymbol{\Xi}$) as follows:

$$\begin{aligned}
\min_r \overline{F}(r, \boldsymbol{\Xi}) &= -\sum_{l=1}^{L} \sum_{m=1}^{M} \left( \log \phi_{l,m}(V_{l,m}, \Xi_{l,m}) + \frac{\sigma^2}{2} g^2(V_{l,m}, \Xi_{l,m}) \right) \\
&\quad - \tfrac{1}{2\sigma^2} \left\langle \| \check{\boldsymbol{V}} - \boldsymbol{B}\boldsymbol{A}^\top \|_{\mathrm{Fro}}^2 \right\rangle_{r_A(\boldsymbol{A}) r_B(\boldsymbol{B})} + \text{const.} \\
&= -\sum_{l=1}^{L} \sum_{m=1}^{M} \left( \log \phi_{l,m}(V_{l,m}, \Xi_{l,m}) + \frac{\sigma^2}{2} g^2(V_{l,m}, \Xi_{l,m}) \right) \\
&\quad - \tfrac{1}{2\sigma^2} \| \check{\boldsymbol{V}} - \widehat{\boldsymbol{B}}\widehat{\boldsymbol{A}}^\top \|_{\mathrm{Fro}}^2 + \text{const.} \\
&= -\sum_{l=1}^{L} \sum_{m=1}^{M} \left( \log \phi_{l,m}(V_{l,m}, \Xi_{l,m}) + \frac{\sigma^2}{2} g^2(V_{l,m}, \Xi_{l,m}) \right) \\
&\quad - \tfrac{1}{2\sigma^2} \textstyle\sum_{l=1}^{L} \sum_{m=1}^{M} \left( \check{V}_{l,m} - \widehat{U}_{l,m} \right)^2 + \text{const.},
\end{aligned} \tag{9.49}$$

where

$$\widehat{\boldsymbol{U}} = \widehat{\boldsymbol{B}}\widehat{\boldsymbol{A}}^\top.$$

The second-to-last equation in Eq. (9.43), together with Eq. (9.45), implies that

$$\log \underline{\phi}_{l,m}(\widehat{U}_{l,m}, \Xi_{l,m}) = \log \phi_{l,m}(V_{l,m}, \Xi_{l,m}) + \tfrac{\sigma^2}{2} g^2(V_{l,m}, \Xi_{l,m}) - \tfrac{1}{2\sigma^2} \left( \check{V}_{l,m} - \widehat{U}_{l,m} \right)^2,$$

with which Eq. (9.49) is written as

$$\min_r \overline{F}(r, \boldsymbol{\Xi}) = -\textstyle\sum_{l=1}^{L} \sum_{m=1}^{M} \log \underline{\phi}_{l,m}(\widehat{U}_{l,m}, \Xi_{l,m}) + \text{const.} \tag{9.50}$$

**Algorithm 19** Iterative singular value shrinkage algorithm for nonconjugate MF (with missing entries).

1: Set the noise variance $\sigma^2$ with which the condition (9.35) tightly holds, and initialize the variational parameters to $\boldsymbol{\Xi} = \mathbf{0}_{(L,M)}$.
2: Compute $\check{\boldsymbol{V}}$ by Eq. (9.45).
3: Compute the VB posterior (9.46) by applying the global solver (Algorithm 15) with $\check{\boldsymbol{V}}$ substituted for $\boldsymbol{V}$.
4: Update $\boldsymbol{\Xi}$ by Eq. (9.51).
5: Iterate Steps 2 through 4 until convergence.

Since $\log \underline{\phi}_{l,m}(u, \xi)$ is the quadratic upper-bound (the right-hand side in Eq. (9.42)) of $-\log \phi_{l,m}(v, u)$, which is tight at $u = \widehat{U}_{l,m}$ when $\xi = \widehat{U}_{l,m}$, the minimizer of Eq. (9.50) with respect to $\boldsymbol{\Xi}$ is given by

$$\widehat{\boldsymbol{\Xi}} \equiv \underset{\boldsymbol{\Xi}}{\mathrm{argmin}} \min_{r} \overline{F}(r, \boldsymbol{\Xi}) = \widehat{\boldsymbol{U}}. \tag{9.51}$$

In summary, to solve the joint minimization problem (9.38), we can iteratively update $r$ and $\boldsymbol{\Xi}$. The update of $r$ can be performed by the global solver (Algorithm 15) with the observed matrix $\boldsymbol{V}$ replaced with $\check{\boldsymbol{V}}$, defined by Eq. (9.45). The update of $\boldsymbol{\Xi}$ is simply performed by Eq. (9.51). Algorithm 19 summarizes this procedure, where $\mathbf{0}_{(d_1,d_2)}$ denotes the $d_1 \times d_2$ matrix with all entries equal to zero. Seeger and Bouchard (2012) empirically showed that this *iterative singular value shrinkage* algorithm significantly outperforms the MAP solution at comparable computational costs. They also proposed an efficient way to perform SVD when $\boldsymbol{V}$ is huge but sparsely observed, based on the techniques proposed by Tomioka et al. (2010).

### 9.4.3 Examples of Nonconjugate MF

In this subsection, we introduce a few examples of model likelihood $\phi_{l,m}(v, u)$, which satisfy Condition (9.35), and give the corresponding derivatives of the negative log likelihood.

#### Isotropic Gaussian MF with Missing Entries

If we let

$$\phi_{l,m}(v, u) = \begin{cases} \mathrm{Gauss}_1(v; u, \sigma^2) & \text{if } (l, m) \in \Lambda, \\ 1 & \text{otherwise}, \end{cases} \tag{9.52}$$

where $\Lambda$ denotes the set of observed entries, the model distribution (9.32) corresponds to the model distribution (3.44) of MF with missing entries. The first and the second derivatives of the negative log likelihood are given as follows:

$$-\frac{\partial \log \phi_{l,m}(v,u)}{\partial u} = \begin{cases} \frac{1}{\sigma^2}(u-v) & \text{if } (l,m) \in \Lambda, \\ 0 & \text{otherwise,} \end{cases}$$

$$-\frac{\partial^2 \log \phi_{l,m}(v,u)}{\partial u^2} = \begin{cases} \frac{1}{\sigma^2} & \text{if } (l,m) \in \Lambda, \\ 0 & \text{otherwise.} \end{cases} \tag{9.53}$$

**Bernoulli MF with Sigmoid Parameterization**

The *Bernoulli distribution* with *sigmoid* parameterization is suitable for binary observations, i.e., $\boldsymbol{V} \in \{0,1\}^{L \times M}$:

$$\phi_{l,m}(v,u) = \begin{cases} \frac{e^{vu}}{1+e^u} & \text{if } (l,m) \in \Lambda, \\ 1 & \text{otherwise.} \end{cases} \tag{9.54}$$

The first and the second derivatives are given as follows:

$$-\frac{\partial \log \phi_{l,m}(v,u)}{\partial u} = \begin{cases} \frac{1}{1+e^{-u}} - v & \text{if } (l,m) \in \Lambda, \\ 0 & \text{otherwise,} \end{cases}$$

$$-\frac{\partial^2 \log \phi_{l,m}(v,u)}{\partial u^2} = \begin{cases} \frac{1}{(1+e^{-u})(1+e^{u})} & \text{if } (l,m) \in \Lambda, \\ 0 & \text{otherwise.} \end{cases} \tag{9.55}$$

It holds that

$$-\frac{\partial^2 \log \phi_{l,m}(v,u)}{\partial u^2} \leq \frac{1}{4},$$

and therefore, the noise variance should be set to $\sigma^2 = 4$, which satisfies the condition (9.35). Figure 9.7 was depicted for this model.

**Poisson MF**

The *Poisson distribution* is suitable for count data, i.e., $\boldsymbol{V} \in \{0,1,2,\ldots\}^{L \times M}$:

$$\phi_{l,m}(v,u) = \begin{cases} \lambda^v(u)e^{-\lambda(u)} & \text{if } (l,m) \in \Lambda, \\ 1 & \text{otherwise,} \end{cases} \tag{9.56}$$

where $\lambda(u)$ is the *link function*. Since a common choice $\lambda(u) = e^u$ for the link function gives unbounded curvature for large $u$, Seeger and Bouchard (2012) proposed to use another link function $\lambda(u) = \log(1 + e^u)$. The first derivative is given as follows:

$$-\frac{\partial \log \phi_{l,m}(v, u)}{\partial u} = \begin{cases} \frac{1}{1+e^{-u}} \left(1 - \frac{v}{\lambda(u)}\right) & \text{if } (l, m) \in \Lambda, \\ 0 & \text{otherwise.} \end{cases} \tag{9.57}$$

It was confirmed that the second derivative is upper-bounded as

$$-\frac{\partial^2 \log \phi_{l,m}(v, u)}{\partial u^2} \leq \frac{1}{4} + 0.17v,$$

and therefore, the noise variance should be set to

$$\sigma^2 = \frac{1}{1/4 + 0.17 \max_{l,m} V_{l,m}}.$$

Since the bound can be loose if some of the entries $V_{l,m}$ of the observed matrix are huge compared to the others, overly large counts should be clipped.

# 10

# Global Solver for Low-Rank Subspace Clustering

The nonasymptotic theory, described in Chapter 6, for fully observed matrix factorization (MF) has been extended to other bilinear models. In this chapter, we introduce exact and approximate global variational Bayesian (VB) solvers (Nakajima et al., 2013c) for low-rank subspace clustering (LRSC).

## 10.1 Problem Description

The LRSC model, introduced in Section 3.4, is defined as

$$p(\boldsymbol{V}|\boldsymbol{A}', \boldsymbol{B}') \propto \exp\left(-\frac{1}{2\sigma^2}\left\|\boldsymbol{V} - \boldsymbol{V}\boldsymbol{B}'\boldsymbol{A}'^{\top}\right\|_{\mathrm{Fro}}^2\right), \tag{10.1}$$

$$p(\boldsymbol{A}') \propto \exp\left(-\frac{1}{2}\mathrm{tr}(\boldsymbol{A}'\boldsymbol{C}_A^{-1}\boldsymbol{A}'^{\top})\right), \tag{10.2}$$

$$p(\boldsymbol{B}') \propto \exp\left(-\frac{1}{2}\mathrm{tr}(\boldsymbol{B}'\boldsymbol{C}_B^{-1}\boldsymbol{B}'^{\top})\right), \tag{10.3}$$

where $\boldsymbol{V} \in \mathbb{R}^{L\times M}$ is an observation matrix, and $\boldsymbol{A}' \in \mathbb{R}^{M\times H}$ and $\boldsymbol{B}' \in \mathbb{R}^{M\times H}$ for $H \leq \min(L, M)$ are the parameters to be estimated. Note that in this chapter we denote the *original* parameters $\boldsymbol{A}'$ and $\boldsymbol{B}'$ with *primes* for convenience. We assume that hyperparameters

$$\boldsymbol{C}_A = \mathbf{Diag}(c_{a_1}^2, \ldots, c_{a_H}^2), \qquad \boldsymbol{C}_B = \mathbf{Diag}(c_{b_1}^2, \ldots, c_{b_H}^2),$$

are diagonal and positive definite. The LRSC model is similar to MF. The only difference is that the product $\boldsymbol{B}'\boldsymbol{A}'^{\top}$ of the parameters is further multiplied by $\boldsymbol{V}$ in Eq. (10.1). Accordingly, we can hope that similar analysis could be applied to LRSC, providing a global solver for LRSC.

We first transform the parameters as

$$A \leftarrow \boldsymbol{\Omega}_V^{\mathrm{right}\top} A', \qquad B \leftarrow \boldsymbol{\Omega}_V^{\mathrm{right}\top} B', \qquad \text{where} \qquad V = \boldsymbol{\Omega}_V^{\mathrm{left}} \boldsymbol{\Gamma}_V \boldsymbol{\Omega}_V^{\mathrm{right}\top} \tag{10.4}$$

is the singular value decomposition (SVD) of $V$. Here, $\boldsymbol{\Omega}_V^{\mathrm{left}} \in \mathbb{R}^{L\times L}$ and $\boldsymbol{\Omega}_V^{\mathrm{right}} \in \mathbb{R}^{M\times M}$ are orthogonal matrices, and $\boldsymbol{\Gamma}_V \in \mathbb{R}^{L\times M}$ is a (possibly nonsquare) diagonal matrix with nonnegative diagonal entries aligned in nonincreasing order, i.e., $\gamma_1 \geq \gamma_2 \geq \cdots \geq \gamma_{\min(L,M)}$. After this transformation, the LRSC model (10.1) through (10.3) is rewritten as

$$p(\boldsymbol{\Gamma}_V|A, B) \propto \exp\left(-\frac{1}{2\sigma^2}\left\|\boldsymbol{\Gamma}_V - \boldsymbol{\Gamma}_V B A^\top\right\|_{\mathrm{Fro}}^2\right), \tag{10.5}$$

$$p(A) \propto \exp\left(-\frac{1}{2}\mathrm{tr}(A C_A^{-1} A^\top)\right), \tag{10.6}$$

$$p(B) \propto \exp\left(-\frac{1}{2}\mathrm{tr}(B C_B^{-1} B^\top)\right). \tag{10.7}$$

The transformation (10.4) does not affect much the derivation of the VB learning algorithm. The following summarizes the result obtained in Section 3.4 with the transformed parameters $A$ and $B$. The solution of the VB learning problem,

$$\widehat{r} = \underset{r}{\mathrm{argmin}}\, F(r) \quad \text{s.t.} \quad r(A, B) = r_A(A) r_B(B), \qquad \text{where} \tag{10.8}$$

$$F = \left\langle \log \frac{r_A(A) r_B(B)}{p(\boldsymbol{\Gamma}_V|A, B) p(A) p(B)} \right\rangle_{r_A(A) r_B(B)},$$

has the following form:

$$r(A) \propto \exp\left(-\frac{\mathrm{tr}\left((A - \widehat{A})\widehat{\boldsymbol{\Sigma}}_A^{-1}(A - \widehat{A})^\top\right)}{2}\right),$$

$$r(B) \propto \exp\left(-\frac{(\breve{b} - \widehat{\breve{b}})^\top \widehat{\breve{\boldsymbol{\Sigma}}}_B^{-1} (\breve{b} - \widehat{\breve{b}})}{2}\right), \tag{10.9}$$

for $\breve{b} = \mathbf{vec}(B) \in \mathbb{R}^{MH}$, and the free energy can be explicitly written as

$$\begin{aligned} 2F = {} & LM \log(2\pi\sigma^2) + \frac{\left\|\boldsymbol{\Gamma}_V - \boldsymbol{\Gamma}_V \widehat{B}\widehat{A}^\top\right\|_{\mathrm{Fro}}^2}{\sigma^2} + M \log \frac{\det(C_A)}{\det(\widehat{\boldsymbol{\Sigma}}_A)} + \log \frac{\det(C_B \otimes I_M)}{\det(\widehat{\breve{\boldsymbol{\Sigma}}}_B)} \\ & - 2MH + \mathrm{tr}\left\{C_A^{-1}\left(\widehat{A}^\top \widehat{A} + M\widehat{\boldsymbol{\Sigma}}_A\right)\right\} + \mathrm{tr}\left\{C_B^{-1} \widehat{B}^\top \widehat{B}\right\} + \mathrm{tr}\left\{(C_B^{-1} \otimes I_M)\widehat{\breve{\boldsymbol{\Sigma}}}_B\right\} \\ & + \mathrm{tr}\left\{\sigma^{-2} \boldsymbol{\Gamma}_V^\top \boldsymbol{\Gamma}_V \left(-\widehat{B}\widehat{A}^\top \widehat{A}\widehat{B}^\top + \left\langle B(\widehat{A}^\top \widehat{A} + M\widehat{\boldsymbol{\Sigma}}_A) B^\top \right\rangle_{r(B)}\right)\right\}. \end{aligned} \tag{10.10}$$

Therefore, the variational parameters $(\widehat{\boldsymbol{A}}, \widehat{\boldsymbol{B}}, \widehat{\boldsymbol{\Sigma}}_A, \widehat{\breve{\boldsymbol{\Sigma}}}_B)$ can be obtained by solving the following problem:

$$\text{Given } \boldsymbol{C}_A, \boldsymbol{C}_B \in \mathbb{D}^H_{++}, \sigma^2 \in \mathbb{R}_{++},$$

$$\min_{(\widehat{\boldsymbol{A}}, \widehat{\boldsymbol{B}}, \widehat{\boldsymbol{\Sigma}}_A, \widehat{\breve{\boldsymbol{\Sigma}}}_B)} F, \tag{10.11}$$

$$\text{s.t. } \widehat{\boldsymbol{A}}, \widehat{\boldsymbol{B}} \in \mathbb{R}^{M \times H}, \widehat{\boldsymbol{\Sigma}}_A \in \mathbb{S}^H_{++}, \widehat{\breve{\boldsymbol{\Sigma}}}_B \in \mathbb{S}^{MH}_{++}. \tag{10.12}$$

The stationary conditions with respect to the variational parameters are given by

$$\widehat{\boldsymbol{A}} = \frac{1}{\sigma^2} \boldsymbol{\Gamma}_V^\top \boldsymbol{\Gamma}_V \widehat{\boldsymbol{B}} \widehat{\boldsymbol{\Sigma}}_A, \tag{10.13}$$

$$\widehat{\boldsymbol{\Sigma}}_A = \sigma^2 \left( \left\langle \boldsymbol{B}^\top \boldsymbol{\Gamma}_V^\top \boldsymbol{\Gamma}_V \boldsymbol{B} \right\rangle_{r(\boldsymbol{B})} + \sigma^2 \boldsymbol{C}_A^{-1} \right)^{-1}, \tag{10.14}$$

$$\widehat{\breve{\boldsymbol{b}}} = \frac{\widehat{\breve{\boldsymbol{\Sigma}}}_B}{\sigma^2} \mathbf{vec} \left( \boldsymbol{\Gamma}_V^\top \boldsymbol{\Gamma}_V \widehat{\boldsymbol{A}} \right), \tag{10.15}$$

$$\widehat{\breve{\boldsymbol{\Sigma}}}_B = \sigma^2 \left( (\widehat{\boldsymbol{A}}^\top \widehat{\boldsymbol{A}} + M \widehat{\boldsymbol{\Sigma}}_A) \otimes \boldsymbol{\Gamma}_V^\top \boldsymbol{\Gamma}_V + \sigma^2 (\boldsymbol{C}_B^{-1} \otimes \boldsymbol{I}_M) \right)^{-1}. \tag{10.16}$$

For empirical VB (EVB) learning, we solve the problem,

$$\text{Given } \sigma^2 \in \mathbb{R}_{++},$$

$$\min_{(\widehat{\boldsymbol{A}}, \widehat{\boldsymbol{B}}, \widehat{\boldsymbol{\Sigma}}_A, \widehat{\breve{\boldsymbol{\Sigma}}}_B, \boldsymbol{C}_A, \boldsymbol{C}_B)} F \tag{10.17}$$

$$\text{subject to } \widehat{\boldsymbol{A}}, \widehat{\boldsymbol{B}} \in \mathbb{R}^{M \times H}, \widehat{\boldsymbol{\Sigma}}_A \in \mathbb{S}^H_{++}, \widehat{\breve{\boldsymbol{\Sigma}}}_B \in \mathbb{S}^{MH}_{++}, \boldsymbol{C}_A, \boldsymbol{C}_B \in \mathbb{D}^H_{++}, \tag{10.18}$$

for which the stationary conditions with respect to the hyperparameters are given by

$$c_{a_h}^2 = \left\| \widehat{\boldsymbol{a}}_h \right\|^2 / M + \left( \widehat{\boldsymbol{\Sigma}}_A \right)_{h,h}, \tag{10.19}$$

$$c_{b_h}^2 = \left( \left\| \widehat{\boldsymbol{b}}_h \right\|^2 + \mathrm{tr} \left( \widehat{\boldsymbol{\Sigma}}_B^{(h,h)} \right) \right) / M, \tag{10.20}$$

$$\widehat{\sigma}^2 = \frac{\mathrm{tr} \left( \boldsymbol{\Gamma}_V^\top \boldsymbol{\Gamma}_V \left( \boldsymbol{I}_M - 2 \widehat{\boldsymbol{B}} \widehat{\boldsymbol{A}}^\top + \left\langle \boldsymbol{B} (\widehat{\boldsymbol{A}}^\top \widehat{\boldsymbol{A}} + M \widehat{\boldsymbol{\Sigma}}_A) \boldsymbol{B}^\top \right\rangle_{r(\boldsymbol{B})} \right) \right)}{LM}. \tag{10.21}$$

In deriving the global VB solution of fully observed MF in Chapter 6, the following two facts were essential. First, a large portion of the degrees of freedom of the *original* variational parameters are irrelevant (see Section 6.3), and the optimization problem can be decomposed into subproblems, each of which has only a small number of unknown variables. Second, the stationary conditions of each subproblem is written as a *polynomial system* (a set of polynomial equations). These two facts also apply to the LRSC model, which allows us to derive an *exact* global VB solver (EGVBS). However, each of the decomposed subproblems still has too many unknowns whose number is proportional to the problem size, and therefore EGVBS is still computationally demanding for typical problem sizes. As an alternative, we also derive an approximate global VB solver (AGVBS) by imposing an additional constraint, which allows further decomposition of the problem into subproblems with a constant number of unknowns.

In this chapter, we first find irrelevant degrees of freedom of the variational parameters and decompose the VB learning problem. Then we derive EGVBS and AGVBS and empirically show their usefulness.

## 10.2 Conditions for VB Solutions

Let $J$ ($\leq \min(L, M)$) be the rank of the observed matrix $\boldsymbol{V}$. For simplicity, we assume that no pair of positive singular values of $\boldsymbol{V}$ coincide with each other, i.e.,

$$\gamma_1 > \gamma_2 > \cdots > \gamma_J > 0.$$

This holds with probability 1 if $\boldsymbol{V}$ is contaminated with Gaussian noise, as the LRSC model (10.1) assumes. Since $(\boldsymbol{\Gamma}_V^\top \boldsymbol{\Gamma}_V)_{m,m'}$ is zero for $m > J$ or $m' > J$, Eqs. (10.13) and (10.15) imply that

$$\widehat{A}_{m,h} = \widehat{B}_{m,h} = 0 \quad \text{for} \quad m > J. \tag{10.22}$$

Similarly to Lemma 6.1 for the fully observed MF, we can prove the following lemma:

**Lemma 10.1** *Any local solution of the problem* (10.11) *is a stationary point of the free energy* (10.10).

*Proof* Since

$$\left\| \boldsymbol{\Gamma}_V - \boldsymbol{\Gamma}_V \widehat{\boldsymbol{B}} \widehat{\boldsymbol{A}}^\top \right\|_{\mathrm{Fro}}^2 \geq 0,$$

and

$$\begin{aligned}
&\mathrm{tr}\left\{ \boldsymbol{\Gamma}_V^\top \boldsymbol{\Gamma}_V \left( -\widehat{\boldsymbol{B}} \widehat{\boldsymbol{A}}^\top \widehat{\boldsymbol{A}} \widehat{\boldsymbol{B}}^\top + \left\langle \boldsymbol{B} (\widehat{\boldsymbol{A}}^\top \widehat{\boldsymbol{A}} + M \widehat{\boldsymbol{\Sigma}}_A) \boldsymbol{B}^\top \right\rangle_{r(\boldsymbol{B})} \right) \right\} \\
&\qquad = M \cdot \mathrm{tr}\left\{ \boldsymbol{\Gamma}_V^\top \boldsymbol{\Gamma}_V \left\langle \boldsymbol{B} \widehat{\boldsymbol{\Sigma}}_A \boldsymbol{B}^\top \right\rangle_{r(\boldsymbol{B})} \right\} \geq 0,
\end{aligned}$$

the free energy (10.10) is lower-bounded as

$$2F \geq -M \log\det\left(\widehat{\Sigma}_A\right) - \log\det\left(\breve{\Sigma}_B\right)$$
$$+ \mathrm{tr}\left\{C_A^{-1}\left(\widehat{A}^\top\widehat{A} + M\widehat{\Sigma}_A\right)\right\} + \mathrm{tr}\left\{C_B^{-1}\widehat{B}^\top\widehat{B}\right\} + \mathrm{tr}\left\{(C_B^{-1} \otimes I_M)\breve{\Sigma}_B\right\} + \tau, \tag{10.23}$$

where $\tau$ is a finite constant. The right-hand side of Eq. (10.23) diverges to $+\infty$ if any entry of $\widehat{A}$ or $\widehat{B}$ goes to $+\infty$ or $-\infty$. Also it diverges if any eigenvalue of $\widehat{\Sigma}_A$ or $\breve{\Sigma}_B$ goes to $+0$ or $\infty$. This implies that no local solution exists on the boundary of (the closure of) the domain (10.12). Since the free energy is differentiable in the domain (10.12), any local minimizer is a stationary point.

For any (diagonalized) observed matrix $\Gamma_V$, the free energy (10.10) can be finite, for example, at $\widehat{A} = \mathbf{0}_{M,H}$, $\widehat{B} = \mathbf{0}_{M,H}$, $\widehat{\Sigma}_A = I_H$, and $\breve{\Sigma}_B = I_{MH}$. Therefore, at least one minimizer always exists, which completes the proof of Lemma 10.1. □

Lemma 10.1 implies that Eqs. (10.13) through (10.16) hold at any local solution.

## 10.3 Irrelevant Degrees of Freedom

Also similarly to Theorem 6.4, we have the following theorem:

**Theorem 10.2** *When $C_A C_B$ is nondegenerate (i.e., $c_{a_h}c_{b_h} > c_{a_{h'}}c_{b_{h'}}$ for any pair $h < h'$), $(\widehat{A}, \widehat{B}, \widehat{\Sigma}_A, \breve{\Sigma}_B)$ are diagonal for any solution of the problem* (10.11). *When $C_A C_B$ is degenerate, any solution has an* equivalent *solution with diagonal $(\widehat{A}, \widehat{B}, \widehat{\Sigma}_A, \breve{\Sigma}_B)$.*

Theorem 10.2 significantly reduces the complexity of the optimization problem, and furthermore makes the problem separable, as seen in Section 10.5.

## 10.4 Proof of Theorem 10.2

Similarly to Section 6.4, we separately consider the following three cases:

**Case 1** When no pair of diagonal entries of $C_A C_B$ coincide.
**Case 2** When all diagonal entries of $C_A C_B$ coincide.
**Case 3** When (not all but) some pairs of diagonal entries of $C_A C_B$ coincide.

### 10.4.1 Diagonality Implied by Optimality

We can prove the following lemma, which is an extension of Lemma 6.2.

**Lemma 10.3** *Let $\boldsymbol{\Gamma}, \boldsymbol{\Omega}, \boldsymbol{\Phi} \in \mathbb{R}^{H \times H}$ be a nondegenerate diagonal matrix, an orthogonal matrix, and a symmetric matrix, respectively. Let $\{\boldsymbol{\Lambda}^{(k)}, \boldsymbol{\Lambda}'^{(k)} \in \mathbb{R}^{H \times H}; k = 1, \ldots, K\}$ be arbitrary diagonal matrices, and $\{\boldsymbol{\Psi}^{(k')} \in \mathbb{R}^{H \times H}; k' = 1, \ldots, K'\}$ be arbitrary symmetric matrices. If*

$$G(\boldsymbol{\Omega}) = \mathrm{tr}\left\{\boldsymbol{\Gamma}\boldsymbol{\Omega}\boldsymbol{\Phi}\boldsymbol{\Omega}^\top + \sum_{k=1}^{K} \boldsymbol{\Lambda}^{(k)}\boldsymbol{\Omega}\boldsymbol{\Lambda}'^{(k)}\boldsymbol{\Omega}^\top + \sum_{k'=1}^{K'} \boldsymbol{\Omega}\boldsymbol{\Psi}^{(k')}\right\} \tag{10.24}$$

*is minimized or maximized (as a function of $\boldsymbol{\Omega}$, given $\boldsymbol{\Gamma}, \boldsymbol{\Phi}, \{\boldsymbol{\Lambda}^{(k)}, \boldsymbol{\Lambda}'^{(k)}\}, \{\boldsymbol{\Psi}^{(k')}\}$) when $\boldsymbol{\Omega} = \boldsymbol{I}_H$, then $\boldsymbol{\Phi}$ is diagonal. Here, $K$ and $K'$ can be any natural numbers including $K = 0$ and $K' = 0$ (when the second and the third terms, respectively, do not exist).*

*Proof* Let

$$\boldsymbol{\Phi} = \boldsymbol{\Omega}'\boldsymbol{\Gamma}'\boldsymbol{\Omega}'^\top \tag{10.25}$$

be the eigenvalue decomposition of $\boldsymbol{\Phi}$. Let $\boldsymbol{\gamma}, \boldsymbol{\gamma}', \{\boldsymbol{\lambda}^{(k)}\}, \{\boldsymbol{\lambda}'^{(k)}\}$ be the vectors consisting of the diagonal entries of $\boldsymbol{\Gamma}, \boldsymbol{\Gamma}', \{\boldsymbol{\Lambda}^{(k)}\}, \{\boldsymbol{\Lambda}'^{(k)}\}$, respectively, i.e.,

$$\boldsymbol{\Gamma} = \mathbf{Diag}(\boldsymbol{\gamma}), \quad \boldsymbol{\Gamma}' = \mathbf{Diag}(\boldsymbol{\gamma}'), \quad \boldsymbol{\Lambda}^{(k)} = \mathbf{Diag}(\boldsymbol{\lambda}^{(k)}), \quad \boldsymbol{\Lambda}'^{(k)} = \mathbf{Diag}(\boldsymbol{\lambda}'^{(k)}).$$

Then, Eq. (10.24) can be written as

$$\begin{aligned} G(\boldsymbol{\Omega}) &= \mathrm{tr}\left\{\boldsymbol{\Gamma}\boldsymbol{\Omega}\boldsymbol{\Phi}\boldsymbol{\Omega}^\top + \sum_{k=1}^{K} \boldsymbol{\Lambda}^{(k)}\boldsymbol{\Omega}\boldsymbol{\Lambda}'^{(k)}\boldsymbol{\Omega}^\top + \sum_{k'=1}^{K'} \boldsymbol{\Omega}\boldsymbol{\Psi}^{(k')}\right\} \\ &= \boldsymbol{\gamma}^\top\boldsymbol{Q}\boldsymbol{\gamma}' + \sum_{k=1}^{K} \boldsymbol{\lambda}^{(k)\top}\boldsymbol{R}\boldsymbol{\lambda}'^{(k)} + \sum_{k'=1}^{K'} \mathrm{tr}\left\{\boldsymbol{\Omega}\boldsymbol{\Psi}^{(k')}\right\}, \end{aligned} \tag{10.26}$$

where

$$\boldsymbol{Q} = (\boldsymbol{\Omega}\boldsymbol{\Omega}') \odot (\boldsymbol{\Omega}\boldsymbol{\Omega}'), \qquad \boldsymbol{R} = \boldsymbol{\Omega} \odot \boldsymbol{\Omega}.$$

Here, $\odot$ denotes the Hadamard product.

Using this expression, we will prove that $\boldsymbol{\Phi}$ is diagonal if $\boldsymbol{\Omega} = \boldsymbol{I}_H$ minimizes or maximizes Eq. (10.26). Let us consider a bilateral perturbation $\boldsymbol{\Omega} = \boldsymbol{\Delta}$ such that the $2 \times 2$ matrix $\boldsymbol{\Delta}_{(h,h')}$ for $h \neq h'$ consisting of the $h$th and the $h'$th columns and rows form a $2 \times 2$ orthogonal matrix,

$$\boldsymbol{\Delta}_{(h,h')} = \begin{pmatrix} \cos\theta & -\sin\theta \\ \sin\theta & \cos\theta \end{pmatrix},$$

and the remaining entries coincide with those of the identity matrix. Then, the elements of $\boldsymbol{Q}$ become

$$Q_{i,j} = \begin{cases} (\Omega'_{h,j}\cos\theta - \Omega'_{h',j}\sin\theta)^2 & \text{if } i = h, \\ (\Omega'_{h,j}\sin\theta + \Omega'_{h',j}\cos\theta)^2 & \text{if } i = h', \\ \Omega'^2_{i,j} & \text{otherwise,} \end{cases}$$

and Eq. (10.26) can be written as a function of $\theta$ as follows:

$$\begin{aligned} G(\theta) &= \sum_{j=1}^{H} \left\{ \gamma_h (\Omega'_{h,j}\cos\theta - \Omega'_{h',j}\sin\theta)^2 + \gamma_{h'} (\Omega'_{h,j}\sin\theta + \Omega'_{h',j}\cos\theta)^2 \right\} \gamma'_j \\ &+ \sum_{k=1}^{K} \begin{pmatrix} \lambda_h^{(k')} & \lambda_{h'}^{(k')} \end{pmatrix} \begin{pmatrix} \cos^2\theta & \sin^2\theta \\ \sin^2\theta & \cos^2\theta \end{pmatrix} \begin{pmatrix} \lambda_h^{(k')} \\ \lambda_{h'}^{(k')} \end{pmatrix} \\ &+ \sum_{k'=1}^{K'} \left( \Psi_{h,h}^{(k')}\cos\theta - \Psi_{h',h}^{(k')}\sin\theta + \Psi_{h,h'}^{(k')}\sin\theta + \Psi_{h',h'}^{(k')}\cos\theta \right) + \text{const.} \end{aligned} \tag{10.27}$$

Since Eq. (10.27) is differentiable at $\theta = 0$, our assumption that Eq. (10.26) is minimized or maximized when $\boldsymbol{\Omega} = \boldsymbol{I}_H$ requires that $\theta = 0$ is a stationary point of Eq. (10.27) for any $h \neq h'$. Therefore, it holds that

$$\begin{aligned} 0 = \left.\frac{\partial G}{\partial \theta}\right|_{\theta=0} &= \Bigg[ 2\sum_j \left\{ \gamma_h (\Omega'_{h,j}\cos\theta - \Omega'_{h',j}\sin\theta)(-\Omega'_{h,j}\sin\theta - \Omega'_{h',j}\cos\theta) \right. \\ &\quad \left. + \gamma_{h'} (\Omega'_{h,j}\sin\theta + \Omega'_{h',j}\cos\theta)(\Omega'_{h,j}\cos\theta - \Omega'_{h',j}\sin\theta) \right\} \gamma'_j \\ &\quad + \sum_{k'=1}^{K'} \left( -\Psi_{h,h}^{(k')}\sin\theta - \Psi_{h',h}^{(k')}\cos\theta + \Psi_{h,h'}^{(k')}\cos\theta - \Psi_{h',h'}^{(k')}\sin\theta \right) \Bigg] \Bigg|_{\theta=0} \\ &= 2(\gamma_{h'} - \gamma_h) \sum_j \Omega'_{h,j}\gamma'_j\Omega'_{h',j} + \sum_{k'=1}^{K'} \left( \Psi_{h,h'}^{(k')} - \Psi_{h',h}^{(k')} \right) \\ &= 2(\gamma_{h'} - \gamma_h)\Phi_{h,h'}. \end{aligned} \tag{10.28}$$

In the last equation, we used Eq. (10.25) and the assumption that $\{\boldsymbol{\Psi}^{(k')}\}$ are symmetric. Since we assume that $\boldsymbol{\Gamma}$ is nondegenerate ($\gamma_h \neq \gamma_{h'}$ for $h \neq h'$), Eq. (10.28) implies that $\boldsymbol{\Phi}$ is diagonal, which completes the proof of Lemma 10.3. □

### 10.4.2 Proof for Case 1

Assume that $(\boldsymbol{A}^*, \boldsymbol{B}^*, \boldsymbol{\Sigma}_A^*, \breve{\boldsymbol{\Sigma}}_B^*)$ is a minimizer, and consider the following variation defined with an arbitrary $H \times H$ orthogonal matrix $\boldsymbol{\Omega}_1$:

$$\widehat{\boldsymbol{A}} = \boldsymbol{A}^* \boldsymbol{C}_B^{1/2} \boldsymbol{\Omega}_1^\top \boldsymbol{C}_B^{-1/2}, \tag{10.29}$$

$$\widehat{\boldsymbol{B}} = \boldsymbol{B}^* \boldsymbol{C}_B^{-1/2} \boldsymbol{\Omega}_1^\top \boldsymbol{C}_B^{1/2}, \tag{10.30}$$
$$\widehat{\boldsymbol{\Sigma}}_A = \boldsymbol{C}_B^{-1/2} \boldsymbol{\Omega}_1 \boldsymbol{C}_B^{1/2} \boldsymbol{\Sigma}_A^* \boldsymbol{C}_B^{1/2} \boldsymbol{\Omega}_1^\top \boldsymbol{C}_B^{-1/2}, \tag{10.31}$$
$$\widehat{\breve{\boldsymbol{\Sigma}}}_B = (\boldsymbol{C}_B^{1/2} \boldsymbol{\Omega}_1 \boldsymbol{C}_B^{-1/2} \otimes \boldsymbol{I}_M) \breve{\boldsymbol{\Sigma}}_B^* (\boldsymbol{C}_B^{-1/2} \boldsymbol{\Omega}_1^\top \boldsymbol{C}_B^{1/2} \otimes \boldsymbol{I}_M). \tag{10.32}$$

Then the free energy (10.10) can be written as a function of $\boldsymbol{\Omega}_1$:

$$2F(\boldsymbol{\Omega}_1) = \mathrm{tr}\left\{ (\boldsymbol{C}_A^{-1} \boldsymbol{C}_B^{-1} \boldsymbol{\Omega}_1 \boldsymbol{C}_B^{1/2} \left( \boldsymbol{A}^{*\top} \boldsymbol{A}^* + M \boldsymbol{\Sigma}_A^* \right) \boldsymbol{C}_B^{1/2} \boldsymbol{\Omega}_1^\top \right\} + \text{const.} \tag{10.33}$$

Since Eq. (10.33) is minimized when $\boldsymbol{\Omega}_1 = \boldsymbol{I}_H$ by assumption, Lemma 10.3 implies that

$$\boldsymbol{C}_B^{1/2} \left( \boldsymbol{A}^{*\top} \boldsymbol{A}^* + M \boldsymbol{\Sigma}_A^* \right) \boldsymbol{C}_B^{1/2}$$

is diagonal. Therefore,

$$\boldsymbol{\Phi}_1 = \boldsymbol{A}^{*\top} \boldsymbol{A}^* + M \boldsymbol{\Sigma}_A^* \tag{10.34}$$

is diagonal, with which Eq. (10.16) implies that $\breve{\boldsymbol{\Sigma}}_B^*$ is diagonal.

Since we have proved the diagonality of $\breve{\boldsymbol{\Sigma}}_B^*$, the expectations in Eqs. (10.10) and (10.14), respectively, can be expressed in the following simple forms at the solution $(\widehat{\boldsymbol{A}}, \widehat{\boldsymbol{B}}, \widehat{\boldsymbol{\Sigma}}_A, \widehat{\breve{\boldsymbol{\Sigma}}}_B) = (\boldsymbol{A}^*, \boldsymbol{B}^*, \boldsymbol{\Sigma}_A^*, \breve{\boldsymbol{\Sigma}}_B^*)$:

$$\left\langle \boldsymbol{B} \left( \widehat{\boldsymbol{A}}^\top \widehat{\boldsymbol{A}} + M \widehat{\boldsymbol{\Sigma}}_A \right) \boldsymbol{B}^\top \right\rangle_{r_B(\boldsymbol{B})} = \widehat{\boldsymbol{B}} \left( \widehat{\boldsymbol{A}}^\top \widehat{\boldsymbol{A}} + M \widehat{\boldsymbol{\Sigma}}_A \right) \widehat{\boldsymbol{B}}^\top + \boldsymbol{\Xi}_{\boldsymbol{\Phi}_1}, \tag{10.35}$$
$$\left\langle \boldsymbol{B}^\top \boldsymbol{\Gamma}_V^\top \boldsymbol{\Gamma}_V \boldsymbol{B} \right\rangle_{r_B(\boldsymbol{B})} = \widehat{\boldsymbol{B}}^\top \boldsymbol{\Gamma}_V^\top \boldsymbol{\Gamma}_V \widehat{\boldsymbol{B}} + \boldsymbol{\Xi}_{\boldsymbol{\Gamma}_V}, \tag{10.36}$$

where $\boldsymbol{\Xi}_{\boldsymbol{\Gamma}_V} \in \mathbb{R}^{H \times H}$ and $\boldsymbol{\Xi}_{\boldsymbol{\Phi}_1} \in \mathbb{R}^{M \times M}$ are diagonal matrices with their entries given by

$$(\boldsymbol{\Xi}_{\boldsymbol{\Phi}_1})_{m,m} = \sum_{h=1}^{H} \left( \widehat{\boldsymbol{A}}^\top \widehat{\boldsymbol{A}} + M \widehat{\boldsymbol{\Sigma}}_A \right)_{h,h} \widehat{\sigma}_{B_{m,h}}^2,$$
$$(\boldsymbol{\Xi}_{\boldsymbol{\Gamma}_V})_{h,h} = \sum_{m=1}^{M} \gamma_m^2 \widehat{\sigma}_{B_{m,h}}^2.$$

Here $\{\widehat{\sigma}_{B_{m,h}}^2\}$ are the diagonal entries of $\widehat{\breve{\boldsymbol{\Sigma}}}_B$ such that

$$\widehat{\breve{\boldsymbol{\Sigma}}}_B = \mathbf{Diag}((\widehat{\sigma}_{B_{1,1}}^2, \ldots, \widehat{\sigma}_{B_{M,1}}^2), (\widehat{\sigma}_{B_{1,2}}^2, \ldots, \widehat{\sigma}_{B_{M,2}}^2), \ldots\ldots, (\widehat{\sigma}_{B_{1,H}}^2, \ldots, \widehat{\sigma}_{B_{M,H}}^2)).$$

Next consider the following variation defined with an $M \times M$ matrix $\boldsymbol{\Omega}_2$ such that the upper-left $J \times J$ submatrix is an arbitrary orthogonal matrix and the other entries are zero:

$$\widehat{\boldsymbol{A}} = \boldsymbol{\Omega}_2^\top \boldsymbol{A}^*,$$
$$\widehat{\boldsymbol{B}} = \boldsymbol{\Omega}_2^\top \boldsymbol{B}^*.$$

Then, by using Eq. (10.35), the free energy (10.10) is written as

$$2F(\boldsymbol{\Omega}_2) = \frac{1}{\sigma^2}\mathrm{tr}\left\{\boldsymbol{\Gamma}_V^\top \boldsymbol{\Gamma}_V \boldsymbol{\Omega}_2^\top \left(-2\boldsymbol{B}^*\boldsymbol{A}^{*\top} + \boldsymbol{B}^*\left(\boldsymbol{A}^{*\top}\boldsymbol{A}^* + M\widehat{\boldsymbol{\Sigma}}_A\right)\boldsymbol{B}^{*\top}\right)\boldsymbol{\Omega}_2\right\} + \text{const.} \tag{10.37}$$

Applying Lemma 10.3 to the upper-left $J \times J$ submatrix in the trace, and then using Eq. (10.22), we find that

$$\boldsymbol{\Phi}_2 = -2\boldsymbol{B}^*\boldsymbol{A}^{*\top} + \boldsymbol{B}^*\left(\boldsymbol{A}^{*\top}\boldsymbol{A}^* + M\widehat{\boldsymbol{\Sigma}}_A\right)\boldsymbol{B}^{*\top} \tag{10.38}$$

is diagonal. Eq. (10.38) also implies that $\boldsymbol{B}^*\boldsymbol{A}^{*\top}$ is symmetric.

Consider the following variation defined with an $M \times M$ matrix $\boldsymbol{\Omega}_3$ such that the upper-left $J \times J$ submatrix is an arbitrary orthogonal matrix and the other entries are zero:

$$\widehat{\boldsymbol{B}} = \boldsymbol{\Omega}_3^\top \boldsymbol{B}^*.$$

Then the free energy is written as

$$2F(\boldsymbol{\Omega}_3) = \frac{1}{\sigma^2}\mathrm{tr}\left\{\boldsymbol{\Gamma}_V^\top \boldsymbol{\Gamma}_V \boldsymbol{\Omega}_3^\top \left(-2\boldsymbol{B}^*\boldsymbol{A}^{*\top}\right)\right.$$
$$\left. + \boldsymbol{\Gamma}_V^\top \boldsymbol{\Gamma}_V \boldsymbol{\Omega}_3^\top \left(\boldsymbol{B}^*\left(\boldsymbol{A}^{*\top}\boldsymbol{A}^* + M\widehat{\boldsymbol{\Sigma}}_A\right)\boldsymbol{B}^{*\top}\right)\boldsymbol{\Omega}_3\right\} + \text{const.} \tag{10.39}$$

Applying Lemma 10.3 to the upper-left $J \times J$ submatrix in the trace, we find that

$$\boldsymbol{\Phi}_3 = \boldsymbol{B}^*\left(\boldsymbol{A}^{*\top}\boldsymbol{A}^* + M\widehat{\boldsymbol{\Sigma}}_A\right)\boldsymbol{B}^{*\top} \tag{10.40}$$

is diagonal. Since Eqs. (10.34) and (10.40) are diagonal, $\boldsymbol{B}^*$ is diagonal. Consequently, Eq. (10.14) combined with Eq. (10.36) implies that $\boldsymbol{A}^*$ and $\boldsymbol{\Sigma}_A^*$ are diagonal.

Thus we proved that the solution for $(\widehat{\boldsymbol{A}}, \widehat{\boldsymbol{B}}, \widehat{\boldsymbol{\Sigma}}_A, \widehat{\breve{\boldsymbol{\Sigma}}}_B)$ are diagonal, provided that $\boldsymbol{C}_A\boldsymbol{C}_B$ is nondegenerate.

### 10.4.3 Proof for Case 2

When $\boldsymbol{C}_A\boldsymbol{C}_B$ is degenerate, there are multiple *equivalent* solutions giving the same free energy (10.10) and the output $\widehat{\boldsymbol{B}\boldsymbol{A}}^\top$. In the following, we show that one of the equivalent solutions has diagonal $(\boldsymbol{A}^*, \boldsymbol{B}^*, \boldsymbol{\Sigma}_A^*, \breve{\boldsymbol{\Sigma}}_B^*)$.

Assume that $\boldsymbol{C}_A\boldsymbol{C}_B = c^2\boldsymbol{I}_H$ for some $c^2 \in \mathbb{R}_{++}$. In this case, the free energy (10.10) is invariant with respect to $\boldsymbol{\Omega}_1$ under the transformation (10.29) through (10.32). Let us focus on the solution with diagonal $\widehat{\breve{\boldsymbol{\Sigma}}}_B$, which can be obtained by the transform (10.29) through (10.32) with a certain $\boldsymbol{\Omega}_1$ from any solution satisfying Eq. (10.16). Then we can show, in the same way as in the nondegenerate case, that Eqs. (10.34), (10.38), and (10.40) are diagonal. This proves the existence of a solution such that $(\boldsymbol{A}^*, \boldsymbol{B}^*, \boldsymbol{\Sigma}_A^*, \breve{\boldsymbol{\Sigma}}_B^*)$ are diagonal.

### 10.4.4 Proof for Case 3

When $c_{a_h}c_{b_h} = c_{a_{h'}}c_{b_{h'}}$ for (not all but) some pairs $h \neq h'$, we can show that $\widehat{\boldsymbol{\Sigma}}_A$ and $\widehat{\breve{\boldsymbol{\Sigma}}}_B$ are block diagonal where the blocks correspond to the groups sharing the same $c_{a_h}c_{b_h}$. In each block, multiple equivalent solutions exist, one of which is a solution such that $(\boldsymbol{A}^*, \boldsymbol{B}^*, \boldsymbol{\Sigma}_A^*, \breve{\boldsymbol{\Sigma}}_B^*)$ are diagonal.

This completes the proof of Theorem 10.2. □

## 10.5 Exact Global VB Solver (EGVBS)

Theorem 10.2 allows us to focus on the solutions such that $(\widehat{\boldsymbol{A}}, \widehat{\boldsymbol{B}}, \widehat{\boldsymbol{\Sigma}}_A, \widehat{\breve{\boldsymbol{\Sigma}}}_B)$ are diagonal. Accordingly, we express the solution of the VB learning problem (10.11) with diagonal entries, i.e.,

$$\widehat{\boldsymbol{A}} = \mathbf{Diag}_{M,H}(\widehat{a}_1, \ldots, \widehat{a}_H), \tag{10.41}$$

$$\widehat{\boldsymbol{B}} = \mathbf{Diag}_{M,H}(\widehat{b}_1, \ldots, \widehat{b}_H), \tag{10.42}$$

$$\widehat{\boldsymbol{\Sigma}}_A = \mathbf{Diag}(\widehat{\sigma}^2_{a_1}, \ldots, \widehat{\sigma}^2_{a_H}), \tag{10.43}$$

$$\widehat{\breve{\boldsymbol{\Sigma}}}_B = \mathbf{Diag}((\widehat{\sigma}^2_{B_{1,1}}, \ldots, \widehat{\sigma}^2_{B_{M,1}}), (\widehat{\sigma}^2_{B_{1,2}}, \ldots, \widehat{\sigma}^2_{B_{M,2}}), \ldots\ldots, (\widehat{\sigma}^2_{B_{1,H}}, \ldots, \widehat{\sigma}^2_{B_{M,H}})), \tag{10.44}$$

where $\mathbf{Diag}_{D_1,D_2}(\cdot)$ denotes the $D_1 \times D_2$ diagonal matrix with the specified diagonal entries. Remember that $J$ $(\leq \min(L, M))$ is the rank of the observed matrix $\boldsymbol{V}$, and $\{\gamma_m\}$ are the singular values arranged in nonincreasing order. Without loss of generality, we assume that $\widehat{a}_h, \widehat{b}_h \in \mathbb{R}_+$ for all $h = 1, \ldots, H$.

We can easily obtain the following theorem:

**Theorem 10.4** *Any local solution of the VB learning problem* (10.11) *satisfies, for all* $h = 1, \ldots, H$,

$$\widehat{a}_h = \frac{\gamma_h^2}{\sigma^2}\widehat{b}_h\widehat{\sigma}^2_{a_h}, \tag{10.45}$$

$$\widehat{\sigma}^2_{a_h} = \sigma^2\left(\gamma_h^2\widehat{b}_h^2 + \sum_{m=1}^{J}\gamma_m^2\widehat{\sigma}^2_{B_{m,h}} + \frac{\sigma^2}{c^2_{a_h}}\right)^{-1}, \tag{10.46}$$

$$\widehat{b}_h = \frac{\gamma_h^2}{\sigma^2}\widehat{a}_h\widehat{\sigma}^2_{B_{h,h}}, \tag{10.47}$$

$$\widehat{\sigma}^2_{B_{m,h}} = \begin{cases} \sigma^2 \left( \gamma_m^2 \left( \widehat{a}_h^2 + M \widehat{\sigma}^2_{a_h} \right) + \frac{\sigma^2}{c^2_{b_h}} \right)^{-1} & (\textit{for } m = 1, \ldots, J), \\ c^2_{b_h} & (\textit{for } m = J+1, \ldots, M), \end{cases} \tag{10.48}$$

*and has the free energy given by*

$$2F = LM \log(2\pi\sigma^2) + \frac{\sum_{h=1}^{J} \gamma_h^2}{\sigma^2} + \sum_{h=1}^{H} 2F_h, \qquad \textit{where} \tag{10.49}$$

$$2F_h = M \log \frac{c^2_{a_h}}{\widehat{\sigma}^2_{a_h}} + \sum_{m=1}^{J} \log \frac{c^2_{b_h}}{\widehat{\sigma}^2_{B_{m,h}}} - (M+J) + \frac{\widehat{a}_h^2 + M\widehat{\sigma}^2_{a_h}}{c^2_{a_h}} + \frac{\widehat{b}_h^2 + \sum_{m=1}^{J} \widehat{\sigma}^2_{B_{m,h}}}{c^2_{b_h}}$$
$$+ \frac{1}{\sigma^2} \left\{ \gamma_h^2 \left( -2\widehat{a}_h \widehat{b}_h + \widehat{b}_h^2 (\widehat{a}_h^2 + M\widehat{\sigma}^2_{a_h}) \right) + \sum_{m=1}^{J} \gamma_m^2 \widehat{\sigma}^2_{B_{m,h}} (\widehat{a}_h^2 + M\widehat{\sigma}^2_{a_h}) \right\}. \tag{10.50}$$

*Proof* By substituting the diagonal expression, Eqs. (10.41) through (10.44), into the free energy (10.10), we have

$$2F = LM \log(2\pi\sigma^2) + M \sum_{h=1}^{H} \log \frac{c^2_{a_h}}{\widehat{\sigma}^2_{a_h}} + \sum_{m=1}^{M} \sum_{h=1}^{H} \log \frac{c^2_{b_h}}{\widehat{\sigma}^2_{B_{m,h}}} + \frac{\sum_{h=1}^{M} \gamma_h^2}{\sigma^2} - 2MH$$
$$+ \sum_{h=1}^{H} \left\{ \frac{1}{c^2_{a_h}} \left( \widehat{a}_h^2 + M\widehat{\sigma}^2_{a_h} \right) + \frac{1}{c^2_{b_h}} \left( \widehat{b}_h^2 + \sum_{m=1}^{M} \widehat{\sigma}^2_{B_{m,h}} \right) \right\}$$
$$+ \frac{1}{\sigma^2} \sum_{h=1}^{H} \left\{ \gamma_h^2 \left( -2\widehat{a}_h \widehat{b}_h + \widehat{b}_h^2 (\widehat{a}_h^2 + M\widehat{\sigma}^2_{a_h}) \right) + \sum_{m=1}^{J} \gamma_m^2 \widehat{\sigma}^2_{B_{m,h}} (\widehat{a}_h^2 + M\widehat{\sigma}^2_{a_h}) \right\}. \tag{10.51}$$

Eqs. (10.45) through (10.48) are obtained as the stationary conditions of Eq. (10.51) that any solution satisfies, according to Lemma 10.1. By substituting Eq. (10.48) for $m = J+1, \ldots, M$ into Eq. (10.51), we obtain Eq. (10.49). □

For EVB learning, where the prior covariances $\boldsymbol{C}_A, \boldsymbol{C}_B$ are also estimated, we have the following theorem:

**Theorem 10.5** *Any local solution of the EVB learning problem* (10.17) *satisfies the following. For each* $h = 1, \ldots, H$, $(\widehat{a}_h, \widehat{b}_h, \widehat{\sigma}^2_{a_h}, \{\widehat{\sigma}^2_{B_{m,h}}\}_{m=1}^{M}, c^2_{a_h}, c^2_{b_h})$ *is either a (positive) stationary point that satisfies Eqs.* (10.45) *through* (10.48) *and*

$$c^2_{a_h} = \widehat{a}_h^2 / M + \widehat{\sigma}^2_{a_h}, \tag{10.52}$$

$$c^2_{b_h} = \left( \widehat{b}_h^2 + \sum_{m=1}^{J} \widehat{\sigma}^2_{B_{m,h}} \right) / J, \tag{10.53}$$

*or the null local solution defined by*

$$\widehat{a}_h = \widehat{b}_h = 0, \quad \widehat{\sigma}^2_{a_h} = c^2_{a_h} \to +0, \quad \widehat{\sigma}^2_{B_{m,h}} = c^2_{b_h} \to +0 \quad (\textit{for } m = 1, \ldots, M), \tag{10.54}$$

*of which the contribution* (10.50) *to the free energy is*

$$F_h \to +0. \tag{10.55}$$

*The total free energy is given by Eq.* (10.49).

*Proof* Considering the derivatives of Eq. (10.51) with respect to $c^2_{a_h}$ and $c^2_{b_h}$, we have

$$2Mc^2_{a_h} = \widehat{a}^2_h + M\widehat{\sigma}^2_{a_h}, \tag{10.56}$$

$$2Mc^2_{b_h} = \widehat{b}^2_h + \sum_{m=1}^{M} \widehat{\sigma}^2_{B_{m,h}}, \tag{10.57}$$

as stationary conditions. By using Eq. (10.48), we can easily obtain Eqs. (10.52) and (10.53).

Unlike in VB learning, where Lemma 10.1 guarantees that any local solution is a stationary point, there exist nonstationary local solutions in EVB learning. We can confirm that, along any path such that

$$\widehat{a}_h, \widehat{b}_h = 0, \quad \widehat{\sigma}^2_{a_h}, \widehat{\sigma}^2_{B_{m,h}}, c^2_{a_h}, c^2_{b_h} \to +0$$
$$\text{with } \beta_a = \frac{\widehat{\sigma}^2_{a_h}}{c^2_{a_h}} \text{ and } \beta_b = \frac{\widehat{\sigma}^2_{B_{m,h}}}{c^2_{b_h}} \text{ kept constant}, \tag{10.58}$$

the free energy contribution (10.50) from the $h$th component decreases monotonically. Among the possible paths, $\beta_a = \beta_b = 1$ gives the lowest free energy (10.55). □

Based on Theorem 10.5, we can obtain the following corollary for the global solution.

**Corollary 10.6** *The global solution of the EVB learning problem* (10.17) *can be found in the following way. For each $h = 1, \ldots, H$, find all stationary points that satisfy Eqs.* (10.45) *through* (10.48), (10.52), *and* (10.53), *and choose the one giving the minimum free energy contribution $\underline{F}_h$. The chosen stationary point is the global solution if $\underline{F}_h < 0$. Otherwise (including the case where no stationary point exists), the null local solution* (10.54) *is global.*

*Proof* For each $h = 1, \ldots, H$, any candidate for a local solution is a stationary point or the null local solution. Therefore, if the minimum free energy contribution over all stationary points is negative, i.e., $\underline{F}_h < 0$, the corresponding stationary point is the global minimizer. With this fact, Corollary 10.6 is a straightforward deduction from Theorem 10.5. □

Taking account of the trivial relations $c_{b_h}^2 = \widehat{\sigma}_{B_{m,h}}^2$ for $m > J$, the stationary conditions consisting of Eqs. (10.45) through (10.48), (10.52), and (10.53) for each $h$ can be seen as a *polynomial system*, a set of polynomial equations, with $5 + J$ unknown variables, $\left(\widehat{a}_h, \widehat{b}_h, \widehat{\sigma}_{a_h}^2, \{\widehat{\sigma}_{B_{m,h}}^2\}_{m=1}^J, c_{a_h}^2, c_{b_h}^2\right)$. Thus, Theorem 10.5 has decomposed the original problem with $O(M^2H^2)$ unknown variables, for which the stationary conditions are given by Eqs. (10.13) through (10.16), (10.19), and (10.20), into $H$ subproblems with $O(J)$ unknown variables each.

Fortunately, there is a reliable numerical method to solve a polynomial system, called the *homotopy method* or *continuation method* (Drexler, 1978; Garcia and Zangwill, 1979; Gunji et al., 2004; Lee et al., 2008). It provides all isolated solutions to a system of $n$ polynomials $\boldsymbol{f}(\boldsymbol{x}) \equiv (f_1(\boldsymbol{x}), \ldots, f_n(\boldsymbol{x})) = \boldsymbol{0}$ by defining a smooth set of homotopy systems with a parameter $t \in [0, 1]$, i.e., $\boldsymbol{g}(\boldsymbol{x}, t) \equiv (g_1(\boldsymbol{x}, t), g_2(\boldsymbol{x}, t), \ldots, g_n(\boldsymbol{x}, t)) = \boldsymbol{0}$ such that one can continuously trace the solution path from the easiest ($t = 0$) to the target ($t = 1$). For empirical evaluation, which will be given in Section 10.8, we use HOM4PS-2.0 (Lee et al., 2008), one of the most successful polynomial system solvers.

With the homotopy method in hand, Corollary 10.6 allows us to solve the EVB learning problem (10.17) in the following way, which we call the *exact global VB solver (EGVBS)*. For each $h = 1, \ldots, H$, we first find all stationary points that satisfy the polynomial system, Eqs. (10.45) through (10.48), (10.52), and (10.53). After that, we discard the prohibitive solutions with complex numbers or negative variances, and then select the stationary point giving the minimum free energy contribution $F_h$, defined by Eq. (10.50). The global solution is the selected stationary point if it satisfies $F_h < 0$; otherwise, the null local solution (10.54) is the global solution. Algorithm 20 summarizes the procedure of EGVBS. When the noise variance $\sigma^2$ is unknown, we conduct a naive one-dimensional search to minimize the total free energy (10.49), with EGVBS applied for every candidate value of $\sigma^2$.

It is straightforward to modify Algorithm 20 to solve the VB learning problem (10.11), where the prior covariances $\boldsymbol{C}_A, \boldsymbol{C}_B$ are given. In this case, we should solve the polynomial system (10.45) through (10.48) in Step 3, and skip Step 6 since all local solutions are stationary points.

## 10.6 Approximate Global VB Solver (AGVBS)

Theorems 10.4 and 10.5 significantly reduced the complexity of the optimization problem. However, EGVBS is still not applicable to data with typical

**Algorithm 20** Exact global VB solver (EGVBS) for LRSC.

1: Compute the SVD of $\boldsymbol{V} = \boldsymbol{\Omega}_V^{\text{left}} \boldsymbol{\Gamma}_V \boldsymbol{\Omega}_V^{\text{right}\top}$.
2: **for** $h = 1$ to $H$ **do**
3: Find all solutions of the polynomial system, Eqs. (10.45) through (10.48), (10.52), and (10.53) by the homotopy method.
4: Discard prohibitive solutions with complex numbers or negative variances.
5: Select the stationary point giving the smallest $F_h$ (defined by Eq. (10.50)).
6: The global solution for the $h$th component is the selected stationary point if it satisfies $F_h < 0$; otherwise, the null local solution (10.54) is the global solution.
7: **end for**
8: Compute $\widehat{\boldsymbol{U}} = \boldsymbol{\Omega}_V^{\text{right}} \widehat{\boldsymbol{B}} \widehat{\boldsymbol{A}}^\top \boldsymbol{\Omega}_V^{\text{right}\top}$.
9: Apply spectral clustering with the affinity matrix equal to $\text{abs}(\widehat{\boldsymbol{U}}) + \text{abs}(\widehat{\boldsymbol{U}}^\top)$.

problem sizes. This is because the homotopy method is not guaranteed to find all solutions in polynomial time in $J$, when the polynomial system involves $O(J)$ unknown variables.

The following simple trick further reduces the complexity and leads to an efficient approximate solver. Let us impose an additional constraint that $\gamma_m^2 \widehat{\sigma}_{B_{m,h}}^2$ are constant over $m = 1, \ldots, J$, i.e.,

$$\gamma_m^2 \widehat{\sigma}_{B_{m,h}}^2 = \overline{\widehat{\sigma}}_{b_h}^2 \quad \text{for} \quad m = 1, \ldots, J. \tag{10.59}$$

Under this constraint, the stationary conditions for the six unknowns $(\widehat{a}_h, \widehat{b}_h, \widehat{\sigma}_{a_h}^2, \overline{\widehat{\sigma}}_{b_h}^2, c_{a_h}^2, c_{b_h}^2)$ (for each $h$) become similar to the stationary conditions for fully observed MF, which allows us to obtain the following theorem:

**Theorem 10.7** *Under the constraint* (10.59), *any stationary point of the free energy* (10.50) *for each h satisfies the following polynomial equation for a single variable* $\widehat{\gamma}_h \in \mathbb{R}$:

$$\xi_6 \widehat{\gamma}_h^6 + \xi_5 \widehat{\gamma}_h^5 + \xi_4 \widehat{\gamma}_h^4 + \xi_3 \widehat{\gamma}_h^3 + \xi_2 \widehat{\gamma}_h^2 + \xi_1 \widehat{\gamma}_h + \xi_0 = 0, \tag{10.60}$$

*where*

$$\xi_6 = \frac{\phi_h^2}{\gamma_h^2}, \tag{10.61}$$

$$\xi_5 = -2\frac{\phi_h^2 M \sigma^2}{\gamma_h^3} + \frac{2\phi_h}{\gamma_h}, \tag{10.62}$$

$$\xi_4 = \frac{\phi_h^2 M^2 \sigma^4}{\gamma_h^4} - \frac{2\phi_h(2M-J)\sigma^2}{\gamma_h^2} + 1 + \frac{\phi_h^2(M\sigma^2-\gamma_h^2)}{\gamma_h^2}, \tag{10.63}$$

$$\xi_3 = \frac{2\phi_h M(M-J)\sigma^4}{\gamma_h^3} - \frac{2(M-J)\sigma^2}{\gamma_h} + \frac{\phi_h((M+J)\sigma^2-\gamma_h^2)}{\gamma_h} - \frac{\phi_h^2 M\sigma^2(M\sigma^2-\gamma_h^2)}{\gamma_h^3} + \frac{\phi_h(M\sigma^2-\gamma_h^2)}{\gamma_h}, \tag{10.64}$$

$$\xi_2 = \frac{(M-J)^2\sigma^4}{\gamma_h^2} - \frac{\phi_h M\sigma^2((M+J)\sigma^2-\gamma_h^2)}{\gamma_h^2} + ((M+J)\sigma^2 - \gamma_h^2) - \frac{\phi_h(M-J)\sigma^2(M\sigma^2-\gamma_h^2)}{\gamma_h^2}, \tag{10.65}$$

$$\xi_1 = -\frac{(M-J)\sigma^2((M+J)\sigma^2-\gamma_h^2)}{\gamma_h} + \frac{\phi_h MJ\sigma^4}{\gamma_h}, \tag{10.66}$$

$$\xi_0 = MJ\sigma^4. \tag{10.67}$$

*Here* $\phi_h = 1 - \frac{\gamma_h^2}{\underline{\gamma}^2}$ *for* $\underline{\gamma}^2 = (\sum_{m=1}^{J} \gamma_m^{-2}/J)^{-1}$. *For each real solution* $\widehat{\widehat{\gamma}}_h$ *such that*

$$\widehat{\gamma}_h = \widehat{\widehat{\gamma}}_h + \gamma_h - \frac{M\sigma^2}{\gamma_h}, \tag{10.68}$$

$$\widehat{\kappa}_h = \gamma_h^2 - (M+J)\sigma^2 - \left(M\sigma^2 - \gamma_h^2\right)\phi_h \frac{\widehat{\widehat{\gamma}}_h}{\gamma_h}, \tag{10.69}$$

$$\widehat{\tau}_h = \frac{1}{2MJ}\left(\widehat{\kappa}_h + \sqrt{\widehat{\kappa}_h^2 - 4MJ\sigma^4\left(1 + \phi_h \frac{\widehat{\widehat{\gamma}}_h}{\gamma_h}\right)}\right), \tag{10.70}$$

$$\widehat{\delta}_h = \frac{\sigma^2}{\sqrt{\widehat{\tau}_h}}\left(\gamma_h - \frac{M\sigma^2}{\gamma_h} - \widehat{\gamma}_h\right)^{-1}, \tag{10.71}$$

*are real and positive, there exists the corresponding stationary point given by*

$$\left(\widehat{a}_h, \widehat{b}_h, \widehat{\sigma}^2_{a_h}, \overline{\widehat{\sigma}}^2_{b_h}, c^2_{a_h}, c^2_{b_h}\right) = \left(\sqrt{\widehat{\gamma}_h\widehat{\delta}_h}, \frac{\sqrt{\widehat{\gamma}_h/\widehat{\delta}_h}}{\gamma_h}, \frac{\sigma^2\widehat{\delta}_h}{\gamma_h}, \frac{\sigma^2}{\gamma_h\widehat{\delta}_h - \phi_h\frac{\sigma^2}{\sqrt{\widehat{\tau}_h}}}, \sqrt{\widehat{\tau}_h}, \frac{\sqrt{\widehat{\tau}_h}}{\gamma_h^2}\right). \tag{10.72}$$

Given the noise variance $\sigma^2$, computing the coefficients (10.61) through (10.67) is straightforward. Theorem 10.7 implies that the following algorithm, which we call the AGVBS, provides the global solution of the EVB learning problem (10.17) under the additional constraint (10.59). After computing the SVD of the observed matrix $\boldsymbol{V}$, AGVBS first finds all real solutions of the sixth-order polynomial equation (10.60) by using, e.g., the "`roots`" command in MATLAB®, for each $h$. Then, it discards the prohibitive solutions such that any of Eqs. (10.68) through (10.71) gives a complex or negative number. For each of the retained solutions, AGVBS computes the corresponding stationary point by Eq. (10.72), along with the free energy contribution $F_h$ by Eq. (10.50). Here, Eq. (10.59) is used for retrieving the original posterior variances $\{\widehat{\sigma}^2_{B_{m,h}}\}_{m=1}^{J}$ for $\boldsymbol{B}$. Finally, AGVBS selects the stationary point giving the minimum free energy contribution $\underline{F}_h$. The global solution is the selected stationary point if it satisfies $\underline{F}_h < 0$; otherwise, the null local solution (10.54) is the global solution. Algorithm 21 summarizes the procedure of AGVBS.

**Algorithm 21** Approximate global VB solver (AGVBS) for LRSC.

1: Compute the SVD of $\boldsymbol{V} = \boldsymbol{\Omega}_V^{\text{left}} \boldsymbol{\Gamma}_V \boldsymbol{\Omega}_V^{\text{right}\top}$.
2: **for** $h = 1$ to $H$ **do**
3: Find all real solutions of the sixth-order polynomial equation (10.60).
4: Discard prohibitive solutions such that any of Eqs. (10.68) through (10.71) gives a complex or negative number.
5: Compute the corresponding stationary point by Eq. (10.72) and its free energy contribution $F_h$ by Eq. (10.50) for each of the retained solutions.
6: Select the stationary point giving the minimum free energy contribution $\underline{F}_h$.
7: The global solution for the $h$th component is the selected stationary point if it satisfies $\underline{F}_h < 0$; otherwise, the null local solution (10.54) is the global solution.
8: **end for**
9: Compute $\widehat{\boldsymbol{U}} = \boldsymbol{\Omega}_V^{\text{right}} \widehat{\boldsymbol{B}\boldsymbol{A}}^\top \boldsymbol{\Omega}_V^{\text{right}\top}$.
10: Apply spectral clustering with the affinity matrix equal to $\text{abs}(\widehat{\boldsymbol{U}}) + \text{abs}(\widehat{\boldsymbol{U}}^\top)$.

As in EGVBS, a naive one-dimensional search is conducted when the noise variance $\sigma^2$ is unknown.

In Section 10.8, we show that AGVBS is practically a good alternative to the Kronecker product covariance approximation (KPCA), an approximate EVB algorithm for LRSC under the Kronecker product covariance constraint (see Section 3.4.2), in terms of accuracy and computation time.

## 10.7 Proof of Theorem 10.7

Let us rescale $\widehat{b}_h$ and $c_{b_h}^2$ as follows:

$$\overline{\widehat{b}}_h = \gamma_h \widehat{b}_h, \qquad \overline{c}_{b_h}^2 = \gamma_h^2 c_{b_h}^2. \tag{10.73}$$

By substituting Eqs. (10.59) and (10.73) into Eq. (10.50), we have

$$2F_h = M \log \frac{c_{a_h}^2}{\widehat{\sigma}_{a_h}^2} + J \log \frac{\overline{c}_{b_h}^2}{\overline{\widehat{\sigma}}_{b_h}^2} + \frac{1}{c_{a_h}^2}\left(\widehat{a}_h^2 + M\widehat{\sigma}_{a_h}^2\right) + \frac{1}{\overline{c}_{b_h}^2}\left(\overline{\widehat{b}}_h^2 + J\frac{\gamma_h^2}{\underline{\gamma}^2}\overline{\widehat{\sigma}}_{b_h}^2\right)$$
$$+ \frac{1}{\sigma^2}\left(-2\gamma_h \widehat{a}_h \overline{\widehat{b}}_h + (\widehat{a}_h^2 + M\widehat{\sigma}_{a_h}^2)(\overline{\widehat{b}}_h^2 + J\overline{\widehat{\sigma}}_{b_h}^2)\right) - (M+J) + \sum_{m=1}^{J} \log \frac{\gamma_m^2}{\gamma_h^2}, \tag{10.74}$$

where

$$\underline{\gamma}^2 = \left(\sum_{m=1}^{J} \gamma_m^{-2}/J\right)^{-1}.$$

Ignoring the last two constant terms, we find that Eq. (10.74) is in almost the same form as the free energy of fully observed MF for a $J \times M$ observed matrix (see Eq. (6.43)). Only the difference is in the fourth term: $J\overline{\overline{\sigma}}^2_{b_h}$ is multiplied by $\frac{\gamma_h^2}{\underline{\gamma}^2}$. Note that, as in MF, the free energy (10.74) is invariant under the following transformation:

$$\left\{\left(\widehat{a}_h, \overline{\overline{b}}_h, \widehat{\sigma}^2_{a_h}, \overline{\overline{\sigma}}^2_{b_h}, c^2_{a_h}, \overline{c}^2_{b_h}\right)\right\} \to \left\{\left(s_h\widehat{a}_h, s_h^{-1}\overline{\overline{b}}_h, s_h^2\widehat{\sigma}^2_{a_h}, s_h^{-2}\overline{\overline{\sigma}}^2_{b_h}, s_h^2 c^2_{a_h}, s_h^{-2}\overline{c}^2_{b_h}\right)\right\}$$

for any $\{s_h \neq 0; h = 1, \ldots, H\}$. Accordingly, we fix the ratio between $c_{a_h}$ and $\overline{c}_{b_h}$ to $c_{a_h}/\overline{c}_{b_h} = 1$ without loss of generality.

By differentiating the free energy (10.74) with respect to $\widehat{a}_h, \widehat{\sigma}^2_{a_h}, \overline{\overline{b}}_h, \overline{\sigma}^2_{b_h}, c^2_{a_h}$, and $\overline{c}^2_{b_h}$, respectively, we obtain the following stationary conditions:

$$\widehat{a}_h = \frac{1}{\sigma^2}\gamma_h\overline{\overline{b}}_h\widehat{\sigma}^2_{a_h}, \tag{10.75}$$

$$\widehat{\sigma}^2_{a_h} = \sigma^2\left(\overline{\overline{b}}^2_h + J\overline{\overline{\sigma}}^2_{b_h} + \frac{\sigma^2}{c^2_{a_h}}\right)^{-1}, \tag{10.76}$$

$$\overline{\overline{b}}_h = \gamma_h\widehat{a}_h\left(\widehat{a}^2_h + M\widehat{\sigma}^2_{a_h} + \frac{\sigma^2}{\overline{c}^2_{b_h}}\right)^{-1}, \tag{10.77}$$

$$\overline{\sigma}^2_{b_h} = \sigma^2\left(\widehat{a}^2_h + M\widehat{\sigma}^2_{a_h} + \frac{\sigma^2\gamma_h^2}{\overline{c}^2_{b_h}\underline{\gamma}^2}\right)^{-1}, \tag{10.78}$$

$$c^2_{a_h} = \widehat{a}^2_h/M + \widehat{\sigma}^2_{a_h}, \tag{10.79}$$

$$\overline{c}^2_{b_h} = \overline{\overline{b}}^2_h/J + \frac{\gamma_h^2}{\underline{\gamma}^2}\overline{\overline{\sigma}}^2_{b_h}. \tag{10.80}$$

Note that, unlike the case of fully observed MF, $\boldsymbol{A}$ and $\boldsymbol{B}$ are not symmetric, which makes analysis more involved. Apparently, if $\widehat{a}_h = 0$ or $\overline{\overline{b}}_h = 0$, the null solution (10.54) gives the minimum $F_h \to +0$ of the free energy (10.74). In the following, we identify the positive stationary points such that $\widehat{a}_h, \overline{\overline{b}}_h > 0$. To this end, we derive a polynomial equation with a single unknown variable from the stationary conditions (10.75) through (10.80). Let

$$\widehat{\gamma}_h = \widehat{a}_h\overline{\overline{b}}_h, \tag{10.81}$$

$$\widehat{\delta}_h = \widehat{a}_h/\overline{\overline{b}}_h. \tag{10.82}$$

From Eqs. (10.75) through (10.78), we obtain

$$\gamma_h^2 = \left(\widehat{a}_h^2 + M\widehat{\sigma}_{a_h}^2 + \frac{\sigma^2}{\overline{c}_{b_h}^2}\right)\left(\overline{\widehat{b}}_h^2 + J\overline{\widehat{\sigma}}_{b_h}^2 + \frac{\sigma^2}{c_{a_h}^2}\right), \tag{10.83}$$

$$\gamma_h \widehat{\delta}_h^{-1} = \left(\overline{\widehat{b}}_h^2 + J\overline{\widehat{\sigma}}_{b_h}^2 + \frac{\sigma^2}{c_{a_h}^2}\right), \tag{10.84}$$

$$\gamma_h \widehat{\delta}_h = \left(\widehat{a}_h^2 + M\widehat{\sigma}_{a_h}^2 + \frac{\sigma^2}{\overline{c}_{b_h}^2}\right). \tag{10.85}$$

Substituting Eq. (10.84) into Eq. (10.76) gives

$$\widehat{\sigma}_{a_h}^2 = \frac{\sigma^2 \widehat{\delta}_h}{\gamma_h}. \tag{10.86}$$

Substituting Eq. (10.85) into Eq. (10.78) gives

$$\overline{\widehat{\sigma}}_{b_h}^2 = \frac{\sigma^2}{\gamma_h \widehat{\delta}_h - \phi_h \frac{\sigma^2}{\overline{c}_{b_h}^2}}, \tag{10.87}$$

where

$$\phi_h = 1 - \frac{\gamma_h^2}{\underline{\gamma}^2}.$$

Thus, the variances $\widehat{\sigma}_{a_h}^2$ and $\overline{\widehat{\sigma}}_{b_h}^2$ have been written as functions of $\widehat{\delta}_h$ and $\overline{c}_{b_h}^2$.

Substituting Eqs. (10.86) and (10.87) into Eq. (10.78) gives

$$\frac{\sigma^2}{\gamma_h \widehat{\delta}_h - \phi_h \frac{\sigma^2}{\overline{c}_{b_h}^2}} \left(\widehat{a}_h^2 + M\frac{\sigma^2 \widehat{\delta}_h}{\gamma_h} + \frac{\sigma^2 \gamma_h^2}{\overline{c}_{b_h}^2 \underline{\gamma}^2}\right) = \sigma^2,$$

and therefore

$$\widehat{\gamma}_h + \frac{M\sigma^2}{\gamma_h} - \gamma_h + \frac{\sigma^2}{\overline{c}_{b_h}^2}\widehat{\delta}_h^{-1} = 0.$$

Solving the preceding equation with respect to $\widehat{\delta}_h^{-1}$ gives

$$\widehat{\delta}_h^{-1} = \frac{\overline{c}_{b_h}^2}{\sigma^2}\left(\gamma_h - \frac{M\sigma^2}{\gamma_h} - \widehat{\gamma}_h\right). \tag{10.88}$$

Thus, we have obtained an expression of $\widehat{\delta}_h$ as a function of $\widehat{\gamma}_h$ and $\overline{c}_{b_h}^2$.

Substituting Eqs. (10.86) and (10.87) into Eq. (10.76) gives

$$\frac{\sigma^2 \widehat{\delta}_h}{\gamma_h}\left(\overline{\widehat{b}}_h^2 + J\frac{\sigma^2}{\gamma_h \widehat{\delta}_h - \phi_h \frac{\sigma^2}{\overline{c}_{b_h}^2}} + \frac{\sigma^2}{c_{a_h}^2}\right) = \sigma^2.$$

Rearranging the previous equation with respect to $\widehat{\delta}_h^{-1}$ gives

$$(\gamma_h - \widehat{\gamma}_h)\frac{\phi_h\sigma^2}{\overline{c}_{b_h}^2\gamma_h}\widehat{\delta}_h^{-2} + \left(\widehat{\gamma}_h + \frac{J\sigma^2}{\gamma_h} - \gamma_h - \frac{\phi_h\sigma^4}{c_{a_h}^2\overline{c}_{b_h}^2\gamma_h}\right)\widehat{\delta}_h^{-1} + \frac{\sigma^2}{c_{a_h}^2} = 0. \quad (10.89)$$

Substituting Eq. (10.88) into Eq. (10.89), we have

$$\frac{\phi_h}{\gamma_h}(\widehat{\gamma}_h - \gamma_h)\left(\widehat{\gamma}_h - \left(\gamma_h - \frac{M\sigma^2}{\gamma_h}\right)\right)^2 - \frac{\sigma^4}{c_{a_h}^2\overline{c}_{b_h}^2}$$
$$+\left(\widehat{\gamma}_h - \left(\gamma_h - \frac{J\sigma^2}{\gamma_h} + \frac{\phi_h\sigma^4}{c_{a_h}^2\overline{c}_{b_h}^2\gamma_h}\right)\right)\left(\widehat{\gamma}_h - \left(\gamma_h - \frac{M\sigma^2}{\gamma_h}\right)\right) = 0. \quad (10.90)$$

Thus we have derived an equation that includes only two unknown variables, $\widehat{\gamma}_h$ and $c_{a_h}^2\overline{c}_{b_h}^2$.

Next we will obtain another equation that includes only $\widehat{\gamma}_h$ and $c_{a_h}^2\overline{c}_{b_h}^2$. Substituting Eqs. (10.79) and (10.80) into Eq. (10.83), we have

$$\gamma_h^2 = \left(Mc_{a_h}^2 + \frac{\sigma^2}{\overline{c}_{b_h}^2}\right)\left(J\overline{c}_{b_h}^2 + J\phi_h\overline{\overline{\sigma}}_{b_h}^2 + \frac{\sigma^2}{c_{a_h}^2}\right). \quad (10.91)$$

Substituting Eq. (10.88) into Eq. (10.87) gives

$$\overline{\overline{\sigma}}_{b_h}^2 = \frac{\overline{c}_{b_h}^2\left(\gamma_h - \frac{M\sigma^2}{\gamma_h} - \widehat{\gamma}_h\right)}{\gamma_h - \phi_h\left(\gamma_h - \frac{M\sigma^2}{\gamma_h} - \widehat{\gamma}_h\right)}. \quad (10.92)$$

Substituting Eq. (10.92) into Eq. (10.91) gives

$$\gamma_h^2 = MJc_{a_h}^2\overline{c}_{b_h}^2 + (M+J)\sigma^2 + \frac{\sigma^4}{c_{a_h}^2\overline{c}_{b_h}^2} + J\phi_h\frac{\left(Mc_{a_h}^2\overline{c}_{b_h}^2 + \sigma^2\right)\left(\gamma_h - \frac{M\sigma^2}{\gamma_h} - \widehat{\gamma}_h\right)}{\gamma_h - \phi_h\left(\gamma_h - \frac{M\sigma^2}{\gamma_h} - \widehat{\gamma}_h\right)}.$$

Rearranging the preceding equation with respect to $c_{a_h}^2\overline{c}_{b_h}^2$, we have

$$MJc_{a_h}^4\overline{c}_{b_h}^4 + \left((M+J)\sigma^2 - \gamma_h^2 + \left(M\sigma^2 - \gamma_h^2\right)\phi_h\frac{\widehat{\widehat{\gamma}}_h}{\gamma_h}\right)c_{a_h}^2\overline{c}_{b_h}^2 + \sigma^4\left(1 + \phi_h\frac{\widehat{\widehat{\gamma}}_h}{\gamma_h}\right) = 0, \quad (10.93)$$

where

$$\widehat{\widehat{\gamma}}_h = \widehat{\gamma}_h - \left(\gamma_h - \frac{M\sigma^2}{\gamma_h}\right). \quad (10.94)$$

The solution of Eq. (10.93) with respect to $c_{a_h}^2\overline{c}_{b_h}^2$ is given by

$$c_{a_h}^2\overline{c}_{b_h}^2 = \frac{\widehat{\kappa}_h + \sqrt{\widehat{\kappa}_h^2 - 4MJ\sigma^4\left(1 + \phi_h\frac{\widehat{\widehat{\gamma}}_h}{\gamma_h}\right)}}{2MJ}, \quad (10.95)$$

where

$$\widehat{\kappa}_h = \gamma_h^2 - (M+J)\sigma^2 - \left(M\sigma^2 - \gamma_h^2\right)\phi_h \frac{\widehat{\widehat{\gamma}}_h}{\gamma_h}. \tag{10.96}$$

By using Eq. (10.94), Eq. (10.90) can be rewritten as

$$\frac{1}{\gamma_h}\phi_h\left(\widehat{\widehat{\gamma}}_h - \frac{M\sigma^2}{\gamma_h}\right)\widehat{\widehat{\gamma}}_h^2 + \left(\widehat{\widehat{\gamma}}_h - \frac{(M-J)\sigma^2}{\gamma_h}\right)\widehat{\widehat{\gamma}}_h - \left(\frac{1}{\gamma_h}\phi_h\widehat{\widehat{\gamma}}_h + 1\right)\frac{\sigma^4}{c_{a_h}^2 \bar{c}_{b_h}^2} = 0. \tag{10.97}$$

Thus, we have obtained two equations, Eqs. (10.95) and (10.97), that relate two unknown variables, $\widehat{\widehat{\gamma}}_h$ (or $\widehat{\gamma}_h$) and $c_{a_h}^2 \bar{c}_{b_h}^2$. Substituting Eq. (10.95) into Eq. (10.97) gives a polynomial equation involving only a single unknown variable $\widehat{\widehat{\gamma}}_h$. With some algebra, we obtain Eq. (10.60).

Let

$$\widehat{\tau}_h = c_{a_h}^2 \bar{c}_{b_h}^2. \tag{10.98}$$

Since we fixed the arbitrary ratio to $c_{a_h}^2 / \bar{c}_{b_h}^2 = 1$, we have

$$c_{a_h}^2 = \sqrt{\widehat{\tau}_h}, \tag{10.99}$$

$$\bar{c}_{b_h}^2 = \sqrt{\widehat{\tau}_h}. \tag{10.100}$$

Some solutions of Eq. (10.60) have no corresponding points in the problem domain (10.18). Assume that a solution $\widehat{\widehat{\gamma}}_h$ is associated with a point in the domain. Then $\widehat{\gamma}_h$ is given by Eq. (10.94), which is real and positive by its definition (10.81). $\widehat{\tau}_h$ is defined and given, respectively, by Eqs. (10.98) and (10.95), which is real and positive. $\widehat{\kappa}_h$, defined by Eq. (10.96), is also real and positive, since $\widehat{\tau}_h$ cannot be real and positive otherwise. $\widehat{\delta}_h$ is given by Eq. (10.88), which is real and positive by its definition (10.82). Finally, remembering the variable change (10.73), we can obtain Eq. (10.72) from Eqs. (10.81), (10.82), (10.86), (10.87), (10.99), and (10.100), which completes the proof of Theorem 10.7. □

## 10.8 Empirical Evaluation

In this section, we empirically compare the global solvers, EGVBS (Algorithm 20) and AGVBS (Algorithm 21), with the standard iterative algorithm (Algorithm 4 in Section 3.4.2) and its approximation (Algorithm 5 in Section 3.4.2), which we here call the standard VB (SVB) iteration and the KPCA iteration, respectively. We assume that the prior covariances $(\boldsymbol{C}_A, \boldsymbol{C}_B)$ and the noise variance $\sigma^2$ are unknown and estimated from observation. We use the full-rank model (i.e., $H = \min(L, M)$), and expect EVB learning to automatically find the true rank without any parameter tuning.

**Artificial Data Experiment**

We first conducted an experiment with a small artificial data set ("artificial small"), on which the exact algorithms, i.e., EGVBS and the SVB iteration, are computationally tractable. Through this experiment, we can assess the accuracy of the efficient approximate solvers, i.e., AGVBS and the KPCA iteration. We randomly created $M = 75$ samples in the $L = 10$ dimensional space. We assumed $K = 2$ clusters: $M^{(1)*} = 50$ samples lie in a $H^{(1)*} = 3$-dimensional subspace, and the other $M^{(2)*} = 25$ samples lie in a $H^{(2)*} = 1$-dimensional subspace. For each cluster $k$, we independently drew $M^{(k)*}$ samples from $\mathrm{Gauss}_{H^{(k)*}}(\mathbf{0}, 10 \cdot \boldsymbol{I}_{H^{(k)*}})$, and projected them onto the observed $L$-dimensional space by $\boldsymbol{R}^{(k)} \in \mathbb{R}^{L \times H^{(k)*}}$, each entry of which follows $R^{(k)}_{l,h} \sim \mathrm{Gauss}_1(0, 1)$. Thus, we obtained a noiseless matrix $\boldsymbol{V}^{(k)*} \in \mathbb{R}^{L \times M^{(k)*}}$ for the $k$th cluster. Concatenating all clusters, $\boldsymbol{V}^* = (\boldsymbol{V}^{(1)*}, \ldots, \boldsymbol{V}^{(K)*})$, and adding random noise subject to $\mathrm{Gauss}_1(0, 1)$ to each entry gave an artificial observed matrix $\boldsymbol{V} \in \mathbb{R}^{L \times M}$, where $M = \sum_{k=1}^{K} M^{(k)*} = 75$. The *true* rank of $\boldsymbol{V}^*$ is given by $H^* = \min(\sum_{k=1}^{K} H^{(k)*}, L, M) = 4$. Note that $H^*$ is different from the rank of the observed matrix $\boldsymbol{V}$, which is almost surely equal to $J = \min(L, M)$ $(= 10)$ under the Gaussian noise.

Figure 10.1 shows the free energy, the computation time, and the estimated rank of $\widehat{\boldsymbol{U}} = \widehat{\boldsymbol{B}}' \widehat{\boldsymbol{A}}'^{\top}$ over iterations. For the iterative methods, we show the results of 10 trials starting from different random initializations. We can see that AGVBS gives almost the same free energy as the exact methods (EGVBS and the SVB iteration). The exact methods require large computation costs: EGVBS took 621 sec to obtain the global solution, and the SVB iteration took $\sim$ 100 sec to achieve almost the same free energy. On the other hand, the approximate methods are much faster: AGVBS took less than 1 sec, and the KPCA iteration took ~10 sec. Since the KPCA iteration had not converged after 250 iterations, we continued its computation until 2,500 iterations, and found that it sometimes converges to a local solution with a significantly higher free energy than the other methods. EGVBS, AGVBS, and the SVB iteration successfully found the *true* rank $H^* = 4$, while the KPCA iteration sometimes failed to find it. This difference is actually reflected to the clustering error, i.e., the misclassification rate with all possible cluster correspondences taken into account, after spectral clustering (Shi and Malik, 2000) is performed: 1.3% for EGVBS, AGVBS, and the SVB iteration, and 2.4% for the KPCA iteration.

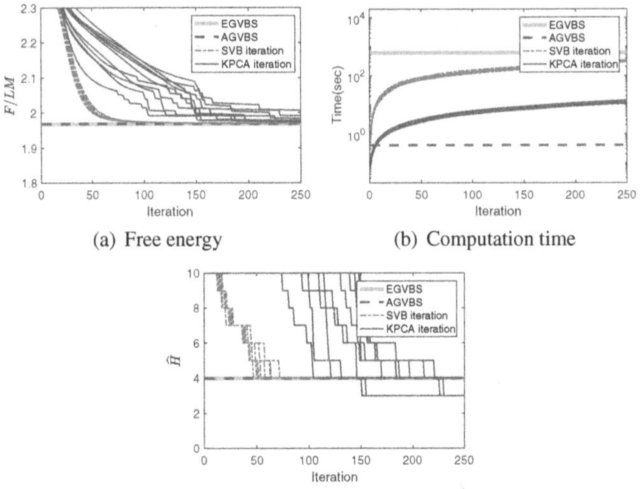


(a) Free energy

(b) Computation time

(c) Estimated rank

Figure 10.1 Results on the “artificial small” data set ($L = 10, M = 75, H^* = 4$). The clustering errors were 1.3% for EGVBS, AGVBS, and the SVB iteration, and 2.4% for the KPCA iteration.

Next we conducted the same experiment with a larger artificial data set (“artificial large”) ($L = 50, K = 4, (M^{(1)*}, \ldots, M^{(K)*}) = (100, 50, 50, 25)$, $(H^{(1)*}, \ldots, H^{(K)*}) = (2, 1, 1, 1)$), on which EGVBS and the SVB iteration are computationally intractable. Figure 10.2 shows the results with AGVBS and the KPCA iteration. The advantage in computation time is clear: AGVBS only took ~0.1 sec, while the KPCA iteration took more than 100 sec. The clustering errors were 4.0% for AGVBS and 11.2% for the KPCA iteration.

**Benchmark Data Experiment**

Finally, we applied AGVBS and the KPCA iteration to the *Hopkins 155 motion* database (Tron and Vidal, 2007). In this data set, each sample corresponds to the trajectory of a point in a video, and clustering the trajectories amounts to finding a set of rigid bodies. Figure 10.3 shows the results on the “1R2RC” ($L = 59, M = 459$) sequence.[1] We see that AGVBS gave a lower free energy with much less computation time than the KPCA iteration. Figure 10.4 shows the clustering errors on the first 20 sequences, which implies that AGVBS generally outperforms the KPCA iteration. Figure 10.4 also shows the results

[1] Peaks in the free energy curves are due to pruning. As noted in Section 3.1.1, the free energy can increase right after pruning happens, but immediately gets lower than the free energy before pruning.

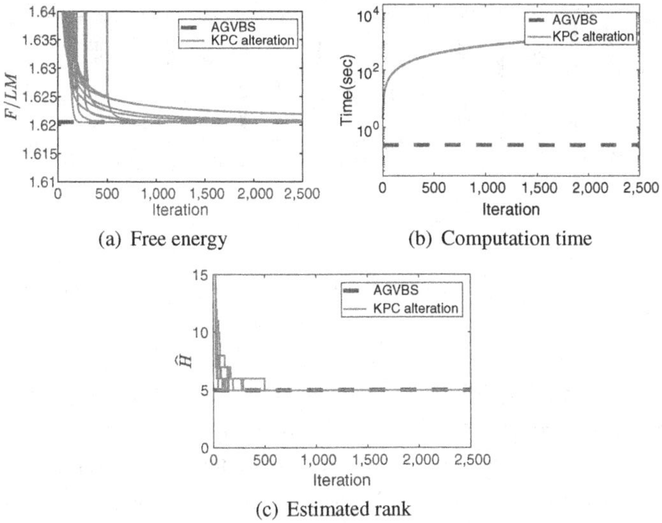


(a) Free energy (b) Computation time

(c) Estimated rank

Figure 10.2 Results on the "artificial large" data set ($L = 50, M = 225, H^* = 5$). The clustering errors were 4.0% for AGVBS and 11.2% for the KPCA iteration.

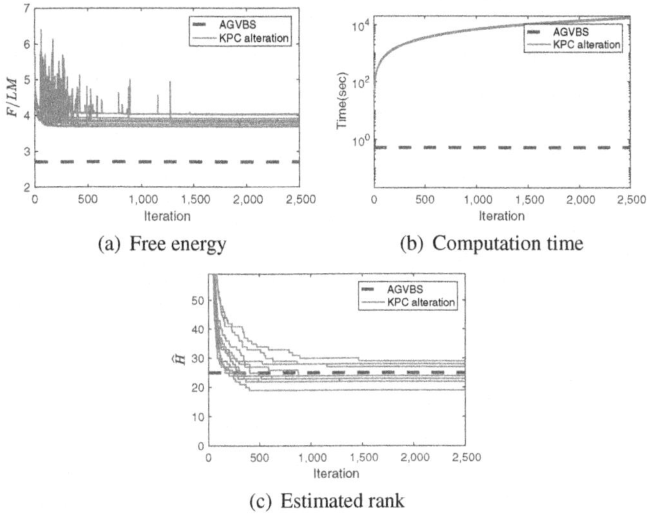


(a) Free energy (b) Computation time

(c) Estimated rank

Figure 10.3 Results on the "1R2RC" sequence ($L = 59, M = 459$) of the Hopkins 155 motion database. Peaks in the free energy curves are due to pruning. The clustering errors are shown in Figure 10.4.

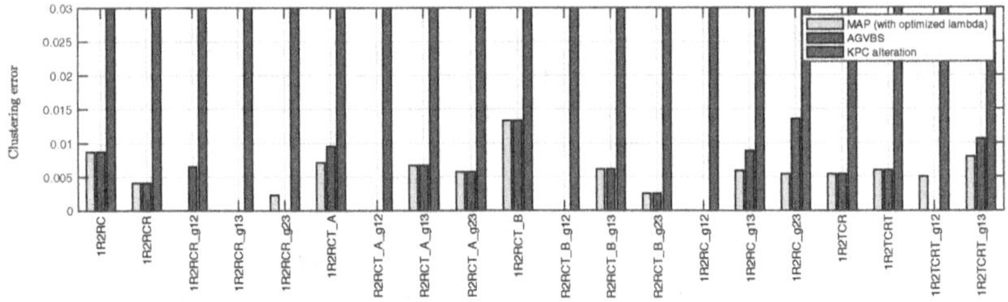


Figure 10.4 Clustering errors on the first 20 sequences of the Hopkins 155 data set.

by MAP learning (Eq. (3.87) in Section 3.4) with the tuning parameter $\lambda$ optimized over the 20 sequences (i.e., we performed MAP learning with different values for $\lambda$, and selected the one giving the lowest average clustering error). We see that AGVBS performs comparably to MAP learning with optimized $\lambda$, which implies that EVB learning estimates the hyperparameters and the noise variance reasonably well.

# 11

# Efficient Solver for Sparse Additive Matrix Factorization

In this chapter, we introduce an efficient variational Bayesian (VB) solver (Nakajima et al., 2013b) for sparse additive matrix factorization (SAMF), where the global VB solver, derived in Chapter 9, for fully observed MF is used as a subroutine.

## 11.1 Problem Description

The SAMF model, introduced in Section 3.5, is defined as

$$p(\boldsymbol{V}|\boldsymbol{\Theta}) \propto \exp\left(-\frac{1}{2\sigma^2}\left\|\boldsymbol{V} - \sum_{s=1}^{S}\boldsymbol{U}^{(s)}\right\|_{\mathrm{Fro}}^2\right), \tag{11.1}$$

$$p(\{\boldsymbol{\Theta}_A^{(s)}\}_{s=1}^S) \propto \exp\left(-\frac{1}{2}\sum_{s=1}^{S}\sum_{k=1}^{K^{(s)}}\mathrm{tr}\left(\boldsymbol{A}^{(k,s)}\boldsymbol{C}_A^{(k,s)-1}\boldsymbol{A}^{(k,s)\top}\right)\right), \tag{11.2}$$

$$p(\{\boldsymbol{\Theta}_B^{(s)}\}_{s=1}^S) \propto \exp\left(-\frac{1}{2}\sum_{s=1}^{S}\sum_{k=1}^{K^{(s)}}\mathrm{tr}\left(\boldsymbol{B}^{(k,s)}\boldsymbol{C}_B^{(k,s)-1}\boldsymbol{B}^{(k,s)\top}\right)\right), \tag{11.3}$$

where

$$\boldsymbol{U}^{(s)} = \boldsymbol{G}(\{\boldsymbol{B}^{(k,s)}\boldsymbol{A}^{(k,s)\top}\}_{k=1}^{K^{(s)}}; \mathcal{X}^{(s)}) \tag{11.4}$$

is the $s$th sparse matrix factorization (SMF) term. Here $\boldsymbol{G}(\cdot;\mathcal{X}) : \mathbb{R}^{\prod_{k=1}^{K}(L'^{(k)}\times M'^{(k)})} \mapsto \mathbb{R}^{L\times M}$ maps the partitioned-and-rearranged (PR) matrices $\{\boldsymbol{U}'^{(k)}\}_{k=1}^K$ to the target matrix $\boldsymbol{U} \in \mathbb{R}^{L\times M}$, based on the one-to-one map $\mathcal{X} : (k, l', m') \mapsto (l, m)$ from the indices of the entries in $\{\boldsymbol{U}'^{(k)}\}_{k=1}^K$ to the indices of the entries in $\boldsymbol{U}$ such that

$$\left(\boldsymbol{G}(\{\boldsymbol{U}'^{(k)}\}_{k=1}^K; \mathcal{X})\right)_{l,m} = U_{l,m} = U_{\mathcal{X}(k,l',m')} = U'^{(k)}_{l',m'}. \tag{11.5}$$

The prior covariances of $\boldsymbol{A}^{(k,s)}$ and $\boldsymbol{B}^{(k,s)}$ are assumed to be diagonal and positive-definite:

$$\boldsymbol{C}_A^{(k,s)} = \mathbf{Diag}(c_{a_1}^{(k,s)2}, \ldots, c_{a_H}^{(k,s)2}),$$
$$\boldsymbol{C}_B^{(k,s)} = \mathbf{Diag}(c_{b_1}^{(k,s)2}, \ldots, c_{b_H}^{(k,s)2}),$$

and $\boldsymbol{\Theta}$ summarizes the parameters as follows:

$$\boldsymbol{\Theta} = \{\boldsymbol{\Theta}_A^{(s)}, \boldsymbol{\Theta}_B^{(s)}\}_{s=1}^S, \text{ where } \boldsymbol{\Theta}_A^{(s)} = \{\boldsymbol{A}^{(k,s)}\}_{k=1}^{K^{(s)}}, \boldsymbol{\Theta}_B^{(s)} = \{\boldsymbol{B}^{(k,s)}\}_{k=1}^{K^{(s)}}.$$

Under the independence constraint,

$$r(\boldsymbol{\Theta}) = \prod_{s=1}^{S} r_A^{(s)}(\boldsymbol{\Theta}_A^{(s)}) r_B^{(s)}(\boldsymbol{\Theta}_B^{(s)}), \tag{11.6}$$

the VB posterior minimizing the free energy can be written as

$$\begin{aligned} r(\boldsymbol{\Theta}) &= \prod_{s=1}^{S}\prod_{k=1}^{K^{(s)}} \Bigg( \mathrm{MGauss}_{M'^{(k,s)},H'^{(k,s)}}(\boldsymbol{A}^{(k,s)}; \widehat{\boldsymbol{A}}^{(k,s)}, \widehat{\boldsymbol{\Sigma}}_A^{(k,s)}) \\ &\qquad \cdot \mathrm{MGauss}_{L'^{(k,s)},H'^{(k,s)}}(\boldsymbol{B}^{(k,s)}; \widehat{\boldsymbol{B}}^{(k,s)}, \widehat{\boldsymbol{\Sigma}}_B^{(k,s)}) \Bigg) \\ &= \prod_{s=1}^{S}\prod_{k=1}^{K^{(s)}} \Bigg( \prod_{m'=1}^{M'^{(k,s)}} \mathrm{Gauss}_{H'^{(k,s)}}(\widetilde{\boldsymbol{a}}_{m'}^{(k,s)}; \widehat{\widetilde{\boldsymbol{a}}}_{m'}^{(k,s)}, \widehat{\boldsymbol{\Sigma}}_A^{(k,s)}) \\ &\qquad \cdot \prod_{l'=1}^{L'^{(k,s)}} \mathrm{Gauss}_{H'^{(k,s)}}(\widetilde{\boldsymbol{b}}_{l'}^{(k,s)}; \widehat{\widetilde{\boldsymbol{b}}}_{l'}^{(k,s)}, \widehat{\boldsymbol{\Sigma}}_B^{(k,s)}) \Bigg). \end{aligned} \tag{11.7}$$

The free energy can be explicitly written as

$$\begin{aligned} 2F &= LM\log(2\pi\sigma^2) + \frac{\|\boldsymbol{V}\|_{\mathrm{Fro}}^2}{\sigma^2} \\ &+ \sum_{s=1}^{S}\sum_{k=1}^{K^{(s)}} \left( M'^{(k,s)} \log \frac{\det\left(\boldsymbol{C}_A^{(k,s)}\right)}{\det\left(\widehat{\boldsymbol{\Sigma}}_A^{(k,s)}\right)} + L'^{(k,s)} \log \frac{\det\left(\boldsymbol{C}_B^{(k,s)}\right)}{\det\left(\widehat{\boldsymbol{\Sigma}}_B^{(k,s)}\right)} \right) \\ &+ \sum_{s=1}^{S}\sum_{k=1}^{K^{(S)}} \mathrm{tr}\Big\{ \boldsymbol{C}_A^{(k,s)-1} (\widehat{\boldsymbol{A}}^{(k,s)\top}\widehat{\boldsymbol{A}}^{(k,s)} + M'^{(k,s)}\widehat{\boldsymbol{\Sigma}}_A^{(k,s)}) \\ &+ \boldsymbol{C}_B^{(k,s)-1} (\widehat{\boldsymbol{B}}^{(k,s)\top}\widehat{\boldsymbol{B}}^{(k,s)} + L'^{(k,s)}\widehat{\boldsymbol{\Sigma}}_B^{(k,s)}) \Big\} \\ &+ \frac{1}{\sigma^2}\mathrm{tr}\left\{ -2\boldsymbol{V}^\top \left( \sum_{s=1}^{S} \boldsymbol{G}(\{\widehat{\boldsymbol{B}}^{(k,s)}\widehat{\boldsymbol{A}}^{(k,s)\top}\}_{k=1}^{K^{(s)}}; \mathcal{X}^{(s)}) \right) \right. \end{aligned}$$

$$+2\sum_{s=1}^{S}\sum_{s'=s+1}^{S} G^{\top}(\{\widehat{B}^{(k,s)}\widehat{A}^{(k,s)\top}\}_{k=1}^{K^{(s)}};\mathcal{X}^{(s)})G(\{\widehat{B}^{(k,s')}\widehat{A}^{(k,s')\top}\}_{k=1}^{K^{(s')}};\mathcal{X}^{(s')})\Bigg\}$$

$$+\frac{1}{\sigma^2}\sum_{s=1}^{S}\sum_{k=1}^{K^{(S)}}\mathrm{tr}\left\{(\widehat{A}^{(k,s)\top}\widehat{A}^{(k,s)}+M'^{(k,s)}\widehat{\Sigma}_A^{(k,s)})(\widehat{B}^{(k,s)\top}\widehat{B}^{(k,s)}+L'^{(k,s)}\widehat{\Sigma}_B^{(k,s)})\right\}$$

$$-\sum_{s=1}^{S}\sum_{k=1}^{K^{(S)}}(L'^{(k,s)}+M'^{(k,s)})H'^{(k,s)}, \tag{11.8}$$

of which the stationary conditions are given by

$$\widehat{A}^{(k,s)} = \sigma^{-2}Z'^{(k,s)\top}\widehat{B}^{(k,s)}\widehat{\Sigma}_A^{(k,s)}, \tag{11.9}$$

$$\widehat{\Sigma}_A^{(k,s)} = \sigma^2\left(\widehat{B}^{(k,s)\top}\widehat{B}^{(k,s)}+L'^{(k,s)}\widehat{\Sigma}_B^{(k,s)}+\sigma^2 C_A^{(k,s)-1}\right)^{-1}, \tag{11.10}$$

$$\widehat{B}^{(k,s)} = \sigma^{-2}Z'^{(k,s)}\widehat{A}^{(k,s)}\widehat{\Sigma}_B^{(k,s)}, \tag{11.11}$$

$$\widehat{\Sigma}_B^{(k,s)} = \sigma^2\left(\widehat{A}^{(k,s)\top}\widehat{A}^{(k,s)}+M'^{(k,s)}\widehat{\Sigma}_A^{(k,s)}+\sigma^2 C_B^{(k,s)-1}\right)^{-1}. \tag{11.12}$$

Here $Z'^{(k,s)} \in \mathbb{R}^{L'^{(k,s)}\times M'^{(k,s)}}$ is defined as

$$Z'^{(k,s)}_{l',m'} = Z^{(s)}_{\mathcal{X}^{(s)}(k,l',m')}, \quad \text{where} \quad Z^{(s)} = V - \sum_{s'\neq s}\widehat{U}^{(s)}. \tag{11.13}$$

The stationary conditions for the hyperparameters $\{C_A^{(k,s)}, C_B^{(k,s)}\}_{k=1,s=1}^{K^{(s)}\ S}, \sigma^2$ are given as

$$c_{a_h}^{(k,s)2} = \left\|\widehat{a}_h^{(k,s)}\right\|^2/M'^{(k,s)} + (\widehat{\Sigma}_A^{(k,s)})_{hh}, \tag{11.14}$$

$$c_{b_h}^{(k,s)2} = \left\|\widehat{b}_h^{(k,s)}\right\|^2/L'^{(k,s)} + (\widehat{\Sigma}_B^{(k,s)})_{hh}, \tag{11.15}$$

$$\sigma^2 = \frac{1}{LM}\Bigg\{\|V\|_{\mathrm{Fro}}^2 - 2\sum_{s=1}^{S}\mathrm{tr}\left(\widehat{U}^{(s)\top}\left(V-\sum_{s'=s+1}^{S}\widehat{U}^{(s')}\right)\right)$$

$$+\sum_{s=1}^{S}\sum_{k=1}^{K^{(s)}}\mathrm{tr}\Big((\widehat{A}^{(k,s)\top}\widehat{A}^{(k,s)}+M'^{(k,s)}\widehat{\Sigma}_{\mathrm{A}}^{(k,s)})\cdot(\widehat{B}^{(k,s)\top}\widehat{B}^{(k,s)}+L'^{(k,s)}\widehat{\Sigma}_{\mathrm{B}}^{(k,s)})\Big)\Bigg\}. \tag{11.16}$$

The standard VB algorithm (Algorithm 6 in Section 3.5) iteratively applies Eqs. (11.9) through (11.12) and (11.14) through (11.16) until convergence.

## 11.2 Efficient Algorithm for SAMF

In this section, we derive a more efficient algorithm than the standard VB algorithm. We first present a theorem that reduces a partial SAMF problem to the (fully observed) MF problem, which can be solved analytically. Then we describe the algorithm that solves the entire SAMF problem.

### 11.2.1 Reduction of the Partial SAMF Problem to the MF Problem

Let us denote the mean of $\boldsymbol{U}^{(s)}$, defined in Eq. (11.4), over the VB posterior by

$$\widehat{\boldsymbol{U}}^{(s)} = \left\langle \boldsymbol{U}^{(s)} \right\rangle_{r_A^{(s)}(\boldsymbol{\Theta}_A^{(s)}) r_B^{(s)}(\boldsymbol{\Theta}_B^{(s)})} = \boldsymbol{G}\left( \left\{ \widehat{\boldsymbol{B}}^{(k,s)} \widehat{\boldsymbol{A}}^{(k,s)\top} \right\}_{k=1}^{K^{(s)}} ; \mathcal{X}^{(s)} \right). \tag{11.17}$$

Then we obtain the following theorem:

**Theorem 11.1** *Given* $\{\widehat{\boldsymbol{U}}^{(s')}\}_{s' \neq s}$ *and the noise variance* $\sigma^2$, *the VB posterior of* $(\boldsymbol{\Theta}_A^{(s)}, \boldsymbol{\Theta}_B^{(s)}) = \{\boldsymbol{A}^{(k,s)}, \boldsymbol{B}^{(k,s)}\}_{k=1}^{K^{(s)}}$ *coincides with the VB posterior of the following MF model:*

$$p(\boldsymbol{Z}'^{(k,s)} | \boldsymbol{A}^{(k,s)}, \boldsymbol{B}^{(k,s)}) \propto \exp\left( -\frac{1}{2\sigma^2} \left\| \boldsymbol{Z}'^{(k,s)} - \boldsymbol{B}^{(k,s)} \boldsymbol{A}^{(k,s)\top} \right\|_{\mathrm{Fro}}^2 \right), \tag{11.18}$$

$$p(\boldsymbol{A}^{(k,s)}) \propto \exp\left( -\frac{1}{2} \mathrm{tr}\left( \boldsymbol{A}^{(k,s)} \boldsymbol{C}_A^{(k,s)-1} \boldsymbol{A}^{(k,s)\top} \right) \right), \tag{11.19}$$

$$p(\boldsymbol{B}^{(k,s)}) \propto \exp\left( -\frac{1}{2} \mathrm{tr}\left( \boldsymbol{B}^{(k,s)} \boldsymbol{C}_B^{(k,s)-1} \boldsymbol{B}^{(k,s)\top} \right) \right), \tag{11.20}$$

*for each* $k = 1, \ldots, K^{(s)}$. *Here,* $\boldsymbol{Z}'^{(k,s)} \in \mathbb{R}^{L'^{(k,s)} \times M'^{(k,s)}}$ *is defined by Eq.* (11.13).

*Proof* Given $\{\widehat{\boldsymbol{U}}^{(s)}\}_{s' \neq s} = \{\{\widehat{\boldsymbol{B}}^{(k,s')} \widehat{\boldsymbol{A}}^{(k,s')\top}\}_{k=1}^{K^{(s')}}\}_{s' \neq s}$ and $\sigma^2$ as fixed constants, the free energy (11.8) can be written as a function of $\{\widehat{\boldsymbol{A}}^{(k,s)}, \widehat{\boldsymbol{B}}^{(k,s)}, \widehat{\boldsymbol{\Sigma}}_A^{(k,s)}, \widehat{\boldsymbol{\Sigma}}_B^{(k,s)}, \boldsymbol{C}_A^{(k,s)}, \boldsymbol{C}_B^{(k,s)}\}_{k=1}^{K^{(s)}}$ as follows:

$$2F^{(s)}\left( \{\widehat{\boldsymbol{A}}^{(k,s)}, \widehat{\boldsymbol{B}}^{(k,s)}, \widehat{\boldsymbol{\Sigma}}_A^{(k,s)}, \widehat{\boldsymbol{\Sigma}}_B^{(k,s)}, \boldsymbol{C}_A^{(k,s)}, \boldsymbol{C}_B^{(k,s)}\}_{k=1}^{K^{(s)}} \right) = \sum_{k=1}^{K^{(s)}} 2F^{(k,s)} + \text{const.}, \tag{11.21}$$

where

$$2F^{(k,s)} = M'^{(k,s)} \log \frac{\det\left(\boldsymbol{C}_A^{(k,s)}\right)}{\det\left(\widehat{\boldsymbol{\Sigma}}_A^{(k,s)}\right)} + L'^{(k,s)} \log \frac{\det\left(\boldsymbol{C}_B^{(k,s)}\right)}{\det\left(\widehat{\boldsymbol{\Sigma}}_B^{(k,s)}\right)}$$
$$+ \mathrm{tr}\left\{ \boldsymbol{C}_A^{(k,s)-1} (\widehat{\boldsymbol{A}}^{(k,s)\top} \widehat{\boldsymbol{A}}^{(k,s)} + M'^{(k,s)} \widehat{\boldsymbol{\Sigma}}_A^{(k,s)}) + \boldsymbol{C}_B^{(k,s)-1} (\widehat{\boldsymbol{B}}^{(k,s)\top} \widehat{\boldsymbol{B}}^{(k,s)} + L'^{(k,s)} \widehat{\boldsymbol{\Sigma}}_B^{(k,s)}) \right\}$$

$$+\frac{1}{\sigma^2}\mathrm{tr}\Big\{-2\widehat{\boldsymbol{A}}^{(k,s)\top}\boldsymbol{Z}'^{(k,s)\top}\widehat{\boldsymbol{B}}^{(k,s)}$$
$$+(\widehat{\boldsymbol{A}}^{(k,s)\top}\widehat{\boldsymbol{A}}^{(k,s)}+M'^{(k,s)}\widehat{\boldsymbol{\Sigma}}_A^{(k,s)})(\widehat{\boldsymbol{B}}^{(k,s)\top}\widehat{\boldsymbol{B}}^{(k,s)}+L'^{(k,s)}\widehat{\boldsymbol{\Sigma}}_B^{(k,s)})\Big\}. \qquad (11.22)$$

Eq. (11.22) coincides with the free energy of the fully observed matrix factorization model (11.18) through (11.20) up to a constant (see Eq. (3.23) with $\boldsymbol{Z}'$ substituted for $\boldsymbol{V}$). Therefore, the VB solution is the same. □

Eq. (11.13) relates the entries of $\boldsymbol{Z}^{(s)} \in \mathbb{R}^{L\times M}$ to the entries of $\{\boldsymbol{Z}'^{(k,s)} \in \mathbb{R}^{L'^{(k,s)}\times M'^{(k,s)}}\}_{k=1}^{K^{(s)}}$ by using the map $\mathcal{X}^{(s)} : (k, l', m') \mapsto (l, m)$ (see Eq. (11.5) and Figure 3.3).

### 11.2.2 Mean Update Algorithm

Theorem 11.1 states that a partial problem of SAMF—finding the posterior of $(\boldsymbol{A}^{(k,s)}, \boldsymbol{B}^{(k,s)})$ for each $k = 1, \ldots, K(s)$ given $\{\widehat{\boldsymbol{U}}^{(s')}\}_{s'\neq s}$ and $\sigma^2$—can be solved by the global solver for the fully observed MF model. Specifically, we use Algorithm 16, introduced in Chapter 9, for estimating each SMF term $\widehat{\boldsymbol{U}}^{(s)}$ in turn. We use Eq. (11.16) for updating the noise variance $\sigma^2$. The whole procedure, called the *mean update (MU) algorithm* (Nakajima et al., 2013b), is summarized in Algorithm 22, where $\boldsymbol{0}_{(d_1,d_2)}$ denotes the $d_1 \times d_2$ matrix with all entries equal to zero.

The MU algorithm is similar in spirit to the *backfitting algorithm* (Hastie and Tibshirani, 1986; D'Souza et al., 2004), where each additive term is updated to fit a dummy target. In the MU algorithm, $\boldsymbol{Z}^{(s)}$ defined in Eq. (11.13) corresponds to the dummy target. Although the MU algorithm globally solves a partial problem in each step, its joint global optimality over the entire

---

**Algorithm 22** Mean update (MU) algorithm for VB SAMF.

1: Initialize: $\widehat{\boldsymbol{U}}^{(s)} \leftarrow \boldsymbol{0}_{(L,M)}$ for $s = 1, \ldots, S$, $\sigma^2 \leftarrow \|\boldsymbol{V}\|_{\mathrm{Fro}}^2/(LM)$.
2: **for** $s = 1$ to $S$ **do**
3: Compute $\boldsymbol{Z}'^{(k,s)} \in \mathbb{R}^{L'^{(k,s)}\times M'^{(k,s)}}$ by Eq. (11.13).
4: For each partition $k = 1, \ldots, K^{(s)}$, compute the solution $\boldsymbol{U}'^{(k,s)} = \boldsymbol{B}^{(k,s)}\boldsymbol{A}^{(k,s)\top}$ for the fully observed MF by Algorithm 16 with $\boldsymbol{Z}'^{(k,s)}$ as the observed matrix.
5: $\widehat{\boldsymbol{U}}^{(s)} \leftarrow \boldsymbol{G}(\{\widehat{\boldsymbol{B}}^{(k,s)}\widehat{\boldsymbol{A}}^{(k,s)\top}\}_{k=1}^{K^{(s)}}; \mathcal{X}^{(s)})$.
6: **end for**
7: Update $\sigma^2$ by Eq. (11.16).
8: Repeat 2 to 7 until convergence.

---

parameter space is not guaranteed. Nevertheless, experimental results in Section 11.3 show that the MU algorithm performs well in practice.

When Algorithm 16 is applied to the dummy target matrix $Z'^{(k,s)} \in \mathbb{R}^{L'^{(k,s)} \times M'^{(k,s)}}$ in Step 4, singular value decomposition is required, which dominates the computation time. However, for many practical SMF terms, including the rowwise (3.114), the columnwise (3.115), and the elementwise (3.116) terms (as well as the segmentwise term, which will be defined for a video application in Section 11.3.2), $Z'^{(k,s)}$ is a vector or scalar, i.e., $L'^{(k,s)} = 1$ or $M'^{(k,s)} = 1$ holds. In such cases, the singular value and the singular vectors are given simply by

$$\gamma_1^{(k,s)} = \|Z'^{(k,s)}\|, \quad \omega_{a_1}^{(k,s)} = Z'^{(k,s)}/\|Z'^{(k,s)}\|, \quad \omega_{b_1}^{(k,s)} = 1 \qquad \text{if } L'^{(k,s)} = 1,$$
$$\gamma_1^{(k,s)} = \|Z'^{(k,s)}\|, \quad \omega_{a_1}^{(k,s)} = 1, \quad \omega_{b_1}^{(k,s)} = Z'^{(k,s)}/\|Z'^{(k,s)}\| \qquad \text{if } M'^{(k,s)} = 1.$$

## 11.3 Experimental Results

In this section, we experimentally show good performance of the MU algorithm (Algorithm 22) over the standard VB algorithm (Algorithm 6 in Section 3.5). We also demonstrate advantages of SAMF in its flexibility in a real-world application.

### 11.3.1 Mean Update vs. Standard VB

We compare the algorithms under the following model:

$$V = U^{\mathrm{LRCE}} + \mathcal{E},$$

where

$$U^{\mathrm{LRCE}} = \sum_{s=1}^{4} U^{(s)} = U^{\text{low-rank}} + U^{\text{row}} + U^{\text{column}} + U^{\text{element}}. \tag{11.23}$$

Here, "LRCE" stands for the sum of the low-rank, rowwise, columnwise, and elementwise terms, each of which is defined in Eqs. (3.113) through (3.116). We call this model "LRCE"-SAMF. As explained in Section 3.5, "LRCE"-SAMF may be used to separate the clean signal $U^{\text{low-rank}}$ from a possible rowwise sparse component (constantly broken sensors), a columnwise sparse component (accidental disturbances affecting all sensors), and an elementwise sparse component (randomly distributed spiky noise). We also evaluate "LCE"-SAMF, "LRE"-SAMF, and "LE"-SAMF, which can be regarded as generalizations of robust PCA (Candès et al., 2011; Ding et al., 2011; Babacan et al., 2012b). Note that "LE"-SAMF corresponds to an SAMF counterpart of robust PCA.

First, we conducted an experiment with artificial data. We assume the empirical VB scenario with unknown noise variance, i.e., all hyperparameters, $\{C_A^{(k,s)}, C_B^{(k,s)}\}_{k=1,s=1}^{K^{(s)}\ S}$, and $\sigma^2$, are estimated from observations. We use the full-rank model ($H = \min(L, M)$) for the low-rank term $U^{\text{low-rank}}$, and expect the model-induced regularization (MIR) effect (see Chapter 7) to find the true rank of $U^{\text{low-rank}}$, as well as the nonzero entries in $U^{\text{row}}$, $U^{\text{column}}$, and $U^{\text{element}}$.

We created an artificial data set with the data matrix size $L = 40$ and $M = 100$, and the rank $H^* = 10$ for a *true* low-rank matrix $U^{\text{low-rank}*} = B^* A^{*\top}$. Each entry in $A^* \in \mathbb{R}^{L \times H^*}$ and $B^* \in \mathbb{R}^{L \times H^*}$ was drawn from $\text{Gauss}_1(0, 1)$. A *true* rowwise (columnwise) part $U^{\text{row}*}$ ($U^{\text{column}*}$) was created by first randomly selecting $\rho L$ rows ($\rho M$ columns) for $\rho = 0.05$, and then adding a noise subject to $\text{Gauss}_M(\mathbf{0}, \zeta I_M)$ ($\text{Gauss}_L(\mathbf{0}, \zeta I_L)$) for $\zeta = 100$ to each of the selected rows (columns). A *true* elementwise part $U^{\text{element}*}$ was similarly created by first selecting $\rho LM$ entries and then adding a noise subject to $\text{Gauss}_1(0, \zeta)$ to each of the selected entries. Finally, an observed matrix $V$ was created by adding a noise subject to $\text{Gauss}_1(0, 1)$ to each entry of the sum $U^{\text{LRCE}*}$ of the aforementioned four *true* matrices.

For the standard VB algorithm, we initialized the variational parameters and the hyperparameters in the following way: the mean parameters, $\{\widehat{A}^{(k,s)}, \widehat{B}^{(k,s)}\}_{k=1,s=1}^{K^{(s)}\ S}$, were randomly created so that each entry follows $\text{Gauss}_1(0, 1)$; the covariances, $\{\widehat{\Sigma}_A^{(k,s)}, \widehat{\Sigma}_B^{(k,s)}\}_{k=1,s=1}^{K^{(s)}\ S}$ and $\{C_A^{(k,s)}, C_B^{(k,s)}\}_{k=1,s=1}^{K^{(s)}\ S}$, were set to be identity; and the noise variance was set to $\sigma^2 = 1$. Note that we rescaled $V$ so that $\|V\|_{\text{Fro}}^2 / (LM) = 1$, before starting iteration. We ran the standard VB algorithm 10 times, starting from different initial points, and each trial is plotted by a solid line (labeled as "Standard(iniRan)") in Figure 11.1.

Initialization for the MU algorithm is simple: we simply set $\widehat{U}^{(s)} = \mathbf{0}_{(L,M)}$ for $s = 1, \ldots, S$, and $\sigma^2 = 1$. Initialization of all other variables is not needed. Furthermore, we empirically observed that the initial value for $\sigma^2$ does not affect the result much, unless it is too small. Actually, initializing $\sigma^2$ to a large value is not harmful in the MU algorithm, because it is set to an adequate value after the first iteration with the mean parameters kept $\widehat{U}^{(s)} = \mathbf{0}_{(L,M)}$. The performance of the MU algorithm is plotted by the dashed line in Figure 11.1.

Figures 11.1(a) through 11.1(c) show the free energy, the computation time, and the estimated rank, respectively, over iterations, and Figure 11.1(d) shows the reconstruction errors after 250 iterations. The reconstruction errors consist of the *overall* error $\left\|\widehat{U}^{\text{LRCE}} - U^{\text{LRCE}*}\right\|_{\text{Fro}}^2 / (LM)$, and the four componentwise

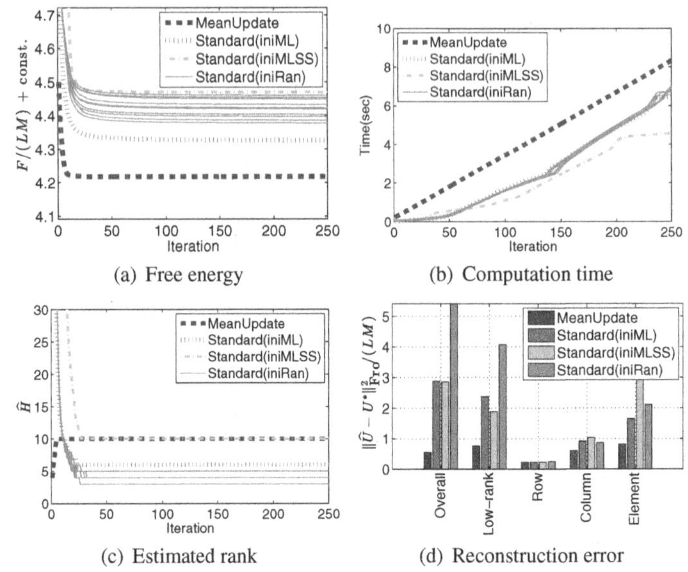


(a) Free energy (b) Computation time

(c) Estimated rank (d) Reconstruction error

Figure 11.1 Experimental results of "LRCE"-SAMF on an artificial data set ($L = 40, M = 100, H^* = 10, \rho = 0.05$).

errors $\left\| \widehat{U}^{(s)} - U^{(s)*} \right\|_{\mathrm{Fro}}^2 /(LM)$. The graphs show that the MU algorithm, whose iteration is computationally slightly more expensive than the standard VB algorithm, immediately converges to a local minimum with the free energy substantially lower than the standard VB algorithm. The estimated rank agrees with the true rank $\widehat{H} = H^* = 10$, while all 10 trials of the standard VB algorithm failed to estimate the true rank. It is also observed that the MU algorithm well reconstructs each of the four terms.

We can slightly improve the performance of the standard VB algorithm by adopting different initialization schemes. The line labeled as "Standard(iniML)" in Figure 11.1 indicates the maximum likelihood (ML) initialization, i.e, $(\widehat{\boldsymbol{a}}_h^{(k,s)}, \widehat{\boldsymbol{b}}_h^{(k,s)}) = (\gamma_h^{(k,s)1/2} \boldsymbol{\omega}_{a_h}^{(k,s)}, \gamma_h^{(k,s)1/2} \boldsymbol{\omega}_{b_h}^{(k,s)})$. Here, $\gamma_h^{(k,s)}$ is the $h$th largest singular value of the $(k, s)$th PR matrix $\boldsymbol{V}'^{(k,s)}$ of $\boldsymbol{V}$ (such that $V_{l',m'}'^{(k,s)} = V_{\mathcal{X}^{(s)}(k,l',m')}$), and $\boldsymbol{\omega}_{a_h}^{(k,s)}$ and $\boldsymbol{\omega}_{b_h}^{(k,s)}$ are the associated right and left singular vectors. Also, we empirically found that starting from small $\sigma^2$ alleviates the local minimum problem. The line labeled as "Standard (iniMLSS)" indicates the ML initialization with $\sigma^2 = 0.0001$. We can see that this scheme successfully recovered the true rank. However, it still performs substantially worse than the MU algorithm in terms of the free energy and the reconstruction error.

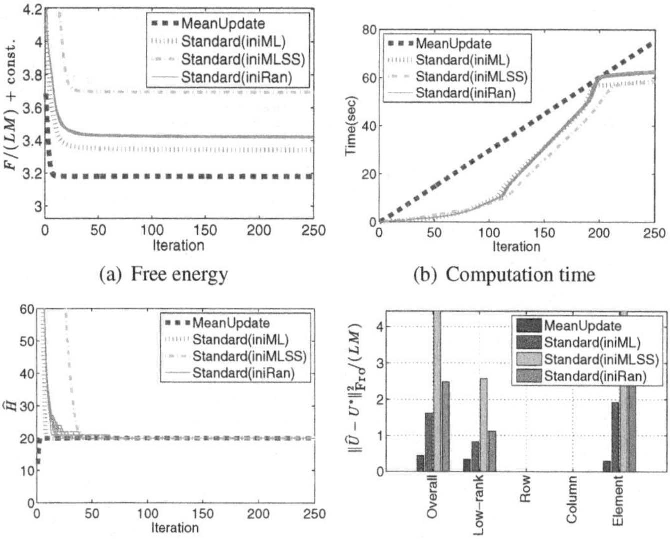


(a) Free energy (b) Computation time

(c) Estimated rank (d) Reconstruction error

Figure 11.2 Experimental results of "LE"-SAMF on an artificial data set ($L = 100, M = 300, H^* = 20, \rho = 0.1$).

Figure 11.2 shows results of "LE"-SAMF on another artificial data set with $L = 100$, $M = 300$, $H^* = 20$, and $\rho = 0.1$. We see that the MU algorithm performs better than the standard VB algorithm. We also tested various SAMF models including "LCE"-SAMF, "LRE"-SAMF, and "LE"-SAMF under different settings for $M$, $L$, $H^*$, and $\rho$, and empirically found that the MU algorithm generally gives a better solution with lower free energy and smaller reconstruction errors than the standard VB algorithm.

Next, we conducted experiments on several data sets from the *UCI repository* (Asuncion and Newman, 2007). Since we do not know the *true* model of those data sets, we only focus on the achieved free energy. Figure 11.3 shows the free energy after convergence in "LRCE"-SAMF, "LCE"-SAMF, "LRE"-SAMF, and "LE"-SAMF. For better comparison, a constant is added so that the free energy achieved by the MU algorithm is zero. We can see a clear advantage of the MU algorithm over the standard VB algorithm.

### 11.3.2 Real-World Application

Finally, we demonstrate the usefulness of the flexibility of SAMF in a foreground (FG)/background (BG) video separation problem (Figure 3.5 in

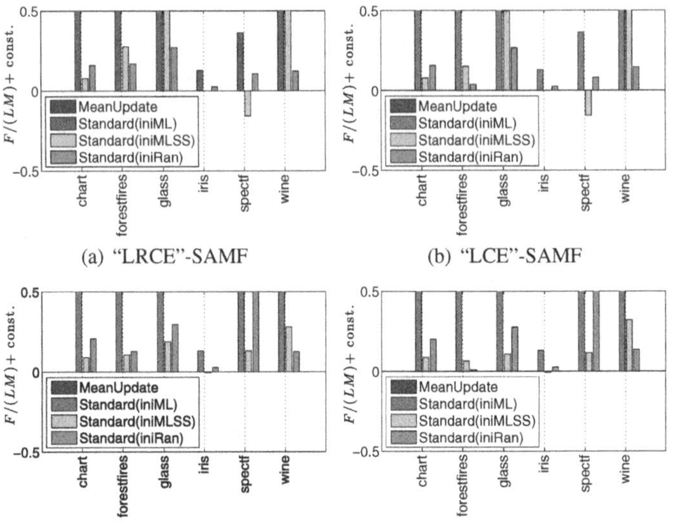


(a) "LRCE"-SAMF (b) "LCE"-SAMF

(c) "LRE"-SAMF (d) "LE"-SAMF

Figure 11.3 Experimental results on benchmark data sets. For better comparison, a constant is added so that the free energy achieved by the MU algorithm is zero (therefore, the bar for "MeanUpdate" is invisible).

Section 3.5). Candès et al. (2011) formed the observed matrix $\boldsymbol{V}$ by stacking all pixels in each frame into each column (Figure 3.6), and applied the *robust PCA* (with "LE"-terms)—the low-rank term captures the *static* BG and the elementwise (or pixelwise) term captures the *moving* FG, e.g., people walking through. SAMF can be seen as an extension of the VB variant of robust PCA (Babacan et al., 2012b). Accordingly, we use "LE"-SAMF,

$$\boldsymbol{V} = \boldsymbol{U}^{\text{low-rank}} + \boldsymbol{U}^{\text{element}} + \boldsymbol{\mathcal{E}},$$

as a baseline method for comparison.

The SAMF framework enables a fine-tuned design for the FG term. Assuming that pixels in an image segment with similar intensity values tend to share the same label (i.e., FG or BG), we formed a segmentwise sparse SMF term: $\boldsymbol{U}'^{(k)}$ for each $k$ is a column vector consisting of all pixels in each segment. We produced an oversegmented image from each frame by using the *efficient graph-based segmentation (EGS)* algorithm (Felzenszwalb and Huttenlocher, 2004), and substituted the segmentwise sparse term for the FG term (see Figure 3.7):

$$\boldsymbol{V} = \boldsymbol{U}^{\text{low-rank}} + \boldsymbol{U}^{\text{segment}} + \boldsymbol{\mathcal{E}}.$$

We call this model *segmentation-based SAMF (sSAMF)*. Note that EGS is computationally very efficient: it takes less than 0.05 sec on a usual laptop to segment a $192 \times 144$ gray image. EGS has several tuning parameters, and the obtained segmentation is sensitive to some of them. However, we confirmed that sSAMF performs similarly with visually different segmentations obtained over a wide range of tuning parameters (see the detailed information in the section "Segmentation Algorithm"). Therefore, careful parameter tuning of EGS is not necessary for our purpose.

We compared sSAMF with "LE"-SAMF on the "WalkByShop1front" video from the *Caviar data set*.[1] Thanks to the Bayesian framework, all unknown parameters (except the ones for segmentation) are estimated from the data, and therefore no manual parameter tuning is required. For both models ("LE"-SAMF and sSAMF), we used the MU algorithm, which was shown in Section 11.3.1 to be practically more reliable than the standard VB algorithm. The original video consists of 2,360 frames, each of which is a color image with $384 \times 288$ pixels. We resized each image into $192 \times 144$ pixels, averaged over the color channels, and subsampled every 15 frames (the frame IDs are $0, 15, 30, \ldots, 2355$). Thus, $\boldsymbol{V}$ is of the size of 27,684 [pixels] $\times$ 158 [frames]. We evaluated "LE"-SAMF and sSAMF on this video, and found that both models perform well (although "LE"-SAMF failed in a few frames).

In order to contrast between the two models more clearly, we created a more *difficult* video by subsampling every five frames from 1,501 to 2,000 (the frame IDs are $1501, 1506, \ldots, 1996$ and $\boldsymbol{V}$ is of the size of 27,684 [pixels] $\times$ 100 [frames]). Since more people walked through in this period, FG/BG separation is more challenging.

Figure 11.4(a) shows one of the original frames. This is a difficult snap shot, because a person stayed at the same position for a while, which confuses separation—any object in the FG pixels is assumed to be *moving*. Figures 11.4(c) and 11.4(d) show the BG and the FG terms, respectively, obtained by "LE"-SAMF. We can see that "LE"-SAMF failed to separate the person from BG (the person is partly captured in the BG term). On the other hand, Figures 11.4(e) and 11.4(f) show the BG and the FG terms obtained by sSAMF based on the segmented image shown in Figure 11.4(b). We can see that sSAMF successfully separated the person from BG in this difficult frame. A careful look at the legs of the person reveals how segmentation helps separation—the legs form a single segment in Figure 11.4(b), and the segmentwise sparse

[1] The EC Funded CAVIAR project/IST 2001 37540, found at URL: http://homepages.inf.ed.ac.uk/rbf/CAVIAR/.

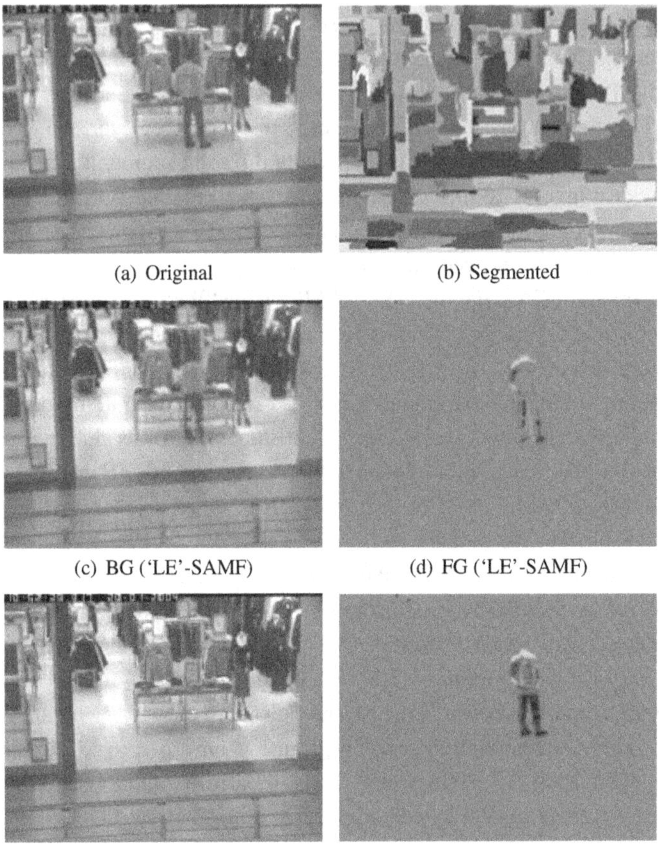

(a) Original (b) Segmented

(c) BG ('LE'-SAMF) (d) FG ('LE'-SAMF)

(e) BG (sSAMF) (f) FG (sSAMF)

Figure 11.4 "LE"-SAMF vs. segmentation-based SAMF in FG/BG video separation.

term (Figure 11.4(f)) captured all pixels on the legs, while the pixelwise sparse term (Figure 11.4(d)) captured only a part of those pixels. We observed that, in all frames of the *difficult* video, as well as the *easier* one, sSAMF gave good separation, while "LE"-SAMF failed in several frames.

For reference, we applied the convex formulation of robust PCA (Candès et al., 2011), which solves the following minimization problem by the inexact augmented Lagrange multiplier (ALM) algorithm (Lin et al., 2009):

$$\min_{\boldsymbol{U}^{\mathrm{BG}}, \boldsymbol{U}^{\mathrm{FG}}} \|\boldsymbol{U}^{\mathrm{BG}}\|_{\mathrm{tr}} + \lambda \|\boldsymbol{U}^{\mathrm{FG}}\|_1 \qquad \text{s.t.} \qquad \boldsymbol{V} = \boldsymbol{U}^{\mathrm{BG}} + \boldsymbol{U}^{\mathrm{FG}}, \tag{11.24}$$

(a) BG (ALM $\lambda = 0.001$) (b) FG (ALM $\lambda = 0.001$)

(c) BG (ALM $\lambda = 0.005$) (d) FG (ALM $\lambda = 0.05$)

(e) BG (ALM $\lambda = 0.025$) (f) FG (ALM $\lambda = 0.025$)

Figure 11.5 FG/BG video separation by the convex formulation of robust PCA (11.24) for $\lambda = 0.001$ (top row), $\lambda = 0.005$ (middle row), and $\lambda = 0.025$ (bottom row).

where $\|\cdot\|_{\mathrm{tr}}$ and $\|\cdot\|_1$ denote the *trace norm* and the $\ell_1$*-norm* of a matrix, respectively. Figure 11.5 shows the obtained BG and FG terms of the same frame as that in Figure 11.4 with $\lambda = 0.001, 0.005, 0.025$. We see that the performance strongly depends on the parameter value of $\lambda$, and that sSAMF gives an almost identical result (bottom row in Figure 11.4) to the best ALM result with $\lambda = 0.005$ (middle row in Figure 11.5) without any manual parameter tuning.

In the following subsections, we give detailed information on the segmentation algorithm and the computation time.

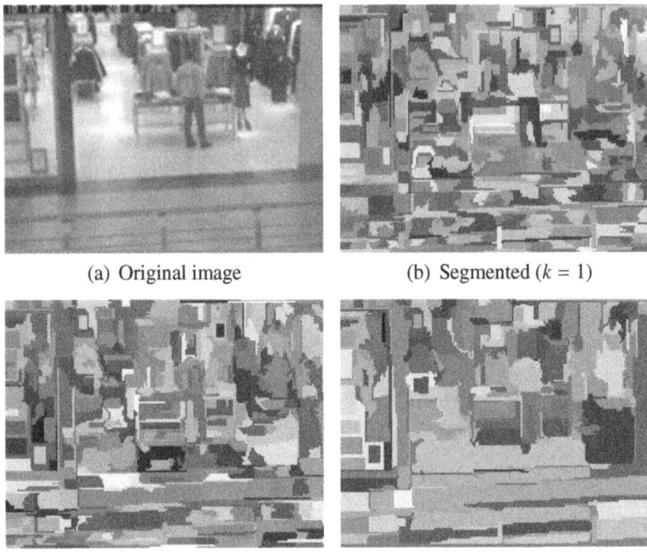

(a) Original image (b) Segmented ($k = 1$)

(c) Segmented ($k = 10$) (d) Segmented ($k = 100$)

Figure 11.6 Segmented images by the efficient graph-based segmentation (EGS) algorithm with different $k$ values. They are visually different, but with all these segmentations, sSAMF gave almost identical FB/BG separations. The original image (a) is the same frame as the one in Figure 11.4.

### Segmentation Algorithm

For the EGS algorithm (Felzenszwalb and Huttenlocher, 2004), we used the code publicly available from the authors' homepage.[2] EGS has three tuning parameters: *sigma*, the smoothing parameter; $k$, the threshold parameter; and *minc*, the minimum segment size. Among them, $k$ dominantly determines the typical size of segments (larger $k$ leads to larger segments). To obtain oversegmented images for sSAMF in our experiment, we chose $k = 50$, and the other parameters are set to $sigma = 0.5$ and $minc = 20$ as recommended by the authors. We also tested other parameter setting, and observed that FG/BG separation by sSAMF performed almost equally for $1 \leq k \leq 100$, despite the visual variation of segmented images (see Figure 11.6). Overall, we empirically observed that the performance of sSAMF is not very sensitive to the selection of segmented images, unless it is highly undersegmented.

[2] www.cs.brown.edu/~pff/

**Computation Time**

The computation time for segmentation by EGS was less than 10 sec (for 100 frames). Forming the one-to-one map $\mathcal{X}$ took more than 80 sec (which is expected to be improved). In total, sSAMF took 600 sec on a Linux machine with Xeon X5570 (2.93GHz), while "LE"-SAMF took 700 sec. This slight reduction in computation time comes from the reduction in the number $K$ of partitions for the FG term, and hence the number of calculations of partial analytic solutions.

# 12
# MAP and Partially Bayesian Learning

Variational Bayesian (VB) learning generally offers a tractable approximation of Bayesian learning, and efficient iterative local search algorithms were derived for many practical models based on the conditional conjugacy (see Part II). However, in some applications, VB learning is still computationally too costly. In such cases, cruder approximation methods, where all or some of the parameters are point-estimated, with potentially less computation cost, are attractive alternatives. For example, Chu and Ghahramani (2009) applied partially Bayesian (PB) learning (introduced in Section 2.2.2), where the core tensor is integrated out and the factor matrices are point-estimated, to Tucker factorization (TF) (Carroll and Chang, 1970; Harshman, 1970; Tucker, 1996; Kolda and Bader, 2009). Mørup and Hansen (2009) applied the maximum a posteriori (MAP) learning to TF with the empirical Bayesian procedure, i.e., the hyperparameters are also estimated from observations. Their proposed empirical MAP learning, which only requires the same order of computation costs as the plain alternating least squares algorithm (Kolda and Bader, 2009), showed its model selection capability through the automatic relevance determination (ARD) property.

Motivated by the empirical success, we have analyzed PB learning and MAP learning and their empirical Bayesian variants (Nakajima et al., 2011; Nakajima and Sugiyama, 2014), which this chapter introduces. Focusing on fully observed matrix factorization (MF), we first analyze the global and local solutions of MAP learning and PB learning and their empirical Bayesian variants. This analysis theoretically reveals similarities and dissimilarities to VB learning. After that, we discuss more general cases, including MF with missing entries and TF.

## 12.1 Theoretical Analysis in Fully Observed MF

In this section, we formulate MAP learning and PB learning in the free energy minimization framework (Section 2.1.1) and derive analytic-form solutions.

### 12.1.1 Problem Description

The model likelihood and the prior of the MF model are given by

$$p(\boldsymbol{V}|\boldsymbol{A},\boldsymbol{B}) \propto \exp\left(-\frac{1}{2\sigma^2}\left\|\boldsymbol{V}-\boldsymbol{B}\boldsymbol{A}^\top\right\|_{\mathrm{Fro}}^2\right), \tag{12.1}$$

$$p(\boldsymbol{A}) \propto \exp\left(-\frac{1}{2}\mathrm{tr}\left(\boldsymbol{A}\boldsymbol{C}_A^{-1}\boldsymbol{A}^\top\right)\right), \tag{12.2}$$

$$p(\boldsymbol{B}) \propto \exp\left(-\frac{1}{2}\mathrm{tr}\left(\boldsymbol{B}\boldsymbol{C}_B^{-1}\boldsymbol{B}^\top\right)\right), \tag{12.3}$$

where the prior covariance matrices are restricted to be diagonal:

$$\boldsymbol{C}_A = \mathbf{Diag}(c_{a_1}^2,\ldots,c_{a_H}^2),$$
$$\boldsymbol{C}_B = \mathbf{Diag}(c_{b_1}^2,\ldots,c_{b_H}^2),$$

for $c_{a_h}, c_{b_h} > 0, h = 1,\ldots,H$. Without loss of generality, we assume that the diagonal entries of the product $\boldsymbol{C}_A\boldsymbol{C}_B$ are arranged in nonincreasing order, i.e., $c_{a_h}c_{b_h} \geq c_{a_{h'}}c_{b_{h'}}$ for any pair $h < h'$.

As in Section 2.2.2, we treat MAP learning and PB learning as special cases of VB learning in the free energy minimization framework. The Bayes posterior is given by

$$p(\boldsymbol{A},\boldsymbol{B}|\boldsymbol{V}) = \frac{p(\boldsymbol{V}|\boldsymbol{A},\boldsymbol{B})p(\boldsymbol{A})p(\boldsymbol{B})}{p(\boldsymbol{V})}, \tag{12.4}$$

which is intractable for the MF model (12.1) through (12.3). Accordingly, we approximate it by

$$\widehat{r} = \underset{r}{\mathrm{argmin}} F(r) \quad \text{s.t.} \quad r(\boldsymbol{A},\boldsymbol{B}) \in \mathcal{G}, \tag{12.5}$$

where $\mathcal{G}$ specifies the constraint on the approximate posterior. $F(r)$ is the free energy, defined as

$$F(r) = \left\langle \log\frac{r(\boldsymbol{A},\boldsymbol{B})}{p(\boldsymbol{V}|\boldsymbol{A},\boldsymbol{B})p(\boldsymbol{A})p(\boldsymbol{B})}\right\rangle_{r(\boldsymbol{A},\boldsymbol{B})} \tag{12.6}$$
$$= \left\langle \log\frac{r(\boldsymbol{A},\boldsymbol{B})}{p(\boldsymbol{A},\boldsymbol{B}|\boldsymbol{V})}\right\rangle_{r(\boldsymbol{A},\boldsymbol{B})} - \log p(\boldsymbol{V}),$$

which is a monotonic function of the KL divergence $\left\langle \log \frac{r(\boldsymbol{A},\boldsymbol{B})}{p(\boldsymbol{A},\boldsymbol{B}|V)} \right\rangle_{r(\boldsymbol{A},\boldsymbol{B})}$ to the Bayes posterior.

**Constraints for MAP Learning and PB Learning**

MAP learning finds the mode of the posterior distribution, which amounts to approximating the posterior with the Dirac delta function. Accordingly, solving the problem (12.5) with the following constraint gives the MAP solution:

$$r^{\mathrm{MAP}}(\boldsymbol{A},\boldsymbol{B}) = \delta(\boldsymbol{A};\widehat{\boldsymbol{A}})\delta(\boldsymbol{B};\widehat{\boldsymbol{B}}), \tag{12.7}$$

where $\delta(\boldsymbol{\mu};\widehat{\boldsymbol{\mu}})$ denotes the (*pseudo-*)Dirac delta function located at $\widehat{\boldsymbol{\mu}}$.[1]

Under the MAP constraint (12.7), the free energy (12.6) is written as

$$\begin{aligned} F^{\mathrm{MAP}}(\widehat{\boldsymbol{A}},\widehat{\boldsymbol{B}}) &= \left\langle \log \frac{\delta(\boldsymbol{A};\widehat{\boldsymbol{A}})\delta(\boldsymbol{B};\widehat{\boldsymbol{B}})}{p(V|\boldsymbol{A},\boldsymbol{B})p(\boldsymbol{A})p(\boldsymbol{B})} \right\rangle_{\delta(\boldsymbol{A};\widehat{\boldsymbol{A}})\delta(\boldsymbol{B};\widehat{\boldsymbol{B}})} \\ &= -\log p(V|\widehat{\boldsymbol{A}},\widehat{\boldsymbol{B}})p(\widehat{\boldsymbol{A}})p(\widehat{\boldsymbol{B}}) + \chi_A + \chi_B, \end{aligned} \tag{12.8}$$

where

$$\chi_A = \left\langle \log \delta(\boldsymbol{A};\widehat{\boldsymbol{A}}) \right\rangle_{\delta(\boldsymbol{A};\widehat{\boldsymbol{A}})}, \qquad \chi_B = \left\langle \log \delta(\boldsymbol{B};\widehat{\boldsymbol{B}}) \right\rangle_{\delta(\boldsymbol{B};\widehat{\boldsymbol{B}})} \tag{12.9}$$

are the negative entropies of the pseudo-Dirac delta functions.

PB learning is a strategy to *analytically* integrate out as many parameters as possible, and the rest are point-estimated. In the MF model, a natural choice is to integrate $\boldsymbol{A}$ out and point-estimate $\boldsymbol{B}$, which we call PB-A learning, or to integrate $\boldsymbol{B}$ out and point-estimate $\boldsymbol{A}$, which we call PB-B learning. Their solutions can be obtained by solving the problem (12.5) with the following constraints, respectively:

$$r^{\mathrm{PB\text{-}A}}(\boldsymbol{A},\boldsymbol{B}) = r_A^{\mathrm{PB}}(\boldsymbol{A})\delta(\boldsymbol{B};\widehat{\boldsymbol{B}}), \tag{12.10}$$

$$r^{\mathrm{PB\text{-}B}}(\boldsymbol{A},\boldsymbol{B}) = \delta(\boldsymbol{A};\widehat{\boldsymbol{A}})r_B^{\mathrm{PB}}(\boldsymbol{B}). \tag{12.11}$$

Under the PB-A constraint (12.10), the free energy (12.6) is written as

$$\begin{aligned} F^{\mathrm{PB\text{-}A}}(r_A^{\mathrm{PB}},\widehat{\boldsymbol{B}}) &= \left\langle \log \frac{r_A^{\mathrm{PB}}(\boldsymbol{A})}{p(V|\boldsymbol{A},\widehat{\boldsymbol{B}})p(\boldsymbol{A})p(\widehat{\boldsymbol{B}})} \right\rangle_{r_A^{\mathrm{PB}}(\boldsymbol{A})} + \chi_B \\ &= \left\langle \log \frac{r_A^{\mathrm{PB}}(\boldsymbol{A})}{p(\boldsymbol{A}|V,\widehat{\boldsymbol{B}})} \right\rangle_{r_A^{\mathrm{PB}}(\boldsymbol{A})} - \log p(V|\widehat{\boldsymbol{B}})p(\widehat{\boldsymbol{B}}) + \chi_B, \end{aligned} \tag{12.12}$$

[1] By the *pseudo*-Dirac delta function, we mean an extremely localized density function, e.g., $\delta(\boldsymbol{A};\widehat{\boldsymbol{A}}) \propto \exp\left(-\|\boldsymbol{A}-\widehat{\boldsymbol{A}}\|_{\mathrm{Fro}}^2/(2\varepsilon^2)\right)$ with a very small but strictly positive variance $\varepsilon^2 > 0$, such that its tail effect can be ignored, while its negative entropy $\chi_A = \langle \log \delta(\boldsymbol{A};\widehat{\boldsymbol{A}}) \rangle_{\delta(\boldsymbol{A};\widehat{\boldsymbol{A}})}$ remains finite.

where

$$p(\boldsymbol{A}|\boldsymbol{V}, \widehat{\boldsymbol{B}}) = \frac{p(\boldsymbol{V}|\boldsymbol{A}, \widehat{\boldsymbol{B}})p(\boldsymbol{A})}{p(\boldsymbol{V}|\widehat{\boldsymbol{B}})}, \tag{12.13}$$

and

$$p(\boldsymbol{V}|\widehat{\boldsymbol{B}}) = \left\langle p(\boldsymbol{V}|\boldsymbol{A}, \widehat{\boldsymbol{B}}) \right\rangle_{p(\boldsymbol{A})} \tag{12.14}$$

are the posterior distribution with respect to $\boldsymbol{A}$ (given $\widehat{\boldsymbol{B}}$) and the marginal distribution, respectively. Note that Eq. (12.12) is a functional of $r_A^{\mathrm{PB}}$ and $\widehat{\boldsymbol{B}}$, and $\chi_B$ is a constant with respect to them.

Since only the first term depends on $r_A^{\mathrm{PB}}$, on which no restriction is imposed, Eq. (12.12) is minimized when

$$r_A^{\mathrm{PB}}(\boldsymbol{A}) = p(\boldsymbol{A}|\boldsymbol{V}, \widehat{\boldsymbol{B}}) \tag{12.15}$$

for any $\widehat{\boldsymbol{B}}$. With Eq. (12.15), the first term in Eq. (12.12) vanishes, and thus the estimator for $\widehat{\boldsymbol{B}}$ is given by

$$\widehat{\boldsymbol{B}}^{\mathrm{PB-A}} = \underset{\widehat{\boldsymbol{B}}}{\mathrm{argmin}}\, \acute{F}^{\mathrm{PB-A}}(\widehat{\boldsymbol{B}}), \tag{12.16}$$

where

$$\acute{F}^{\mathrm{PB-A}}(\widehat{\boldsymbol{B}}) \equiv \min_{r_A^{\mathrm{PB}}} F^{\mathrm{PB-A}}(r_A^{\mathrm{PB}}, \widehat{\boldsymbol{B}}) = -\log p(\boldsymbol{V}|\widehat{\boldsymbol{B}})p(\widehat{\boldsymbol{B}}) + \chi_B. \tag{12.17}$$

The process to compute $\acute{F}^{\mathrm{PB-A}}(\widehat{\boldsymbol{B}})$ in Eq. (12.17) corresponds to integrating $\boldsymbol{A}$ out based on the conditional conjugacy. The probabilistic PCA, introduced in Section 3.1.2, was originally proposed with PB-A learning (Tipping and Bishop, 1999).

In the same way, we can obtain the approximate posterior under the PB-B constraint (12.11) as follows:

$$r_B^{\mathrm{PB}}(\boldsymbol{B}) = p(\boldsymbol{B}|\boldsymbol{V}, \widehat{\boldsymbol{A}}), \tag{12.18}$$

$$\widehat{\boldsymbol{A}}^{\mathrm{PB-B}} = \underset{\widehat{\boldsymbol{A}}}{\mathrm{argmin}}\, \acute{F}^{\mathrm{PB-B}}(\widehat{\boldsymbol{A}}), \tag{12.19}$$

where

$$p(\boldsymbol{B}|\boldsymbol{V}, \widehat{\boldsymbol{A}}) = \frac{p(\boldsymbol{V}|\widehat{\boldsymbol{A}}, \boldsymbol{B})p(\boldsymbol{B})}{p(\boldsymbol{V}|\widehat{\boldsymbol{A}})}, \tag{12.20}$$

$$p(\boldsymbol{V}|\widehat{\boldsymbol{A}}) = \left\langle p(\boldsymbol{V}|\widehat{\boldsymbol{A}}, \boldsymbol{B}) \right\rangle_{p(\boldsymbol{B})}, \tag{12.21}$$

$$\acute{F}^{\mathrm{PB-B}}(\widehat{\boldsymbol{A}}) \equiv \min_{r_B^{\mathrm{PB}}} F^{\mathrm{PB-B}}(r_B^{\mathrm{PB}}, \widehat{\boldsymbol{A}}) = -\log p(\boldsymbol{V}|\widehat{\boldsymbol{A}})p(\widehat{\boldsymbol{A}}) + \chi_A. \tag{12.22}$$

We define PB learning as one of PB-A learning and PB-B learning giving a lower free energy. Namely,

$$r^{\mathrm{PB}}(\boldsymbol{A},\boldsymbol{B}) = \begin{cases} r^{\mathrm{PB\text{-}A}}(\boldsymbol{A},\boldsymbol{B}) & \text{if } \min F^{\mathrm{PB\text{-}A}}(r_A^{\mathrm{PB}}, \widehat{\boldsymbol{B}}) \leq \min F^{\mathrm{PB\text{-}B}}(r_B^{\mathrm{PB}}, \widehat{\boldsymbol{A}}), \\ r^{\mathrm{PB\text{-}B}}(\boldsymbol{A},\boldsymbol{B}) & \text{otherwise.} \end{cases}$$

**Free Energies for MAP Learning and PB Learning**

Apparently, the constraint (12.7) for MAP learning and the constraints (12.10) and (12.11) for PB learning forces independence between $\boldsymbol{A}$ and $\boldsymbol{B}$, and therefore, they are stronger than the independence constraint

$$r^{\mathrm{VB}}(\boldsymbol{A},\boldsymbol{B}) = r_A^{\mathrm{VB}}(\boldsymbol{A})r_B^{\mathrm{VB}}(\boldsymbol{B}) \tag{12.23}$$

for VB learning. In Chapter 6, we showed that the VB posterior under the independence constraint (12.23) is in the following Gaussian form:

$$r_A(\boldsymbol{A}) = \mathrm{MGauss}_{M,H}(\boldsymbol{A};\widehat{\boldsymbol{A}},\boldsymbol{I}_M \otimes \widehat{\boldsymbol{\Sigma}}_A) \propto \exp\left(-\frac{\mathrm{tr}\left((\boldsymbol{A}-\widehat{\boldsymbol{A}})\widehat{\boldsymbol{\Sigma}}_A^{-1}(\boldsymbol{A}-\widehat{\boldsymbol{A}})^\top\right)}{2}\right), \tag{12.24}$$

$$r_B(\boldsymbol{B}) = \mathrm{MGauss}_{L,H}(\boldsymbol{B};\widehat{\boldsymbol{B}},\boldsymbol{I}_L \otimes \widehat{\boldsymbol{\Sigma}}_B) \propto \exp\left(-\frac{\mathrm{tr}\left((\boldsymbol{B}-\widehat{\boldsymbol{B}})\widehat{\boldsymbol{\Sigma}}_B^{-1}(\boldsymbol{B}-\widehat{\boldsymbol{B}})^\top\right)}{2}\right), \tag{12.25}$$

where the posterior covariances, $\widehat{\boldsymbol{\Sigma}}_A$ and $\widehat{\boldsymbol{\Sigma}}_B$, are diagonal. Furthermore, Eqs. (12.24) and (12.25) can be the pseudo-Dirac delta functions by setting the posterior covariances to $\widehat{\boldsymbol{\Sigma}}_A = \varepsilon^2\boldsymbol{I}_H$ and $\widehat{\boldsymbol{\Sigma}}_B = \varepsilon^2\boldsymbol{I}_H$, respectively, for a very small $\varepsilon^2 > 0$.

Consequently, the MAP and the PB solutions can be obtained by minimizing the free energy for VB learning with posterior covariances clipped to $\varepsilon^2\boldsymbol{I}_H$, according to the corresponding constraint. Namely, we start from the free energy expression (6.42) for VB learning, i.e.,

$$2F = LM\log(2\pi\sigma^2) + \frac{\sum_{h=1}^{L}\gamma_h^2}{\sigma^2} + \sum_{h=1}^{H} 2F_h, \tag{12.26}$$

where

$$\begin{aligned} 2F_h = {} & M\log\frac{c_{a_h}^2}{\widehat{\sigma}_{a_h}^2} + L\log\frac{c_{b_h}^2}{\widehat{\sigma}_{b_h}^2} + \frac{\widehat{a}_h^2 + M\widehat{\sigma}_{a_h}^2}{c_{a_h}^2} + \frac{\widehat{b}_h^2 + L\widehat{\sigma}_{b_h}^2}{c_{b_h}^2} \\ & - (L+M) + \frac{-2\widehat{a}_h\widehat{b}_h\gamma_h + \left(\widehat{a}_h^2 + M\widehat{\sigma}_{a_h}^2\right)\left(\widehat{b}_h^2 + L\widehat{\sigma}_{b_h}^2\right)}{\sigma^2}, \end{aligned} \tag{12.27}$$

and set

$$\widehat{\sigma}_{a_h}^2 = \varepsilon^2 \qquad h = 1,\ldots,H, \tag{12.28}$$

for MAP learning and PB-B learning, and

$$\widehat{\sigma}^2_{b_h} = \varepsilon^2 \qquad h = 1, \ldots, H, \tag{12.29}$$

for MAP learning and PB-A learning. Here,

$$\widehat{\boldsymbol{A}} = \left(\widehat{a}_1 \boldsymbol{\omega}_{a_1}, \ldots, \widehat{a}_H \boldsymbol{\omega}_{a_H}\right),$$
$$\widehat{\boldsymbol{B}} = \left(\widehat{b}_1 \boldsymbol{\omega}_{b_1}, \ldots, \widehat{b}_H \boldsymbol{\omega}_{b_H}\right),$$
$$\widehat{\boldsymbol{\Sigma}}_A = \mathbf{Diag}\left(\widehat{\sigma}^2_{a_1}, \ldots, \widehat{\sigma}^2_{a_H}\right),$$
$$\widehat{\boldsymbol{\Sigma}}_B = \mathbf{Diag}\left(\widehat{\sigma}^2_{b_1}, \ldots, \widehat{\sigma}^2_{b_H}\right),$$

and

$$\boldsymbol{V} = \sum_{h=1}^{L} \gamma_h \boldsymbol{\omega}_{b_h} \boldsymbol{\omega}_{a_h}^\top$$

is the singular value decomposition (SVD) of $\boldsymbol{V}$.

Thus, the free energies for MAP learning, PB-A learning, and PB-B learning are given by Eq. (12.26) for

$$2F_h^{\mathrm{MAP}} = M \log c^2_{a_h} + L \log c^2_{b_h} + \frac{\widehat{a}_h^2}{c^2_{a_h}} + \frac{\widehat{b}_h^2}{c^2_{b_h}} + \frac{-2\widehat{a}_h \widehat{b}_h \gamma_h + \widehat{a}_h^2 \widehat{b}_h^2}{\sigma^2}$$
$$- (L + M) + (L + M)\chi, \tag{12.30}$$

$$2F_h^{\mathrm{PB\text{-}A}} = M \log \frac{c^2_{a_h}}{\widehat{\sigma}^2_{a_h}} + L \log c^2_{b_h} + \frac{\widehat{a}_h^2 + M\widehat{\sigma}^2_{a_h}}{c^2_{a_h}} + \frac{\widehat{b}_h^2}{c^2_{b_h}} + \frac{-2\widehat{a}_h \widehat{b}_h \gamma_h + \left(\widehat{a}_h^2 + M\widehat{\sigma}^2_{a_h}\right)\widehat{b}_h^2}{\sigma^2}$$
$$- (L + M) + L\chi, \tag{12.31}$$

$$2F_h^{\mathrm{PB\text{-}B}} = M \log c^2_{a_h} + L \log \frac{c^2_{b_h}}{\widehat{\sigma}^2_{b_h}} + \frac{\widehat{a}_h^2}{c^2_{a_h}} + \frac{\widehat{b}_h^2 + L\widehat{\sigma}^2_{b_h}}{c^2_{b_h}} + \frac{-2\widehat{a}_h \widehat{b}_h \gamma_h + \widehat{a}_h^2\left(\widehat{b}_h^2 + L\widehat{\sigma}^2_{b_h}\right)}{\sigma^2}$$
$$- (L + M) + M\chi, \tag{12.32}$$

respectively, where

$$\chi = -\log \varepsilon^2 \tag{12.33}$$

is a large positive constant corresponding to the negative entropy of the one-dimensional pseudo-Dirac delta function.

As in VB learning, the free energy (12.26) is separable for each singular component as long as the noise variance $\sigma^2$ is treated as a constant. Therefore, the variational parameters $(\widehat{a}_h, \widehat{b}_h, \widehat{\sigma}^2_{a_h}, \widehat{\sigma}^2_{b_h})$ and the prior covariances $(c^2_{a_h}, c^2_{b_h})$ for the $h$th component can be estimated by minimizing $F_h$.

### 12.1.2 Global Solutions

In this section, we derive the global minimizers of the free energies, (12.30) through (12.32), and analyze their behavior.

#### Global MAP and PB Solutions

By minimizing the MAP free energy (12.30), we can obtain the global solution for MAP learning, given the hyperparameters $\boldsymbol{C}_A, \boldsymbol{C}_B, \sigma^2$ treated as fixed constants. Let

$$\boldsymbol{U} = \boldsymbol{B}\boldsymbol{A}^\top$$

be the target low-rank matrix.

**Theorem 12.1** *Given $\boldsymbol{C}_A, \boldsymbol{C}_B \in \mathbb{D}^H_{++}$, and $\sigma^2 \in \mathbb{R}_{++}$, the MAP solution of the MF model* (12.1) *through* (12.3) *is given by*

$$\widehat{\boldsymbol{U}}^{\mathrm{MAP}} = \sum_{h=1}^{H} \widehat{\gamma}_h^{\mathrm{MAP}} \boldsymbol{\omega}_{b_h} \boldsymbol{\omega}_{a_h}^\top, \qquad \textit{where} \qquad \widehat{\gamma}_h^{\mathrm{MAP}} = \begin{cases} \breve{\gamma}_h^{\mathrm{MAP}} & \textit{if } \gamma_h \geq \underline{\gamma}_h^{\mathrm{MAP}}, \\ 0 & \textit{otherwise}, \end{cases} \tag{12.34}$$

*where*

$$\underline{\gamma}_h^{\mathrm{MAP}} = \frac{\sigma^2}{c_{a_h} c_{b_h}}, \tag{12.35}$$

$$\breve{\gamma}_h^{\mathrm{MAP}} = \gamma_h - \frac{\sigma^2}{c_{a_h} c_{b_h}}. \tag{12.36}$$

*Proof* Eq. (12.30) can be written as a function of $\widehat{a}_h$ and $\widehat{b}_h$ as

$$\begin{aligned} 2F_h^{\mathrm{MAP}} &= \frac{\widehat{a}_h^2}{c_{a_h}^2} + \frac{\widehat{b}_h^2}{c_{b_h}^2} + \frac{-2\widehat{a}_h\widehat{b}_h\gamma_h + \widehat{a}_h^2\widehat{b}_h^2}{\sigma^2} + \text{const.} \\ &= \left(\frac{\widehat{a}_h}{c_{a_h}} - \frac{\widehat{b}_h}{c_{b_h}}\right)^2 + \frac{\left(\widehat{a}_h\widehat{b}_h - \left(\gamma_h - \frac{\sigma^2}{c_{a_h}c_{b_h}}\right)\right)^2}{\sigma^2} + \text{const.} \end{aligned} \tag{12.37}$$

Noting that the first two terms are nonnegative, we find that, if $\gamma_h > \underline{\gamma}_h^{\mathrm{MAP}}$, Eq. (12.37) is minimized when

$$\breve{\gamma}_h^{\mathrm{MAP}} \equiv \widehat{a}_h\widehat{b}_h = \gamma_h - \frac{\sigma^2}{c_{a_h}c_{b_h}}, \tag{12.38}$$

$$\widehat{\delta}_h^{\mathrm{MAP}} \equiv \frac{\widehat{a}_h}{\widehat{b}_h} = \frac{c_{a_h}}{c_{b_h}}. \tag{12.39}$$

Otherwise, it is minimized when

$$\widehat{a}_h = 0,$$
$$\widehat{b}_h = 0,$$

which completes the proof. □

Eqs. (12.38) and (12.39) immediately lead to the following corollary:

**Corollary 12.2** *The MAP posterior is given by*

$$r^{\mathrm{MAP}}(\boldsymbol{A}, \boldsymbol{B}) = \prod_{h=1}^{H} \delta(\boldsymbol{a}_h; \widehat{a}_h \boldsymbol{\omega}_{a_h}) \prod_{h=1}^{H} \delta(\boldsymbol{b}_h; \widehat{b}_h \boldsymbol{\omega}_{b_h}), \tag{12.40}$$

*with the following estimators: if* $\gamma_h > \underline{\gamma}_h^{\mathrm{MAP}}$,

$$\widehat{a}_h = \pm\sqrt{\breve{\gamma}_h^{\mathrm{MAP}} \widehat{\delta}_h^{\mathrm{MAP}}}, \quad \widehat{b}_h = \pm\sqrt{\frac{\breve{\gamma}_h^{\mathrm{MAP}}}{\widehat{\delta}_h^{\mathrm{MAP}}}}, \tag{12.41}$$

$$\textit{where} \qquad \widehat{\delta}_h^{\mathrm{MAP}} \left( \equiv \frac{\widehat{a}_h}{\widehat{b}_h} \right) = \frac{c_{a_h}}{c_{b_h}}, \tag{12.42}$$

*and otherwise*

$$\widehat{a}_h = 0, \quad \widehat{b}_h = 0. \tag{12.43}$$

Similarly, by minimizing the PB-A free energy (12.31) and the PB-B free energy (12.32) and comparing them, we can obtain the global solution for PB learning:

**Theorem 12.3** *Given* $\boldsymbol{C}_A, \boldsymbol{C}_B \in \mathbb{D}_{++}^H$, *and* $\sigma^2 \in \mathbb{R}_{++}$, *the PB solution of the MF model* (12.1) *through* (12.3) *is given by*

$$\widehat{\boldsymbol{U}}^{\mathrm{PB}} = \sum_{h=1}^{H} \widehat{\gamma}_h^{\mathrm{PB}} \boldsymbol{\omega}_{b_h} \boldsymbol{\omega}_{a_h}^\top, \qquad \textit{where} \qquad \widehat{\gamma}_h^{\mathrm{PB}} = \begin{cases} \breve{\gamma}_h^{\mathrm{PB}} & \textit{if } \gamma_h \geq \underline{\gamma}_h^{\mathrm{PB}}, \\ 0 & \textit{otherwise}, \end{cases} \tag{12.44}$$

*where*

$$\underline{\gamma}_h^{\mathrm{PB}} = \sigma \sqrt{\max(L, M) + \frac{\sigma^2}{c_{a_h}^2 c_{b_h}^2}}, \tag{12.45}$$

$$\breve{\gamma}_h^{\mathrm{PB}} = \left( 1 - \frac{\sigma^2 \left( \max(L, M) + \sqrt{\max(L, M)^2 + 4 \frac{\gamma_h^2}{c_{a_h}^2 c_{b_h}^2}} \right)}{2\gamma_h^2} \right) \gamma_h. \tag{12.46}$$

**Corollary 12.4** *The PB posterior is given by*

$$r^{\mathrm{PB}}(\boldsymbol{A},\boldsymbol{B}) = \begin{cases} r^{\mathrm{PB\text{-}A}}(\boldsymbol{A},\boldsymbol{B}) & \text{if } L < M, \\ r^{\mathrm{PB\text{-}A}}(\boldsymbol{A},\boldsymbol{B}) \text{ or } r^{\mathrm{PB\text{-}B}}(\boldsymbol{A},\boldsymbol{B}) & \text{if } L = M, \\ r^{\mathrm{PB\text{-}B}}(\boldsymbol{A},\boldsymbol{B}) & \text{if } L > M, \end{cases} \tag{12.47}$$

*where $r^{\mathrm{PB\text{-}A}}(\boldsymbol{A},\boldsymbol{B})$ and $r^{\mathrm{PB\text{-}B}}(\boldsymbol{A},\boldsymbol{B})$ are the PB-A posterior and the PB-B posterior, respectively, given as follows. The PB-A posterior is given by*

$$r^{\mathrm{PB\text{-}A}}(\boldsymbol{A},\boldsymbol{B}) = \prod_{h=1}^{H} \mathrm{Gauss}_M(\boldsymbol{a}_h; \widehat{a}_h \boldsymbol{\omega}_{a_h}, \widehat{\sigma}_{a_h}^2 \boldsymbol{I}_M) \prod_{h=1}^{H} \delta(\boldsymbol{b}_h; \widehat{b}_h \boldsymbol{\omega}_{b_h}), \tag{12.48}$$

*with the following estimators: if*

$$\gamma_h > \underline{\gamma}_h^{\mathrm{PB\text{-}A}} \equiv \sigma \sqrt{M + \frac{\sigma^2}{c_{a_h}^2 c_{b_h}^2}}, \tag{12.49}$$

*then*

$$\widehat{a}_h = \pm\sqrt{\breve{\gamma}_h^{\mathrm{PB\text{-}A}} \widehat{\delta}_h^{\mathrm{PB\text{-}A}}}, \quad \widehat{b}_h = \pm\sqrt{\frac{\breve{\gamma}_h^{\mathrm{PB\text{-}A}}}{\widehat{\delta}_h^{\mathrm{PB\text{-}A}}}}, \quad \widehat{\sigma}_{a_h}^2 = \frac{\sigma^2}{\breve{\gamma}_h^{\mathrm{PB\text{-}A}} / \widehat{\delta}_h^{\mathrm{PB\text{-}A}} + \sigma^2 / c_{a_h}^2}, \tag{12.50}$$

*where*

$$\breve{\gamma}_h^{\mathrm{PB\text{-}A}} \left(\equiv \widehat{a}_h \widehat{b}_h\right) = \left(1 - \frac{\sigma^2\left(M + \sqrt{M^2 + 4\frac{\gamma_h^2}{c_{a_h}^2 c_{b_h}^2}}\right)}{2\gamma_h^2}\right)\gamma_h, \tag{12.51}$$

$$\widehat{\delta}_h^{\mathrm{PB\text{-}A}} \left(\equiv \frac{\widehat{a}_h}{\widehat{b}_h}\right) = \frac{c_{a_h}^2\left(M + \sqrt{M^2 + 4\frac{\gamma_h^2}{c_{a_h}^2 c_{b_h}^2}}\right)}{2\gamma_h}, \tag{12.52}$$

*and otherwise*

$$\widehat{a}_h = 0, \quad \widehat{b}_h = 0, \quad \widehat{\sigma}_{a_h}^2 = c_{a_h}^2. \tag{12.53}$$

*The PB-B posterior is given by*

$$r^{\mathrm{PB\text{-}B}}(\boldsymbol{A},\boldsymbol{B}) = \prod_{h=1}^{H} \delta(\boldsymbol{a}_h; \widehat{a}_h \boldsymbol{\omega}_{a_h}) \prod_{h=1}^{H} \mathrm{Gauss}_L(\boldsymbol{b}_h; \widehat{b}_h \boldsymbol{\omega}_{b_h}, \widehat{\sigma}_{b_h}^2 \boldsymbol{I}_L), \tag{12.54}$$

*with the following estimators: if*

$$\gamma_h > \underline{\gamma}_h^{\mathrm{PB\text{-}B}} \equiv \sigma \sqrt{L + \frac{\sigma^2}{c_{a_h}^2 c_{b_h}^2}}, \tag{12.55}$$

*then*

$$\widehat{a}_h = \pm\sqrt{\breve{\gamma}_h^{\mathrm{PB-B}}\widehat{\delta}_h^{\mathrm{PB-B}}}, \quad \widehat{b}_h = \pm\sqrt{\frac{\breve{\gamma}_h^{\mathrm{PB-B}}}{\widehat{\delta}_h^{\mathrm{PB-B}}}}, \quad \widehat{\sigma}^2_{b_h} = \frac{\sigma^2}{\breve{\gamma}_h^{\mathrm{PB-B}}\widehat{\delta}_h^{\mathrm{PB-B}} + \sigma^2/c^2_{b_h}}, \tag{12.56}$$

*where*

$$\breve{\gamma}_h^{\mathrm{PB-B}}\left(\equiv \widehat{a}_h\widehat{b}_h\right) = \left(1 - \frac{\sigma^2\left(L+\sqrt{L^2+4\frac{\gamma_h^2}{c^2_{a_h}c^2_{b_h}}}\right)}{2\gamma_h^2}\right)\gamma_h, \tag{12.57}$$

$$\widehat{\delta}_h^{\mathrm{PB-B}}\left(\equiv \frac{\widehat{a}_h}{\widehat{b}_h}\right) = \left(\frac{c^2_{b_h}\left(L+\sqrt{L^2+4\frac{\gamma_h^2}{c^2_{a_h}c^2_{b_h}}}\right)}{2\gamma_h}\right)^{-1}, \tag{12.58}$$

*and otherwise*

$$\widehat{a}_h = 0, \quad \widehat{b}_h = 0, \quad \widehat{\sigma}^2_{b_h} = c^2_{b_h}. \tag{12.59}$$

Note that, when $L = M$, the choice from the PB-A and the PB-B posteriors depends on the prior covariances $\boldsymbol{C}_A$ and $\boldsymbol{C}_B$. However, as long as the estimator for the target low-rank matrix $\boldsymbol{U}$ is concerned, the choice does not matter, as Theorem 12.3 states. This is because

$$\underline{\gamma}_h^{\mathrm{PB-A}} = \underline{\gamma}_h^{\mathrm{PB-B}} \quad \text{and} \quad \breve{\gamma}_h^{\mathrm{PB-A}} = \breve{\gamma}_h^{\mathrm{PB-B}} \quad \text{when} \quad L = M,$$

as Corollary 12.4 implies.

**Proofs of Theorem 12.3 and Corollary 12.4**

We first derive the PB-A solution by minimizing the corresponding free energy (12.31):

$$2F_h^{\mathrm{PB-A}} = M\log\frac{c^2_{a_h}}{\widehat{\sigma}^2_{a_h}} + L\log c^2_{b_h} + \frac{\widehat{a}_h^2+M\widehat{\sigma}^2_{a_h}}{c^2_{a_h}} + \frac{\widehat{b}_h^2}{c^2_{b_h}} + \frac{-2\widehat{a}_h\widehat{b}_h\gamma_h+\left(\widehat{a}_h^2+M\widehat{\sigma}^2_{a_h}\right)\widehat{b}_h^2}{\sigma^2}$$
$$- (L+M) + L\chi. \tag{12.60}$$

As a function of $\widehat{a}_h$ (treating $\widehat{b}_h$ and $\widehat{\sigma}^2_{a_h}$ as fixed constants), Eq. (12.60) can be written as

$$2F_h^{\mathrm{PB-A}}(\widehat{a}_h) = \frac{\widehat{a}_h^2}{c^2_{a_h}} + \frac{-2\widehat{a}_h\widehat{b}_h\gamma_h + \widehat{a}_h^2\widehat{b}_h^2}{\sigma^2} + \text{const.}$$

$$= \frac{\widehat{b}_h^2 + \sigma^2/c_{a_h}^2}{\sigma^2}\left(\widehat{a}_h^2 - 2\frac{\widehat{b}_h \gamma_h}{\widehat{b}_h^2 + \sigma^2/c_{a_h}^2}\widehat{a}_h\right) + \text{const.}$$
$$= \frac{\widehat{b}_h^2 + \sigma^2/c_{a_h}^2}{\sigma^2}\left(\widehat{a}_h - \frac{\widehat{b}_h \gamma_h}{\widehat{b}_h^2 + \sigma^2/c_{a_h}^2}\right)^2 + \text{const.},$$

which is minimized when

$$\widehat{a}_h = \frac{\widehat{b}_h \gamma_h}{\widehat{b}_h^2 + \sigma^2/c_{a_h}^2}. \tag{12.61}$$

As a function of $\widehat{\sigma}_{a_h}^2$ (treating $\widehat{a}_h$ and $\widehat{b}_h$ as fixed constants), Eq. (12.60) can be written as

$$2F_h^{\text{PB-A}}(\widehat{\sigma}_{a_h}^2) = M\left(-\log \widehat{\sigma}_{a_h}^2 + \left(\tfrac{1}{c_{a_h}^2} + \tfrac{\widehat{b}_h^2}{\sigma^2}\right)\widehat{\sigma}_{a_h}^2\right) + \text{const.}$$
$$= M\left(-\log\left(\tfrac{1}{c_{a_h}^2} + \tfrac{\widehat{b}_h^2}{\sigma^2}\right)\widehat{\sigma}_{a_h}^2 + \left(\tfrac{1}{c_{a_h}^2} + \tfrac{\widehat{b}_h^2}{\sigma^2}\right)\widehat{\sigma}_{a_h}^2\right) + \text{const.},$$

which is minimized when

$$\widehat{\sigma}_{a_h}^2 = \left(\frac{1}{c_{a_h}^2} + \frac{\widehat{b}_h^2}{\sigma^2}\right)^{-1} = \frac{\sigma^2}{\widehat{b}_h^2 + \sigma^2/c_{a_h}^2}. \tag{12.62}$$

Therefore, substituting Eqs. (12.61) and (12.62) into Eq. (12.60) gives the free energy with $\widehat{a}_h$ and $\widehat{\sigma}_{a_h}^2$ already optimized:

$$2\acute{F}_h^{\text{PB-A}} = \min_{\widehat{a}_h, \widehat{\sigma}_{a_h}^2} 2F_h^{\text{PB-A}}$$
$$= -M \log \widehat{\sigma}_{a_h}^2 + \tfrac{\widehat{a}_h^2 + M\widehat{\sigma}_{a_h}^2}{\widehat{\sigma}_{a_h}^2} + \tfrac{\widehat{b}_h^2}{c_{b_h}^2} - \tfrac{2\widehat{a}_h \widehat{b}_h \gamma_h}{\sigma^2} + \text{const.}$$
$$= M \log(\widehat{b}_h^2 + \sigma^2/c_{a_h}^2) - \tfrac{\widehat{b}_h^2 \gamma_h^2}{\sigma^2(\widehat{b}_h^2 + \sigma^2/c_{a_h}^2)} + \tfrac{\widehat{b}_h^2}{c_{b_h}^2} + \text{const.}$$
$$= M \log(\widehat{b}_h^2 + \sigma^2/c_{a_h}^2) + \left(\tfrac{\gamma_h^2}{\sigma^2} - \tfrac{\widehat{b}_h^2 \gamma_h^2}{\sigma^2(\widehat{b}_h^2 + \sigma^2/c_{a_h}^2)}\right) + \left(\tfrac{\widehat{b}_h^2}{c_{b_h}^2} + \tfrac{\sigma^2}{c_{a_h}^2 c_{b_h}^2}\right) + \text{const.}$$
$$= M \log(\widehat{b}_h^2 + \sigma^2/c_{a_h}^2) + \tfrac{\gamma_h^2}{c_{a_h}^2}(\widehat{b}_h^2 + \sigma^2/c_{a_h}^2)^{-1} + \tfrac{1}{c_{b_h}^2}(\widehat{b}_h^2 + \sigma^2/c_{a_h}^2) + \text{const.} \tag{12.63}$$

In the second last equation, we added some constants so that the minimizer can be found by the following lemma:

**Lemma 12.5** *The function*

$$f(x) = \xi_{\log} \log x + \xi_{-1} x^{-1} + \xi_1 x$$

*of $x > 0$ for positive coefficients $\xi_{\log}, \xi_{-1}, \xi_1 > 0$ is strictly* quasiconvex,[2] *and minimized at*

$$\widehat{x} = \frac{-\xi_{\log} + \sqrt{\xi_{\log}^2 + 4\xi_1\xi_{-1}}}{2\xi_1}.$$

*Proof* $f(x)$ is differentiable in $x > 0$, and its first derivative is

$$\begin{aligned}\frac{\partial f}{\partial x} &= \xi_{\log} x^{-1} - \xi_{-1} x^{-2} + \xi_1 \\ &= x^{-2}\left(\xi_1 x^2 + \xi_{\log} x - \xi_{-1}\right) \\ &= x^{-2}\left(x + \frac{\xi_{\log} + \sqrt{\xi_{\log}^2 + 4\xi_1\xi_{-1}}}{2\xi_1}\right)\left(x - \frac{-\xi_{\log} + \sqrt{\xi_{\log}^2 + 4\xi_1\xi_{-1}}}{2\xi_1}\right).\end{aligned}$$

Since the first two factors are positive, we find that $f(x)$ is strictly decreasing for $0 < x < \widehat{x}$, and strictly increasing for $x > \widehat{x}$, which proves the lemma. □

By applying Lemma 12.5 to Eq. (12.63) with $x = \widehat{b_h^2} + \sigma^2/c_{a_h}^2$, we find that $\acute{F}_h^{\mathrm{PB-A}}$ is strictly quasiconvex and minimized when

$$\widehat{b_h^2} + \sigma^2/c_{a_h}^2 = \frac{c_{b_h}^2\left(-M + \sqrt{M^2 + \frac{4\gamma_h^2}{c_{a_h}^2 c_{b_h}^2}}\right)}{2}. \tag{12.64}$$

Since $\widehat{b_h^2}$ is nonnegative, the minimizer of the free energy (12.63) is given by

$$\widehat{b_h^2} = \max\left\{0, \frac{c_{b_h}^2\left(-\left(M + \frac{2\sigma^2}{c_{a_h}^2 c_{b_h}^2}\right) + \sqrt{M^2 + \frac{4\gamma_h^2}{c_{a_h}^2 c_{b_h}^2}}\right)}{2}\right\}. \tag{12.65}$$

Apparently, Eq. (12.65) is positive when

$$\left(M + \frac{2\sigma^2}{c_{a_h}^2 c_{b_h}^2}\right)^2 < M^2 + \frac{4\gamma_h^2}{c_{a_h}^2 c_{b_h}^2},$$

which leads to the thresholding condition:

$$\gamma_h > \underline{\gamma}_h^{\mathrm{PB-A}}.$$

By using Eqs. (12.61), (12.62), (12.65), and

$$\left(\widehat{b_h^2} + \sigma^2/c_{a_h}^2\right)^{-1} = \begin{cases} \frac{c_{a_h}^2\left(M + \sqrt{M^2 + \frac{4\gamma_h^2}{c_{a_h}^2 c_{b_h}^2}}\right)}{2\gamma_h^2} & \text{if } \gamma_h > \underline{\gamma}_h^{\mathrm{PB-A}}, \\ \frac{c_{a_h}^2}{\sigma^2} & \text{otherwise,} \end{cases} \tag{12.66}$$

[2] The definition of *quasiconvexity* is given in footnote 1 in Section 8.1.

derived from Eqs. (12.64) and (12.65), we obtain

$$\widehat{\gamma}_h^{\mathrm{PB-A}} \equiv \widehat{a}_h \widehat{b}_h = \frac{\widehat{b}_h^2 \gamma_h}{\widehat{b}_h^2 + \sigma^2/c_{a_h}^2}$$
$$= \left(1 - \frac{\sigma^2/c_{a_h}^2}{\widehat{b}_h^2 + \sigma^2/c_{a_h}^2}\right)\gamma_h$$
$$= \begin{cases} \left(1 - \frac{\sigma^2\left(M + \sqrt{M^2 + 4\frac{\gamma_h^2}{c_{a_h}^2 c_{b_h}^2}}\right)}{2\gamma_h^2}\right)\gamma_h & \text{if } \gamma_h > \underline{\gamma}_h^{\mathrm{PB-A}}, \\ 0 & \text{otherwise,} \end{cases}$$

$$\widehat{\delta}_h^{\mathrm{PB-A}} \equiv \frac{\widehat{a}_h}{\widehat{b}_h} = \frac{\gamma_h}{\widehat{b}_h^2 + \sigma^2/c_{a_h}^2}$$
$$= \frac{c_{a_h}^2\left(M + \sqrt{M^2 + 4\frac{\gamma_h^2}{c_{a_h}^2 c_{b_h}^2}}\right)}{2\gamma_h} \qquad \text{for } \gamma_h > \underline{\gamma}_h^{\mathrm{PB-A}},$$

$$\widehat{\sigma}_{a_h}^2 = \frac{\sigma^2}{\widehat{b}_h^2 + \sigma^2/c_{a_h}^2}$$
$$= \begin{cases} \frac{\sigma^2}{\widehat{\gamma}_h^{\mathrm{PB-A}}/\widehat{\delta}_h^{\mathrm{PB-A}} + \sigma^2/c_{a_h}^2} & \text{if } \gamma_h > \underline{\gamma}_h^{\mathrm{PB-A}}, \\ c_{a_h}^2 & \text{otherwise.} \end{cases}$$

Thus, we have obtained the PB-A posterior (12.48) specified by Eqs. (12.49) through (12.53).

In exactly the same way, we can obtain the PB-B posterior (12.54) specified by Eqs. (12.55) through (12.59), by minimizing the free energy (12.32) for PB-B learning.

Finally, for the choice from the PB-A posterior and the PB-B posterior, we can easily prove the following lemma:

**Lemma 12.6** *It holds that*

$$\min_{\widehat{a}_h, \widehat{b}_h, \widehat{\sigma}_{a_h}^2} F_h^{\mathrm{PB-A}} < \min_{\widehat{a}_h, \widehat{b}_h, \widehat{\sigma}_{b_h}^2} F_h^{\mathrm{PB-B}} \qquad \text{if} \qquad L < M,$$
$$\min_{\widehat{a}_h, \widehat{b}_h, \widehat{\sigma}_{a_h}^2} F_h^{\mathrm{PB-A}} > \min_{\widehat{a}_h, \widehat{b}_h, \widehat{\sigma}_{b_h}^2} F_h^{\mathrm{PB-B}} \qquad \text{if} \qquad L > M.$$

*Proof* When comparing the PB-A free energy (12.31) and the PB-B free energy (12.32), the last terms are dominant since we assume that the negative entropy $\chi$, defined by Eq. (12.33), of the one-dimensional pseudo-Dirac delta function is finite but arbitrarily large. Then, comparing the last terms, each of which is proportional to the number of parameters point-estimated, of Eqs. (12.31) and (12.32) proves Lemma 12.6. □

Combining Lemma 12.6 with the PB-A posterior and the PB-B posterior obtained before, we have Corollary 12.4. Theorem 12.3 is a direct consequence of Corollary 12.4. □

**Comparison between MAP, PB, and VB Solutions**

Here we compare the MAP solution (Theorem 12.1), the PB solution (Theorem 12.3), and the VB solution (Theorem 6.7 in Chapter 6). For all methods, the solution is a shrinkage estimator applied to each singular value, i.e., in the following form for $\widehat{\gamma}_h \leq \gamma_h$:

$$\widehat{U} = \sum_{h=1}^{H} \widehat{\gamma}_h \omega_{b_h} \omega_{a_h}^{\top}, \qquad \text{where} \qquad \widehat{\gamma}_h = \begin{cases} \breve{\gamma}_h & \text{if } \gamma_h \geq \underline{\gamma}_h, \\ 0 & \text{otherwise.} \end{cases} \tag{12.67}$$

When the prior is flat, i.e., $c_{a_h} c_{b_h} \to \infty$, the truncation threshold $\underline{\gamma}_h$ and the shrinkage factor $\breve{\gamma}_h$ are simplified as

$$\lim_{c_{a_h} c_{b_h} \to \infty} \underline{\gamma}_h^{\mathrm{MAP}} = 0, \qquad \lim_{c_{a_h} c_{b_h} \to \infty} \breve{\gamma}_h^{\mathrm{MAP}} = \gamma_h, \tag{12.68}$$

$$\lim_{c_{a_h} c_{b_h} \to \infty} \underline{\gamma}_h^{\mathrm{PB}} = \sigma \sqrt{\max(L, M)}, \qquad \lim_{c_{a_h} c_{b_h} \to \infty} \breve{\gamma}_h^{\mathrm{PB}} = \left(1 - \frac{\max(L,M)\sigma^2}{\gamma_h}\right) \gamma_h, \tag{12.69}$$

$$\lim_{c_{a_h} c_{b_h} \to \infty} \underline{\gamma}_h^{\mathrm{VB}} = \sigma \sqrt{\max(L, M)}, \qquad \lim_{c_{a_h} c_{b_h} \to \infty} \breve{\gamma}_h^{\mathrm{VB}} = \left(1 - \frac{\max(L,M)\sigma^2}{\gamma_h}\right) \gamma_h, \tag{12.70}$$

and therefore the estimators $\widehat{\gamma}_h$ can be written as

$$\widehat{\gamma}_h^{\mathrm{MAP}} = \gamma_h, \tag{12.71}$$

$$\widehat{\gamma}_h^{\mathrm{PB}} = \max\left(0, \left(1 - \frac{\max(L,M)\sigma^2}{\gamma_h}\right) \gamma_h\right), \tag{12.72}$$

$$\widehat{\gamma}_h^{\mathrm{VB}} = \max\left(0, \left(1 - \frac{\max(L,M)\sigma^2}{\gamma_h}\right) \gamma_h\right). \tag{12.73}$$

As expected, the MAP estimator (12.71) coincides with the maximum likelihood (ML) estimator when the prior is flat. On the other hand, the PB estimator (12.72) and the VB estimator (12.73) do not converge to the ML estimator. Interestingly, the PB estimator and the VB estimator coincide with each other, and they are in the form of the *positive-part James–Stein (PJS) estimator* (James and Stein, 1961; Efron and Morris, 1973) applied to each singular component (see Appendix A for a short introduction to the James–Stein estimator). The reason why the VB estimator is shrunken even with the flat prior was explained in Chapter 7 in terms of model-induced regularization

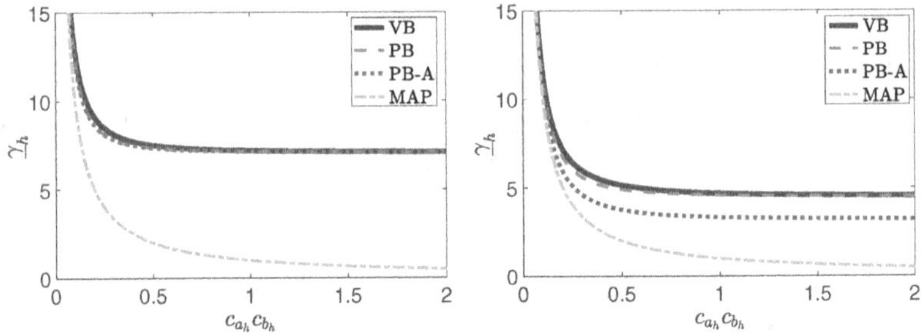


(a) $L = 20, M = 50$ (b) $L = 20, M = 10$

Figure 12.1 Truncation thresholds, $\underline{\gamma}_h^{\mathrm{VB}}$, $\underline{\gamma}_h^{\mathrm{PB}}$, $\underline{\gamma}_h^{\mathrm{PB-A}}$, and $\underline{\gamma}_h^{\mathrm{MAP}}$ as functions of the product $c_{a_h}c_{b_h}$ of the prior covariances. The noise variance is set to $\sigma^2 = 1$.

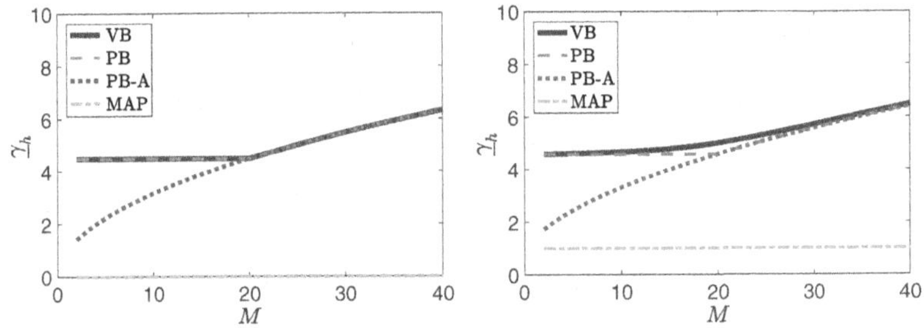


(a) $L = 20, c_{a_h}c_{b_h} \to \infty$ (b) $L = 20, c_{a_h}c_{b_h} = 1$

Figure 12.2 Truncation thresholds as functions of $M$.

(MIR) enhanced by phase transitions. Eq. (12.72) implies that PB learning—a cruder approximation to Bayesian learning—shares the same property as VB learning.

To investigate the dimensionality selection behavior, we depict the truncation thresholds of VB learning, PB learning, PB-A learning, and MAP learning as functions of the product $c_{a_h}c_{b_h}$ of the prior covariances in Figure 12.1. The left panel is for the case with $L = 20, M = 50$, and the right panel is for the case with $L = 20, M = 10$. PB-A learning corresponds to PB learning with the predetermined marginalized and the point-estimated spaces as in Tipping and Bishop (1999) and Chu and Ghahramani (2009), i.e., the matrix $\boldsymbol{A}$ is always marginalized out and $\boldsymbol{B}$ is point-estimated regardless of the dimensionality. We see in Figure 12.1 that PB learning and VB learning show similar dimensionality selection behaviors, while PB-A learning behaves differently when $L > M$.

Figure 12.2 shows the truncation thresholds as functions of $M$ for $L = 20$. With the flat prior $c_{a_h}c_{b_h} \to \infty$ (left panel), the PB and the VB solutions agree with each other, as Eqs. (12.69) and (12.70) imply. The PB-A solution is also identical to them when $M \geq L$. However, its behavior changes at $M = L$: the truncation threshold of PB-A learning smoothly goes down as $M$ decreases, while those of PB learning and VB learning make a sudden turn and become constant. The right panel is for the case with a nonflat prior ($c_{a_h} c_{b_h} = 1$), which shows similar tendency to the case with the flat prior.

A question is which behavior is more desirable, a sudden turn in the threshold curve in VB/PB learning, or the smooth behavior in PB-A learning? We argue that the behavior of VB/PB learning is more desirable for the following reason. Let us consider the case where no *true* signal exists, i.e., the true rank is $H^* = 0$. In this case, we merely observe pure noise, $\boldsymbol{V} = \boldsymbol{\mathcal{E}}$, and the average of the squared singular values of $\boldsymbol{V}$ over all components is given by

$$\frac{\left\langle \mathrm{tr}(\boldsymbol{\mathcal{E}}\boldsymbol{\mathcal{E}}^\top) \right\rangle_{\mathrm{MGauss}_{L,M}(\boldsymbol{\mathcal{E}};\boldsymbol{0}_{L,M},\sigma^2 \boldsymbol{I}_L \otimes \boldsymbol{I}_M)}}{\min(L, M)} = \sigma^2 \max(L, M). \qquad (12.74)$$

Comparing Eq. (12.74) with Eqs. (12.70) and (12.69), we find that VB learning and PB learning always discard the components with singular values no greater than the average noise contribution (note here that Eqs. (12.70) and (12.69) give the thresholds for the flat prior $c_{a_h} c_{b_h} \to \infty$, and the thresholds increase as $c_{a_h} c_{b_h}$ decreases). The sudden turn in the threshold curve actually follows the behavior of the average noise contribution (12.74) to the singular values. On the other hand, PB-A learning does not necessarily discard such noise-dominant components, and can strongly overfit the noise when $L \gg M$.

Figure 12.3 shows the estimators $\widehat{\gamma}_h$ by VB learning, PB learning, PB-A learning, and MAP learning for $c_{a_h} c_{b_h} = 1$, as functions of the observed

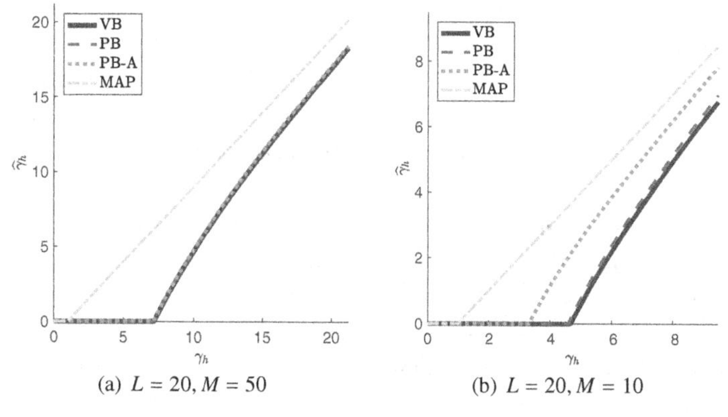


Figure 12.3 Behavior of VB, PB, PB-A, and MAP estimators (the vertical axis) for $c_{a_h} c_{b_h} = 1$, when the singular value $\gamma_h$ (the horizontal axis) is observed. The noise variance is set to $\sigma^2 = 1$.

singular value $\gamma_h$. We can see that the PB estimator behaves similarly to the VB estimator, while the MAP estimator behaves significantly differently. The right panel shows that PB-A learning also behaves differently from VB learning when $L > M$, which implies that the choice between the PB-A posterior and the PB-B posterior based on the free energy is essential to accurately approximate VB learning.

Actually, the coincidence between the VB solution (12.70) and the PB solution (12.69) with the flat prior can be seen as a natural consequence from the similarity in the posterior shape. From Theorem 6.7 and Corollary 6.8, we can derive the following corollary:

**Corollary 12.7** *Assume that, when we make the prior flat* $c_{a_h} c_{b_h} \to \infty$, $c_{a_h}$ *and* $c_{b_h}$ *go to infinity in the same order, i.e.,* $c_{a_h}/c_{b_h} = \Theta(1)$.[3] *Then, the following hold for the variances of the VB posterior* (6.40)*: when* $L < M$,

$$\lim_{c_{a_h} c_{b_h} \to \infty} \widehat{\sigma}^2_{a_h} = \infty, \qquad \lim_{c_{a_h} c_{b_h} \to \infty} \widehat{\sigma}^2_{b_h} = 0, \tag{12.75}$$

*and when* $L > M$,

$$\lim_{c_{a_h} c_{b_h} \to \infty} \widehat{\sigma}^2_{a_h} = 0, \qquad \lim_{c_{a_h} c_{b_h} \to \infty} \widehat{\sigma}^2_{b_h} = \infty. \tag{12.76}$$

*Proof* Assume first that $L < M$. When $\gamma_h > \underline{\gamma}_h^{\mathrm{VB}}$, Eq. (6.52) gives $\lim_{c_{a_h} c_{b_h} \to \infty} \widehat{\delta}_h^{\mathrm{VB}} = \infty$, and therefore Eq. (6.51) gives Eq. (12.75). When $\gamma_h \le \underline{\gamma}_h^{\mathrm{VB}}$, Eq. (6.54) gives $\widehat{\zeta}_h^{\mathrm{VB}} = \sigma^2/M - \Theta(c_{a_h}^{-2} c_{b_h}^{-2})$ as $c_{a_h} c_{b_h} \to \infty$, and therefore Eq. (6.53) gives Eq. (12.75).

When $L > M$, Theorem 6.7 and Corollary 6.8 hold for $\boldsymbol{V} \leftarrow \boldsymbol{V}^\top$, meaning that the VB posterior is obtained by exchanging the variational parameters for $\boldsymbol{A}$ and those for $\boldsymbol{B}$. Thus, we obtain Eq. (12.76), and complete the proof. □

Corollary 12.7 implies that, with the flat prior $c_{a_h} c_{b_h} \to \infty$, the shape of the VB posterior is similar to the shape of the PB posterior: they extend in the space of $\boldsymbol{A}$ when $M > L$, and extend in the space of $\boldsymbol{B}$ when $M < L$. Therefore, it is no wonder that the solutions coincide with each other.

### Global Empirical MAP and Empirical PB Solutions

Next, we investigate the empirical Bayesian variants of MAP learning and PB learning, where the hyperparameters $\boldsymbol{C}_A$ and $\boldsymbol{C}_B$ are also estimated from observations. Note that the noise variance $\sigma^2$ is still considered as a fixed constant (noise variance estimation will be discussed in Section 12.1.6).

[3] $\Theta(f(x))$ is a positive function such that $\limsup_{x\to\infty} |\Theta(f(x))/f(x)| < \infty$ and $\liminf_{x\to\infty} |\Theta(f(x))/f(x)| > 0$.

Let us first consider the MAP free energy (12.30):

$$2F_h^{\text{MAP}} = M \log c_{a_h}^2 + L \log c_{b_h}^2 + \frac{\widehat{a}_h^2}{c_{a_h}^2} + \frac{\widehat{b}_h^2}{c_{b_h}^2} + \frac{-2\widehat{a}_h\widehat{b}_h\gamma_h + \widehat{a}_h^2\widehat{b}_h^2}{\sigma^2}$$
$$- (L + M) + (L + M)\chi.$$

We can make the MAP free energy arbitrarily small, i.e., $F_h^{\text{MAP}} \to -\infty$ by setting $c_{a_h}^2, c_{b_h}^2 \to +0$ with the variational parameters set to the corresponding solution, i.e., $\widehat{a}_h = \widehat{b}_h = 0$ (see Corollary 12.2). Therefore, the global solution of *empirical MAP (EMAP) learning* is given by

$$\widehat{a}_h = 0, \quad \widehat{b}_h = 0, \quad c_{a_h}^2 \to +0, \quad c_{b_h}^2 \to +0, \quad \text{for } h = 1, \ldots, H,$$

which results in the following theorem:

**Theorem 12.8** *The global solution of EMAP learning is* $\widehat{\gamma}_h^{\text{EMAP}} \left(\equiv \widehat{a}_h\widehat{b}_h\right) = 0$ *for all* $h = 1, \ldots, H$*, regardless of observations.*

The same happens in *empirical PB (EPB) learning*. The PB-A free energy (12.31),

$$2F_h^{\text{PB-A}} = M \log \frac{c_{a_h}^2}{\widehat{\sigma}_{a_h}^2} + L \log c_{b_h}^2 + \frac{\widehat{a}_h^2 + M\widehat{\sigma}_{a_h}^2}{c_{a_h}^2} + \frac{\widehat{b}_h^2}{c_{b_h}^2} + \frac{-2\widehat{a}_h\widehat{b}_h\gamma_h + \left(\widehat{a}_h^2 + M\widehat{\sigma}_{a_h}^2\right)\widehat{b}_h^2}{\sigma^2}$$
$$- (L + M) + L\chi,$$

can be arbitrarily small, i.e., $F_h^{\text{PB-A}} \to -\infty$ by setting $c_{a_h}^2, c_{b_h}^2 \to +0$ with the variational parameters set to the corresponding solution $\widehat{a}_h = \widehat{b}_h = 0, \widehat{\sigma}_{a_h}^2 = c_{a_h}^2$ (see Corollary 12.4). Also, the PB-B free energy (12.32),

$$2F_h^{\text{PB-B}} = M \log c_{a_h}^2 + L \log \frac{c_{b_h}^2}{\widehat{\sigma}_{b_h}^2} + \frac{\widehat{a}_h^2}{c_{a_h}^2} + \frac{\widehat{b}_h^2 + L\widehat{\sigma}_{b_h}^2}{c_{b_h}^2} + \frac{-2\widehat{a}_h\widehat{b}_h\gamma_h + \widehat{a}_h^2\left(\widehat{b}_h^2 + L\widehat{\sigma}_{b_h}^2\right)}{\sigma^2}$$
$$- (L + M) + M\chi,$$

can be arbitrarily small, i.e., $F_h^{\text{PB-B}} \to -\infty$ by setting $c_{a_h}^2, c_{b_h}^2 \to +0$ with the variational parameters set to the corresponding solution $\widehat{a}_h = \widehat{b}_h = 0, \widehat{\sigma}_{b_h}^2 = c_{b_h}^2$. Thus, we have the following theorem:

**Theorem 12.9** *The global solution of EPB learning is* $\widehat{\gamma}_h^{\text{EPB}} \left(\equiv \widehat{a}_h\widehat{b}_h\right) = 0$ *for all* $h = 1, \ldots, H$*, regardless of observations.*

Theorems 12.8 and 12.9 imply that empirical Bayesian variants of MAP learning and PB learning give useless trivial estimators. This happens because the posterior variances of the parameters to be point-estimated are fixed to a small value, so that the posteriors form the pseudo-Dirac delta functions. In VB learning, if we set $c_{a_h}c_{b_h}$ to a small value, the posterior variances, $\widehat{\sigma}_{a_h}^2$ and $\widehat{\sigma}_{b_h}^2$, get small accordingly, so that the third and the fourth terms in Eq. (12.27) do not diverge to $+\infty$. As a result, the first and the second terms in Eq. (12.27) remain finite. On the other hand, in MAP learning and PB learning, at least one of the posterior variances, $\widehat{\sigma}^2_{a_h}$ and $\widehat{\sigma}^2_{b_h}$, is treated as a constant and cannot be adjusted to the corresponding prior covariance when it is set to be small. This makes the free energy lower-unbounded. Actually, if we lower-bound the prior covariances as $c^2_{a_h}, c^2_{b_h} \geq \varepsilon^2$ with the same $\varepsilon^2$ as the one we used for defining the variances (12.28) and (12.29) of the pseudo-Dirac delta functions and their entropy (12.33), the MAP and the PB free energies, $F_h^{\mathrm{MAP}}$, $F_h^{\mathrm{PB-A}}$, and $F_h^{\mathrm{PB-B}}$, are also lower-bounded by zero, as the VB free energy, $F_h^{\mathrm{VB}}$.

### 12.1.3 Local Solutions

The analysis in Section 12.1.2 might seem contradictory with the reported results in Mørup and Hansen (2009), where EMAP showed good performance with the ARD property in TF—since the free energies in MF and TF are similar to each other, they should share the same issue of the lower-unboundedness. In the following, we elucidate that this apparent contradiction is because of the local solutions in EMAP learning and EPB learning that behave similarly to the nontrivial positive solution of EVB learning. Actually, EMAP learning and EPB learning can behave similarly to EVB learning when the free energy is minimized by local search.

#### Local EMAP and EPB Solutions

Here we conduct more detailed analysis of the free energies for EMAP learning and EPB learning, and clarify the behavior of their local minima. To make the free energy always comparable (finite), we slightly modify the problem. Specifically, we solve the following problem:

$$\text{Given} \quad \sigma^2 \in \mathbb{R}_{++},$$
$$\min_{r, \{c^2_{a_h}, c^2_{b_h}\}_{h=1}^H} \quad F, \tag{12.77}$$
$$\text{s.t.} \quad c_{a_h} c_{b_h} \geq \varepsilon^2, \quad c_{a_h}/c_{b_h} = 1 \quad \text{for } h = 1, \ldots, H,$$
$$\text{and} \quad \begin{cases} r(\boldsymbol{A}, \boldsymbol{B}) = \delta(\boldsymbol{A}; \widehat{\boldsymbol{A}})\delta(\boldsymbol{B}; \widehat{\boldsymbol{B}}) & \text{(for EMAP learning)}, \\ r(\boldsymbol{A}, \boldsymbol{B}) = r_A(\boldsymbol{A})\delta(\boldsymbol{B}; \widehat{\boldsymbol{B}}) & \text{(for EPB-A learning)}, \\ r(\boldsymbol{A}, \boldsymbol{B}) = \delta(\boldsymbol{A}; \widehat{\boldsymbol{A}}) r_B(\boldsymbol{B}) & \text{(for EPB-B learning)}, \\ r(\boldsymbol{A}, \boldsymbol{B}) = r_A(\boldsymbol{A}) r_B(\boldsymbol{B}) & \text{(for EVB learning)}, \end{cases}$$

where the free energy $F$ is defined by Eq. (12.6), and the pseudo-Dirac delta function is defined as Gaussian with an arbitrarily small but positive variance $\varepsilon^2 > 0$:

$$\delta(\boldsymbol{A}; \widehat{\boldsymbol{A}}) = \mathrm{MGauss}_{M,H}(\boldsymbol{A}; \widehat{\boldsymbol{A}}, \varepsilon^2 \boldsymbol{I}_M \otimes \boldsymbol{I}_H) \propto \exp\left(-\frac{\|\boldsymbol{A}-\widehat{\boldsymbol{A}}\|^2_{\mathrm{Fro}}}{2\varepsilon^2}\right),$$

$$\delta(\boldsymbol{B}; \widehat{\boldsymbol{B}}) = \mathrm{MGauss}_{L,H}(\boldsymbol{B}; \widehat{\boldsymbol{B}}, \varepsilon^2 \boldsymbol{I}_L \otimes \boldsymbol{I}_H) \propto \exp\left(-\frac{\|\boldsymbol{B}-\widehat{\boldsymbol{B}}\|^2_{\mathrm{Fro}}}{2\varepsilon^2}\right).$$

Note that, in Eq. (12.77), we lower-bounded the product $c_{a_h} c_{b_h}$ of the prior covariances and fixed the ratio $c_{a_h}/c_{b_h}$. We added the constraint for EVB learning for comparison.

Following the discussion in Section 12.1.1, we can write the posterior as

$$r(\boldsymbol{A}, \boldsymbol{B}) = r_A(\boldsymbol{A}) r_B(\boldsymbol{B}), \qquad \text{where}$$

$$r_A(\boldsymbol{A}) = \mathrm{MGauss}_{M,H}(\boldsymbol{A}; \widehat{\boldsymbol{A}}, \boldsymbol{I}_M \otimes \widehat{\boldsymbol{\Sigma}}_A) \propto \exp\left(-\frac{\mathrm{tr}\left((\boldsymbol{A}-\widehat{\boldsymbol{A}})\widehat{\boldsymbol{\Sigma}}_A^{-1}(\boldsymbol{A}-\widehat{\boldsymbol{A}})^\top\right)}{2}\right),$$

$$r_B(\boldsymbol{B}) = \mathrm{MGauss}_{L,H}(\boldsymbol{B}; \widehat{\boldsymbol{B}}, \boldsymbol{I}_L \otimes \widehat{\boldsymbol{\Sigma}}_B) \propto \exp\left(-\frac{\mathrm{tr}\left((\boldsymbol{B}-\widehat{\boldsymbol{B}})\widehat{\boldsymbol{\Sigma}}_B^{-1}(\boldsymbol{B}-\widehat{\boldsymbol{B}})^\top\right)}{2}\right),$$

for

$$\widehat{\boldsymbol{A}} = \left(\widehat{a}_1 \boldsymbol{\omega}_{a_1}, \ldots, \widehat{a}_H \boldsymbol{\omega}_{a_H}\right),$$
$$\widehat{\boldsymbol{B}} = \left(\widehat{b}_1 \boldsymbol{\omega}_{b_1}, \ldots, \widehat{b}_H \boldsymbol{\omega}_{b_H}\right),$$
$$\widehat{\boldsymbol{\Sigma}}_A = \mathbf{Diag}\left(\widehat{\sigma}^2_{a_1}, \ldots, \widehat{\sigma}^2_{a_H}\right),$$
$$\widehat{\boldsymbol{\Sigma}}_B = \mathbf{Diag}\left(\widehat{\sigma}^2_{b_1}, \ldots, \widehat{\sigma}^2_{b_H}\right),$$

and the variational parameters $\{\widehat{a}_h, \widehat{b}_h, \widehat{\sigma}^2_{a_h}, \widehat{\sigma}^2_{b_h}\}_{h=1}^H$ are the solution of the following problem:

$$\text{Given} \quad \sigma^2 \in \mathbb{R}_{++},$$

$$\min_{\{\widehat{a}_h, \widehat{b}_h, \widehat{\sigma}^2_{a_h}, \widehat{\sigma}^2_{b_h}, c^2_{a_h}, c^2_{b_h}\}_{h=1}^H} F, \tag{12.78}$$

$$\text{s.t.} \quad \widehat{a}_h, \widehat{b}_h \in \mathbb{R}, \quad c_{a_h} c_{b_h} \geq \varepsilon^2, \quad c_{a_h}/c_{b_h} = 1,$$

$$\text{and} \begin{cases} \widehat{\sigma}^2_{a_h} = \varepsilon^2, & \widehat{\sigma}^2_{b_h} = \varepsilon^2 \quad \text{(for EMAP learning)}, \\ \widehat{\sigma}^2_{a_h} \geq \varepsilon^2, & \widehat{\sigma}^2_{b_h} = \varepsilon^2 \quad \text{(for EPB-A learning)}, \\ \widehat{\sigma}^2_{a_h} = \varepsilon^2, & \widehat{\sigma}^2_{b_h} \geq \varepsilon^2 \quad \text{(for EPB-B learning)}, \\ \widehat{\sigma}^2_{a_h} \geq \varepsilon^2, & \widehat{\sigma}^2_{b_h} \geq \varepsilon^2 \quad \text{(for EVB learning)}, \end{cases} \tag{12.79}$$

$$\text{for } h = 1, \ldots, H,$$

where the free energy $F$ is explicitly written by Eqs. (12.26) and (12.27), that is,

$$2F = LM\log(2\pi\sigma^2) + \frac{\sum_{h=1}^{\min(L,M)}\gamma_h^2}{\sigma^2} + \sum_{h=1}^{H} 2F_h, \tag{12.80}$$

where

$$\begin{aligned} 2F_h = {} & M\log\frac{c_{a_h}^2}{\widehat{\sigma}_{a_h}^2} + L\log\frac{c_{b_h}^2}{\widehat{\sigma}_{b_h}^2} + \frac{\widehat{a}_h^2 + M\widehat{\sigma}_{a_h}^2}{c_{a_h}^2} + \frac{\widehat{b}_h^2 + L\widehat{\sigma}_{b_h}^2}{c_{b_h}^2} \\ & - (L+M) + \frac{-2\widehat{a}_h\widehat{b}_h\gamma_h + \left(\widehat{a}_h^2 + M\widehat{\sigma}_{a_h}^2\right)\left(\widehat{b}_h^2 + L\widehat{\sigma}_{b_h}^2\right)}{\sigma^2}. \end{aligned} \tag{12.81}$$

By substituting the MAP solution (Corollary 12.2) and the PB solution (Corollary 12.4), respectively, into Eq. (12.81), we can write the free energy as a function of the product $c_{a_h}c_{b_h}$ of the prior covariances. We have the following lemmas (the proofs are given in Sections 12.1.4 and 12.1.5, respectively):

**Lemma 12.10** *In EMAP learning, the free energy* (12.81) *can be written as a function of* $c_{a_h}c_{b_h}$ *as follows:*

$$\begin{aligned} 2\acute{F}_h^{\mathrm{MAP}} &= \min_{\widehat{a}_h,\widehat{b}_h\in\mathbb{R},\ \widehat{\sigma}_{a_h}^2=\widehat{\sigma}_{b_h}^2=\varepsilon^2} 2F_h \\ &= \begin{cases} (L+M)\left(\log c_{a_h}c_{b_h} + \frac{\varepsilon^2}{c_{a_h}c_{b_h}} - 1 + \chi\right) & \text{for } \varepsilon^2 \le c_{a_h}c_{b_h} \le \frac{\sigma^2}{\gamma_h}, \\ (L+M)(\log c_{a_h}c_{b_h} - 1 + \chi) - \sigma^{-2}\left(\gamma_h - \frac{\sigma^2}{c_{a_h}c_{b_h}}\right)^2 & \text{for } c_{a_h}c_{b_h} > \frac{\sigma^2}{\gamma_h}. \end{cases} \end{aligned} \tag{12.82}$$

**Lemma 12.11** *In EPB learning, the free energy* (12.81) *can be written as a function of* $c_{a_h}c_{b_h}$ *as follows: if* $\gamma_h > \sigma\sqrt{\max(L,M)}$,

$$\begin{aligned} 2\acute{F}_h^{\mathrm{PB}} &= \min\left\{2\acute{F}_h^{\mathrm{PB-A}},\ 2\acute{F}_h^{\mathrm{PB-B}}\right\} \\ &= \min\left\{\min_{\widehat{a}_h,\widehat{b}_h\in\mathbb{R},\ \widehat{\sigma}_{a_h}^2\ge\varepsilon^2,\ \widehat{\sigma}_{b_h}^2=\varepsilon^2} 2F_h,\ \min_{\widehat{a}_h,\widehat{b}_h\in\mathbb{R},\ \widehat{\sigma}_{a_h}^2=\varepsilon^2,\ \widehat{\sigma}_{b_h}^2\ge\varepsilon^2} 2F_h\right\} \\ &= \begin{cases} \min(L,M)\left(\log c_{a_h}c_{b_h} + \frac{\varepsilon^2}{c_{a_h}c_{b_h}} - 1 + \chi\right) \\ \qquad\qquad \text{for } \varepsilon^2 \le c_{a_h}c_{b_h} \le \frac{\sigma^2}{\sqrt{\gamma_h^2 - \max(L,M)\sigma^2}}, \\ \frac{\min(L,M)+2\max(L,M)}{2}\log c_{a_h}^2c_{b_h}^2 + \sqrt{\max(L,M)^2 + \frac{4\gamma_h^2}{c_{a_h}^2c_{b_h}^2}} - \frac{\sigma^2}{c_{a_h}^2c_{b_h}^2} \\ \quad + \max(L,M)\log\left(-\max(L,M) + \sqrt{\max(L,M)^2 + \frac{4\gamma_h^2}{c_{a_h}^2c_{b_h}^2}}\right) \\ \quad - \frac{\gamma_h^2}{\sigma^2} - \max(L,M)\log(2\sigma^2) + \min(L,M)(\chi - 1) \\ \qquad\qquad \text{for } c_{a_h}c_{b_h} > \frac{\sigma^2}{\sqrt{\gamma_h^2 - \max(L,M)\sigma^2}}, \end{cases} \end{aligned} \tag{12.83}$$

*and otherwise,*

$$2\acute{F}_h^{\mathrm{PB}} = \min(L, M)\left(\log c_{a_h} c_{b_h} + \frac{\varepsilon^2}{c_{a_h} c_{b_h}} - 1 + \chi\right). \tag{12.84}$$

By minimizing the EMAP free energy (12.82) and the EPB free energy (12.83), respectively, with respect to the product $c_{a_h} c_{b_h}$ of the prior covariances, we obtain the following theorems (the proofs are given also in Sections 12.1.4 and 12.1.5, respectively):

**Theorem 12.12** *In EMAP learning, the free energy* (12.81) *has the global minimum such that*

$$\widehat{\gamma}_h^{\mathrm{EMAP}} \left(\equiv \widehat{a}_h \widehat{b}_h\right) = 0.$$

*It has a nontrivial local minimum such that*

$$\widehat{\gamma}_h^{\mathrm{local-EMAP}} \left(\equiv \widehat{a}_h \widehat{b}_h\right) = \breve{\gamma}_h^{\mathrm{local-EMAP}} \qquad \textit{if and only if} \qquad \gamma_h > \underline{\gamma}^{\mathrm{local-EMAP}},$$

*where*

$$\underline{\gamma}^{\mathrm{local-EMAP}} = \sigma \sqrt{2(L + M)}, \tag{12.85}$$

$$\breve{\gamma}_h^{\mathrm{local-EMAP}} = \frac{1}{2}\left(\gamma_h + \sqrt{\gamma_h^2 - 2\sigma^2(L + M)}\right). \tag{12.86}$$

**Theorem 12.13** *In EPB learning, the free energy* (12.81) *has the global minimum such that*

$$\widehat{\gamma}_h^{\mathrm{EPB}} \left(\equiv \widehat{a}_h \widehat{b}_h\right) = 0.$$

*It has a non-trivial local minimum such that*

$$\widehat{\gamma}_h^{\mathrm{local-EPB}} \left(\equiv \widehat{a}_h \widehat{b}_h\right) = \breve{\gamma}_h^{\mathrm{local-EPB}} \qquad \textit{if and only if} \qquad \gamma_h > \underline{\gamma}^{\mathrm{local-EPB}},$$

*where*

$$\underline{\gamma}^{\mathrm{local-EPB}} = \sigma \sqrt{L + M + \sqrt{2LM + \min(L, M)^2}}, \tag{12.87}$$

$$\breve{\gamma}_h^{\mathrm{local-EPB}} = \frac{\gamma_h}{2}\left(1 + \frac{-\max(L,M)\sigma^2 + \sqrt{\gamma_h^4 - 2(L+M)\sigma^2\gamma_h^2 + \min(L,M)^2\sigma^4}}{\gamma_h^2}\right). \tag{12.88}$$

Figure 12.4 shows the free energy (normalized by $LM$) as a function of $c_{a_h} c_{b_h}$ for EMAP learning (given in Lemma 12.10), EPB learning (given in Lemma 12.11), and EVB learning, defined by

$$2\acute{F}_h^{\mathrm{VB}} = \min_{\widehat{a}_h, \widehat{b}_h \in \mathbb{R},\ \widehat{\sigma}_{a_h}^2, \widehat{\sigma}_{b_h}^2 \geq \varepsilon^2} 2F_h.$$

For EMAP learing and EPB learning, we ignored some constants (e.g., the entropy terms proportional to $\chi$) to make the *shapes* of the free energies

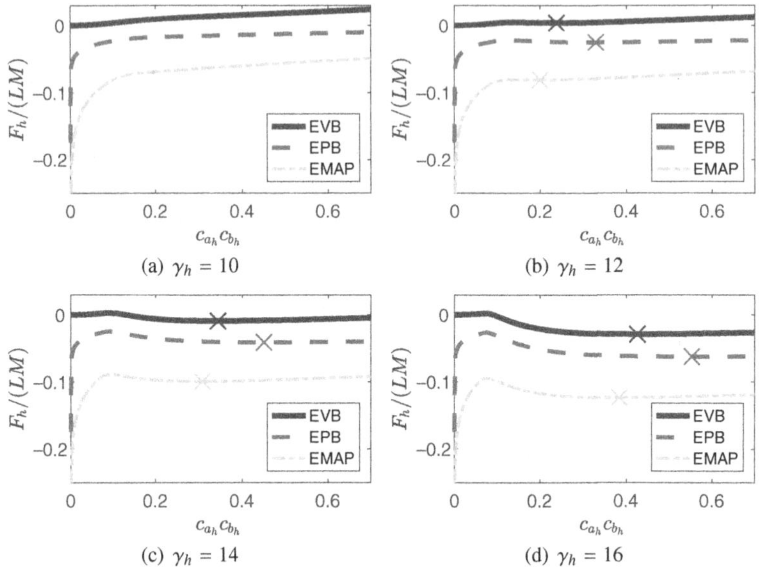


Figure 12.4 Free energy dependence on $c_{a_h}c_{b_h}$, where $L = 20, M = 50$. Crosses indicate nontrivial local minima.

comparable. We can see deep pits at $c_{a_h}c_{b_h} \to +0$ in EMAP and EPB free energies, which correspond to the global solutions. However, we also see nontrivial local minima, which behave similarly to the nontrivial local solution for VB learning. Namely, nontrivial local minima of EMAP, EPB, and EVB free energies appear at locations similar to each other when the observed singular value $\gamma_h$ exceeds the thresholds given by Eqs. (12.85), (12.87), and (6.127), respectively.

The deep pit at $c_{a_h}c_{b_h} \to +0$ is essential when we stick to the global solution. The VB free energy does not have such a pit, which enables consistent inference based on the free energy minimization principle. However, as long as we rely on local search, the pit at the origin is not essential in practice. Assume that a nontrivial local minimum exists, and we perform local search only once. Then, whether local search for EMAP learning or EPB learning converges to the trivial global solution or the nontrivial local solution simply depends on the initialization. Note that the same applies also to EVB learning, for which local search is not guaranteed to converge to the global solution. This is because of the multimodality of the VB free energy, which can be seen in Figure 12.4.

One might wonder if some hyperpriors on $c_{a_h}^2$ and $c_{b_h}^2$ could fill the deep pits at $c_{a_h}c_{b_h} \to +0$ in the EMAP and the EPB free energies, so that the nontrivial local solutions are global when some reasonable conditions hold. However, when we rely on the ARD property for model selection, hyperpriors should be almost noninformative. With such an almost noninformative hyperprior, e.g., the inverge-Gamma, $p(c_{a_h}^2, c_{b_h}^2) \propto (c_{a_h}^2 c_{b_h}^2)^{1.001} + 0.001/(c_{a_h}^2 c_{b_h}^2)$, which was used in Bishop (1999b), deep pits still exist very close to the origin, which keep the global EMAP and EPB estimators useless.

**Comparison between Local-EMAP, Local-EPB, and EVB Solutions**
Let us observe the behavior of local solutions. We define the *local-EMAP estimator* and the *local-EPB estimator*, respectively, by

$$\widehat{U}^{\text{local-EMAP}} = \sum_{h=1}^{H} \widehat{\gamma}_h^{\text{local-EMAP}} \omega_{b_h} \omega_{a_h}^\top,$$
$$\text{where} \quad \widehat{\gamma}_h^{\text{local-EMAP}} = \begin{cases} \breve{\gamma}_h^{\text{local-EMAP}} & \text{if } \gamma_h \geq \underline{\gamma}^{\text{local-EMAP}}, \\ 0 & \text{otherwise,} \end{cases} \tag{12.89}$$

$$\widehat{U}^{\text{local-EPB}} = \sum_{h=1}^{H} \widehat{\gamma}_h^{\text{local-EPB}} \omega_{b_h} \omega_{a_h}^\top,$$
$$\text{where} \quad \widehat{\gamma}_h^{\text{local-EPB}} = \begin{cases} \breve{\gamma}_h^{\text{local-EPB}} & \text{if } \gamma_h \geq \underline{\gamma}^{\text{local-EPB}}, \\ 0 & \text{otherwise,} \end{cases} \tag{12.90}$$

following the definition of the local-EVB estimator (6.131) in Chapter 6. In the following, we assume that local search algorithms for EMAP learning and EPB learning find these solutions.

Define the normalized (by the average noise contribution (12.74)) singular values:

$$\gamma'_h = \frac{\gamma_h}{\sqrt{\max(L, M)\sigma^2}}.$$

We also define normalized versions of the estimator, the truncation threshold, and the shrinkage factor as

$$\widehat{\gamma}'_h = \frac{\widehat{\gamma}_h}{\sqrt{\max(L, M)\sigma^2}}, \quad \underline{\gamma}'_h = \frac{\underline{\gamma}_h}{\sqrt{\max(L, M)\sigma^2}}, \quad \breve{\gamma}'_h = \frac{\breve{\gamma}_h}{\sqrt{\max(L, M)\sigma^2}}. \tag{12.91}$$

Then the normalized truncation thresholds and the normalized shrinkage factors can be written as functions of $\alpha = \min(L, M)/\max(L, M)$ as follows:

$$\underline{\gamma}'^{\text{EVB}} = \sigma \sqrt{1 + \alpha + \sqrt{\alpha}\left(\underline{\kappa} + \tfrac{1}{\underline{\kappa}}\right)}, \tag{12.92}$$

$$\breve{\gamma}_h'^{\text{EVB}} = \frac{\gamma'_h}{2}\left(1 - \frac{(1+\alpha)\sigma^2}{\gamma_h'^2} + \sqrt{\left(1 - \frac{(1+\alpha)\sigma^2}{\gamma_h'^2}\right)^2 - \frac{4\alpha\sigma^4}{\gamma_h'^4}}\right), \tag{12.93}$$

$$\underline{\gamma}'^{\text{local-EPB}} = \sigma\sqrt{1 + \alpha + \sqrt{2\alpha + \alpha^2}}, \tag{12.94}$$

$$\breve{\gamma}_h'^{\text{local-EPB}} = \frac{\gamma_h'}{2}\left(1 + \frac{-\sigma^2 + \sqrt{\gamma_h'^4 - 2(1+\alpha)\sigma^2\gamma_h'^2 + \sigma^4}}{\gamma_h'^2}\right), \tag{12.95}$$

$$\underline{\gamma}'^{\text{local-EMAP}} = \sigma\sqrt{2(1 + \alpha)}, \tag{12.96}$$

$$\breve{\gamma}_h'^{\text{local-EMAP}} = \frac{1}{2}\left(\gamma_h' + \sqrt{\gamma_h'^2 - 2\sigma^2(1 + \alpha)}\right). \tag{12.97}$$

Note that $\underline{\kappa}$ is also a function of $\alpha$.

Figure 12.5 compares the normalized versions of the (global) EVB estimator $\widehat{\gamma}_h'^{\text{EVB}}$, the local-EPB estimator $\widehat{\gamma}_h'^{\text{local-EPB}}$, and the local-EMAP estimator $\widehat{\gamma}_h'^{\text{local-EMAP}}$. We can observe similar behaviors of those three empirical Bayesian estimators. This is in contrast to the nonempirical Bayesian estimators shown in Figure 12.3, where the PB estimator behaves similarly to the VB estimator, while the MAP estimator behaves differently.

Figure 12.6 compares the normalized versions of the EVB truncation threshold (12.92), the local-EPB truncation threshold (12.94), and the local-EMAP truncation threshold (12.96). We can see that those thresholds behave similarly. However, we can find an essential difference of the local-EPB threshold from the EVB and the local-EMAP thresholds: it holds that, for any $\alpha$,

$$\underline{\gamma}'^{\text{local-EPB}} < \overline{\gamma}'^{\text{MPUL}} \leq \underline{\gamma}'^{\text{EVB}}, \underline{\gamma}'^{\text{local-EMAP}}, \tag{12.98}$$

$$\text{where} \qquad \overline{\gamma}'^{\text{MPUL}} = \frac{\overline{\gamma}^{\text{MPUL}}}{\sqrt{\max(L, M)\sigma^2}} = (1 + \sqrt{\alpha})$$

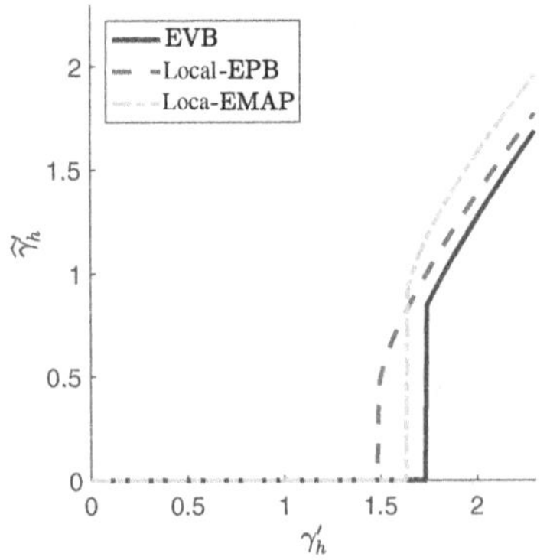


Figure 12.5 Behavior of (global) EVB, the local-EPB, and the local-EMAP estimators for $\alpha = \min(L, M)/\max(L, M) = 1/3$.

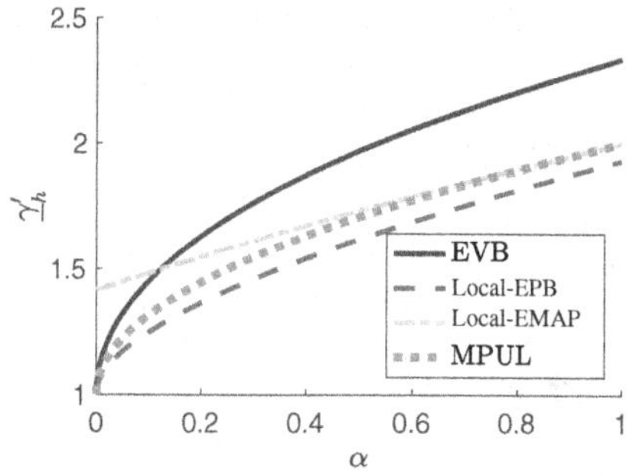


Figure 12.6 Truncation thresholds.

is the normalized version of the Marčenko–Pastur upper limit (MPUL) (Eq. (8.41) in Chapter 8), which is also shown in Figure 12.6.

As discussed in Section 8.4.1, the MPUL is the largest singular value of an $L \times M$ zero-mean independent random matrix in the *large-scale limit* where the matrix size $(L, M)$ goes to infinity with fixed ratio $\alpha = \min(L, M)/\max(L, M)$. In other words, the MPUL corresponds to the minimum observed singular value detectable (or distinguishable from noise) by any dimensionality reduction method. The inequalities (12.98) say that local-EPB threshold is always smaller than the MPUL, while the EVB threshold and the local-EMAP threshold are never smaller than the MPUL. This implies that, for a large-scale observed matrix, the EVB estimator and the local-EMAP estimator discard the singular components dominated by noise, while the local-EPB estimator retains some of them.

### 12.1.4 Proofs of Lemma 12.10 and Theorem 12.12

By substituting the MAP solution, given by Corollary 12.2, we can write the free energy (12.81) as follows: for $\varepsilon^2 \leq c_{a_h} c_{b_h} \leq \frac{\sigma^2}{\gamma_h}$,

$$
\begin{aligned}
2\acute{F}_h^{\text{MAP}} &= \min_{\widehat{a}_h, \widehat{b}_h \in \mathbb{R},\ \widehat{\sigma}^2_{a_h} = \widehat{\sigma}^2_{b_h} = \varepsilon^2} 2F_h \\
&= M \log c_{a_h}^2 + L \log c_{b_h}^2 \\
&\quad + \left( \frac{M}{c_{a_h}^2} + \frac{L}{c_{b_h}^2} \right) \varepsilon^2 - (L + M) + (L + M)\chi \\
&= M \log c_{a_h}^2 + L \log c_{b_h}^2 + \left( \frac{M}{c_{a_h}^2} + \frac{L}{c_{b_h}^2} \right) \varepsilon^2 \\
&\quad - (L + M) + (L + M)\chi, \qquad (12.99)
\end{aligned}
$$

and for $c_{a_h} c_{b_h} > \frac{\sigma^2}{\gamma_h}$,

$$\begin{aligned}
2\acute{F}_h^{\mathrm{MAP}} &= \min_{\widehat{a}_h, \widehat{b}_h \in \mathbb{R},\ \widehat{\sigma}^2_{a_h} = \widehat{\sigma}^2_{b_h} = \varepsilon^2} 2F_h \\
&= M \log c_{a_h}^2 + L \log c_{b_h}^2 + \left(\gamma_h - \frac{\sigma^2}{c_{a_h} c_{b_h}}\right)\left(\frac{2}{c_{a_h} c_{b_h}} + \frac{-2\gamma_h + \gamma_h - \frac{\sigma^2}{c_{a_h} c_{b_h}}}{\sigma^2}\right) \\
&\quad + \left(\frac{M}{c_{a_h}^2} + \frac{L}{c_{b_h}^2}\right)\varepsilon^2 - (L+M) + (L+M)\chi \\
&= M \log c_{a_h}^2 + L \log c_{b_h}^2 - \sigma^{-2}\left(\gamma_h - \frac{\sigma^2}{c_{a_h} c_{b_h}}\right)^2 \\
&\quad - (L+M) + (L+M)\chi. \qquad (12.100)
\end{aligned}$$

In the second-to-last equation in Eq. (12.100), we ignored the fourth term because $c_{a_h} c_{b_h} > \frac{\sigma^2}{\gamma_h}$ implies $c_{a_h}^2, c_{b_h}^2 \gg \varepsilon^2$ (with an arbitrarily high probability depending on $\varepsilon^2$). By fixing the ratio to $c_{a_h}/c_{b_h} = 1$, we obtain Eq. (12.82), which proves Lemma 12.10.

Now we minimize the free energy (12.82) with respect to $c_{a_h} c_{b_h}$, and find nontrivial local solutions. The free energy is continuous in the domain $\varepsilon^2 \leq c_{a_h} c_{b_h} < \infty$, and differentiable except at $c_{a_h} c_{b_h} = \frac{\sigma^2}{\gamma_h}$. The derivative is given by

$$\begin{aligned}
\frac{\partial 2\acute{F}_h^{\mathrm{MAP}}}{\partial (c_{a_h} c_{b_h})} &= \begin{cases} (L+M)\left(\frac{1}{c_{a_h} c_{b_h}} - \frac{\varepsilon^2}{c_{a_h}^2 c_{b_h}^2}\right) & \text{for } \varepsilon^2 \leq c_{a_h} c_{b_h} \leq \frac{\sigma^2}{\gamma_h} \\ \left(\frac{L+M}{c_{a_h} c_{b_h}} + 2\sigma^{-2}\left(\gamma_h - \frac{\sigma^2}{c_{a_h} c_{b_h}}\right)\frac{\sigma^2}{c_{a_h}^2 c_{b_h}^2}\right) & \text{for } c_{a_h} c_{b_h} > \frac{\sigma^2}{\gamma_h} \end{cases} \\
&= \begin{cases} \frac{L+M}{c_{a_h}^2 c_{b_h}^2}\left(c_{a_h} c_{b_h} - \varepsilon^2\right) & \text{for } \varepsilon^2 \leq c_{a_h} c_{b_h} \leq \frac{\sigma^2}{\gamma_h}, \\ \frac{1}{c_{a_h}^3 c_{b_h}^3}\left((L+M) c_{a_h}^2 c_{b_h}^2 + 2\gamma_h c_{a_h} c_{b_h} - 2\sigma^2\right) & \text{for } c_{a_h} c_{b_h} > \frac{\sigma^2}{\gamma_h}. \end{cases} \qquad (12.101)
\end{aligned}$$

Eq. (12.101) implies that the free energy $\acute{F}_h^{\mathrm{MAP}}$ is increasing for $\varepsilon^2 \leq c_{a_h} c_{b_h} \leq \frac{\sigma^2}{\gamma_h}$, and that it is increasing at $c_{a_h} c_{b_h} = \frac{\sigma^2}{\gamma_h}$ and at $c_{a_h} c_{b_h} \to +\infty$.

In the region of $\frac{\sigma^2}{\gamma_h} < c_{a_h} c_{b_h} < +\infty$, the free energy has stationary points if and only if

$$\gamma_h \geq \sigma\sqrt{2(L+M)}\left(\equiv \underline{\gamma}^{\text{local-EMAP}}\right), \qquad (12.102)$$

because the derivative can be factorized (with real factors if and only if the condition (12.102) holds) as

$$\frac{\partial 2\acute{F}_h^{\mathrm{MAP}}}{\partial (c_{a_h} c_{b_h})} = \frac{L+M}{c_{a_h}^3 c_{b_h}^3}\left(c_{a_h} c_{b_h} - \acute{c}_{a_h} \acute{c}_{b_h}\right)\left(c_{a_h} c_{b_h} - \check{c}_{a_h} \check{c}_{b_h}\right),$$

where

$$\acute{c}_{a_h}\acute{c}_{b_h} = \frac{\gamma_h - \sqrt{\gamma_h^2 - 2\sigma^2(L+M)}}{L+M}, \tag{12.103}$$

$$\check{c}_{a_h}\check{c}_{b_h} = \frac{\gamma_h + \sqrt{\gamma_h^2 - 2\sigma^2(L+M)}}{L+M}. \tag{12.104}$$

Summarizing the preceding discussion, we have the following lemma:

**Lemma 12.14** *If $\gamma_h \leq \underline{\gamma}^{\text{local-EMAP}}$, the EMAP free energy $\acute{F}_h^{\text{MAP}}$, defined by Eq.* (12.82), *is increasing for $c_{a_h}c_{b_h} > \varepsilon^2$, and therefore minimized at $c_{a_h}c_{b_h} = \varepsilon^2$. If $\gamma_h > \underline{\gamma}^{\text{local-EMAP}}$,*

$$\acute{F}_h^{\text{MAP}} \text{ is } \begin{cases} \text{increasing} & \text{for} \quad \varepsilon^2 < c_{a_h}c_{b_h} < \acute{c}_{a_h}\acute{c}_{b_h}, \\ \text{decreasing} & \text{for} \quad \acute{c}_{a_h}\acute{c}_{b_h} < c_{a_h}c_{b_h} < \check{c}_{a_h}\check{c}_{b_h}, \\ \text{increasing} & \text{for} \quad \check{c}_{a_h}\check{c}_{b_h} < c_{a_h}c_{b_h} < +\infty, \end{cases}$$

*and therefore has two (local) minima at $c_{a_h}c_{b_h} = \varepsilon^2$ and at $c_{a_h}c_{b_h} = \check{c}_{a_h}\check{c}_{b_h}$. Here $\acute{c}_{a_h}\acute{c}_{b_h}$ and $\check{c}_{a_h}\check{c}_{b_h}$ are defined by Eqs.* (12.103) *and* (12.104), *respectively.*

When $\gamma_h > \underline{\gamma}^{\text{local-EMAP}}$, the EMAP free energy (12.82) at the local minima is

$$2\acute{F}_h^{\text{MAP}} = \begin{cases} 0 & \text{at } c_{a_h}c_{b_h} = \varepsilon^2, \\ (L+M)\left(\log \check{c}_{a_h}\check{c}_{b_h} - 1 + \chi\right) - \sigma^{-2}\left(\gamma_h - \frac{\sigma^2}{\check{c}_{a_h}\check{c}_{b_h}}\right)^2 & \text{at } c_{a_h}c_{b_h} = \check{c}_{a_h}\check{c}_{b_h}, \end{cases}$$

respectively. Since we assume that $\chi = -\log \varepsilon^2$ is an arbitrarily large constant ($\varepsilon^2 > 0$ is arbitrarily small), $c_{a_h}c_{b_h} = \varepsilon^2$ is always the global minimum.

Substituting Eq. (12.104) into Eq. (12.36) gives Eq. (12.86), which completes the proof of Theorem 12.12. □

### 12.1.5 Proofs of Lemma 12.11 and Theorem 12.13

We first analyze the free energy for EPB-A learning. From Eq. (12.49), we have

$$c_{a_h}c_{b_h} = \frac{\sigma^2}{\sqrt{(\underline{\gamma}_h^{\text{PB-A}})^2 - M\sigma^2}}.$$

Therefore, if

$$\gamma_h \leq \sigma\sqrt{M},$$

there exists only the null solution (12.53) for any $c_{a_h} c_{b_h} > 0$, and therefore the free energy (12.81) is given by

$$2\acute{F}_h^{\mathrm{PB-A}} = \min_{\widehat{a}_h, \widehat{b}_h \in \mathbb{R},\ \widehat{\sigma}^2_{a_h} \geq \varepsilon^2,\ \widehat{\sigma}^2_{b_h} = \varepsilon^2} 2F_h$$
$$= L\left(\log c^2_{b_h} + \frac{\varepsilon^2}{c^2_{b_h}} - 1 + \chi\right). \tag{12.105}$$

In the following, we consider the case where

$$\gamma_h > \sigma\sqrt{M}. \tag{12.106}$$

For $\varepsilon^2 \leq c_{a_h} c_{b_h} \leq \frac{\sigma^2}{\sqrt{\gamma_h^2 - M\sigma^2}}$, there still exists only the null solution (12.53) with the free energy given by Eq. (12.105). The positive solution (12.50) appears for $c_{a_h} c_{b_h} > \frac{\sigma^2}{\sqrt{\gamma_h^2 - M\sigma^2}}$ with the free energy given by

$$2\acute{F}_h^{\mathrm{PB-A}} = \min_{\widehat{a}_h, \widehat{b}_h \in \mathbb{R},\ \widehat{\sigma}^2_{a_h} \geq \varepsilon^2,\ \widehat{\sigma}^2_{b_h} = \varepsilon^2} 2F_h$$
$$= \min_{\widehat{a}_h, \widehat{b}_h \in \mathbb{R},\ \widehat{\sigma}^2_{a_h} \geq \varepsilon^2} \left\{ M\log\frac{c^2_{a_h}}{\widehat{\sigma}^2_{a_h}} + L\log c^2_{b_h} + \frac{\widehat{b}_h^2}{c^2_{b_h}} \right.$$
$$\left. - \frac{2\widehat{a}_h\widehat{b}_h\gamma_h}{\sigma^2} + \left(\widehat{a}_h^2 + M\widehat{\sigma}^2_{a_h}\right)\left(\frac{\widehat{b}_h^2}{\sigma^2} + \frac{1}{c^2_{a_h}}\right)\right\} - (L+M) + L\chi$$
$$= \min_{\widehat{a}_h, \widehat{b}_h \in \mathbb{R},\ \widehat{\sigma}^2_{a_h} \geq \varepsilon^2} \left\{ M\log\frac{\widehat{b}_h^2 + \sigma^2/c^2_{a_h}}{\sigma^2} + \frac{\widehat{b}_h^2}{c^2_{b_h}} - \frac{\widehat{b}_h^2\gamma_h}{\widehat{b}_h^2 + \sigma^2/c^2_{a_h}}\frac{2\gamma_h}{\sigma^2} + \frac{\widehat{a}_h^2 + M\widehat{\sigma}^2_{a_h}}{\widehat{\sigma}^2_{a_h}}\right\}$$
$$+ M\log c^2_{a_h} + L\log c^2_{b_h} - (L+M) + L\chi$$
$$= \min_{\widehat{b}_h \in \mathbb{R}} \left\{ M\log\left(\widehat{b}_h^2 + \sigma^2/c^2_{a_h}\right) + \frac{\widehat{b}_h^2 + \sigma^2/c^2_{a_h}}{c^2_{b_h}} - \frac{\widehat{b}_h^2\gamma_h^2}{\sigma^2(\widehat{b}_h^2 + \sigma^2/c^2_{a_h})}\right\}$$
$$- \frac{\sigma^2}{c^2_{a_h}c^2_{b_h}} + M\log c^2_{a_h} + L\log c^2_{b_h} - M\log\sigma^2 - L + L\chi$$
$$= \min_{\widehat{b}_h \in \mathbb{R}} \left\{ M\log\left(\widehat{b}_h^2 + \sigma^2/c^2_{a_h}\right) + \frac{\widehat{b}_h^2 + \sigma^2/c^2_{a_h}}{c^2_{b_h}} + \frac{\gamma_h^2}{c^2_{a_h}(\widehat{b}_h^2 + \sigma^2/c^2_{a_h})}\right\}$$
$$- \frac{\gamma_h^2}{\sigma^2} - \frac{\sigma^2}{c^2_{a_h}c^2_{b_h}} + M\log c^2_{a_h} + L\log c^2_{b_h} - M\log\sigma^2 - L + L\chi. \tag{12.107}$$

Here we used the conditions (12.61) and (12.62) for the PB-A solution. By substituting the other conditions (12.64) and (12.66) into Eq. (12.107), we have

$$2\acute{F}_h^{\text{PB-A}} = M\log\left(-M + \sqrt{M^2 + \frac{4\gamma_h^2}{c_{a_h}^2 c_{b_h}^2}}\right) + \frac{-M + \sqrt{M^2 + \frac{4\gamma_h^2}{c_{a_h}^2 c_{b_h}^2}}}{2} + \frac{M + \sqrt{M^2 + \frac{4\gamma_h^2}{c_{a_h}^2 c_{b_h}^2}}}{2}$$
$$- \frac{\gamma_h^2}{\sigma^2} - \frac{\sigma^2}{c_{a_h}^2 c_{b_h}^2} + M\log c_{a_h}^2 + (L+M)\log c_{b_h}^2 - M\log(2\sigma^2) - L + L\chi$$
$$= M\log\left(-M + \sqrt{M^2 + \frac{4\gamma_h^2}{c_{a_h}^2 c_{b_h}^2}}\right) + \sqrt{M^2 + \frac{4\gamma_h^2}{c_{a_h}^2 c_{b_h}^2}} - \frac{\sigma^2}{c_{a_h}^2 c_{b_h}^2}$$
$$- \frac{\gamma_h^2}{\sigma^2} + M\log c_{a_h}^2 + (L+M)\log c_{b_h}^2 - M\log(2\sigma^2) - L + L\chi. \quad (12.108)$$

The PB-B free energy can be derived in exactly the same way, and the result is symmetric to the PB-A free energy. Namely, if

$$\gamma_h \leq \sigma\sqrt{L},$$

there exists only the null solution (12.59) for any $c_{a_h} c_{b_h} > 0$ with the free energy given by

$$2\acute{F}_h^{\text{PB-B}} = \min_{\widehat{a}_h, \widehat{b}_h \in \mathbb{R},\ \widehat{\sigma}_{a_h}^2 = \varepsilon^2,\ \widehat{\sigma}_{b_h}^2 \geq \varepsilon^2} 2F_h$$
$$= M\left(\log c_{a_h}^2 + \frac{\varepsilon^2}{c_{a_h}^2} - 1 + \chi\right). \quad (12.109)$$

Assume that

$$\gamma_h > \sigma\sqrt{L}.$$

For $\varepsilon^2 \leq c_{a_h} c_{b_h} \leq \frac{\sigma^2}{\sqrt{\gamma_h^2 - L\sigma^2}}$, there still exists only the null solution (12.59) with the free energy given by Eq. (12.109). The positive solution (12.56) appears for $c_{a_h} c_{b_h} > \frac{\sigma^2}{\sqrt{\gamma_h^2 - L\sigma^2}}$ with the free energy given by

$$2\acute{F}_h^{\text{PB-B}} = \min_{\widehat{a}_h, \widehat{b}_h \in \mathbb{R},\ \widehat{\sigma}_{a_h}^2 = \varepsilon^2,\ \widehat{\sigma}_{b_h}^2 \geq \varepsilon^2} 2F_h$$
$$= L\log\left(-L + \sqrt{L^2 + \frac{4\gamma_h^2}{c_{a_h}^2 c_{b_h}^2}}\right) + \sqrt{L^2 + \frac{4\gamma_h^2}{c_{a_h}^2 c_{b_h}^2}} - \frac{\sigma^2}{c_{a_h}^2 c_{b_h}^2}$$
$$- \frac{\gamma_h^2}{\sigma^2} + (L+M)\log c_{a_h}^2 + L\log c_{b_h}^2 - L\log(2\sigma^2) - M + M\chi. \quad (12.110)$$

By fixing the ratio between the prior covariances to $c_{a_h}/c_{b_h} = 1$ in Eqs. (12.105) and (12.108) through (12.110), and taking the posterior choice in Eq. (12.47) into account, we obtain Eqs. (12.83) and (12.84), which prove Lemma 12.11.

Let us minimize the free energy with respect to $c_{a_h}c_{b_h}$. When

$$\gamma_h \leq \sigma\sqrt{\max(L, M)},$$

the free energy is given by Eq. (12.84), and its derivative is given by

$$\frac{\partial 2\acute{F}_h^{\mathrm{PB}}}{\partial(c_{a_h}c_{b_h})} = \frac{\min(L, M)}{c_{a_h}^2 c_{b_h}^2}\left(c_{a_h}c_{b_h} - \varepsilon^2\right).$$

This implies that the free energy $\acute{F}_h^{\mathrm{PB}}$ is increasing for $\varepsilon^2 < c_{a_h}c_{b_h} < \infty$, and therefore minimized at $c_{a_h}c_{b_h} = \varepsilon^2$.

Assume that

$$\gamma_h > \sigma\sqrt{\max(L, M)}.$$

In this case, the free energy is given by Eq. (12.83), which is continuous in the domain $\varepsilon^2 \leq c_{a_h}c_{b_h} < \infty$, and differentiable except at $c_{a_h}c_{b_h} = \frac{\sigma^2}{\sqrt{\gamma_h^2 - \max(L,M)\sigma^2}}$. Although the continuity is not very obvious, one can verify it by checking the value at $c_{a_h}c_{b_h} = \frac{\sigma^2}{\sqrt{\gamma_h^2 - \max(L,M)\sigma^2}}$ for each case in Eq. (12.83). The continuity is also expected from the fact that the PB solution is continuous at the threshold $\gamma_h = \underline{\gamma}_h^{\mathrm{PB}}$, i.e., the positive solution (Eq. (12.50) for PB-A and Eq. (12.56) for PB-B) converges to the null solution (Eq. (12.53) for PB-A and Eq. (12.59) for PB-B) when $\gamma_h \to \underline{\gamma}_h^{\mathrm{PB-A}} + 0$.

The free energy (12.83) is the same as Eq. (12.84) for $\varepsilon^2 \leq c_{a_h}c_{b_h} \leq \frac{\sigma^2}{\sqrt{\gamma_h^2 - \max(L,M)\sigma^2}}$, and therefore increasing in $\varepsilon^2 < c_{a_h}c_{b_h} \leq \frac{\sigma^2}{\sqrt{\gamma_h^2 - \max(L,M)\sigma^2}}$. For $c_{a_h}c_{b_h} > \frac{\sigma^2}{\sqrt{\gamma_h^2 - \max(L,M)\sigma^2}}$, the derivative of the free energy with respect to $c_{a_h}^2 c_{b_h}^2$ is given by

$$\begin{aligned}
\frac{\partial 2\acute{F}_h^{\mathrm{PB}}}{\partial(c_{a_h}^2 c_{b_h}^2)} &= \frac{\min(L,M)+2\max(L,M)}{2c_{a_h}^2 c_{b_h}^2} - \frac{4\gamma_h^2}{2c_{a_h}^4 c_{b_h}^4\sqrt{\max(L,M)^2 + \frac{4\gamma_h^2}{c_{a_h}^2 c_{b_h}^2}}} + \frac{\sigma^2}{c_{a_h}^4 c_{b_h}^4} \\
&\quad - \frac{4\max(L,M)\gamma_h^2}{2c_{a_h}^4 c_{b_h}^4\sqrt{\max(L,M)^2 + \frac{4\gamma_h^2}{c_{a_h}^2 c_{b_h}^2}}\left(-\max(L,M) + \sqrt{\max(L,M)^2 + \frac{4\gamma_h^2}{c_{a_h}^2 c_{b_h}^2}}\right)} \\
&= \frac{\min(L,M)+2\max(L,M)}{2c_{a_h}^2 c_{b_h}^2} + \frac{\sigma^2}{c_{a_h}^4 c_{b_h}^4} - \frac{4\gamma_h^2}{2c_{a_h}^4 c_{b_h}^4\left(-\max(L,M) + \sqrt{\max(L,M)^2 + \frac{4\gamma_h^2}{c_{a_h}^2 c_{b_h}^2}}\right)} \\
&= \frac{1}{2c_{a_h}^2 c_{b_h}^2}\left(L + M + 2\frac{\sigma^2}{c_{a_h}^2 c_{b_h}^2} - \sqrt{\max(L,M)^2 + \frac{4\gamma_h^2}{c_{a_h}^2 c_{b_h}^2}}\right) \\
&= \frac{1}{2c_{a_h}^4 c_{b_h}^4}\left((L+M)c_{a_h}^2 c_{b_h}^2 + 2\sigma^2 - c_{a_h}c_{b_h}\sqrt{\max(L,M)^2 c_{a_h}^2 c_{b_h}^2 + 4\gamma_h^2}\right),
\end{aligned} \tag{12.111}$$

which has the same sign as

$$\tau(c_{a_h}^2 c_{b_h}^2) = \left\{(L+M)c_{a_h}^2 c_{b_h}^2 + 2\sigma^2\right\}^2 - \left\{c_{a_h} c_{b_h} \sqrt{\max(L,M)^2 c_{a_h}^2 c_{b_h}^2 + 4\gamma_h^2}\right\}^2$$
$$= \left(2LM + \min(L,M)^2\right) c_{a_h}^4 c_{b_h}^4 - 4\left(\gamma_h^2 - \sigma^2(L+M)\right) c_{a_h}^2 c_{b_h}^2 + 4\sigma^4. \tag{12.112}$$

Eq. (12.112) is a quadratic function of $c_{a_h}^2 c_{b_h}^2$, being positive at $c_{a_h}^2 c_{b_h}^2 \to +0$ and at $c_{a_h}^2 c_{b_h}^2 \to +\infty$. The free energy has stationary points if and only if

$$\gamma_h \geq \sigma \sqrt{L + M + \sqrt{2LM + \min(L,M)^2}} \left(\equiv \underline{\gamma}^{\text{local-EPB}}\right), \tag{12.113}$$

because $\tau(c_{a_h}^2 c_{b_h}^2)$, which has the same sign as the derivative of the free energy, can be factorized (with real factors if and only if the condition (12.113) holds) as

$$\tau(c_{a_h}^2 c_{b_h}^2) = \left(2LM + \min(L,M)^2\right)\left(c_{a_h}^2 c_{b_h}^2 - \acute{c}_{a_h}^2 \acute{c}_{b_h}^2\right)\left(c_{a_h}^2 c_{b_h}^2 - \breve{c}_{a_h}^2 \breve{c}_{b_h}^2\right),$$

where

$$\acute{c}_{a_h}^2 \acute{c}_{b_h}^2 = 2 \cdot \frac{(\gamma_h^2 - \sigma^2(L+M)) - \sqrt{(\gamma_h^2 - \sigma^2(L+M))^2 - (2LM + \min(L,M)^2)\sigma^4}}{2LM + \min(L,M)^2}, \tag{12.114}$$

$$\breve{c}_{a_h}^2 \breve{c}_{b_h}^2 = 2 \cdot \frac{(\gamma_h^2 - \sigma^2(L+M)) + \sqrt{(\gamma_h^2 - \sigma^2(L+M))^2 - (2LM + \min(L,M)^2)\sigma^4}}{2LM + \min(L,M)^2}. \tag{12.115}$$

Summarizing the preceding discussion, we have the following lemma:

**Lemma 12.15** *If* $\gamma_h \leq \underline{\gamma}^{\text{local-EPB}}$, *the EPB free energy* $\acute{F}_h^{\text{PB}}$, *defined by Eqs.* (12.83) *and* (12.84), *is increasing for* $c_{a_h} c_{b_h} > \varepsilon^2$, *and therefore minimized at* $c_{a_h} c_{b_h} = \varepsilon^2$. *If* $\gamma_h > \underline{\gamma}^{\text{local-EPB}}$,

$$\acute{F}_h^{\text{PB}} \textit{ is } \begin{cases} \textit{increasing} & \textit{for} \quad \varepsilon^2 < c_{a_h} c_{b_h} < \acute{c}_{a_h} \acute{c}_{b_h}, \\ \textit{decreasing} & \textit{for} \quad \acute{c}_{a_h} \acute{c}_{b_h} < c_{a_h} c_{b_h} < \breve{c}_{a_h} \breve{c}_{b_h}, \\ \textit{increasing} & \textit{for} \quad \breve{c}_{a_h} \breve{c}_{b_h} < c_{a_h} c_{b_h} < +\infty, \end{cases}$$

*and therefore has two (local) minima at* $c_{a_h} c_{b_h} = \varepsilon^2$ *and at* $c_{a_h} c_{b_h} = \breve{c}_{a_h} \breve{c}_{b_h}$. *Here,* $\acute{c}_{a_h} \acute{c}_{b_h}$ *and* $\breve{c}_{a_h} \breve{c}_{b_h}$ *are defined by Eqs.* (12.114) *and* (12.115), *respectively.*

When $\gamma_h > \underline{\gamma}^{\text{local-EPB}}$, the EPB free energy (12.83) at the null local solution $c_{a_h} c_{b_h} = \varepsilon^2$ is $2\acute{F}_h^{\text{PB}} = 0$, while the EPB free energy at the positive local solution $c_{a_h} c_{b_h} = \breve{c}_{a_h} \breve{c}_{b_h}$ contains the term $\min(L,M)\chi$ with $\chi = -\log \varepsilon^2$ assumed to be arbitrarily large. Therefore, the null solution is always the global minimum.

Substituting Eq. (12.115) into Eq. (12.46) gives Eq. (12.88), which completes the proof of Theorem 12.13. □

### 12.1.6 Noise Variance Estimation

The noise variance $\sigma^2$ is unknown in many practical applications. In VB learning, minimizing the free energy (12.26) with respect also to $\sigma^2$ gives a reasonable estimator, with which perfect dimensionality recovery was proven in Chapter 8. Here, we investigate whether MAP learning and PB learning offer good noise variance estimators.

We first consider the nonempirical Bayesian variants where the prior covariances $\boldsymbol{C}_A, \boldsymbol{C}_B$ are treated as given constants. By using Lemma 12.10, we can write the MAP free energy with the variational parameters optimized, as a function of $\sigma^2$, as follows:

$$
\begin{aligned}
2\acute{F}^{\mathrm{MAP}} &= LM\log(2\pi\sigma^2) + \frac{\sum_{h=1}^{\min(L,M)}\gamma_h^2}{\sigma^2} + \sum_{h=1}^{H} 2\acute{F}_h^{\mathrm{MAP}} \\
&= LM\log(2\pi\sigma^2) + \frac{\sum_{h=1}^{\min(L,M)}\gamma_h^2}{\sigma^2} \\
&\quad + \sum_{h=1}^{\min(H,\overline{H})}\left\{(L+M)\left(\log c_{a_h}c_{b_h} - 1 + \chi\right) - \sigma^{-2}\left(\gamma_h - \frac{\sigma^2}{c_{a_h}c_{b_h}}\right)^2\right\} \\
&\quad + \sum_{h=\min(H,\overline{H})+1}^{\min(L,M)}(L+M)\left(\log c_{a_h}c_{b_h} + \frac{\varepsilon^2}{c_{a_h}c_{b_h}} - 1 + \chi\right) \\
&= LM\log\sigma^2 + \frac{\sum_{h=\min(H,\overline{H})+1}^{\min(L,M)}\gamma_h^2}{\sigma^2} + \sum_{h=1}^{\min(H,\overline{H})}\left(\frac{2\gamma_h}{c_{a_h}c_{b_h}} - \frac{\sigma^2}{c_{a_h}^2c_{b_h}^2}\right) \\
&\quad + LM\log(2\pi) + \sum_{h=1}^{\min(L,M)}(L+M)\left(\log c_{a_h}c_{b_h} + \frac{\varepsilon^2}{c_{a_h}c_{b_h}} - 1 + \chi\right) \\
&= LM\log\sigma^2 + \frac{\sum_{h=\min(H,\overline{H})+1}^{\min(L,M)}\gamma_h^2}{\sigma^2} + \sum_{h=1}^{\min(H,\overline{H})}\left(\frac{2\gamma_h}{c_{a_h}c_{b_h}} - \frac{\sigma^2}{c_{a_h}^2c_{b_h}^2}\right) + \text{const.}
\end{aligned}
\tag{12.116}
$$

for

$$
\underline{\sigma}_{\overline{H}+1}^{2\,\mathrm{MAP}} \le \sigma^2 \le \underline{\sigma}_{\overline{H}}^{2\,\mathrm{MAP}}, \tag{12.117}
$$

where

$$
\underline{\sigma}_h^{2\,\mathrm{MAP}} = \begin{cases} \infty & \text{for } h = 0, \\ c_{a_h}c_{b_h}\gamma_h & \text{for } h = 1,\ldots,\min(L,M), \\ 0 & \text{for } h = \min(L,M)+1. \end{cases} \tag{12.118}
$$

Assume that we use the full-rank model $H = \min(L, M)$, and expect the ARD property to find the correct rank. Under this setting, the free energy (12.116) can be arbitrarily small for $\sigma^2 \to +0$, because the first term diverges to $-\infty$, and the second term is equal to zero for $0\ (= \underline{\sigma}_{\min(L,M)+1}^{2\,\mathrm{MAP}}) \le \sigma^2 \le c_{a_{\min(L,M)}}c_{b_{\min(L,M)}}\gamma_{\min(L,M)} (= \underline{\sigma}_{\min(L,M)}^{2\,\mathrm{MAP}})$ (note that $\gamma_{\min(L,M)} > 0$ with probability 1). This leads to the following lemma:

**Lemma 12.16** *Assume that* $H = \min(L, M)$ *and* $C_A, C_B$ *are given as constants. Then the MAP free energy with respect to* $\sigma^2$ *is (globally) minimized at*

$$\widehat{\sigma}^{2\,MAP} \to +0.$$

The PB free energy behaves differently. By using Lemma 12.11, we can write the PB free energy with the variational parameters optimized, as a function of $\sigma^2$, as follows:

$$\begin{aligned}
2\acute{F}^{\mathrm{PB}} &= LM\log(2\pi\sigma^2) + \frac{\sum_{h=1}^{\min(L,M)}\gamma_h^2}{\sigma^2} + \sum_{h=1}^{H} 2\acute{F}_h^{\mathrm{PB}} \\
&= LM\log(2\pi\sigma^2) + \frac{\sum_{h=1}^{\min(L,M)}\gamma_h^2}{\sigma^2} \\
&\quad + \sum_{h=1}^{\min(H,\overline{H})}\left\{\frac{\min(L,M)+2\max(L,M)}{2}\log c_{a_h}^2 c_{b_h}^2 + \sqrt{\max(L,M)^2 + \frac{4\gamma_h^2}{c_{a_h}^2 c_{b_h}^2}}\right. \\
&\quad - \frac{\sigma^2}{c_{a_h}^2 c_{b_h}^2} + \max(L,M)\log\left(-\max(L,M) + \sqrt{\max(L,M)^2 + \frac{4\gamma_h^2}{c_{a_h}^2 c_{b_h}^2}}\right) \\
&\quad \left. - \frac{\gamma_h^2}{\sigma^2} - \max(L,M)\log(2\sigma^2) + \min(L,M)(\chi - 1)\right\} \\
&\quad + \sum_{h=\min(H,\overline{H})+1}^{\min(L,M)} \min(L,M)\left(\log c_{a_h}c_{b_h} + \frac{\varepsilon^2}{c_{a_h}c_{b_h}} - 1 + \chi\right) \\
&= (\min(L,M) - \min(H,\overline{H}))\max(L,M)\log(2\sigma^2) + \frac{\sum_{h=\min(H,\overline{H})+1}^{\min(L,M)}\gamma_h^2}{\sigma^2} \\
&\quad + \sum_{h=1}^{\min(H,\overline{H})}\left\{\max(L,M)\log c_{a_h}^2 c_{b_h}^2 + \sqrt{\max(L,M)^2 + \frac{4\gamma_h^2}{c_{a_h}^2 c_{b_h}^2}}\right. \\
&\quad \left. - \frac{\sigma^2}{c_{a_h}^2 c_{b_h}^2} + \max(L,M)\log\left(-\max(L,M) + \sqrt{\max(L,M)^2 + \frac{4\gamma_h^2}{c_{a_h}^2 c_{b_h}^2}}\right)\right\} \\
&\quad + LM\log(\pi) + \sum_{h=1}^{\min(L,M)}\min(L,M)\left(\log c_{a_h}c_{b_h} - 1 + \chi\right) \\
&= (\min(L,M) - \min(H,\overline{H}))\max(L,M)\log(2\sigma^2) + \frac{\sum_{h=\min(H,\overline{H})+1}^{\min(L,M)}\gamma_h^2}{\sigma^2} \\
&\quad + \sum_{h=1}^{\min(H,\overline{H})}\left\{\max(L,M)\log c_{a_h}^2 c_{b_h}^2 + \sqrt{\max(L,M)^2 + \frac{4\gamma_h^2}{c_{a_h}^2 c_{b_h}^2}}\right. \\
&\quad \left. - \frac{\sigma^2}{c_{a_h}^2 c_{b_h}^2} + \max(L,M)\log\left(-\max(L,M) + \sqrt{\max(L,M)^2 + \frac{4\gamma_h^2}{c_{a_h}^2 c_{b_h}^2}}\right)\right\} \\
&\quad + \mathrm{const.}
\end{aligned} \tag{12.119}$$

for

$$\underline{\sigma}_{\overline{H}+1}^{2\,\mathrm{PB}} \le \sigma^2 \le \underline{\sigma}_{\overline{H}}^{2\,\mathrm{PB}}, \tag{12.120}$$

where

$$\underline{\sigma}_h^{2\ \mathrm{PB}} = \begin{cases} \infty & \text{for } h = 0, \\ \frac{c_{a_h}^2 c_{b_h}^2}{2}\left(-\max(L,M) + \sqrt{\max(L,M)^2 + 4\frac{\gamma_h}{c_{a_h}^2 c_{b_h}^2}}\right) & \\ & \text{for } h = 1, \ldots, \min(L,M), \\ 0 & \text{for } h = \min(L,M)+1. \end{cases} \tag{12.121}$$

We find a remarkable difference between the MAP free energy (12.116) and the PB free energy (12.119): unlike in the MAP free energy, the first log term in the PB free energy disappears for $0 = \underline{\sigma}^{2\ \mathrm{PB}}_{\min(L,M)+1} < \sigma^2 < \underline{\sigma}^{2\ \mathrm{PB}}_{\min(L,M)}$, and therefore, the PB free energy does not diverge to $-\infty$ at $\sigma^2 \to +0$. We can actually prove that the noise variance estimator is lower-bounded by a positive value as follows. The PB free energy (12.121) is continuous, and, for $0 = \underline{\sigma}^{2\ \mathrm{PB}}_{\min(L,M)+1} < \sigma^2 < \underline{\sigma}^{2\ \mathrm{PB}}_{\min(L,M)}$, it can be written as

$$2\acute{F}^{\mathrm{PB}} = -\frac{\sigma^2}{c_{a_h}^2 c_{b_h}^2} + \mathrm{const.},$$

which is monotonically decreasing. This leads to the following lemma:

**Lemma 12.17** *Assume that* $H = \min(L, M)$ *and* $C_A, C_B$ *are given as constants. Then the noise variance estimator in PB learning is lower-bounded by*

$$\begin{aligned}\widehat{\sigma}^{2\ MAP} &\geq \underline{\sigma}^{2\ \mathrm{PB}}_{\min(L,M)} \\ &= \frac{c_{a_{\min(L,M)}}^2 c_{b_{\min(L,M)}}^2}{2}\left(-\max(L,M) + \sqrt{\max(L,M)^2 + 4\frac{\gamma_{\min(L,M)}}{c_{a_{\min(L,M)}}^2 c_{b_{\min(L,M)}}^2}}\right).\end{aligned} \tag{12.122}$$

We numerically investigated the behavior of the noise variance estimator by creating random observed matrices $V = B^* A^{*\top} + \mathcal{E} \in \mathbb{R}^{L\times M}$, and depicting the VB, PB, and MAP free energies as functions of $\sigma^2$ with the variational parameters optimized. Figure 12.7 shows a typical case for $L = 20, M = 50$, $H^* = 2$ with the entries of $A^* \in \mathbb{R}^{M\times H^*}$ and $B^* \in \mathbb{R}^{L\times H^*}$ independently drawn from $\mathrm{Gauss}_1(0, 1^2)$, and the entries of $\mathcal{E} \in \mathbb{R}^{L\times M}$ independently drawn from $\mathrm{Gauss}_1(0, 0.3^2)$. We set the prior covariances to $c_{a_h} c_{b_h} = 1$. As Lemma 12.16 states, the global minimum of the MAP free energy is at $\sigma^2 \to +0$. Since no nontrivial local minimum is observed, local search gives the same trivial solution. On the other hand, the PB free energy has a minimum in the positive region $\sigma^2 > 0$ with probability 1, as Lemma 12.17 states. However, we empirically observed that PB learning tends to underestimate the noise

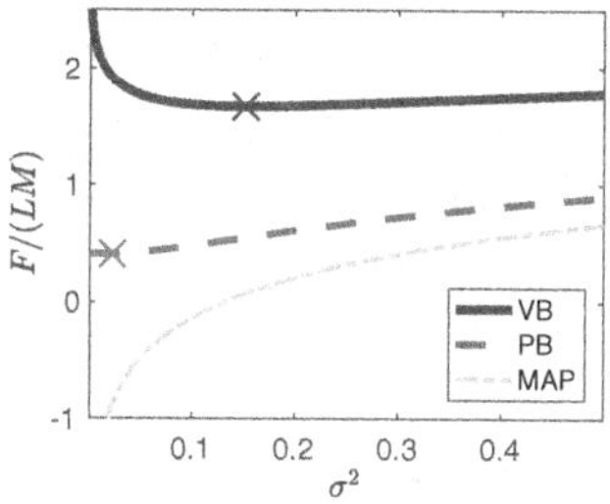


Figure 12.7 Free energy dependence on $\sigma^2$. Crosses indicate nontrivial minima.

variance, as in Figure 12.7. Therefore, we cannot expect that the noise variance estimation works well in PB learning, either.

The situation is more complicated in the empirical Bayesian variants. Since the global EMAP estimator and the global EPB estimator, given any $\sigma^2 > 0$, are the null solution, the joint global optimization over all variational parameters and the hyperparameters results in $\sigma^2 = \sum_{h=1}^{\min(L,M)} \gamma_h^2/(LM)$, regardless of observations—all observed signals are considered to be noise. If we adopt nontrivial local solutions as estimators, i.e., the local-EMAP estimator and the local-EPB estimator, the free energies are not continuous anymore as functions of $\sigma^2$, because of the energy jump by the entropy factor $\chi$ of the pseudo-Dirac delta function. Even in that case, if we globally minimize the free energies with respect to $\sigma^2$, the estimator contains no nontrivial local solution, because the null solutions cancel all entropy factors. As such, no reasonable way to estimate the noise variance has been found in EMAP learning and in EPB learning.

In the previous work on the TF model with PB learning (Chu and Ghahramani, 2009) and with EMAP learning (Mørup and Hansen, 2009), the noise variance was treated as a given constant. This was perhaps because the noise variance estimation failed, which is consistent with the preceding discussion.

## 12.2 More General Cases

Although extending the analysis for fully observed MF to more general cases is not easy in general, some basic properties can be shown. Specifically, this section shows that the global solutions for EMAP learning and EPB learning are also trivial and useless in the MF model with missing entries and in the TF model. Nevertheless, we experimentally show in Section 12.3 that local search for EMAP learning and EPB learning provides estimators that behave similarly to the EVB estimator.

### 12.2.1 Matrix Factorization with Missing Entries

The MF model with missing entries was introduced in Section 3.2. There, the likelihood (12.1) is replaced with

$$p(V|A,B) \propto \exp\left(-\frac{1}{2\sigma^2}\left\|\mathcal{P}_\Lambda(V) - \mathcal{P}_\Lambda\left(BA^\top\right)\right\|_{\mathrm{Fro}}^2\right), \tag{12.123}$$

where $\Lambda$ denotes the set of observed indices, and

$$(\mathcal{P}_\Lambda(V))_{l,m} = \begin{cases} V_{l,m} & \text{if } (l,m) \in \Lambda, \\ 0 & \text{otherwise.} \end{cases}$$

The VB free energy is explicitly written as

$$\begin{aligned} 2F = {} & \#(\Lambda) \cdot \log(2\pi\sigma^2) + M \log\det(C_A) + L \log\det(C_B) \\ & - \sum_{m=1}^{M} \log\det\left(\widehat{\Sigma}_{A,m}\right) - \sum_{l=1}^{L} \log\det\left(\widehat{\Sigma}_{B,l}\right) - (L+M)H \\ & + \mathrm{tr}\left\{C_A^{-1}\left(\widehat{A}^\top\widehat{A} + \sum_{m=1}^{M}\widehat{\Sigma}_{A,m}\right) + C_B^{-1}\left(\widehat{B}^\top\widehat{B} + \sum_{l=1}^{L}\widehat{\Sigma}_{B,l}\right)\right\} \\ & + \sigma^{-2}\sum_{(l,m)\in\Lambda}\left(V_{l,m} - 2V_{l,m}\widehat{\widetilde{a}}_m^\top\widehat{\widetilde{b}}_l + \mathrm{tr}\left\{\left(\widehat{\widetilde{a}}_m\widehat{\widetilde{a}}_m^\top + \widehat{\Sigma}_{A,m}\right)\left(\widehat{\widetilde{b}}_l\widehat{\widetilde{b}}_l^\top + \widehat{\Sigma}_{B,l}\right)\right\}\right), \end{aligned} \tag{12.124}$$

where $\#(\Lambda)$ denotes the number of observed entries.

We define the EMAP learning problem and the EPB learning problem by Eq. (12.77) with the free energy given by Eq. (12.124). The following holds:

**Lemma 12.18** *The global solutions of EMAP learning and EPB learning for the MF model with missing entries, i.e., Eqs.* (12.123), (12.2), *and* (12.3), *are* $\widehat{U}^{\mathrm{EMAP}} = \widehat{U}^{\mathrm{EPB}} = \widehat{B}\widehat{A}^\top = \mathbf{0}_{(L,M)}$, *regardless of observations.*

*Proof* The posterior covariance for $A$ is clipped to $\widehat{\Sigma}_{A,m} = \varepsilon^2 I_H$ in EPB-B learning, while the posterior covariance for $B$ is clipped to $\widehat{\Sigma}_{B,m} = \varepsilon^2 I_H$ in EPB-A learning. In either case, one can make the second or the third term in the free energy (12.124) arbitrarily small to cancel the fourth or the fifth term by setting $C_A = \varepsilon^2 I_H$ or $C_B = \varepsilon^2 I_H$. Then, because of the terms in the third line of Eq. (12.124), which come from the prior distributions, it holds that $\widehat{A} \to \mathbf{0}_{(M,H)}$ or $\widehat{B} \to \mathbf{0}_{(M,H)}$ for $\varepsilon^2 \to +0$, which results in $\widehat{U}^{\mathrm{EPB}} = \widehat{B}\widehat{A}^\top \to \mathbf{0}_{(L,M)}$. In EMAP learning, both posterior covariances are clipped to $\widehat{\Sigma}_{A,m} = \widehat{\Sigma}_{B,m} = \varepsilon^2 I_H$. By the same argument as for EPB learning, we can show that $\widehat{U}^{\mathrm{EMAP}} = \widehat{B}\widehat{A}^\top \to \mathbf{0}_{(L,M)}$, which completes the proof. □

### 12.2.2 Tucker Factorization

The TF model was introduced in Section 3.3.1. The likelihood and the priors are given by

$$p(\mathcal{V}|\mathcal{G},\{A^{(n)}\}) \propto \exp\left(-\frac{\left\|\mathcal{V}-\mathcal{G}\times_1 A^{(1)}\cdots\times_N A^{(N)}\right\|^2}{2\sigma^2}\right), \tag{12.125}$$

$$p(\mathcal{G}) \propto \exp\left(-\frac{\mathbf{vec}(\mathcal{G})^\top\left(C_{G^{(N)}}\otimes\cdots\otimes C_{G^{(1)}}\right)^{-1}\mathbf{vec}(\mathcal{G})}{2}\right), \tag{12.126}$$

$$p(\{A^{(n)}\}) \propto \exp\left(-\frac{\sum_{n=1}^N \mathrm{tr}(A^{(n)}C_{A^{(n)}}^{-1}A^{(n)\top})}{2}\right), \tag{12.127}$$

where $\otimes$ and $\mathbf{vec}(\cdot)$ denote the *Kronecker product* and the *vectorization operator*, respectively. $\{C_{G^{(n)}}\}$ and $\{C_{A^{(n)}}\}$ are the prior covariances restricted to be diagonal, i.e.,

$$C_{G^{(n)}} = \mathbf{Diag}\left(c^2_{g_1^{(n)}},\ldots,c^2_{g^{(n)}_{H^{(n)}}}\right),$$

$$C_{A^{(n)}} = \mathbf{Diag}\left(c^2_{a_1^{(n)}},\ldots,c^2_{a^{(n)}_{H^{(n)}}}\right).$$

We denote $\breve{C}_G = C_{G^{(N)}}\otimes\cdots\otimes C_{G^{(1)}}$.

The VB free energy is explicitly written as

$$\begin{aligned}
2F =& \left(\prod_{n=1}^N M^{(n)}\right)\log(2\pi\sigma^2) + \log\det\left(\breve{C}_G\right) + \sum_{n=1}^N M^{(n)}\log\det\left(C_{A^{(n)}}\right)\\
&- \log\det\left(\widehat{\breve{\Sigma}}_G\right) - \sum_{n=1}^N M^{(n)}\log\det\left(\widehat{\Sigma}_{A^{(n)}}\right)\\
&+ \frac{\|\mathcal{V}\|^2}{\sigma^2} - \prod_{n=1}^N H^{(n)} - \prod_{n=1}^N (M^{(n)}H^{(n)})\\
&+ \mathrm{tr}\left(\breve{C}_G^{-1}(\widehat{\breve{g}}\widehat{\breve{g}}^\top + \widehat{\breve{\Sigma}}_G)\right) + \sum_{n=1}^N \mathrm{tr}\left(C_{A^{(n)}}^{-1}(\widehat{A}^{(n)\top}\widehat{A}^{(n)} + M^{(n)}\widehat{\Sigma}_{A^{(n)}})\right)\\
&- \frac{2}{\sigma^2}\breve{v}^\top(\widehat{A}^{(N)}\otimes\cdots\otimes\widehat{A}^{(1)})\widehat{\breve{g}}\\
&+ \frac{1}{\sigma^2}\mathrm{tr}\left\{\left((\widehat{A}^{(N)\top}\widehat{A}^{(N)} + M^{(N)}\widehat{\Sigma}_{A^{(N)}})\otimes\cdots\otimes(\widehat{A}^{(1)\top}\widehat{A}^{(1)} + M^{(1)}\widehat{\Sigma}_{A^{(1)}})\right)\right.\\
&\qquad\left.\cdot(\widehat{\breve{g}}\widehat{\breve{g}}^\top + \widehat{\breve{\Sigma}}_G)\right\}.
\end{aligned} \tag{12.128}$$

In the TF model, we refer as PB-G learning to the approximate Bayesian method where the posteriors for the factor matrices $\{A^{(N)}\}$ are approximated by the pseudo-Dirac delta function, and as PB-A learning to the one where the posterior for the core tensor $\mathcal{G}$ is approximated by the pseudo-Dirac delta function. PB learning chooses the one giving a lower free energy from PB-G learning and PB-A learning. MAP learning approximates both posteriors by the pseudo-Dirac delta function. Note that the approach by Chu and Ghahramani (2009) corresponds to PB-G learning with the prior covariances fixed to $\boldsymbol{C}_{G^{(n)}} = \boldsymbol{C}_{A^{(n)}} = \boldsymbol{I}_{H^{(n)}}$ for $n = 1, \ldots, N$, while the approach, called *ARD Tucker*, by Mørup and Hansen (2009) corresponds to EMAP learning with the prior covariances estimated from observations. In both approaches, the noise variance $\sigma^2$ was treated as a given constant.

Again the global solutions of EMAP learning and EPB learning are trivial and useless.

**Lemma 12.19** *The global solutions of EMAP learning and EPB learning for the TF model, i.e., Eqs.* (12.125) *through* (12.127), *are* $\widehat{\mathcal{U}}^{\mathrm{EMAP}} = \widehat{\mathcal{U}}^{\mathrm{EPB}} = \widehat{\mathcal{G}} \times_1 \widehat{\boldsymbol{A}}^{(1)} \cdots \times_N \widehat{\boldsymbol{A}}^{(N)} = \boldsymbol{0}_{(M^{(1)}, \ldots, M^{(N)})}$, *regardless of observations.*

*Proof* The posterior covariance for $\mathcal{G}$ is clipped to $\breve{\boldsymbol{\Sigma}}_G = \varepsilon^2 \boldsymbol{I}_{\prod_{n=1}^N H^{(n)}}$ in EPB-A learning, while the posterior covariances for $\{\boldsymbol{A}^{(n)}\}$ are clipped to $\{\widehat{\boldsymbol{\Sigma}}_{A^{(n)}} = \varepsilon^2 \boldsymbol{I}_{H^{(n)}}\}$ in EPB-G learning. In either case, one can make the second or the third term in the free energy (12.128) arbitrarily small to cancel the fourth or the fifth term by setting $\{\boldsymbol{C}_{G^{(n)}} = \varepsilon^2 \boldsymbol{I}_{H^{(n)}}\}$ or $\{\boldsymbol{C}_{A^{(n)}} = \varepsilon^2 \boldsymbol{I}_{H^{(n)}}\}$. Then, because of the terms in the fourth line of Eq. (12.128), which come from the prior distributions, it holds that $\widehat{\mathcal{G}} \to \boldsymbol{0}_{(H^{(1)}, \ldots, H^{(N)})}$ or $\{\widehat{\boldsymbol{A}}^{(N)} \to \boldsymbol{0}_{(M^{(N)}, H^{(N)})}\}$ for $\varepsilon^2 \to +0$, which results in $\widehat{\mathcal{U}}^{\mathrm{EPB}} = \boldsymbol{0}_{(M^{(1)}, \ldots, M^{(N)})}$. In EMAP learning, both posterior covariances are clipped to $\breve{\boldsymbol{\Sigma}}_G = \varepsilon^2 \boldsymbol{I}_{\prod_{n=1}^N H^{(n)}}$ and $\{\widehat{\boldsymbol{\Sigma}}_{A^{(n)}} = \varepsilon^2 \boldsymbol{I}_{H^{(n)}}\}$, respectively. By the same argument as for EPB learning, we can show that $\widehat{\mathcal{U}}^{\mathrm{EMAP}} = \boldsymbol{0}_{(M^{(1)}, \ldots, M^{(N)})}$, which completes the proof. □

## 12.3 Experimental Results

In this section, we experimentally investigate the behavior of EMAP learning and EPB learning, in comparison with EVB learning. We start from the fully observed MF model, where we can assess how often local search finds the nontrivial local solution (derived in Section 12.1.3) rather than the global null solution. After that, we conduct experiments in collaborative filtering, where the MF model with missing entries is used, and in TF.

For local search, we adopted the standard iterative algorithm. The standard iterative algorithm for EVB learning has been derived in Chapter 3. The standard iterative algorithms for EPB learning and EMAP learning, which can be derived simply by setting the derivatives of the corresponding free energies with respect to the unknown parameters to zero, similarly apply the stationary conditions in turn to update unknown parameters. For initialization, the entries of the mean parameters, e.g., $\widehat{\boldsymbol{A}}, \widehat{\boldsymbol{B}}$, and $\widehat{\mathcal{G}}$, were drawn from $\mathrm{Gauss}_1(0, 1^2)$, while the covariance parameters were set to the identity, e.g., $\widehat{\breve{\boldsymbol{\Sigma}}}_G = \boldsymbol{I}_{\prod_{n=1}^N H^{(n)}}, \widehat{\boldsymbol{\Sigma}}_{A^{(n)}} = \boldsymbol{C}_{G^{(n)}} = \boldsymbol{C}_{A^{(n)}} = \boldsymbol{I}_{H^{(n)}}$. We used this initialization scheme through all experiments in this section.

### 12.3.1 Fully Observed MF

We first conducted an experiment on an artificial (*Artificial1*) data set, which was generated as follows. We randomly generated *true* matrices $\boldsymbol{A}^* \in \mathbb{R}^{M \times H^*}$ and $\boldsymbol{B}^* \in \mathbb{R}^{L \times H^*}$ such that each entry of $\boldsymbol{A}^*$ and $\boldsymbol{B}^*$ follows $\mathrm{Gauss}_1(0, 1)$. An observed matrix $\boldsymbol{V} \in \mathbb{R}^{L \times M}$ was created by adding a noise subject to $\mathrm{Gauss}_1(0, 1)$ to each entry of $\boldsymbol{B}^* \boldsymbol{A}^{*\top}$. Figures 12.8 through 12.10 show the free energy and the estimated rank over iterations in EVB learning, local-EPB

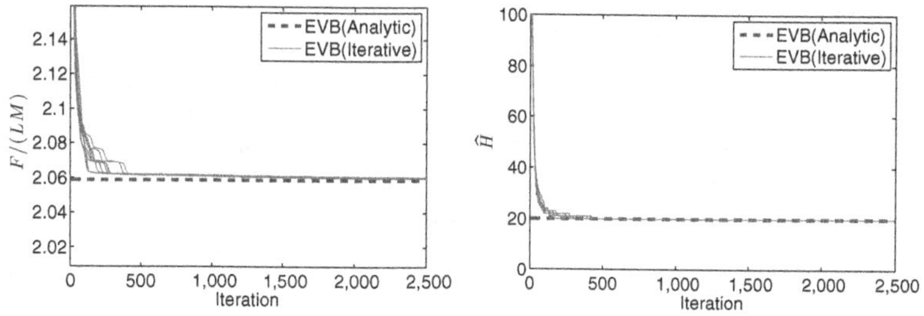


(a) Free energy (b) Estimated rank

Figure 12.8 EVB learning on *Artificial1* ($L = 100, M = 300, H^* = 20$).

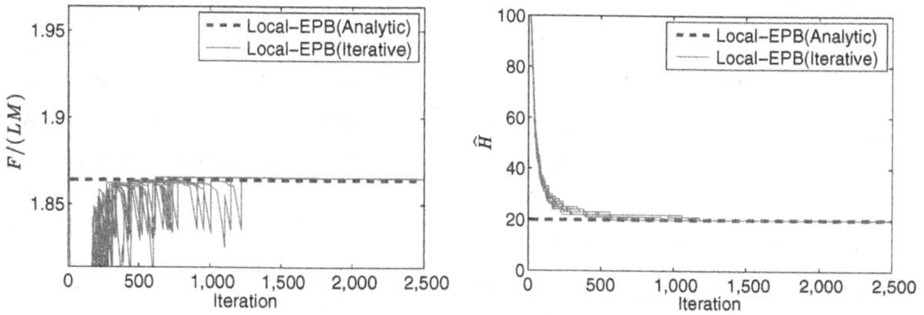


(a) Free energy (b) Estimated rank

Figure 12.9 Local-EPB learning on *Artificial1*.

Table 12.1 *Estimated rank in fully observed MF experiments.*

| | | | | $\widehat{H}^{\text{EVB}}$ | | $\widehat{H}^{\text{local-EPB}}$ | | $\widehat{H}^{\text{local-EMAP}}$ | |
|---|---|---|---|---|---|---|---|---|---|
| Data set | $M$ | $L$ | $H^*$ | Analytic | Iterative | Analytic | Iterative | Analytic | Iterative |
| *Artificial1* | 300 | 100 | 20 | 20 | 20 (100%) | 20 | 20 (100%) | 20 | 20 (100%) |
| *Artificial2* | 500 | 400 | 5 | 5 | 5 (100%) | 8 | 8 (90%)<br>9 (10%) | 5 | 5 (100%) |
| *Chart* | 600 | 60 | – | 2 | 2 (100%) | 2 | 2 (100%) | 2 | 2 (100%) |
| *Glass* | 214 | 9 | – | 1 | 1 (100%) | 1 | 1 (100%) | 1 | 1 (100%) |
| *Optical Digits* | 5,620 | 64 | – | 10 | 10 (100%) | 10 | 10 (100%) | 6 | 6 (100%) |
| *Satellite* | 6,435 | 36 | – | 2 | 2 (100%) | 2 | 2 (100%) | 1 | 1 (100%) |

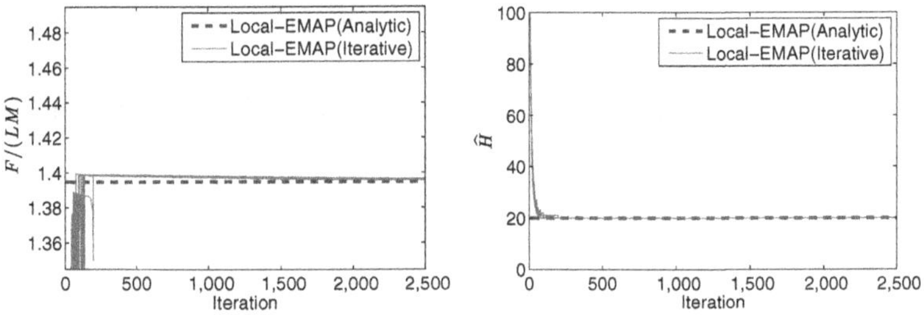


(a) Free energy (b) Estimated rank

Figure 12.10 Local-EMAP learning on *Artificial1*.

learning, and local-EMAP learning, respectively, on the *Artificial1* data set with the data matrix size $L = 100$ and $M = 300$, and the true rank $H^* = 20$. The noise variance was assumed to be known, i.e., it was set to $\sigma^2 = 1$. We performed iterative local search 10 times, starting from different initial points, and each trial is plotted by a solid curve in the figures. The results computed by the analytic-form solutions for EVB learning (Theorem 6.13), local-EPB learning (Theorem 12.13), and local-EMAP learning (Theorem 12.12) were plotted as dashed lines. We can observe that iterative local search for EPB learning and EMAP learning tends to successfully find the nontrivial local solutions, although they are not global solutions.

We also conducted experiments on another artificial data set and benchmark data sets. The results are summarized in Table 12.1. *Artificial2* was created in the same way as *Artificial1*, but with $L = 400, M = 500$, and $H^* = 5$. The benchmark data sets were collected from the UCI repository (Asuncion and

Newman, 2007), on which we set the noise variance under the assumption that the signal to noise ratio is 0 db, following Mørup and Hansen (2009).

In the table, the estimated ranks by the analytic-form solution and by iterative local search are shown. The percentages for iterative local search indicate the frequencies over 10 trials. We observe the following: first, iterative local search tends to estimate the same rank as the analytic-form (local) solution; and second, the estimated rank tends to be consistent among EVB learning, local-EPB learning, and local-EMAP learning. Furthermore, on the artificial data sets, where the true rank is known, the rank is correctly estimated in most of the cases. Exceptions are *Artificial2*, where local-EPB learning overestimates the rank, and *Optical Digits* and *Satellite*, where local-EMAP learning estimates a smaller rank than the others. These phenomena can be explained by the theoretical implications in Section 12.1.3: in *Artificial2*, the ratio $\xi = H^* / \min(L, M) = 5/400$ between the true rank and the possible largest rank is small, which means that most of the singular components consist of noise. In such a case, local-EPB learning with its truncation threshold lower than MPUL tends to retain components purely consisting of noise (see Figure 12.6). In *Optical Digits* and *Satellite*, $\alpha$ (= 64/5620 for *Optical Digits* and = 36/6435 for *Satellite*) is extremely small, and therefore local-EMAP learning with its higher truncation threshold tends to discard more components than the others, as Figure 12.6 implies.

### 12.3.2 Collaborative Filtering

Next we conducted experiments in the collaborative filtering (CF) scenario, where the observed matrix has missing entries to be predicted by the MF model.

We generated an artificial (*ArtificialCF*) data set in the same way as the fully observed case for $L = 2{,}000, M = 5{,}000, H^* = 5$, and then masked 99% of the entries as missing values. We applied EVB learning, local-EPB learning, and local-EMAP learning to the MF model with missing entries, i.e., Eqs. (12.123), (12.2), and (12.3).[4] Figure 12.11 shows the estimated rank and the *generalization error* over iterations for 10 trials, where the generalization error is defined as $\mathrm{GE} = \|\mathcal{P}_{\Lambda'}(V) - \mathcal{P}_{\Lambda'}(\widehat{BA^\top})\|^2_{\mathrm{Fro}} / (\#(\Lambda')\sigma^2)$ for $\Lambda'$ being the set of test indices.

[4] Here we solve the EVB learning problem, the EPB learning problem, and the EMAP learning problem, respectively, by the standard iterative algorithms. However, we refer to the last two methods as local-EPB learning and local-EMAP learning, since we expect the local search algorithm to find not the global null solution but the nontrivial local solution.

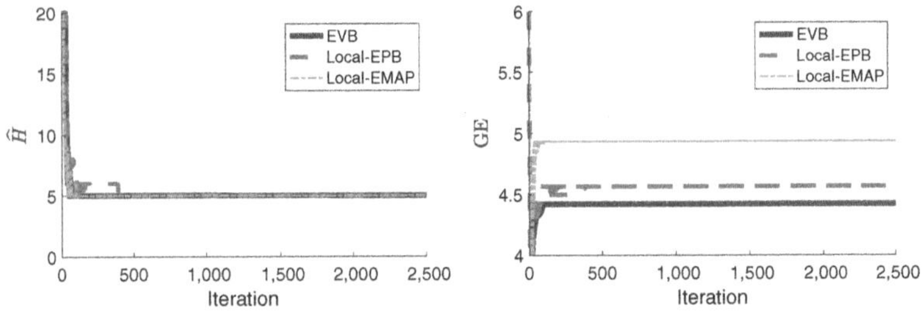


(a) Estimated rank (b) Generalization error

Figure 12.11 CF result on *ArtificialCF* ($L$ = 2,000, $M$ = 5,000, $H^*$ = 5 with 99% missing ratio).

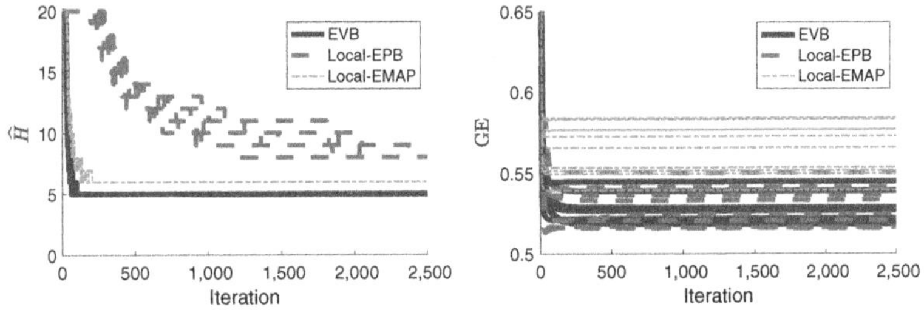


(a) Estimated rank (b) Generalization error

Figure 12.12 CF result on *MovieLens* ($L$ = 943, $M$ = 1,682 with 99% missing ratio).

We also conducted an experiment on the *MovieLens* data sets (with $L = 943$, $M = 1{,}682$).[5] We randomly divided the observed entries into training entries and test entries, so that 99% of the entries are missing in the training phase. The test entries are used to evaluate the generalization error. Figure 12.12 shows the result in the same format as Figure 12.11.

We see that, on both data sets, local-EMAP learning tends to estimate a similar rank to EVB learning, while local-EPB learning tends to estimate a larger rank—a similar tendency to the fully observed case. In terms of the generalization error, local-EPB learning performs comparably to EVB learning, while local-EMAP learning performs slightly worse.

### 12.3.3 Tensor Factorization

Finally, we conducted experiments on TF. We created an artificial (*ArtificialTF*) data set, following Mørup and Hansen (2009): we drew a three-mode

[5] www.grouplens.org/

Table 12.2 *Estimated rank (effective size of core tensor) in TF experiments.*

| Data set | $M$ | $H^*$ | $\widehat{H}^{\mathrm{EVB}}$ | $\widehat{H}^{\mathrm{local\text{-}EPB}}$ | $\widehat{H}^{\mathrm{local\text{-}EMAP}}$ | $\widehat{H}^{\mathrm{ARD\text{-}Tucker}}$ |
|---|---|---|---|---|---|---|
| *ArtificialTF* | (30, 40, 50) | (3, 4, 5) | (3, 4, 5): 100% | (3, 4, 5): 100% | (3, 4, 5): 90%<br>(3, 7, 5): 10% | (3, 4, 5): 100% |
| *FIA* | (12, 100, 89) | (3, 6, 4) | (3, 5, 3): 100% | (3, 5, 3): 100% | (3, 5, 2): 50%<br>(4, 5, 2): 20%<br>(5, 4, 2): 10%<br>(4, 4, 2): 10%<br>(8, 5, 2): 10% | (3, 4, 2): 70%<br>(3, 5, 2): 10%<br>(3, 7, 2): 10%<br>(10, 4, 3): 10% |

random tensor of the size $(M^{(1)}, M^{(1)}, M^{(1)}) = (30, 40, 50)$ with the signal components $(H^{(1)*}, H^{(2)*}, H^{(3)*}) = (3, 4, 5)$. The noise is added so that the signal-to-noise ratio is 0 db. We also used the *Flow Injection Analysis* (*FIA*) data set.[6] Table 12.2 shows the estimated rank with frequencies over 10 trials. Here we also show the results by ARD Tucker with the ridge prior (Mørup and Hansen, 2009), performed with the code provided by the authors. Local-EMAP learning and ARD Tucker minimize exactly the same objective, and the slightly different results come from the differences in the local search algorithm (standard iterative vs. gradient descent) and in the initialization scheme.

We generally observe that all learning methods provide reasonable results, although local-EMAP learning, as well as ARD Tucker, is less stable than the others.

[6] www.models.kvl.dk/datasets

# Part IV

# Asymptotic Theory

# 13
# Asymptotic Learning Theory

Part IV is dedicated to asymptotic theory of variational Bayesian (VB) learning. In this part, "asymptotic limit" always means the limit when the number $N$ of training samples goes to infinity. The main goal of asymptotic learning theory is to clarify the behavior of some statistics, e.g., the generalization error, the training error, and the Bayes free energy, which indicate how fast a learning machine can be trained as a function of the number of training samples and how the trained machine is biased to the training samples by overfitting. This provides the mathematical foundation of information criteria for model selection—a task to choose the degree of freedom of a statistical model based on observed training data. We can also evaluate the approximation accuracy of VB learning to full Bayesian learning in terms of the free energy, i.e., the gap between the VB free energy and the Bayes free energy, which corresponds to the tightness of the evidence lower-bound (ELBO) (see Section 2.1.1). In this first chapter of Part IV, we give an overview of asymptotic learning theory as the background for the subsequent chapters.

## 13.1 Statistical Learning Machines

A statistical learning machine consists of two fundamental components, a statistical model and a learning algorithm (Figure 13.1). The statistical model is denoted by a probabilistic distribution depending on some unknown parameters, and the learning algorithm estimates the unknown parameters from observed training samples. Before introducing asymptotic learning theory, we categorize statistical learning machines based on the model and the learning algorithm.

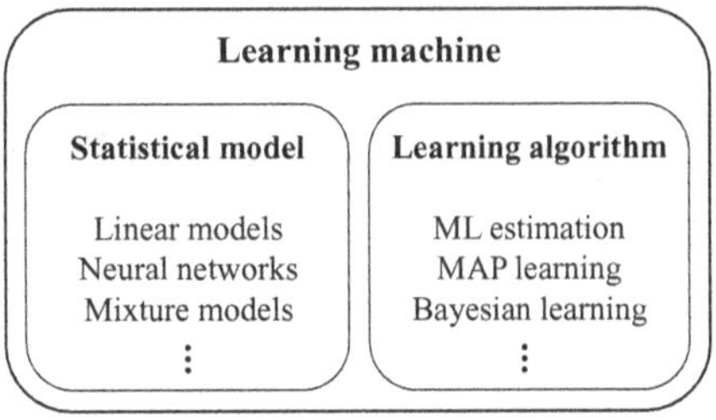


Figure 13.1 A statistical learning machine consists of a statistical model and a learning algorithm.

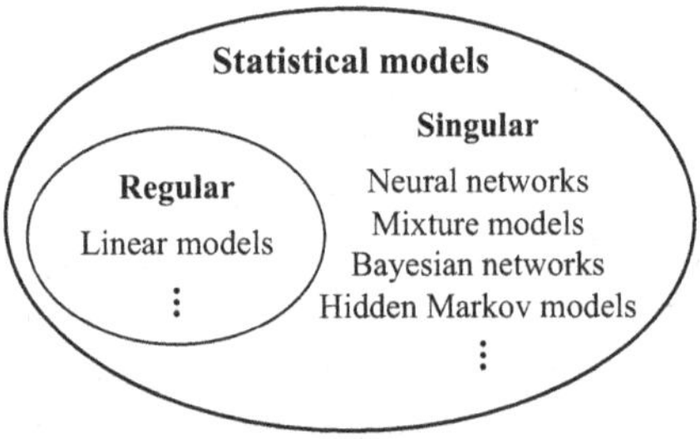


Figure 13.2 Statistical models are classified into regular models and singular models.

### 13.1.1 Statistical Models—Regular and Singular

We classify the statistical models into two classes, the *regular models* and the *singular models* (Figure 13.2). The regular models are identifiable (Definition 7.4 in Section 7.3.1), i.e.,

$$p(\boldsymbol{x}|\boldsymbol{w}_1) = p(\boldsymbol{x}|\boldsymbol{w}_2) \iff \boldsymbol{w}_1 = \boldsymbol{w}_2 \qquad \text{for any } \boldsymbol{w}_1, \boldsymbol{w}_2 \in \mathcal{W}, \tag{13.1}$$

and do not have *singularities* in the parameter space, i.e., the *Fisher information*

$$\mathbb{S}_{+}^{D} \ni \boldsymbol{F}(\boldsymbol{w}) = \int \frac{\partial \log p(\boldsymbol{x}|\boldsymbol{w})}{\partial \boldsymbol{w}} \left( \frac{\partial \log p(\boldsymbol{x}|\boldsymbol{w})}{\partial \boldsymbol{w}} \right)^{\top} p(\boldsymbol{x}|\boldsymbol{w}) d\boldsymbol{x} \tag{13.2}$$

is nonsingular (or full-rank) for any $\boldsymbol{w} \in \mathcal{W}$.

With a few additional assumptions, the regular models were analyzed under the *regularity conditions* (Section 13.4.1), which lead to the *asymptotic normality* of the distribution of the maximum likelihood (ML) estimator, and the asymptotic normality of the Bayes posterior distribution (Cramer,

1949; Sakamoto et al., 1986; van der Vaart, 1998). Based on those asymptotic normalities, a unified theory was established, clarifying the asymptotic behavior of generalization properties, which are common over all regular models, and over all reasonable learning algorithms, including ML learning, maximum a posteriori (MAP) learning, and Bayesian learning, as will be seen in Section 13.4.

On the other hand, analyzing singular models requires specific techniques for different models and different learning algorithms, and it was revealed that the asymptotic behavior of generalization properties depends on the model and the algorithm (Hartigan, 1985; Bickel and Chernoff, 1993; Takemura and Kuriki, 1997; Kuriki and Takemura, 2001; Amari et al., 2002; Hagiwara, 2002; Fukumizu, 2003; Watanabe, 2009). This is because the true parameter is at a singular point when the model size is larger than necessary to express the true distribution, and, in such cases, singularities affect the distribution of the ML estimator, as well as the Bayes posterior distribution even in the asymptotic limit. Consequently, the asymptotic normality, on which the regular learning theory relies, does not hold in singular models.

### 13.1.2 Learning Algorithms—Point Estimation and Bayesian Learning

When analyzing singular models, we also classify learning algorithms into two classes, point estimation and Bayesian learning (Figure 13.3). The point estimation methods, including ML learning and MAP learning, choose a single model (i.e., a single point in the parameter space) that maximizes a certain criterion such as the likelihood or the posterior probability, while Bayesian learning methods use an *ensemble* of models over the posterior distribution or its approximation.

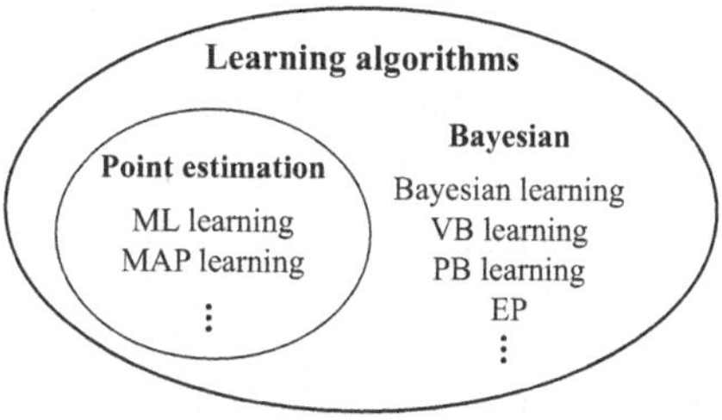


Figure 13.3 Learning algorithms are classified into point-estimation and Bayesian learning.

Unlike in the regular models, point estimation and Bayesian learning show different learning behavior in singular models. This is because how singularities affect the learning property depends on the learning methods. For example, as discussed in Chapter 7, strong nonuniformity of the density of the volume element leads to model-induced regularization (MIR) in Bayesian learning, while it does not affect point-estimation methods.

## 13.2 Basic Tools for Asymptotic Analysis

Here we introduce basic tools for asymptotic analysis.

### 13.2.1 Central Limit Theorem

Asymptotic learning theory heavily relies on the *central limit theorem.*

**Theorem 13.1** *(Central limit theorem) (van der Vaart, 1998) Let* $\{\boldsymbol{x}^{(1)}, \ldots, \boldsymbol{x}^{(N)}\}$ *be N i.i.d. samples from an arbitrary distribution with finite mean* $\boldsymbol{\mu} \in \mathbb{R}^D$ *and finite covariance* $\boldsymbol{\Sigma} \in \mathbb{S}^D_{++}$, *and let* $\overline{\boldsymbol{x}} = N^{-1} \sum_{n=1}^{N} \boldsymbol{x}^{(n)}$ *be their average. Then, the distribution of* $\boldsymbol{z} = \sqrt{N}(\overline{\boldsymbol{x}} - \boldsymbol{\mu})$ *converges to the Gaussian distribution with mean zero and covariance* $\boldsymbol{\Sigma}$,[1] *i.e.,*

$$p(\boldsymbol{z}) \to \mathrm{Gauss}_D(\boldsymbol{z}; \boldsymbol{0}, \boldsymbol{\Sigma}) \qquad as \qquad N \to \infty. \tag{13.3}$$

Intuitively, Eq. (13.3) can be interpreted as

$$p(\overline{\boldsymbol{x}}) \to \mathrm{Gauss}_D\left(\overline{\boldsymbol{x}}; \boldsymbol{\mu}, N^{-1}\boldsymbol{\Sigma}\right) \qquad \text{as} \qquad N \to \infty, \tag{13.4}$$

implying that the distribution of the average $\overline{\boldsymbol{x}}$ of i.i.d. random variables converges to the Gaussian distribution with mean $\boldsymbol{\mu}$ and covariance $N^{-1}\boldsymbol{\Sigma}$.

The central limit theorem implies the (weak) *law of large numbers*,[2] i.e., for any $\varepsilon > 0$,

$$\lim_{N \to \infty} \mathrm{Prob}\left(\|\overline{\boldsymbol{x}} - \boldsymbol{\mu}\| > \varepsilon\right) = 0. \tag{13.5}$$

### 13.2.2 Asymptotic Notation

We use the following *asymptotic notation*, a.k.a, *Bachmann–Landau notation*, to express the order of functions when the number $N$ of samples goes to infinity:

[1] We consider weak topology in the space of distributions, i.e., $p(\boldsymbol{x})$ is identified with $r(\boldsymbol{x})$ if $\langle f(\boldsymbol{x})\rangle_{p(\boldsymbol{x})} = \langle f(\boldsymbol{x})\rangle_{r(\boldsymbol{x})}$ for any bounded continuous function $f(x)$. Convergence (of a random variable $\boldsymbol{x}$) in this sense is called *convergence in distribution, weak convergence*, or *convergence in law*, and denoted as $p(\boldsymbol{x}) \to r(\boldsymbol{x})$ or $\boldsymbol{x} \rightsquigarrow r(\boldsymbol{x})$ (van der Vaart, 1998).

[2] Convergence $\boldsymbol{x} \to \boldsymbol{\mu}$ in the sense that $\lim_{N\to\infty} \mathrm{Prob}\left(\|\boldsymbol{x} - \boldsymbol{\mu}\| > \varepsilon\right) = 0, \forall \varepsilon > 0$ is called *convergence in probability*.

$$O(f(N)) : \text{A function such that} \limsup_{N\to\infty} |O(f(N))/f(N)| < \infty,$$

$$o(f(N)) : \text{A function such that} \lim_{N\to\infty} o(f(N))/f(N) = 0,$$

$$\Omega(f(N)) : \text{A function such that} \liminf_{N\to\infty} |\Omega(f(N))/f(N)| > 0,$$

$$\omega(f(N)) : \text{A function such that} \lim_{N\to\infty} |\omega(f(N))/f(N)| = \infty,$$

$$\Theta(f(N)) : \text{A function such that} \limsup_{N\to\infty} |\Theta(f(N))/f(N)| < \infty$$

$$\text{and} \liminf_{N\to\infty} |\Theta(f(N))/f(N)| > 0.$$

Intuitively, as a function of $N$, $O(f(N))$ is a function of no greater order than $f(N)$, $o(f(N))$ is a function of less order than $f(N)$, $\Omega(f(N))$ is a function of no less order than $f(N)$, $\omega(f(N))$ is a function of greater order than $f(N)$, and $\Theta(f(N))$ is a function of the same order as $f(N)$. One thing we need to be careful of is that the upper-bounding notations, $O$ and $o$, preserve after addition and subtraction, while lower-bounding notations, $\Omega$ and $\omega$, as well as the both-sides-bounding notation $\Theta$, do not necessarily preserve. For example, if $g_1(N) = \Theta(f(N))$ and $g_2(N) = \Theta(f(N))$ then $g_1(N) + g_2(N) = O(f(N))$, while it can happen that $g_1(N) + g_2(N) \neq \Theta(f(N))$ since the leading terms of $g_1(N)$ and $g_2(N)$ can coincide with each other with opposite signs and be canceled out.

For random variables, we use their probabilistic versions, $O_{\mathrm{p}}, o_{\mathrm{p}}, \Omega_{\mathrm{p}}, \omega_{\mathrm{p}}$, and $\Theta_{\mathrm{p}}$, for which the corresponding conditions hold in probability. For example, for i.i.d. samples $\{x^{(n)}\}_{n=1}^N$ from $\mathrm{Gauss}_1(x; 0, 1^2)$, we can say that

$$x^{(n)} = \Theta_{\mathrm{p}}(1),$$

$$\overline{x} = \frac{1}{N}\sum_{n=1}^{N} x^{(n)} = \Theta_{\mathrm{p}}(N^{-1/2}),$$

$$\overline{x^2} = \frac{1}{N}\sum_{n=1}^{N} \left(x^{(n)}\right)^2 = 1 + \Theta_{\mathrm{p}}(N^{-1/2}).$$

Note that the second and the third equations are consequences from the central limit theorem (Theorem 13.1) applied to the samples $\{x^{(n)}\}$ that follow the Gaussian distribution, and to the samples $\{(x^{(n)})^2\}$ that follow the chi-squared distribution, respectively.

In this book, we express asymptotic approximation mostly by using asymptotic notation. To this end, we sometimes need to translate convergence of a random variable into an equation with asymptotic notation. Let $\boldsymbol{x}$ be a random variable depending on $N$, $r(\boldsymbol{x})$ be a distribution with finite mean and covariance, and $f(\boldsymbol{x})$ be an arbitrary bounded continuous function. Then the following hold:

- If $p(\boldsymbol{x}) \to r(\boldsymbol{x})$, i.e., the distribution of $\boldsymbol{x}$ converges to $r(\boldsymbol{x})$, then

$$\boldsymbol{x} = O_{\mathrm{p}}(1) \qquad \text{and} \qquad \langle f(\boldsymbol{x})\rangle_{p(\boldsymbol{x})} = \langle f(\boldsymbol{x})\rangle_{r(\boldsymbol{x})} \left(1 + o(1)\right).$$

- If $\lim_{N\to\infty} \mathrm{Prob}\left(\|\boldsymbol{x} - \boldsymbol{y}\| > \varepsilon\right) = 0$ for any $\varepsilon > 0$, then

$$\boldsymbol{x} = \boldsymbol{y} + o_{\mathrm{p}}(1).$$

For example, the central limit theorem (13.3) implies that

$$\overline{\boldsymbol{x}} = \boldsymbol{\mu} + O_{\mathrm{p}}(N^{-1/2}),$$
$$\left\langle (\overline{\boldsymbol{x}} - \boldsymbol{\mu})(\overline{\boldsymbol{x}} - \boldsymbol{\mu})^{\top} \right\rangle_{p(\overline{\boldsymbol{x}})} = N^{-1}\boldsymbol{\Sigma} + o(N^{-1}),$$

while the law of large numbers (13.5) implies that

$$\overline{\boldsymbol{x}} = \boldsymbol{\mu} + o_{\mathrm{p}}(1).$$

## 13.3 Target Quantities

Here we introduce target quantities to be analyzed in asymptotic learning theory.

### 13.3.1 Generalization Error and Training Error

Consider a statistical model $p(\boldsymbol{x}|\boldsymbol{w})$, where $\boldsymbol{x} \in \mathbb{R}^M$ is an observed random variable and $\boldsymbol{w} \in \mathbb{R}^D$ is a parameter to be estimated. Let $\boldsymbol{X} = (\boldsymbol{x}^{(1)}, \ldots, \boldsymbol{x}^{(N)})^{\top} \in \mathbb{R}^{N\times M}$ be $N$ i.i.d. training samples taken from the true distribution $q(\boldsymbol{x})$. We assume *realizability*—the true distribution can be exactly expressed by the statistical model, i.e., $\exists \boldsymbol{w}^*$ s.t. $q(\boldsymbol{x}) = p(\boldsymbol{x}|\boldsymbol{w}^*)$, where $\boldsymbol{w}^*$ is called the true parameter.

Learning algorithms estimate the parameter value $\boldsymbol{w}$ or its posterior distribution given the training data $\mathcal{D} = \boldsymbol{X}$, and provide the predictive distribution $p(\boldsymbol{x}|\boldsymbol{X})$ for a new sample $\boldsymbol{x}$. For example, ML learning provides the predictive distribution given by

$$p^{\mathrm{ML}}(\boldsymbol{x}|\boldsymbol{X}) = p(\boldsymbol{x}|\widehat{\boldsymbol{w}}^{\mathrm{ML}}), \tag{13.6}$$

where

$$\widehat{\boldsymbol{w}}^{\mathrm{ML}} = \underset{\boldsymbol{w}}{\mathrm{argmax}}\, p(\boldsymbol{X}|\boldsymbol{w}) = \underset{\boldsymbol{w}}{\mathrm{argmax}} \left( \prod_{n=1}^{N} p(\boldsymbol{x}^{(n)}|\boldsymbol{w}) \right) \tag{13.7}$$

is the ML estimator, while Bayesian learning provides the predictive distribution given by

$$p^{\mathrm{Bayes}}(\boldsymbol{x}|\boldsymbol{X}) = \langle p(\boldsymbol{x}|\boldsymbol{w})\rangle_{p(\boldsymbol{w}|\boldsymbol{X})} = \int p(\boldsymbol{x}|\boldsymbol{w})p(\boldsymbol{w}|\boldsymbol{X})d\boldsymbol{w}, \tag{13.8}$$

where

$$p(\boldsymbol{w}|\boldsymbol{X}) = \frac{p(\boldsymbol{X}|\boldsymbol{w})p(\boldsymbol{w})}{p(\boldsymbol{X})} = \frac{p(\boldsymbol{X}|\boldsymbol{w})p(\boldsymbol{w})}{\int p(\boldsymbol{X}|\boldsymbol{w})p(\boldsymbol{w})d\boldsymbol{w}} \tag{13.9}$$

is the posterior distribution (see Section 1.1).

The *generalization error*, a criterion of generalization performance, is defined as the *Kullback–Leibler (KL) divergence* of the predictive distribution from the true distribution:

$$\mathrm{GE}(\boldsymbol{X}) = \int q(\boldsymbol{x})\log\frac{q(\boldsymbol{x})}{p(\boldsymbol{x}|\boldsymbol{X})}d\boldsymbol{x}. \tag{13.10}$$

Its *empirical* variant,

$$\mathrm{TE}(\boldsymbol{X}) = \frac{1}{N}\sum_{n=1}^{N}\log\frac{q(\boldsymbol{x}^{(n)})}{p(\boldsymbol{x}^{(n)}|\boldsymbol{X})}, \tag{13.11}$$

is called the *training error*, which is often used as an estimator of the generalization error. Note that, for the ML predictive distribution (13.6),

$$-N\cdot\mathrm{TE}^{\mathrm{ML}}(\boldsymbol{X}) = \sum_{n=1}^{N}\log\frac{p(\boldsymbol{x}^{(n)}|\widehat{\boldsymbol{w}}^{\mathrm{ML}})}{q(\boldsymbol{x}^{(n)})} \tag{13.12}$$

corresponds to the *log-likelihood ratio*, an important statistic for statistical test, when the null hypothesis is true.

The generalization error (13.10) and the training error (13.11) are random variables that depend on realization of the training data $\boldsymbol{X}$. Taking the average over the distribution of training samples, we define deterministic quantities,

$$\overline{\mathrm{GE}}(N) = \langle \mathrm{GE}(\boldsymbol{X})\rangle_{q(\boldsymbol{X})}, \tag{13.13}$$

$$\overline{\mathrm{TE}}(N) = \langle \mathrm{TE}(\boldsymbol{X})\rangle_{q(\boldsymbol{X})}, \tag{13.14}$$

which are called the *average generalization error* and the *average training error*, respectively. Here $\langle\cdot\rangle_{q(\boldsymbol{X})}$ denotes the expectation value over the distribution of $N$ training samples. The average generalization error and the average training error are scalar functions of the number $N$ of samples, and represent generalization performance of a learning machine consisting of a statistical model and a learning algorithm. The optimality of Bayesian learning is proven in terms of the average generalization error (see Appendix D).

If a learning algorithm can successfully estimate the true parameter $\boldsymbol{w}^*$ with reasonably small error, the average generalization error and the average training error converge to zero with the rate $\Theta(N^{-1})$ in the asymptotic limit.[3] One of the main goals of asymptotic learning theory is to identify or bound the coefficients of their leading terms, i.e., $\lambda$ and $\nu$ in the following asymptotic expansions:

$$\overline{\mathrm{GE}}(N) = \lambda N^{-1} + o(N^{-1}), \tag{13.15}$$

$$\overline{\mathrm{TE}}(N) = \nu N^{-1} + o(N^{-1}). \tag{13.16}$$

We call $\lambda$ and $\nu$ the *generalization coefficient* and the *training coefficient*, respectively.

### 13.3.2 Bayes Free Energy

The marginal likelihood (defined by Eq. (1.6) in Chapter 1),

$$p(\boldsymbol{X}) = \int p(\boldsymbol{X}|\boldsymbol{w})p(\boldsymbol{w})d\boldsymbol{w} = \int p(\boldsymbol{w}) \prod_{n=1}^{N} p(\boldsymbol{x}^{(n)}|\boldsymbol{w})d\boldsymbol{w}, \tag{13.17}$$

is also an important quantity in Bayesian learning. As explained in Section 1.1.3, the marginal likelihood can be regarded as the likelihood of an *ensemble* of models—the set of model distributions with the parameters subject to the prior distribution. Following the concept of the "likelihood" in statistics, we can say that the ensemble of models giving the highest marginal likelihood is most likely. Therefore, we can perform model selection by maximizing the marginal likelihood (Efron and Morris, 1973; Schwarz, 1978; Akaike, 1980; MacKay, 1992; Watanabe, 2009). Maximizing the marginal likelihood (13.17) amounts to minimizing the *Bayes free energy*, defined by Eq. (1.60):

$$F^{\mathrm{Bayes}}(\boldsymbol{X}) = -\log p(\boldsymbol{X}). \tag{13.18}$$

The Bayes free energy is a random variable depending on the training samples $\boldsymbol{X}$, and is of the order of $\Theta_{\mathrm{p}}(N)$. However, the dominating part comes from the entropy of the true distribution, and does not depend on the statistical model nor the learning algorithm. In statistical learning theory, we therefore analyze the behavior of the *relative Bayes free energy*,

$$\widetilde{F}^{\mathrm{Bayes}}(\boldsymbol{X}) = \log \frac{q(\boldsymbol{X})}{p(\boldsymbol{X})} = F^{\mathrm{Bayes}}(\boldsymbol{X}) - NS_N(\boldsymbol{X}), \tag{13.19}$$

[3] This holds if the estimator achieves a mean squared error in the same order as the *Cramér–Rao lower-bound*, i.e., $\left\langle \|\widehat{\boldsymbol{w}} - \boldsymbol{w}^*\|^2 \right\rangle_{q(\boldsymbol{X})} \geq N^{-1}\mathrm{tr}\left(\boldsymbol{F}^{-1}(\boldsymbol{w}^*)\right)$, where $\boldsymbol{F}$ is the Fisher information (13.2) at $\boldsymbol{w}^*$. The Cramér–Rao lower-bound holds for any unbiased estimator under the regularity conditions.

where

$$S_N(\boldsymbol{X}) = -\frac{1}{N}\sum_{n=1}^{N} \log q(\boldsymbol{x}^{(n)}) \tag{13.20}$$

is the *empirical entropy*. The negative of the relative Bayes free energy,

$$-\widetilde{F}^{\text{Bayes}}(\boldsymbol{X}) = \log \frac{\int p(\boldsymbol{w}) \prod_{n=1}^{N} p(\boldsymbol{x}^{(n)}|\boldsymbol{w}) d\boldsymbol{w}}{\prod_{n=1}^{N} q(\boldsymbol{x}^{(n)})}, \tag{13.21}$$

can be seen as an *ensemble* version of the log-likelihood ratio—the logarithm of the ratio between the *marginal likelihood* (alternative hypothesis) and the true likelihood (null hypothesis).

When the prior $p(\boldsymbol{w})$ is positive around the true parameter $\boldsymbol{w}^*$, the relative Bayes free energy (13.19) is known to be of the order of $\Theta(\log N)$ and can be asymptotically expanded as follows:

$$\widetilde{F}^{\text{Bayes}}(\boldsymbol{X}) = \lambda'^{\text{Bayes}} \log N + o_{\text{p}}(\log N), \tag{13.22}$$

where the coefficient of the leading term $\lambda'^{\text{Bayes}}$ is called the Bayes *free energy coefficient*. Note that, although the relative Bayes free energy is a random variable depending on realization of the training data $\boldsymbol{X}$, the leading term in Eq. (13.22) is deterministic.

Let us define the average relative Bayes free energy over the distribution of training samples:

$$\overline{F}^{\text{Bayes}}(N) = \left\langle \widetilde{F}^{\text{Bayes}}(\boldsymbol{X}) \right\rangle_{q(\boldsymbol{X})} = \left\langle \log \frac{\prod_{n=1}^{N} q(\boldsymbol{x}^{(n)})}{\int p(\boldsymbol{w}) \prod_{n=1}^{N} p(\boldsymbol{x}^{(n)}|\boldsymbol{w}) d\boldsymbol{w}} \right\rangle_{q(\boldsymbol{X})}. \tag{13.23}$$

An interesting and useful relation can be found between the average Bayes generalization error and the average relative Bayes free energy (Levin et al., 1990):

$$\begin{aligned}
\overline{\text{GE}}^{\text{Bayes}}(N) &= \left\langle \int q(\boldsymbol{x}) \log \frac{q(\boldsymbol{x})}{p^{\text{Bayes}}(\boldsymbol{x}|\boldsymbol{X})} d\boldsymbol{x} \right\rangle_{q(\boldsymbol{X})} \\
&= \left\langle \int q(\boldsymbol{x}) \log \frac{q(\boldsymbol{x})}{\int p(\boldsymbol{x}|\boldsymbol{w}) p(\boldsymbol{w}|\boldsymbol{X}) d\boldsymbol{w}} d\boldsymbol{x} \right\rangle_{q(\boldsymbol{X})} \\
&= \left\langle \int q(\boldsymbol{x}) \log \frac{q(\boldsymbol{x})}{\int p(\boldsymbol{x}|\boldsymbol{w}) p(\boldsymbol{X}|\boldsymbol{w}) p(\boldsymbol{w}) d\boldsymbol{w}} d\boldsymbol{x} \right\rangle_{q(\boldsymbol{X})} \\
&\quad - \left\langle \int q(\boldsymbol{x}) \log \frac{1}{\int p(\boldsymbol{X}|\boldsymbol{w}) p(\boldsymbol{w}) d\boldsymbol{w}} d\boldsymbol{x} \right\rangle_{q(\boldsymbol{X})} \\
&= \left\langle \log \frac{q(\boldsymbol{x}) q(\boldsymbol{X})}{\int p(\boldsymbol{x}|\boldsymbol{w}) p(\boldsymbol{X}|\boldsymbol{w}) p(\boldsymbol{w}) d\boldsymbol{w}} \right\rangle_{q(\boldsymbol{x}) q(\boldsymbol{X})} - \left\langle \log \frac{q(\boldsymbol{X})}{\int p(\boldsymbol{X}|\boldsymbol{w}) p(\boldsymbol{w}) d\boldsymbol{w}} \right\rangle_{q(\boldsymbol{X})} \\
&= \overline{F}^{\text{Bayes}}(N+1) - \overline{F}^{\text{Bayes}}(N).
\end{aligned} \tag{13.24}$$

The relation (13.24) combined with the asymptotic expansions, Eqs. (13.15) and (13.22), implies that the Bayes generalization coefficient and the Bayes free energy coefficient coincide with each other, i.e.,

$$\lambda'^{\text{Bayes}} = \lambda^{\text{Bayes}}. \tag{13.25}$$

Importantly, this relation holds for any statistical model, regardless of being regular or singular.

### 13.3.3 Target Quantities under Conditional Modeling

Many statistical models are for the regression or classification setting, where the model distribution $p(\boldsymbol{y}|\boldsymbol{x}, \boldsymbol{w})$ is the distribution of an output $\boldsymbol{y} \in \mathbb{R}^L$ conditional on an input $\boldsymbol{x} \in \mathbb{R}^M$ and an unknown parameter $\boldsymbol{w} \in \mathbb{R}^D$. The input is assumed to be given for all samples including the future test samples. Let $\mathcal{D} = \{(\boldsymbol{x}^{(1)}, \boldsymbol{y}^{(1)}), \ldots, (\boldsymbol{x}^{(N)}, \boldsymbol{y}^{(N)})\}$ be $N$ i.i.d. training samples drawn from the true joint distribution $q(\boldsymbol{x}, \boldsymbol{y}) = q(\boldsymbol{y}|\boldsymbol{x})q(\boldsymbol{x})$. As noted in Example 1.2 in Chapter 1, we can proceed with most computations without knowing the input distribution $q(\boldsymbol{x})$.

Let $\boldsymbol{X} = (\boldsymbol{x}^{(1)}, \ldots, \boldsymbol{x}^{(N)})^\top \in \mathbb{R}^{N \times M}$ and $\boldsymbol{Y} = (\boldsymbol{y}^{(1)}, \ldots, \boldsymbol{y}^{(N)})^\top \in \mathbb{R}^{N \times L}$ separately summarize the inputs and the outputs in the training data. The predictive distribution, given as a conditional distribution on a new input $\boldsymbol{x}$ as well as the whole training samples $\mathcal{D} = (\boldsymbol{X}, \boldsymbol{Y})$, can usually be computed without any information on $q(\boldsymbol{X})$. For example, the ML predictive distribution is given as

$$p^{\text{ML}}(\boldsymbol{y}|\boldsymbol{x}, \mathcal{D}) = p(\boldsymbol{y}|\boldsymbol{x}, \widehat{\boldsymbol{w}}^{\text{ML}}), \tag{13.26}$$

where

$$\widehat{\boldsymbol{w}}^{\text{ML}} = \underset{\boldsymbol{w}}{\text{argmax}}\, p(\boldsymbol{Y}|\boldsymbol{X}, \boldsymbol{w}) \cdot q(\boldsymbol{X}) = \underset{\boldsymbol{w}}{\text{argmax}} \left( \prod_{n=1}^{N} p(\boldsymbol{y}^{(n)}|\boldsymbol{x}^{(n)}, \boldsymbol{w}) \right) \tag{13.27}$$

is the ML estimator, while the Bayes predictive distribution is given as

$$p^{\text{Bayes}}(\boldsymbol{y}|\boldsymbol{x}, \mathcal{D}) = \langle p(\boldsymbol{y}|\boldsymbol{x}, \boldsymbol{w}) \rangle_{p(\boldsymbol{w}|\boldsymbol{X}, \boldsymbol{Y})} = \int p(\boldsymbol{y}|\boldsymbol{x}, \boldsymbol{w}) p(\boldsymbol{w}|\boldsymbol{X}, \boldsymbol{Y}) d\boldsymbol{w}, \tag{13.28}$$

where

$$p(\boldsymbol{w}|\boldsymbol{X}, \boldsymbol{Y}) = \frac{p(\boldsymbol{Y}|\boldsymbol{X}, \boldsymbol{w})p(\boldsymbol{w}) \cdot q(\boldsymbol{X})}{\int p(\boldsymbol{Y}|\boldsymbol{X}, \boldsymbol{w})p(\boldsymbol{w})d\boldsymbol{w} \cdot q(\boldsymbol{X})} = \frac{p(\boldsymbol{Y}|\boldsymbol{X}, \boldsymbol{w})p(\boldsymbol{w})}{\int p(\boldsymbol{Y}|\boldsymbol{X}, \boldsymbol{w})p(\boldsymbol{w})d\boldsymbol{w}} \tag{13.29}$$

is the Bayes posterior distribution. Here $(\boldsymbol{x}, \boldsymbol{y})$ is a new input–output sample pair, assumed to be drawn from the true distribution $q(\boldsymbol{y}|\boldsymbol{x})q(\boldsymbol{x})$.

The generalization error (13.10), the training error (13.11), and the relative Bayes free energy (13.19) can be expressed as follows:

$$\mathrm{GE}(\mathcal{D}) = \left\langle \log \frac{q(y|\boldsymbol{x})q(\boldsymbol{x})}{p(y|\boldsymbol{x},\mathcal{D})q(\boldsymbol{x})} \right\rangle_{q(y|\boldsymbol{x})q(\boldsymbol{x})} = \left\langle \log \frac{q(y|\boldsymbol{x})}{p(y|\boldsymbol{x},\mathcal{D})} \right\rangle_{q(y|\boldsymbol{x})q(\boldsymbol{x})}, \quad (13.30)$$

$$\mathrm{TE}(\mathcal{D}) = \frac{1}{N}\sum_{n=1}^{N} \log \frac{q(y^{(n)}|\boldsymbol{x}^{(n)})q(\boldsymbol{x}^{(n)})}{p(y^{(n)}|\boldsymbol{x}^{(n)},\mathcal{D})q(\boldsymbol{x}^{(n)})} = \frac{1}{N}\sum_{n=1}^{N} \log \frac{q(y^{(n)}|\boldsymbol{x}^{(n)})}{p(y^{(n)}|\boldsymbol{x}^{(n)},\mathcal{D})}, \quad (13.31)$$

$$\widetilde{F}^{\mathrm{Bayes}}(\mathcal{D}) = \log \frac{q(\boldsymbol{Y}|\boldsymbol{X})q(\boldsymbol{X})}{p(\boldsymbol{Y}|\boldsymbol{X})q(\boldsymbol{X})} = F^{\mathrm{Bayes}}(\boldsymbol{Y}|\boldsymbol{X}) - NS_N(\boldsymbol{Y}|\boldsymbol{X}), \quad (13.32)$$

where

$$F^{\mathrm{Bayes}}(\boldsymbol{Y}|\boldsymbol{X}) = \log \int p(\boldsymbol{w}) \prod_{n=1}^{N} p(y^{(n)}|\boldsymbol{x}^{(n)},\boldsymbol{w}) d\boldsymbol{w}, \quad (13.33)$$

$$S_N(\boldsymbol{Y}|\boldsymbol{X}) = -\frac{1}{N}\sum_{n=1}^{N} \log q(y^{(n)}|\boldsymbol{x}^{(n)}). \quad (13.34)$$

We can see that the input distribution $q(\boldsymbol{x})$ cancels out in most of the preceding equations, and therefore Eqs. (13.30) through (13.34) can be computed without considering $q(\boldsymbol{x})$. Note that in Eq. (13.30), $q(\boldsymbol{x})$ remains the distribution over which the expectation is taken. However, it is necessary only formally, and the expectation value does not depend on $q(\boldsymbol{x})$ (as long as the regularity conditions hold). The same applies to the average generalization error (13.13), the average training error (13.14), and the average relative Bayes free energy (13.23), where the expectation $\langle\cdot\rangle_{q(\boldsymbol{Y}|\boldsymbol{X})q(\boldsymbol{X})}$ over the distribution of the training samples is taken.

## 13.4 Asymptotic Learning Theory for Regular Models

In this section, we introduce the *regular learning theory*, which generally holds under the regularity conditions.

### 13.4.1 Regularity Conditions

The *regularity conditions* are defined for the statistical model $p(\boldsymbol{x}|\boldsymbol{w})$ parameterized by a finite-dimensional parameter vector $\boldsymbol{w} \in \mathcal{W} \subseteq \mathbb{R}^D$, and the true distribution $q(\boldsymbol{x})$. We include conditions for the prior distribution $p(\boldsymbol{w})$, which are necessary for analyzing MAP learning and Bayesian learning. There are variations, and we here introduce a (rough) simple set.

(i) The statistical model $p(\boldsymbol{x}|\boldsymbol{w})$ is differentiable (as many times as necessary) with respect to the parameter $\boldsymbol{w} \in \mathcal{W}$ for any $\boldsymbol{x}$, and the differential operator and the integral operator are commutable.

(ii) The statistical model $p(\boldsymbol{x}|\boldsymbol{w})$ is *identifiable*, i.e., Eq. (13.1) holds, and the Fisher information (13.2) is nonsingular (full-rank) at any $\boldsymbol{w} \in \mathcal{W}$.

(iii) The support of $p(\boldsymbol{x}|\boldsymbol{w})$, i.e., $\{\boldsymbol{x} \in \mathcal{X}; p(\boldsymbol{x}|\boldsymbol{w}) > 0\}$, is common for all $\boldsymbol{w} \in \mathcal{W}$.

(iv) The true distribution is *realizable* by the statistical model, i.e., $\exists \boldsymbol{w}^*$ s.t. $q(\boldsymbol{x}) = p(\boldsymbol{x}|\boldsymbol{w}^*)$, and the true parameter $\boldsymbol{w}^*$ is an interior point of the domain $\mathcal{W}$.

(v) The prior $p(\boldsymbol{w})$ is twice differentiable and bounded as $0 < p(\boldsymbol{w}) < \infty$ at any $\boldsymbol{w} \in \mathcal{W}$.

Note that the first three conditions are on the model distribution $p(\boldsymbol{x}|\boldsymbol{w})$, the fourth is on the true distribution $q(\boldsymbol{x})$, and the fifth is on the prior distribution $p(\boldsymbol{w})$.

An important consequence of the regularity conditions is that the log-likelihood can be Taylor-expanded about any $\overline{\boldsymbol{w}} \in \mathcal{W}$:

$$\log p(\boldsymbol{x}|\boldsymbol{w}) = \log p(\boldsymbol{x}|\overline{\boldsymbol{w}}) + (\boldsymbol{w} - \overline{\boldsymbol{w}})^\top \left. \frac{\partial \log p(\boldsymbol{x}|\boldsymbol{w})}{\partial \boldsymbol{w}} \right|_{\boldsymbol{w}=\overline{\boldsymbol{w}}} + \frac{1}{2}(\boldsymbol{w} - \overline{\boldsymbol{w}})^\top \left. \frac{\partial^2 \log p(\boldsymbol{x}|\boldsymbol{w})}{\partial \boldsymbol{w} \partial \boldsymbol{w}^\top} \right|_{\boldsymbol{w}=\overline{\boldsymbol{w}}} (\boldsymbol{w} - \overline{\boldsymbol{w}}) + O(\|\boldsymbol{w} - \overline{\boldsymbol{w}}\|^3). \quad (13.35)$$

### 13.4.2 Consistency and Asymptotic Normality

We first show *consistency* and *asymptotic normality*, which hold in ML learning, MAP learning, and Bayesian learning.

**Consistency of ML Estimator**

The ML estimator is defined by

$$\widehat{\boldsymbol{w}}^{\mathrm{ML}} = \operatorname*{argmax}_{\boldsymbol{w}} \log \left( \prod_{n=1}^{N} p(\boldsymbol{x}^{(n)}|\boldsymbol{w}) \right) = \operatorname*{argmax}_{\boldsymbol{w}} L_N(\boldsymbol{w}), \quad (13.36)$$

where

$$L_N(\boldsymbol{w}) = \frac{1}{N} \sum_{n=1}^{N} \log p(\boldsymbol{x}^{(n)}|\boldsymbol{w}). \quad (13.37)$$

By the law of large numbers (13.5), it holds that

$$L_N(\boldsymbol{w}) = L^*(\boldsymbol{w}) + o_{\mathrm{p}}(1), \quad (13.38)$$

where

$$L^*(\boldsymbol{w}) = \langle \log p(\boldsymbol{x}|\boldsymbol{w}) \rangle_{p(\boldsymbol{x}|\boldsymbol{w}^*)} . \tag{13.39}$$

Identifiability of the statistical model guarantees that

$$\boldsymbol{w}^* = \underset{\boldsymbol{w}}{\operatorname{argmax}}\, L^*(\boldsymbol{w}) \tag{13.40}$$

is the unique maximizer. Eqs. (13.36), (13.38), and (13.40), imply the consistency of the ML estimator, i.e.,

$$\widehat{\boldsymbol{w}}^{\mathrm{ML}} = \boldsymbol{w}^* + o_{\mathrm{p}}(1). \tag{13.41}$$

**Asymptotic Normality of the ML Estimator**

Since the gradient $\partial L_N(\boldsymbol{w})/\partial \boldsymbol{w}$ is differentiable, the *mean value theorem*[4] guarantees that there exists $\acute{\boldsymbol{w}} \in [\min(\widehat{\boldsymbol{w}}^{\mathrm{ML}}, \boldsymbol{w}^*), \max(\widehat{\boldsymbol{w}}^{\mathrm{ML}}, \boldsymbol{w}^*)]^D$ (where $\min(\cdot)$ and $\max(\cdot)$ operate elementwise) such that

$$\left.\frac{\partial L_N(\boldsymbol{w})}{\partial \boldsymbol{w}}\right|_{\boldsymbol{w}=\widehat{\boldsymbol{w}}^{\mathrm{ML}}} = \left.\frac{\partial L_N(\boldsymbol{w})}{\partial \boldsymbol{w}}\right|_{\boldsymbol{w}=\boldsymbol{w}^*} + \left.\frac{\partial^2 L_N(\boldsymbol{w})}{\partial \boldsymbol{w}\partial \boldsymbol{w}^\top}\right|_{\boldsymbol{w}=\acute{\boldsymbol{w}}} (\widehat{\boldsymbol{w}}^{\mathrm{ML}} - \boldsymbol{w}^*). \tag{13.42}$$

By the definition (13.36) of the ML estimator and the differentiability of $L_N(\boldsymbol{w})$, the left-hand side of Eq. (13.42) is equal to zero, i.e.,

$$\left.\frac{\partial L_N(\boldsymbol{w})}{\partial \boldsymbol{w}}\right|_{\boldsymbol{w}=\widehat{\boldsymbol{w}}^{\mathrm{ML}}} = \boldsymbol{0}. \tag{13.43}$$

The first term in the right-hand side of Eq. (13.42) can be written as

$$\left.\frac{\partial L_N(\boldsymbol{w})}{\partial \boldsymbol{w}}\right|_{\boldsymbol{w}=\boldsymbol{w}^*} = \frac{1}{N}\sum_{n=1}^{N} \left.\frac{\partial \log p(\boldsymbol{x}^{(n)}|\boldsymbol{w})}{\partial \boldsymbol{w}}\right|_{\boldsymbol{w}=\boldsymbol{w}^*} . \tag{13.44}$$

Since Eq. (13.40) and the differentiability of $L^*(\boldsymbol{w})$ imply that

$$\left.\frac{\partial L^*(\boldsymbol{w})}{\partial \boldsymbol{w}}\right|_{\boldsymbol{w}=\boldsymbol{w}^*} = \left\langle \left.\frac{\partial \log p(\boldsymbol{x}|\boldsymbol{w})}{\partial \boldsymbol{w}}\right|_{\boldsymbol{w}=\boldsymbol{w}^*} \right\rangle_{p(\boldsymbol{x}|\boldsymbol{w}^*)} = \boldsymbol{0}, \tag{13.45}$$

the right-hand side of Eq. (13.44) is the average over $N$ i.i.d. samples of the random variable

$$\left.\frac{\partial \log p(\boldsymbol{x}^{(n)}|\boldsymbol{w})}{\partial \boldsymbol{w}}\right|_{\boldsymbol{w}=\boldsymbol{w}^*},$$

[4] The mean value theorem states that, for a differentiable function $f : [a, b] \mapsto \mathbb{R}$,

$$\exists c \in [a, b] \qquad \text{s.t.} \qquad \left.\frac{df(x)}{dx}\right|_{x=c} = \frac{f(b)-f(a)}{b-a}.$$

which follows a distribution with zero mean (Eq. (13.45)) and the covariance given by the Fisher information (13.2) at $\boldsymbol{w} = \boldsymbol{w}^*$, i.e.,

$$\boldsymbol{F}(\boldsymbol{w}^*) = \left\langle \left.\frac{\partial \log p(\boldsymbol{x}|\boldsymbol{w})}{\partial \boldsymbol{w}}\right|_{\boldsymbol{w}=\boldsymbol{w}^*} \left.\frac{\partial \log p(\boldsymbol{x}|\boldsymbol{w})}{\partial \boldsymbol{w}}\right|_{\boldsymbol{w}=\boldsymbol{w}^*}^{\top} \right\rangle_{p(\boldsymbol{x}|\boldsymbol{w}^*)}. \tag{13.46}$$

Therefore, according to the central limit theorem (Theorem 13.1), the distribution of the first term in the right-hand side of Eq. (13.42) converges to

$$p\left(\left.\frac{\partial L_N(\boldsymbol{w})}{\partial \boldsymbol{w}}\right|_{\boldsymbol{w}=\boldsymbol{w}^*}\right) \to \mathrm{Gauss}_D\left(\left.\frac{\partial L_N(\boldsymbol{w})}{\partial \boldsymbol{w}}\right|_{\boldsymbol{w}=\boldsymbol{w}^*}; \boldsymbol{0}, N^{-1}\boldsymbol{F}(\boldsymbol{w}^*)\right). \tag{13.47}$$

The coefficient of the second term in the right-hand side of Eq. (13.42) satisfies

$$\left.\frac{\partial^2 L_N(\boldsymbol{w})}{\partial \boldsymbol{w} \partial \boldsymbol{w}}\right|_{\boldsymbol{w}=\acute{\boldsymbol{w}}} = \left.\frac{\partial^2 L^*(\boldsymbol{w})}{\partial \boldsymbol{w} \partial \boldsymbol{w}^{\top}}\right|_{\boldsymbol{w}=\boldsymbol{w}^*} + o_{\mathrm{p}}(1), \tag{13.48}$$

because of the law of large numbers and the consistency of the ML estimator, i.e., $[\min(\widehat{\boldsymbol{w}}^{\mathrm{ML}}, \boldsymbol{w}^*), \max(\widehat{\boldsymbol{w}}^{\mathrm{ML}}, \boldsymbol{w}^*)] \ni \acute{\boldsymbol{w}} \to \boldsymbol{w}^*$ since $\widehat{\boldsymbol{w}}^{\mathrm{ML}} \to \boldsymbol{w}^*$. Furthermore, the following relation holds under the regularity conditions (see Appendix B.2):

$$\left.\frac{\partial^2 L^*(\boldsymbol{w})}{\partial \boldsymbol{w} \partial \boldsymbol{w}^{\top}}\right|_{\boldsymbol{w}=\boldsymbol{w}^*} = \left\langle \left.\frac{\partial^2 \log p(\boldsymbol{x}|\boldsymbol{w})}{\partial \boldsymbol{w} \partial \boldsymbol{w}^{\top}}\right|_{\boldsymbol{w}=\boldsymbol{w}^*} \right\rangle_{p(\boldsymbol{x}|\boldsymbol{w}^*)} = -\boldsymbol{F}(\boldsymbol{w}^*). \tag{13.49}$$

Substituting Eqs. (13.43), (13.48), and (13.49) into Eq. (13.42) gives

$$\left(\boldsymbol{F}(\boldsymbol{w}^*) + o_{\mathrm{p}}(1)\right)(\widehat{\boldsymbol{w}}^{\mathrm{ML}} - \boldsymbol{w}^*) = \left.\frac{\partial L_N(\boldsymbol{w})}{\partial \boldsymbol{w}}\right|_{\boldsymbol{w}=\boldsymbol{w}^*}. \tag{13.50}$$

Since the Fisher information is assumed to be invertible, Eq. (13.47) leads to the following theorem:

**Theorem 13.2** *(Asymptotic normality of ML estimator) Under the regularity conditions, the distribution of* $\boldsymbol{v}^{\mathrm{ML}} = \sqrt{N}(\widehat{\boldsymbol{w}}^{\mathrm{ML}} - \boldsymbol{w}^*)$ *converges to*

$$p\left(\boldsymbol{v}^{\mathrm{ML}}\right) \to \mathrm{Gauss}_D\left(\boldsymbol{v}^{\mathrm{ML}}; \boldsymbol{0}, \boldsymbol{F}^{-1}(\boldsymbol{w}^*)\right) \quad \text{as} \quad N \to \infty. \tag{13.51}$$

Theorem 13.2 implies that

$$\widehat{\boldsymbol{w}}^{\mathrm{ML}} = \boldsymbol{w}^* + O_{\mathrm{p}}(N^{-1/2}). \tag{13.52}$$

### 13.4.3 Asymptotic Normality of the Bayes Posterior

The Bayes posterior can be written as follows:

$$p(\boldsymbol{w}|\boldsymbol{X}) = \frac{\exp\left(NL_N(\boldsymbol{w}) + \log p(\boldsymbol{w})\right)}{\int \exp\left(NL_N(\boldsymbol{w}) + \log p(\boldsymbol{w})\right) d\boldsymbol{w}}. \tag{13.53}$$

In the asymptotic limit, the factor $\exp(NL_N(\boldsymbol{w}))$ dominates the numerator, and the probability mass concentrates around the peak of $L_N(\boldsymbol{w})$—the ML estimator $\widehat{\boldsymbol{w}}^{\mathrm{ML}}$. The Taylor expansion of $L_N(\boldsymbol{w})$ about $\widehat{\boldsymbol{w}}^{\mathrm{ML}}$ gives

$$\begin{aligned} L_N(\boldsymbol{w}) &\approx L_N(\widehat{\boldsymbol{w}}^{\mathrm{ML}}) + (\boldsymbol{w} - \widehat{\boldsymbol{w}}^{\mathrm{ML}})^\top \left.\frac{\partial L_N(\boldsymbol{w})}{\partial \boldsymbol{w}}\right|_{\boldsymbol{w}=\widehat{\boldsymbol{w}}^{\mathrm{ML}}} \\ &\quad + \frac{1}{2}(\boldsymbol{w} - \widehat{\boldsymbol{w}}^{\mathrm{ML}})^\top \left.\frac{\partial^2 L_N(\boldsymbol{w})}{\partial \boldsymbol{w} \partial \boldsymbol{w}^\top}\right|_{\boldsymbol{w}=\widehat{\boldsymbol{w}}^{\mathrm{ML}}} (\boldsymbol{w} - \widehat{\boldsymbol{w}}^{\mathrm{ML}}) \\ &\approx L_N(\widehat{\boldsymbol{w}}^{\mathrm{ML}}) - \frac{1}{2}(\boldsymbol{w} - \widehat{\boldsymbol{w}}^{\mathrm{ML}})^\top \boldsymbol{F}(\boldsymbol{w}^*)(\boldsymbol{w} - \widehat{\boldsymbol{w}}^{\mathrm{ML}}), \end{aligned} \tag{13.54}$$

where we used Eq. (13.43) and

$$\left.\frac{\partial^2 L_N(\boldsymbol{w})}{\partial \boldsymbol{w} \partial \boldsymbol{w}^\top}\right|_{\boldsymbol{w}=\widehat{\boldsymbol{w}}^{\mathrm{ML}}} = -\boldsymbol{F}(\boldsymbol{w}^*) + o_{\mathrm{p}}(1), \tag{13.55}$$

which is implied by the law of large numbers and the consistency of the ML estimator. Eqs. (13.53) and (13.54) imply that the Bayes posterior can be approximated by Gaussian in the asymptotic limit:

$$p(\boldsymbol{w}|\boldsymbol{X}) \approx \mathrm{Gauss}_D\left(\boldsymbol{w}; \widehat{\boldsymbol{w}}^{\mathrm{ML}}, N^{-1}\boldsymbol{F}^{-1}(\boldsymbol{w}^*)\right).$$

The following theorem was derived with more accurate discussion.

**Theorem 13.3** *(Asymptotic normality of the Bayes posterior) (van der Vaart, 1998) Under the regularity conditions, the (rescaled) Bayes posterior distribution* $p(\boldsymbol{v}|\boldsymbol{X})$ *where* $\boldsymbol{v} = \sqrt{N}(\boldsymbol{w} - \boldsymbol{w}^*)$ *converges to*

$$p(\boldsymbol{v}|\boldsymbol{X}) \to \mathrm{Gauss}_D\left(\boldsymbol{v}; \boldsymbol{v}^{\mathrm{ML}}, \boldsymbol{F}^{-1}(\boldsymbol{w}^*)\right) \quad \textit{as} \quad N \to \infty, \tag{13.56}$$

*where* $\boldsymbol{v}^{\mathrm{ML}} = \sqrt{N}(\widehat{\boldsymbol{w}}^{\mathrm{ML}} - \boldsymbol{w}^*)$.

Theorem 13.3 implies that

$$\widehat{\boldsymbol{w}}^{\mathrm{MAP}} = \widehat{\boldsymbol{w}}^{\mathrm{ML}} + o_{\mathrm{p}}(N^{-1/2}), \tag{13.57}$$

$$\widehat{\boldsymbol{w}}^{\mathrm{Bayes}} = \langle \boldsymbol{w} \rangle_{p(\boldsymbol{w}|\boldsymbol{X})} = \widehat{\boldsymbol{w}}^{\mathrm{ML}} + o_{\mathrm{p}}(N^{-1/2}), \tag{13.58}$$

which prove the consistency of the MAP estimator and the Bayesian estimator.

### 13.4.4 Generalization Properties

Now we analyze the generalization error and the training error in ML learning, MAP learning, and Bayesian learning, as well as the Bayes free energy. After that, we introduce information criteria for model selection, which were developed based on the asymptotic behavior of those quantities.

### 13.4.5 ML Learning

The generalization error of ML learning can be written as

$$\begin{aligned}\mathrm{GE}^{\mathrm{ML}}_{\mathrm{Regular}}(\boldsymbol{X}) &= \left\langle \log \frac{p(\boldsymbol{x}|\boldsymbol{w}^*)}{p(\boldsymbol{x}|\widehat{\boldsymbol{w}}^{\mathrm{ML}})} \right\rangle_{p(\boldsymbol{x}|\boldsymbol{w}^*)} \\ &= L^*(\boldsymbol{w}^*) - L^*(\widehat{\boldsymbol{w}}^{\mathrm{ML}}) \end{aligned} \tag{13.59}$$

with $L^*(\boldsymbol{w})$ defined by Eq. (13.39). The Taylor expansion of the second term of Eq. (13.59) about the true parameter $\boldsymbol{w}^*$ gives

$$\begin{aligned} L^*(\widehat{\boldsymbol{w}}^{\mathrm{ML}}) &= L^*(\boldsymbol{w}^*) + (\widehat{\boldsymbol{w}}^{\mathrm{ML}} - \boldsymbol{w}^*)^\top \left. \frac{\partial L^*(\boldsymbol{w})}{\partial \boldsymbol{w}} \right|_{\boldsymbol{w}=\boldsymbol{w}^*} \\ &\quad + \frac{1}{2}(\widehat{\boldsymbol{w}}^{\mathrm{ML}} - \boldsymbol{w}^*)^\top \left. \frac{\partial^2 L^*(\boldsymbol{w})}{\partial \boldsymbol{w} \partial \boldsymbol{w}^\top} \right|_{\boldsymbol{w}=\boldsymbol{w}^*} (\widehat{\boldsymbol{w}}^{\mathrm{ML}} - \boldsymbol{w}^*) + O(\|\widehat{\boldsymbol{w}}^{\mathrm{ML}} - \boldsymbol{w}^*\|^3) \\ &= L^*(\boldsymbol{w}^*) - \frac{1}{2}(\widehat{\boldsymbol{w}}^{\mathrm{ML}} - \boldsymbol{w}^*)^\top \boldsymbol{F}(\boldsymbol{w}^*)(\widehat{\boldsymbol{w}}^{\mathrm{ML}} - \boldsymbol{w}^*) + O(\|\widehat{\boldsymbol{w}}^{\mathrm{ML}} - \boldsymbol{w}^*\|^3), \end{aligned} \tag{13.60}$$

where we used Eqs. (13.45) and (13.49) in the last equality. Substituting Eq. (13.60) into Eq. (13.59) gives

$$\mathrm{GE}^{\mathrm{ML}}_{\mathrm{Regular}}(\boldsymbol{X}) = \frac{1}{2}(\widehat{\boldsymbol{w}}^{\mathrm{ML}} - \boldsymbol{w}^*)^\top \boldsymbol{F}(\boldsymbol{w}^*)(\widehat{\boldsymbol{w}}^{\mathrm{ML}} - \boldsymbol{w}^*) + O(\|\widehat{\boldsymbol{w}}^{\mathrm{ML}} - \boldsymbol{w}^*\|^3). \tag{13.61}$$

The asymptotic normality (Theorem 13.2) of the ML estimator implies that

$$\sqrt{N}\boldsymbol{F}^{\frac{1}{2}}(\boldsymbol{w}^*)(\widehat{\boldsymbol{w}}^{\mathrm{ML}} - \boldsymbol{w}^*) \rightsquigarrow \mathrm{Gauss}_D(\boldsymbol{0}, \boldsymbol{I}_D), \tag{13.62}$$

and that

$$O(\|\widehat{\boldsymbol{w}}^{\mathrm{ML}} - \boldsymbol{w}^*\|^3) = O_{\mathrm{p}}(N^{-3/2}). \tag{13.63}$$

Eq. (13.62) implies that the distribution of $s = N(\widehat{\boldsymbol{w}}^{\mathrm{ML}} - \boldsymbol{w}^*)^\top \boldsymbol{F}(\boldsymbol{w}^*)(\widehat{\boldsymbol{w}}^{\mathrm{ML}} - \boldsymbol{w}^*)$ converges to the *chi-squared distribution* with $D$ degrees of freedom:[5]

$$p(s) \to \chi^2(s; D), \tag{13.64}$$

and therefore,

$$N\left\langle (\widehat{\boldsymbol{w}}^{\mathrm{ML}} - \boldsymbol{w}^*)^\top \boldsymbol{F}(\boldsymbol{w}^*)(\widehat{\boldsymbol{w}}^{\mathrm{ML}} - \boldsymbol{w}^*) \right\rangle_{q(\boldsymbol{X})} = D + o(1). \tag{13.65}$$

Eqs. (13.61), (13.63), and (13.65) lead to the following theorem:

[5] The chi-squared distribution with $D$ degrees of freedom is the distribution of the sum of the squares of $D$ i.i.d. samples drawn from $\mathrm{Gauss}_1(0, 1^2)$. It is actually a special case of the Gamma distribution, and it holds that $\chi^2(x; D) = \mathrm{Gamma}(x; D/2, 1/2)$. The mean and the variance are equal to $D$ and $2D$, respectively.

**Theorem 13.4** *The average generalization error of ML learning in the regular models can be asymptotically expanded as*

$$\overline{\mathrm{GE}}^{\mathrm{ML}}_{\mathrm{Regular}}(N) = \left\langle \mathrm{GE}^{\mathrm{ML}}_{\mathrm{Regular}}(\boldsymbol{X}) \right\rangle_{q(\boldsymbol{X})} = \lambda^{\mathrm{ML}}_{\mathrm{Regular}} N^{-1} + o(N^{-1}), \tag{13.66}$$

*where the generalization coefficient is given by*

$$2\lambda^{\mathrm{ML}}_{\mathrm{Regular}} = D. \tag{13.67}$$

Interestingly, the leading term of the generalization error only depends on the parameter dimension or the degree of freedom of the statistical model.

The training error of ML learning can be analyzed in a similar fashion. It can be written as

$$\begin{aligned} \mathrm{TE}^{\mathrm{ML}}_{\mathrm{Regular}}(\boldsymbol{X}) &= N^{-1} \sum_{n=1}^{N} \log \frac{p(\boldsymbol{x}^{(n)}|\boldsymbol{w}^*)}{p(\boldsymbol{x}^{(n)}|\widehat{\boldsymbol{w}}^{\mathrm{ML}})} \\ &= L_N(\boldsymbol{w}^*) - L_N(\widehat{\boldsymbol{w}}^{\mathrm{ML}}) \end{aligned} \tag{13.68}$$

with $L_N(\boldsymbol{w})$ defined by Eq. (13.37). The Taylor expansion of the first term of Eq. (13.68) about the ML estimator $\widehat{\boldsymbol{w}}^{\mathrm{ML}}$ gives

$$\begin{aligned} L_N(\boldsymbol{w}^*) &= L_N(\widehat{\boldsymbol{w}}^{\mathrm{ML}}) + (\boldsymbol{w}^* - \widehat{\boldsymbol{w}}^{\mathrm{ML}})^\top \left.\frac{\partial L_N(\boldsymbol{w})}{\partial \boldsymbol{w}}\right|_{\boldsymbol{w}=\widehat{\boldsymbol{w}}^{\mathrm{ML}}} \\ &\quad + \frac{1}{2}(\boldsymbol{w}^* - \widehat{\boldsymbol{w}}^{\mathrm{ML}})^\top \left.\frac{\partial^2 L_N(\boldsymbol{w})}{\partial \boldsymbol{w} \partial \boldsymbol{w}^\top}\right|_{\boldsymbol{w}=\widehat{\boldsymbol{w}}^{\mathrm{ML}}} (\boldsymbol{w}^* - \widehat{\boldsymbol{w}}^{\mathrm{ML}}) + O(\|\boldsymbol{w}^* - \widehat{\boldsymbol{w}}^{\mathrm{ML}}\|^3) \\ &= L_N(\widehat{\boldsymbol{w}}^{\mathrm{ML}}) - \frac{1}{2}(\boldsymbol{w}^* - \widehat{\boldsymbol{w}}^{\mathrm{ML}})^\top \left(\boldsymbol{F}(\boldsymbol{w}^*) + o_{\mathrm{p}}(1)\right)(\boldsymbol{w}^* - \widehat{\boldsymbol{w}}^{\mathrm{ML}}) \\ &\quad + O(\|\boldsymbol{w}^* - \widehat{\boldsymbol{w}}^{\mathrm{ML}}\|^3), \end{aligned} \tag{13.69}$$

where we used Eqs. (13.43) and (13.55). Substituting Eq. (13.69) into Eq. (13.68) and applying Eq. (13.52), we have

$$\mathrm{TE}^{\mathrm{ML}}_{\mathrm{Regular}}(\boldsymbol{X}) = -\frac{1}{2}(\widehat{\boldsymbol{w}}^{\mathrm{ML}} - \boldsymbol{w}^*)^\top \boldsymbol{F}(\boldsymbol{w}^*)(\widehat{\boldsymbol{w}}^{\mathrm{ML}} - \boldsymbol{w}^*) + o_{\mathrm{p}}(N^{-1}). \tag{13.70}$$

Thus, Eq. (13.70) together with Eq. (13.65) gives the following theorem:

**Theorem 13.5** *The average training error of ML learning in the regular models can be asymptotically expanded as*

$$\overline{\mathrm{TE}}^{\mathrm{ML}}_{\mathrm{Regular}}(N) = \left\langle \mathrm{TE}^{\mathrm{ML}}_{\mathrm{Regular}}(\boldsymbol{X}) \right\rangle_{q(\boldsymbol{X})} = \nu^{\mathrm{ML}}_{\mathrm{Regular}} N^{-1} + o(N^{-1}), \tag{13.71}$$

*where the training coefficient is given by*

$$2\nu^{\mathrm{ML}}_{\mathrm{Regular}} = -D. \tag{13.72}$$

Comparing Theorems 13.4 and 13.5, we see that the generalization coefficient and the training coefficient are antisymmetric with each other:

$$\lambda^{\mathrm{ML}}_{\mathrm{Regular}} = -\nu^{\mathrm{ML}}_{\mathrm{Regular}}.$$

### 13.4.6 MAP Learning

We first prove the following theorem:

**Theorem 13.6** *For any (point-) estimator such that*

$$\widehat{\boldsymbol{w}} = \widehat{\boldsymbol{w}}^{\mathrm{ML}} + o_{\mathrm{p}}(N^{-1/2}), \tag{13.73}$$

*it holds that*

$$\mathrm{GE}^{\widehat{\boldsymbol{w}}}_{\mathrm{Regular}}(\boldsymbol{X}) = \left\langle \log \frac{p(\boldsymbol{x}|\boldsymbol{w}^*)}{p(\boldsymbol{x}|\widehat{\boldsymbol{w}})} \right\rangle_{p(\boldsymbol{x}|\boldsymbol{w}^*)} = \mathrm{GE}^{\mathrm{ML}}_{\mathrm{Regular}}(\boldsymbol{X}) + o_{\mathrm{p}}(N^{-1}), \tag{13.74}$$

$$\mathrm{TE}^{\widehat{\boldsymbol{w}}}_{\mathrm{Regular}}(\boldsymbol{X}) = N^{-1} \sum_{n=1}^{N} \log \frac{p(\boldsymbol{x}^{(n)}|\boldsymbol{w}^*)}{p(\boldsymbol{x}^{(n)}|\widehat{\boldsymbol{w}})} = \mathrm{TE}^{\mathrm{ML}}_{\mathrm{Regular}}(\boldsymbol{X}) + o_{\mathrm{p}}(N^{-1}). \tag{13.75}$$

*Proof* The generalization error of the estimator $\widehat{\boldsymbol{w}}$ can be written as

$$\begin{aligned}
\mathrm{GE}^{\widehat{\boldsymbol{w}}}_{\mathrm{Regular}}(\boldsymbol{X}) &= \left\langle \log \frac{p(\boldsymbol{x}|\boldsymbol{w}^*)}{p(\boldsymbol{x}|\widehat{\boldsymbol{w}})} \right\rangle_{p(\boldsymbol{x}|\boldsymbol{w}^*)} \\
&= L^*(\boldsymbol{w}^*) - L^*(\widehat{\boldsymbol{w}}) \\
&= \mathrm{GE}^{\mathrm{ML}}_{\mathrm{Regular}}(\boldsymbol{X}) + \left( L^*(\widehat{\boldsymbol{w}}^{\mathrm{ML}}) - L^*(\widehat{\boldsymbol{w}}) \right),
\end{aligned} \tag{13.76}$$

where the second term can be expanded as

$$\begin{aligned}
L^*(\widehat{\boldsymbol{w}}^{\mathrm{ML}}) - L^*(\widehat{\boldsymbol{w}}) &= -(\widehat{\boldsymbol{w}} - \widehat{\boldsymbol{w}}^{\mathrm{ML}})^\top \left. \frac{\partial L^*(\boldsymbol{w})}{\partial \boldsymbol{w}} \right|_{\boldsymbol{w}=\widehat{\boldsymbol{w}}^{\mathrm{ML}}} \\
&\quad - \frac{1}{2} (\widehat{\boldsymbol{w}} - \widehat{\boldsymbol{w}}^{\mathrm{ML}})^\top \left. \frac{\partial^2 L^*(\boldsymbol{w})}{\partial \boldsymbol{w} \partial \boldsymbol{w}^\top} \right|_{\boldsymbol{w}=\widehat{\boldsymbol{w}}^{\mathrm{ML}}} (\widehat{\boldsymbol{w}} - \widehat{\boldsymbol{w}}^{\mathrm{ML}}) + O(\|\widehat{\boldsymbol{w}} - \widehat{\boldsymbol{w}}^{\mathrm{ML}}\|^3).
\end{aligned} \tag{13.77}$$

Eqs. (13.45) and (13.52) (with the differentiability of $\partial L^*(\boldsymbol{w})/\partial \boldsymbol{w}$) imply that

$$\left. \frac{\partial L^*(\boldsymbol{w})}{\partial \boldsymbol{w}} \right|_{\boldsymbol{w}=\widehat{\boldsymbol{w}}^{\mathrm{ML}}} = \left. \frac{\partial L^*(\boldsymbol{w})}{\partial \boldsymbol{w}} \right|_{\boldsymbol{w}=\boldsymbol{w}^*+O_{\mathrm{p}}(N^{-1/2})} = O_{\mathrm{p}}(N^{-1/2}),$$

with which Eqs. (13.73) and (13.77) lead to

$$L^*(\widehat{\boldsymbol{w}}^{\mathrm{ML}}) - L^*(\widehat{\boldsymbol{w}}) = o_{\mathrm{p}}(N^{-1}).$$

Substituting the preceding into Eq. (13.76) gives Eq. (13.74).

Similarly, the training error of the estimator $\widehat{w}$ can be written as

$$\begin{aligned}
\mathrm{TE}^{\widehat{w}}_{\mathrm{Regular}}(X) &= N^{-1}\sum_{n=1}^{N}\log\frac{p(x^{(n)}|w^*)}{p(x^{(n)}|\widehat{w})} \\
&= L_N(w^*) - L_N(\widehat{w}) \\
&= \mathrm{TE}^{\mathrm{ML}}_{\mathrm{Regular}}(X) + \left(L_N(\widehat{w}^{\mathrm{ML}}) - L_N(\widehat{w})\right),
\end{aligned} \tag{13.78}$$

where the second term can be expanded as

$$\begin{aligned}
L_N(\widehat{w}^{\mathrm{ML}}) - L_N(\widehat{w}) &= -(\widehat{w}-\widehat{w}^{\mathrm{ML}})^\top \left.\frac{\partial L_N(w)}{\partial w}\right|_{w=\widehat{w}^{\mathrm{ML}}} \\
&\quad - \frac{1}{2}(\widehat{w}-\widehat{w}^{\mathrm{ML}})^\top \left.\frac{\partial^2 L_N(w)}{\partial w \partial w^\top}\right|_{w=\widehat{w}^{\mathrm{ML}}} (\widehat{w}-\widehat{w}^{\mathrm{ML}}) + O(\|\widehat{w}-\widehat{w}^{\mathrm{ML}}\|^3).
\end{aligned} \tag{13.79}$$

Eqs. (13.43), (13.73), and (13.79) lead to

$$L_N(\widehat{w}^{\mathrm{ML}}) - L_N(\widehat{w}) = o_{\mathrm{p}}(N^{-1}).$$

Substituting the preceding into Eq. (13.78) gives Eq. (13.75), which completes the proof. □

Since the MAP estimator satisfies the condition (13.73) of Theorem 13.6 (see Eq. (13.57)), we obtain the following corollaries:

**Corollary 13.7** *The average generalization error of MAP learning in the regular models can be asymptotically expanded as*

$$\overline{\mathrm{GE}}^{\mathrm{MAP}}_{\mathrm{Regular}}(N) = \left\langle \mathrm{GE}^{\mathrm{MAP}}_{\mathrm{Regular}}(X)\right\rangle_{q(X)} = \lambda^{\mathrm{MAP}}_{\mathrm{Regular}} N^{-1} + o(N^{-1}), \tag{13.80}$$

*where the generalization coefficient is given by*

$$2\lambda^{\mathrm{MAP}}_{\mathrm{Regular}} = D. \tag{13.81}$$

**Corollary 13.8** *The average training error of MAP learning in the regular models can be asymptotically expanded as*

$$\overline{\mathrm{TE}}^{\mathrm{MAP}}_{\mathrm{Regular}}(N) = \left\langle \mathrm{TE}^{\mathrm{MAP}}_{\mathrm{Regular}}(X)\right\rangle_{q(X)} = \nu^{\mathrm{MAP}}_{\mathrm{Regular}} N^{-1} + o(N^{-1}), \tag{13.82}$$

*where the training coefficient is given by*

$$2\nu^{\mathrm{MAP}}_{\mathrm{Regular}} = -D. \tag{13.83}$$

### 13.4.7 Bayesian Learning

Eq. (13.58) and Theorem 13.6 imply that the Bayesian estimator also gives the same generalization and training coefficients as ML learning, if the *plug-in* predictive distribution $p(\boldsymbol{x}|\widehat{\boldsymbol{w}}^{\mathrm{Bayes}})$, i.e., the model distribution with the Bayesian parameter plugged-in, is used for prediction. We can show that the proper Bayesian procedure with the predictive distribution $p(\boldsymbol{x}|\boldsymbol{X}) = \langle p(\boldsymbol{x}|\boldsymbol{w})\rangle_{p(\boldsymbol{w}|\boldsymbol{X})}$ also gives the same generalization and training coefficients.

We first prove the following theorem:

**Theorem 13.9** *Let $r(\boldsymbol{w})$ be a (possibly approximate posterior) distribution of the parameter, of which the mean and the covariance satisfy the following:*

$$\widehat{\boldsymbol{w}} = \langle \boldsymbol{w}\rangle_{r(\boldsymbol{w})} = \boldsymbol{w}^* + O_{\mathrm{p}}(N^{-1/2}), \tag{13.84}$$

$$\widehat{\boldsymbol{\Sigma}}_w = \left\langle \left(\boldsymbol{w} - \langle \boldsymbol{w}\rangle_{r(\boldsymbol{w})}\right)\left(\boldsymbol{w} - \langle \boldsymbol{w}\rangle_{r(\boldsymbol{w})}\right)^\top \right\rangle_{r(\boldsymbol{w})} = O_{\mathrm{p}}(N^{-1}). \tag{13.85}$$

*Then the generalization error and the training error of the predictive distribution $p(\boldsymbol{x}|\boldsymbol{X}) = \langle p(\boldsymbol{x}|\boldsymbol{w})\rangle_{r(\boldsymbol{w})}$ satisfy*

$$\mathrm{GE}^{r}_{\mathrm{Regular}}(\boldsymbol{X}) = \left\langle \log \frac{p(\boldsymbol{x}|\boldsymbol{w}^*)}{\langle p(\boldsymbol{x}|\boldsymbol{w})\rangle_{r(\boldsymbol{w})}} \right\rangle_{p(\boldsymbol{x}|\boldsymbol{w}^*)} = \mathrm{GE}^{\widehat{\boldsymbol{w}}}_{\mathrm{Regular}}(\boldsymbol{X}) + o_{\mathrm{p}}(N^{-1}), \tag{13.86}$$

$$\mathrm{TE}^{r}_{\mathrm{Regular}}(\boldsymbol{X}) = N^{-1}\sum_{n=1}^{N} \log \frac{p(\boldsymbol{x}^{(n)}|\boldsymbol{w}^*)}{\langle p(\boldsymbol{x}^{(n)}|\boldsymbol{w})\rangle_{r(\boldsymbol{w})}} = \mathrm{TE}^{\widehat{\boldsymbol{w}}}_{\mathrm{Regular}}(\boldsymbol{X}) + o_{\mathrm{p}}(N^{-1}), \tag{13.87}$$

*where $\mathrm{GE}^{\widehat{\boldsymbol{w}}}_{\mathrm{Regular}}(\boldsymbol{X})$ and $\mathrm{TE}^{\widehat{\boldsymbol{w}}}_{\mathrm{Regular}}(\boldsymbol{X})$ are, respectively, the generalization error and the training error of the point estimator $\widehat{\boldsymbol{w}}$ (defined in Theorem 13.6).*

*Proof* The predictive distribution can be expressed as

$$\begin{aligned}
\langle p(\boldsymbol{x}|\boldsymbol{w})\rangle_{r(\boldsymbol{w})} &= \langle \exp\left(\log p(\boldsymbol{x}|\boldsymbol{w})\right)\rangle_{r(\boldsymbol{w})} \\
&= \Bigg\langle \exp\Bigg(\log p(\boldsymbol{x}|\widehat{\boldsymbol{w}}) + (\boldsymbol{w} - \widehat{\boldsymbol{w}})^\top \left.\frac{\partial \log p(\boldsymbol{x}|\boldsymbol{w}')}{\partial \boldsymbol{w}'}\right|_{\boldsymbol{w}'=\widehat{\boldsymbol{w}}} \\
&\quad + \frac{1}{2}(\boldsymbol{w} - \widehat{\boldsymbol{w}})^\top \left.\frac{\partial^2 \log p(\boldsymbol{x}|\boldsymbol{w}')}{\partial \boldsymbol{w}' \partial \boldsymbol{w}'^\top}\right|_{\boldsymbol{w}'=\widehat{\boldsymbol{w}}} (\boldsymbol{w} - \widehat{\boldsymbol{w}}) + O\left(\|\boldsymbol{w} - \widehat{\boldsymbol{w}}\|^3\right)\Bigg)\Bigg\rangle_{r(\boldsymbol{w})} \\
&= p(\boldsymbol{x}|\widehat{\boldsymbol{w}}) \cdot \Bigg\langle\Bigg( 1 + (\boldsymbol{w} - \widehat{\boldsymbol{w}})^\top \left.\frac{\partial \log p(\boldsymbol{x}|\boldsymbol{w}')}{\partial \boldsymbol{w}'}\right|_{\boldsymbol{w}'=\widehat{\boldsymbol{w}}} \\
&\quad + \frac{1}{2}(\boldsymbol{w} - \widehat{\boldsymbol{w}})^\top \left.\frac{\partial \log p(\boldsymbol{x}|\boldsymbol{w}')}{\partial \boldsymbol{w}'}\right|_{\boldsymbol{w}'=\widehat{\boldsymbol{w}}} \left.\frac{\partial \log p(\boldsymbol{x}|\boldsymbol{w}')}{\partial \boldsymbol{w}'}\right|^\top_{\boldsymbol{w}'=\widehat{\boldsymbol{w}}} (\boldsymbol{w} - \widehat{\boldsymbol{w}}) \\
&\quad + \frac{1}{2}(\boldsymbol{w} - \widehat{\boldsymbol{w}})^\top \left.\frac{\partial^2 \log p(\boldsymbol{x}|\boldsymbol{w}')}{\partial \boldsymbol{w}' \partial \boldsymbol{w}'^\top}\right|_{\boldsymbol{w}'=\widehat{\boldsymbol{w}}} (\boldsymbol{w} - \widehat{\boldsymbol{w}}) + O\left(\|\boldsymbol{w} - \widehat{\boldsymbol{w}}\|^3\right)\Bigg)\Bigg\rangle_{r(\boldsymbol{w})}.
\end{aligned}$$

Here we first expanded $\log p(\boldsymbol{x}|\boldsymbol{w})$ about $\widehat{\boldsymbol{w}}$, and then expanded the exponential function (with $\exp(z) = 1 + z + z^2/2 + O(z^3)$).

Using the conditions (13.84) and (13.85) on $r(\boldsymbol{w})$, we have

$$\langle p(\boldsymbol{x}|\boldsymbol{w})\rangle_{r(\boldsymbol{w})} = p(\boldsymbol{x}|\widehat{\boldsymbol{w}}) \cdot \left(1 + \tfrac{1}{2}\mathrm{tr}\left(\widehat{\boldsymbol{\Sigma}}_w \boldsymbol{\Phi}(\boldsymbol{x};\widehat{\boldsymbol{w}})\right) + O_{\mathrm{p}}\left(N^{-3/2}\right)\right), \tag{13.88}$$

where

$$\boldsymbol{\Phi}(\boldsymbol{x};\widehat{\boldsymbol{w}}) = \left.\frac{\partial \log p(\boldsymbol{x}|\boldsymbol{w})}{\partial \boldsymbol{w}}\right|_{\boldsymbol{w}=\widehat{\boldsymbol{w}}} \left.\frac{\partial \log p(\boldsymbol{x}|\boldsymbol{w})}{\partial \boldsymbol{w}}\right|_{\boldsymbol{w}=\widehat{\boldsymbol{w}}}^{\top} + \left.\frac{\partial^2 \log p(\boldsymbol{x}|\boldsymbol{w})}{\partial \boldsymbol{w}' \partial \boldsymbol{w}'^{\top}}\right|_{\boldsymbol{w}=\widehat{\boldsymbol{w}}}. \tag{13.89}$$

Therefore,

$$\begin{aligned}\left\langle \log \frac{\langle p(\boldsymbol{x}|\boldsymbol{w})\rangle_{r(\boldsymbol{w})}}{p(\boldsymbol{x}|\widehat{\boldsymbol{w}})}\right\rangle_{p(\boldsymbol{x}|\boldsymbol{w}^*)} &= \left\langle \log\left(1 + \tfrac{1}{2}\mathrm{tr}\left(\widehat{\boldsymbol{\Sigma}}_w \boldsymbol{\Phi}(\boldsymbol{x};\widehat{\boldsymbol{w}})\right) + O_{\mathrm{p}}\left(N^{-3/2}\right)\right)\right\rangle_{p(\boldsymbol{x}|\boldsymbol{w}^*)} \\ &= \tfrac{1}{2}\left\langle \mathrm{tr}\left(\widehat{\boldsymbol{\Sigma}}_w \boldsymbol{\Phi}(\boldsymbol{x};\widehat{\boldsymbol{w}})\right)\right\rangle_{p(\boldsymbol{x}|\boldsymbol{w}^*)} + O_{\mathrm{p}}\left(N^{-3/2}\right). \end{aligned} \tag{13.90}$$

Here we expanded the logarithm function (with $\log(1+z) = z + O(z^2)$), using the condition (13.85) on the covariance, i.e., $\widehat{\boldsymbol{\Sigma}}_w = O_{\mathrm{p}}(N^{-1})$.

The condition (13.84) on the mean, i.e., $\widehat{\boldsymbol{w}} = \boldsymbol{w}^* + O_{\mathrm{p}}(N^{-1/2})$, implies that

$$\begin{aligned}\langle \boldsymbol{\Phi}(\boldsymbol{x};\widehat{\boldsymbol{w}})\rangle_{p(\boldsymbol{x}|\boldsymbol{w}^*)} &= \langle \boldsymbol{\Phi}(\boldsymbol{x};\boldsymbol{w}^*)\rangle_{p(\boldsymbol{x}|\boldsymbol{w}^*)} + O_{\mathrm{p}}(N^{-1/2}) \\ &= \boldsymbol{F}(\boldsymbol{w}^*) - \boldsymbol{F}(\boldsymbol{w}^*) + O_{\mathrm{p}}(N^{-1/2}) \\ &= O_{\mathrm{p}}(N^{-1/2}), \end{aligned} \tag{13.91}$$

where we used the definition of the Fisher information (13.46) and its equivalent expression (13.49) (under the regularity conditions). Eqs. (13.90) and (13.91) together with the condition (13.85) give

$$\left\langle \log \frac{\langle p(\boldsymbol{x}|\boldsymbol{w})\rangle_{r(\boldsymbol{w})}}{p(\boldsymbol{x}|\widehat{\boldsymbol{w}})}\right\rangle_{p(\boldsymbol{x}|\boldsymbol{w}^*)} = O_{\mathrm{p}}(N^{-3/2}),$$

which results in Eq. (13.86).

Similarly, by using the expression (13.88) of the predictive distribution, we have

$$\begin{aligned} N^{-1}\sum_{n=1}^{N} \log \frac{\langle p(\boldsymbol{x}^{(n)}|\boldsymbol{w})\rangle_{r(\boldsymbol{w})}}{p(\boldsymbol{x}^{(n)}|\widehat{\boldsymbol{w}})} &= N^{-1}\sum_{n=1}^{N} \log\left(1 + \tfrac{1}{2}\mathrm{tr}\left(\widehat{\boldsymbol{\Sigma}}_w \boldsymbol{\Phi}(\boldsymbol{x}^{(n)};\widehat{\boldsymbol{w}})\right) + O_{\mathrm{p}}\left(N^{-3/2}\right)\right) \\ &= \frac{1}{2}N^{-1}\sum_{n=1}^{N} \mathrm{tr}\left(\widehat{\boldsymbol{\Sigma}}_w \boldsymbol{\Phi}(\boldsymbol{x}^{(n)};\widehat{\boldsymbol{w}})\right) + O_{\mathrm{p}}\left(N^{-3/2}\right). \end{aligned} \tag{13.92}$$

The law of large numbers (13.5) and Eq. (13.91) lead to

$$\begin{aligned} N^{-1}\sum_{n=1}^{N} \boldsymbol{\Phi}(\boldsymbol{x}^{(n)};\widehat{\boldsymbol{w}}) &= \langle \boldsymbol{\Phi}(\boldsymbol{x};\widehat{\boldsymbol{w}})\rangle_{p(\boldsymbol{x}|\boldsymbol{w}^*)} + o_{\mathrm{p}}(1) \\ &= o_{\mathrm{p}}(1). \end{aligned}$$

Substituting the preceding and the condition (13.85) into Eq. (13.92) gives

$$N^{-1} \sum_{n=1}^{N} \log \frac{\left\langle p(\boldsymbol{x}^{(n)}|\boldsymbol{w}) \right\rangle_{r(\boldsymbol{w})}}{p(\boldsymbol{x}^{(n)}|\widehat{\boldsymbol{w}})} = o_{\mathrm{p}}\left(N^{-1}\right),$$

which results in Eq. (13.87). This completes the proof. □

The asymptotic normality of the Bayes posterior (Theorem 13.3), combined with the asymptotic normality of the ML estimator (Theorem 13.2), guarantees that the conditions (13.84) and (13.85) of Theorem 13.9 hold in Bayesian learning, which leads to the following corollaries:

**Corollary 13.10** *The average generalization error of Bayesian learning in the regular models can be asymptotically expanded as*

$$\overline{\mathrm{GE}}^{\mathrm{Bayes}}_{\mathrm{Regular}}(N) = \left\langle \mathrm{GE}^{\mathrm{Bayes}}_{\mathrm{Regular}}(\boldsymbol{X}) \right\rangle_{q(\boldsymbol{X})} = \lambda^{\mathrm{Bayes}}_{\mathrm{Regular}} N^{-1} + o(N^{-1}), \tag{13.93}$$

*where the generalization coefficient is given by*

$$2\lambda^{\mathrm{Bayes}}_{\mathrm{Regular}} = D. \tag{13.94}$$

**Corollary 13.11** *The average training error of Bayesian learning in the regular models can be asymptotically expanded as*

$$\overline{\mathrm{TE}}^{\mathrm{Bayes}}_{\mathrm{Regular}}(N) = \left\langle \mathrm{TE}^{\mathrm{Bayes}}_{\mathrm{Regular}}(\boldsymbol{X}) \right\rangle_{q(\boldsymbol{X})} = \nu^{\mathrm{Bayes}}_{\mathrm{Regular}} N^{-1} + o(N^{-1}), \tag{13.95}$$

*where the training coefficient is given by*

$$2\nu^{\mathrm{Bayes}}_{\mathrm{Regular}} = -D. \tag{13.96}$$

Asymptotic behavior of the Bayes free energy (13.18) was also analyzed (Schwarz, 1978; Watanabe, 2009). The Bayes free energy can be written as

$$\begin{aligned} F^{\mathrm{Bayes}}(\boldsymbol{X}) &= -\log p(\boldsymbol{X}) \\ &= -\log \int p(\boldsymbol{w}) \prod_{n=1}^{N} p(\boldsymbol{x}^{(n)}|\boldsymbol{w}) d\boldsymbol{w} \\ &= -\log \int \exp\left(N L_N(\boldsymbol{w}) + \log p(\boldsymbol{w})\right) d\boldsymbol{w}, \end{aligned}$$

where the factor $\exp\left(N L_N(\boldsymbol{w})\right)$ dominates in the asymptotic limit. By using the Taylor expansion

$$L_N(\boldsymbol{w}) = L_N(\widehat{\boldsymbol{w}}^{\mathrm{ML}}) + (\boldsymbol{w} - \widehat{\boldsymbol{w}}^{\mathrm{ML}})^\top \left.\frac{\partial L_N(\boldsymbol{w})}{\partial \boldsymbol{w}}\right|_{\boldsymbol{w}=\widehat{\boldsymbol{w}}^{\mathrm{ML}}}$$

$$+\frac{1}{2}(w-\widehat{w}^{\mathrm{ML}})^\top \left.\frac{\partial^2 L_N(w)}{\partial w \partial w^\top}\right|_{w=\widehat{w}^{\mathrm{ML}}}(w-\widehat{w}^{\mathrm{ML}})+O\left(\|w-\widehat{w}^{\mathrm{ML}}\|^3\right)$$

$$= L_N(\widehat{w}^{\mathrm{ML}})-\frac{1}{2}(w-\widehat{w}^{\mathrm{ML}})^\top\left(F(w^*)+o_{\mathrm{p}}(1)\right)(w-\widehat{w}^{\mathrm{ML}})+O\left(\|w-\widehat{w}^{\mathrm{ML}}\|^3\right),$$

we can approximate the Bayes free energy as follows:

$$\begin{aligned}
F^{\mathrm{Bayes}}(X) &\approx -NL_N(\widehat{w}^{\mathrm{ML}}) - \log\int \exp\Bigg(-\frac{N}{2}(w-\widehat{w}^{\mathrm{ML}})^\top F(w^*)(w-\widehat{w}^{\mathrm{ML}}) \\
&\qquad\qquad + \log p(w)\Bigg)dw \\
&= -NL_N(\widehat{w}^{\mathrm{ML}}) - \log\int \exp\Bigg(-\frac{1}{2}v^\top F(w^*)v \\
&\qquad\qquad + \log p(\widehat{w}^{\mathrm{ML}}+N^{-1/2}v)\Bigg)\frac{dv}{N^{D/2}} \\
&= -NL_N(\widehat{w}^{\mathrm{ML}}) + \frac{D}{2}\log N + O_{\mathrm{p}}(1). \qquad (13.97)
\end{aligned}$$

where $v = \sqrt{N}(w-\widehat{w}^{\mathrm{ML}})$ is a rescaled parameter, on which the integration was performed with $dv = N^{D/2}dw$.

Therefore, the relative Bayes free energy (13.19) can be written as

$$\begin{aligned}
\widetilde{F}^{\mathrm{Bayes}}(X) &= F^{\mathrm{Bayes}}(X) + NL_N(w^*) \\
&\approx \frac{D}{2}\log N + N\left(L_N(w^*) - L_N(\widehat{w}^{\mathrm{ML}})\right) + O_{\mathrm{p}}(1). \qquad (13.98)
\end{aligned}$$

Here we used $S_N(X) = -L_N(w^*)$, which can be confirmed by their definitions (13.20) and (13.37). The second term in Eq. (13.98) is of the order of $O_{\mathrm{p}}(1)$, because Eqs. (13.68), (13.70), and (13.52) imply that

$$L_N(w^*) - L_N(\widehat{w}^{\mathrm{ML}}) = \mathrm{TE}^{\mathrm{ML}}_{\mathrm{Regular}}(X) = O_{\mathrm{p}}(N^{-1}).$$

The following theorem was obtained with more rigorous discussion.

**Theorem 13.12** *(Watanabe, 2009) The relative Bayes free energy for the regular models can be asymptotically expanded as*

$$\widetilde{F}^{\mathrm{Bayes}}_{\mathrm{Regular}}(X) = F^{\mathrm{Bayes}}(X) - NS_N(X) = \lambda'^{\mathrm{Bayes}}_{\mathrm{Regular}}\log N + O_{\mathrm{p}}(1), \qquad (13.99)$$

*where the Bayes free energy coefficient is given by*

$$2\lambda'^{\mathrm{Bayes}}_{\mathrm{Regular}} = D. \qquad (13.100)$$

Note that Corollary 13.10 and Theorem 13.12 are consistent with Eq. (13.25), which holds for any statistical model.

### 13.4.8 Information Criteria

We have seen that the leading terms of the generalization error, the training error, and the relative Bayes free energy are proportional to the parameter dimension. Those results imply that how much a regular statistical model *overfits* training data mainly depends on the degrees of freedom of statistical models. Based on this insight, various *information criteria* were proposed for *model selection*.

Let us first recapitulate the model selection problem. Consider a $(D-1)$-degree polynomial regression model for one-dimensional input $t$ and output $y$:

$$y = \sum_{d=1}^{D} w_d t^{d-1} + \varepsilon,$$

where $\varepsilon$ denotes a noise. This model can be written as

$$y = \boldsymbol{w}^\top \boldsymbol{x} + \varepsilon,$$

where $\boldsymbol{w} \in \mathbb{R}^D$ is a parameter vector, and $\boldsymbol{x} = (1, t, t^2, \ldots, t^{D-1})^\top$ is a transformed input vector. Suppose that the true distribution can be realized *just* with a $(D^* - 1)$-degree polynomial:

$$y = \sum_{d=1}^{D^*} w_d^* t^{d-1} + \varepsilon = \boldsymbol{w}^{*\top} \boldsymbol{x}' + \varepsilon,$$

where $\boldsymbol{w}^* \in \mathbb{R}^{D^*}$ is the *true* parameter vector, and $\boldsymbol{x}' = (1, t, t^2, \ldots, t^{D^*-1})^\top$.[6]

If we train a $(D-1)$-degree polynomial model for $D < D^*$, we expect poor generalization performance because the true distribution is not realizable, i.e., the model is too simple to express the true distribution. On the other hand, it was observed that if we train a model such that $D \gg D^*$, the generalization performance is also not optimal, because the unnecessarily high degree terms cause overfitting. Accordingly, finding an appropriate degree $D$ of freedom, based on the observed data, is an important task, which is known as a model selection problem.

It would be a good strategy if we could choose $D$, which minimizes the generalization error (13.30). Ignoring the terms that do not depend on the model (or $D$), the generalization error can be written as

$$\mathrm{GE}(\mathcal{D}) = -\int q(\boldsymbol{x}) q(y|\boldsymbol{x}) \log p(y|\boldsymbol{x}, \mathcal{D}) d\boldsymbol{x} dy + \text{const.} \tag{13.101}$$

[6] By "just," we mean that $w_{D^*}^* \neq 0$, and therefore the true distribution is not realizable with any $(D-1)$-degree polynomial for $D < D^*$.

Unfortunately, we cannot directly evaluate Eq. (13.101), since the true distribution $q(y|\boldsymbol{x})$ is inaccessible. Instead, the training error (13.31),

$$\mathrm{TE}(\mathcal{D}) = -\frac{1}{N}\sum_{n=1}^{N} \log p(y^{(n)}|\boldsymbol{x}^{(n)}, \mathcal{D}) + \text{const.}, \tag{13.102}$$

is often used as an estimator for the generalization error. Although Eq. (13.102) is accessible, the training error is known to be a biased estimator for the generalization error (13.101). In fact, the training error does not reflect the negative effect of redundancy of the statistical model, and tends to be monotonically decreasing as the parameter dimension $D$ increases.

*Akaike's information criterion (AIC)* (Akaike, 1974),

$$\mathrm{AIC} = -2\sum_{n=1}^{N} \log p(y^{(n)}|\boldsymbol{x}^{(n)}, \widehat{\boldsymbol{w}}^{\mathrm{ML}}) + 2D, \tag{13.103}$$

was proposed as an estimator for the generalization error of ML learning with bias correction. Theorems 13.4 and 13.5 provide the bias between the generalization error and the training error as follows:

$$\begin{aligned}\left\langle \mathrm{GE}^{\mathrm{ML}}_{\mathrm{Regular}}(\mathcal{D}) - \mathrm{TE}^{\mathrm{ML}}_{\mathrm{Regular}}(\mathcal{D})\right\rangle_{q(\mathcal{D})} &= \overline{\mathrm{GE}}^{\mathrm{ML}}_{\mathrm{Regular}}(N) - \overline{\mathrm{TE}}^{\mathrm{ML}}_{\mathrm{Regular}}(N)\\ &= \frac{\lambda^{\mathrm{ML}}_{\mathrm{Regular}} - \nu^{\mathrm{ML}}_{\mathrm{Regular}}}{N} + o(N^{-1})\\ &= \frac{D}{N} + o(N^{-1}).\end{aligned} \tag{13.104}$$

Therefore, it holds that

$$\begin{aligned}&\mathrm{TE}^{\mathrm{ML}}_{\mathrm{Regular}}(\mathcal{D}) + \left\langle \mathrm{GE}^{\mathrm{ML}}_{\mathrm{Regular}}(\mathcal{D}) - \mathrm{TE}^{\mathrm{ML}}_{\mathrm{Regular}}(\mathcal{D})\right\rangle_{q(\mathcal{D})}\\ &\quad = \mathrm{TE}^{\mathrm{ML}}_{\mathrm{Regular}}(\mathcal{D}) + \frac{D}{N} + o(N^{-1})\\ &\quad = \frac{\mathrm{AIC}}{2N} - S_N(\boldsymbol{Y}|\boldsymbol{X}) + o(N^{-1}),\end{aligned} \tag{13.105}$$

where $S_N(\boldsymbol{Y}|\boldsymbol{X})$ is the (conditional) empirical entropy (13.34). Since the empirical entropy $S_N(\boldsymbol{Y}|\boldsymbol{X})$ does not depend on the model, Eq. (13.105) implies that minimizing AIC amounts to minimizing an asymptotically unbiased estimator for the generalization error.

Another strategy for model selection is to minimize an approximation to the Bayes free energy (13.33). Instead of performing integration for computing the Bayes free energy, Schwarz (1978) proposed to minimize the *Bayesian information criterion (BIC)*:

$$\mathrm{BIC} = \mathrm{MDL} = -2\sum_{n=1}^{N} \log p(\boldsymbol{y}^{(n)}|\boldsymbol{x}^{(n)}, \widehat{\boldsymbol{w}}^{\mathrm{ML}}) + D \log N. \tag{13.106}$$

Interestingly, an equivalent criterion, called the *minimum description length (MDL)* (Rissanen, 1986), was derived in the context of information theory in communication. The relation between BIC and the Bayes free energy can be directly found from the approximation (13.97), i.e., it holds that

$$\begin{aligned} F^{\mathrm{Bayes}}(\boldsymbol{Y}|\boldsymbol{X}) &\approx -NL_N(\widehat{\boldsymbol{w}}^{\mathrm{ML}}) + \frac{D}{2}\log N + O_{\mathrm{p}}(1) \\ &= \frac{\mathrm{BIC}}{2} + O_{\mathrm{p}}(1), \end{aligned} \tag{13.107}$$

and therefore minimizing BIC amounts to minimizing an approximation to the Bayes free energy.

The first terms of AIC (13.103) and BIC (13.106) are the *maximum log-likelihood*—the log-likelihood at the ML estimator—multiplied by −2. The second terms, called penalty terms, penalize high model complexity, which explicitly work as *Occam's razor* (MacKay, 1992) to prune off irrelevant degrees of freedom of the statistical model. AIC, BIC, and MDL are easily computable and have shown their usefulness in many applications. However, their derivations rely on the fact that the generalization coefficient, the training coefficient, and the free energy coefficient depend only on the parameter dimension *under the regularity conditions*. Actually, it has been revealed that, in singular models, those coefficients depend not only on the parameter dimension $D$ but also on the true distribution.

## 13.5 Asymptotic Learning Theory for Singular Models

Many popular statistical models do not satisfy the regularity conditions. For example, neural networks, matrix factorization, mixture models, hidden Markov models, and Bayesian networks are all unidentifiable and have singularities, where the Fisher information is singular, in the parameter space.

As discussed in Chapter 7, the true parameter is on a singular point when the true distribution is realizable with a model with parameter dimension smaller than the used model, i.e., when the model has redundant components for expressing the true distribution. In such cases, the likelihood cannot be Taylor-expanded about the true parameter, and the asymptotic normality does not hold.

Consequently, the regular learning theory, described in Section 13.4, cannot be applied to singular models.

In this section, we first give intuition on how singularities affect generalization properties, and then introduce asymptotic theoretical results on ML learning and Bayesian learning. After that, we give an overview of asymptotic theory of VB learning, which will be described in detail in the subsequent chapters.

### 13.5.1 Effect of Singularities

Two types of effects of singularities have been observed, which will be detailed in the following subsections.

#### Basis Selection Effect

Consider a regression model for one-dimensional input $x \in [-10, 10]$ and output $y \in \mathbb{R}$ with $H$ *radial basis function (RBF)* units:

$$p(y|x, \boldsymbol{a}, \boldsymbol{b}, \boldsymbol{c}) = \frac{1}{\sqrt{2\pi\sigma^2}} \exp\left(-\frac{1}{2\sigma^2}(y - f(x; \boldsymbol{a}, \boldsymbol{b}, \boldsymbol{c}))^2\right), \tag{13.108}$$

$$\text{where} \qquad f(x; \boldsymbol{a}, \boldsymbol{b}, \boldsymbol{c}) = \sum_{h=1}^{H} \rho_h\left(x; a_h, b_h, c_h^2\right). \tag{13.109}$$

Each RBF unit in Eq. (13.109) is a weighted Gaussian RBF,

$$\rho_h\left(x; a_h, b_h, c_h^2\right) = a_h \cdot \mathrm{Gauss}_1\left(x; b_h, c_h^2\right) = \frac{a_h}{\sqrt{2\pi c_h^2}} \exp\left(-\frac{(x - b_h)^2}{2c_h^2}\right),$$

controlled by a weight parameter $a_h \in \mathbb{R}$, a mean parameter $b_h \in \mathbb{R}$, and a scale parameter $c_h^2 \in \mathbb{R}_{++}$. Treating the noise variance $\sigma^2$ in Eq. (13.108) as a known constant, the parameters to be estimated are summarized as $\boldsymbol{w} = (\boldsymbol{a}^\top, \boldsymbol{b}^\top, \boldsymbol{c}^\top)^\top \in \mathbb{R}^{3H}$, where $\boldsymbol{a} = (a_1, \ldots, a_H)^\top \in \mathbb{R}^H$, $\boldsymbol{b} = (b_1, \ldots, b_H)^\top \in \mathbb{R}^H$, and $\boldsymbol{c} = (c_1^2, \ldots, c_H^2)^\top \in \mathbb{R}_{++}^H$. Figure 13.4(a) shows an example of the RBF regression function (13.109) for $H = 2$.

Apparently, the model (13.108) is unidentifiable, and has singularities—since $\rho_h(x; 0, b_h, c_h^2) = 0$ for any $b_h \in \mathbb{R}, c_h^2 \in \mathbb{R}_{++}$, the $(b_h, c_h^2)$ half-space at $a_h = 0$ is an unidentifiable set, on which the Fisher information is singular (see Figure 13.5).[7] Accordingly, we call the model (13.108) a singular RBF regression model, of which the parameter dimension is equal to $D_{\text{sin-RBF}} = 3H$.

[7] More unidentifiable sets can exist, depending on the other RBF units. See Section 7.3.1 for details on identifiability.

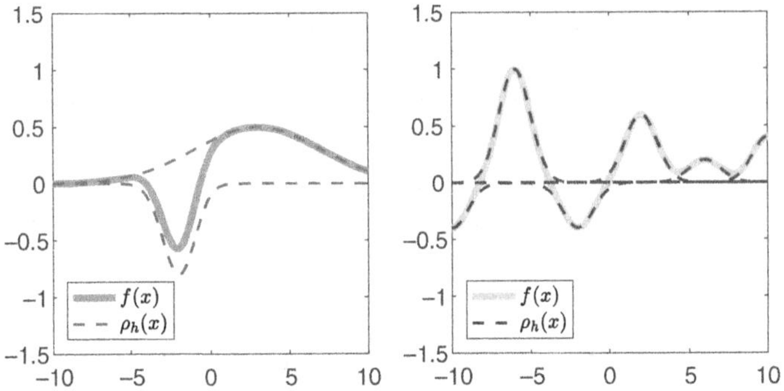


(a) Singular RBF with $H = 2$ units. (b) Regular RBF with $H = 6$ units.

Figure 13.4 Examples (solid curves) of the singular RBF regression function (13.109) and the regular RBF regression function (13.111). Each RBF unit $\rho_h(x)$ is depicted as a dashed curve.

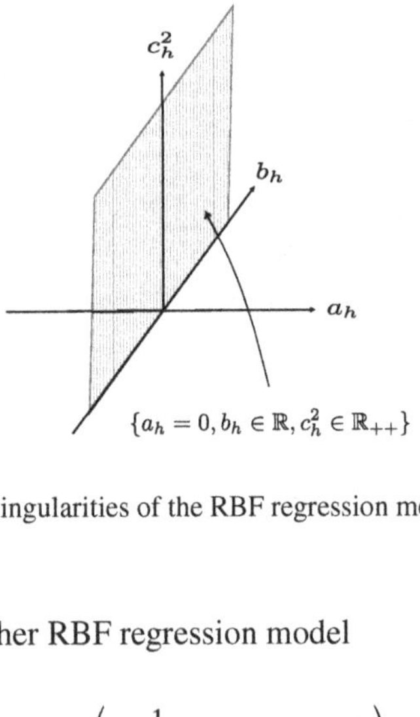


Figure 13.5 Singularities of the RBF regression model (13.108).

Let us consider another RBF regression model

$$p(y|x, \boldsymbol{a}) = \frac{1}{\sqrt{2\pi\sigma^2}} \exp\left(-\frac{1}{2\sigma^2}(y - f(x; \boldsymbol{a}))^2\right), \tag{13.110}$$

where
$$f(x; \boldsymbol{a}) = \sum_{h=1}^{H} \breve{\rho}_h(x; a_h) = \sum_{h=1}^{H} \rho_h\left(x; a_h, \breve{b}_h, \breve{c}_h^2\right). \tag{13.111}$$

Unlike the singular RBF model (13.108), we here treat the mean parameters $\breve{\boldsymbol{b}} = (\breve{b}_1, \ldots, \breve{b}_H)^\top$ and the scale parameters $\breve{\boldsymbol{c}} = (\breve{c}_1^2, \ldots, \breve{c}_H^2)^\top$ as fixed constants, and only estimate the weight parameters $\boldsymbol{a} = (a_1, \ldots, a_H)^\top \in \mathbb{R}^H$.

Let us set the means and the scales as follows, so that the model covers the input domain $[-10, 10]$:

$$\breve{b}_h = -10 + 20 \cdot \frac{h-1}{H-1}, \tag{13.112}$$

$$\breve{c}_h^2 = 1. \tag{13.113}$$

Figure 13.4(b) shows an example of the RBF regression function (13.111) for $H = 6$. Clearly, it holds that $\breve{\rho}_h(x; a_h) \neq \breve{\rho}_h(x; a'_h)$ if $a_h \neq a'_h$, and therefore the model is identifiable. The other regularity conditions (summarized in Section 13.4.1) on the model distribution $p(y|x, \boldsymbol{a})$ are also satisfied. Accordingly, we call the model (13.110) a regular RBF regression model, of which the parameter dimension is equal to $D_{\text{reg-RBF}} = H$.

Now we investigate difference in learning behavior between the singular RBF model (13.108) and the regular RBF model (13.110). Figure 13.6 shows trained regression functions (by ML learning) from $N = 50$ samples (shown as crosses) generated from the regression model,

$$q(y|x) = \frac{1}{\sqrt{2\pi\sigma^2}} \exp\left(-\frac{1}{2\sigma^2}(y - f^*(x))^2\right), \tag{13.114}$$

with the following true functions:

(i) *poly*: Polynomial function $f^*(x) = -0.002x^3$.
(ii) *cos*: Cosine function $f^*(x) = \cos(0.5x)$.
(iii) *tanh*: Tangent hyperbolic function $f^*(x) = \tanh(-0.5x)$.
(iv) *sin-sig*: Sine times sigmoid function $f^*(x) = \sin(x) \cdot \frac{1}{1+e^{-x}}$.
(v) *sin-alg*: Sine function aligned for the regular model $f^*(x) = \sin(2\pi \frac{9}{70} x)$.
(vi) *rbf*: Single RBF function $f^*(x) = \rho_1(x; 3, -10, 1)$.

The noise variance is set to $\sigma^2 = 0.01$, and assumed to be known. We set the number of RBF units to $H = 2$ for the singular model, and $H = 6$ for the regular model, so that both models have the same degrees of freedom, $D_{\text{sin-RBF}} = D_{\text{reg-RBF}} = 6$.

In Figure 13.6, we can observe the following: the singular RBF model can flexibly fit functions in different shapes (a) through (d), unless the function has too many peaks (e); the regular RBF model is not as flexible as the singular RBF model (a) through (d), unless the peaks and valleys match the predefined means of the RBF units (e). Actually, the frequency of *sin-alg* is aligned so that the peaks and the valleys match Eq. (13.112). These observations are quantitatively supported by the generalization error and the training error shown in Figure 13.7, leaving us an impression that the singular RBF model

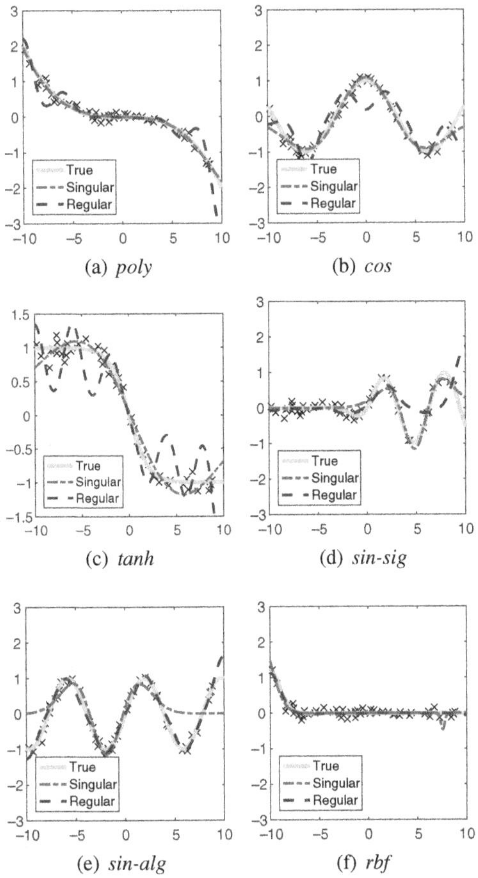


Figure 13.6 Trained regression functions by the singular RBF model (13.108) with $H = 2$ RBF units, and the regular RBF model (13.110) with $H = 6$ RBF units.

with two modifiable basis functions is more flexible than the regular RBF model with six prefixed basis functions.

However, *flexibility* is granted at the risk of overfitting to noise, which can be observed in Figure 13.6(f). We can see that the true RBF function at $x = -10$ is captured by both models. However, the singular RBF model shows a small valley around $x = 8$, which is a consequence of overfitting to sample noise. Figure 13.7 also shows that, in the *rbf* case, the singular RBF model gives lower training error and higher generalization error than the regular RBF model—typical behavior when overfitting occurs. This overfitting tendency is reflected to the generalization and the training coefficients.

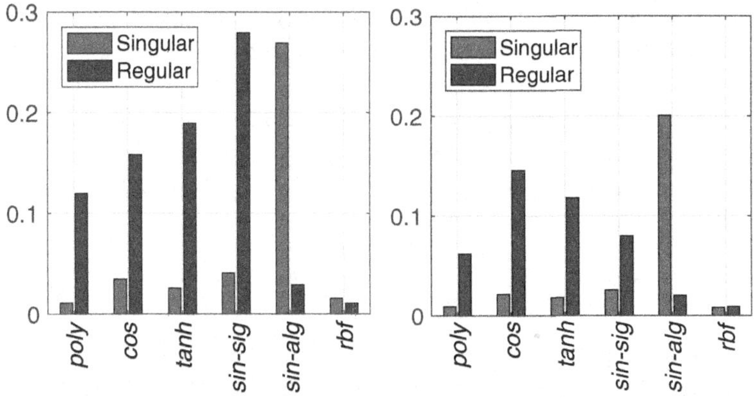


(a) Generalization error (b) Training error

Figure 13.7 The generalization error and the training error by the singular RBF model and the regular RBF model.

Apparently, if the true function is realizable, i.e., $\exists \boldsymbol{w}^* \text{s.t.} f^*(x) = f(x; \boldsymbol{w}^*)$, the true distribution (13.114) is realizable by the RBF regression model (Eq. (13.108) or (13.110)). In the examples (a) through (e) in Figure 13.7, the true function is not realizable by the RBF regression model. In such cases, the generalization error and the training error do not converge to zero, and it holds that $\overline{\mathrm{GE}}(N) = \Theta(1)$ and $\overline{\mathrm{TE}}(N) = \Theta(1)$ for the best learning algorithm. On the other hand, the true function (f) consists of a single RBF unit, and furthermore its mean and variance match those of the first unit of the regular RBF model (see Eqs. (13.112) and (13.113)). Accordingly, the true function (f) and therefore the true distribution (13.114) in the example (f) are realizable, i.e., $\exists \boldsymbol{w}^*, \text{s.t.} q(y|x) = p(y|x, \boldsymbol{w}^*)$, by both of the singular RBF model (13.108) and the regular RBF model (13.110).

When the true parameter $\boldsymbol{w}^*$ exists, the generalization error converges to zero, and, for any reasonable learning algorithm, the average generalization error and the average training error can be asymptotically expanded as Eqs. (13.15) and (13.16):

$$\overline{\mathrm{GE}}(N) = \lambda N^{-1} + o(N^{-1}),$$
$$\overline{\mathrm{TE}}(N) = \nu N^{-1} + o(N^{-1}).$$

Since the regular RBF model (13.110) is regular, its generalization coefficient and the training coefficient are given by

$$2\lambda_{\text{reg-RBF}} = -2\nu_{\text{reg-RBF}} = D = H_{\text{reg-RBF}} \tag{13.115}$$

for ML learning, MAP learning, and Bayesian learning (under the additional regularity conditions on the prior). On the other hand, the generalization coefficient and the training coefficient for the singular RBF model (13.108) are unknown and can be significantly different from the regular models. As will be introduced in Section 13.5.3, the generalization coefficients of ML learning and MAP learning for various singular models have been clarified, and all results that have been found so far satisfy

$$2\lambda^{\mathrm{ML}}_{\mathrm{Singuler}} \geq D, \qquad 2\lambda^{\mathrm{MAP}}_{\mathrm{Singuler}} \geq D, \tag{13.116}$$

where $D$ is the parameter dimensionality. By comparing Eq. (13.116) with Eq. (13.115) (or Eqs. (13.67) and (13.81)), we find that the ML and the MAP generalization coefficients per single model parameter in singular models are larger than those in the regular models, which implies that singular models tend to overfit more than the regular models.

We can explain this phenomenon as an effect of the neighborhood structure around singularities. Recall the example (f), where the singular RBF model and the regular RBF model learn the true distribution $f^*(x) = \rho_1(x; 3, -10, 1)$. For the singular RBF model, $\boldsymbol{w}^*_{\mathrm{sin-RBF}} = (a_1, a_2, b_1, b_2, c_1^2, c_2^2) = (3, 0, -10, *, 1, *)$, where $*$ allows any value in the domain, are possible true parameters, while, for the regular RBF model, $\boldsymbol{w}^*_{\mathrm{sin-RBF}} = (a_1, a_2, \ldots, a_6) = (3, 0, \ldots, 0)$ is the unique true parameter. Figure 13.8(a) shows the space of the three parameters $(a_2, b_2, c_2^2)$ of the second RBF unit of the singular RBF model, in which the true parameter is *on the singularities*. Since the true parameter extends over the two-dimensional half-space $\{(b_2, c_2^2); b_2 \in \mathbb{R}, c_2^2 \in \mathbb{R}_{++}\}$, the neighborhood (shown by small arrows) contains any RBF with adjustable mean and variance. Although the estimated parameter converges to the singularities in the asymptotic limit, ML learning on finite training samples tries to fit the noise, which contaminates the training samples, by *selecting the optimal basis function*, where the optimality is in terms of the training error. On the other hand, Figure 13.8(b) shows the parameter space $(a_h, \breve{b}_h, \breve{c}_h^2)$ for $h = 2, \ldots, 4$. For each $h$, the true distribution corresponds to a single point, indicated by a shadowed circle, and its neighborhood extends only in one direction, i.e., $a_h = 0 \pm \varepsilon$ with a *prefixed* RBF basis specified by the constants $(\breve{b}_h, \breve{c}_h^2)$. Consequently, with the same number of redundant parameters as the singular RBF model, ML learning tries to fit the training noise only with those three basis functions.

Although the probability that the three prefixed basis functions can fit the training noise better than a single flexible basis function is not zero, we would expect that the singular RBF model would likely capture the noise more flexibly than the regular RBF model. This intuition is supported by previous theoretical work that showed Eq. (13.116) in many singular models,

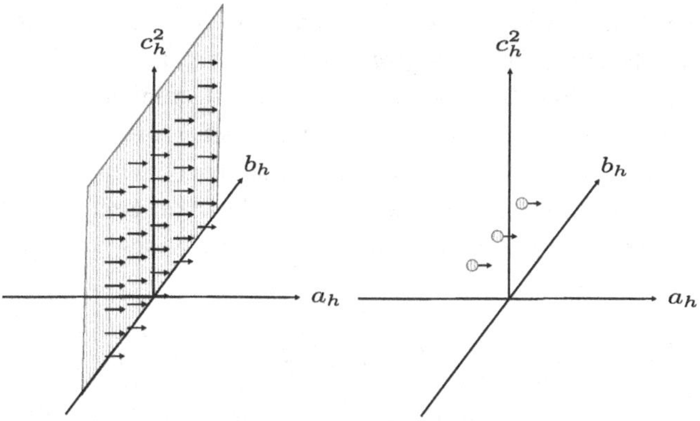


(a) Singular RBF (b) Regular RBF

Figure 13.8 Neighborhood of the true distribution in the *rbf* example. (a) The parameter space of the second ($h = 2$) RBF unit of the singular RBF model. The true parameter is on the singularities, of which the neighborhood contains any RBF with adjustable mean and variance. (b) The parameter space of the second to the fourth ($h = 2, \ldots, 4$) RBF units of the regular RBF model. With the same degrees of freedom as a single singular RBF unit, the neighborhood of the true parameter contains only three different RBF bases with *prefixed* means and variances.

as well as the numerical example in Figure 13.6. We call this phenomenon, i.e., singular models tending to overfit more than regular models, the *basis selection effect*. Although Eq. (13.116) was shown for ML learning and MAP learning, the basis selection effect should occur for any reasonable learning algorithms, including Bayesian learning. However, in Bayesian learning, this effect is canceled by the other effect of singularities, which is explained in the following subsection.

**Integration Effect**

Assume that, in Bayesian learning with a singular model, we adopt a prior distribution $p(\boldsymbol{w})$ bounded as $0 < p(\boldsymbol{w}) < \infty$ at any $\boldsymbol{w} \in \mathcal{W}$. This assumption is the same as one of the regularity conditions in Section 13.4.1. However, this assumption excludes the use of the Jeffreys prior (see Appendix B) and positive mass is distributed over singularities. As discussed in detail in Chapter 7, this prior choice leads to *nonuniformity of the volume element* and favors models with smaller degrees of freedom, if a learning algorithm involving integral computations in the parameter space is adopted. As a result, singularities induce MIR in Bayesian learning and its approximation methods, e.g., VB learning. Importantly, the integration effect does not occur in point estimation methods, including ML learning and MAP learning, since the nonuniformity of the volume element affects the estimator only through *integral computations*. We call this phenomenon the *integration effect* of singularities.

The basis selection effect and the integration effect influence the learning behavior in the opposite way: the former intensifies overfitting, while the latter suppresses it. A question is which is stronger in Bayesian learning. Singular learning theory, which will be introduced in Section 13.5.4, has already answered this question. The following has been shown for *any* singular models:

$$2\lambda^{\text{Bayes}}_{\text{Singuler}} \leq D. \tag{13.117}$$

Comparing Eq. (13.117) with Eq. (13.115) (or Eq. (13.94)), we find that the Bayes generalization coefficient per single model parameter in the singular models is smaller than that in the regular models. Note that the conclusion is opposite to ML learning and MAP learning—singular models overfit training noise more than the regular models in ML learning and MAP learning, while they less overfit in Bayesian learning. Since the basis selection effect should occur in any reasonable learning algorithm, we can interpret Eq. (13.117) as evidence that the integration effect is stronger than the basis selection effect in Bayesian learning.

One might wonder why Bayesian learning is not analyzed with the Jeffreys prior—the parameterization invariant noninformative prior. Actually, the Jeffreys prior, or other prior distribution with zero mass at the singularities, is rarely used in singular models because of the computational reasons: when the computational tractability relies on the (conditional) conjugacy, the Jefferey prior is out of choice in singular models; when some sampling method is used for approximating the Bayes posterior, the diverging outskirts of the Jeffreys prior prevents the sampling sequence to converge. Note that this excludes the empirical Bayesian procedure, where the prior can be collapsed after training. Little is known about the learning behavior of empirical Bayesian learning in singular models, and the asymptotic learning theory part (Part IV) of this book also excludes this case.

In the following subsections, we give a brief summary of theoretical results that revealed learning properties of singular models.

### 13.5.2 Conditions Assumed in Asymptotic Theory for Singular Models

Singular models were analyzed under the following conditions on the true distribution and the prior distribution:

(i) The true distribution is *realizable* by the statistical model, i.e., $\exists w^* \text{ s.t. } q(x) = p(x|w^*)$.
(ii) The prior $p(w)$ is twice differentiable and bounded as $0 < p(w) < \infty$ at any $w \in \mathcal{W}$.

Under the second condition, the prior choice does not affect the generalization coefficient. Accordingly, the results, introduced in the following subsection, for ML learning can be directly applied to MAP learning.

### 13.5.3 ML Learning and MAP Learning

Fukumizu (1999) analyzed the asymptotic behavior of the generalization error of ML learning for the reduced rank regression (RRR) model (3.36), by applying the *random matrix theory* to evaluate the singular value distribution of the ML estimator. Specifically, the large-scale limit where the dimensions of the input and the output are infinitely large was considered, and the exact generalization coefficient was derived. The training coefficient can be obtained in the same way (Nakajima and Watanabe, 2007).

The Gaussian mixture model (GMM) (4.6) has been studied as a prototype of singular models in the case of ML learning. Akaho and Kappen (2000) showed that the generalization error and the training error behave quite differently from regular models. As defined in Eq. (13.12), $-N \cdot \mathrm{TE}^{\mathrm{ML}}(X)$ is the log-likelihood ratio, which asymptotically follows the chi-squared distribution for regular models, while little is known about its behavior for singular models. In fact, it is conjectured for the spherical GMM (4.6) that the log-likelihood ratio diverges to infinity in the order of $\log \log N$ (Hartigan, 1985). For mixture models with discrete components such as binomial mixture models, the asymptotic distribution of the log-likelihood ratio was studied through the distribution of the maximum of the Gaussian random field (Bickel and Chernoff, 1993; Takemura and Kuriki, 1997; Kuriki and Takemura, 2001).

Based on the idea of locally conic parameterization (Dacunha-Castelle and Gassiat, 1997), the asymptotic behaviors of the log-likelihood ratio in some singular models were analyzed. For some mixture models with continuous components, including GMMs, it can be proved that the log-likelihood ratio diverges to infinity as $N \to \infty$. In neural networks, it is known that the log-likelihood ratio diverges in the order of $\log N$ when there are at least two redundant hidden units (Fukumizu, 2003; Hagiwara and Fukumizu, 2008).

In all previous works, the obtained generalization coefficient or its equivalent satisfies Eq. (13.116).

### 13.5.4 Singular Learning Theory for Bayesian Learning

For analyzing the generalization performance of Bayesian learning, a general approach, called the *singular learning theory (SLT)*, was established, based on the mathematical techniques in algebraic geometry (Watanabe, 2001a, 2009).

The average relative Bayes free energy (13.23),

$$\overline{F}^{\text{Bayes}}(N) = \left\langle \log \frac{\prod_{n=1}^{N} q(\boldsymbol{x}^{(n)})}{\int p(\boldsymbol{w}) \prod_{n=1}^{N} p(\boldsymbol{x}^{(n)}|\boldsymbol{w}) d\boldsymbol{w}} \right\rangle_{q(X)}$$

$$= -\left\langle \log \int \exp\left(-N \log \frac{q(\boldsymbol{x})}{p(\boldsymbol{x}|\boldsymbol{w})}\right) \cdot p(\boldsymbol{w}) d\boldsymbol{w} \right\rangle_{q(\boldsymbol{x})},$$

can be approximated as

$$\overline{F}^{\text{Bayes}}(N) \approx -\log \int \exp\left(-NE(\boldsymbol{w})\right) \cdot p(\boldsymbol{w}) d\boldsymbol{w}, \tag{13.118}$$

where

$$E(\boldsymbol{w}) = \left\langle \log \frac{q(\boldsymbol{x})}{p(\boldsymbol{x}|\boldsymbol{w})} \right\rangle_{q(\boldsymbol{x})} \tag{13.119}$$

is the KL divergence between the true distribution $q(\boldsymbol{x})$ and the model distribution $p(\boldsymbol{x}|\boldsymbol{w})$.[8]

Let us see the KL divergence (13.119) as the energy in physics, and define the *state density* function for the energy value $s > 0$:

$$v(s) = \int \delta(s - E(\boldsymbol{w})) \cdot p(\boldsymbol{w}) d\boldsymbol{w}, \tag{13.120}$$

where $\delta(\cdot)$ is the Dirac delta function (located at the origin). Note that the state density (13.120) and the (approximation to the relative) Bayes free energy (13.118) are connected by the Laplace transform:

$$\overline{F}^{\text{Bayes}}(N) = -\log \int \exp(-s) v\left(\frac{s}{N}\right) \frac{ds}{N}. \tag{13.121}$$

Define furthermore the *zeta function* as the Mellin transform, an extension of the Laplace transform, of the state density (13.120):

$$\zeta(z) = \int s^z v(s) ds = \int E(\boldsymbol{w})^z p(\boldsymbol{w}) d\boldsymbol{w}. \tag{13.122}$$

The zeta function (13.122) is a function of a complex number $z \in \mathbb{C}$, and it was proved that all the poles of $\zeta(z)$ are real, negative, and rational numbers.

[8] It holds that $\overline{F}^{\text{Bayes}}(N) = -\log \int \exp\left(-NE(\boldsymbol{w})\right) \cdot p(\boldsymbol{w}) d\boldsymbol{w} + O(1)$ if the support of the prior is compact (Watanabe, 2001a, 2009).

By using the relations through Laplace/Mellin transform among the free energy (13.118), the state density (13.120), and the zeta function (13.122), Watanabe (2001a) proved the following theorem:

**Theorem 13.13** *(Watanabe, 2001a, 2009) Let* $0 > -\lambda_1 > -\lambda_2 > \cdots$ *be the sequence of the poles of the zeta function* (13.122) *in the decreasing order, and* $m_1, m_2, \ldots$ *be the corresponding orders of the poles. Then the average relative Bayes free energy* (13.119) *can be asymptotically expanded as*

$$\overline{F}^{\text{Bayes}}(N) = \lambda_1 \log N - (m_1 - 1) \log\log N + O(1). \tag{13.123}$$

Let $c(N) = \overline{F}^{\text{Bayes}}(N) - \lambda_1 \log N + (m_1 - 1)\log\log N$ be the $O(1)$ term in Eq. (13.123). The relation (13.24) between the generalization error and the free energy leads to the following corollary:

**Corollary 13.14** *(Watanabe, 2001a, 2009) If* $c(N+1) - c(N) = o\left(\frac{1}{N \log N}\right)$, *the average generalization error* (13.13) *can be asymptotically expanded as*

$$\overline{\text{GE}}^{\text{Bayes}}(N) = \frac{\lambda_1}{N} - \frac{m_1 - 1}{N \log N} + o\left(\frac{1}{N \log N}\right). \tag{13.124}$$

To sum up, finding the maximum pole $\lambda_1$ of the zeta function $\zeta(z)$ gives the Bayes free energy coefficient

$$\lambda'^{\text{Bayes}} = \lambda_1,$$

which is equal to the Bayes generalization coefficient

$$\lambda^{\text{Bayes}} = \lambda_1.$$

Note that Theorem 13.13 and Corollary 13.14 hold both for regular and singular models. As discussed in Section 7.3.2, MIR (or the integration effect of singularities) is caused by strong nonuniformity of the density of the volume element. Since the state density (13.120) reflects the strength of the nonuniformity, one can see that finding the maximum pole of $\zeta(z)$ amounts to finding the strength of the nonuniformity at the most concentrated point.

Some general inequalities were proven (Watanabe, 2001b, 2009):

- If the prior is positive at any singular point, i.e., $p(\boldsymbol{w}) > 0, \ \forall \boldsymbol{w} \in \{\boldsymbol{w}; \det(\boldsymbol{F}(\boldsymbol{w})) = 0\}$, then

$$2\lambda'^{\text{Bayes}} = 2\lambda^{\text{Bayes}} \leq D. \tag{13.125}$$

- If the Jeffreys prior (see Appendiex B.4) is adopted, for which $p(\boldsymbol{w}) = 0$ holds at any singular point, then

$$2\lambda'^{\text{Bayes}} = 2\lambda^{\text{Bayes}} \geq D. \tag{13.126}$$

Some cases have been found where $2\lambda'^{\mathrm{Bayes}} = 2\lambda^{\mathrm{Bayes}}$ are strictly larger than $D$.

These results support the discussion in Section 13.5.1 on the two effects of singularities: Eq. (13.125) implies that the integration effect dominates the basis selection effect in Bayesian learning, and Eq. (13.126) implies that the basis selection effect appears also in Bayesian learning if the integration effect is suppressed by using the Jeffreys prior.

Theorem 13.13 and Corollary 13.14 hold for general statistical models, while they do not immediately tell us learning properties of singular models. This is because finding the maximum pole of the zeta function $\zeta(z)$ is not an easy task, and requires a specific technique in algebraic geometry called the *resolution of singularities*. Good news is that, when any pole larger than $-D/2$ is found, it provides an upper bound of the generalization coefficient and thus guarantees the performance with a tighter bound (Theorem 13.13 implies that the larger the found pole is, the tighter the provided bound is).

For the RRR model (Aoyagi and Watanabe, 2005) and for the GMM (Aoyagi and Nagata, 2012), the maximum pole was found for general cases, and therefore the exact value of the free energy coefficient, as well as the generalization coefficient, was obtained. In other singular models, including neural networks (Watanabe, 2001a), mixture models (Yamazaki and Watanabe, 2003a), hidden Markov models (Yamazaki and Watanabe, 2005), and Bayesian networks (Yamazaki and Watanabe, 2003b; Rusakov and Geiger, 2005), upper-bounds of the free energy coefficient were obtained by finding some poles of the zeta function. An effort has been made to perform the resolution of singularities systematically by using the newton diagram (Yamazaki and Watanabe, 2004).

### 13.5.5 Information Criteria for Singular Models

The information criteria introduced in Section 13.4.8 rely on the learning theory under the regularity conditions. Therefore, although they were sometimes applied for model selection in singular models, their relations to generalization properties, e.g., AIC to the ML generalization error, and BIC to the Bayes free energy, do not generally hold. In the following, we introduce information criteria applicable for general statistical models including the regular and the singular models (Watanabe, 2009, 2010, 2013). They also cover a generalization of Bayesian learning.

Consider a learning method, called *generalized Bayesian learning*, based on the *generalized posterior distribution*,

$$p^{(\beta)}(\boldsymbol{w}|\boldsymbol{X}) = \frac{p(\boldsymbol{w}) \prod_{n=1}^{N} \left\{p(\boldsymbol{x}^{(n)}|\boldsymbol{w})\right\}^{\beta}}{\int p(\boldsymbol{w}) \prod_{n=1}^{N} \{p(\boldsymbol{x}^{(n)}|\boldsymbol{w})\}^{\beta} d\boldsymbol{w}}, \tag{13.127}$$

where $\beta$, called the *inverse temperature parameter*, modifies the importance of the likelihood per training sample. The prediction is made by the *generalized predictive distribution*,

$$p^{(\beta)}(\boldsymbol{x}|\boldsymbol{X}) = \langle p(\boldsymbol{x}|\boldsymbol{w}) \rangle_{p^{(\beta)}(\boldsymbol{w}|\boldsymbol{X})}. \tag{13.128}$$

Generalized Bayesian learning covers both Bayesian learning and ML learning as special cases: when $\beta = 1$, the generalized posterior distribution (13.127) is reduced to the Bayes posterior distribution (13.9), with which the generalized predictive distribution (13.128) gives the Bayes predictive distribution (13.8); As $\beta$ increases, the probability mass of the generalized posterior distribution concentrates around the ML estimator, and, in the limit when $\beta \to \infty$, the generalized predictive distribution converges to the ML predictive distribution (13.6).

Define the following quantities:

$$\mathrm{GL}(\boldsymbol{X}) = -\left\langle \log \int p(\boldsymbol{x}|\boldsymbol{w}) p^{(\beta)}(\boldsymbol{w}|\boldsymbol{X}) d\boldsymbol{w} \right\rangle_{q(\boldsymbol{x})}, \tag{13.129}$$

$$\mathrm{TL}(\boldsymbol{X}) = -\frac{1}{N} \sum_{n=1}^{N} \log \int p(\boldsymbol{x}^{(n)}|\boldsymbol{w}) p^{(\beta)}(\boldsymbol{w}|\boldsymbol{X}) d\boldsymbol{w}, \tag{13.130}$$

$$\mathrm{GGL}(\boldsymbol{X}) = -\left\langle \int (\log p(\boldsymbol{x}|\boldsymbol{w}))\, p^{(\beta)}(\boldsymbol{w}|\boldsymbol{X}) d\boldsymbol{w} \right\rangle_{q(\boldsymbol{x})}, \tag{13.131}$$

$$\mathrm{GTL}(\boldsymbol{X}) = -\frac{1}{N} \sum_{n=1}^{N} \int \left(\log p(\boldsymbol{x}^{(n)}|\boldsymbol{w})\right) p^{(\beta)}(\boldsymbol{w}|\boldsymbol{X}) d\boldsymbol{w}, \tag{13.132}$$

which are called the *Bayes generalization loss*, the *Bayes training loss*, the *Gibbs generalization loss*, and the *Gibbs training loss*, respectively. The generalization error and the training error of generalized Bayesian learning are, respectively, related to the Bayes generalization loss and the Bayes training loss as follows (Watanabe, 2009):

$$\begin{aligned} \mathrm{GE}^{(\beta)}(\boldsymbol{X}) &= \left\langle \log \frac{q(\boldsymbol{x})}{\int p(\boldsymbol{x}|\boldsymbol{w}) p^{(\beta)}(\boldsymbol{w}|\boldsymbol{X}) d\boldsymbol{w}} \right\rangle_{q(\boldsymbol{x})} \\ &= \mathrm{GL}(\boldsymbol{X}) - S, \end{aligned} \tag{13.133}$$

$$\mathrm{TE}^{(\beta)}(\boldsymbol{X}) = \frac{1}{N}\sum_{n=1}^{N}\log\frac{q(\boldsymbol{x}^{(n)})}{\int p(\boldsymbol{x}^{(n)}|\boldsymbol{w})p^{(\beta)}(\boldsymbol{w}|\boldsymbol{X})d\boldsymbol{w}}$$
$$= \mathrm{TL}(\boldsymbol{X}) - S_N(\boldsymbol{X}), \tag{13.134}$$

where

$$S = -\langle\log q(\boldsymbol{x})\rangle_{q(\boldsymbol{x})} \quad \text{and} \quad S_N(\boldsymbol{X}) = -\frac{1}{N}\sum_{n=1}^{N}\log q(\boldsymbol{x}^{(n)}) \tag{13.135}$$

are the *entropy* of the true distribution and its empirical version, respectively.[9] Also, Gibbs counterparts have the following relations:

$$\mathrm{GGE}^{(\beta)}(\boldsymbol{X}) = \left\langle \int\left(\log\frac{q(\boldsymbol{x})}{p(\boldsymbol{x}|\boldsymbol{w})}\right)p^{(\beta)}(\boldsymbol{w}|\boldsymbol{X})d\boldsymbol{w}\right\rangle_{q(\boldsymbol{x})}$$
$$= \mathrm{GGL}(\boldsymbol{X}) - S, \tag{13.136}$$

$$\mathrm{GTE}^{(\beta)}(\boldsymbol{X}) = \frac{1}{N}\sum_{n=1}^{N}\int\left(\log\frac{q(\boldsymbol{x}^{(n)})}{p(\boldsymbol{x}^{(n)}|\boldsymbol{w})}\right)p^{(\beta)}(\boldsymbol{w}|\boldsymbol{X})d\boldsymbol{w}$$
$$= \mathrm{GTL}(\boldsymbol{X}) - S_N(\boldsymbol{X}). \tag{13.137}$$

Here $\mathrm{GGE}^{(\beta)}(\boldsymbol{X})$ and $\mathrm{GTE}^{(\beta)}(\boldsymbol{X})$ are the generalization error and the training error, respectively, of *Gibbs learning*, where prediction is made by $p(\boldsymbol{x}|\boldsymbol{w})$ with its parameter $\boldsymbol{w}$ sampled from the generalized posterior distribution (13.127).

The following relations were proven (Watanabe, 2009):

$$\langle\mathrm{GL}(\boldsymbol{X})\rangle_{q(\boldsymbol{X})} = \langle\mathrm{TL}(\boldsymbol{X})\rangle_{q(\boldsymbol{X})} + 2\beta\left(\langle\mathrm{GTL}(\boldsymbol{X})\rangle_{q(\boldsymbol{X})} - \langle\mathrm{TL}(\boldsymbol{X})\rangle_{q(\boldsymbol{X})}\right) + o(N^{-1}), \tag{13.138}$$

$$\langle\mathrm{GGL}(\boldsymbol{X})\rangle_{q(\boldsymbol{X})} = \langle\mathrm{GTL}(\boldsymbol{X})\rangle_{q(\boldsymbol{X})} + 2\beta\left(\langle\mathrm{GTL}(\boldsymbol{X})\rangle_{q(\boldsymbol{X})} - \langle\mathrm{TL}(\boldsymbol{X})\rangle_{q(\boldsymbol{X})}\right) + o(N^{-1}), \tag{13.139}$$

which imply that asymptotically unbiased estimators for generalization losses (the left-hand sides of Eqs. (13.138) and (13.139)) can be constructed from training losses (the right-hand sides). The aforementioned equations lead to *widely applicable information criteria (WAIC)* (Watanabe, 2009), defined as

$$\mathrm{WAIC}_1 = \mathrm{TL}(\boldsymbol{X}) + 2\beta\left(\mathrm{GTL}(\boldsymbol{X}) - \mathrm{TL}(\boldsymbol{X})\right), \tag{13.140}$$

$$\mathrm{WAIC}_2 = \mathrm{GTL}(\boldsymbol{X}) + 2\beta\left(\mathrm{GTL}(\boldsymbol{X}) - \mathrm{TL}(\boldsymbol{X})\right). \tag{13.141}$$

[9] $S_N(\boldsymbol{X})$ was defined in Eq. (13.20), and it holds that $S = \langle S_N(\boldsymbol{X})\rangle_{q(\boldsymbol{X})}$.

Clearly, $\mathrm{WAIC}_1$ and $\mathrm{WAIC}_2$ are asymptotically unbiased estimators for the Bayes generalization loss $\mathrm{GL}(\boldsymbol{X})$ and the Gibbs generalization loss $\mathrm{GGL}(\boldsymbol{X})$, respectively, and therefore minimizing them amounts to minimizing the Bayes generalization error (13.133) and the Gibbs generalization error (13.136), respectively.

The training losses, $\mathrm{TL}(\boldsymbol{X})$ and $\mathrm{GTL}(\boldsymbol{X})$ , can be computed by, e.g., MCMC sampling (see Sections 2.2.4 and 2.2.5). Let $\boldsymbol{w}^{(1)}, \ldots, \boldsymbol{w}^{(T)}$ be samples drawn from the generalized posterior distribution (13.127). Then we can estimate the training losses by

$$\mathrm{TL}(\boldsymbol{X}) \approx -\frac{1}{N}\sum_{n=1}^{N}\log\left(\frac{1}{T}\sum_{t=1}^{T}p(\boldsymbol{x}^{(n)}|\boldsymbol{w}^{(t)})\right), \tag{13.142}$$

$$\mathrm{GTL}(\boldsymbol{X}) \approx -\frac{1}{N}\sum_{n=1}^{N}\frac{1}{T}\sum_{t=1}^{T}\log p(\boldsymbol{x}^{(n)}|\boldsymbol{w}^{(t)}). \tag{13.143}$$

WAIC can be seen as an extension of AIC, since minimizing it amounts to minimizing an asymptotically unbiased estimator for the generalization error. Indeed, under the regularity conditions, it holds that

$$\lim_{\beta\to\infty} 2\beta\left(\mathrm{GTL}(\boldsymbol{X}) - \mathrm{TL}(\boldsymbol{X})\right) = \frac{D}{N},$$

and therefore

$$\mathrm{WAIC}_1, \mathrm{WAIC}_2 \to \frac{\mathrm{AIC}}{2N} \qquad \text{as} \qquad \beta \to \infty.$$

An extension of BIC was also proposed. The widely applicable Bayesian information criterion (WBIC) (Watanabe, 2013) is defined as

$$\mathrm{WBIC} = -\sum_{n=1}^{N}\int \log p(\boldsymbol{x}^{(n)}|\boldsymbol{w})p^{(\beta=1/\log N)}(\boldsymbol{w}|\boldsymbol{X})d\boldsymbol{w}, \tag{13.144}$$

where $p^{(\beta=1/\log N)}(\boldsymbol{w}|\boldsymbol{X})$ is the generalized posterior distribution (13.127) with the inverse temperature parameter set to $\beta = 1/\log N$. It was shown that

$$F^{\mathrm{Bayes}}(\boldsymbol{X})\left(\equiv -\log\int p(\boldsymbol{w})\prod_{n=1}^{N}p(\boldsymbol{x}^{(n)}|\boldsymbol{w})d\boldsymbol{w}\right) = \mathrm{WBIC} + O_{\mathrm{p}}(\sqrt{\log N}), \tag{13.145}$$

and therefore WBIC can be used as an estimator or approximation for the Bayes free energy (13.18) when $N$ is large. It was also shown that, under the regularity conditions, it holds that

$$\mathrm{WBIC} = \frac{\mathrm{BIC}}{2} + O_{\mathrm{p}}(1).$$

WBIC (13.144) can be estimated, similarly to WAIC, from samples $\boldsymbol{w}^{(1)}_{\beta=1/\log N}, \ldots, \boldsymbol{w}^{(T)}_{\beta=1/\log N}$ drawn from $p^{(\beta=1/\log N)}(\boldsymbol{w}|\boldsymbol{X})$ as

$$\mathrm{WBIC} \approx -\sum_{n=1}^{N} \frac{1}{T} \sum_{t=1}^{T} \log p(\boldsymbol{x}^{(n)}|\boldsymbol{w}^{(t)}_{\beta=1/\log N}). \tag{13.146}$$

Note that evaluating the Bayes free energy (13.18) is much more computationally demanding in general. For example, the *all temperatures method* (Watanabe, 2013) requires posterior samples $\{\boldsymbol{w}^{(t)}_{\beta_j}\}$ for many $0 = \beta_1 < \beta_2 < \cdots < \beta_J = 1$, and estimates the Bayes free energy as

$$F^{\mathrm{Bayes}}(\boldsymbol{X}) \approx -\sum_{j=1}^{J-1} \log \frac{1}{T_j} \sum_{t=1}^{T_j} \exp\left( (\beta_{j+1} - \beta_j) \sum_{n=1}^{N} \log p(\boldsymbol{x}^{(n)}|\boldsymbol{w}^{(t)}_{\beta_j}) \right).$$

## 13.6 Asymptotic Learning Theory for VB Learning

In the rest of Part IV, we describe asymptotic learning theory for VB learning in detail. Here we give an overview of the subsequent chapters.

VB learning is rarely applied to regular models.[10] Actually, if the model (and the prior) satisfies the regularity conditions, Laplace approximation (2.2.1) can give a good approximation to the posterior, because of the asymptotic normality (Theorem 13.3). Accordingly, we focus on singular models when analyzing VB learning.

We are interested in generalization properties of the VB posterior, which is defined as

$$\widehat{r} \equiv \mathop{\mathrm{argmin}}_{r} F(r), \qquad \text{s.t.} \qquad r \in \mathcal{G}, \tag{13.147}$$

where

$$F(r) = \left\langle \log \frac{r(\boldsymbol{w})}{p(\boldsymbol{w}) \prod_{n=1}^{N} p(\boldsymbol{x}^{(n)}|\boldsymbol{w})} \right\rangle_{r(\boldsymbol{w})} = \mathrm{KL}\left(r(\boldsymbol{w}) \| p(\boldsymbol{w}|\boldsymbol{X})\right) + F^{\mathrm{Bayes}}(\boldsymbol{X}) \tag{13.148}$$

is the free energy and $\mathcal{G}$ is the model-specific constraint, imposed for computational tractability, on the approximate posterior.

[10] VB learning is often applied to a linear model with an ARD prior. In such a model, the model likelihood satisfies the regularity conditions, while the prior does not. Actually, the model exhibits characteristics of singular models, since it can be translated to a singular model with a constant prior (see Section 7.5). In Part IV, we only consider the case where the prior is fixed, without any hyperparameter optimized.

With the VB predictive distribution

$$p^{\mathrm{VB}}(\boldsymbol{x}|\boldsymbol{X}) = \langle p(\boldsymbol{x}|\boldsymbol{w}) \rangle_{\widehat{r}(\boldsymbol{w})},$$

the generalization error (13.10) and the training error (13.11) are defined and analyzed.

We also analyze the *VB free energy*,

$$F^{\mathrm{VB}}(\boldsymbol{X}) = F(\widehat{r}) = \min_{r} F(r). \tag{13.149}$$

Since Eq. (13.148) implies that

$$F^{\mathrm{VB}}(\boldsymbol{X}) - F^{\mathrm{Bayes}}(\boldsymbol{X}) = \mathrm{KL}\left(\widehat{r}(\boldsymbol{w}) \| p(\boldsymbol{w}|\boldsymbol{X})\right),$$

comparing the VB free energy and the Bayes free energy reveals how accurately VB learning approximates Bayesian learning.

Similarly to the analysis of Bayesian learning, we investigate the asymptotic behavior of the *relative VB free energy*,

$$\widetilde{F}^{\mathrm{VB}}(\boldsymbol{X}) = F^{\mathrm{VB}}(\boldsymbol{X}) - NS_N(\boldsymbol{X}) = \lambda'^{\mathrm{VB}} \log N + o_{\mathrm{p}}(\log N), \tag{13.150}$$

where $S_N(\boldsymbol{X})$ is the empirical entropy defined in Eq. (13.20), and $\lambda'^{\mathrm{VB}}$ is called the VB free energy coefficient.

Chapter 14 introduces asymptotic VB theory for the RRR model. This model was relatively easily analyzed by using the analytic-form solution for fully observed matrix factorization (Chapter 6), and the exact values of the VB generalization coefficient, the VB training coefficient, and the VB free energy coefficient were derived (Nakajima and Watanabe, 2007). Since generalization properties of ML learning and Bayesian learning have also been clarified (Fukumizu, 1999; Aoyagi and Watanabe, 2005), similarities and dissimilarities among ML (and MAP) learning, Bayesian learning, and VB learning will be discussed.

Chapters 15 through 17 are devoted to asymptotic VB theory for latent variable models. Chapter 15 analyzes the VB free energy of mixture models. The VB free energy coefficients and their dependencies on prior hyperparameters are revealed. Chapter 16 proceeds to such analyses of the VB free energy for other latent variable models, namely, Bayesian networks, hidden Markov models, probabilistic context free grammar, and latent Dirichlet allocation. Chapter 17 provides a formula for general latent variable models, which reduces the asymptotic analysis of the VB free energy to that of the Bayes free energy introduced in Section 13.5.4. Those results will clarify phase transition phenomena with respect to the hyperparameter setting—the shape of the posterior distribution in the asymptotic limit drastically changes when some hyparparameter value exceeds a certain threshold. Such implication suggests to practitioners how to choose hyperparameters.

Note that the relation (13.25) does not necessarily hold for VB learning and other approximate Bayesian methods, since Eq. (13.24) only holds for the exact Bayes predictive distribution. Therefore, unlike Bayesian learning, the asymptotic behavior of the VB free energy does not necessarily inform us of the asymptotic behavior of the VB generalization error. An effort on relating the VB free energy and the VB generalization error is introduced in Chapter 17, although clarifying VB generalization error requires further effort and techniques.

# 14
# Asymptotic VB Theory of Reduced Rank Regression

In this chapter, we introduce asymptotic theory of VB learning in the reduced rank regression (RRR) model (Nakajima and Watanabe, 2007). Among the singular models, the RRR model is one of the simplest, and many aspects of its learning behavior have been clarified. Accordingly, we can discuss similarities and dissimilarities of ML (and MAP) learning, Bayesian learning, and VB learning in terms of generalization error, training error, and free energy. After defining the problem setting, we show theoretical results and summarize insights into VB learning that the analysis on the RRR model provides.

## 14.1 Reduced Rank Regression

RRR (Baldi and Hornik, 1995; Reinsel and Velu, 1998), introduced in Section 3.1.2 as a special case of fully observed matrix factorization, is a regression model with a rank-$H(\leq \min(L, M))$ linear mapping between input $\boldsymbol{x} \in \mathbb{R}^M$ and output $\boldsymbol{y} \in \mathbb{R}^L$:

$$\boldsymbol{y} = \boldsymbol{B}\boldsymbol{A}^\top \boldsymbol{x} + \boldsymbol{\varepsilon}, \tag{14.1}$$

where $\boldsymbol{A} \in \mathbb{R}^{M\times H}$ and $\boldsymbol{B} \in \mathbb{R}^{L\times H}$ are parameters to be estimated, and $\boldsymbol{\varepsilon}$ is observation noise. Assuming Gaussian noise $\boldsymbol{\varepsilon} \sim \mathrm{Gauss}_L(\boldsymbol{0}, \sigma'^2 \boldsymbol{I}_L)$, the model distribution is given as

$$p(\boldsymbol{y}|\boldsymbol{x}, \boldsymbol{A}, \boldsymbol{B}) = \left(2\pi\sigma'^2\right)^{-L/2} \exp\left(-\frac{1}{2\sigma'^2} \left\|\boldsymbol{y} - \boldsymbol{B}\boldsymbol{A}^\top \boldsymbol{x}\right\|^2\right). \tag{14.2}$$

RRR is also called a *linear neural network*, since the three-layer neural network (7.13) is reduced to RRR (14.1) if the activation function $\psi(\cdot)$ is linear (see also Figure 3.1). We assume conditionally conjugate Gaussian priors for the parameters:

$$p(\boldsymbol{A}) \propto \exp\left(-\frac{1}{2}\mathrm{tr}\left(\boldsymbol{A}\boldsymbol{C}_A^{-1}\boldsymbol{A}^\top\right)\right), \quad p(\boldsymbol{B}) \propto \exp\left(-\frac{1}{2}\mathrm{tr}\left(\boldsymbol{B}\boldsymbol{C}_B^{-1}\boldsymbol{B}^\top\right)\right), \tag{14.3}$$

with diagonal convariances $\boldsymbol{C}_A$ and $\boldsymbol{C}_B$:

$$\boldsymbol{C}_A = \mathbf{Diag}(c_{a_1}^2, \ldots, c_{a_H}^2), \qquad \boldsymbol{C}_B = \mathbf{Diag}(c_{b_1}^2, \ldots, c_{b_H}^2),$$

for $c_{a_h}, c_{b_h} > 0, h = 1, \ldots, H$. In the asymptotic analysis, we assume that the hyperparameters $\{c_{a_h}^2, c_{b_h}^2\}, \sigma'^2$ are fixed constants of the order of 1, i.e., $\{c_{a_h}^2, c_{b_h}^2\}, \sigma'^2 \sim \Theta(1)$ when $N \to \infty$.

The degree of freedom of the RRR model is, in general, different from the apparent number, $(M + L)H$, of entries of the parameters $\boldsymbol{A}$ and $\boldsymbol{B}$. This is because of the trivial redundancy in parameterization—the transformation $(\boldsymbol{A}, \boldsymbol{B}) \mapsto (\boldsymbol{A}\boldsymbol{T}^\top, \boldsymbol{B}\boldsymbol{T}^{-1})$ does not change the linear mapping $\boldsymbol{B}\boldsymbol{A}^\top$ for any nonsingular matrix $\boldsymbol{T} \in \mathbb{R}^{H \times H}$. Accordingly, the *essential* parameter dimensionality is counted as

$$D = H(M + L) - H^2. \tag{14.4}$$

Suppose we are given $N$ training samples:

$$\mathcal{D} = \left\{(\boldsymbol{x}^{(n)}, \boldsymbol{y}^{(n)}); \boldsymbol{x}^{(n)} \in \mathbb{R}^M, \boldsymbol{y}^{(n)} \in \mathbb{R}^L, n = 1, \ldots, N\right\}, \tag{14.5}$$

which are independently drawn from the true distribution $q(\boldsymbol{x}, \boldsymbol{y}) = q(\boldsymbol{y}|\boldsymbol{x})q(\boldsymbol{x})$. We also use the matrix forms that summarize the inputs and the outputs separately:

$$\boldsymbol{X} = (\boldsymbol{x}^{(1)}, \ldots, \boldsymbol{x}^{(N)})^\top \in \mathbb{R}^{N \times M}, \qquad \boldsymbol{Y} = (\boldsymbol{y}^{(1)}, \ldots, \boldsymbol{y}^{(N)})^\top \in \mathbb{R}^{N \times L}.$$

We suppose that the data are preprocessed so that the input and the output are centered, i.e.,

$$\frac{1}{N}\sum_{n=1}^{N} \boldsymbol{x}^{(n)} = \boldsymbol{0} \quad \text{and} \quad \frac{1}{N}\sum_{n=1}^{N} \boldsymbol{y}^{(n)} = \boldsymbol{0}, \tag{14.6}$$

and the input is *prewhitened* (Hyvärinen et al., 2001), i.e.,

$$\frac{1}{N}\sum_{n=1}^{N} \boldsymbol{x}^{(n)}\boldsymbol{x}^{(n)\top} = \frac{1}{N}\boldsymbol{X}^\top\boldsymbol{X} = \boldsymbol{I}_M. \tag{14.7}$$

The likelihood of the RRR model (14.1) on the training samples $\mathcal{D} = (\boldsymbol{X}, \boldsymbol{Y})$ is expressed as

$$p(\boldsymbol{Y}|\boldsymbol{X}, \boldsymbol{A}, \boldsymbol{B}) \propto \exp\left(-\frac{1}{2\sigma'^2}\sum_{n=1}^{N}\left\|\boldsymbol{y}^{(n)} - \boldsymbol{B}\boldsymbol{A}^\top\boldsymbol{x}^{(n)}\right\|^2\right). \tag{14.8}$$

As shown in Section 3.1.2, the logarithm of the likelihood (14.8) can be written, as a function of the parameters, as follows:

$$\log p(\boldsymbol{Y}|\boldsymbol{X}, \boldsymbol{A}, \boldsymbol{B}) = -\frac{N}{2\sigma'^2} \left\| \boldsymbol{V} - \boldsymbol{B}\boldsymbol{A}^\top \right\|_{\mathrm{Fro}}^2 + \text{const.}, \tag{14.9}$$

where

$$\boldsymbol{V} = \frac{1}{N} \sum_{n=1}^{N} \boldsymbol{y}^{(n)} \boldsymbol{x}^{(n)\top} = \frac{1}{N} \boldsymbol{Y}^\top \boldsymbol{X}. \tag{14.10}$$

Note that, unlike in Section 3.1.2, we here do not use the rescaled noise variance $\sigma^2 = \sigma'^2/N$, in order to make the dependence on the number $N$ of samples clear for asymptotic analysis. Because the log-likelihood (14.9) is in the same form as that of the fully observed matrix factorization (MF), we can use the global VB solution, derived in Chapter 6, of the MF model for analyzing VB learning in the RRR model.

### 14.1.1 VB Learning

VB learning solves the following problem:

$$\widehat{r} = \underset{r}{\operatorname{argmin}}\, F(r) \quad \text{s.t.} \quad r(\boldsymbol{A}, \boldsymbol{B}) = r_A(\boldsymbol{A}) r_B(\boldsymbol{B}), \tag{14.11}$$

where

$$F = \left\langle \log \frac{r_A(\boldsymbol{A}) r_B(\boldsymbol{B})}{p(\boldsymbol{Y}|\boldsymbol{X}, \boldsymbol{A}, \boldsymbol{B}) p(\boldsymbol{A}) p(\boldsymbol{B})} \right\rangle_{r_A(\boldsymbol{A}) r_B(\boldsymbol{B})}$$

is the free energy. As derived in Section 3.1, the solution to the problem (14.11) is in the following forms:

$$r_A(\boldsymbol{A}) = \mathrm{MGauss}_{M,H}(\boldsymbol{A}; \widehat{\boldsymbol{A}}, \boldsymbol{I}_M \otimes \widehat{\boldsymbol{\Sigma}}_A) \propto \exp\left( -\frac{\mathrm{tr}\left( (\boldsymbol{A} - \widehat{\boldsymbol{A}}) \widehat{\boldsymbol{\Sigma}}_A^{-1} (\boldsymbol{A} - \widehat{\boldsymbol{A}})^\top \right)}{2} \right), \tag{14.12}$$

$$r_B(\boldsymbol{B}) = \mathrm{MGauss}_{L,H}(\boldsymbol{B}; \widehat{\boldsymbol{B}}, \boldsymbol{I}_L \otimes \widehat{\boldsymbol{\Sigma}}_B) \propto \exp\left( -\frac{\mathrm{tr}\left( (\boldsymbol{B} - \widehat{\boldsymbol{B}}) \widehat{\boldsymbol{\Sigma}}_B^{-1} (\boldsymbol{B} - \widehat{\boldsymbol{B}})^\top \right)}{2} \right). \tag{14.13}$$

With the variational parameters $(\widehat{\boldsymbol{A}}, \widehat{\boldsymbol{\Sigma}}_A, \widehat{\boldsymbol{B}}, \widehat{\boldsymbol{\Sigma}}_B)$, the free energy can be explicitly written as

$$2F = NL\log(2\pi\sigma'^2) + \frac{\sum_{n=1}^{N}\left\|\boldsymbol{y}^{(n)}\right\|^2 - N\|\boldsymbol{V}\|_{\mathrm{Fro}}^2}{\sigma'^2}$$
$$+ \frac{N\left\|\boldsymbol{V} - \widehat{\boldsymbol{B}}\widehat{\boldsymbol{A}}^\top\right\|_{\mathrm{Fro}}^2}{\sigma'^2} + M\log\frac{\det(\boldsymbol{C}_A)}{\det\left(\widehat{\boldsymbol{\Sigma}}_A\right)} + L\log\frac{\det(\boldsymbol{C}_B)}{\det\left(\widehat{\boldsymbol{\Sigma}}_B\right)}$$
$$- (L+M)H + \mathrm{tr}\left\{\boldsymbol{C}_A^{-1}\left(\widehat{\boldsymbol{A}}^\top\widehat{\boldsymbol{A}} + M\widehat{\boldsymbol{\Sigma}}_A\right) + \boldsymbol{C}_B^{-1}\left(\widehat{\boldsymbol{B}}^\top\widehat{\boldsymbol{B}} + L\widehat{\boldsymbol{\Sigma}}_B\right)\right.$$
$$\left. + N\sigma'^{-2}\left(-\widehat{\boldsymbol{A}}^\top\widehat{\boldsymbol{A}}\widehat{\boldsymbol{B}}^\top\widehat{\boldsymbol{B}} + \left(\widehat{\boldsymbol{A}}^\top\widehat{\boldsymbol{A}} + M\widehat{\boldsymbol{\Sigma}}_A\right)\left(\widehat{\boldsymbol{B}}^\top\widehat{\boldsymbol{B}} + L\widehat{\boldsymbol{\Sigma}}_B\right)\right)\right\}. \quad (14.14)$$

We can further apply Corollary 6.6, which states that the VB learning problem (14.11) is decomposable in the following way. Let

$$\boldsymbol{V} = \sum_{h=1}^{L} \gamma_h \boldsymbol{\omega}_{b_h}\boldsymbol{\omega}_{a_h}^\top \quad (14.15)$$

be the singular value decomposition (SVD) of $\boldsymbol{V}$ (defined in Eq. (14.10)), where $\gamma_h$ ($\geq 0$) is the $h$th largest singular value, and $\boldsymbol{\omega}_{a_h}$ and $\boldsymbol{\omega}_{b_h}$ are the associated right and left singular vectors. Then the solution (or its equivalent) of the variational parameters $\widehat{\boldsymbol{A}} = (\widehat{\boldsymbol{a}}_1, \ldots, \widehat{\boldsymbol{a}}_H), \widehat{\boldsymbol{B}} = (\widehat{\boldsymbol{b}}_1, \ldots, \widehat{\boldsymbol{b}}_H), \widehat{\boldsymbol{\Sigma}}_A, \widehat{\boldsymbol{\Sigma}}_B$, which minimizes the free energy (14.14), can be expressed as follows:

$$\widehat{\boldsymbol{a}}_h = \widehat{a}_h\boldsymbol{\omega}_{a_h}, \qquad \widehat{\boldsymbol{b}}_h = \widehat{b}_h\boldsymbol{\omega}_{b_h},$$
$$\widehat{\boldsymbol{\Sigma}}_A = \mathbf{Diag}\left(\widehat{\sigma}_{a_1}^2, \ldots, \widehat{\sigma}_{a_H}^2\right), \qquad \widehat{\boldsymbol{\Sigma}}_B = \mathbf{Diag}\left(\widehat{\sigma}_{b_1}^2, \ldots, \widehat{\sigma}_{b_H}^2\right),$$

where $\{\widehat{a}_h, \widehat{b}_h \in \mathbb{R}, \widehat{\sigma}_{a_h}^2, \widehat{\sigma}_{b_h}^2 \in \mathbb{R}_{++}\}_{h=1}^H$ are a new set of variational parameters. Thus, the VB posteriors (14.12) and (14.13) can be written as

$$r_A(\boldsymbol{A}) = \prod_{h=1}^{H} \mathrm{Gauss}_M(\boldsymbol{a}_h; \widehat{a}_h\boldsymbol{\omega}_{a_h}, \widehat{\sigma}_{a_h}^2\boldsymbol{I}_M), \quad (14.16)$$

$$r_B(\boldsymbol{B}) = \prod_{h=1}^{H} \mathrm{Gauss}_L(\boldsymbol{b}_h; \widehat{b}_h\boldsymbol{\omega}_{b_h}, \widehat{\sigma}_{b_h}^2\boldsymbol{I}_L), \quad (14.17)$$

with $\{\widehat{a}_h, \widehat{b}_h, \widehat{\sigma}_{a_h}^2, \widehat{\sigma}_{b_h}^2\}_{h=1}^H$ that are the solution of the following minimization problem:

$$\text{Given} \quad \sigma'^2 \in \mathbb{R}_{++}, \quad \{c_{a_h}^2, c_{b_h}^2 \in \mathbb{R}_{++}\}_{h=1}^H,$$
$$\min_{\{\widehat{a}_h, \widehat{b}_h, \widehat{\sigma}_{a_h}^2, \widehat{\sigma}_{b_h}^2\}_{h=1}^H} F \quad (14.18)$$
$$\text{s.t.} \quad \{\widehat{a}_h, \widehat{b}_h \in \mathbb{R}, \quad \widehat{\sigma}_{a_h}^2, \widehat{\sigma}_{b_h}^2 \in \mathbb{R}_{++}\}_{h=1}^H.$$

Here $F$ is the free energy (14.14), which can be decomposed as

$$2F = NL\log(2\pi\sigma'^2) + \frac{\sum_{n=1}^{N}\left\|\boldsymbol{y}^{(n)}\right\|^2}{\sigma'^2} + \sum_{h=1}^{H} 2F_h, \tag{14.19}$$

where
$$2F_h = M\log\frac{c_{a_h}^2}{\widehat{\sigma}_{a_h}^2} + L\log\frac{c_{b_h}^2}{\widehat{\sigma}_{b_h}^2} + \frac{\widehat{a}_h^2 + M\widehat{\sigma}_{a_h}^2}{c_{a_h}^2} + \frac{\widehat{b}_h^2 + L\widehat{\sigma}_{b_h}^2}{c_{b_h}^2}$$
$$- (L+M) + \frac{N}{\sigma'^2}\left(-2\widehat{a}_h\widehat{b}_h\gamma_h + \left(\widehat{a}_h^2 + M\widehat{\sigma}_{a_h}^2\right)\left(\widehat{b}_h^2 + L\widehat{\sigma}_{b_h}^2\right)\right). \tag{14.20}$$

### 14.1.2 VB Solution

Let us derive an asymptotic-form VB solution from the nonasymptotic global VB solution, derived in Section 6. Theorem 6.7 leads to the following theorem:

**Theorem 14.1** *The VB estimator* $\widehat{\boldsymbol{U}}^{\mathrm{VB}} \equiv \langle \boldsymbol{B}\boldsymbol{A}^\top\rangle_{r_A(\boldsymbol{A})r_B(\boldsymbol{B})}$ *for the linear mapping of the RRR model* (14.2) *and* (14.3) *can be written as*

$$\widehat{\boldsymbol{U}}^{\mathrm{VB}} = \widehat{\boldsymbol{B}}\widehat{\boldsymbol{A}}^\top = \sum_{h=1}^{H}\widehat{\gamma}_h^{\mathrm{VB}}\boldsymbol{\omega}_{b_h}\boldsymbol{\omega}_{a_h}^\top, \quad \textit{where} \quad \widehat{\gamma}_h^{\mathrm{VB}} = \max\left(0, \breve{\gamma}_h^{\mathrm{VB}}\right) \tag{14.21}$$

*for*

$$\breve{\gamma}_h^{\mathrm{VB}} = \gamma_h\left(1 - \frac{\max(L,M)\sigma'^2}{N\gamma_h^2}\right) + O_{\mathrm{p}}(N^{-1}). \tag{14.22}$$

*For each component h,* $\widehat{\gamma}_h^{\mathrm{VB}} > 0$ *if and only if* $\gamma_h > \underline{\gamma}_h^{\mathrm{VB}}$ *for*

$$\underline{\gamma}_h^{\mathrm{VB}} = \sigma'\sqrt{\frac{\max(L,M)}{N}} + O(N^{-1}). \tag{14.23}$$

*Proof* Noting that Theorem 6.7 gives the VB solution for either $\boldsymbol{V}$ or $\boldsymbol{V}^\top \in \mathbb{R}^{L\times M}$ that satisfies $L \leq M$, that the shrinkage estimator $\breve{\gamma}_h^{\mathrm{VB}}$ (given by Eq. (6.50)) is an increasing function of $\gamma_h$, and that $\breve{\gamma}_h^{\mathrm{VB}} = 0$ when $\gamma_h$ is equal to the threshold $\underline{\gamma}_h^{\mathrm{VB}}$ (given by Eq. (6.49)), we have Eq. (14.21) with

$$\breve{\gamma}_h^{\mathrm{VB}} = \gamma_h\left(1 - \frac{\sigma'^2}{2N\gamma_h^2}\left(L + M + \sqrt{(M-L)^2 + \frac{4\gamma_h^2}{c_{a_h}^2 c_{b_h}^2}}\right)\right)$$
$$= \gamma_h\left(1 - \frac{\sigma'^2}{2N\gamma_h^2}\left(L + M + \sqrt{(M-L)^2 + O(\gamma_h^2)}\right)\right)$$

$$
\begin{aligned}
&= \begin{cases} \gamma_h \left(1 - \frac{\sigma'^2}{2N\gamma_h^2} \left(L + M + \max(L, M) - \min(L, M) + O(\gamma_h^2)\right)\right) & (\text{if } L \neq M) \\ \gamma_h \left(1 - \frac{\sigma'^2}{2N\gamma_h^2} \left(L + M + O(\gamma_h)\right)\right) & (\text{if } L = M) \end{cases} \\
&= \gamma_h \left(1 - \frac{\max(L, M)\sigma'^2}{N\gamma_h^2} \left(1 + O_{\mathrm{p}}(\gamma_h)\right)\right) \qquad (14.24) \\
&= \gamma_h \left(1 - \frac{\max(L, M)\sigma'^2}{N\gamma_h^2}\right) + O_{\mathrm{p}}(N^{-1}),
\end{aligned}
$$

and

$$
\begin{aligned}
\underline{\gamma}_h^{\mathrm{VB}} &= \frac{\sigma'}{\sqrt{N}} \sqrt{\frac{(L + M)}{2} + \frac{\sigma'^2}{2Nc_{a_h}^2 c_{b_h}^2} + \sqrt{\left(\frac{(L + M)}{2} + \frac{\sigma'^2}{2Nc_{a_h}^2 c_{b_h}^2}\right)^2 - LM}} \\
&= \frac{\sigma'}{\sqrt{N}} \sqrt{\frac{(L + M)}{2} + \frac{\sigma'^2}{2Nc_{a_h}^2 c_{b_h}^2} + \sqrt{\left(\frac{\max(L, M) - \min(L, M)}{2}\right)^2 + O(N^{-1})}} \\
&= \begin{cases} \frac{\sigma'}{\sqrt{N}} \sqrt{\frac{(L+M)}{2} + \frac{\sigma'^2}{2Nc_{a_h}^2 c_{b_h}^2} + \frac{\max(L,M) - \min(L,M)}{2} + O(N^{-1})} & (\text{if } L \neq M) \\ \frac{\sigma'}{\sqrt{N}} \sqrt{\frac{(L+M)}{2} + \frac{\sigma'^2}{2Nc_{a_h}^2 c_{b_h}^2} + O(N^{-1/2})} & (\text{if } L = M) \end{cases} \\
&= \frac{\sigma'}{\sqrt{N}} \sqrt{\max(L, M)} + O(N^{-1}),
\end{aligned}
$$

which completes the proof. Note that we used $\gamma_h = O_{\mathrm{p}}(1)$ to get Eq. (14.24). □

Theorem 14.1 states that the VB estimator converges to the positive-part James–Stein (PJS) estimator (see Appendix A)—the same solution (Corollary 7.1) as the nonasymptotic MF solution with the flat prior. This is natural because the influence from the constant prior disappears in the asymptotic limit, making MAP learning converge to ML learning.

Corollary 6.8 leads to the following corollary:

**Corollary 14.2** *The VB posterior of the RRR model* (14.2) *and* (14.3) *is given by Eqs. (14.16) and* (14.17) *with the variational parameters given as follows: if* $\gamma_h > \underline{\gamma}_h^{\mathrm{VB}}$,

$$
\widehat{a}_h = \pm \sqrt{\breve{\gamma}_h^{\mathrm{VB}} \widehat{\delta}_h^{\mathrm{VB}}}, \quad \widehat{b}_h = \pm \sqrt{\frac{\breve{\gamma}_h^{\mathrm{VB}}}{\widehat{\delta}_h^{\mathrm{VB}}}}, \quad \widehat{\sigma}_{a_h}^2 = \frac{\sigma'^2 \widehat{\delta}_h^{\mathrm{VB}}}{N\gamma_h}, \quad \widehat{\sigma}_{b_h}^2 = \frac{\sigma'^2}{N\gamma_h \widehat{\delta}_h^{\mathrm{VB}}}, \tag{14.25}
$$

$$\textit{where} \quad \widehat{\delta}_h^{\mathrm{VB}} \left( \equiv \frac{\widehat{a}_h}{\widehat{b}_h} \right) = \begin{cases} \frac{(\max(L,M)-\min(L,M))c_{a_h}}{\gamma_h} + O_{\mathrm{p}}(1) & (\textit{if } L \le M), \\ \left( \frac{(\max(L,M)-\min(L,M))c_{b_h}}{\gamma_h} + O_{\mathrm{p}}(1) \right)^{-1} & (\textit{if } L > M), \end{cases} \tag{14.26}$$

*and otherwise,*

$$\widehat{a}_h = 0, \quad \widehat{b}_h = 0, \quad \widehat{\sigma}^2_{a_h} = c^2_{a_h}\left(1 - \frac{NL\widehat{\zeta}_h^{\mathrm{VB}}}{\sigma'^2}\right), \quad \widehat{\sigma}^2_{b_h} = c^2_{b_h}\left(1 - \frac{NM\widehat{\zeta}_h^{\mathrm{VB}}}{\sigma'^2}\right), \tag{14.27}$$

$$\textit{where} \quad \widehat{\zeta}_h^{\mathrm{VB}} \left( \equiv \widehat{\sigma}^2_{a_h}\widehat{\sigma}^2_{b_h} \right) = \begin{cases} \frac{\min(L,M)\sigma'^2}{NLM} + \Theta(N^{-2}), & (\textit{if } L \ne M), \\ \frac{\min(L,M)\sigma'^2}{NLM} + \Theta(N^{-3/2}), & (\textit{if } L = M). \end{cases} \tag{14.28}$$

*Proof* Noting that Corollary 6.8 gives the VB posterior for either $V$ or $V^\top \in \mathbb{R}^{L\times M}$ that satisfies $L \le M$, we have Eq. (14.25) with

$$\begin{aligned} \widehat{\delta}_h^{\mathrm{VB}} &= \begin{cases} \frac{Nc_{a_h}}{\sigma'^2}\left(\gamma_h - \breve{\gamma}_h^{\mathrm{VB}} - \frac{L\sigma'^2}{N\gamma_h}\right) & (\text{if } L \le M) \\ \left(\frac{Nc_{b_h}}{\sigma'^2}\left(\gamma_h - \breve{\gamma}_h^{\mathrm{VB}} - \frac{M\sigma'^2}{N\gamma_h}\right)\right)^{-1} & (\text{if } L > M) \end{cases} \\ &= \begin{cases} \frac{(\max(L,M)-\min(L,M))c_{a_h}}{\gamma_h} + O_{\mathrm{p}}(1) & (\text{if } L \le M), \\ \left(\frac{(\max(L,M)-\min(L,M))c_{b_h}}{\gamma_h} + O_{\mathrm{p}}(1)\right)^{-1} & (\text{if } L > M), \end{cases} \end{aligned}$$

when $\gamma_h > \underline{\gamma}_h^{\mathrm{VB}}$, and Eq. (14.27) with

$$\begin{aligned} \widehat{\zeta}_h^{\mathrm{VB}} &= \frac{\sigma'^2}{2NLM}\left\{L + M + \frac{\sigma'^2}{Nc^2_{a_h}c^2_{b_h}} - \sqrt{\left(L + M + \frac{\sigma'^2}{Nc^2_{a_h}c^2_{b_h}}\right)^2 - 4LM}\right\} \\ &= \frac{\sigma'^2}{2NLM}\left\{L + M + \frac{\sigma'^2}{Nc^2_{a_h}c^2_{b_h}}\right. \\ &\quad \left. - \sqrt{(L+M)^2 + 2(L+M)\frac{\sigma'^2}{Nc^2_{a_h}c^2_{b_h}} + \left(\frac{\sigma'^2}{Nc^2_{a_h}c^2_{b_h}}\right)^2 - 4LM}\right\} \\ &= \frac{\sigma'^2}{2NLM}\left\{L + M + \frac{\sigma'^2}{Nc^2_{a_h}c^2_{b_h}}\right. \\ &\quad \left. - \sqrt{(\max(L,M) - \min(L,M))^2 + 2(L+M)\frac{\sigma'^2}{Nc^2_{a_h}c^2_{b_h}} + \left(\frac{\sigma'^2}{Nc^2_{a_h}c^2_{b_h}}\right)^2}\right\} \end{aligned}$$

$$= \begin{cases} \frac{\sigma'^2}{2NLM}\left\{L + M + \frac{\sigma'^2}{Nc_{a_h}^2 c_{b_h}^2} - (\max(L,M) - \min(L,M)) \right. \\ \qquad \left. \cdot\left(1 + \frac{L+M}{(\max(L,M)-\min(L,M))^2}\frac{\sigma'^2}{Nc_{a_h}^2 c_{b_h}^2}\right) + O(N^{-2})\right\} & (\text{if } L \neq M) \\ \frac{\sigma'^2}{2NLM}\left\{L + M + \frac{\sigma'^2}{Nc_{a_h}^2 c_{b_h}^2} \right. \\ \qquad \left. - \sqrt{2(L+M)\left(\frac{\sigma'^2}{Nc_{a_h}^2 c_{b_h}^2}\right) + \left(\frac{\sigma'^2}{Nc_{a_h}^2 c_{b_h}^2}\right)^2}\right\} & (\text{if } L = M) \end{cases}$$

$$= \begin{cases} \frac{\sigma'^2}{2NLM}\left(2\min(L,M) + \Theta(N^{-1})\right) & (\text{if } L \neq M) \\ \frac{\sigma'^2}{2NLM}\left(2\min(L,M) + \Theta(N^{-1/2})\right) & (\text{if } L = M) \end{cases}$$

$$= \begin{cases} \frac{\min(L,M)\sigma'^2}{NLM} + \Theta(N^{-2}) & (\text{if } L \neq M), \\ \frac{\min(L,M)\sigma'^2}{NLM} + \Theta(N^{-3/2}) & (\text{if } L = M), \end{cases}$$

when $\gamma_h \leq \underline{\gamma}_h^{\mathrm{VB}}$. This completes the proof. □

From Corollary 14.2, we can evaluate the orders of the optimal variational parameters in the asymptotic limit, which will be used when the VB free energy is analyzed.

**Corollary 14.3** *The orders of the optimal variational parameters, given by Eq.* (14.25) *or Eq.* (14.27)*, are as follows: if* $\gamma_h > \underline{\gamma}_h^{\mathrm{VB}} (= \Theta(N^{-1/2}))$,

$$\begin{aligned} &\widehat{a}_h = \Theta_{\mathrm{p}}(1), && \widehat{b}_h = \Theta_{\mathrm{p}}(\gamma_h), && \widehat{\sigma}_{a_h}^2 = \Theta_{\mathrm{p}}(N^{-1}\gamma_h^{-2}), && \widehat{\sigma}_{b_h}^2 = \Theta_{\mathrm{p}}(N^{-1}) && (\textit{if } L < M), \\ &\widehat{a}_h = \Theta_{\mathrm{p}}(\gamma_h^{1/2}), && \widehat{b}_h = \Theta_{\mathrm{p}}(\gamma_h^{1/2}), && \widehat{\sigma}_{a_h}^2 = \Theta_{\mathrm{p}}(N^{-1}\gamma_h^{-1}), && \widehat{\sigma}_{b_h}^2 = \Theta_{\mathrm{p}}(N^{-1}\gamma_h^{-1}) && (\textit{if } L = M), \\ &\widehat{a}_h = \Theta_{\mathrm{p}}(\gamma_h), && \widehat{b}_h = \Theta_{\mathrm{p}}(1), && \widehat{\sigma}_{a_h}^2 = \Theta_{\mathrm{p}}(N^{-1}), && \widehat{\sigma}_{b_h}^2 = \Theta_{\mathrm{p}}(N^{-1}\gamma_h^{-2}) && (\textit{if } L > M), \end{aligned} \tag{14.29}$$

*and otherwise,*

$$\begin{aligned} &\widehat{a}_h = 0, && \widehat{b}_h = 0, && \widehat{\sigma}_{a_h}^2 = \Theta(1), && \widehat{\sigma}_{b_h}^2 = \Theta(N^{-1}) && (\textit{if } L < M), \\ &\widehat{a}_h = 0, && \widehat{b}_h = 0, && \widehat{\sigma}_{a_h}^2 = \Theta(N^{-1/2}), && \widehat{\sigma}_{b_h}^2 = \Theta(N^{-1/2}) && (\textit{if } L = M), \\ &\widehat{a}_h = 0, && \widehat{b}_h = 0, && \widehat{\sigma}_{a_h}^2 = \Theta(N^{-1}), && \widehat{\sigma}_{b_h}^2 = \Theta(1) && (\textit{if } L > M). \end{aligned} \tag{14.30}$$

*Proof* Eqs. (14.22) and (14.23) give

$$\underline{\gamma}_h^{\mathrm{VB}} = \Theta(N^{-1/2}), \qquad \breve{\gamma}_h^{\mathrm{VB}} = \Theta(\gamma_h),$$

and Eq. (14.26) gives

$$\widehat{\delta}_h^{\mathrm{VB}} = \begin{cases} \Theta_{\mathrm{p}}(\gamma_h^{-1}) & (\text{if } L < M), \\ \Theta_{\mathrm{p}}(1) & (\text{if } L = M), \\ \Theta_{\mathrm{p}}(\gamma_h) & (\text{if } L > M). \end{cases}$$

Substituting the preceding into Eq. (14.25) gives Eq. (14.29), and substituting Eq. (14.28) into Eq. (14.27) gives Eq. (14.30), which complete the proof. □

Corollary 14.3 implies that the posterior probability mass does not necessarily converge to a single point, for example, $\widehat{\sigma}_{a_h}^2 = \Theta(1)$ if $\gamma_h < \underline{\gamma}_h^{\mathrm{VB}}$ and $L < M$. This is typical behavior of singular models with *nonidentifiability*. On the other hand, the probability mass of the linear mapping $\boldsymbol{U} = \boldsymbol{B}\boldsymbol{A}^\top$ converges to a single point.

**Corollary 14.4** *It holds that*

$$\left\langle \left\| \boldsymbol{B}\boldsymbol{A}^\top - \widehat{\boldsymbol{B}\boldsymbol{A}^\top} \right\|_{\mathrm{Fro}}^2 \right\rangle_{r_A(\boldsymbol{A}) r_B(\boldsymbol{B})} = O_{\mathrm{p}}(N^{-1}).$$

*Proof* We have

$$\begin{aligned}
\left\langle \left\| \boldsymbol{B}\boldsymbol{A}^\top - \widehat{\boldsymbol{B}\boldsymbol{A}^\top} \right\|_{\mathrm{Fro}}^2 \right\rangle_{r_A(\boldsymbol{A}) r_B(\boldsymbol{B})} &= \mathrm{tr}\left\langle \left(\boldsymbol{B}\boldsymbol{A}^\top - \widehat{\boldsymbol{B}\boldsymbol{A}^\top}\right)^\top \left(\boldsymbol{B}\boldsymbol{A}^\top - \widehat{\boldsymbol{B}\boldsymbol{A}^\top}\right) \right\rangle_{r_A(\boldsymbol{A}) r_B(\boldsymbol{B})} \\
&= \mathrm{tr}\left\langle \boldsymbol{A}\boldsymbol{B}^\top\boldsymbol{B}\boldsymbol{A}^\top - 2\boldsymbol{A}\boldsymbol{B}^\top\widehat{\boldsymbol{B}\boldsymbol{A}^\top} + \widehat{\boldsymbol{A}\boldsymbol{B}^\top}\widehat{\boldsymbol{B}\boldsymbol{A}^\top} \right\rangle_{r_A(\boldsymbol{A}) r_B(\boldsymbol{B})} \\
&= \mathrm{tr}\left( \left\langle \boldsymbol{A}^\top\boldsymbol{A}\boldsymbol{B}^\top\boldsymbol{B} \right\rangle_{r_A(\boldsymbol{A}) r_B(\boldsymbol{B})} - \widehat{\boldsymbol{A}}^\top\widehat{\boldsymbol{A}\boldsymbol{B}}^\top\widehat{\boldsymbol{B}} \right) \\
&= \mathrm{tr}\left( \left(\widehat{\boldsymbol{A}}^\top\widehat{\boldsymbol{A}} + M\widehat{\boldsymbol{\Sigma}}_A\right)\left(\widehat{\boldsymbol{B}}^\top\widehat{\boldsymbol{B}} + L\widehat{\boldsymbol{\Sigma}}_B\right) - \widehat{\boldsymbol{A}}^\top\widehat{\boldsymbol{A}\boldsymbol{B}}^\top\widehat{\boldsymbol{B}} \right) \\
&= \sum_{h=1}^{H} \left( \left(\widehat{a}_h^2 + M\widehat{\sigma}_{a_h}^2\right)\left(\widehat{b}_h^2 + L\widehat{\sigma}_{b_h}^2\right) - \widehat{a}_h^2\widehat{b}_h^2 \right) \\
&= \sum_{h=1}^{H} \left( L\widehat{a}_h^2\widehat{\sigma}_{b_h}^2 + M\widehat{b}_h^2\widehat{\sigma}_{a_h}^2 + LM\widehat{\sigma}_{a_h}^2\widehat{\sigma}_{b_h}^2 \right).
\end{aligned} \tag{14.31}$$

Corollary 14.3 guarantees that all terms in Eq. (14.31) are of the order of $\Theta_{\mathrm{p}}(N^{-1})$ for any $L$, $M$, and $\{\gamma_h\}$, which completes the proof. □

Now we derive an asymptotic form of the VB predictive distribution,

$$p\left(\boldsymbol{y}|\boldsymbol{x}, \boldsymbol{X}, \boldsymbol{Y}\right) = \left\langle p(\boldsymbol{y}|\boldsymbol{x}, \boldsymbol{A}, \boldsymbol{B}) \right\rangle_{r_A(\boldsymbol{A}) r_B(\boldsymbol{B})}. \tag{14.32}$$

From Corollary 14.4, we expect that the predictive distribution is not very far from the *plug-in* VB predictive distribution (see Section 1.1.3):

$$p(\boldsymbol{y}|\boldsymbol{x}, \widehat{\boldsymbol{A}}, \widehat{\boldsymbol{B}}) = \mathrm{Gauss}_L(\boldsymbol{y}; \widehat{\boldsymbol{B}\boldsymbol{A}}^\top \boldsymbol{x}, \sigma'^2 \boldsymbol{I}_L). \tag{14.33}$$

Indeed, we will show in the next section that both predictive distributions (14.32) and (14.33) give the same generalization and training coefficients. This justifies the use of the *plug-in* VB predictive distribution, which is easy to compute from the optimal variational parameters.

By expanding the VB predictive distribution around the plug-in VB predictive distribution, we have the following theorem:

**Theorem 14.5** *The VB predictive distribution* (14.32) *of the RRR model* (14.2) *and* (14.3) *can be written as*

$$p(\boldsymbol{y}|\boldsymbol{x}, \boldsymbol{X}, \boldsymbol{Y}) = \mathrm{Gauss}_L(\boldsymbol{y}; \boldsymbol{\Psi}\widehat{\boldsymbol{B}\boldsymbol{A}}^\top \boldsymbol{x}, \sigma'^2 \boldsymbol{\Psi}) + O_{\mathrm{p}}(N^{-3/2}) \tag{14.34}$$

*for* $\boldsymbol{\Psi} = \boldsymbol{I}_L + O_{\mathrm{p}}(N^{-1})$.

*Proof* The VB predictive distribution can be written as follows:

$$\begin{aligned} p(\boldsymbol{y}|\boldsymbol{x}, \boldsymbol{X}, \boldsymbol{Y}) &= \langle p(\boldsymbol{y}|\boldsymbol{x}, \boldsymbol{A}, \boldsymbol{B})\rangle_{r_A(\boldsymbol{A})r_B(\boldsymbol{B})} \\ &= p(\boldsymbol{y}|\boldsymbol{x}, \widehat{\boldsymbol{A}}, \widehat{\boldsymbol{B}}) \left\langle \frac{p(\boldsymbol{y}|\boldsymbol{x}, \boldsymbol{A}, \boldsymbol{B})}{p(\boldsymbol{y}|\boldsymbol{x}, \widehat{\boldsymbol{A}}, \widehat{\boldsymbol{B}})} \right\rangle_{r_A(\boldsymbol{A})r_B(\boldsymbol{B})} \\ &= p(\boldsymbol{y}|\boldsymbol{x}, \widehat{\boldsymbol{A}}, \widehat{\boldsymbol{B}}) \left\langle \exp\left( -\frac{\|\boldsymbol{y} - \boldsymbol{B}\boldsymbol{A}^\top \boldsymbol{x}\|^2 - \|\boldsymbol{y} - \widehat{\boldsymbol{B}\boldsymbol{A}}^\top \boldsymbol{x}\|^2}{2\sigma'^2} \right) \right\rangle_{r_A(\boldsymbol{A})r_B(\boldsymbol{B})} \\ &= p(\boldsymbol{y}|\boldsymbol{x}, \widehat{\boldsymbol{A}}, \widehat{\boldsymbol{B}}) \left\langle \exp\left( -\frac{\left(\boldsymbol{y} - \boldsymbol{B}\boldsymbol{A}^\top \boldsymbol{x} + (\boldsymbol{y} - \widehat{\boldsymbol{B}\boldsymbol{A}}^\top \boldsymbol{x})\right)^\top \left(\boldsymbol{y} - \boldsymbol{B}\boldsymbol{A}^\top \boldsymbol{x} - (\boldsymbol{y} - \widehat{\boldsymbol{B}\boldsymbol{A}}^\top \boldsymbol{x})\right)}{2\sigma'^2} \right) \right\rangle_{r_A(\boldsymbol{A})r_B(\boldsymbol{B})} \\ &= p(\boldsymbol{y}|\boldsymbol{x}, \widehat{\boldsymbol{A}}, \widehat{\boldsymbol{B}}) \left\langle \exp\left( \frac{\left(\boldsymbol{y} - (\boldsymbol{B}\boldsymbol{A}^\top - \widehat{\boldsymbol{B}\boldsymbol{A}}^\top)\boldsymbol{x}\right)^\top (\boldsymbol{B}\boldsymbol{A}^\top - \widehat{\boldsymbol{B}\boldsymbol{A}}^\top)\boldsymbol{x}}{\sigma'^2} \right) \right\rangle_{r_A(\boldsymbol{A})r_B(\boldsymbol{B})}. \end{aligned} \tag{14.35}$$

Corollary 14.4 implies that the exponent in Eq. (14.35) is of the order of $N^{-1/2}$, i.e.,

$$\phi \equiv \frac{\left(\boldsymbol{y} - (\boldsymbol{B}\boldsymbol{A}^\top - \widehat{\boldsymbol{B}\boldsymbol{A}}^\top)\boldsymbol{x}\right)^\top (\boldsymbol{B}\boldsymbol{A}^\top - \widehat{\boldsymbol{B}\boldsymbol{A}}^\top)\boldsymbol{x}}{\sigma'^2} = O_{\mathrm{p}}(N^{-1/2}). \tag{14.36}$$

By applying the Taylor expansion of the exponential function to Eq. (14.35), we obtain an asymptotic expansion of the predictive distribution around the plug-in predictive distribution:

$$p(\boldsymbol{y}|\boldsymbol{x}, \boldsymbol{X}, \boldsymbol{Y}) = p(\boldsymbol{y}|\boldsymbol{x}, \widehat{\boldsymbol{A}}, \widehat{\boldsymbol{B}}) \left( 1 + \langle \phi \rangle_{r_A(\boldsymbol{A})r_B(\boldsymbol{B})} + \frac{1}{2} \left\langle \phi^2 \right\rangle_{r_A(\boldsymbol{A})r_B(\boldsymbol{B})} + O_{\mathrm{p}}(N^{-3/2}) \right).$$

Focusing on the dependence on the random variable $\boldsymbol{y}$, we can identify the function form of the predictive distribution as follows:

$$
\begin{aligned}
p(\boldsymbol{y}|\boldsymbol{x}, \boldsymbol{X}, \boldsymbol{Y}) &\propto \exp\Bigg( - \frac{\left\|\boldsymbol{y}-\widehat{\boldsymbol{B}\boldsymbol{A}}^\top \boldsymbol{x}\right\|^2}{2\sigma'^2} \\
&\qquad + \log\left(1 + \langle\phi\rangle_{r_A(\boldsymbol{A})r_B(\boldsymbol{B})} + \tfrac{1}{2}\left\langle\phi^2\right\rangle_{r_A(\boldsymbol{A})r_B(\boldsymbol{B})} + O_{\mathrm{p}}(N^{-3/2})\right)\Bigg) \\
&= \exp\Bigg( - \frac{\left\|\boldsymbol{y}-\widehat{\boldsymbol{B}\boldsymbol{A}}^\top \boldsymbol{x}\right\|^2}{2\sigma'^2} + \langle\phi\rangle_{r_A(\boldsymbol{A})r_B(\boldsymbol{B})} + \tfrac{1}{2}\left\langle\phi^2\right\rangle_{r_A(\boldsymbol{A})r_B(\boldsymbol{B})} \\
&\qquad - \tfrac{1}{2}\langle\phi\rangle^2_{r_A(\boldsymbol{A})r_B(\boldsymbol{B})} + O_{\mathrm{p}}(N^{-3/2})\Bigg) \\
&\propto \exp\left( - \frac{\left\|\boldsymbol{y}-\widehat{\boldsymbol{B}\boldsymbol{A}}^\top \boldsymbol{x}\right\|^2}{2\sigma'^2} + \tfrac{1}{2}\left\langle\phi^2\right\rangle_{r_A(\boldsymbol{A})r_B(\boldsymbol{B})} + O_{\mathrm{p}}(N^{-3/2})\right) \\
&\propto \exp\left( - \frac{\left\|\boldsymbol{y}-\widehat{\boldsymbol{B}\boldsymbol{A}}^\top \boldsymbol{x}\right\|^2 - \boldsymbol{y}^\top \boldsymbol{\Psi}_1 \boldsymbol{y}}{2\sigma'^2} + O_{\mathrm{p}}(N^{-3/2})\right) \\
&\propto \exp\left( - \frac{\|\boldsymbol{y}\|^2 - 2\boldsymbol{y}^\top \widehat{\boldsymbol{B}\boldsymbol{A}}^\top \boldsymbol{x} - \boldsymbol{y}^\top \boldsymbol{\Psi}_1 \boldsymbol{y}}{2\sigma'^2} + O_{\mathrm{p}}(N^{-3/2})\right) \\
&\propto \exp\left( - \frac{\left(\boldsymbol{y}-\boldsymbol{\Psi}\widehat{\boldsymbol{B}\boldsymbol{A}}^\top \boldsymbol{x}\right)^\top \boldsymbol{\Psi}^{-1} \left(\boldsymbol{y}-\boldsymbol{\Psi}\widehat{\boldsymbol{B}\boldsymbol{A}}^\top \boldsymbol{x}\right)}{2\sigma'^2} + O_{\mathrm{p}}(N^{-3/2})\right),
\end{aligned} \tag{14.37}
$$

where

$$
\boldsymbol{\Psi} = (\boldsymbol{I}_L - \boldsymbol{\Psi}_1)^{-1}, \tag{14.38}
$$

$$
\boldsymbol{\Psi}_1 = \left\langle \frac{(\boldsymbol{B}\boldsymbol{A}^\top - \widehat{\boldsymbol{B}\boldsymbol{A}}^\top)\boldsymbol{x}\boldsymbol{x}^\top (\boldsymbol{B}\boldsymbol{A}^\top - \widehat{\boldsymbol{B}\boldsymbol{A}}^\top)^\top}{\sigma'^2} \right\rangle_{r_A(\boldsymbol{A})r_B(\boldsymbol{B})}. \tag{14.39}
$$

Here we used

$$
\begin{aligned}
\langle\phi\rangle_{r_A(\boldsymbol{A})r_B(\boldsymbol{B})} &= \left\langle \frac{\left(\boldsymbol{y}-(\boldsymbol{B}\boldsymbol{A}^\top - \widehat{\boldsymbol{B}\boldsymbol{A}}^\top)\boldsymbol{x}\right)^\top (\boldsymbol{B}\boldsymbol{A}^\top - \widehat{\boldsymbol{B}\boldsymbol{A}}^\top)\boldsymbol{x}}{\sigma'^2} \right\rangle_{r_A(\boldsymbol{A})r_B(\boldsymbol{B})} \\
&= \left\langle \frac{\left\|(\boldsymbol{B}\boldsymbol{A}^\top - \widehat{\boldsymbol{B}\boldsymbol{A}}^\top)\boldsymbol{x}\right\|^2}{\sigma'^2} \right\rangle_{r_A(\boldsymbol{A})r_B(\boldsymbol{B})} \\
&= \mathrm{const.}, \\
\left\langle\phi^2\right\rangle_{r_A(\boldsymbol{A})r_B(\boldsymbol{B})} &= \left\langle \frac{\left(\boldsymbol{y}-(\boldsymbol{B}\boldsymbol{A}^\top - \widehat{\boldsymbol{B}\boldsymbol{A}}^\top)\boldsymbol{x}\right)^\top (\boldsymbol{B}\boldsymbol{A}^\top - \widehat{\boldsymbol{B}\boldsymbol{A}}^\top)\boldsymbol{x}\boldsymbol{x}^\top (\boldsymbol{B}\boldsymbol{A}^\top - \widehat{\boldsymbol{B}\boldsymbol{A}}^\top)^\top \left(\boldsymbol{y}-(\boldsymbol{B}\boldsymbol{A}^\top - \widehat{\boldsymbol{B}\boldsymbol{A}}^\top)\boldsymbol{x}\right)}{\sigma'^4} \right\rangle_{r_A(\boldsymbol{A})r_B(\boldsymbol{B})} \\
&= \frac{\boldsymbol{y}^\top \boldsymbol{\Psi}_1 \boldsymbol{y}}{\sigma'^2} + O_{\mathrm{p}}(N^{-3/2}).
\end{aligned}
$$

Eq. (14.39) implies that $\boldsymbol{\Psi}_1$ is symmetric and $\boldsymbol{\Psi}_1 = O_{\mathrm{p}}(N^{-1})$. Therefore, $\boldsymbol{\Psi}$, defined by Eq. (14.38), is symmetric, positive definite, and can be written as $\boldsymbol{\Psi} = \boldsymbol{I}_L + O_p(N^{-1})$. The function form of Eq. (14.37) implies that the VB predictive distribution converges to the Gaussian distribution in the asymptotic limit, and we thus have

$$p(\boldsymbol{y}|\boldsymbol{x}, X, Y) = \frac{\exp\left(-\frac{\left(\boldsymbol{y}-\boldsymbol{\Psi}\widehat{\boldsymbol{B}\boldsymbol{A}}^\top \boldsymbol{x}\right)^\top \boldsymbol{\Psi}^{-1}\left(\boldsymbol{y}-\boldsymbol{\Psi}\widehat{\boldsymbol{B}\boldsymbol{A}}^\top \boldsymbol{x}\right)}{2\sigma'^2} + O_p(N^{-3/2})\right)}{\int \exp\left(-\frac{\left(\boldsymbol{y}-\boldsymbol{\Psi}\widehat{\boldsymbol{B}\boldsymbol{A}}^\top \boldsymbol{x}\right)^\top \boldsymbol{\Psi}^{-1}\left(\boldsymbol{y}-\boldsymbol{\Psi}\widehat{\boldsymbol{B}\boldsymbol{A}}^\top \boldsymbol{x}\right)}{2\sigma'^2} + O_p(N^{-3/2})\right) d\boldsymbol{y}}$$

$$= \frac{\exp\left(-\frac{\left(\boldsymbol{y}-\boldsymbol{\Psi}\widehat{\boldsymbol{B}\boldsymbol{A}}^\top \boldsymbol{x}\right)^\top \boldsymbol{\Psi}^{-1}\left(\boldsymbol{y}-\boldsymbol{\Psi}\widehat{\boldsymbol{B}\boldsymbol{A}}^\top \boldsymbol{x}\right)}{2\sigma'^2}\right)(1+O_p(N^{-3/2}))}{\int \exp\left(-\frac{\left(\boldsymbol{y}-\boldsymbol{\Psi}\widehat{\boldsymbol{B}\boldsymbol{A}}^\top \boldsymbol{x}\right)^\top \boldsymbol{\Psi}^{-1}\left(\boldsymbol{y}-\boldsymbol{\Psi}\widehat{\boldsymbol{B}\boldsymbol{A}}^\top \boldsymbol{x}\right)}{2\sigma'^2}\right)(1+O_p(N^{-3/2}))d\boldsymbol{y}}$$

$$= \frac{1}{(2\pi\sigma'^2)^{L/2}\det(\boldsymbol{\Psi})^{1/2}} \exp\left(-\frac{\left(\boldsymbol{y}-\boldsymbol{\Psi}\widehat{\boldsymbol{B}\boldsymbol{A}}^\top \boldsymbol{x}\right)^\top \boldsymbol{\Psi}^{-1}\left(\boldsymbol{y}-\boldsymbol{\Psi}\widehat{\boldsymbol{B}\boldsymbol{A}}^\top \boldsymbol{x}\right)}{2\sigma'^2}\right) + O_p(N^{-3/2}),$$

which completes the proof. □

## 14.2 Generalization Properties

Let us analyze generalization properties of VB learning based on the posterior distribution and the predictive distribution, derived in Section 14.1.2.

### 14.2.1 Assumption on True Distribution

We assume that the true distribution can be expressed by the model distribution with the *true* parameter $\boldsymbol{A}^*$ and $\boldsymbol{B}^*$ with their rank $H^*$:

$$q(\boldsymbol{y}|\boldsymbol{x}) = \mathrm{Gauss}_L\left(\boldsymbol{y}, \boldsymbol{B}^*\boldsymbol{A}^{*\top}\boldsymbol{x}, \sigma'^2\boldsymbol{I}_L\right)$$
$$= \left(2\pi\sigma'^2\right)^{-L/2} \exp\left(-\frac{\left\|\boldsymbol{y} - \boldsymbol{B}^*\boldsymbol{A}^{*\top}\boldsymbol{x}\right\|^2}{2\sigma'^2}\right). \tag{14.40}$$

Let

$$\boldsymbol{U}^* \equiv \boldsymbol{B}^*\boldsymbol{A}^{*\top} = \sum_{h=1}^{\min(L,M)} \gamma_h^* \boldsymbol{\omega}_{b_h}^* \boldsymbol{\omega}_{a_h}^{*\top} \tag{14.41}$$

be the SVD of the true linear mapping $\boldsymbol{B}^*\boldsymbol{A}^{*\top}$, where $\gamma_h^*$ ($\geq 0$) is the $h$th largest singular value, and $\boldsymbol{\omega}_{a_h}^*$ and $\boldsymbol{\omega}_{b_h}^*$ are the associated right and left singular vectors. The assumption that the true linear mapping has rank $H^*$ amounts to

$$\gamma_h^* = \begin{cases} \Theta(1) & \text{for} \quad h = 1, \ldots, H^*, \\ 0 & \text{for} \quad h = H^* + 1, \ldots, \min(L, M). \end{cases} \tag{14.42}$$

### 14.2.2 Consistency of VB Estimator

Since the training samples are drawn from the true distribution (14.40), the central limit theorem (Theorem 13.1) guarantees the following:

$$\boldsymbol{V}\left(\equiv \frac{1}{N}\sum_{n=1}^{N}\boldsymbol{y}^{(n)}\boldsymbol{x}^{(n)\top}\right) = \frac{1}{N}\sum_{n=1}^{N}\left(\boldsymbol{B}^{*}\boldsymbol{A}^{*\top}\boldsymbol{x}^{(n)} + \boldsymbol{\varepsilon}^{(n)}\right)\boldsymbol{x}^{(n)\top}$$
$$= \boldsymbol{B}^{*}\boldsymbol{A}^{*\top} + O_{\mathrm{p}}(N^{-1/2}), \tag{14.43}$$

$$\left\langle \boldsymbol{x}\boldsymbol{x}^{\top}\right\rangle_{q(\boldsymbol{x})} = \frac{1}{N}\sum_{n=1}^{N}\boldsymbol{x}^{(n)}\boldsymbol{x}^{(n)\top} + O_{\mathrm{p}}(N^{-1/2})$$
$$= \boldsymbol{I}_M + O_{\mathrm{p}}(N^{-1/2}). \tag{14.44}$$

Here we used the assumption (14.7) that the input is prewhitened. Eq. (14.43) is consistent with Eq. (14.9), which implies that the distribution of $\boldsymbol{V}$ is given by

$$q(\boldsymbol{V}) = \mathrm{MGauss}_{L,M}\left(\boldsymbol{V}; \boldsymbol{B}^{*}\boldsymbol{A}^{*\top}, \frac{\sigma'^2}{N}\boldsymbol{I}_L \otimes \boldsymbol{I}_M\right), \tag{14.45}$$

and therefore

$$\langle \boldsymbol{V}\rangle_{q(\boldsymbol{X},\boldsymbol{Y})} = \langle \boldsymbol{V}\rangle_{q(\boldsymbol{V})} = \boldsymbol{B}^{*}\boldsymbol{A}^{*\top}, \tag{14.46}$$

and for each $(l, m)$,

$$\left\langle \left\| V_{l,m} - \left(\boldsymbol{B}^{*}\boldsymbol{A}^{*\top}\right)_{l,m}\right\|_{\mathrm{Fro}}^{2}\right\rangle_{q(\boldsymbol{X},\boldsymbol{Y})} = \frac{\sigma'^2}{N}. \tag{14.47}$$

Eq. (14.43) implies that

$$\gamma_h = \gamma_h^{*} + O_{\mathrm{p}}(N^{-1/2}), \tag{14.48}$$

where $\gamma_h$ is the $h$th largest singular value of $\boldsymbol{V}$ (see Eq. (14.15)). Eq. (14.45) also implies that, for any $h$,

$$\sum_{h':\gamma_{h'}^{*}=\gamma_h^{*}} \left\langle \gamma_{h'}\boldsymbol{\omega}_{b_{h'}}\boldsymbol{\omega}_{a_{h'}}^{\top}\right\rangle_{q(\boldsymbol{X},\boldsymbol{Y})} = \sum_{h':\gamma_{h'}^{*}=\gamma_h^{*}} \gamma_{h'}^{*}\boldsymbol{\omega}_{b_{h'}}^{*}\boldsymbol{\omega}_{a_{h'}}^{*\top}, \tag{14.49}$$

$$\sum_{h':\gamma_{h'}^{*}=\gamma_h^{*}} \gamma_{h'}\boldsymbol{\omega}_{b_{h'}}\boldsymbol{\omega}_{a_{h'}}^{\top} = \sum_{h':\gamma_{h'}^{*}=\gamma_h^{*}} \gamma_{h'}^{*}\boldsymbol{\omega}_{b_{h'}}^{*}\boldsymbol{\omega}_{a_{h'}}^{*\top} + O_{\mathrm{p}}(N^{-1/2}), \tag{14.50}$$

where $\sum_{h':\gamma_{h'}^{*}=\gamma_h^{*}}$ denotes the sum over all $h'$ such that $\gamma_{h'}^{*} = \gamma_h^{*}$. Eq. (14.50) implies that for any nonzero and nondegenerate singular component $h$ (i.e., $\gamma_h^{*} > 0$ and $\gamma_h^{*} \neq \gamma_{h'}^{*} \forall h' \neq h$), it holds that

$$\omega_{a_h} = \omega^*_{a_h} + O_p(N^{-1/2}),$$
$$\omega_{b_h} = \omega^*_{b_h} + O_p(N^{-1/2}).$$

Eq. (14.9) implies that the ML estimator is given by

$$\left(\widehat{BA^\top}\right)^{\text{ML}} = \sum_{h=1}^{H} \gamma_h \omega_{b_h} \omega_{a_h}^\top. \tag{14.51}$$

Therefore, Eq. (14.43) guarantees the convergence of the ML estimator to the true linear mapping $B^* A^{*\top}$ when $H \geq H^*$.

**Lemma 14.6** *(Consistency of ML estimator in RRR) It holds that*

$$\left(\widehat{BA^\top}\right)^{\text{ML}} - B^* A^{*\top} = \begin{cases} \Theta(1) & \textit{if} \quad H < H^*, \\ O_p(N^{-1/2}) & \textit{if} \quad H \geq H^*. \end{cases}$$

We can also show the convergence of the VB estimator:

**Lemma 14.7** *(Consistency of VB estimator in RRR) It holds that*

$$\widehat{BA^\top} - B^* A^{*\top} = \begin{cases} \Theta(1) & \textit{if} \quad H < H^*, \\ O_p(N^{-1/2}) & \textit{if} \quad H \geq H^*. \end{cases}$$

*Proof* The case where $H < H^*$ is trivial because the rank $H$ matrix $\widehat{BA^\top}$ can never converge to the rank $H^*$ matrix $B^* A^{*\top}$. Assume that $H \geq H^*$. Theorem 14.1 implies that, when $\gamma_h > \underline{\gamma}_h^{\text{VB}} (= \Theta(N^{-1/2}))$,

$$\widehat{\gamma}_h^{\text{VB}} = \breve{\gamma}_h^{\text{VB}} = \gamma_h \left( 1 - \frac{\max(L, M)\sigma'^2}{N\gamma_h^2} \right) + O_p(N^{-1}) = \gamma_h + O_p(N^{-1/2}),$$

and otherwise

$$\widehat{\gamma}_h^{\text{VB}} = 0.$$

Since $\gamma_h = O_p(N^{-1/2})$ for $h = H^* + 1, \ldots, \min(L, M)$, the preceding two equations lead to

$$\widehat{BA^\top} = \sum_{h=1}^{H} \widehat{\gamma}_h^{\text{VB}} \omega_{b_h} \omega_{a_h}^\top = \sum_{h=1}^{\min(L,M)} \gamma_h \omega_{b_h} \omega_{a_h}^\top + O_p(N^{-1/2}) = V + O_p(N^{-1/2}). \tag{14.52}$$

Substituting Eq. (14.43) into Eq. (14.52) completes the proof. □

### 14.2.3 Generalization Error

Now we analyze the asymptotic behavior of the generalization error. We first show the asymptotic equivalence between the VB predictive distribution, given by Theorem 14.5, and the plug-in VB predictive distribution (14.33)—both give the same leading term of the generalization error with $O_p(N^{-3/2})$ difference. To this end, we use the following lemma:

**Lemma 14.8** *For any three sets of Gaussian parameters* $(\mu^*, \Sigma^*)$, $(\widehat{\mu}, \widehat{\Sigma})$, $(\acute{\mu}, \acute{\Sigma})$ *such that*

$$\widehat{\mu} = \mu^* + O_p(N^{-1/2}), \qquad \widehat{\Sigma} = \Sigma^* + O_p(N^{-1/2}), \tag{14.53}$$

$$\acute{\mu} = \widehat{\mu} + O_p(N^{-1}), \qquad \acute{\Sigma} = \widehat{\Sigma} + O_p(N^{-1}), \tag{14.54}$$

*it holds that*

$$\left\langle \log \frac{\mathrm{Gauss}_L\left(y; \acute{\mu}, \acute{\Sigma}\right) + O_p(N^{-3/2})}{\mathrm{Gauss}_L\left(y; \widehat{\mu}, \widehat{\Sigma}\right)} \right\rangle_{\mathrm{Gauss}_L(y;\mu^*,\Sigma^*)} = O_p(N^{-3/2}). \tag{14.55}$$

*Proof* The (twice of the) left-hand side of Eq. (14.55) can be written as

$$\begin{aligned}
\psi_1 &\equiv 2\left\langle \log \frac{\mathrm{Gauss}_L\left(y; \acute{\mu}, \acute{\Sigma}\right) + O_p(N^{-3/2})}{\mathrm{Gauss}_L\left(y; \widehat{\mu}, \widehat{\Sigma}\right)} \right\rangle_{\mathrm{Gauss}_L(y;\mu^*,\Sigma^*)} \\
&= \left\langle \log \frac{\det(\widehat{\Sigma})}{\det(\acute{\Sigma})} - (y - \acute{\mu})^\top \acute{\Sigma}^{-1} (y - \acute{\mu}) + (y - \widehat{\mu})^\top \widehat{\Sigma}^{-1} (y - \widehat{\mu}) \right\rangle_{\mathrm{Gauss}_L(y;\mu^*,\Sigma^*)} \\
&\quad + O_p(N^{-3/2}) \\
&= -\log\det\left(\Sigma^* \widehat{\Sigma}^{-1} \Sigma^{*-1} \acute{\Sigma}\right) \\
&\quad - \left\langle (y - \mu^* - (\acute{\mu} - \mu^*))^\top \acute{\Sigma}^{-1} (y - \mu^* - (\acute{\mu} - \mu^*)) \right\rangle_{\mathrm{Gauss}_L(y;\mu^*,\Sigma^*)} \\
&\quad + \left\langle (y - \mu^* - (\widehat{\mu} - \mu^*))^\top \widehat{\Sigma}^{-1} (y - \mu^* - (\widehat{\mu} - \mu^*)) \right\rangle_{\mathrm{Gauss}_L(y;\mu^*,\Sigma^*)} + O_p(N^{-3/2}) \\
&= \mathrm{tr}\left(\log\left(\Sigma^* \acute{\Sigma}^{-1}\right) - \log\left(\Sigma^* \widehat{\Sigma}^{-1}\right)\right) - \mathrm{tr}\left(\Sigma^* \acute{\Sigma}^{-1}\right) - (\acute{\mu} - \mu^*)^\top \acute{\Sigma}^{-1} (\acute{\mu} - \mu^*) \\
&\quad + \mathrm{tr}\left(\Sigma^* \widehat{\Sigma}^{-1}\right) + (\widehat{\mu} - \mu^*)^\top \widehat{\Sigma}^{-1} (\widehat{\mu} - \mu^*) + O_p(N^{-3/2}).
\end{aligned}$$

By using Eqs. (14.53) and (14.54) and the Taylor expansion of the logarithmic function, we have

$$\begin{aligned}
\psi_1 &= \mathrm{tr}\Bigg(\left(\Sigma^* \acute{\Sigma}^{-1} - I_L\right) - \frac{\left(\Sigma^* \acute{\Sigma}^{-1} - I_L\right)^\top \left(\Sigma^* \acute{\Sigma}^{-1} - I_L\right)}{2} \\
&\quad - \left(\Sigma^* \widehat{\Sigma}^{-1} - I_L\right) + \frac{\left(\Sigma^* \widehat{\Sigma}^{-1} - I_L\right)^\top \left(\Sigma^* \widehat{\Sigma}^{-1} - I_L\right)}{2}\Bigg) \\
&\quad - \mathrm{tr}\left(\Sigma^* \acute{\Sigma}^{-1}\right) - (\widehat{\mu} - \mu^*)^\top \widehat{\Sigma}^{-1} (\widehat{\mu} - \mu^*)
\end{aligned}$$

$$+ \operatorname{tr}\left(\boldsymbol{\Sigma}^* \widehat{\boldsymbol{\Sigma}}^{-1}\right) + (\widehat{\boldsymbol{\mu}} - \boldsymbol{\mu}^*)^\top \widehat{\boldsymbol{\Sigma}}^{-1} (\widehat{\boldsymbol{\mu}} - \boldsymbol{\mu}^*) + O_{\mathrm{p}}(N^{-3/2})$$
$$= O_{\mathrm{p}}(N^{-3/2}),$$

which completes the proof. □

Given a test input $\boldsymbol{x}$, Lemma 14.8 can be applied to the true distribution (14.40), the plug-in VB predictive distribution (14.33), and the predictive distribution (14.34) when $H \geq H^*$, where

$$\begin{aligned}
\boldsymbol{\mu}^* &= \boldsymbol{B}^* \boldsymbol{A}^{*\top} \boldsymbol{x}, & \boldsymbol{\Sigma}^* &= \sigma'^2 \boldsymbol{I}_L, \\
\widehat{\boldsymbol{\mu}} &= \widehat{\boldsymbol{B}\boldsymbol{A}}^\top \boldsymbol{x} = \boldsymbol{\mu}^* + O_{\mathrm{p}}(N^{-1/2}), & \widehat{\boldsymbol{\Sigma}} &= \sigma'^2 \boldsymbol{I}_L = \boldsymbol{\Sigma}^*, \\
\acute{\boldsymbol{\mu}} &= \boldsymbol{\Psi} \widehat{\boldsymbol{B}\boldsymbol{A}}^\top \boldsymbol{x} = \widehat{\boldsymbol{\mu}} + O_{\mathrm{p}}(N^{-1}), & \acute{\boldsymbol{\Sigma}} &= \sigma'^2 \boldsymbol{\Psi} = \widehat{\boldsymbol{\Sigma}} + O_{\mathrm{p}}(N^{-1}),
\end{aligned}$$

for $\boldsymbol{\Psi} = \boldsymbol{I}_L + O_{\mathrm{p}}(N^{-1})$. Here, Lemma 14.7 was used in the equation for $\widehat{\boldsymbol{\mu}}$. Thus, we have the following corollary:

**Corollary 14.9** *When $H \geq H^*$, it holds that*

$$\left\langle \log \frac{p(\boldsymbol{y}|\boldsymbol{x}, \boldsymbol{X}, \boldsymbol{Y})}{p(\boldsymbol{y}|\boldsymbol{x}, \widehat{\boldsymbol{A}}, \widehat{\boldsymbol{B}})} \right\rangle_{q(\boldsymbol{y}|\boldsymbol{x})} = O_{\mathrm{p}}(N^{-3/2}),$$

*and therefore the difference between the generalization error* (13.30) *of the VB predictive distribution* (14.34) *and the generalization error of the plug-in VB predictive distribution* (14.33) *is of the order of $N^{-3/2}$, i.e.,*

$$\begin{aligned}
\mathrm{GE}(\mathcal{D}) &= \left\langle \log \frac{q(\boldsymbol{y}|\boldsymbol{x})}{p(\boldsymbol{y}|\boldsymbol{x}, \boldsymbol{X}, \boldsymbol{Y})} \right\rangle_{q(\boldsymbol{y}|\boldsymbol{x})q(\boldsymbol{x})} \\
&= \left\langle \log \frac{q(\boldsymbol{y}|\boldsymbol{x})}{p(\boldsymbol{y}|\boldsymbol{x}, \widehat{\boldsymbol{A}}, \widehat{\boldsymbol{B}})} \right\rangle_{q(\boldsymbol{y}|\boldsymbol{x})q(\boldsymbol{x})} + O_{\mathrm{p}}(N^{-3/2}).
\end{aligned}$$

Corollary 14.9 leads to the following theorem:

**Theorem 14.10** *The generalization error of the RRR model is written as*

$$\mathrm{GE}(\mathcal{D}) = \begin{cases} \Theta(1) & \textit{if} \quad H < H^*, \\ \dfrac{\left\| \widehat{\boldsymbol{B}\boldsymbol{A}}^\top - \boldsymbol{B}^* \boldsymbol{A}^{*\top} \right\|_{\mathrm{Fro}}^2}{2\sigma'^2} + O_{\mathrm{p}}(N^{-3/2}) & \textit{if} \quad H \geq H^*. \end{cases} \tag{14.56}$$

*Proof* When $H < H^*$, Theorem 14.5 implies that

$$\begin{aligned}
\mathrm{GE}(\mathcal{D}) &= \left\langle \log \frac{q(\boldsymbol{y}|\boldsymbol{x})}{p(\boldsymbol{y}|\boldsymbol{x}, \widehat{\boldsymbol{A}}, \widehat{\boldsymbol{B}})} \right\rangle_{q(\boldsymbol{y}|\boldsymbol{x})q(\boldsymbol{x})} + O_{\mathrm{p}}(N^{-1}) \\
&= \frac{\left\| \widehat{\boldsymbol{B}\boldsymbol{A}}^\top - \boldsymbol{B}^* \boldsymbol{A}^{*\top} \right\|_{\mathrm{Fro}}^2}{2\sigma'^2} + O_{\mathrm{p}}(N^{-1}).
\end{aligned}$$

With Lemma 14.7, we have $\mathrm{GE}(\mathcal{D}) = \Theta(1)$. When $H \geq H^*$, we have

$$\mathrm{GE}(\mathcal{D}) = \left\langle \log \frac{q(y|x)}{p(y|x, \widehat{A}, \widehat{B})} \right\rangle_{q(y|x)q(x)} + O_p(N^{-3/2})$$

$$= \left\langle - \frac{\|y - B^* A^{*\top} x\|^2 - \|y - \widehat{B}\widehat{A}^\top x\|^2}{2\sigma'^2} \right\rangle_{q(y|x)q(x)} + O_p(N^{-3/2})$$

$$= \left\langle - \frac{\|y - B^* A^{*\top} x\|^2 - \left\|y - B^* A^{*\top} x - \left(\widehat{B}\widehat{A}^\top - B^* A^{*\top}\right) x\right\|^2}{2\sigma'^2} \right\rangle_{q(y|x)q(x)} + O_p(N^{-3/2})$$

$$= \left\langle \frac{\left\|\left(\widehat{B}\widehat{A}^\top - B^* A^{*\top}\right) x\right\|^2}{2\sigma'^2} \right\rangle_{q(x)} + O_p(N^{-3/2})$$

$$= \left\langle \frac{\mathrm{tr}\left\{\left(\widehat{B}\widehat{A}^\top - B^* A^{*\top}\right) x x^\top \left(\widehat{B}\widehat{A}^\top - B^* A^{*\top}\right)^\top\right\}}{2\sigma'^2} \right\rangle_{q(x)} + O_p(N^{-3/2}).$$

By using Eq. (14.44) and Lemma 14.7, we obtain Eq. (14.56), which completes the proof. □

Next we compute the average generalization error (13.13) over the distribution of training samples. As Theorem 14.10 states, the generalization error never converges to zero if $H < H^*$, since a rank $H^*$ matrix cannot be well approximated by a rank $H$ matrix. Accordingly, we hereafter focus on the case where $H \geq H^*$. By $\mathrm{Wishart}_D(V, \nu)$ we denote the $D$-dimensional Wishart distribution with scale matrix $V$ and degree of freedom $\nu$. Then we have the following theorem:

**Theorem 14.11** *The average generalization error of the RRR model for $H \geq H^*$ is asymptotically expanded as*

$$\overline{\mathrm{GE}}(N) = \langle \mathrm{GE}(\mathcal{D}) \rangle_{q(\mathcal{D})} = \lambda^{\mathrm{VB}} N^{-1} + O(N^{-3/2}),$$

*where the generalization coefficient is given by*

$$2\lambda^{\mathrm{VB}} = (H^*(L+M) - H^{*2}) + \left\langle \sum_{h=1}^{H-H^*} \theta\left(\gamma_h'^2 > \max(L, M)\right)\left(1 - \frac{\max(L,M)}{\gamma_h'^2}\right)^2 \gamma_h'^2 \right\rangle_{q(W)}. \tag{14.57}$$

*Here $\gamma_h'^2$ is the hth largest eigenvalue of a random matrix $W \in \mathbb{S}_+^{\min(L,M)}$ subject to $\mathrm{Wishart}_{\min(L,M)-H^*}(I_{\min(L,M)-H^*}, \max(L, M) - H^*)$, and $\theta(\cdot)$ is the indicator function such that $\theta(\text{condition}) = 1$ if the condition is true and $\theta(\text{condition}) = 0$ otherwise.*

*Proof* Theorem 14.1 and Eqs. (14.42) and (14.48) imply that

$$\widehat{\gamma}_h^{\mathrm{VB}} = \begin{cases} \gamma_h + O_{\mathrm{p}}(N^{-1}) = O_{\mathrm{p}}(1) & \text{for } h = 1, \ldots, H^*, \\ \max\left(0, \gamma_h \left(1 - \frac{\max(L,M)\sigma'^2}{N\gamma_h^2}\right)\right) + O_{\mathrm{p}}(N^{-1}) = O_{\mathrm{p}}(N^{-1/2}) & \\ & \text{for } h = H^* + 1, \ldots, H. \end{cases} \tag{14.58}$$

Therefore, we have

$$\begin{aligned} \left\| \widehat{B}\widehat{A}^\top - B^* A^{*\top} \right\|_{\mathrm{Fro}}^2 &= \left\| \sum_{h=1}^{H} \widehat{\gamma}_h^{\mathrm{VB}} \omega_{b_h} \omega_{a_h}^\top - \sum_{h=1}^{\min(L,M)} \gamma_h^* \omega_{b_h}^* \omega_{a_h}^{*\top} \right\|_{\mathrm{Fro}}^2 \\ &= \left\| \sum_{h=1}^{H^*} \left( \gamma_h \omega_{b_h} \omega_{a_h}^\top - \gamma_h^* \omega_{b_h}^* \omega_{a_h}^{*\top} \right) + \sum_{h=H^*+1}^{H} \widehat{\gamma}_h^{\mathrm{VB}} \omega_{b_h} \omega_{a_h}^\top + O_{\mathrm{p}}(N^{-1}) \right\|_{\mathrm{Fro}}^2 \\ &= \left\| \sum_{h=1}^{H^*} \left( \gamma_h \omega_{b_h} \omega_{a_h}^\top - \gamma_h^* \omega_{b_h}^* \omega_{a_h}^{*\top} \right) + \sum_{h=H^*+1}^{H} \widehat{\gamma}_h^{\mathrm{VB}} \omega_{b_h} \omega_{a_h}^\top \right\|_{\mathrm{Fro}}^2 + O_{\mathrm{p}}(N^{-3/2}) \\ &= \left\| V - B^* A^* + \sum_{h=H^*+1}^{H} \widehat{\gamma}_h^{\mathrm{VB}} \omega_{b_h} \omega_{a_h}^\top - \sum_{h=H^*+1}^{\min(L,M)} \gamma_h \omega_{b_h} \omega_{a_h}^\top \right\|_{\mathrm{Fro}}^2 \\ &\qquad + O_{\mathrm{p}}(N^{-3/2}). \end{aligned}$$

Here, in order to get the third equation, we used the fact that the first two terms in the norm in the second equation are of the order of $O_{\mathrm{p}}(N^{-1/2})$. The expectation over the distribution of training samples is given by

$$\begin{aligned} &\left\langle \left\| \widehat{B}\widehat{A}^\top - B^* A^{*\top} \right\|_{\mathrm{Fro}}^2 \right\rangle_{q(\mathcal{D})} \\ &= \left\langle \left\| V - B^* A^* + \sum_{h=H^*+1}^{H} \widehat{\gamma}_h^{\mathrm{VB}} \omega_{b_h} \omega_{a_h}^\top - \sum_{h=H^*+1}^{\min(L,M)} \gamma_h \omega_{b_h} \omega_{a_h}^\top \right\|_{\mathrm{Fro}}^2 \right\rangle_{q(\mathcal{D})} \\ &\qquad + O(N^{-3/2}) \\ &= \left\langle \| V - B^* A^* \|_{\mathrm{Fro}}^2 \right\rangle_{q(\mathcal{D})} \\ &\quad + 2 \left\langle (V - B^* A^*)^\top \left( \sum_{h=H^*+1}^{H} \widehat{\gamma}_h^{\mathrm{VB}} \omega_{b_h} \omega_{a_h}^\top - \sum_{h=H^*+1}^{\min(L,M)} \gamma_h \omega_{b_h} \omega_{a_h}^\top \right) \right\rangle_{q(\mathcal{D})} \\ &\quad + \left\langle \sum_{h=H^*+1}^{H} (\widehat{\gamma}_h^{\mathrm{VB}})^2 - 2 \sum_{h=H^*+1}^{H} \gamma_h \widehat{\gamma}_h^{\mathrm{VB}} + \sum_{h=H^*+1}^{\min(L,M)} \gamma_h^2 \right\rangle_{q(\mathcal{D})} \\ &\qquad + O(N^{-3/2}) \\ &= \left\langle \| V - B^* A^* \|_{\mathrm{Fro}}^2 \right\rangle_{q(\mathcal{D})} + 2 \left\langle \sum_{h=H^*+1}^{H} \gamma_h \widehat{\gamma}_h^{\mathrm{VB}} - \sum_{h=H^*+1}^{\min(L,M)} \gamma_h^2 \right\rangle_{q(\mathcal{D})} \\ &\quad + \left\langle \sum_{h=H^*+1}^{H} (\widehat{\gamma}_h^{\mathrm{VB}})^2 - 2 \sum_{h=H^*+1}^{H} \gamma_h \widehat{\gamma}_h^{\mathrm{VB}} + \sum_{h=H^*+1}^{\min(L,M)} \gamma_h^2 \right\rangle_{q(\mathcal{D})} + O(N^{-3/2}) \\ &= \left\langle \| V - B^* A^* \|_{\mathrm{Fro}}^2 \right\rangle_{q(\mathcal{D})} - \left\langle \sum_{h=H^*+1}^{\min(L,M)} \gamma_h^2 \right\rangle_{q(\mathcal{D})} \\ &\quad + \left\langle \sum_{h=H^*+1}^{H} (\widehat{\gamma}_h^{\mathrm{VB}})^2 \right\rangle_{q(\mathcal{D})} + O(N^{-3/2}). \end{aligned} \tag{14.59}$$

Here we used Eq. (14.49) and the orthonormality of the singular vectors.

Eq. (14.45) implies that the first term in Eq. (14.59) is equal to

$$\left\langle \|\boldsymbol{V} - \boldsymbol{B}^* \boldsymbol{A}^*\|^2_{\mathrm{Fro}} \right\rangle_{q(\mathcal{D})} = LM \frac{\sigma'^2}{N}. \tag{14.60}$$

The redundant components $\{\gamma_h \boldsymbol{\omega}_{b_h} \boldsymbol{\omega}_{a_h}^\top\}_{h=H^*+1}^{\min(L,M)}$ are zero-mean (see Eq. (14.49)) Gaussian matrices capturing the Gaussian noise in the orthogonal space to the necessary components $\{\gamma_h \boldsymbol{\omega}_{b_h} \boldsymbol{\omega}_{a_h}^\top\}_{h=1}^{H^*}$. Therefore, the distribution of the corresponding singular values $\{\gamma_h\}_{h=H^*+1}^{\min(L,M)}$ coincides with the distribution of the singular values of $\boldsymbol{V}' \in \mathbb{R}^{(\min(L,M)-H^*)\times(\max(L,M)-H^*)}$ subject to

$$q(\boldsymbol{V}') = \mathrm{MGauss}_{\min(L,M)-H^*,\max(L,M)-H^*} \left( \boldsymbol{V}'; \boldsymbol{0}_{\min(L,M)-H^*,\max(L,M)-H^*}, \frac{\sigma'^2}{N} \boldsymbol{I}_{\min(L,M)-H^*} \otimes \boldsymbol{I}_{\max(L,M)-H^*} \right). \tag{14.61}$$

This leads to

$$\left\langle \sum_{h=H^*+1}^{\min(L,M)} \gamma_h^2 \right\rangle_{q(\mathcal{D})} = (L - H^*)(M - H^*) \frac{\sigma'^2}{N}. \tag{14.62}$$

Let $\{\gamma'_h\}_{h=1}^{\min(L,M)-H^*}$ be the singular values of $\frac{\sqrt{N}}{\sigma'} \boldsymbol{V}'$. Then, $\{\gamma'^2_h\}_{h=1}^{\min(L,M)-H^*}$ are the eigenvalues of $\boldsymbol{W} = \frac{N}{\sigma'^2} \boldsymbol{V}' \boldsymbol{V}'^\top$, which is subject to $\mathrm{Wishart}_{\min(L,M)-H^*}$ $(\boldsymbol{I}_{\min(L,M)-H^*}, \max(L,M) - H^*)$. By substituting Eqs. (14.60), (14.62), and (14.58) into Eq. (14.59), we have

$$\begin{aligned}
&\left\langle \left\| \widehat{\boldsymbol{B}\boldsymbol{A}^\top} - \boldsymbol{B}^* \boldsymbol{A}^{*\top} \right\|^2_{\mathrm{Fro}} \right\rangle_{q(\mathcal{D})} \\
&= \frac{\sigma'^2}{N} \{LM - (L - H^*)(M - H^*)\} \\
&\quad + \left\langle \textstyle\sum_{h=H^*+1}^{H} \left\{ \max\left(0, \gamma_h \left(1 - \frac{\max(L,M)\sigma'^2}{N\gamma_h^2}\right)\right)\right\}^2 \right\rangle_{q(\mathcal{D})} + O(N^{-3/2}) \\
&= \frac{\sigma'^2}{N} \left\{ (H^*(L+M) - H^{*2}) + \left\langle \textstyle\sum_{h=1}^{H-H^*} \left\{ \max\left(0, 1 - \frac{\max(L,M)}{\gamma'^2_h}\right)\right\}^2 \gamma'^2_h \right\rangle_{q(\boldsymbol{W})} \right\} \\
&\quad + O(N^{-3/2}).
\end{aligned}$$

Substituting the preceding into Eq. (14.56) completes the proof. □

The first and the second terms in Eq. (14.57) correspond to the contribution from the necessary components $h = 1, \ldots, H^*$ and the contribution from the redundant components $h = H^* + 1, \ldots, H$, respectively. If we focus on the parameter space of the first $H^*$ components, i.e., $\{\boldsymbol{a}_h, \boldsymbol{b}_h\}_{h=1}^{H^*}$, the true linear mapping $\{\boldsymbol{a}^*_h, \boldsymbol{b}^*_h\}_{h=1}^{H^*}$ lies at an (essentially) nonsingular point (after removing the trivial $H^{*2}$ redundancy). Therefore, as the regular learning theory states, the contribution from the necessary components is equal to the (essential) degree of freedom (see Eq. (14.4)) of the RRR model for $H = H^*$. On the other hand, the regular learning theory cannot be applied to the redundant components $\{\boldsymbol{a}_h, \boldsymbol{b}_h\}_{h=H^*+1}^{H}$ since the true parameter is on the *singularities* $\{\boldsymbol{a}_h = \boldsymbol{0}\} \cup \{\boldsymbol{b}_h = \boldsymbol{0}\}$, making the second term different from the degree of freedom of the redundant parameters.

Assuming that $L$ and $M$ are large, we can approximate the second term in Eq. (14.57) by using the *random matrix theory* (see Section 8.4.1). Consider the large-scale limit when $L, M, H, H^*$ go to infinity with the same ratio, so that

$$\alpha = \frac{\min(L, M) - H^*}{\max(L, M) - H^*}, \tag{14.63}$$

$$\beta = \frac{H - H^*}{\min(L, M) - H^*}, \tag{14.64}$$

$$\kappa = \frac{\max(L, M)}{\max(L, M) - H^*} \tag{14.65}$$

are constant. Then Marčenko–Pastur law (Proposition 8.11) states that the empirical distribution of the eigenvalues $\{y_1, \ldots, y_{\min(L,M)-H^*}\}$ of the random matrix $\frac{NV'V'^{\top}}{(\max(L,M)-H^*)\sigma'^2} \sim \mathrm{Wishart}_{\min(L,M)-H^*}(\boldsymbol{I}_{\min(L,M)-H^*}, 1)$ almost surely converges to

$$p(y) \to p^{\mathrm{MP}}(y) \equiv \frac{\sqrt{(y-\underline{y})(\overline{y}-y)}}{2\pi\alpha y}\theta\left(\underline{y} < y < \overline{y}\right), \tag{14.66}$$

$$\text{where} \qquad \overline{y} = (1 + \sqrt{\alpha})^2, \qquad \underline{y} = (1 - \sqrt{\alpha})^2. \tag{14.67}$$

Let

$$(2\pi\alpha)^{-1} J'_s(u) = \int_u^{\infty} y^s p(y) dy \tag{14.68}$$

be the $s$th order (incomplete) moment of the Marčenko–Pastur distribution (14.66) with the lower bound $u$ of the integration range. Then, the second term of Eq. (14.57) can be written as

$$\begin{aligned}
&\left\langle \sum_{h=1}^{H-H^*} \theta(\gamma_h'^2 > \max(L, M)) \left(1 - \tfrac{\max(L,M)}{\gamma_h'^2}\right)^2 \gamma_h'^2 \right\rangle_{q(\boldsymbol{W})} \\
&\quad \to (\min(L, M) - H^*)(\max(L, M) - H^*) \int_{u_\beta}^{\infty} \theta(y > \kappa) \left(1 - \tfrac{\kappa}{y}\right)^2 y p(y) dy \\
&\quad = (\min(L, M) - H^*)(\max(L, M) - H^*) \int_{\max(\kappa, u_\beta)}^{\infty} \left(y - 2\kappa + \kappa^2 y^{-1}\right) p(y) dy \\
&\quad = \frac{(\min(L, M) - H^*)(\max(L, M) - H^*)}{2\pi\alpha} \left(J'_1(\acute{u}) - 2\kappa J'_0(\acute{u}) + \kappa^2 J'_{-1}(\acute{u})\right),
\end{aligned}$$

where $u_\beta$ is the $\beta$-percentile point of $p(y)$, i.e.,

$$\beta = \int_{u_\beta}^{\infty} p(y)dy = (2\pi\alpha)^{-1} J_0'(u_\beta), \tag{14.69}$$

and

$$\acute{u} = \max(\kappa, u_\beta). \tag{14.70}$$

Using the transformation $z = \left(y - (\underline{y} + \overline{y})/2\right)/(2\sqrt{\alpha})$, we can derive analytic forms of the moments (14.68) and thus obtain the following theorem:

**Theorem 14.12** *The VB generalization coefficient of the RRR model in the large-scale limit is given by*

$$\begin{aligned} 2\lambda^{\mathrm{VB}} \to\ & (H^*(L+M) - H^{*2}) \\ & + \frac{(\min(L,M) - H^*)(\max(L,M) - H^*)}{2\pi\alpha} \left\{ J_1(\acute{z}) - 2\kappa J_0(\acute{z}) + \kappa^2 J_{-1}(\acute{z}) \right\}, \end{aligned} \tag{14.71}$$

*where*

$$J_1(z) = 2\alpha(-z\sqrt{1-z^2} + \cos^{-1} z),$$

$$J_0(z) = -2\sqrt{\alpha}\sqrt{1-z^2} + (1+\alpha)\cos^{-1} z - (1-\alpha)\cos^{-1}\frac{\sqrt{\alpha}(1+\alpha)z + 2\alpha}{2\alpha z + \sqrt{\alpha}(1+\alpha)},$$

$$J_{-1}(z) = \begin{cases} 2\sqrt{\alpha}\frac{\sqrt{1-z^2}}{2\sqrt{\alpha}z+1+\alpha} - \cos^{-1} z + \frac{1+\alpha}{1-\alpha}\cos^{-1}\frac{\sqrt{\alpha}(1+\alpha)z+2\alpha}{2\alpha z+\sqrt{\alpha}(1+\alpha)} & (0 < \alpha < 1), \\ 2\sqrt{\frac{1-z}{1+z}} - \cos^{-1} z & (\alpha = 1), \end{cases}$$

*and* $\acute{z} = \max\left((\kappa - (1+\alpha))/2\sqrt{\alpha},\, J_0^{-1}(2\pi\alpha\beta)\right)$. *Here* $J_s^{-1}(\cdot)$ *denotes the inverse function of* $J_s(z)$.

Theorem 14.12 allows us to compare the generalization error of VB learning with those of ML (MAP) learning and Bayesian learning in Section 14.2.6.

### 14.2.4 Training Error

The training error can be analyzed in a similar way to the generalization error. We first prove the following lemma:

**Lemma 14.13** *Let* $\overline{\boldsymbol{U}} \in \mathbb{R}^{L,M}$ *and* $\overline{\boldsymbol{\Sigma}} \in \mathbb{S}_{+}^{L}$ *be the ML estimators of the linear regression model* $\boldsymbol{y} = \boldsymbol{U}\boldsymbol{x} + \boldsymbol{\varepsilon}$ *with Gaussian noise* $\boldsymbol{\varepsilon} \sim \mathrm{Gauss}(\boldsymbol{0}, \boldsymbol{\Sigma})$. *For any two sets of parameters* $(\widehat{\boldsymbol{U}}, \widehat{\boldsymbol{\Sigma}}), (\acute{\boldsymbol{U}}, \acute{\boldsymbol{\Sigma}})$ *such that*

$$\widehat{\boldsymbol{U}} = \overline{\boldsymbol{U}} + O_{\mathrm{p}}(N^{-1/2}), \qquad \widehat{\boldsymbol{\Sigma}} = \overline{\boldsymbol{\Sigma}} + O_{\mathrm{p}}(N^{-1/2}), \tag{14.72}$$

$$\acute{\boldsymbol{U}} = \widehat{\boldsymbol{U}} + O_{\mathrm{p}}(N^{-1}), \qquad \acute{\boldsymbol{\Sigma}} = \widehat{\boldsymbol{\Sigma}} + O_{\mathrm{p}}(N^{-1}), \tag{14.73}$$

*it holds that*

$$\frac{1}{N}\sum_{n=1}^{N}\log\frac{\mathrm{Gauss}_L\left(\boldsymbol{y}^{(n)};\acute{\boldsymbol{U}}\boldsymbol{x}^{(n)},\acute{\boldsymbol{\Sigma}}\right)+O_{\mathrm{p}}(N^{-3/2})}{\mathrm{Gauss}_L\left(\boldsymbol{y}^{(n)};\widehat{\boldsymbol{U}}\boldsymbol{x}^{(n)},\widehat{\boldsymbol{\Sigma}}\right)}=O_{\mathrm{p}}(N^{-3/2}). \tag{14.74}$$

*Proof* The (twice of the) left-hand side of Eq. (14.74) can be written as

$$\begin{aligned}
\psi_2 &\equiv \frac{2}{N}\sum_{n=1}^{N}\log\frac{\mathrm{Gauss}_L\left(\boldsymbol{y}^{(n)};\acute{\boldsymbol{U}}\boldsymbol{x}^{(n)},\acute{\boldsymbol{\Sigma}}\right)+O_{\mathrm{p}}(N^{-3/2})}{\mathrm{Gauss}_L\left(\boldsymbol{y}^{(n)};\widehat{\boldsymbol{U}}\boldsymbol{x}^{(n)},\widehat{\boldsymbol{\Sigma}}\right)}\\
&=\log\frac{\det\left(\widehat{\boldsymbol{\Sigma}}\right)}{\det\left(\acute{\boldsymbol{\Sigma}}\right)}+\frac{1}{N}\sum_{n=1}^{N}\Big(-(\boldsymbol{y}^{(n)}-\acute{\boldsymbol{U}}\boldsymbol{x}^{(n)})^{\top}\acute{\boldsymbol{\Sigma}}^{-1}(\boldsymbol{y}^{(n)}-\acute{\boldsymbol{U}}\boldsymbol{x}^{(n)})\\
&\quad+(\boldsymbol{y}^{(n)}-\widehat{\boldsymbol{U}}\boldsymbol{x}^{(n)})^{\top}\widehat{\boldsymbol{\Sigma}}^{-1}(\boldsymbol{y}^{(n)}-\widehat{\boldsymbol{U}}\boldsymbol{x}^{(n)})\Big)+O_{\mathrm{p}}(N^{-3/2})\\
&=-\log\det\left(\overline{\boldsymbol{\Sigma}}\widehat{\boldsymbol{\Sigma}}^{-1}\overline{\boldsymbol{\Sigma}}^{-1}\acute{\boldsymbol{\Sigma}}\right)+O_{\mathrm{p}}(N^{-3/2})\\
&\quad-\frac{1}{N}\sum_{n=1}^{N}\left(\boldsymbol{y}^{(n)}-\overline{\boldsymbol{U}}\boldsymbol{x}^{(n)}-(\acute{\boldsymbol{U}}-\overline{\boldsymbol{U}})\boldsymbol{x}^{(n)}\right)^{\top}\acute{\boldsymbol{\Sigma}}^{-1}\left(\boldsymbol{y}^{(n)}-\overline{\boldsymbol{U}}\boldsymbol{x}^{(n)}-(\acute{\boldsymbol{U}}-\overline{\boldsymbol{U}})\boldsymbol{x}^{(n)}\right)\\
&\quad+\frac{1}{N}\sum_{n=1}^{N}\left(\boldsymbol{y}^{(n)}-\overline{\boldsymbol{U}}\boldsymbol{x}^{(n)}-(\widehat{\boldsymbol{U}}-\overline{\boldsymbol{U}})\boldsymbol{x}^{(n)}\right)^{\top}\widehat{\boldsymbol{\Sigma}}^{-1}\left(\boldsymbol{y}^{(n)}-\overline{\boldsymbol{U}}\boldsymbol{x}^{(n)}-(\widehat{\boldsymbol{U}}-\overline{\boldsymbol{U}})\boldsymbol{x}^{(n)}\right)\\
&=\mathrm{tr}\left(\log\left(\overline{\boldsymbol{\Sigma}}\acute{\boldsymbol{\Sigma}}^{-1}\right)-\log\left(\overline{\boldsymbol{\Sigma}}\widehat{\boldsymbol{\Sigma}}^{-1}\right)\right)+O_{\mathrm{p}}(N^{-3/2})\\
&\quad-\mathrm{tr}\left(\overline{\boldsymbol{\Sigma}}\acute{\boldsymbol{\Sigma}}^{-1}\right)-\frac{1}{N}\sum_{n=1}^{N}\left((\acute{\boldsymbol{U}}-\overline{\boldsymbol{U}})\boldsymbol{x}^{(n)}\right)^{\top}\acute{\boldsymbol{\Sigma}}^{-1}\left((\acute{\boldsymbol{U}}-\overline{\boldsymbol{U}})\boldsymbol{x}^{(n)}\right)\\
&\quad+\mathrm{tr}\left(\overline{\boldsymbol{\Sigma}}\widehat{\boldsymbol{\Sigma}}^{-1}\right)+\frac{1}{N}\sum_{n=1}^{N}\left((\widehat{\boldsymbol{U}}-\overline{\boldsymbol{U}})\boldsymbol{x}^{(n)}\right)^{\top}\widehat{\boldsymbol{\Sigma}}^{-1}\left((\widehat{\boldsymbol{U}}-\overline{\boldsymbol{U}})\boldsymbol{x}^{(n)}\right).
\end{aligned}$$

By using Eqs. (14.72) and (14.73) and the Taylor expansion of the logarithmic function, we have

$$\begin{aligned}
\psi_2 &= \mathrm{tr}\left(\left(\overline{\boldsymbol{\Sigma}}\acute{\boldsymbol{\Sigma}}^{-1}-\boldsymbol{I}_L\right)-\frac{\left(\overline{\boldsymbol{\Sigma}}\acute{\boldsymbol{\Sigma}}^{-1}-\boldsymbol{I}_L\right)^{\top}\left(\overline{\boldsymbol{\Sigma}}\acute{\boldsymbol{\Sigma}}^{-1}-\boldsymbol{I}_L\right)}{2}-\left(\overline{\boldsymbol{\Sigma}}\widehat{\boldsymbol{\Sigma}}^{-1}-\boldsymbol{I}_L\right)+\frac{\left(\overline{\boldsymbol{\Sigma}}\widehat{\boldsymbol{\Sigma}}^{-1}-\boldsymbol{I}_L\right)^{\top}\left(\overline{\boldsymbol{\Sigma}}\widehat{\boldsymbol{\Sigma}}^{-1}-\boldsymbol{I}_L\right)}{2}\right)\\
&\quad-\mathrm{tr}\left(\overline{\boldsymbol{\Sigma}}\acute{\boldsymbol{\Sigma}}^{-1}\right)-\frac{1}{N}\sum_{n=1}^{N}\left((\widehat{\boldsymbol{U}}-\overline{\boldsymbol{U}})\boldsymbol{x}^{(n)}\right)^{\top}\acute{\boldsymbol{\Sigma}}^{-1}\left((\widehat{\boldsymbol{U}}-\overline{\boldsymbol{U}})\boldsymbol{x}^{(n)}\right)\\
&\quad+\mathrm{tr}\left(\overline{\boldsymbol{\Sigma}}\widehat{\boldsymbol{\Sigma}}^{-1}\right)+\frac{1}{N}\sum_{n=1}^{N}\left((\widehat{\boldsymbol{U}}-\overline{\boldsymbol{U}})\boldsymbol{x}^{(n)}\right)^{\top}\widehat{\boldsymbol{\Sigma}}^{-1}\left((\widehat{\boldsymbol{U}}-\overline{\boldsymbol{U}})\boldsymbol{x}^{(n)}\right)+O_{\mathrm{p}}(N^{-3/2})\\
&=O_{\mathrm{p}}(N^{-3/2}),
\end{aligned}$$

which completes the proof. □

When $H \geq H^*$, Lemma 14.13 can be applied to the plug-in VB predictive distribution (14.33) and the VB predictive distribution (14.34), where

$$\begin{aligned}
\overline{U} &= V, & \overline{\Sigma} &= \frac{\sigma'^2}{N}\sum_{n=1}^{N}\left(y^{(n)} - Vx^{(n)}\right)\left(y^{(n)} - Vx^{(n)}\right)^\top \\
& & &= \sigma'^2 I_L + O_p(N^{-1/2}), \\
\widehat{U} &= \widehat{BA}^\top = \overline{U} + O_p(N^{-1/2}), & \widehat{\Sigma} &= \sigma'^2 I_L = \overline{\Sigma} + O_p(N^{-1/2}), \\
\acute{U} &= \Psi\widehat{BA}^\top = \widehat{U} + O_p(N^{-1}), & \acute{\Sigma} &= \sigma'^2 \Psi = \widehat{\Sigma} + O_p(N^{-1}),
\end{aligned}$$

for $\Psi = I_L + O_p(N^{-1})$. Here Eq. (14.43) and Lemma 14.7 were used in the equation for $\widehat{U}$. Thus, we have the following corollary:

**Corollary 14.14** *When $H \geq H^*$, it holds that*

$$\frac{1}{N}\sum_{n=1}^{N}\log\frac{p\left(y^{(n)}|x^{(n)}, X, Y\right)}{p\left(y^{(n)}|x^{(n)}, \widehat{A}, \widehat{B}\right)} = O_p(N^{-3/2}),$$

*and therefore the difference between the training error* (13.31) *of the VB predictive distribution* (14.34) *and the training error of the plug-in VB predictive distribution* (14.33) *is of the order of $N^{-3/2}$, i.e.,*

$$\begin{aligned}
\mathrm{TE}(\mathcal{D}) &= \frac{1}{N}\sum_{n=1}^{N}\log\frac{q(y^{(n)}|x^{(n)})}{p(y^{(n)}|x^{(n)}, X, Y)} \\
&= \frac{1}{N}\sum_{n=1}^{N}\log\frac{q(y^{(n)}|x^{(n)})}{p(y^{(n)}|x^{(n)}, \widehat{A}, \widehat{B})} + O_p(N^{-3/2}).
\end{aligned}$$

Corollary 14.14 leads to the following theorem:

**Theorem 14.15** *The training error of the RRR model is written as*

$$\mathrm{TE}(\mathcal{D}) = \begin{cases} \Theta(1) & \text{if} \quad H < H^*, \\ \frac{\left\|V - \widehat{BA}^\top\right\|_{\mathrm{Fro}}^2 - \left\|V - B^* A^{*\top}\right\|_{\mathrm{Fro}}^2}{2\sigma'^2} + O_p(N^{-3/2}) & \text{if} \quad H \geq H^*. \end{cases} \tag{14.75}$$

*Proof* When $H < H^*$, Theorem 14.5 implies that

$$\begin{aligned}
\mathrm{TE}(\mathcal{D}) &= \frac{1}{N}\sum_{n=1}^{N}\log\frac{q(y^{(n)}|x^{(n)})}{p(y^{(n)}|x^{(n)}, \widehat{A}, \widehat{B})} + O_p(N^{-1}) \\
&= \frac{\left\|V - \widehat{BA}^\top\right\|_{\mathrm{Fro}}^2 - \left\|V - B^* A^{*\top}\right\|_{\mathrm{Fro}}^2}{2\sigma'^2} + O_p(N^{-1}).
\end{aligned}$$

With Lemma 14.7, we have $\mathrm{TE}(\mathcal{D}) = \Theta(1)$. When $H \geq H^*$, we have

$$\begin{aligned}
\mathrm{TE}(\mathcal{D}) &= -\frac{1}{N}\sum_{n=1}^{N} \frac{\|\boldsymbol{y}^{(n)} - \boldsymbol{B}^*\boldsymbol{A}^{*\top}\boldsymbol{x}^{(n)}\|^2 - \left\|\boldsymbol{y}^{(n)} - \widehat{\boldsymbol{B}\boldsymbol{A}}^\top \boldsymbol{x}^{(n)}\right\|^2}{2\sigma'^2} + O_{\mathrm{p}}(N^{-3/2}) \\
&= -\frac{1}{N}\sum_{n=1}^{N} \frac{\|\boldsymbol{y}^{(n)} - \boldsymbol{V}\boldsymbol{x}^{(n)} - (\boldsymbol{B}^*\boldsymbol{A}^{*\top} - \boldsymbol{V})\boldsymbol{x}^{(n)}\|^2 - \left\|\boldsymbol{y}^{(n)} - \boldsymbol{V}\boldsymbol{x}^{(n)} - \left(\widehat{\boldsymbol{B}\boldsymbol{A}}^\top - \boldsymbol{V}\right)\boldsymbol{x}^{(n)}\right\|^2}{2\sigma'^2} + O_{\mathrm{p}}(N^{-3/2}) \\
&= -\frac{1}{N}\sum_{n=1}^{N} \frac{\|(\boldsymbol{B}^*\boldsymbol{A}^{*\top} - \boldsymbol{V})\boldsymbol{x}^{(n)}\|^2 - \left\|\left(\widehat{\boldsymbol{B}\boldsymbol{A}}^\top - \boldsymbol{V}\right)\boldsymbol{x}^{(n)}\right\|^2}{2\sigma'^2} + O_{\mathrm{p}}(N^{-3/2}) \\
&= -\frac{1}{N}\sum_{n=1}^{N} \frac{\mathrm{tr}\left\{(\boldsymbol{B}^*\boldsymbol{A}^{*\top} - \boldsymbol{V})\boldsymbol{x}^{(n)}\boldsymbol{x}^{(n)\top}(\boldsymbol{B}^*\boldsymbol{A}^{*\top} - \boldsymbol{V})^\top - \left(\widehat{\boldsymbol{B}\boldsymbol{A}}^\top - \boldsymbol{V}\right)\boldsymbol{x}^{(n)}\boldsymbol{x}^{(n)\top}\left(\widehat{\boldsymbol{B}\boldsymbol{A}}^\top - \boldsymbol{V}\right)^\top\right\}}{2\sigma'^2} \\
&\quad + O_{\mathrm{p}}(N^{-3/2}).
\end{aligned}$$

By using the prewhitening condition (14.7) and Lemma 14.7, we obtain Eq. (14.75), which completes the proof. □

Now we can derive an asymptotic form of the average training error:

**Theorem 14.16** *The average training error of the RRR model for $H \geq H^*$ is asymptotically expanded as*

$$\overline{\mathrm{TE}}(N) = \langle \mathrm{TE}(\mathcal{D}) \rangle_{q(\mathcal{D})} = \nu^{\mathrm{VB}} N^{-1} + O(N^{-3/2}),$$

*where the training coefficient is given by*

$$\begin{aligned}
2\nu^{\mathrm{VB}} &= -(H^*(L+M) - H^{*2}) \\
&\quad - \left\langle \sum_{h=1}^{H-H^*} \theta\left(\gamma_h'^2 > \max(L,M)\right)\left(1 - \frac{\max(L,M)}{\gamma_h'^2}\right)\left(1 + \frac{\max(L,M)}{\gamma_h'^2}\right)\gamma_h'^2 \right\rangle_{q(\boldsymbol{W})}.
\end{aligned} \tag{14.76}$$

*Here $\gamma_h'^2$ is the hth largest eigenvalue of a random matrix $\boldsymbol{W} \in \mathbb{S}_+^{\min(L,M)}$ subject to* $\mathrm{Wishart}_{\min(L,M)-H^*}(\boldsymbol{I}_{\min(L,M)-H^*}, \max(L,M) - H^*)$.

*Proof* From Eq. (14.58), we have

$$\begin{aligned}
\left\|\widehat{\boldsymbol{B}\boldsymbol{A}}^\top - \boldsymbol{V}\right\|_{\mathrm{Fro}}^2 &= \left\|\sum_{h=1}^{H} \widehat{\gamma}_h^{\mathrm{VB}}\boldsymbol{\omega}_{b_h}\boldsymbol{\omega}_{a_h}^\top - \sum_{h=1}^{\min(L,M)} \gamma_h \boldsymbol{\omega}_{b_h}\boldsymbol{\omega}_{a_h}^\top\right\|_{\mathrm{Fro}}^2 \\
&= \left\|\sum_{h=H^*+1}^{H} (\widehat{\gamma}_h^{\mathrm{VB}} - \gamma_h)\boldsymbol{\omega}_{b_h}\boldsymbol{\omega}_{a_h}^\top - \sum_{h=H+1}^{\min(L,M)} \gamma_h \boldsymbol{\omega}_{b_h}\boldsymbol{\omega}_{a_h}^\top + O_{\mathrm{p}}(N^{-1})\right\|_{\mathrm{Fro}}^2 \\
&= \left\|\sum_{h=H^*+1}^{H} (\widehat{\gamma}_h^{\mathrm{VB}} - \gamma_h)\boldsymbol{\omega}_{b_h}\boldsymbol{\omega}_{a_h}^\top - \sum_{h=H+1}^{\min(L,M)} \gamma_h \boldsymbol{\omega}_{b_h}\boldsymbol{\omega}_{a_h}^\top\right\|_{\mathrm{Fro}}^2 + O_{\mathrm{p}}(N^{-3/2}) \\
&= \sum_{h=H^*+1}^{H} (\widehat{\gamma}_h^{\mathrm{VB}} - \gamma_h)^2 + \sum_{h=H+1}^{\min(L,M)} \gamma_h^2 + O_{\mathrm{p}}(N^{-3/2}) \\
&= \sum_{h=H^*+1}^{H} \left(\max\left(0, \gamma_h\left(1 - \frac{\max(L,M)\sigma'^2}{N\gamma_h^2}\right)\right) - \gamma_h\right)^2 \\
&\quad + \sum_{h=H+1}^{\min(L,M)} \gamma_h^2 + O_{\mathrm{p}}(N^{-3/2})
\end{aligned}$$

$$= -\sum_{h=H^*+1}^{H} \theta\left(\gamma_h^2 > \frac{\max(L,M)\sigma'^2}{N}\right) \cdot \left(\gamma_h - \frac{\max(L,M)\sigma'^2}{N\gamma_h}\right)\left(\gamma_h + \frac{\max(L,M)\sigma'^2}{N\gamma_h}\right)$$
$$+ \sum_{h=H^*+1}^{\min(L,M)} \gamma_h^2 + O_p(N^{-3/2}). \qquad (14.77)$$

By using Eqs. (14.60), (14.62) and (14.77), we have

$$\left\langle \left\| V - \widehat{BA}^\top \right\|_{\text{Fro}}^2 - \left\| V - B^* A^{*\top} \right\|_{\text{Fro}}^2 \right\rangle_{q(\mathcal{D})}$$
$$= -\frac{\sigma'^2}{N}\Big\{(H^*(L+M) - H^{*2})$$
$$+ \sum_{h=H^*+1}^{H} \theta\left(\gamma_h^2 > \frac{\max(L,M)\sigma'^2}{N}\right) \cdot \left(\gamma_h - \frac{\max(L,M)\sigma'^2}{N\gamma_h}\right)\left(\gamma_h + \frac{\max(L,M)\sigma'^2}{N\gamma_h}\right)\Big\}$$
$$+ O_p(N^{-3/2}).$$

Thus, by introducing the singular values $\{\gamma'_h\}_{h=1}^{\min(L,M)-H^*}$ of $\frac{\sqrt{N}}{\sigma'} V'$, where $V'$ is a random matrix subject to Eq. (14.61), and using Theorem 14.15, we obtain Eq. (14.76), which completes the proof. □

Finally, we apply the Marčenko–Pastur law (Proposition 8.11) for evaluating the second term in Eq. (14.76). In the large-scale limit when $L, M, H, H^*$ go to infinity with the same ratio, so that Eqs. (14.63) through (14.65) are constant, we have

$$\left\langle \sum_{h=1}^{H-H^*} \theta\left(\gamma_h'^2 > \max(L,M)\right)\left(1 - \frac{\max(L,M)}{\gamma_h'^2}\right)\left(1 + \frac{\max(L,M)}{\gamma_h'^2}\right)\gamma_h'^2 \right\rangle_{q(W)}$$
$$\to (\min(L,M) - H^*)(\max(L,M) - H^*) \int_{u_\beta}^{\infty} \theta(y > \kappa)\left(1 - \tfrac{\kappa}{y}\right)\left(1 + \tfrac{\kappa}{y}\right) y p(y) dy$$
$$= (\min(L,M) - H^*)(\max(L,M) - H^*) \int_{\max(\kappa,u_\beta)}^{\infty} \left(y - \kappa^2 y^{-1}\right) p(y) dy$$
$$= \frac{(\min(L,M) - H^*)(\max(L,M) - H^*)}{2\pi\alpha} \left(J'_1(\acute{u}) - \kappa^2 J'_{-1}(\acute{u})\right),$$

where $J'_s(u)$, $\beta$, and $\acute{u}$ are defined in Eqs. (14.68), (14.69), and (14.70), respectively. Thus, the transformation $z = \left(y - (\underline{y} + \overline{y})/2\right)/(2\sqrt{\alpha})$ gives the following theorem:

**Theorem 14.17** *The VB training coefficient of the RRR model in the large scale limit is given by*

$$2\nu^{\text{VB}} \to -(H^*(L+M) - H^{*2})$$
$$- \frac{(\min(L,M) - H^*)(\max(L,M) - H^*)}{2\pi\alpha} \left\{J_1(\acute{z}) - \kappa^2 J_{-1}(\acute{z})\right\}, \qquad (14.78)$$

*where $J_1(z)$, $J_{-1}(z)$, and $\acute{z}$ are defined in Theorem 14.12.*

### 14.2.5 Free Energy

The VB free energy can be analyzed relatively easily based on the orders of the variational parameters, given by Corollary 14.3:

**Theorem 14.18** *The relative VB free energy* (13.150) *of the RRR model for* $H \geq H^*$ *is asymptotically expanded as*

$$\widetilde{F}^{\mathrm{VB}}(\mathcal{D}) = F^{\mathrm{VB}}(\boldsymbol{Y}|\boldsymbol{X}) - NS_N(\boldsymbol{Y}|\boldsymbol{X}) = \lambda'^{\mathrm{VB}} \log N + O_{\mathrm{p}}(1), \tag{14.79}$$

*where the free energy coefficient is given by*

$$2\lambda'^{\mathrm{VB}} = H^*(L+M) + (H-H^*)\min(L,M). \tag{14.80}$$

*Proof* The VB free energy for the RRR model is given by Eq. (14.14), and the empirical entropy is given by

$$\begin{aligned}
2S_N(\boldsymbol{Y}|\boldsymbol{X}) &= -\frac{2}{N}\sum_{n=1}^{N} \log q(\boldsymbol{y}^{(n)}|\boldsymbol{x}^{(n)}) \\
&= L\log(2\pi\sigma'^2) + \frac{\sum_{n=1}^{N}\left\|\boldsymbol{y}^{(n)} - \boldsymbol{B}^*\boldsymbol{A}^{*\top}\boldsymbol{x}^{(n)}\right\|^2}{N\sigma'^2} \\
&= L\log(2\pi\sigma'^2) + \frac{\frac{1}{N}\sum_{n=1}^{N}\left\|\boldsymbol{y}^{(n)}\right\|^2 - 2\mathrm{tr}(\boldsymbol{V}^\top\boldsymbol{B}^*\boldsymbol{A}^{*\top}) + \left\|\boldsymbol{B}^*\boldsymbol{A}^{*\top}\right\|_{\mathrm{Fro}}^2}{\sigma'^2} \\
&= L\log(2\pi\sigma'^2) + \frac{\frac{1}{N}\sum_{n=1}^{N}\left\|\boldsymbol{y}^{(n)}\right\|^2 + \left\|\boldsymbol{V} - \boldsymbol{B}^*\boldsymbol{A}^{*\top}\right\|_{\mathrm{Fro}}^2 - \|\boldsymbol{V}\|_{\mathrm{Fro}}^2}{\sigma'^2}.
\end{aligned}$$

Therefore, the relative VB free energy (14.79) is given as

$$\begin{aligned}
2\widetilde{F}^{\mathrm{VB}}(\mathcal{D}) &= N\cdot\frac{\left\|\boldsymbol{V}-\widehat{\boldsymbol{B}}\widehat{\boldsymbol{A}}^\top\right\|_{\mathrm{Fro}}^2 - \left\|\boldsymbol{V}-\boldsymbol{B}^*\boldsymbol{A}^{*\top}\right\|_{\mathrm{Fro}}^2}{\sigma'^2} + M\log\frac{\det(\boldsymbol{C}_A)}{\det\left(\widehat{\boldsymbol{\Sigma}}_A\right)} + L\log\frac{\det(\boldsymbol{C}_B)}{\det\left(\widehat{\boldsymbol{\Sigma}}_B\right)} \\
&\quad - (L+M)H + \mathrm{tr}\Big\{\boldsymbol{C}_A^{-1}\left(\widehat{\boldsymbol{A}}^\top\widehat{\boldsymbol{A}} + M\widehat{\boldsymbol{\Sigma}}_A\right) + \boldsymbol{C}_B^{-1}\left(\widehat{\boldsymbol{B}}^\top\widehat{\boldsymbol{B}} + L\widehat{\boldsymbol{\Sigma}}_B\right) \\
&\quad + N\sigma'^{-2}\left(-\widehat{\boldsymbol{A}}^\top\widehat{\boldsymbol{A}}\widehat{\boldsymbol{B}}^\top\widehat{\boldsymbol{B}} + \left(\widehat{\boldsymbol{A}}^\top\widehat{\boldsymbol{A}} + M\widehat{\boldsymbol{\Sigma}}_A\right)\left(\widehat{\boldsymbol{B}}^\top\widehat{\boldsymbol{B}} + L\widehat{\boldsymbol{\Sigma}}_B\right)\right)\Big\}.
\end{aligned} \tag{14.81}$$

Eqs. (14.52) and (14.43) imply that the first term in Eq. (14.81) is $O_{\mathrm{p}}(1)$. Corollary 14.3 with Eqs. (14.42) and (14.48) implies that, for $h = 1, \ldots, H^*$,

$$\widehat{a}_h = \Theta_{\mathrm{p}}(1), \qquad \widehat{b}_h = \Theta_{\mathrm{p}}(1), \qquad \widehat{\sigma}_{a_h}^2 = \Theta_{\mathrm{p}}(N^{-1}), \qquad \widehat{\sigma}_{b_h}^2 = \Theta_{\mathrm{p}}(N^{-1}),$$

and, for $h = H^* + 1, \ldots, H$,

$$\begin{aligned}
&\widehat{a}_h = O_{\mathrm{p}}(1), && \widehat{b}_h = O_{\mathrm{p}}(N^{-1/2}), && \widehat{\sigma}_{a_h}^2 = \Theta_{\mathrm{p}}(1), && \widehat{\sigma}_{b_h}^2 = \Theta_{\mathrm{p}}(N^{-1}), && (\text{if } L < M), \\
&\widehat{a}_h = O_{\mathrm{p}}(N^{-1/4}), && \widehat{b}_h = O_{\mathrm{p}}(N^{-1/4}), && \widehat{\sigma}_{a_h}^2 = \Theta_{\mathrm{p}}(N^{-1/2}), && \widehat{\sigma}_{b_h}^2 = \Theta_{\mathrm{p}}(N^{-1/2}), && (\text{if } L = M), \\
&\widehat{a}_h = O_{\mathrm{p}}(N^{-1/2}), && \widehat{b}_h = O_{\mathrm{p}}(1), && \widehat{\sigma}_{a_h}^2 = \Theta_{\mathrm{p}}(N^{-1}), && \widehat{\sigma}_{b_h}^2 = \Theta_{\mathrm{p}}(1), && (\text{if } L > M).
\end{aligned}$$

These results imply that the most terms in Eq. (14.81) are $O_{\mathrm{p}}(1)$, and we thus have

$$2\widetilde{F}^{\mathrm{VB}}(\mathcal{D}) = M \log \frac{\det(\boldsymbol{C}_A)}{\det\left(\widehat{\boldsymbol{\Sigma}}_A\right)} + L \log \frac{\det(\boldsymbol{C}_B)}{\det\left(\widehat{\boldsymbol{\Sigma}}_B\right)} + O_{\mathrm{p}}(1)$$
$$= M \log \prod_{h=1}^{H} \widehat{\sigma}_{a_h}^{-2} + L \log \prod_{h=1}^{H} \widehat{\sigma}_{b_h}^{-2} + O_{\mathrm{p}}(1)$$
$$= \{H^*(L+M) + (H - H^*)\min(L, M)\} \log N + O_{\mathrm{p}}(1),$$

which completes the proof. □

Clearly from the proof, the first term and the second term in Eq. (14.80) correspond to the contribution from the necessary components, $h = 1, \ldots, H^*$, and the contribution from the redundant components, $h = H^*, \ldots, H$, respectively. A remark is that the contribution from the necessary components contains the trivial redundancy, i.e., it is $H^*(L + M)$ instead of $H^*(L + M) - H^{*2}$. This is because the independence between $\boldsymbol{A}$ and $\boldsymbol{B}$ prevents the VB posterior distribution from extending along the trivial redundancy.

### 14.2.6 Comparison with Other Learning Algorithms

Theorems 14.12, 14.17, and 14.18 allow us to compute the generalization, the training, and the free energy coefficients of VB learning. We can now compare those properties with those of ML learning and Bayesian learning, which have been clarified for the RRR model. Note that MAP learning with a smooth and finite prior (e.g., the Gaussian prior (14.3)) with fixed hyperparameters is asymptotically equivalent to ML learning, and has the same generalization and training coefficients.

**ML Learning**

The generalization error of ML learning in the RRR model was analyzed (Fukumizu, 1999), based on the Marčenko–Pastur law (Proposition 8.11). Let $\gamma_h'^2$ be the $h$th largest eigenvalue of a random matrix $\boldsymbol{W} \in \mathbb{S}_+^{\min(L,M)}$ subject to $\mathrm{Wishart}_{\min(L,M)-H^*}(\boldsymbol{I}_{\min(L,M)-H^*}, \max(L, M) - H^*)$.

**Theorem 14.19** *(Fukumizu, 1999) The average ML generalization error of the RRR model for $H \geq H^*$ is asymptotically expanded as*

$$\overline{\mathrm{GE}}^{\mathrm{ML}}(N) = \lambda^{\mathrm{ML}} N^{-1} + O(N^{-3/2}),$$

*where the generalization coefficient is given by*

$$2\lambda^{\mathrm{ML}} = (H^*(L+M) - H^{*2}) + \left\langle \sum_{h=1}^{H-H^*} \gamma_h'^2 \right\rangle_{q(\boldsymbol{W})}. \tag{14.82}$$

**Theorem 14.20** *(Fukumizu, 1999) The ML generalization coefficient of the RRR model in the large-scale limit is given by*

$$2\lambda^{\text{ML}} \to (H^*(L+M) - H^{*2}) + \frac{(\min(L,M) - H^*)(\max(L,M) - H^*)}{2\pi\alpha} J_1(\acute{z}), \qquad (14.83)$$

*where $J_1(\cdot)$ and $\acute{z}$ are defined in Theorem 14.12.*

Actually, Theorems 14.11 and 14.12 were derived by extending Theorems 14.19 and 14.20 to VB learning. We can derive Theorems 14.19 and 14.20 in the same way as VB learning by replacing the VB estimator (14.58) with the ML estimator $\widehat{\gamma}_h^{\text{ML}} = \gamma_h$.

The training error can be similarly analyzed.

**Theorem 14.21** *The average ML training error of the RRR model for $H \geq H^*$ is asymptotically expanded as*

$$\overline{\text{TE}}^{\text{ML}}(N) = \nu^{\text{ML}} N^{-1} + O(N^{-3/2}),$$

*where the training coefficient is given by*

$$2\nu^{\text{ML}} = -(H^*(L+M) - H^{*2}) - \left\langle \sum_{h=1}^{H-H^*} \gamma_h'^2 \right\rangle_{q(\mathbf{W})}. \qquad (14.84)$$

**Theorem 14.22** *The ML training coefficient of the RRR model in the large-scale limit is given by*

$$2\nu^{\text{ML}} \to -(H^*(L+M) - H^{*2}) - \frac{(\min(L,M) - H^*)(\max(L,M) - H^*)}{2\pi\alpha} J_1(\acute{z}), \qquad (14.85)$$

*where $J_1(\cdot)$ and $\acute{z}$ are defined in Theorem 14.12.*

A note is that Theorems 14.19 and 14.21 imply that the generalization coefficient and the training coefficient are antisymmetric in ML learning, i.e., $\lambda^{\text{ML}} = -\nu^{\text{ML}}$, while they are not antisymmetric in VB learning, i.e., $\lambda^{\text{VB}} \neq -\nu^{\text{VB}}$ (see Theorems 14.11 and 14.16).

**Bayesian Learning**

The Bayes free energy in the RRR model was clarified based on the singular learning theory (see Section 13.5.4).

**Theorem 14.23** *(Aoyagi and Watanabe, 2005) The relative Bayes free energy* (13.32) *in the RRR model is asymptotically expanded as*

$$\widetilde{F}^{\text{Bayes}}(\mathcal{D}) = F^{\text{Bayes}}(\boldsymbol{Y}|\boldsymbol{X}) - NS_N(\boldsymbol{Y}|\boldsymbol{X})$$
$$= \lambda'^{\text{Bayes}} \log N - (m-1)\log\log N + O_{\text{p}}(1),$$

*where the free energy coefficient, as well as the coefficient of the second leading term, is given as follows:*

(i) *When* $L + H^* \leq M + H$, $M + H^* \leq L + H$, *and* $H^* + H \leq L + M$:

(a) *If* $L + M + H + H^*$ *is even, then* $m = 1$ *and*

$$2\lambda'^{\text{Bayes}} = \frac{-(H^* + H)^2 - (L - M)^2 + 2(H^* + H)(L + M)}{4}.$$

(b) *If* $L + M + H + H^*$ *is odd, then* $m = 2$ *and*

$$2\lambda'^{\text{Bayes}} = \frac{-(H^* + H)^2 - (L - M)^2 + 2(H^* + H)(L + M) + 1}{4}.$$

(ii) *When* $M + H < L + H^*$, *then* $m = 1$ *and*

$$2\lambda'^{\text{Bayes}} = HM - HH^* + LH^*.$$

(iii) *When* $L + H < M + H^*$, *then* $m = 1$ *and*

$$2\lambda'^{\text{Bayes}} = HL - HH^* + MH^*.$$

(iv) *When* $L + M < H + H^*$, *then* $m = 1$ *and*

$$2\lambda'^{\text{Bayes}} = LM.$$

Theorem 14.23 immediately informs us of the asymptotic behavior of the Bayes generalization error, based on Corollary 13.14.

**Theorem 14.24** *(Aoyagi and Watanabe, 2005) The Bayes generalization error of the RRR model for* $H \geq H^*$ *is asymptotically expanded as*

$$\overline{\text{GE}}^{\text{Bayes}}(N) = \lambda^{\text{Bayes}} N^{-1} - (m-1)(N \log N)^{-1} + o\left((N\log N)^{-1}\right),$$

*where* $\lambda^{\text{Bayes}} = \lambda'^{\text{Bayes}}$ *and* $m$ *are given in Theorem 14.23.*

Unfortunately, the Bayes training error has not been clarified yet.

### Numerical Comparison

Let us visually compare the theoretically clarified generalization properties. Figures 14.1 through 14.4 show the generalization coefficients and the training coefficients of the RRR model under the following settings:

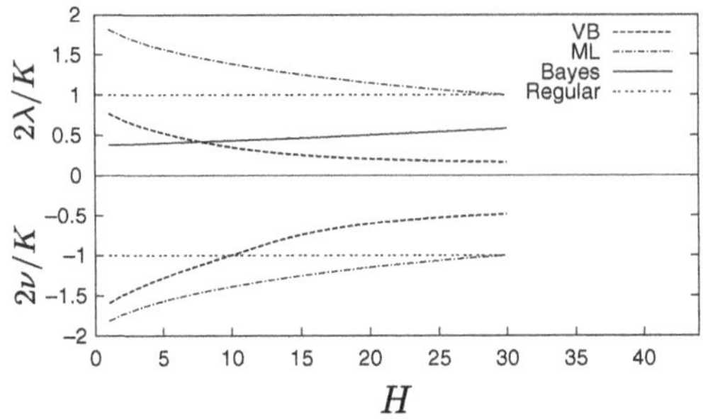


Figure 14.1 The generalization coefficients (in the positive vertical region) and the training coefficients (in the negative vertical region) of VB learning, ML learning, and Bayesian learning in the RRR model with $\max(L, M) = 50$, $\min(L, M) = 30$, $H = 1, \ldots, 30$, and $H^* = 0$.

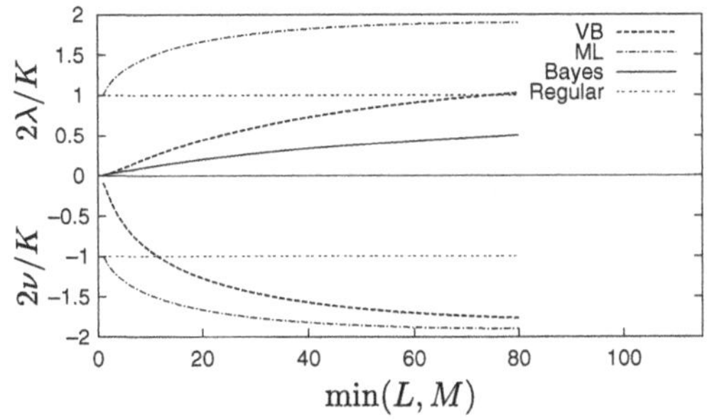


Figure 14.2 The generalization coefficients and the training coefficients ($\max(L, M) = 80$, $\min(L, M) = 1, \ldots, 80$, $H = 1$, and $H^* = 0$).

(i) $\max(L, M) = 50, \min(L, M) = 30, H = 1, \ldots, 30$ (horizontal axis), $H^* = 0$,
(ii) $\max(L, M) = 80, \min(L, M) = 1, \ldots, 80$ (horizontal axis), $H = 1, H^* = 0$,
(iii) $L = M = 80, H = 1, \ldots, 80$ (horizontal axis), $H^* = 0$,
(iv) $\max(L, M) = 50, \min(L, M) = 30, H = 20, H^* = 1, \ldots, 20$ (horizontal axis).

The vertical axis indicates the coefficient normalized by the half of the *essential* parameter dimension $D$, given by Eq. (14.4). The curves in the positive vertical region correspond to the generalization coefficients of VB learning, ML learning, and Bayesian learning, while the curves in the negative vertical region correspond to the training coefficients. As a guide, we depicted the lines $2\lambda/D = 1$ and $2\nu/D = -1$, which correspond to the generalization and the training coefficients (by ML learning and Bayesian learning) of the regular

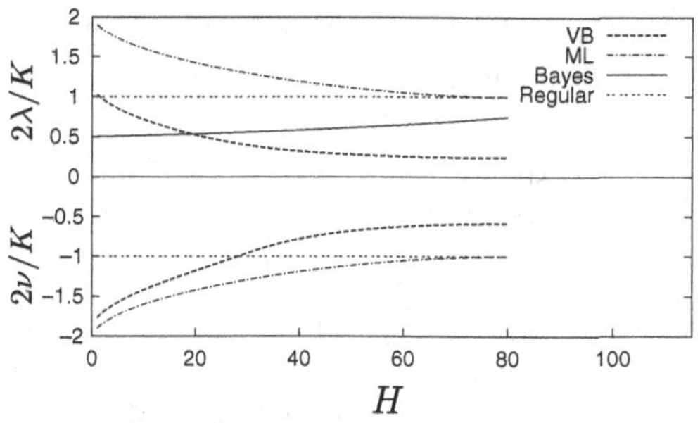


Figure 14.3 The generalization coefficients and the training coefficients ($L = M = 80$, $H = 1, \ldots, 80$, and $H^* = 0$).

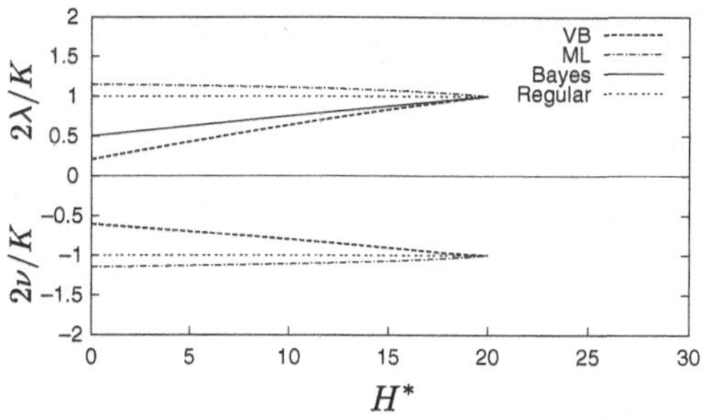


Figure 14.4 The generalization coefficients and the training coefficients ($\max(L, M) = 50$, $\min(L, M) = 30$, $H = 20$, and $H^* = 1, \ldots, 20$).

models with the same parameter dimensionality. The curves for ML learning and VB learning were computed under the large-scale approximation, i.e., by using Theorems 14.12, 14.17, 14.20, and 14.22.[1]

We see in Figures 14.1 through 14.4 that VB learning generally provides comparable generalization performance to Bayesian learning. However, significant differences are also observed. For example, we see in Figure 14.1 that VB learning provides much worse generalization performance than Bayesian learning when $H \ll \min(L, M)$, and much better performance when $H \sim \min(L, M)$.

Another finding is that, in Figures 14.1 and 14.3, the VB generalization coefficient depends on $H$ similarly to the ML generalization coefficient. Moreover, we see that, when $\min(L, M) = 80$ in Figure 14.2 and when $H = 1$ in Figure 14.3, the VB generalization coefficient slightly exceeds the line

[1] We confirmed that numerical computation with Theorems 14.11, 14.16, 14.19, and 14.21 gives visually indistinguishable results.

$2\lambda/D = 1$—the VB generalization coefficient per parameter dimension can be larger than that in the regular models, which never happens for the Bayes generalization coefficient (see Eq. (13.125)).

Finally, Figure 14.4 shows that, for this particular RRR model with $\max(L, M) = 50$, $\min(L, M) = 30$, and $H = 20$, VB learning always gives smaller generalization error than Bayesian learning in the asymptotic limit, regardless of the true rank $H^*$. This might be seen contradictory with the proven optimality of Bayesian learning—Bayesian learning is never dominated by any other method (see Appendix D for the optimality of Bayesin learning and Appendix A for the definition of the term "domination"). We further discuss this issue by considering *subtle true singular values* in Section 14.2.7.

Next we compare the VB free energy with the Bayes free energy, by using Theorems 14.18 and 14.23. Figures 14.5 through 14.8 show the free energy

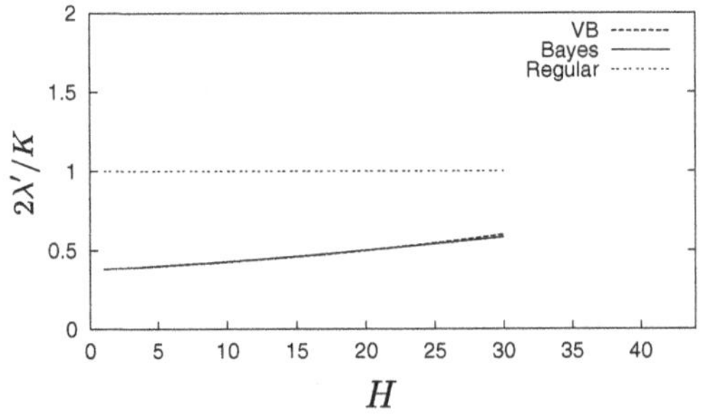


Figure 14.5 Free energy coefficients ($\max(L, M) = 50$, $\min(L, M) = 30$, $H = 1, \ldots, 30$, and $H^* = 0$). The VB and the Bayes free energy coefficients are almost overlapped.

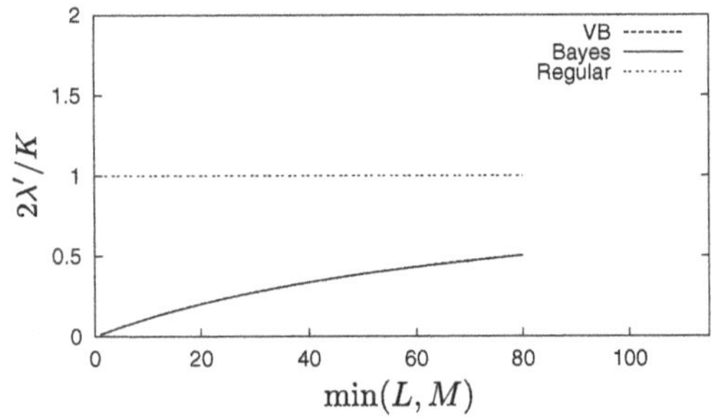


Figure 14.6 Free energy coefficients ($\max(L, M) = 80$, $\min(L, M) = 1, \ldots, 80$, $H = 1$, and $H^* = 0$). The VB and the Bayes free energy coefficients are almost overlapped.

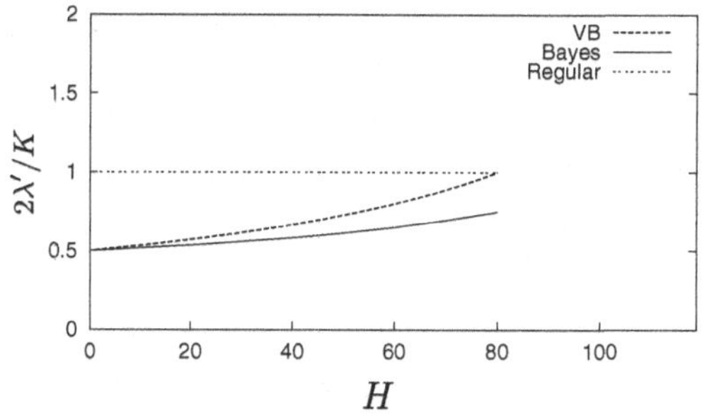


Figure 14.7 Free energy coefficients ($L = M = 80$, $H = 1, \ldots, 80$, and $H^* = 0$).

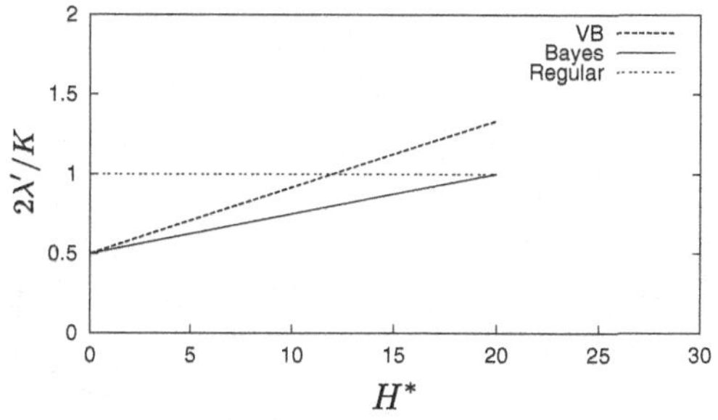


Figure 14.8 Free energy coefficients ($\max(L, M) = 50$, $\min(L, M) = 30$, $H = 20$, and $H^* = 1, \ldots, 20$).

coefficients of the RRR model with the same setting as Figures 14.1 through 14.4, respectively. As for the generalization and the training coefficients, the vertical axis indicates the free energy coefficient normalized by the half of the *essential* parameter dimensionality $D$, given by Eq. (14.4). The curves correspond to the VB free energy coefficient (Theorem 14.18), the Bayes free energy coefficient (Theorem 14.23), and the Bayes free energy coefficient $2\lambda'^{\mathrm{Bayes}}_{\mathrm{Regular}} = D$ of the regular models with the same parameter dimensionality. We find that the VB free energy almost coincides with the Bayes free energy in Figures 14.5 and 14.6, while the VB free energy is much larger than the Bayes free energy in Figures 14.7 and 14.8.

Since the gap between the VB free energy and the Bayes free energy indicates how well the VB posterior approximates the Bayes posterior in terms of the KL divergence (see Section 13.6), our observation is not exactly what we would expect. For example, we see in Figure 14.1 that the generalization performance of VB learning is significantly different from Bayesian learning (when $H \ll \min(L, M)$ and when $H \sim \min(L, M)$), while the free energies in

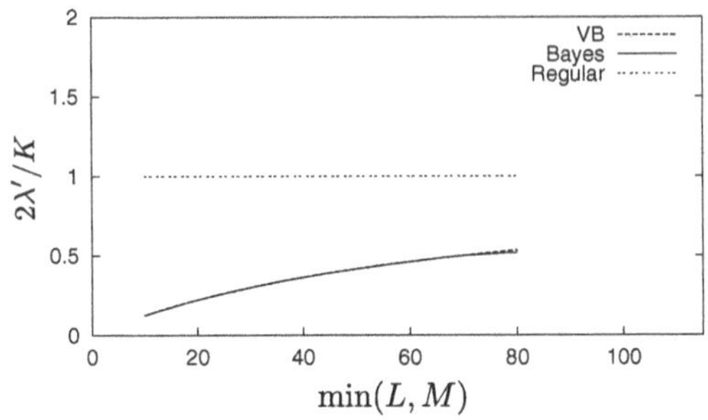


Figure 14.9 Free energy coefficients ($\max(L, M) = 80$, $\min(L, M) = 10, \ldots, 80$, $H = 10$, and $H^* = 0$).

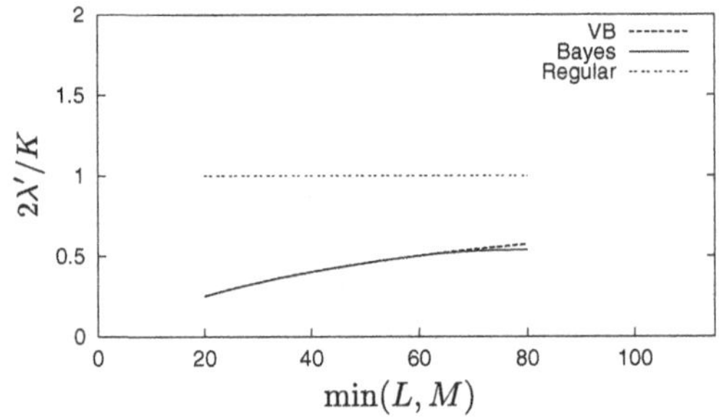


Figure 14.10 Free energy coefficients ($\max(L, M) = 80$, $\min(L, M) = 20, \ldots, 80$, $H = 20$, and $H^* = 0$).

Figure 14.5 imply that the VB posterior well approximates the Bayes posterior. Also, by comparing Figures 14.3 and 14.7, we observe that, when $H \ll \min(L, M)$, VB learning provides much worse generalization performance than Bayesian learning, while the VB free energy well approximates the Bayes free energy; and that, when $H \sim \min(L, M)$, VB learning provides much better generalization performance, while the VB free energy is significantly larger than the Bayes free energy. Further investigation is required to understand the relation between the generalization performance and the gap between the VB and the Bayes free energies.

Figures 14.9 through 14.11 show similar cases to Figure 14.6 but for different ranks $H = 10, 20, 40$, respectively. From Figures 14.5 through 14.11, we conclude that, in general, the VB free energy behaves similarly to the Bayes free energy when $L$ and $M$ are significantly different from each other or $H \ll \min(L, M)$. In Figure 14.8, the VB free energy behaves strangely and poorly approximates the Bayes free energy when $H^*$ is large. This is because

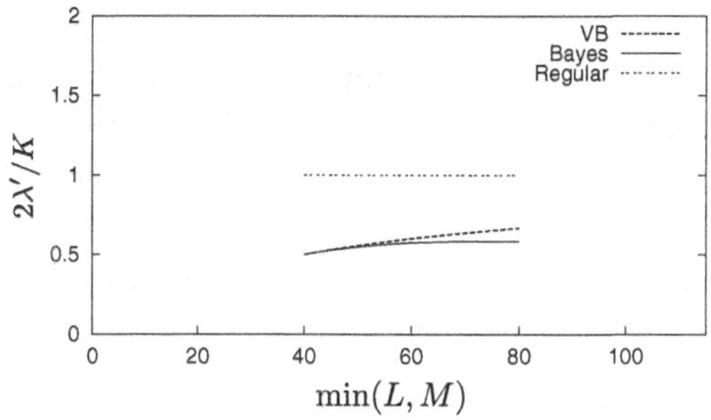


Figure 14.11 Free energy coefficients ($\max(L, M) = 80$, $\min(L, M) = 40, \ldots, 80$, $H = 40$, and $H^* = 0$).

of the trivial redundancy of the RRR model, of which VB learning with the independence constraint cannot make use to reduce the free energy (see the remark in the last paragraph of Section 14.2.5).

### 14.2.7 Analysis with Subtle True Singular Values

Here we conduct an additional analysis to explain the seemingly contradictory observation in Figure 14.4—in the RRR model with $\max(L, M) = 50$, $\min(L, M) = 30$, $H = 20$, VB learning always gives smaller generalization error than Bayesian learning, regardless of the true rank $H^*$. We show that this does not mean the domination by VB learning over Bayesian learning, which was proven to be never dominated by any other method (see Appendix D).

**Distinct and Subtle Signal Assumptions**

The contradictory observation was due to the assumption (14.42) on the true singular values:

$$\gamma_h^* = \begin{cases} \Theta(1) & \text{for} \quad h = 1, \ldots, H^*, \\ 0 & \text{for} \quad h = H^* + 1, \ldots, \min(L, M), \end{cases} \tag{14.86}$$

which we call the *distinct signal assumption*. This assumption seems to cover any true linear mapping $\boldsymbol{B}^*\boldsymbol{A}^{*\top} = \sum_{h=1}^{H} \gamma_h^* \omega_{b_h}^* \omega_{a_h}^*$ by classifying all singular components such that $\gamma_h^* > 0$ to the necessary components $h = 1, \ldots, H^*$, and the other components such that $\gamma_h^* = 0$ to the redundant components $h = H^* + 1, \ldots, \min(L, M)$. However, in the asymptotic limit, the assumption (14.86) *implicitly* prohibits the existence of true singular values in the same order as the noise contribution, i.e., $\gamma_h^* = \Theta_p(N^{-1/2})$. In other words, the distinct signal assumption (14.86) considers all true singular values to be either *infinitely* larger than the noise or exactly equal to zero. As a result, asymptotic analysis under the distinct signal assumption reflects only the *overfitting* tendency of a learning machine, and ignores the *underfitting* tendency, which happens when the signal is not clearly separable from the noise. Since overfitting and underfitting are in the trade-off relation, it is important to investigate both tendencies when generalization performance is analyzed.

To relax the restriction discussed previously, we replace the assumption (14.42) with

$$\gamma_h^* = \begin{cases} \Theta(1) & \text{for} \quad h = 1, \ldots, H^*, \\ O(N^{-1/2}) & \text{for} \quad h = H^* + 1, \ldots, \min(L, M), \end{cases} \tag{14.87}$$

which we call the *subtle signal assumption*, in the following analysis (Watanabe and Amari, 2003; Nakajima and Watanabe, 2007). Note that, with the assumption (14.87), we do not intend to analyze the case where the true singular values depend on $N$. Rather, we assume realistic situations where the number of necessary components $H^*$ depends on $N$. Let us keep in mind the following two points, which are usually true when we analyze real-world data:

- The number $N$ of samples is always finite.
  Asymptotic theory is not to investigate what happens when $N \to \infty$, but to approximate the situation where $N$ is finite but large.
- It rarely happens that real-world data can be *exactly* expressed by a low-rank model.
  Statistical models are supposed to be simpler than the real-world data generation process, but expected to approximate it with certain accuracy, and the accuracy depends on the noise level and the number of samples.

Then we expect that, for most real-world data, it holds that $\gamma_h^* > 0$ for all $h = 1, \ldots, \min(L, M)$, but, given finite $N$, some of the true singular values are comparable to the noise contribution $\gamma_h^* = \Theta(N^{-1/2})$, and some others are negligible $\gamma_h^* = o(N^{-1/2})$. The subtle signal assumption (14.87) covers such realistic situations.

### Generalization Error under Subtle Signal Assumption

Replacing the distinct signal assumption (14.86) with the subtle signal assumption (14.87) does not affect the discussion up to Theorem 14.10, i.e., Theorems 14.1, 14.5, and 14.10, Lemmas 14.6 through 14.8, and their corollaries still hold. Instead of Theorem 14.11, we have the following theorem:

**Theorem 14.25** *Under the subtle signal assumption* (14.87), *the average generalization error of the RRR model for* $H \geq H^*$ *is asymptotically expanded as*

$$\overline{\mathrm{GE}}(N) = \lambda^{\mathrm{VB}} N^{-1} + O(N^{-3/2}),$$

*where the generalization coefficient is given by*

$$\begin{aligned}2\lambda^{\mathrm{VB}} &= (H^*(L+M) - H^{*2}) + \tfrac{N}{\sigma'^2} \textstyle\sum_{h=H^*+1}^{\min(L,M)} \gamma_h^{*2} \\ &+ \Big\langle \textstyle\sum_{h=1}^{H-H^*} \theta\left(\gamma_h''^2 > \max(L,M)\right) \\ &\quad \cdot \left\{ \left(1 - \tfrac{\max(L,M)}{\gamma_h''^2}\right)^2 \gamma_h''^2 - 2\left(1 - \tfrac{\max(L,M)}{\gamma_h''^2}\right) \gamma_h'' \omega_{b_h}''^\top V''^* \omega_{a_h}'' \right\} \Big\rangle_{q(V'')} .\end{aligned} \tag{14.88}$$

*Here,*

$$V'' = \sum_{h=1}^{\min(L,M)-H^*} \gamma_h'' \omega_{b_h}'' \omega_{a_h}''^\top \tag{14.89}$$

*is the SVD of a random matrix* $V'' \in \mathbb{R}^{(\min(L,M)-H^*)\times(\max(L,M)-H^*)}$ *subject to*

$$q(V'') = \mathrm{MGauss}_{\min(L,M)-H^*,\max(L,M)-H^*}\left(V''; V''^*, I_{\min(L,M)-H^*} \otimes I_{\max(L,M)-H^*}\right), \tag{14.90}$$

*and* $V''^* \in \mathbb{R}^{(\min(L,M)-H^*)\times(\max(L,M)-H^*)}$ *is a (nonsquare) diagonal matrix with the diagonal entries given by* $V''_{h,h} = \frac{\sqrt{N}}{\sigma'}\gamma^*_{H^*+h}$ *for* $h = 1, \ldots, \min(L,M) - H^*$.

*Proof* From Eq. (14.58), we have

$$\begin{aligned}\left\|\widehat{B}\widehat{A}^\top - B^* A^{*\top}\right\|_{\mathrm{Fro}}^2 &= \left\|\textstyle\sum_{h=1}^H \widehat{\gamma}_h^{\mathrm{VB}} \omega_{b_h} \omega_{a_h}^\top - B^* A^{*\top}\right\|_{\mathrm{Fro}}^2 \\ &= \left\|\textstyle\sum_{h=1}^{H^*} \gamma_h \omega_{b_h} \omega_{a_h}^\top - B^* A^{*\top} + \textstyle\sum_{h=H^*+1}^H \widehat{\gamma}_h^{\mathrm{VB}} \omega_{b_h} \omega_{a_h}^\top\right\|_{\mathrm{Fro}}^2 + O_{\mathrm{p}}(N^{-3/2}) \\ &= \left\|V - B^* A^* + \textstyle\sum_{h=H^*+1}^H \widehat{\gamma}_h^{\mathrm{VB}} \omega_{b_h} \omega_{a_h}^\top - \textstyle\sum_{h=H^*+1}^{\min(L,M)} \gamma_h \omega_{b_h} \omega_{a_h}^\top\right\|_{\mathrm{Fro}}^2 \\ &\quad + O_{\mathrm{p}}(N^{-3/2}),\end{aligned}$$

and therefore,

$$\begin{aligned}&\left\langle\left\|\widehat{B}\widehat{A}^\top - B^* A^{*\top}\right\|_{\mathrm{Fro}}^2\right\rangle_{q(\mathcal{D})} \\ &= \left\langle\left\|V - B^* A^* + \textstyle\sum_{h=H^*+1}^H \widehat{\gamma}_h^{\mathrm{VB}} \omega_{b_h} \omega_{a_h}^\top - \textstyle\sum_{h=H^*+1}^{\min(L,M)} \gamma_h \omega_{b_h} \omega_{a_h}^\top\right\|_{\mathrm{Fro}}^2\right\rangle_{q(\mathcal{D})} \\ &\quad + O(N^{-3/2}) \\ &= \left\langle\|V - B^* A^*\|_{\mathrm{Fro}}^2\right\rangle_{q(\mathcal{D})} \\ &\quad + 2\left\langle (V - B^* A^*)^\top \left(\textstyle\sum_{h=H^*+1}^H \widehat{\gamma}_h^{\mathrm{VB}} \omega_{b_h} \omega_{a_h}^\top - \textstyle\sum_{h=H^*+1}^{\min(L,M)} \gamma_h \omega_{b_h} \omega_{a_h}^\top\right)\right\rangle_{q(\mathcal{D})}\end{aligned}$$

$$+\left\langle \sum_{h=H^*+1}^{H} (\widehat{\gamma}_h^{\mathrm{VB}})^2 - 2\sum_{h=H^*+1}^{H} \gamma_h \widehat{\gamma}_h^{\mathrm{VB}} + \sum_{h=H^*+1}^{\min(L,M)} \gamma_h^2 \right\rangle_{q(\mathcal{D})} + O(N^{-3/2})$$

$$= \left\langle \|V - B^* A^*\|_{\mathrm{Fro}}^2 \right\rangle_{q(\mathcal{D})}$$

$$+ 2\left\langle \sum_{h=H^*+1}^{H} (\gamma_h \omega_{b_h} \omega_{a_h}^\top - \gamma_h^* \omega_{b_h}^* \omega_{a_h}^{*\top})^\top \widehat{\gamma}_h^{\mathrm{VB}} \omega_{b_h} \omega_{a_h}^\top \right.$$

$$\left. - \sum_{h=H^*+1}^{\min(L,M)} (\gamma_h \omega_{b_h} \omega_{a_h}^\top - \gamma_h^* \omega_{b_h}^* \omega_{a_h}^{*\top})^\top \gamma_h \omega_{b_h} \omega_{a_h}^\top \right\rangle_{q(\mathcal{D})}$$

$$+ \left\langle \sum_{h=H^*+1}^{H} (\widehat{\gamma}_h^{\mathrm{VB}})^2 - 2\sum_{h=H^*+1}^{H} \gamma_h \widehat{\gamma}_h^{\mathrm{VB}} + \sum_{h=H^*+1}^{\min(L,M)} \gamma_h^2 \right\rangle_{q(\mathcal{D})}$$

$$+ O(N^{-3/2})$$

$$= \left\langle \|V - B^* A^*\|_{\mathrm{Fro}}^2 \right\rangle_{q(\mathcal{D})} - 2\left\langle \sum_{h=H^*+1}^{H} (\gamma_h^* \omega_{b_h}^* \omega_{a_h}^{*\top})^\top \widehat{\gamma}_h^{\mathrm{VB}} \omega_{b_h} \omega_{a_h}^\top \right\rangle_{q(\mathcal{D})}$$

$$+ 2\left\langle \sum_{h=H^*+1}^{\min(L,M)} (\gamma_h^* \omega_{b_h}^* \omega_{a_h}^{*\top})^\top \gamma_h \omega_{b_h} \omega_{a_h}^\top \right\rangle_{q(\mathcal{D})}$$

$$+ \left\langle \sum_{h=H^*+1}^{H} (\widehat{\gamma}_h^{\mathrm{VB}})^2 \right\rangle_{q(\mathcal{D})} - \left\langle \sum_{h=H^*+1}^{\min(L,M)} \gamma_h^2 \right\rangle_{q(\mathcal{D})} + O(N^{-3/2})$$

$$= \left\langle \|V - B^* A^*\|_{\mathrm{Fro}}^2 \right\rangle_{q(\mathcal{D})} - \left\langle \sum_{h=H^*+1}^{\min(L,M)} \left\| \gamma_h \omega_{b_h} \omega_{a_h}^\top - \gamma_h^* \omega_{b_h}^* \omega_{a_h}^{*\top} \right\|_{\mathrm{Fro}}^2 \right\rangle_{q(\mathcal{D})}$$

$$- 2\left\langle \sum_{h=H^*+1}^{H} (\gamma_h^* \omega_{b_h}^* \omega_{a_h}^{*\top})^\top \widehat{\gamma}_h^{\mathrm{VB}} \omega_{b_h} \omega_{a_h}^\top \right\rangle_{q(\mathcal{D})}$$

$$+ \left\langle \sum_{h=H^*+1}^{H} (\widehat{\gamma}_h^{\mathrm{VB}})^2 \right\rangle_{q(\mathcal{D})} + \sum_{h=H^*+1}^{\min(L,M)} \gamma_h^{*2} + O(N^{-3/2})$$

$$= \frac{\sigma'^2}{N} \left( LM - (L - H^*)(M - H^*) + \tfrac{N}{\sigma'^2} \sum_{h=H^*+1}^{\min(L,M)} \gamma_h^{*2} \right)$$

$$+ \left\langle \sum_{h=H^*+1}^{H} (\widehat{\gamma}_h^{\mathrm{VB}})^2 \right\rangle_{q(\mathcal{D})} - 2\left\langle \sum_{h=H^*+1}^{H} (\gamma_h^* \omega_{b_h}^* \omega_{a_h}^{*\top})^\top \widehat{\gamma}_h^{\mathrm{VB}} \omega_{b_h} \omega_{a_h}^\top \right\rangle_{q(\mathcal{D})}$$

$$+ O(N^{-3/2}). \tag{14.91}$$

In the orthogonal space to the *distinctly* necessary components $\{\gamma_h, \omega_{a_h}, \omega_{b_h}\}_{h=1}^{H^*}$, the distribution of $\{\gamma_h, \omega_{a_h}, \omega_{b_h}\}_{h=H^*+1}^{\min(L,M)}$ coincides with the distribution of $\{\frac{\sigma'^2}{\sqrt{N}}\gamma_h'', \omega_{a_h}'', \omega_{b_h}''\}_{h=1}^{\min(L,M)-H^*}$, defined in Eq. (14.89), with $V''^*$ as the true matrix for subtle or the redundant components, $h = H^* + 1, \ldots, \min(L, M)$. By using Eq. (14.58), we thus have

$$\left\langle \left\| \widehat{B}\widehat{A}^\top - B^* A^{*\top} \right\|_{\mathrm{Fro}}^2 \right\rangle_{q(\mathcal{D})}$$

$$= \frac{\sigma'^2}{N} \left( (H^*(L + M) - H^{*2}) + \tfrac{N}{\sigma'^2} \sum_{h=H^*+1}^{\min(L,M)} \gamma_h^{*2} \right.$$

$$+ \left\langle \sum_{h=1}^{H-H^*} \theta\left(\gamma_h''^2 > \max(L, M)\right) \right.$$

$$\cdot\left\{\left(1-\frac{\max(L,M)}{\gamma_h''^2}\right)^2\gamma_h''^2-2\left(1-\frac{\max(L,M)}{\gamma_h''^2}\right)\gamma_h''\omega_{b_h}''^{\top}V''^{*}\omega_{a_h}''\right\}\Bigg\rangle_{q(V'')}\Bigg)$$

$$+\,O(N^{-3/2}),$$

which completes the proof. □

**Training Error under Subtle Signal Assumption**

The training error can be analyzed more easily.

**Theorem 14.26** *Under the subtle signal assumption* (14.87), *the average training error of the RRR model for $H \geq H^*$ is asymptotically expanded as*

$$\overline{\mathrm{TE}}(N) = \nu^{\mathrm{VB}}N^{-1} + O(N^{-3/2}),$$

*where the training coefficient is given by*

$$2\nu^{\mathrm{VB}} = -\left(H^*(L+M)-H^{*2}\right)+\tfrac{N}{\sigma'^2}\textstyle\sum_{h=H^*+1}^{\min(L,M)}\gamma_h^{*2}$$
$$+\left\langle\textstyle\sum_{h=1}^{H-H^*}\theta\left(\gamma_h''^2>\max(L,M)\right)\cdot\left(1-\frac{\max(L,M)}{\gamma_h''^2}\right)\left(1+\frac{\max(L,M)}{\gamma_h''^2}\right)\gamma_h''^2\right\rangle_{q(V'')}. \tag{14.92}$$

*Here $V''$ and $\{\gamma_h''\}$ are defined in Theorem 14.25.*

*Proof* Theorem 14.15 and Eq. (14.77) still hold under the assumption (14.87). Therefore,

$$\left\|V-\widehat{B}\widehat{A}^{\top}\right\|_{\mathrm{Fro}}^2$$
$$=-\textstyle\sum_{h=H^*+1}^{H}\theta\left(\gamma_h^2>\frac{\max(L,M)\sigma'^2}{N}\right)\cdot\left(\gamma_h-\frac{\max(L,M)\sigma'^2}{N\gamma_h}\right)\left(\gamma_h+\frac{\max(L,M)\sigma'^2}{N\gamma_h}\right)$$
$$+\textstyle\sum_{h=H^*+1}^{\min(L,M)}(\gamma_h-\gamma_h^*)^2+\sum_{h=H^*+1}^{\min(L,M)}\gamma_h^{*2}+O_{\mathrm{p}}(N^{-3/2}),$$

and

$$\left\langle\left\|V-\widehat{B}\widehat{A}^{\top}\right\|_{\mathrm{Fro}}^2-\left\|V-B^*A^{*\top}\right\|_{\mathrm{Fro}}^2\right\rangle_{q(\mathcal{D})}$$
$$=-\frac{\sigma'^2}{N}\left\{\left(H^*(L+M)-H^{*2}\right)-\frac{N}{\sigma'^2}\sum_{h=H^*+1}^{\min(L,M)}\gamma_h^{*2}\right.$$
$$\left.+\textstyle\sum_{h=H^*+1}^{H}\theta\left(\gamma_h^2>\frac{\max(L,M)\sigma'^2}{N}\right)\cdot\left(\gamma_h-\frac{\max(L,M)\sigma'^2}{N\gamma_h}\right)\left(\gamma_h+\frac{\max(L,M)\sigma'^2}{N\gamma_h}\right)\right\}$$
$$+\,O_{\mathrm{p}}(N^{-3/2}).$$

Substituting the preceding equation into Eq. (14.75) and using the random matrix $V''$ and its singular values $\{\gamma_h''\}$, defined in Theorem 14.25, we obtain Eq. (14.92). □

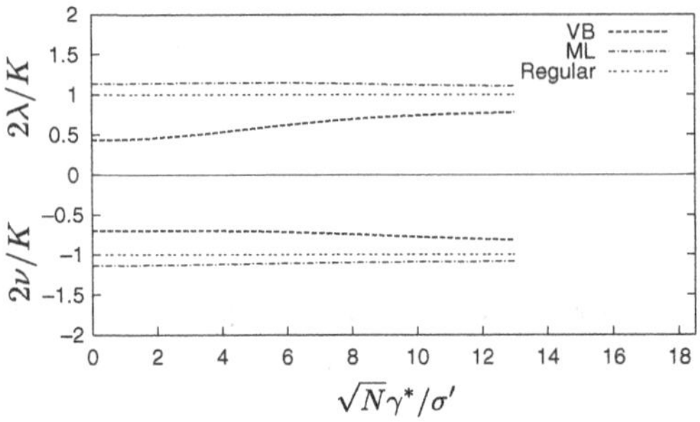


Figure 14.12 The generalization coefficients and the training coefficients under the subtle signal assumption (14.87) in the RRR model with $\max(L, M) = 50$, $\min(L, M) = 30$, $H = 20$, and $H^* = 5$.

## Comparison with Other Learning Algorithms

Figure 14.12 shows the generalization coefficients and the training coefficients, computed by using Theorems 14.25 and 14.26, respectively, as functions of a rescaled subtle true singular value $\sqrt{N}\gamma_h^*/\sigma'$. The considered RRR model is with $\max(L, M) = 50$, $\min(L, M) = 30$, and $H = 20$, and the true linear mapping is assumed to consist of $H^* = 5$ *distinctly* necessary components ($\gamma_h^* = \Theta(1)$ for $h = 1, \ldots, 5$), 10 *subtle* components ($\gamma_h^* = \Theta(N^{-1/2})$ for $h = 6, \ldots, 15$), and the other five null components ($\gamma_h^* = 0$ for $h = 16, \ldots, 20$). The subtle singular values are assumed to be identical, $\gamma_h^* = \gamma^*$ for $h = 6, \ldots, 15$, and the horizontal axis indicates $\sqrt{N}\gamma^*/\sigma'$. The generalization coefficients and the training coefficients of ML learning can be derived in the same way as Theorems 14.25 and 14.26 with the VB estimator $\widehat{\gamma}_h^{\mathrm{VB}}$ replaced with the ML estimator $\widehat{\gamma}_h^{\mathrm{ML}} = \gamma_h$. Unfortunately, the generalization error nor the training error of Bayesian learning under the subtle signal assumption for the general RRR model has not been clarified.

Only in the case where $L = H = 1$, the Bayes generalization error under the subtle signal assumption has been analyzed.

**Theorem 14.27** *(Watanabe and Amari, 2003) The Bayes generalization error of the RRR model with $M \geq 2, L = H = 1$ under the assumption that the true mapping is $b^*\boldsymbol{a}^* = O(N^{-1/2})$ is asymptotically expanded as*

$$\overline{\mathrm{GE}}^{\mathrm{Bayes}}(N) = \lambda^{\mathrm{Bayes}} N^{-1} + o(N^{-1}),$$

*where the generalization coefficient is given by*

$$2\lambda^{\mathrm{Bayes}} = 1 + \left\langle \left( \left\| \frac{\sqrt{N}}{\sigma'} b^* \boldsymbol{a}^* \right\|^2 + \frac{\sqrt{N}}{\sigma'} b^* \boldsymbol{a}^{*\top} \boldsymbol{v} \right) \frac{\Phi_M(\boldsymbol{v})}{\Phi_{M-2}(\boldsymbol{v})} \right\rangle_{q(\boldsymbol{v})}. \tag{14.93}$$

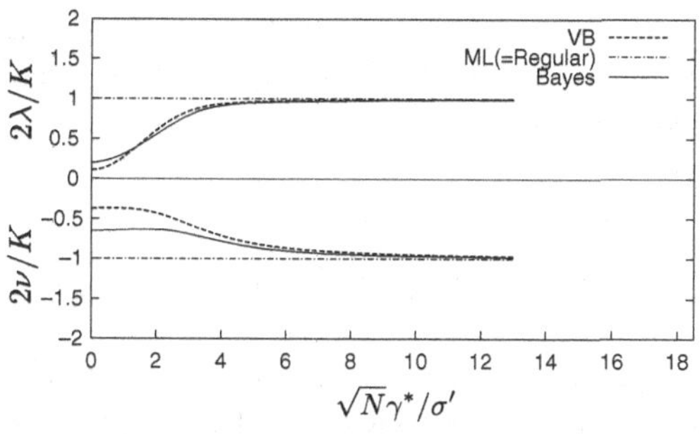


Figure 14.13 The generalization coefficients and the training coefficients under the subtle signal assumption (14.87) in the RRR model with $M = 5$, $L = H = 1$, and $H^* = 0$.

*Here,*

$$\Phi_M(\boldsymbol{v}) = \int_0^{\pi/2} \sin^M \theta \exp\left(-\frac{1}{2}\left\|\frac{\sqrt{N}}{\sigma'} b^* \boldsymbol{a}^* + \boldsymbol{v}\right\|^2 \sin^2\theta\right) d\theta,$$

*and* $\boldsymbol{v} \in \mathbb{R}^M$ *is a random vector subject to* $q(\boldsymbol{v}) = \mathrm{Gauss}_M(\boldsymbol{v}; \boldsymbol{0}, \boldsymbol{I}_M)$.

Figure 14.13 compares the generalization coefficients when $M = 5, L = H = 1$, and $H^* = 0$, where the horizontal axis corresponds to a rescaled subtle true singular value $\sqrt{N}\gamma^*/\sigma' = \sqrt{N}\,\|b^*\boldsymbol{a}^*\|\,/\sigma'$.[2] We see that the generalization error of VB learning is smaller than that of Bayesian learning when $\sqrt{N}\gamma^*/\sigma' = 0$, and identical when $\sqrt{N}\gamma^*/\sigma' \to \infty$. This means that, under the distinct signal assumption (14.86), which considers only the case where $\gamma^* = 0$ (i.e., $H^* = 0$) or $\gamma^* = \Theta(1)$ (i.e., $H^* = 1$), VB learning always performs better than Bayesian learning. However, we can see in Figure 14.13 that, when $\sqrt{N}\gamma^*/\sigma' \approx 3$, Bayesian learning outperforms VB learning. Figure 14.13 simply implies that VB learning is more strongly regularized than Bayesian learning, or in other words, VB learning tends to underfit subtle signals such that $\gamma^* = \Theta(N^{-1/2})$, while Bayesian learning tends to overfit noise.

Knowing the proved optimality of Bayesian learning (Appendix D), we would expect that the same happens in Figure 14.12, where the limits $\sqrt{N}\gamma^*/\sigma' = 0$ and $\sqrt{N}\gamma^*/\sigma' \to \infty$ correspond to the cases with $H^* = 5$ and $H^* = 15$, respectively, in Figure 14.4 under the distinct signal assumption

[2] When $L = H = 1$, the parameter transformation $b\boldsymbol{a} \to \boldsymbol{w}$ makes the RRR model identifiable, and therefore the ML generalization coefficient is identical to that of the regular models. This is the reason why only the *integration effect* or *model induced-regularization* was observed in the one-dimensional matrix factorization model in Section 7.2 and Section 7.3.3. The *basis selection effect* appears only when a singular model cannot be equivalently transformed to a regular model (see the discussion in Section 14.3).

(14.86). Namely, if we could depict the Bayes generalization coefficient in Figure 14.12, there should be some interval where Bayesian learning outperforms VB learning.

## 14.3 Insights into VB Learning

In this chapter, we analyzed the generalization error, the training error, and the free energy of VB learning in the RRR model, and derived their asymptotic forms. We also introduced theoretical results providing those properties for ML learning and Bayesian learning. As mentioned in Section 13.5, the RRR model is the only singular model of which those three properties have been theoretically clarified for ML learning, Bayesian learning, and VB learning. Accordingly, we here summarize our observations, and discuss effects of singularities in VB learning and other learning algorithms.

(i) In the RRR model, the *basis selection effect*, explained in Section 13.5.1, appears as a selection bias of largest singular values of a zero-mean random matrix.

Theorem 14.19 gives an asymptotic expansion of the ML generalization error. The second term in the generalization coefficient (14.82) is the expectation of the square of the $(H - H^*)$ largest singular values of a random matrix $\frac{\sqrt{N}}{\sigma'}\boldsymbol{V}'$, where $\boldsymbol{V}'$ is subject to the zero-mean Gaussian (14.61). This corresponds to the effect of basis selection: ML learning chooses the singular components that best fit the observation noise. With the full-rank model, i.e., $H = \min(L, M)$, the second term in the generalization coefficient (14.82) is equal to

$$\left\langle \sum_{h=1}^{\min(L,M)-H^*} \gamma_h'^2 \right\rangle_{q(\boldsymbol{W})} = (\min(L, M) - H^*)(\max(L, M) - H^*)$$
$$= (L - H^*)(M - H^*),$$

and therefore the generalization coefficient becomes

$$2\lambda^{\mathrm{ML}} = (H^*(L + M) - H^{*2}) + (L - H^*)(M - H^*) = LM$$
$$= D,$$

which is the same as the generalization coefficient of the regular models. Indeed, the full-rank RRR model is equivalently transformed to a regular model by $\boldsymbol{B}\boldsymbol{A}^\top \rightarrow \boldsymbol{U}$, where the domain for $\boldsymbol{U}$ is the whole $\mathbb{R}^{L\times M}$ space (no low-rank restriction is imposed to $\boldsymbol{U}$). In this case, no *basis selection* occurs because all possible bases are supposed to be used.

(ii) In the RRR model, the integration effect, explained in Section 13.5.1, appears as the James–Stein (JS) type shrinkage.

This was shown in Theorem 14.1 in the asymptotic limit: the VB estimator converges to the positive-part JS estimator operated on each singular component separately. By comparing Theorems 14.11 and 14.19, we see that ML learning and VB learning differ from each other by the factor $\theta\left(\gamma_h'^2 > \max(L, M)\right)\left(1 - \frac{\max(L,M)}{\gamma_h'^2}\right)^2$, which comes from the postive-part JS shrinkage. Unlike the basis selection effect, the integration effect appears even if the model can be equivalently transformed to a regular model—the full-rank RRR model (with $H = \min(L, M)$) is still affected by the singularities. The relation between VB learning and the JS shrinkage estimator was also observed in nonasymptotic analysis in Chapter 7, where model-induced regularization (MIR) was illustrated as a consequence of the integration effect, by focusing on the one-dimensional matrix factorization model. Note that basis selection effect does not appear in the one-dimensional matrix factorization model (where $L = M = H$), because it can be equivalently transformed to a regular model.

(iii) VB learning shows similarity both to ML learning and Bayesian learning.

Figures 14.1 through 14.4 generally show that VB learning is regularized as much as Bayesian learning, while its dependence on the model size ($H, L, M$, etc.) is more like ML learning. Unlike Bayesian learning, the integration effect does not always dominate the basis selection effect in VB learning—a good property of Bayesian learning, $2\lambda^{\mathrm{Bayes}} \leq D$, does not necessarily hold in VB learning, e.g., we observe that $2\lambda^{\mathrm{VB}} > D$ at $\min(L, M) = 80$ in Figure 14.2, and $H = 1$ in Figure 14.3.

(iv) In VB learning, the relation between the generalization error and the free energy is not as simple as in Bayesian learning.

In Bayesian learning, the generalization coefficient and the free energy coefficient coincide with each other, i.e., $\lambda^{\mathrm{Bayes}} = \lambda'^{\mathrm{Bayes}}$. This property does not hold in VB learning even approximately, as seen by comparing Figures 14.1 through 14.4 and Figures 14.5 through 14.8. In many cases, the VB free energy well approximates the Bayes free energy, while the VB generalization error significantly differs from the Bayes generalization error. Research on the relation between the free energy and the generalization error in VB learning is ongoing (see Section 17.4).

(v) MIR in VB learning can be stronger than that in Bayesian learning.

By definition, the VB free energy is never less than the Bayes free energy, and therefore it holds that $\lambda'^{\mathrm{VB}} \geq \lambda'^{\mathrm{Bayes}}$. On the other hand, such a relation does not hold for the generalization error, i.e., $\lambda^{\mathrm{VB}}$ can be larger or less than $\lambda^{\mathrm{Bayes}}$. However, even if $\lambda^{\mathrm{VB}}$ is less than or equal to $\lambda^{\mathrm{Bayes}}$ for any true rank $H^*$ in some RRR model, it does not mean the domination of VB learning over Bayesian learning. Since the optimality of Bayesian learning was proved (see Appendix D), $\lambda^{\mathrm{VB}} < \lambda^{\mathrm{Bayes}}$ simply means that VB learning is more strongly regularized than Bayesian learning, or in other words, VB learning tends to underfit small signals while Bayesian learning tends to overfit noise. Extending the analysis under the subtle signal assumption (14.87) to the general RRR model would clarify this point.

(vi) The generalization error depends on the dimensionality in an interesting way.

The shrinkage factor is governed by $\max(L, M)$ and independent of $\min(L, M)$ in the asymptotic limit (see Theorem 14.1). This is because the shrinkage is caused by the VB posterior extending into the parameter space with larger dimensional space ($M$-dimensional input space or $L$-dimensional output space) for the redundant components, as seen in Corollary 14.3. This choice was made by maximizing the entropy of the VB posterior distribution when the free energy is minimized. Consequently, when $L \neq M$, the shape of the VB posterior in the asymptotic limit is similar to the partially Bayesian learning, where the posterior of $\boldsymbol{A}$ or $\boldsymbol{B}$ is approximated by the Dirac delta function (see Chapter 12). On the other hand, increase of the smaller dimensionality $\min(L, M)$ broadens the variety of basis selection: as mentioned in (i), the basis selection effect in the RRR model occurs by the redundant components selecting the $(H - H^*)$ largest singular components, and $(\min(L, M) - H^*)$ corresponds to the dimensionality that the basis functions can span. This phenomenon can be seen in Figure 8.3—the Marčenko–Pastur distribution is diverse when $\alpha = (\min(L, M) - H^*)/(\max(L, M) - H^*)$ is large. We can conclude that a large $\max(L, M)$ enhances the integration effect, leading to strong regularization, while a large $\min(L, M)$ enhances the basis selection effect, leading to overfitting. As a result, VB learning tends to be strongly regularized when $L \ll M$ or $L \gg M$, and tends to overfit when $L \approx M$.

# 15
# Asymptotic VB Theory of Mixture Models

In this chapter, we discuss the asymptotic behavior of the VB free energy of mixture models, for which VB learning algorithms were introduced in Sections 4.1.1 and 4.1.2. We first prepare basic lemmas commonly used in this and the following chapters.

## 15.1 Basic Lemmas

Consider the *latent variable model* expressed as

$$p(\mathcal{D}|\boldsymbol{w}) = \sum_{\mathcal{H}} p(\mathcal{D}, \mathcal{H}|\boldsymbol{w}).$$

In this chapter, we analyze the VB free energy, which is the minimum of the free energy under the constraint,

$$r(\boldsymbol{w}, \mathcal{H}) = r_w(\boldsymbol{w}) r_{\mathcal{H}}(\mathcal{H}), \tag{15.1}$$

i.e.,

$$F^{\mathrm{VB}}(\mathcal{D}) = \min_{r_w(\boldsymbol{w}), r_{\mathcal{H}}(\mathcal{H})} F(r), \tag{15.2}$$

where

$$\begin{aligned} F(r) &= \left\langle \log \frac{r_w(\boldsymbol{w}) r_{\mathcal{H}}(\mathcal{H})}{p(\boldsymbol{w}, \mathcal{H}, \mathcal{D})} \right\rangle_{r_w(\boldsymbol{w}) r_{\mathcal{H}}(\mathcal{H})} \\ &= F^{\mathrm{Bayes}}(\mathcal{D}) + \mathrm{KL}\left( r_w(\boldsymbol{w}) r_{\mathcal{H}}(\mathcal{H}) \| p(\boldsymbol{w}, \mathcal{H}|\mathcal{D}) \right). \end{aligned} \tag{15.3}$$

Here,

$$F^{\mathrm{Bayes}}(\mathcal{D}) = -\log p(\mathcal{D}) = -\log \sum_{\mathcal{H}} p(\mathcal{D}, \mathcal{H}) = -\log \sum_{\mathcal{H}} \int p(\mathcal{D}, \mathcal{H}, \boldsymbol{w}) d\boldsymbol{w}$$

is the Bayes free energy. Recall that the stationary condition of the free energy yields

$$r_w(\boldsymbol{w}) = \frac{1}{C_w} p(\boldsymbol{w}) \exp \langle \log p(\mathcal{D}, \mathcal{H}|\boldsymbol{w}) \rangle_{r_{\mathcal{H}}(\mathcal{H})}, \tag{15.4}$$

$$r_{\mathcal{H}}(\mathcal{H}) = \frac{1}{C_{\mathcal{H}}} \exp \langle \log p(\mathcal{D}, \mathcal{H}|\boldsymbol{w}) \rangle_{r_w(\boldsymbol{w})}. \tag{15.5}$$

For the minimizer $\widehat{r}_w(\boldsymbol{w})$ of $F(r)$, let

$$\widehat{\boldsymbol{w}} = \langle \boldsymbol{w} \rangle_{\widehat{r}_w(\boldsymbol{w})} \tag{15.6}$$

be the VB estimator.

The following lemma shows that the free energy is decomposed into the sum of two terms.

**Lemma 15.1** *It holds that*

$$F^{\mathrm{VB}}(\mathcal{D}) = \min_{r_w(\boldsymbol{w})} \{R + Q\}, \tag{15.7}$$

*where*

$$R = \mathrm{KL}(r_w(\boldsymbol{w}) \| p(\boldsymbol{w})),$$
$$Q = -\log C_{\mathcal{H}},$$

*for* $C_{\mathcal{H}} = \sum_{\mathcal{H}} \exp \langle \log p(\mathcal{D}, \mathcal{H}|\boldsymbol{w}) \rangle_{r_w(\boldsymbol{w})}$.

*Proof* From the restriction of the VB approximation in Eq. (15.1), $F(r)$ can be divided into two terms,

$$F(r) = \left\langle \log \frac{r_w(\boldsymbol{w})}{p(\boldsymbol{w})} \right\rangle_{r_w(\boldsymbol{w})} + \left\langle \log \frac{r_{\mathcal{H}}(\mathcal{H})}{p(\mathcal{D}, \mathcal{H}|\boldsymbol{w})} \right\rangle_{r_w(\boldsymbol{w}) r_{\mathcal{H}}(\mathcal{H})}.$$

Since the optimal VB posteriors satisfy Eqs. (15.4) and (15.5), if the VB posterior $r_{\mathcal{H}}(\mathcal{H})$ is optimized, then

$$\left\langle \log \frac{r_{\mathcal{H}}(\mathcal{H})}{p(\mathcal{D}, \mathcal{H}|\boldsymbol{w})} \right\rangle_{r_w(\boldsymbol{w}) r_{\mathcal{H}}(\mathcal{H})} = -\log C_{\mathcal{H}}$$

holds. Thus, we obtain Eq. (15.7). □

The free energies of mixture models and other latent variable models involve the di-gamma function $\Psi(x)$ and the log-gamma function $\log \Gamma(x)$ (see, e.g., Eq. (4.22)). To analyze the free energy, we will use the inequalities on these functions in the following lemma:

**Lemma 15.2** *(Alzer, 1997) For $x > 0$,*

$$\frac{1}{2x} < \log x - \Psi(x) < \frac{1}{x}, \tag{15.8}$$

*and*

$$0 \leq \log \Gamma(x) - \left\{ \left( x - \frac{1}{2} \right) \log x - x + \frac{1}{2} \log 2\pi \right\} \leq \frac{1}{12x}. \tag{15.9}$$

The inequalities (15.8) ensure that substituting $\log x$ for $\Psi(x)$ only contributes at most additive constant terms to the VB free energy. The substitution for $\log \Gamma(x)$ is given by Eq. (15.9) as well.

For the i.i.d. latent variable models defined as

$$p(\boldsymbol{x}|\boldsymbol{w}) = \sum_{z} p(\boldsymbol{x}, z|\boldsymbol{w}), \tag{15.10}$$

the likelihood for the observed data $\mathcal{D} = \{\boldsymbol{x}^{(n)}\}_{n=1}^N$ and the complete data $\{\mathcal{D}, \mathcal{H}\} = \{\boldsymbol{x}^{(n)}, z^{(n)}\}_{n=1}^N$ is given by

$$p(\mathcal{D}|\boldsymbol{w}) = \prod_{n=1}^{N} p(\boldsymbol{x}^{(n)}|\boldsymbol{w}),$$

$$p(\mathcal{D}, \mathcal{H}|\boldsymbol{w}) = \prod_{n=1}^{N} p(\boldsymbol{x}^{(n)}, z^{(n)}|\boldsymbol{w}),$$

respectively. In the asymptotic analysis of the free energy for such a model, when the free energy is minimized, the second term in Eq. (15.7), $Q = -\log C_{\mathcal{H}}$, is proved to be close to $N$ times the empirical entropy (13.20),

$$S_N(\mathcal{D}) = -\frac{1}{N} \sum_{n=1}^{N} \log p(\boldsymbol{x}^{(n)}|\boldsymbol{w}^*), \tag{15.11}$$

where $\boldsymbol{w}^*$ is the true parameter generating the data. Thus, the first term in Eq. (15.7) shows the asymptotic behavior of the VB free energy, which is analyzed with the inequalities in Lemma 15.2.

Let

$$\widetilde{Q} = Q - NS_N(\mathcal{D}). \tag{15.12}$$

It follows from Jensen's inequality that

$$\begin{aligned} \widetilde{Q} &= \log p(\mathcal{D}|\boldsymbol{w}^*) - \log \sum_{\mathcal{H}} \exp \langle \log p(\mathcal{D}, \mathcal{H}|\boldsymbol{w}) \rangle_{r_w(\boldsymbol{w})} \\ &\geq \log \frac{p(\mathcal{D}|\boldsymbol{w}^*)}{\langle p(\mathcal{D}|\boldsymbol{w}) \rangle_{r_w(\boldsymbol{w})}} \\ &\geq NE_N(\widehat{\boldsymbol{w}}^{\mathrm{ML}}), \end{aligned} \tag{15.13}$$

where $\widehat{w}^{\mathrm{ML}}$ is the *maximum likelihood (ML) estimator*, and

$$E_N(w) = L_N(w^*) - L_N(w) = \frac{1}{N}\sum_{n=1}^{N} \log \frac{p(x^{(n)}|w^*)}{p(x^{(n)}|w)} \tag{15.14}$$

is the empirical KL divergence. Note here that $L_N$ is defined in Eq. (13.37), and $E_N(w)$ corresponds to the training error of the plug-in predictive distribution (defined in Eq. (13.75) for regular models) with an estimator $w$.

If the domain of data $\mathcal{X}$ is discrete and with finite cardinality, $\#(\mathcal{X}) = M$, $Q$ in Eq. (15.7) can be analyzed in detail. In such a case, we can assume without loss of generality that $x \in \{e_1, \ldots, e_M\}$, where $e_m$ is the one-of-$M$ representation, i.e., only the $m$th entry is one and the other entries are zeros. Let $\widehat{N}_m$ be the number of output $m$ in the sequence $\mathcal{D}$, i.e., $\widehat{N}_m = \sum_{n=1}^{N} x_m^{(n)}$, and define the strongly $\varepsilon$-typical set $T_\varepsilon^N(p^*)$ with respect to the probability mass function,

$$p^* = (p_1^*, \ldots, p_M^*)^\top = (p(x_1 = 1|w^*), \ldots, p(x_M = 1|w^*))^\top \in \Delta^{M-1}$$

as follows:

$$T_\varepsilon^N(p^*) = \left\{ \mathcal{D} \in \mathcal{X}^N; \left| \frac{\widehat{N}_m}{N} - p_m^* \right| \le \frac{p_m^*}{\log M}\varepsilon, \ \ m = 1, \ldots, M \right\}. \tag{15.15}$$

It is known that the probability that the observed data sequence is not strongly $\varepsilon$-typical is upper-bounded as follows:

**Lemma 15.3** *(Han and Kobayashi, 2007) It holds that*

$$\mathrm{Prob}(\mathcal{D} \notin T_\varepsilon^N(p^*)) \le \frac{\kappa M}{N\varepsilon^2},$$

*where*

$$\kappa = (\log M)^2 \max_{m: p_m^* \neq 0} \frac{1 - p_m^*}{p_m^*}.$$

Let

$$\widehat{p} = (\widehat{p}_1, \ldots, \widehat{p}_M)^\top = \left( \langle p(x_1 = 1|w)\rangle_{r_w(w)}, \ldots, \langle p(x_M = 1|w)\rangle_{r_w(w)} \right)^\top \in \Delta^{M-1}$$

be the probability mass function defined by the predictive distribution $\langle p(x|w)\rangle_{r_w(w)}$ with the VB posterior $r_w(w)$. For any fixed $\delta > 0$, define

$$R_\delta^* = \left\{ \widehat{p} \in \Delta^{M-1}; \mathrm{KL}(p^* \| \widehat{p}) \le \delta \right\}, \tag{15.16}$$

where $\mathrm{KL}(p^*\|\widehat{p}) = \sum_{m=1}^{M} p_m^* \log \frac{p_m^*}{\widehat{p}_m}$. Then the following lemma holds:

**Lemma 15.4** *Suppose that the domain $\mathcal{X}$ is discrete and with finite cardinality, $\#(\mathcal{X}) = M$. For all $\varepsilon > 0$ and $\mathcal{D} \in T^N_\varepsilon(\boldsymbol{p}^*)$, there exists a constant $C > 0$ such that if $\widehat{\boldsymbol{p}} \notin R^*_{C\varepsilon^2}$,*

$$\widetilde{Q} = \Omega_{\mathrm{p}}(N). \tag{15.17}$$

*Furthermore,*

$$\min_{r_w(\boldsymbol{w})} \widetilde{Q} = O_{\mathrm{p}}(1). \tag{15.18}$$

*Proof* From Eq. (15.13), we have

$$\begin{aligned}
\widetilde{Q} &\geq \log \frac{p(\mathcal{D}|\boldsymbol{w}^*)}{\langle p(\mathcal{D}|\boldsymbol{w})\rangle_{r_w(\boldsymbol{w})}} \\
&= N \sum_{m=1}^{M} \frac{\widehat{N}_m}{N} \log \frac{p^*_m}{\widehat{p}_m} \\
&= N \left\{ \mathrm{KL}(\widehat{\boldsymbol{p}}^{\mathrm{ML}} \| \widehat{\boldsymbol{p}}) - \mathrm{KL}(\widehat{\boldsymbol{p}}^{\mathrm{ML}} \| \boldsymbol{p}^*) \right\},
\end{aligned} \tag{15.19}$$

where $\widehat{\boldsymbol{p}}^{\mathrm{ML}} = (\widehat{p}^{\mathrm{ML}}_1, \ldots, \widehat{p}^{\mathrm{ML}}_M)^\top = (\widehat{N}_1/N, \ldots, \widehat{N}_M/N)^\top \in \Delta^{M-1}$ is the type, namely the empirical distribution of $\mathcal{D}$. Thus, if $\mathrm{KL}(\widehat{\boldsymbol{p}}^{\mathrm{ML}} \| \widehat{\boldsymbol{p}}) > \mathrm{KL}(\widehat{\boldsymbol{p}}^{\mathrm{ML}} \| \boldsymbol{p}^*)$, the right-hand side of Eq. (15.19) grows in the order of $N$. If $\mathcal{D} \in T^N_\varepsilon(\boldsymbol{p}^*)$, $\mathrm{KL}(\widehat{\boldsymbol{p}}^{\mathrm{ML}} \| \boldsymbol{p}^*) = O_{\mathrm{p}}(\varepsilon^2)$ since $\mathrm{KL}(\widehat{\boldsymbol{p}}^{\mathrm{ML}} \| \boldsymbol{p}^*)$ is well approximated by a quadratic function of $\widehat{\boldsymbol{p}}^{\mathrm{ML}} - \boldsymbol{p}^*$. To prove the first assertion of the lemma, it suffices to see that $\mathrm{KL}(\boldsymbol{p}^* \| \widehat{\boldsymbol{p}}) \leq C\varepsilon^2$ is equivalent to $\mathrm{KL}(\widehat{\boldsymbol{p}}^{\mathrm{ML}} \| \widehat{\boldsymbol{p}}) \leq C'\varepsilon^2$ for a constant $C' > 0$ if $\mathcal{D} \in T^N_\varepsilon(\boldsymbol{p}^*)$. In fact, we have

$$\mathrm{KL}(\boldsymbol{p}^* \| \widehat{\boldsymbol{p}}) = \mathrm{KL}(\boldsymbol{p}^* \| \widehat{\boldsymbol{p}}^{\mathrm{ML}}) + \mathrm{KL}(\widehat{\boldsymbol{p}}^{\mathrm{ML}} \| \widehat{\boldsymbol{p}}) + \sum_{m=1}^{M} (p^*_m - \widehat{p}^{\mathrm{ML}}_m) \left( \log \widehat{p}^{\mathrm{ML}}_m - \log \widehat{p}_m \right). \tag{15.20}$$

It follows from $\mathcal{D} \in T^N_\varepsilon(\boldsymbol{p}^*)$ that $\mathrm{KL}(\boldsymbol{p}^* \| \widehat{\boldsymbol{p}}^{\mathrm{ML}})/\varepsilon^2$ and $|p^*_m - \widehat{p}^{\mathrm{ML}}_m|/\varepsilon$ are bounded by constants. Then $\mathrm{KL}(\widehat{\boldsymbol{p}}^{\mathrm{ML}} \| \widehat{\boldsymbol{p}}) \leq C'\varepsilon^2$ implies that $|\widehat{p}^{\mathrm{ML}}_m - \widehat{p}_m|/\varepsilon$ is bounded by a constant, and hence all the terms in Eq. (15.20) divided by $\varepsilon^2$ are bounded by constants.

It follows from Eq. (15.19) that

$$\min_{r_w(\boldsymbol{w})} \widetilde{Q} \geq -N \mathrm{KL}(\widehat{\boldsymbol{p}}^{\mathrm{ML}} \| \boldsymbol{p}^*). \tag{15.21}$$

The standard asymptotic theory of the multinomial model implies that twice the right-hand side of Eq. (15.21), with its sign flipped, asymptotically follows the chi-squared distribution with degree of freedom $M - 1$ as discussed in Section 13.4.5. □

This lemma is used for proving the consistency of the VB posterior and evaluating lower-bounds of VB free energy for discrete models in Sections 15.3 and 15.4 and Chapter 16.

## 15.2 Mixture of Gaussians

In this section, we consider the following Gaussian mixture model (GMM) introduced in Section 4.1.1 and give upper- and lower-bounds for the VB free energy (Watanabe and Watanabe, 2004, 2006):

$$p(z|\alpha) = \mathrm{Multinomial}_{K,1}(z;\alpha), \tag{15.22}$$

$$p(\boldsymbol{x}|z,\{\boldsymbol{\mu}_k\}_{k=1}^K) = \prod_{k=1}^K \{\mathrm{Gauss}_M(\boldsymbol{x};\boldsymbol{\mu}_k,\boldsymbol{I}_M)\}^{z_k}, \tag{15.23}$$

$$p(\alpha|\phi) = \mathrm{Dirichlet}_K(\alpha;(\phi,\ldots,\phi)^\top), \tag{15.24}$$

$$p(\boldsymbol{\mu}_k|\boldsymbol{\mu}_0,\xi) = \mathrm{Gauss}_M(\boldsymbol{\mu}_k|\boldsymbol{\mu}_0,(1/\xi)\boldsymbol{I}_M). \tag{15.25}$$

Under the constraint,

$$r(\mathcal{H},\boldsymbol{w}) = r_{\mathcal{H}}(\mathcal{H})r_w(\boldsymbol{w}),$$

the VB posteriors are given as follows:

$$r(\{\boldsymbol{z}^{(n)}\}_{n=1}^N,\alpha,\{\boldsymbol{\mu}_k\}_{k=1}^K) = r_z(\{\boldsymbol{z}^{(n)}\}_{n=1}^N)r_\alpha(\alpha)r_\mu(\{\boldsymbol{\mu}_k\}_{k=1}^K),$$

$$r_z(\{\boldsymbol{z}^{(n)}\}_{n=1}^N) = \prod_{n=1}^N \mathrm{Multinomial}_{K,1}\left(\boldsymbol{z}^{(n)};\widehat{\boldsymbol{z}}^{(n)}\right),$$

$$r_\alpha(\alpha) = \mathrm{Dirichlet}\left(\alpha;(\widehat{\phi}_1,\ldots,\widehat{\phi}_K)^\top\right),$$

$$r_\mu(\{\boldsymbol{\mu}_k\}_{k=1}^K) = \prod_{k=1}^K \mathrm{Gauss}_M\left(\boldsymbol{\mu}_k;\widehat{\boldsymbol{\mu}}_k,\widehat{\sigma}_k^2\boldsymbol{I}_M\right).$$

The variational parameters $\{\boldsymbol{z}^{(n)}\}_{n=1}^N, \{\widehat{\phi}_k\}_{k=1}^K, \{\widehat{\boldsymbol{\mu}}_k,\widehat{\sigma}_k^2\}_{k=1}^K$ minimize the free energy,

$$F = \log\left(\frac{\Gamma(\sum_{k=1}^K\widehat{\phi}_k)}{\prod_{k=1}^K\Gamma(\widehat{\phi}_k)}\right) - \log\left(\frac{\Gamma(K\phi)}{(\Gamma(\phi))^K}\right) - \frac{M}{2}\sum_{k=1}^K\log\left(\xi\widehat{\sigma}_k^2\right) - \frac{KM}{2}$$
$$+\sum_{n=1}^N\sum_{k=1}^K\widehat{z}_k^{(n)}\log\widehat{z}_k^{(n)} + \sum_{k=1}^K\left(\widehat{\phi}_k - \phi - \overline{N}_k\right)\left(\Psi(\widehat{\phi}_k) - \Psi(\textstyle\sum_{k'=1}^K\widehat{\phi}_{k'})\right)$$

$$+\sum_{k=1}^{K}\frac{\xi\left(\|\widehat{\boldsymbol{\mu}}_k-\boldsymbol{\mu}_0\|^2+M\widehat{\sigma}_k^2\right)}{2}+\sum_{k=1}^{K}\frac{\overline{N}_k\left(M\log(2\pi)+M\widehat{\sigma}_k^2\right)}{2}$$
$$+\sum_{k=1}^{K}\frac{\overline{N}_k\|\overline{\boldsymbol{x}}_k-\widehat{\boldsymbol{\mu}}_k\|^2+\sum_{n=1}^{N}\widehat{z}_k^{(n)}\|\boldsymbol{x}^{(n)}-\overline{\boldsymbol{x}}_k\|^2}{2}, \tag{15.26}$$

where

$$\overline{N}_k=\sum_{n=1}^{N}\widehat{z}_k^{(n)}, \tag{15.27}$$

$$\overline{\boldsymbol{x}}_k=\frac{1}{\overline{N}_k}\sum_{n=1}^{N}\boldsymbol{x}^{(n)}\widehat{z}_k^{(n)}. \tag{15.28}$$

The stationary condition of the free energy yields

$$\widehat{z}_k^{(n)}=\frac{\overline{z}_k^{(n)}}{\sum_{k'=1}^{K}\overline{z}_{k'}^{(n)}}, \tag{15.29}$$

$$\widehat{\phi}_k=\overline{N}_k+\phi, \tag{15.30}$$

$$\widehat{\boldsymbol{\mu}}_k=\frac{\overline{N}_k\overline{\boldsymbol{x}}_k+\xi\boldsymbol{\mu}_0}{\overline{N}_k+\xi}, \tag{15.31}$$

$$\widehat{\sigma}_k^2=\frac{1}{\overline{N}_k+\xi}, \tag{15.32}$$

where

$$\overline{z}_k^{(n)}\propto\exp\left(\Psi(\widehat{\phi}_k)-\frac{1}{2}\|\boldsymbol{x}^{(n)}-\widehat{\boldsymbol{\mu}}_k\|^2+M\widehat{\sigma}_k^2\right). \tag{15.33}$$

The following condition is assumed.

**Assumption 15.1** *The true distribution $q(\boldsymbol{x})$ is an $M$-dimensional GMM $p(\boldsymbol{x}|\boldsymbol{w}^*)$, which has $K_0$ components and parameter $\boldsymbol{w}^*=(\boldsymbol{\alpha}^*,\{\boldsymbol{\mu}_k^*\}_{k=1}^{K_0})$:*

$$q(\boldsymbol{x})=p(\boldsymbol{x}|\boldsymbol{w}^*)=\sum_{k=1}^{K_0}\alpha_k^*\mathrm{Gauss}_M(\boldsymbol{x};\boldsymbol{\mu}_k^*,\boldsymbol{I}_M), \tag{15.34}$$

*where $\boldsymbol{x},\boldsymbol{\mu}_k^*\in\mathbb{R}^M$. Suppose that the true distribution can be realized by our model in hand, i.e., $K\geq K_0$ holds.*

Under this condition, we prove the following theorem, which evaluates the relative VB free energy,

$$\widetilde{F}^{\mathrm{VB}}(\mathcal{D})=F^{\mathrm{VB}}(\mathcal{D})-NS_N(\mathcal{D}). \tag{15.35}$$

The proof will appear in the next section.

**Theorem 15.5** *The relative VB free energy of the GMM satisfies*

$$\underline{\lambda}'^{\mathrm{VB}}_{\mathrm{MM}} \log N + NE_N(\widehat{\boldsymbol{w}}) + O_\mathrm{p}(1) \leq \widetilde{F}^{\mathrm{VB}}(\mathcal{D}) \leq \overline{\lambda}'^{\mathrm{VB}}_{\mathrm{MM}} \log N + O_\mathrm{p}(1), \qquad (15.36)$$

*where $E_N$ is the empirical KL divergence* (15.14), *and the coefficients $\underline{\lambda}'^{\mathrm{VB}}_{\mathrm{MM}}$, $\overline{\lambda}'^{\mathrm{VB}}_{\mathrm{MM}}$ are given by*

$$\underline{\lambda}'^{\mathrm{VB}}_{\mathrm{MM}} = \begin{cases} (K-1)\phi + \frac{M}{2} & \left(\phi < \frac{M+1}{2}\right), \\ \frac{MK+K-1}{2} & \left(\phi \geq \frac{M+1}{2}\right), \end{cases} \qquad (15.37)$$

$$\overline{\lambda}'^{\mathrm{VB}}_{\mathrm{MM}} = \begin{cases} (K-K_0)\phi + \frac{MK_0+K_0-1}{2} & \left(\phi < \frac{M+1}{2}\right), \\ \frac{MK+K-1}{2} & \left(\phi \geq \frac{M+1}{2}\right). \end{cases} \qquad (15.38)$$

In this theorem, $E_N(\widehat{\boldsymbol{w}})$ is the training error of the VB estimator. Let $\widehat{\boldsymbol{w}}^{\mathrm{ML}}$ be the ML estimator. Then it immediately follows from Eq. (15.14) that

$$NE_N(\widehat{\boldsymbol{w}}) \geq NE_N(\widehat{\boldsymbol{w}}^{\mathrm{ML}}), \qquad (15.39)$$

where $NE_N(\widehat{\boldsymbol{w}}^{\mathrm{ML}}) = \min_{\boldsymbol{w}} \sum_{n=1}^{N} \log \frac{p(\boldsymbol{x}^{(n)}|\boldsymbol{w}^*)}{p(\boldsymbol{x}^{(n)}|\boldsymbol{w})}$ is the (maximum) log-likelihood ratio statistic with sign inversion. As discussed in Section 13.5.3, it is conjectured for the GMM defined by Eqs. (15.22) and (15.23) that the log-likelihood ratio diverges in the order of $\log\log N$ (Hartigan, 1985). If this conjecture is proved, the statement of the theorem is simplified to

$$\widetilde{F}^{\mathrm{VB}}(\mathcal{D}) = \lambda' \log N + o_\mathrm{p}(\log N),$$

for $\underline{\lambda}'^{\mathrm{VB}}_{\mathrm{MM}} \leq \lambda' \leq \overline{\lambda}'^{\mathrm{VB}}_{\mathrm{MM}}$. Note, however, that even if $NE_N(\widehat{\boldsymbol{w}}^{\mathrm{ML}})$ diverges to minus infinity, Eq. (15.39) does not necessarily mean $NE_N(\widehat{\boldsymbol{w}})$ diverges in the same order. Also note that $NE_N(\widehat{\boldsymbol{w}})$ does not affect the upper-bound in Eq. (15.36).

Since the dimension of the parameter $\boldsymbol{w}$ is $D = MK + K - 1$, the relative Bayes free energy coefficient of regular statistical models, on which the Bayesian information criterion (BIC) (Schwarz, 1978) and the minimum description length (MDL) (Rissanen, 1986) are based, is given by $D/2$. Note that, unlike regular models, the advantage of Bayesian learning for singular models is demonstrated by the asymptotic analysis as seen in Eqs. (13.123), (13.124), and (13.125). Theorem 15.5 claims that the coefficient $\overline{\lambda}'^{\mathrm{VB}}_{\mathrm{MM}}$ of $\log N$ is smaller than $D/2$ when $\phi < (M+1)/2$. This means that the VB free energy $F^{\mathrm{VB}}$ becomes smaller than that of regular models, i.e., $2\lambda'^{\mathrm{VB}} \leq D$ holds.

Theorem 15.5 shows how the hyperparameters affect the learning process. The coefficients $\underline{\lambda}'^{\mathrm{VB}}_{\mathrm{MM}}$ and $\overline{\lambda}'^{\mathrm{VB}}_{\mathrm{MM}}$ in Eqs. (15.37) and (15.38) are divided into two cases. These cases correspond to whether $\phi < \frac{M+1}{2}$ holds, indicating that the influence of the hyperparameter $\phi$ in the prior $p(\alpha|\phi)$ appears depending on the number $M$ of parameters in each component. Let $\widehat{K}$ be the number of components satisfying $\overline{N}_k = \Theta_p(N)$. Then the following corollary follows from the proof of Theorem 15.5.

**Corollary 15.6** *The upper-bound in Eq.* (15.36) *is attained when* $\widehat{K} = K_0$ *if* $\phi < \frac{M+1}{2}$ *and* $\widehat{K} = K$ *if* $\phi \geq \frac{M+1}{2}$.

This corollary implies that the phase transition of the VB posterior occurs at $\phi = \frac{M+1}{2}$, i.e., only when $\phi < \frac{M+1}{2}$, the prior distribution reduces redundant components; otherwise, it uses all the components. The phase transition of the posterior occurs also in Bayesian learning while the phase transition point is different from that of VB learning (Yamazaki and Kaji, 2013).

Theorem 15.5 also implies that the hyperparameter $\phi$ is the only hyperparameter on which the leading term of the VB free energy $F^{\mathrm{VB}}$ depends. This is due to the influence of the hyperparameters on the prior probability density around the true parameters. Consider the case where $K_0 < K$. In this case, for a parameter that gives the true distribution, either of the followings holds: $\alpha_k = 0$ for some $k$ or $\boldsymbol{\mu}_i = \boldsymbol{\mu}_j$ for some pair $(i, j)$. The prior distribution $p(\alpha|\phi)$ given by Eq. (15.24) can drastically change the probability density around the points where $\alpha_k = 0$ for some $k$ by changing the hyperparameter $\phi$ while the prior distribution $p(\boldsymbol{\mu}_k|\boldsymbol{\mu}_0, \xi)$ given by Eq. (15.25) always takes positive values for any values of the hyperparameters $\xi$ and $\boldsymbol{\mu}_0$. While the condition for the prior density $p(\alpha|\phi)$ to diverge at $\alpha_k = 0$ is $\alpha_k < 1$, and hence is independent of $M$, the phase transition point of the VB posterior is $\phi = \frac{M+1}{2}$. As we will see in Section 15.4 for the Bernoulli mixture model, if some of the components are located at the boundary of the parameter space, the leading term of the relative VB free energy depends also on the hyperparameter of the prior for component parameters.

Theorem 15.5 is also extended to the case of the general Dirichlet prior $p(\alpha|\boldsymbol{\phi}) = \mathrm{Dirichlet}_K(\alpha; \boldsymbol{\phi})$, where $\boldsymbol{\phi} = (\phi_1, \ldots, \phi_K)^\top$ is the hyperparameter as follows:

**Theorem 15.7** *(Nakamura and Watanabe, 2014) The relative VB free energy of the GMM satisfies*

$$\sum_{k=1}^{K} \underline{\lambda}_k'^{\mathrm{VB}} \log N + N E_N(\widehat{\boldsymbol{w}}) + O_p(1) \leq \widetilde{F}^{\mathrm{VB}}(\mathcal{D}) \leq \sum_{k=1}^{K} \overline{\lambda}_k'^{\mathrm{VB}} \log N + O_p(1),$$

*where the coefficients* $\underline{\lambda}_k'^{\mathrm{VB}}$, $\overline{\lambda}_k'^{\mathrm{VB}}$ *are given by*

$$\underline{\lambda}_k'^{\mathrm{VB}} = \begin{cases} \phi_k - \frac{1}{2K} & \left(k \neq 1 \ \ and \ \ \phi_k < \frac{M+1}{2}\right), \\ \frac{M+1}{2} - \frac{1}{2K} & \left(k = 1 \ \ or \ \ \phi_k \geq \frac{M+1}{2}\right), \end{cases}$$

$$\overline{\lambda}_k'^{\mathrm{VB}} = \begin{cases} \phi_k - \frac{1}{2K} & \left(k > K_0 \ \ and \ \ \phi_k < \frac{M+1}{2}\right), \\ \frac{M+1}{2} - \frac{1}{2K} & \left(k \leq K_0 \ \ or \ \ \phi_k \geq \frac{M+1}{2}\right). \end{cases}$$

The proof of this theorem is omitted. This theorem implies that the phase transition of the VB posterior of each component occurs at the same transition point $\phi_k = \frac{M+1}{2}$ as Theorem 15.5.

**Proof of Theorem 15.5**

Before proving Theorem 15.5, we show two lemmas where the two terms, $R = \mathrm{KL}(r_w(\boldsymbol{w})\|p(\boldsymbol{w}))$ and $Q = -\log C_{\mathcal{H}}$, in Lemma 15.1 are respectively evaluated.

**Lemma 15.8** *It holds that*

$$\left| R - \left\{ G(\widehat{\boldsymbol{\alpha}}) + \frac{\xi}{2} \sum_{k=1}^{K} \|\widehat{\boldsymbol{\mu}}_k - \boldsymbol{\mu}_0\|^2 \right\} \right| \leq C,$$

*where C is a constant,* $\widehat{\boldsymbol{\mu}}_k = \langle \boldsymbol{\mu}_k \rangle_{r_\mu(\{\boldsymbol{\mu}_k\}_{k=1}^K)} = \frac{\overline{N}_k \overline{\boldsymbol{x}}_k + \xi \boldsymbol{\mu}_0}{\overline{N}_k + \xi}$, *and the function* $G(\widehat{\boldsymbol{\alpha}})$ *of* $\widehat{\boldsymbol{\alpha}} = \left\{ \widehat{\alpha}_k = \langle \alpha_k \rangle_{r_\alpha(\boldsymbol{\alpha})} = \frac{\overline{N}_k + \phi}{N + K\phi} \right\}_{k=1}^{K}$ *is defined by*

$$G(\widehat{\boldsymbol{\alpha}}) = \frac{MK + K - 1}{2} \log N + \left\{ \frac{M+1}{2} - \phi \right\} \sum_{k=1}^{K} \log \widehat{\alpha}_k. \tag{15.40}$$

*Proof* Calculating the KL divergence between the posterior and the prior, we obtain

$$\mathrm{KL}(r_\alpha(\boldsymbol{\alpha})\|p(\boldsymbol{\alpha}|\phi)) = \sum_{k=1}^{K} h(\overline{N}_k) - N\Psi(N + K\phi) + \log \Gamma(N + K\phi) + \log \frac{\Gamma(\phi)^K}{\Gamma(K\phi)}, \tag{15.41}$$

where we use the notation $h(x) = x\Psi(x+\phi) - \log \Gamma(x+\phi)$. Similarly, we obtain

$$\begin{aligned} &\mathrm{KL}(r_\mu(\{\boldsymbol{\mu}_k\}_{k=1}^K)\|p(\{\boldsymbol{\mu}_k\}_{k=1}^K|\boldsymbol{\mu}_0, \xi)) \\ &= \sum_{k=1}^{K} \frac{M}{2} \log \frac{\overline{N}_k + \xi}{\xi} - \frac{KM}{2} + \frac{\xi}{2} \sum_{k=1}^{K} \left\{ \frac{M}{\overline{N}_k + \xi} + \|\widehat{\boldsymbol{\mu}}_k - \boldsymbol{\mu}_0\|^2 \right\}. \end{aligned} \tag{15.42}$$

By using Inequalities (15.8) and (15.9), we obtain

$$-1 + \frac{12\phi - 1}{12(x + \phi)} \leq h(x) + \left(\phi - \frac{1}{2}\right)\log(x + \phi) - x - \phi + \frac{1}{2}\log 2\pi \leq 0. \quad (15.43)$$

Thus, from Eqs. (15.41), (15.42), (15.43), and

$$R = \mathrm{KL}(r_\alpha(\alpha)\|p(\alpha|\phi)) + \mathrm{KL}(r_\mu(\{\mu_k\}_{k=1}^K)\|p(\{\mu_k\}_{k=1}^K|\mu_0, \xi)),$$

it follows that

$$\begin{aligned}
&\left|R - \left\{G(\widehat{\alpha}) + \frac{\xi}{2}\sum_{k=1}^{K}\|\widehat{\mu}_k - \mu_0\|^2\right\}\right| \\
&\leq \frac{MK + K - 1}{2}\log\left(1 + \frac{K\phi}{N}\right) \\
&\quad + (K-1)\left|\phi - \frac{\log 2\pi}{2}\right| + K + \sum_{k=1}^{K}\frac{|12\phi - 1|}{12(\overline{N}_k + \phi)} + \frac{12N + 1}{12(N + K\phi)} \\
&\quad + \left|\log\frac{\Gamma(\phi)^K}{\Gamma(K\phi)}\right| + \left|\sum_{k=1}^{K}\log\frac{\overline{N}_k + \xi}{\overline{N}_k + \phi} - \frac{MK}{2}(1 + \log\xi) + \frac{\xi}{2}\sum_{k=1}^{K}\frac{M}{\overline{N}_k + \xi}\right|.
\end{aligned}$$

The right-hand side of the preceding inequality is bounded by a constant since

$$\frac{1}{N + \xi} < \frac{1}{\overline{N}_k + \xi} < \frac{1}{\xi},$$

and

$$\frac{1}{N + \phi} < \frac{1}{\overline{N}_k + \phi} < \frac{1}{\phi}.$$

□

**Lemma 15.9** *It holds that*

$$\begin{aligned}
Q = -\sum_{n=1}^{N}\log\Bigg(\sum_{k=1}^{K}\frac{1}{\sqrt{2\pi}^M}\exp\Bigg(&\Psi(\overline{N}_k + \phi) - \Psi(N + K\phi) \\
&- \frac{\|x^{(n)} - \widehat{\mu}_k\|^2}{2} - \frac{M}{2}\frac{1}{\overline{N}_k + \xi}\Bigg)\Bigg),
\end{aligned} \quad (15.44)$$

*and*

$$NE_N(\widehat{w}) - \frac{N}{N + K\phi} \leq \widetilde{Q} \leq N\overline{E}_N(\widehat{w}) - \frac{N}{2(N + K\phi)}, \quad (15.45)$$

*where $E_N(\widehat{w})$ is given by Eq. (15.14) and $\overline{E}_N(\widehat{w})$ is defined by*

$$\overline{E}_N(\widehat{w}) = \frac{1}{N}\sum_{n=1}^{N}\log\frac{p(x^{(n)}|w^*)}{\sum_{k=1}^{K}\frac{\widehat{\alpha}_k}{\sqrt{2\pi}^M}\exp\left(-\frac{\|x^{(n)} - \widehat{\mu}_k\|^2}{2} - \frac{M+2}{2(\overline{N}_k + \min\{\phi, \xi\})}\right)}.$$

*Proof*

$$C_{\mathcal{H}} = \prod_{n=1}^{N} \sum_{z^{(n)}} \exp\left\langle \log p(x^{(n)}, z^{(n)}|w) \right\rangle_{r_w(w)}$$
$$= \prod_{n=1}^{N} \sum_{k=1}^{K} \frac{1}{\sqrt{2\pi}^M} \exp\left( \Psi(\overline{N}_k + \phi) - \Psi(N + K\phi) \right.$$
$$\left. - \frac{\|x^{(n)} - \widehat{\mu}_k\|^2}{2} - \frac{M}{2} \frac{1}{\overline{N}_k + \xi} \right).$$

Thus, we have Eq. (15.44).

Using again Inequality (15.8), we obtain

$$Q \leq - \sum_{n=1}^{N} \log\left( \sum_{k=1}^{K} \frac{\widehat{\alpha}_k}{\sqrt{2\pi}^M} \exp\left( -\frac{\|x^{(n)} - \widehat{\mu}_k\|^2}{2} - \frac{M+2}{2(\overline{N}_k + \min\{\phi, \xi\})} \right) \right)$$
$$- \frac{N}{2(N + K\phi)}, \tag{15.46}$$

and

$$Q \geq - \sum_{n=1}^{N} \log\left( \sum_{k=1}^{K} \frac{\widehat{\alpha}_k}{\sqrt{2\pi}^M} \exp\left( -\frac{\|x^{(n)} - \widehat{\mu}_k\|^2}{2} \right) \right) - \frac{N}{N + K\phi},$$

which give upper- and lower-bounds in Eq. (15.45), respectively. □

Now, from the preceding lemmas, we prove Theorem 15.5 by showing upper- and lower-bounds, respectively. First, we show the upper-bound in Eq. (15.36).

From Lemma 15.1, Lemma 15.8, and Lemma 15.9, it follows that

$$F - NS_N(\mathcal{D}) \leq \min_{\widehat{w}} T_N(\widehat{w}) + C, \tag{15.47}$$

where

$$T_N(\widehat{w}) = G(\widehat{\alpha}) + \frac{\xi}{2} \sum_{k=1}^{K} \|\widehat{\mu}_k - \mu_0\|^2 + N\overline{E}_N(\widehat{w}).$$

From Eq. (15.47), it is noted that the function values of $T_N(\widehat{w})$ at specific points of the variational parameter $\widehat{w}$ give upper-bounds of the VB free energy $F^{\mathrm{VB}}(\mathcal{D})$. Hence, let us consider following two cases.

(I) Consider the case where all components, including redundant ones, are used to learn $K_0$ true components, i.e.,

$$\widehat{\alpha}_k = \frac{\alpha_k^* N + \phi}{N + K\phi} \qquad (1 \leq k \leq K_0 - 1),$$

$$\widehat{\alpha}_k = \frac{\alpha^*_{K_0} N/(K - K_0 + 1) + \phi}{N + K\phi} \qquad (K_0 \le k \le K),$$
$$\widehat{\mu}_k = \mu^*_k \qquad (1 \le k \le K_0 - 1),$$
$$\widehat{\mu}_k = \mu^*_{K_0} \qquad (K_0 \le k \le K).$$

Then we obtain

$$\begin{aligned}
&N\overline{E}_N(\widehat{w}) \\
&< \sum_{n=1}^{N} \log p(x^{(n)}|w^*) - \sum_{n=1}^{N} \log \frac{N+\phi}{N+K\phi} \\
&\quad - \sum_{n=1}^{N} \log\left( \sum_{k=1}^{K_0-1} \frac{\alpha^*_k}{\sqrt{2\pi}^M} \exp\left( -\frac{\|x^{(n)} - \mu^*_k\|^2}{2} - \frac{M+2}{2(\alpha^*_k N + \min\{\xi, \phi\})} \right) \right. \\
&\quad \left. + \frac{\alpha^*_{K_0}}{\sqrt{2\pi}^M} \exp\left( -\frac{\|x^{(n)} - \mu^*_k\|^2}{2} - \frac{M+2}{2(\frac{\alpha^*_{K_0}}{K-K_0+1} N + \min\{\xi, \phi\})} \right) \right) \\
&< \sum_{n=1}^{N} \log \frac{\frac{N+K\phi}{N+\phi} p(x^{(n)}|w^*)}{p(x^{(n)}|w^*) \exp\left( -\frac{(M+2)(K-K_0+1)}{2(\min_k\{\alpha^*_k\}N + \min\{\xi,\phi\}(K-K_0+1))} \right)} \\
&< \frac{(K-1)\phi}{N+\phi} + \frac{(M+2)(K-K_0+1)N}{2(\min_k\{\alpha^*_k\}N + \min\{\xi,\phi\}(K-K_0+1))} \\
&\le (K-1)\phi + \left( \frac{M+2}{2} \right) \frac{K-K_0+1}{\min_k\{\alpha^*_k\}},
\end{aligned}$$

where the first inequality follows from $\frac{\alpha^*_k N+\phi}{N+K\phi} > \alpha^*_k \frac{N+\phi}{N+K\phi}$ and the third inequality follows from $\log(1+x) \le x$ for $x > -1$.

It follows that

$$T_N(\widehat{w}) < \frac{MK + K - 1}{2} \log N + C', \tag{15.48}$$

where $C'$ is a constant.

(II) Consider the case where the redundant components are eliminated, i.e.,

$$\widehat{\alpha}_k = \frac{\alpha^*_k N + \phi}{N + K\phi} \qquad (1 \le k \le K_0),$$
$$\widehat{\alpha}_k = \frac{\phi}{N + K\phi} \qquad (K_0 + 1 \le k \le K),$$
$$\widehat{\mu}_k = \mu^*_k \qquad (1 \le k \le K_0),$$
$$\widehat{\mu}_k = \mu_0 \qquad (K_0 + 1 \le k \le K).$$

Then it holds that

$$\begin{aligned}
&N\overline{E}_N(\widehat{\boldsymbol{w}}) \\
&< \sum_{n=1}^{N} \log \frac{p(\boldsymbol{x}^{(n)}|\boldsymbol{w}^*)}{\frac{N+\phi}{N+K\phi}\sum_{k=1}^{K_0}\frac{\alpha_k^*}{\sqrt{2\pi}^M}\exp\left(-\frac{\|\boldsymbol{x}^{(n)}-\boldsymbol{\mu}_k^*\|^2}{2}-\frac{M+2}{2(\alpha_k^*N+\min\{\xi,\phi\})}\right)} \\
&< \frac{(K-1)\phi N}{N+\phi}+\left(\frac{M+2}{2}\right)\frac{N}{\min_k\{\alpha_k^*\}N+\min\{\xi,\phi\}} \\
&\leq (K-1)\phi+\left(\frac{M+2}{2}\right)\frac{1}{\min_k\{\alpha_k^*\}}.
\end{aligned} \tag{15.49}$$

The first inequality follows from $\frac{\alpha_k^* N+\phi}{N+K\phi} > \alpha_k^* \frac{N+\phi}{N+K\phi}$ and

$$\sum_{k=K_0+1}^{K} \frac{\widehat{\alpha}_k}{\sqrt{2\pi}^M}\exp\left(-\frac{\|\boldsymbol{x}^{(n)}-\widehat{\boldsymbol{\mu}}_k\|^2}{2}-\frac{M+2}{2(\overline{N}_k+\min\{\phi,\xi\})}\right) > 0.$$

The second inequality follows from $\log(1+x) \leq x$ for $x > -1$.
It follows that

$$T_N(\widehat{\boldsymbol{w}}) < \left\{(K-K_0)\phi+\frac{MK_0+K_0-1}{2}\right\}\log N + C'', \tag{15.50}$$

where $C''$ is a constant.

From Eqs. (15.47), (15.48), and (15.50), we obtain the upper-bound in Eq. (15.36).

Next we show the lower-bound in Eq. (15.36). It follows from Lemma 15.1, Lemma 15.8, and Lemma 15.9 that

$$F - NS_N(\mathcal{D}) \geq \min_{\widehat{\alpha}}\{G(\widehat{\alpha})\} + N E_N(\widehat{\boldsymbol{w}}) - C - 1. \tag{15.51}$$

If $\phi \geq \frac{M+1}{2}$, then,

$$G(\widehat{\alpha}) \geq \frac{MK+K-1}{2}\log N - \left(\frac{M+1}{2}-\phi\right)K\log K, \tag{15.52}$$

since Jensen's inequality yields that

$$\sum_{k=1}^{K}\log\widehat{\alpha}_k \leq K\log\left(\frac{1}{K}\sum_{k=1}^{K}\widehat{\alpha}_k\right) = K\log\left(\frac{1}{K}\right).$$

If $\phi < \frac{M+1}{2}$, then

$$G(\widehat{\alpha}) \geq \left\{(K-1)\phi + \frac{M}{2}\right\}\log N + \left(\frac{M+1}{2} - \phi\right)(K-1)\log\frac{\phi N}{N+K\phi} + C'''$$
$$\geq \left\{(K-1)\phi + \frac{M}{2}\right\}\log N + \left(\frac{M+1}{2} - \phi\right)(K-1)\log\frac{\phi}{1+K\phi} + C''', \tag{15.53}$$

where $C'''$ is a constant. The first inequality follows since

$$\widehat{\alpha}_k \geq \frac{\phi}{N+K\phi}$$

holds for every $k$, and the constraint

$$\sum_{k=1}^{K} \widehat{\alpha}_k = 1$$

ensures that $|\log \widehat{\alpha}_k|$ is bounded by a constant independent of $N$ for at least one index $k$. From Eqs. (15.51), (15.52), and (15.53) we obtain the lower-bound in Eq. (15.36).

## 15.3 Mixture of Exponential Family Distributions

The previous theorem for the GMM can be generalized to the mixture of *exponential family distributions* (Watanabe and Watanabe, 2005, 2007). The model that we consider is defined by

$$p(z|\alpha) = \mathrm{Multinomial}_{K,1}(z;\alpha), \tag{15.54}$$

$$p(t|z, \{\eta_k\}_{k=1}^K) = \prod_{k=1}^{K}\left\{\exp\left(\eta_k^\top t - A(\eta_k) + B(t)\right)\right\}^{z_k}, \tag{15.55}$$

$$p(\alpha|\phi) = \mathrm{Dirichlet}_K(\alpha;(\phi,\ldots,\phi)^\top), \tag{15.56}$$

$$p(\eta_k|\nu_0,\xi) = \frac{1}{C(\xi,\nu_0)}\exp\left(\xi(\nu_0^\top\eta_k - A(\eta_k))\right). \tag{15.57}$$

As demonstrated in Section 4.1.2, under the constraint, $r(\mathcal{H}, w) = r_{\mathcal{H}}(\mathcal{H})r_w(w)$, the VB posteriors are given as follows:

$$r(\{z^{(n)}\}_{n=1}^N, \alpha, \{\eta_k\}_{k=1}^K) = r_z(\{z^{(n)}\}_{n=1}^N)r_\alpha(\alpha)r_\eta(\{\eta_k\}_{k=1}^K),$$

$$r_z(\{z^{(n)}\}_{n=1}^N) = \prod_{n=1}^{N}\mathrm{Multinomial}_{K,1}\left(z^{(n)};\widehat{z}^{(n)}\right),$$

$$
\begin{aligned}
r_{\alpha}(\boldsymbol{\alpha}) &= \mathrm{Dirichlet}\left(\boldsymbol{\alpha}; (\widehat{\phi}_1, \ldots, \widehat{\phi}_K)^\top\right), \\
r_{\eta}(\{\boldsymbol{\eta}_k\}_{k=1}^K) &= \prod_{k=1}^K \frac{1}{C(\widehat{\xi}_k, \widehat{\boldsymbol{\nu}}_k)} \exp\left(\widehat{\xi}_k(\widehat{\boldsymbol{\nu}}_k^\top \boldsymbol{\eta}_k - A(\boldsymbol{\eta}_k))\right),
\end{aligned}
\tag{15.58}
$$

The variational parameters $\{\widehat{z}^{(n)}\}_{n=1}^N, \{\widehat{\phi}_k\}_{k=1}^K, \{\widehat{\boldsymbol{\nu}}_k, \widehat{\xi}_k\}_{k=1}^K$ minimize the free energy,

$$
\begin{aligned}
F = {} & \log\left(\frac{\Gamma(\sum_{k=1}^K \widehat{\phi}_k)}{\prod_{k=1}^K \Gamma(\widehat{\phi}_k)}\right) - \log\left(\frac{\Gamma(K\phi)}{(\Gamma(\phi))^K}\right) - \sum_{k=1}^K \log C(\widehat{\xi}_k, \widehat{\boldsymbol{\nu}}_k) + K \log C(\xi, \boldsymbol{\nu}_0) \\
& + \sum_{n=1}^N \sum_{k=1}^K \widehat{z}_k^{(n)} \log \widehat{z}_k^{(n)} + \sum_{k=1}^K \left(\widehat{\phi}_k - \phi - \overline{N}_k\right)\left(\Psi(\widehat{\phi}_k) - \Psi(\textstyle\sum_{k'=1}^K \widehat{\phi}_{k'})\right) \\
& + \sum_{k=1}^K \left[\widehat{\boldsymbol{\eta}}_k^\top \left\{\xi\left(\widehat{\boldsymbol{\nu}}_k - \boldsymbol{\nu}_0\right) + \overline{N}_k\left(\widehat{\boldsymbol{\nu}}_k - \overline{\boldsymbol{t}}_k\right)\right\} + \left(\widehat{\xi}_k - \xi - \overline{N}_k\right) \frac{\partial \log C(\widehat{\xi}_k, \widehat{\boldsymbol{\nu}}_k)}{\partial \xi_k}\right] \\
& - \sum_{n=1}^N B(\boldsymbol{t}^{(n)}),
\end{aligned}
\tag{15.59}
$$

where

$$
\begin{aligned}
\overline{N}_k &= \sum_{n=1}^N \widehat{z}_k^{(n)}, \\
\overline{\boldsymbol{t}}_k &= \frac{1}{\overline{N}_k} \sum_{n=1}^N \left\langle z_k^{(n)} \right\rangle_{r_{\mathcal{H}}(\mathcal{H})} \boldsymbol{t}^{(n)}, \\
\widehat{\boldsymbol{\eta}}_k &= \frac{1}{\widehat{\xi}_k} \frac{\partial \log C(\widehat{\xi}_k, \widehat{\boldsymbol{\nu}}_k)}{\partial \boldsymbol{\nu}_k}.
\end{aligned}
$$

The stationary condition of the free energy yields

$$\widehat{z}_k^{(n)} = \frac{\overline{z}_k^{(n)}}{\sum_{k'=1}^K \overline{z}_{k'}^{(n)}}, \tag{15.60}$$

$$\widehat{\alpha}_k = \overline{N}_k + \phi, \tag{15.61}$$

$$\widehat{\boldsymbol{\nu}}_k = \frac{\overline{N}_k \overline{\boldsymbol{t}}_k + \xi \boldsymbol{\nu}_0}{\overline{N}_k + \xi}, \tag{15.62}$$

$$\widehat{\xi}_k = \overline{N}_k + \xi. \tag{15.63}$$

where

$$\overline{z}_k^{(n)} \propto \exp\left(\Psi(\widehat{\phi}_k) + \widehat{\boldsymbol{\eta}}_k^\top \boldsymbol{t}^{(n)} - \langle A(\boldsymbol{\eta}_k) \rangle_{r_{\eta}(\boldsymbol{\eta}_k)}\right). \tag{15.64}$$

We assume the following conditions.

**Assumption 15.2** *The true distribution $q(\boldsymbol{t})$ of sufficient statistics is represented by a mixture of exponential family distributions $p(\boldsymbol{t}|\boldsymbol{w}^*)$, which has $K_0$ components and the parameter $\boldsymbol{w}^* = \{\alpha_k^*, \boldsymbol{\eta}_k^*\}_{k=1}^{K_0}$:*

$$q(\boldsymbol{t}) = p(\boldsymbol{t}|\boldsymbol{w}^*) = \sum_{k=1}^{K_0} \alpha_k^* \exp\left(\boldsymbol{\eta}_k^{*\top}\boldsymbol{t} - A(\boldsymbol{\eta}_k^*) + B(\boldsymbol{t})\right),$$

*where $\boldsymbol{\eta}_k^* \in \mathbb{R}^M$ and $\boldsymbol{\eta}_k^* \neq \boldsymbol{\eta}_{k'}^* (k \neq k')$. Also, assume that the true distribution can be achieved with the model, i.e., $K \geq K_0$ holds.*

**Assumption 15.3** *The prior distribution $p(\{\boldsymbol{\eta}_k\}_{k=1}^K|\nu_0, \xi)$ defined by Eq. (15.57) satisfies $0 < p(\{\boldsymbol{\eta}_k\}_{k=1}^K|\nu_0, \xi) < \infty$.*

**Assumption 15.4** *Regarding the distribution $p(\boldsymbol{t}|\boldsymbol{\eta})$ of each component, the Fisher information matrix*

$$\boldsymbol{F}(\boldsymbol{\eta}) = \frac{\partial^2 A(\boldsymbol{\eta})}{\partial\boldsymbol{\eta}\partial\boldsymbol{\eta}^\top}$$

*satisfies $0 < \det(\boldsymbol{F}(\boldsymbol{\eta})) < +\infty$ for an arbitrary $\boldsymbol{\eta} \in \boldsymbol{H}$. The function $\boldsymbol{\nu}^\top\boldsymbol{\eta} - A(\boldsymbol{\eta})$ has a stationary point at $\widehat{\boldsymbol{\eta}}$ in the interior of $\boldsymbol{H}$ for each $\boldsymbol{\nu} \in \left\{\frac{\partial A(\boldsymbol{\eta})}{\partial \boldsymbol{\eta}}; \boldsymbol{\eta} \in \boldsymbol{H}\right\}$.*

The following theorem will be proven under these conditions. The proof will appear in the next section. Here,

$$S_N(\mathcal{D}) = -\frac{1}{N}\sum_{n=1}^{N} \log p(\boldsymbol{t}^{(n)}|\boldsymbol{w}^*) \tag{15.65}$$

is the empirical entropy.

**Theorem 15.10** *The relative VB free energy of the mixture of exponential family distributions satisfies*

$$\begin{aligned} \underline{\lambda}_{\mathrm{MM}}^{\prime\mathrm{VB}} \log N + NE_N(\widehat{\boldsymbol{w}}) + O_{\mathrm{p}}(1) \leq \widetilde{F}^{\mathrm{VB}}(\mathcal{D}) &= F^{\mathrm{VB}}(\mathcal{D}) - NS_N(\mathcal{D}) \\ &\leq \overline{\lambda}_{\mathrm{MM}}^{\prime\mathrm{VB}} \log N + O_{\mathrm{p}}(1), \end{aligned} \tag{15.66}$$

*where $\underline{\lambda}_{\mathrm{MM}}^{\prime\mathrm{VB}}$ and $\overline{\lambda}_{\mathrm{MM}}^{\prime\mathrm{VB}}$ are given by*

$$\underline{\lambda}_{\mathrm{MM}}^{\prime\mathrm{VB}} = \begin{cases} (K-1)\phi + \frac{M}{2} & (\phi < \frac{M+1}{2}), \\ \frac{MK+K-1}{2} & (\phi \geq \frac{M+1}{2}), \end{cases} \tag{15.67}$$

$$\overline{\lambda}_{\mathrm{MM}}^{\prime\mathrm{VB}} = \begin{cases} (K-K_0)\phi + \frac{MK_0+K_0-1}{2} & (\phi < \frac{M+1}{2}), \\ \frac{MK+K-1}{2} & (\phi \geq \frac{M+1}{2}). \end{cases} \tag{15.68}$$

Again in this theorem,

$$E_N(\widehat{\boldsymbol{w}}) = \frac{1}{N}\sum_{n=1}^{N} \log \frac{p(\boldsymbol{t}^{(n)}|\boldsymbol{w}^*)}{p(\boldsymbol{t}^{(n)}|\widehat{\boldsymbol{w}})} \tag{15.69}$$

is the training error, and

$$NE_N(\widehat{\boldsymbol{w}}) \geq NE_N(\widehat{\boldsymbol{w}}^{\mathrm{ML}}), \tag{15.70}$$

holds for the (maximum) log-likelihood ratio statistic. As discussed in Section 13.5.3, the log-likelihood ratio statistics of some singular models diverge to infinity as $N$ increases. Some known facts about the divergence of the log-likelihood ratio are described in the following examples. Note again that even if $NE_N(\widehat{\boldsymbol{w}}^{\mathrm{ML}})$ diverges to minus infinity, Eq. (15.70) does not necessarily mean $NE_N(\widehat{\boldsymbol{w}})$ diverges in the same order.

If the domain of the sufficient statistics $\boldsymbol{t}$ of the model $p(\boldsymbol{t}|\boldsymbol{w})$ is discrete and finite, we obtain the following theorem by Lemmas 15.3 and 15.4:

**Theorem 15.11** *If the domain of the sufficient statistics* $\boldsymbol{t}$ *is discrete and finite, the relative VB free energy of the mixture of exponential family distributions satisfies*

$$\widetilde{F}^{\mathrm{VB}}(\mathcal{D}) = \overline{\lambda}'^{\mathrm{VB}}_{\mathrm{MM}} \log N + O_{\mathrm{p}}(1), \tag{15.71}$$

*where the coefficient* $\overline{\lambda}'^{\mathrm{VB}}_{\mathrm{MM}}$ *is given by Eq.* (15.68).

The proof of this theorem follows the proof of the preceding theorem.

## Examples

The following are examples where Theorems 15.10 and 15.11 apply.

**Example 1 (Binomial)** Consider a mixture of binomial component distributions. Each component has a one-dimensional parameter $\nu \in [0,\ 1]$:

$$p(x = k|\nu) = \mathrm{Binomial}_T(k; \nu) = \binom{T}{k} \nu^k (1-\nu)^{T-k}, \tag{15.72}$$

where $T$ is the number of Bernoulli trials and $k = 0, 1, 2, \ldots, T$. Hence, $M = 1$ and the natural parameter is given by $\eta = \log \frac{\nu}{1-\nu}$. Theorem 15.11 applies with $M = 1$.

**Example 2 (Gamma)** Consider the gamma component with shape parameter $\alpha > 0$ and scale parameter $\beta > 0$:

$$p(x|\alpha,\beta) = \mathrm{Gamma}(x;\alpha,\beta) = \frac{\beta^{\alpha}}{\Gamma(\alpha)} x^{\alpha-1} \exp(-\beta x), \tag{15.73}$$

where $0 \le x < \infty$. The natural parameter $\boldsymbol{\eta}$ is given by $\eta_1 = \beta$ and $\eta_2 = \alpha - 1$. Hence, Eq. (15.66) holds where $\underline{\lambda}'^{\mathrm{VB}}_{\mathrm{MM}}$ and $\overline{\lambda}'^{\mathrm{VB}}_{\mathrm{MM}}$ are given by Eqs. (15.67) and (15.68) with $M = 2$. When shape parameter $\alpha$ is known, the likelihood ratio in ML learning diverges in the order of $\log\log N$ (Liu et al., 2003). This implies that $NE_N(\widehat{\boldsymbol{w}}) = O_{\mathrm{p}}(\log\log N)$ from Eq. (15.70).

**Example 3 (Gaussian)** Consider the $L$-dimensional Gaussian component with mean $\boldsymbol{\mu}$ and covariance matrix $\boldsymbol{\Sigma}$:

$$p(\boldsymbol{x}|\boldsymbol{\mu},\boldsymbol{\Sigma}) = \mathrm{Gauss}_L(\boldsymbol{x};\boldsymbol{\mu},\boldsymbol{\Sigma}) = \frac{1}{(2\pi)^{L/2}|\boldsymbol{\Sigma}|^{1/2}} \exp\left(-\frac{1}{2}(\boldsymbol{x}-\boldsymbol{\mu})^{\top}\boldsymbol{\Sigma}^{-1}(\boldsymbol{x}-\boldsymbol{\mu})\right).$$

The natural parameter $\boldsymbol{\eta}$ is given by $\boldsymbol{\mu}^T\boldsymbol{\Sigma}^{-1}$ and $\boldsymbol{\Sigma}^{-1}$. These are functions of the elements of $\boldsymbol{\mu}$ and the upper-right half of $\boldsymbol{\Sigma}^{-1}$. Hence, Eq. (15.66) holds where $\underline{\lambda}'^{\mathrm{VB}}_{\mathrm{MM}}$ and $\overline{\lambda}'^{\mathrm{VB}}_{\mathrm{MM}}$ are given by Eqs. (15.67) and (15.68) with $M = L + L(L+1)/2$. If the covariance matrix $\boldsymbol{\Sigma}$ is known and the parameter is restricted to mean $\boldsymbol{\mu}$, it is conjectured that the likelihood ratio in ML learning diverges in the order of $\log\log N$ (Hartigan, 1985). This suggests that the likelihood ratio can diverge in a higher order than $\log\log N$ if the covariance matrices are also estimated.

Other than these examples, Theorems 15.10 and 15.11 apply to mixtures of distributions such as multinomial, Poisson, and Weibull.

## Proof of Theorem 15.10

Here Theorem 15.10 is proved in the same way as Theorem 15.5.

Since the VB posterior satisfies $r_w(\boldsymbol{w}) = r_\alpha(\boldsymbol{\alpha})r_\eta(\{\boldsymbol{\eta}_k\}_{k=1}^K)$, we have

$$\begin{aligned} R &= \mathrm{KL}(r_w(\boldsymbol{w})\|p(\boldsymbol{w})) \\ &= \mathrm{KL}(r_\alpha(\boldsymbol{\alpha})\|p(\boldsymbol{\alpha}|\phi)) + \sum_{k=1}^{K} \mathrm{KL}(r_\eta(\boldsymbol{\eta}_k)\|p(\boldsymbol{\eta}_k|\boldsymbol{\nu}_0,\xi)). \end{aligned} \tag{15.74}$$

The following lemma is used for evaluating $\mathrm{KL}(r_\eta(\boldsymbol{\eta}_k)\|p(\boldsymbol{\eta}_k|\boldsymbol{\nu}_0,\xi))$ in the case of the mixture of exponential family distributions.

**Lemma 15.12** *It holds that*

$$\mathrm{KL}(r_\eta(\boldsymbol{\eta}_k)\|p(\boldsymbol{\eta}_k|\boldsymbol{\nu}_0,\xi)) = \frac{M}{2}\log(\overline{N}_k + \xi) - \log p(\widehat{\boldsymbol{\eta}}_k|\boldsymbol{\nu}_0,\xi) + O_{\mathrm{p}}(1),$$

*where*

$$\widehat{\boldsymbol{\eta}}_k = \langle \boldsymbol{\eta}_k \rangle_{r_\eta(\boldsymbol{\eta}_k)} = \frac{1}{\widehat{\xi}_k} \frac{\partial \log C(\widehat{\xi}_k, \widehat{\boldsymbol{\nu}}_k)}{\partial \widehat{\boldsymbol{\nu}}_k}. \tag{15.75}$$

*Proof* Using the VB posterior, Eq. (15.58), we obtain

$$\mathrm{KL}(r_\eta(\boldsymbol{\eta}_k) \| p(\boldsymbol{\eta}_k | \boldsymbol{\nu}_0, \xi)) = -\log \frac{C(\widehat{\xi}_k, \widehat{\boldsymbol{\nu}}_k)}{C(\xi, \boldsymbol{\nu}_0)} + \overline{N}_k \left\{ \widehat{\boldsymbol{\nu}}_k \langle \boldsymbol{\eta}_k \rangle_{r_\eta(\boldsymbol{\eta}_k)} - \langle A(\boldsymbol{\eta}_k) \rangle_{r_\eta(\boldsymbol{\eta}_k)} \right\}, \tag{15.76}$$

where we used $\widehat{\xi}_k = \overline{N}_k + \xi$. Let us now evaluate the value of $C(\widehat{\xi}_k, \widehat{\boldsymbol{\nu}}_k)$ when $\widehat{\xi}_k$ is sufficiently large. From Assumption 15.4, using the saddle point approximation, we obtain

$$C(\widehat{\xi}_k, \widehat{\boldsymbol{\nu}}_k) = \exp\left(\widehat{\xi}_k \{\widehat{\boldsymbol{\nu}}_k^\top \widetilde{\boldsymbol{\eta}}_k - A(\widetilde{\boldsymbol{\eta}}_k)\}\right) \left(\frac{2\pi}{\widehat{\xi}_k}\right)^{M/2} \sqrt{\det\left(\boldsymbol{F}(\widetilde{\boldsymbol{\eta}}_k)\right)^{-1}} \left\{1 + O_{\mathrm{p}}\left(\frac{1}{\widehat{\xi}_k}\right)\right\}, \tag{15.77}$$

where $\widetilde{\boldsymbol{\eta}}_k$ is the maximizer of the function $\widehat{\boldsymbol{\nu}}^\top \boldsymbol{\eta}_k - A(\boldsymbol{\eta}_k)$, that is,

$$\frac{\partial A(\widetilde{\boldsymbol{\eta}}_k)}{\partial \boldsymbol{\eta}_k} = \widehat{\boldsymbol{\nu}}_k.$$

Therefore, $-\log C(\widehat{\xi}_k, \widehat{\boldsymbol{\nu}}_k)$ is evaluated as

$$-\log C(\widehat{\xi}_k, \widehat{\boldsymbol{\nu}}_k) = \frac{M}{2} \log \frac{\widehat{\xi}_k}{2\pi} + \frac{1}{2} \log \det\left(\boldsymbol{F}(\widetilde{\boldsymbol{\eta}}_k)\right) - \widehat{\xi}_k \left\{\widehat{\boldsymbol{\nu}}_k^\top \widetilde{\boldsymbol{\eta}}_k - A(\widetilde{\boldsymbol{\eta}}_k)\right\} + O_{\mathrm{p}}\left(\frac{1}{\widehat{\xi}_k}\right). \tag{15.78}$$

Applying the saddle point approximation to

$$\boldsymbol{\eta}_k - \widetilde{\boldsymbol{\eta}}_k = \frac{1}{C(\widehat{\xi}_k, \widehat{\boldsymbol{\nu}}_k)} \int (\boldsymbol{\eta}_k - \widetilde{\boldsymbol{\eta}}_k) \exp\left(\widehat{\xi}_k \left\{\widehat{\boldsymbol{\nu}}_k^\top \boldsymbol{\eta}_k - A(\boldsymbol{\eta}_k)\right\}\right) d\boldsymbol{\eta}_k,$$

we obtain

$$\|\widehat{\boldsymbol{\eta}}_k - \widetilde{\boldsymbol{\eta}}_k\| \leq \frac{A'}{\widehat{\xi}_k} + O_{\mathrm{p}}\left(\widehat{\xi}_k^{-3/2}\right), \tag{15.79}$$

where $A'$ is a constant. Since

$$A(\boldsymbol{\eta}_k) - A(\widetilde{\boldsymbol{\eta}}_k) = (\boldsymbol{\eta}_k - \widetilde{\boldsymbol{\eta}}_k)^\top \widehat{\boldsymbol{\nu}}_k + \frac{1}{2} (\boldsymbol{\eta}_k - \widetilde{\boldsymbol{\eta}}_k)^\top \boldsymbol{F}(\widetilde{\boldsymbol{\eta}}_k)(\boldsymbol{\eta}_k - \widetilde{\boldsymbol{\eta}}_k), \tag{15.80}$$

for some point $\overline{\boldsymbol{\eta}}_k$ on the line segment between $\boldsymbol{\eta}_k$ and $\widetilde{\boldsymbol{\eta}}_k$, we have

$$A(\widehat{\boldsymbol{\eta}}_k) - A(\widetilde{\boldsymbol{\eta}}_k) = (\widehat{\boldsymbol{\eta}}_k - \widetilde{\boldsymbol{\eta}}_k)^\top \widehat{\boldsymbol{\nu}}_k + O_{\mathrm{p}}(\widehat{\xi}_k^{-2}), \tag{15.81}$$

and applying the saddle point approximation, we obtain

$$\langle A(\boldsymbol{\eta}_k)\rangle_{r_\eta(\boldsymbol{\eta}_k)} - A(\widetilde{\boldsymbol{\eta}}_k) = (\widehat{\boldsymbol{\eta}}_k - \widetilde{\boldsymbol{\eta}}_k)^\top \widehat{\boldsymbol{\nu}}_k + \frac{M}{2\widehat{\xi}_k} + O_{\mathrm{p}}\left(\widehat{\xi}_k^{-3/2}\right). \tag{15.82}$$

From Eqs. (15.81) and (15.82), we have

$$\langle A(\boldsymbol{\eta}_k)\rangle_{r_\eta(\boldsymbol{\eta}_k)} - A(\bar{\boldsymbol{\eta}}_k) = \frac{M}{2\widehat{\xi}_k} + O_{\mathrm{p}}\left(\widehat{\xi}_k^{-3/2}\right), \tag{15.83}$$

Thus, from Eqs. (15.76), (15.78), (15.81), and (15.82), we obtain the lemma. □

Lemmas 15.8 and 15.9 are substituted by the following lemmas.

**Lemma 15.13** *It holds that*

$$\left| R - G(\widehat{\boldsymbol{\alpha}}) + \sum_{k=1}^{K} \log p(\widehat{\boldsymbol{\eta}}_k | \nu_0, \xi) \right| \leq C, \tag{15.84}$$

*where $C$ is a constant and the function $G(\widehat{\boldsymbol{\alpha}})$ is defined by Eq.* (15.40).

*Proof* From Eqs. (15.41), (15.43), and (15.74) and Lemma 15.12,

$$\left| R - G(\widehat{\boldsymbol{\alpha}}) + \sum_{k=1}^{K} \log p(\widehat{\boldsymbol{\eta}}_k | \nu_0, \xi) \right|$$

is upper-bounded by a constant since

$$\frac{1}{N + \xi} < \frac{1}{\overline{N}_k + \xi} < \frac{1}{\xi}.$$

□

**Lemma 15.14** *It holds that*

$$N E_N(\widehat{\boldsymbol{w}}) + O_{\mathrm{p}}(1) \leq \widetilde{Q} = -\log C_{\mathcal{H}} - N S_N(\mathcal{D}) \leq N \overline{E}_N(\widehat{\boldsymbol{w}}) + O_{\mathrm{p}}(1), \tag{15.85}$$

*where the function $E_N(\boldsymbol{w})$ is defined by Eq.* (15.69) *and*

$$\overline{E}_N(\widehat{\boldsymbol{w}}) = \frac{1}{N} \sum_{n=1}^{N} \log \frac{p(\boldsymbol{t}^{(n)} | \boldsymbol{w}^*)}{\sum_{k=1}^{K} \widehat{\alpha}_k p(\boldsymbol{t}^{(n)} | \widehat{\boldsymbol{\eta}}_k) \exp\left(-\frac{A}{\overline{N}_k + \min\{\phi, \xi\}}\right)},$$

*where $A$ is a constant.*

*Proof*

$$C_{\mathcal{H}} = \prod_{n=1}^{N} \sum_{k=1}^{K} \exp \left\langle \log \alpha_k p(\boldsymbol{t}^{(n)}|\boldsymbol{\eta}_k) \right\rangle_{r_w(\boldsymbol{w})}$$
$$= \prod_{n=1}^{N} \sum_{k=1}^{K} \exp \left( \Psi(\overline{N}_k + \phi) - \Psi(N + K\phi) + \widehat{\boldsymbol{\eta}}_k^{\top} \boldsymbol{t}^{(n)} - \langle A(\boldsymbol{\eta}_k) \rangle_{r_\eta(\boldsymbol{\eta}_k)} + B(\boldsymbol{t}^{(n)}) \right).$$

Again, using the inequalities in Eqs. (15.8) and (15.83), we obtain

$$Q \leq \sum_{n=1}^{N} \log \left( \sum_{k=1}^{K} \widehat{\alpha}_k p(\boldsymbol{t}^{(n)}|\widehat{\boldsymbol{\eta}}_k) \exp \left( -\frac{M+2}{2(\overline{N}_k + \min\{\phi, \xi\})} + O_{\mathrm{p}}\left( \overline{N}_k^{-\frac{3}{2}} \right) \right) \right) + O_{\mathrm{p}}(1),$$
$$Q \geq - \sum_{n=1}^{N} \log \left( \sum_{k=1}^{K} \widehat{\alpha}_k p(\boldsymbol{t}^{(n)}|\widehat{\boldsymbol{\eta}}_k) \right) + O_{\mathrm{p}}(1),$$

which give the upper- and lower-bounds in Eq. (15.85), respectively. □

Since the prior distribution $p(\{\boldsymbol{\eta}_k\}_{k=1}^K|\nu_0, \xi)$ satisfies $0 < p(\{\boldsymbol{\eta}_k\}_{k=1}^K|\nu_0, \xi) < \infty$, from Lemmas 15.13 and 15.14, we complete the proof of Theorem 15.10 in the same way as that of Theorem 15.5.

**Proof of Theorem 15.11**

The upper-bound follows from Theorem 15.10. From Lemmas 15.1 and 15.13 and the boundedness of the prior, we have the following lower-bound:

$$F - NS_N(\mathcal{D}) \geq G(\widehat{\boldsymbol{\alpha}}) + \widetilde{Q} + O_{\mathrm{p}}(1).$$

Lemma 15.4 implies that for $\varepsilon > 0$, if $\mathcal{D} \in T_\varepsilon^N(\boldsymbol{p}^*)$ and $\widehat{\boldsymbol{p}} \notin R^*_{C\varepsilon^2}$ for the constant $C$ in the lemma,

$$\widetilde{Q} = \Omega_{\mathrm{p}}(N).$$

Since $G(\widehat{\boldsymbol{\alpha}}) = O_{\mathrm{p}}(\log N)$, this means that if the free energy is minimized, $\widehat{\boldsymbol{p}} \in R^*_{C\varepsilon^2}$ for sufficiently large $N$, which implies that at least $K_0$ components are active and

$$|\log \widehat{\alpha}_k| = O_{\mathrm{p}}(1)$$

holds for at least $K_0$ components. By minimizing $G(\widehat{\boldsymbol{\alpha}})$ under this constraint and the second assertion of Lemma 15.4, we have

$$F^{\mathrm{VB}}(\mathcal{D}) - NS_N(\mathcal{D}) \geq \overline{\lambda}'^{\mathrm{VB}}_{\mathrm{MM}} \log N + O_{\mathrm{p}}(1),$$

for $\mathcal{D} \in T_\varepsilon^N(\boldsymbol{p}^*)$. Because the probability that the observed data sequence is strongly $\varepsilon$-typical tends to 1 as $N \to \infty$ for any $\varepsilon > 0$ by Lemma 15.3, we obtain the theorem.

## 15.4 Mixture of Bernoulli with Deterministic Components

In the previous sections, we assumed that all true component parameters are in the interior of the parameter space. In this section, we consider the *Bernoulli mixture model* when some components are at the boundary of the parameter space (Kaji et al., 2010).

For an $M$-dimensional binary vector, $\boldsymbol{x} = (x_1, \ldots, x_M)^\top \in \{0, 1\}^M$, we define the Bernoulli distribution with parameter $\boldsymbol{\mu} = (\mu_1, \ldots, \mu_M)^\top$ as

$$\mathrm{Bern}_M(\boldsymbol{x}|\boldsymbol{\mu}) = \prod_{m=1}^{M} \mu_m^{x_m}(1 - \mu_m)^{(1-x_m)}.$$

For each element of $\boldsymbol{\mu}$, its conjugate prior, the Beta distribution, is given by

$$\mathrm{Beta}(\mu; a, b) = \frac{1}{\mathcal{B}(a, b)}\mu^{a-1}(1 - \mu)^{b-1},$$

for $a, b > 0$.

The Bernoulli mixture model that we consider is given by

$$p(\boldsymbol{z}|\boldsymbol{\alpha}) = \mathrm{Multinomial}_{K,1}(\boldsymbol{z}; \boldsymbol{\alpha}), \tag{15.86}$$

$$p(\boldsymbol{x}|\boldsymbol{z}, \{\boldsymbol{\mu}_k\}_{k=1}^K) = \prod_{k=1}^{K} \{\mathrm{Bern}_M(\boldsymbol{x}; \boldsymbol{\mu}_k)\}^{z_k}, \tag{15.87}$$

$$p(\boldsymbol{\alpha}|\phi) = \mathrm{Dirichlet}_K(\boldsymbol{\alpha}; (\phi, \ldots, \phi)^\top), \tag{15.88}$$

$$p(\boldsymbol{\mu}_k|\xi) = \prod_{m=1}^{M} \mathrm{Beta}(\mu_{km}; \xi, \xi), \tag{15.89}$$

where $\phi > 0$ and $\xi > 0$ are hyperparameters. Under the constraint, $r(\mathcal{H}, \boldsymbol{w}) = r_{\mathcal{H}}(\mathcal{H})r_w(\boldsymbol{w})$, the VB posteriors are given as follows:

$$r(\{\boldsymbol{z}^{(n)}\}_{n=1}^N, \boldsymbol{\alpha}, \{\boldsymbol{\mu}_k\}_{k=1}^K) = r_z(\{\boldsymbol{z}^{(n)}\}_{n=1}^N)r_\alpha(\boldsymbol{\alpha})r_\mu(\{\boldsymbol{\mu}_k\}_{k=1}^K),$$

$$r_z(\{\boldsymbol{z}^{(n)}\}_{n=1}^N) = \prod_{n=1}^{N} \mathrm{Multinomial}_{K,1}\left(\boldsymbol{z}^{(n)}; \widehat{\boldsymbol{z}}^{(n)}\right),$$

$$r_\alpha(\boldsymbol{\alpha}) = \mathrm{Dirichlet}\left(\boldsymbol{\alpha}; (\widehat{\phi}_1, \ldots, \widehat{\phi}_K)^\top\right),$$

$$r_\mu(\{\boldsymbol{\mu}_k\}_{k=1}^K) = \prod_{k=1}^{K}\prod_{m=1}^{M} \mathrm{Beta}\left(\mu_{km}; \widehat{a}_{km}, \widehat{b}_{km}\right).$$

The variational parameters $\{z^{(n)}\}_{n=1}^N, \{\widehat{\phi}_k\}_{k=1}^K, \{\{\widehat{a}_{km}\}_{m=1}^M, \{\widehat{b}_{km}\}_{m=1}^M\}_{k=1}^K$ minimize the free energy,

$$F = \sum_{n=1}^{N}\sum_{k=1}^{K} \widehat{z}_k^{(n)} \log \widehat{z}_k^{(n)} + \log\left(\frac{\Gamma(\sum_{k=1}^K \widehat{\phi}_k)}{\prod_{k=1}^K \Gamma(\widehat{\phi}_k)}\right) - \log\left(\frac{\Gamma(K\phi)}{(\Gamma(\phi))^K}\right)$$
$$+ \sum_{k=1}^{K}\left(\widehat{\phi}_k - \phi - \overline{N}_k\right)\left(\Psi(\widehat{\phi}_k) - \Psi(\textstyle\sum_{k'=1}^K \widehat{\phi}_{k'})\right)$$
$$+ \sum_{k=1}^{K}\sum_{m=1}^{M}\left\{\log\left(\frac{\Gamma(\widehat{a}_{km} + \widehat{b}_{km})}{\Gamma(\widehat{a}_{km})\Gamma(\widehat{b}_{km})}\right) - \log\left(\frac{\Gamma(2\xi)}{(\Gamma(\xi))^2}\right)\right.$$
$$+ \left(\widehat{a}_{km} - \xi - \overline{N}_k\overline{x}_{km}\right)\left(\Psi(\widehat{a}_{km}) - \Psi(\widehat{a}_{km} + \widehat{b}_{km})\right)$$
$$\left. + \left(\widehat{b}_{km} - \xi - \overline{N}_k(1 - \overline{x}_{km})\right)\left(\Psi(\widehat{b}_{km}) - \Psi(\widehat{a}_{km} + \widehat{b}_{km})\right)\right\},$$

where

$$\overline{N}_k = \sum_{n=1}^{N} \left\langle z_k^{(n)} \right\rangle_{r_{\mathcal{H}}(\mathcal{H})},$$
$$\overline{x}_{km} = \frac{1}{\overline{N}_k}\sum_{n=1}^{N} \left\langle z_k^{(n)} \right\rangle_{r_{\mathcal{H}}(\mathcal{H})} x_m^{(n)},$$

for $k = 1, \ldots, K$ and $m = 1, \ldots, M$. The stationary condition of the free energy yields

$$\widehat{z}_k^{(n)} = \frac{\overline{z}_k^{(n)}}{\sum_{k'=1}^K \overline{z}_{k'}^{(n)}}, \tag{15.90}$$
$$\widehat{\phi}_k = \overline{N}_k + \phi, \tag{15.91}$$
$$\widehat{a}_{km} = \overline{N}_k\overline{x}_{km} + \xi, \tag{15.92}$$
$$\widehat{b}_{km} = \overline{N}_k(1 - \overline{x}_{km}) + \xi, \tag{15.93}$$

where

$$\overline{z}_k^{(n)} = \exp\left(\Psi(\widehat{\phi}_k) - \Psi(\textstyle\sum_{k'=1}^K \widehat{\phi}_{k'}) + \sum_{m=1}^M \left\{x_m^{(n)}\left(\Psi(\widehat{a}_{km}) - \Psi(\widehat{a}_{km} + \widehat{b}_{km})\right)\right.\right.$$
$$\left.\left. + (1 - x_m^{(n)})\left(\Psi(\widehat{b}_{km}) - \Psi(\widehat{a}_{km} + \widehat{b}_{km})\right)\right\}\right). \tag{15.94}$$

We assume the following condition.

**Assumption 15.5** *For $0 \leq K_1^* \leq K_0^* \leq K$, the true distribution $q(\boldsymbol{x}) = p(\boldsymbol{x}|\boldsymbol{w}^*)$ is represented by $K_0^*$ components and the parameter is given by $\boldsymbol{w}^* = \{\alpha_k^*, \boldsymbol{\mu}_k^*\}_{k=1}^{K_0^*}$:*

$$q(\boldsymbol{x}) = p(\boldsymbol{x}|\boldsymbol{w}^*) = \sum_{k=1}^{K_0^*} \alpha_k^* \mathrm{Bern}_M(\boldsymbol{x}; \boldsymbol{\mu}_k^*),$$

*where*

$$0 < \mu^*_{km} < 1 \quad (1 \leq k \leq K^*_1),$$

$$\mu^*_{km} = 0 \ \ or \ \ 1 \quad (K^*_1 + 1 \leq k \leq K^*_0).$$

*We define* $\Delta K^* = K^*_0 - K^*_1$.

Let $\widehat{K}_0$ be the number of components satisfying $\overline{N}_k/N = \Omega_\mathrm{p}(1)$ and $\widehat{K}_1$ be the number of components satisfying $\overline{x}_{km} = \Omega_\mathrm{p}(1)$ and $1 - \overline{x}_{km} = \Omega_\mathrm{p}(1)$ for all $m = 1, \cdots, M$. Then, for $\Delta\widehat{K} \equiv \widehat{K}_0 - \widehat{K}_1$ components, it holds that $\overline{N}_k/N = \Omega_\mathrm{p}(1)$ and $\overline{x}_{km} = o_\mathrm{p}(1)$ or $1 - \overline{x}_{km} = o_\mathrm{p}(1)$. Hence, the $\widehat{K}_1$ components with $\overline{N}_k/N = \Omega_\mathrm{p}(1)$ and $\overline{x}_{km} = \Omega_\mathrm{p}(1)$ are said to be "nondeterministic" and the $\Delta\widehat{K}$ components are said to be "deterministic," respectively. We have the following theorem.

**Theorem 15.15** *The relative free energy of the Bernoulli mixture model satisfies*

$$\widetilde{F}^{\mathrm{VB}}(\mathcal{D}) = F^{\mathrm{VB}}(\mathcal{D}) - NS_N(\mathcal{D})$$
$$= \left\{\left(\frac{M+1}{2} - \phi\right)\widehat{K}_1 + \left(\frac{1}{2} - \phi + M\xi\right)\Delta\widehat{K} + K\phi - \frac{1}{2}\right\}\log N + \Omega_\mathrm{p}\left(N\widehat{J}\right) + O_\mathrm{p}(1),$$

*where* $\widehat{J} = 1$ *if* $\widehat{K}_1 < K^*_1$ *or* $\Delta\widehat{K} < \Delta K^*$ *and otherwise* $\widehat{J} = 0$.

The proof of Theorem 15.15 is shown after the next theorem. The following theorem claims that the numbers of deterministic and nondeterministic components are essentially determined by the hyperparameters.

**Theorem 15.16** *The estimated numbers of components* $\widehat{K}_0$ *and* $\widehat{K}_1$ *of the Bernoulli mixture model are determined as follows:*

*(1) If* $\frac{M+1}{2} - \phi > 0$ *and* $\frac{1}{2} - \phi + M\xi > 0$, *then* $\widehat{K}_1 = K^*_1$ *and* $\Delta\widehat{K} = \Delta K^*$.
*(2) If* $\frac{M+1}{2} - \phi > 0$ *and* $\frac{1}{2} - \phi + M\xi < 0$, *then* $\widehat{K}_1 = K^*_1$ *and* $\Delta\widehat{K} = K - K^*_1$.
*(3) If* $\frac{M+1}{2} - \phi < 0$ *and* $\frac{1}{2} - \phi + M\xi > 0$, *then* $\widehat{K}_1 = K - \Delta K^*$ *and* $\Delta\widehat{K} = \Delta K^*$.
*(4) If* $\frac{M+1}{2} - \phi < 0$ *and* $\frac{1}{2} - \phi + M\xi < 0$, *and*

(a) *if* $\xi > \frac{1}{2}$, *then* $\widehat{K}_1 = K - \Delta K^*$ *and* $\Delta\widehat{K} = \Delta K^*$.
(b) *if* $\xi < \frac{1}{2}$, *then* $\widehat{K}_1 = K^*_1$ *and* $\Delta\widehat{K} = K - K^*_1$.

*Proof* Minimizing the coefficient of the relative free energy with respect to $\widehat{K}_1$ and $\Delta\widehat{K}$ under the constraint that the true distribution is realizable, i.e., $\widehat{K}_1 \geq K^*_1$ and $\Delta\widehat{K} \geq \Delta K^*$, we obtain the theorem. □

**Proof of Theorem 15.15**

From Lemma 15.1, we first evaluate $R = \mathrm{KL}(r_w(\boldsymbol{w})\|p(\boldsymbol{w}))$. The inequalities of the di-gamma and log-gamma functions in Eqs. (15.8) and (15.9) yield that

$$R = \sum_{k=1}^{K}\left(\frac{1}{2} - \phi\right)\log(\overline{N}_k + \phi) + \left(K\phi - \frac{1}{2}\right)\log(N + K\phi)$$
$$+ \sum_{k=1}^{K}\sum_{m=1}^{M}\left\{\left(\frac{1}{2} - \xi\right)\log(\overline{N}_k\overline{x}_{km} + \xi) + \left(\frac{1}{2} - \xi\right)\log(\overline{N}_k(1 - \overline{x}_{km}) + \xi)\right\}$$
$$+ \sum_{k=1}^{K} M\left(2\xi - \frac{1}{2}\right)\log(\overline{N}_k + 2\xi) + O_{\mathrm{p}}(1).$$

We consider variational parameters in which $\widehat{K}_0$ components are active, i.e., $\overline{N}_k = \Omega_{\mathrm{p}}(N)$. Furthermore, without loss of generality, we can assume that $0 < \overline{x}_{km} < 1$ $(1 \le m \le M)$ for $\widehat{K}_1$ nondeterministic components and $\overline{x}_{km} = O_{\mathrm{p}}(1/N)$ $(1 \le m \le M)$ for $\Delta\widehat{K}$ deterministic components. Putting such variational parameters into the preceding expression, we have the asymptotic form in the theorem.

Lemmas 15.3 and 15.4 imply that if $\widehat{K}_1 < K_1^*$ or $\Delta\widehat{K} < \Delta K^*$, $\widetilde{Q} = -\log C_{\mathcal{H}} - NS_N(\mathcal{D}) = \Omega_{\mathrm{p}}(N)$, and otherwise $\widetilde{Q} = O_{\mathrm{p}}(1)$. Thus, we obtain the theorem.

# 16

# Asymptotic VB Theory of Other Latent Variable Models

In this chapter, we proceed to asymptotic analyses of VB learning in other latent variable models discussed in Section 4.2, namely, Bayesian networks, hidden Markov models, probabilistic context-free grammar, and latent Dirichlet allocation.

## 16.1 Bayesian Networks

In this section, we analyze the VB free energy of the following *Bayesian network* model (Watanabe et al., 2009), introduced in Section 4.2.1:

$$p(\boldsymbol{x}|\boldsymbol{w}) = \sum_{z \in \mathcal{Z}} p(\boldsymbol{x}, z|\boldsymbol{w}), \tag{16.1}$$

$$p(\boldsymbol{x}, z|\boldsymbol{w}) = p(\boldsymbol{x}|\boldsymbol{b}_z) \prod_{k=1}^{K} \prod_{i=1}^{T_k} a_{(k,i)}^{z_{k,i}},$$

$$p(\boldsymbol{x}|\boldsymbol{b}_z) = \prod_{j=1}^{M} \prod_{l=1}^{Y_j} b_{(j,l|z)}^{x_{j,l}},$$

$$p(\boldsymbol{w}) = \left\{ \prod_{k=1}^{K} p(\boldsymbol{a}_k|\phi) \right\} \left\{ \prod_{z \in \mathcal{Z}} \prod_{j=1}^{M} p(\boldsymbol{b}_{j|z}|\xi) \right\},$$

$$p(\boldsymbol{a}_k|\phi) = \mathrm{Dirichlet}_{T_k}\left(\boldsymbol{a}_k; (\phi, \ldots, \phi)^\top\right),$$

$$p(\boldsymbol{b}_{j|z}|\xi) = \mathrm{Dirichlet}_{Y_j}\left(\boldsymbol{b}_{j|z}; (\xi, \ldots, \xi)^\top\right),$$

where $\phi > 0$ and $\xi > 0$ are hyperparameters. Here, $\mathcal{Z} = \{(z_1, \ldots, z_K); z_k \in \{\boldsymbol{e}_i\}_{i=1}^{T_k}, k = 1, \ldots, K\}$, and $z_k \in \{\boldsymbol{e}_i\}_{i=1}^{T_k}$ is the *one-of-K representation*, i.e., $z_{k,i} = 1$ for some $i \in \{1, \ldots, T_k\}$ and $z_{k,j} = 0$ for $j \neq i$. Also, $\boldsymbol{x} = (\boldsymbol{x}_1, \ldots, \boldsymbol{x}_M)$ for $\boldsymbol{x}_j \in \{\boldsymbol{e}_l\}_{l=1}^{Y_j}$. The number of the parameters of this model is

$$D = M_{\mathrm{obs}} \prod_{k=1}^{K} T_k + \sum_{k=1}^{K} (T_k - 1), \tag{16.2}$$

where

$$M_{\mathrm{obs}} = \sum_{j=1}^{M} (Y_j - 1).$$

Under the constraint, $r(\mathcal{H}, \boldsymbol{w}) = r_{\mathcal{H}}(\mathcal{H}) r_w(\boldsymbol{w})$, the VB posteriors are given by

$$r_w(\boldsymbol{w}) = \left\{ \prod_{k=1}^{K} r_a(\boldsymbol{a}_k) \right\} \left\{ \prod_{z \in \mathcal{Z}} \prod_{j=1}^{M} r_b(\boldsymbol{b}_{j|z}) \right\},$$

$$r_a(\boldsymbol{a}_k) = \mathrm{Dirichlet}_{T_k}\left( \boldsymbol{a}_k; \widehat{\boldsymbol{\phi}}_k \right), \tag{16.3}$$

$$r_b(\boldsymbol{b}_{j|z}) = \mathrm{Dirichlet}_{Y_j}\left( \boldsymbol{b}_{j|z}; \widehat{\boldsymbol{\xi}}_{j|z} \right), \tag{16.4}$$

$$r_{\mathcal{H}}(\mathcal{H}) = \prod_{n=1}^{N} r_z(\boldsymbol{z}^{(n)}),$$

where

$$r_z(\boldsymbol{z}^{(n)} = \boldsymbol{z}) \propto \exp\left( \sum_{k=1}^{K} \left\{ \Psi(\widehat{\phi}_{(k,i_k)}) - \Psi\left( \textstyle\sum_{i'_k=1}^{T_k} \widehat{\phi}_{(k,i'_k)} \right) \right\} \right.$$
$$\left. + \sum_{j=1}^{M} \left\{ \Psi(\widehat{\xi}_{(j,l_j^{(n)}|z)}) - \Psi\left( \textstyle\sum_{l'=1}^{Y_j} \widehat{\xi}_{(j,l'|z)} \right) \right\} \right) \tag{16.5}$$

for $\boldsymbol{z} = (\boldsymbol{e}_{i_1}, \ldots, \boldsymbol{e}_{i_K})$ and $\boldsymbol{x}_j^{(n)} = \boldsymbol{e}_{l_j^{(n)}}$. The free energy is given by

$$F = \sum_{k=1}^{K} \left\{ \log\left( \frac{\Gamma(\sum_{i=1}^{T_k} \widehat{\phi}_{(k,i)})}{\prod_{i=1}^{T_k} \Gamma(\widehat{\phi}_{(k,i)})} \right) - \log\left( \frac{\Gamma(T_k \phi)}{(\Gamma(\phi))^{T_k}} \right) \right.$$
$$\left. + \sum_{i=1}^{T_k} \left( \widehat{\phi}_{(k,i)} - \phi - \overline{N}^{z}_{(k,i)} \right) \left( \Psi(\widehat{\phi}_{(k,i)}) - \Psi\left( \textstyle\sum_{i'=1}^{T_k} \widehat{\phi}_{(k,i')} \right) \right) \right\}$$
$$+ \sum_{z \in \mathcal{Z}} \sum_{j=1}^{M} \left\{ \log\left( \frac{\Gamma(\sum_{l=1}^{Y_j} \widehat{\xi}_{(j,l|z)})}{\prod_{l=1}^{Y_j} \Gamma(\widehat{\xi}_{(j,l|z)})} \right) - \log\left( \frac{\Gamma(Y_j \xi)}{(\Gamma(\xi))^{Y_j}} \right) \right.$$
$$\left. + \sum_{l=1}^{Y_j} \left( \widehat{\xi}_{(j,l|z)} - \xi - \overline{N}^{x}_{(j,l|z)} \right) \left( \Psi(\widehat{\xi}_{(j,l|z)}) - \Psi\left( \textstyle\sum_{l'=1}^{Y_j} \widehat{\xi}_{(j,l'|z)} \right) \right) \right\}$$
$$+ \sum_{n=1}^{N} \sum_{z \in \mathcal{Z}} r_z(\boldsymbol{z}^{(n)} = \boldsymbol{z}) \log r_z(\boldsymbol{z}^{(n)} = \boldsymbol{z}),$$

where

$$\overline{N}^z_{(k,i_k)} = \sum_{n=1}^{N} \left\langle z^{(n)}_{k,i_k} \right\rangle_{r_{\mathcal{H}}(\mathcal{H})},$$

$$\overline{N}^x_{(j,l_j|z)} = \sum_{n=1}^{N} x^{(n)}_{j,l_j} r_z(z^{(n)} = z),$$

for

$$r_z(z^{(n)} = (e_{i_1}, \ldots, e_{i_K})) = \left\langle \prod_{k=1}^{K} z^{(n)}_{k,i_k} \right\rangle_{r_{\mathcal{H}}(\mathcal{H})}. \tag{16.6}$$

We assume the following condition:

**Assumption 16.1** *The true distribution $q(x)$ can be expressed by a Bayesian network with $H$ hidden nodes, each of which has $S_k$ states, for $H \leq K$, i.e.,*

$$q(x) = p(x|w^*) = \sum_{z \in \mathcal{Z}^*} p(x, z|w^*) = \sum_{z \in \mathcal{Z}^*} p(x|b^*_z) \prod_{k=1}^{H} \prod_{i=1}^{S_k} \left\{a^*_{(k,i)}\right\}^{z_{k,i}},$$

*where*

$$p(x|b^*_z) = \prod_{j=1}^{M} \prod_{l=1}^{Y_j} \left\{b^*_{(j,l|z)}\right\}^{x_{j,l}}$$

*for $z \in \mathcal{Z}^* = \{(z_1, \ldots, z_H); z_k \in \{e_i\}_{i=1}^{S_k}, k = 1, \ldots, H\}$. The true parameters $w^* = \{\{a^*_k\}_{k=1}^{H}, \{b^*_z\}_{z \in \mathcal{Z}^*}\}$ are given by*

$$a^*_k = \{a^*_{(k,i)}; 1 \leq i \leq S_k\} \quad (k = 1, \ldots, H),$$

$$b^*_z = \{b^*_{j|z}\}_{j=1}^{M} \quad (z \in \mathcal{Z}^*),$$

$$b^*_{j|z} = \{b^*_{(j,l|z)}; 1 \leq l \leq Y_j\} \quad (j = 1, \ldots, M).$$

*For $k > H$, we define $S_k = 1$.*

*The true distribution can be realized by the model, i.e., the model is given by Eq. (16.1), where $T_k \geq S_k$ holds for $k = 1, \ldots, H$. We assume that the true distribution is the smallest in the sense that it cannot be realized by any model with a smaller number of hidden units and with a smaller number of the states of each hidden unit.*

Under this condition, we prove the following theorem, which evaluates the relative VB free energy. The proof will appear in the next section.

**Theorem 16.1** *The relative VB free energy of the Bayesian network model satisfies*

$$\widetilde{F}^{\mathrm{VB}}(\mathcal{D}) = F^{\mathrm{VB}}(\mathcal{D}) - NS_N(\mathcal{D}) = \lambda_{\mathrm{BN}}'^{\mathrm{VB}} \log N + O_{\mathrm{p}}(1),$$

*where*

$$\lambda_{\mathrm{BN}}'^{\mathrm{VB}} = \phi \sum_{k=1}^{K} T_k - \frac{K}{2} + \min_{\{u_k\}} \left\{ \frac{M_{\mathrm{obs}}}{2} \prod_{k=1}^{K} u_k - \left(\phi - \frac{1}{2}\right) \sum_{k=1}^{K} u_k \right\}. \quad (16.7)$$

*The minimum is taken over the set of positive integers* $\{u_k; S_k \le u_k \le T_k\}_{k=1}^{K}$.

If $K = 1$, this is reduced to the case of the naive Bayesian networks whose Bayes free energy or stochastic complexity has been evaluated (Yamazaki and Watanabe, 2003a; Rusakov and Geiger, 2005). Bounds for their VB free energy have also been obtained (Watanabe and Watanabe, 2004, 2005, 2006).

The coefficient $\lambda_{\mathrm{BN}}'^{\mathrm{VB}}$ is given by the solution of the minimization problem in Eq. (16.7). We present a few exemplary cases as corollaries in this section.

By taking $u_k = S_k$ for $1 \le k \le H$ and $u_k = 1$ for $H + 1 \le k \le K$, we obtain the following upper-bound for the VB free energy (Watanabe et al., 2006). This bound is tight if $\phi \le (1 + M_{\mathrm{obs}} \min_{1 \le k \le K}\{S_k\})/2$.

**Corollary 16.2** *It holds that*

$$\widetilde{F}^{\mathrm{VB}}(\mathcal{D}) \le \lambda_{\mathrm{BN}}'^{\mathrm{VB}} \log N + O_{\mathrm{p}}(1), \quad (16.8)$$

*where*

$$\lambda_{\mathrm{BN}}'^{\mathrm{VB}} = \phi \sum_{k=1}^{K} T_k - \phi K + \left(\phi - \frac{1}{2}\right) H + \left(\frac{1}{2} - \phi\right) \sum_{k=1}^{H} S_k + \frac{M_{\mathrm{obs}}}{2} \prod_{k=1}^{H} S_k. \quad (16.9)$$

If $K = H = 2$, that is, the true network and the model both have two hidden nodes, solving the minimization problem in Eq. (16.7) gives the following corollary. Suppose $S_1 \ge S_2$ and $T_1 \ge T_2$.

**Corollary 16.3** *If* $K = H = 2$,

$$\widetilde{F}^{\mathrm{VB}}(\mathcal{D}) = \lambda_{\mathrm{BN}}'^{\mathrm{VB}} \log N + O_{\mathrm{p}}(1), \quad (16.10)$$

*where*

$$\lambda_{\mathrm{BN}}'^{\mathrm{VB}} = \begin{cases} (T_1 - S_1 + T_2 - S_2)\phi + \frac{M_{\mathrm{obs}}}{2} S_1 S_2 + \frac{S_1 + S_2}{2} - 1 & \left(0 < \phi \le \frac{1 + S_2 M_{\mathrm{obs}}}{2}\right), \\ (T_2 - S_2)\phi + \frac{M_{\mathrm{obs}}}{2} T_1 S_2 + \frac{T_1 + S_2}{2} - 1 & \left(\frac{1 + S_2 M_{\mathrm{obs}}}{2} < \phi \le \frac{1 + T_1 M_{\mathrm{obs}}}{2}\right), \\ \frac{M_{\mathrm{obs}}}{2} T_1 T_2 + \frac{T_1 + T_2}{2} - 1 & \left(\frac{1 + T_1 M_{\mathrm{obs}}}{2} < \phi\right). \end{cases} \quad (16.11)$$

The leading term of the relative Bayes free energy of regular statistical models is given by $(D/2)\log N$ (Schwarz, 1978), where $D$ is the number of parameters in Eq. (16.2). Corollary 16.3 claims that the coefficient $\lambda'^{\mathrm{VB}}_{\mathrm{BN}}$ of the leading term is smaller than $D/2$ when $\phi \leq \frac{1+T_1 M_{\mathrm{obs}}}{2}$.

**Proof of Theorem 16.1**

From Lemma 15.1, we can rewrite the free energy as follows:

$$F^{\mathrm{VB}}(\mathcal{D}) = \min_{r_w(\boldsymbol{w})} [R + Q], \tag{16.12}$$

where

$$R = \mathrm{KL}(r_w(\boldsymbol{w}) \| p(\boldsymbol{w})),$$
$$Q = -\log C_{\mathcal{H}} = -\log \sum_{\mathcal{H}} \langle \log p(\mathcal{D}, \mathcal{H}|\boldsymbol{w}) \rangle_{r_w(\boldsymbol{w})}.$$

From Eqs. (16.3), (16.4), and (16.5), we obtain $Q$ and $R$ in Eq. (16.12) as follows:

$$\begin{aligned} Q &= -\sum_{n=1}^{N} \log \sum_{\boldsymbol{z}^{(n)}} \left\langle \log p(\boldsymbol{x}^{(n)}, \boldsymbol{z}^{(n)}|\boldsymbol{w}) \right\rangle_{r_w(\boldsymbol{w})} \\ &= -\sum_{n=1}^{N} \log \left( \sum_{\boldsymbol{z}=(\boldsymbol{e}_{i_1},\ldots,\boldsymbol{e}_{i_K})} \exp\left( \sum_{k=1}^{K} \left\{ \Psi(\overline{N}^{z}_{(k,i_k)} + \phi) - \Psi(N + T_k\phi) \right\} \right.\right. \\ &\quad \left.\left. + \sum_{j=1}^{M} \sum_{l=1}^{Y_j} x^{(n)}_{j,l} \left\{ \Psi(\overline{N}^{x}_{(j,l|\boldsymbol{z})} + \xi) - \Psi(\overline{N}^{x}_{\boldsymbol{z}} + Y_j\xi) \right\} \right) \right), \end{aligned} \tag{16.13}$$

and

$$\begin{aligned} R &= \sum_{k=1}^{K} \mathrm{KL}(r_a(\boldsymbol{a}_k) \| p(\boldsymbol{a}_k|\phi)) + \sum_{\boldsymbol{z}} \sum_{j=1}^{M} \mathrm{KL}(r_b(\boldsymbol{b}_{j|\boldsymbol{z}}) \| p(\boldsymbol{b}_{j|\boldsymbol{z}}|\xi)) \\ &= \sum_{k=1}^{K} \left[ \sum_{i=1}^{T_k} \left\{ \overline{N}^{z}_{(k,i)} \Psi(\overline{N}^{z}_{(k,i)} + \phi) - \log \Gamma(\overline{N}^{z}_{(k,i)} + \phi) \right\} \right. \\ &\quad \left. -N\Psi(N + T_k\phi) + \log\Gamma(N + T_k\phi) + \log \frac{\Gamma(\phi)^{T_k}}{\Gamma(T_k\phi)} \right] \\ &\quad + \sum_{\boldsymbol{z}} \sum_{j=1}^{M} \left[ \sum_{l=1}^{Y_j} \left\{ \overline{N}^{x}_{(j,l|\boldsymbol{z})} \Psi(\overline{N}^{x}_{(j,l|\boldsymbol{z})} + \xi) - \log\Gamma(\overline{N}^{x}_{(j,l|\boldsymbol{z})} + \xi) \right\} \right. \\ &\quad \left. - \overline{N}^{x}_{\boldsymbol{z}} \Psi(\overline{N}^{x}_{\boldsymbol{z}} + Y_j\xi) + \log\Gamma(\overline{N}^{x}_{\boldsymbol{z}} + Y_j\xi) + \log \frac{\Gamma(\xi)^{Y_j}}{\Gamma(Y_j\xi)} \right]. \end{aligned} \tag{16.14}$$

Furthermore, by using the inequalities for the di-gamma and log-gamma functions in Eqs. (15.8) and (15.9), we can bound $Q$ as follows:

$$Q \le -\sum_{n=1}^{N} \log\left( \sum_{z=(e_{i_1},\ldots,e_{i_K})} \exp\left( \sum_{k=1}^{K} \left\{ \log \frac{\overline{N}^{z}_{(k,i_k)} + \phi}{N + T_k\phi} - \frac{1}{\overline{N}^{z}_{(k,i_k)} + \phi} + \frac{1}{2(N + T_k\phi)} \right\} \right.\right.$$
$$\left.\left. + \sum_{j=1}^{M} \sum_{l=1}^{Y_j} x^{(n)}_{j,l} \left\{ \log \frac{\overline{N}^{x}_{(j,l|z)} + \xi}{\overline{N}^{x}_{z} + Y_j\xi} - \frac{1}{\overline{N}^{x}_{(j,l|z)} + \xi} + \frac{1}{2(\overline{N}^{x}_{z} + Y_j\xi)} \right\} \right) \right). \quad (16.15)$$

We can also evaluate $R$ in Eq. (16.12) as follows:

$$R = \sum_{k=1}^{K} \left\{ \left( T_k\phi - \frac{1}{2} \right) \log (N + T_k\phi) \right\} - \sum_{k=1}^{K} \sum_{i=1}^{T_k} \left\{ \left( \phi - \frac{1}{2} \right) \log \left( \overline{N}^{z}_{(k,i)} + \phi \right) \right\}$$
$$+ \sum_{z} \sum_{j=1}^{M} \left\{ \left( Y_j\xi - \frac{1}{2} \right) \log \left( \overline{N}^{x}_{z} + Y_j\xi \right) - \sum_{l=1}^{Y_j} \left( \xi - \frac{1}{2} \right) \log \left( \overline{N}^{x}_{(j,l|z)} + \xi \right) \right\}$$
$$+ O_{\mathrm{p}}(1). \quad (16.16)$$

Since $F^{\mathrm{VB}}(\mathcal{D})$ is given as the minimum value of the function of $\{\overline{N}^{x}_{(j,l|z)}\}$, we can obtain from Eq. (16.12) an upper-bound for $F^{\mathrm{VB}}(\mathcal{D})$ by substituting each $\overline{N}^{x}_{(j,l|z)}$ by any specific value. Therefore, let $u_k$ be a natural number such that $S_k \le u_k \le T_k$ for $k = 1, \ldots, K$ and consider the following $\overline{N}^{x}_{(j,l|z)}$ for each $j$ and $l$:

$$\overline{N}^{x}_{(j,l|\overline{z})} = N b^{*}_{(j,l|\overline{z})} \prod_{k=1}^{K} \overline{a}_{(k,i_k)}, \quad (16.17)$$

where $\overline{z} = (\boldsymbol{e}_{\min\{l_1,S_1\}}, \boldsymbol{e}_{\min\{l_2,S_2\}}, \ldots, \boldsymbol{e}_{\min\{l_H,S_H\}})$ and

$$\overline{a}_{(k,i_k)} = \begin{cases} a^{*}_{(k,i_k)} & (1 \le i_k \le S_k - 1), \\ a^{*}_{(k,S_k)}/(u_k - S_k + 1) & (S_k \le i_k \le u_k), \\ 0 & (\text{otherwise}). \end{cases} \quad (16.18)$$

This corresponds to the case where $u_k$ ($\ge S_k$) states of the $k$th hidden node are active for $k = 1, \ldots, H$. Then we have $\overline{N}^{x}_{z} = N \prod_{k=1}^{K} \overline{a}_{(k,i_k)}$ and $\overline{N}^{z}_{(k,i)} = N\overline{a}_{(k,i)}$. Substituting them into Eq. (16.16) yields

$$R = \left\{ \phi \sum_{k=1}^{K} T_k - \frac{K}{2} + \frac{M_{\mathrm{obs}}}{2} \prod_{k=1}^{K} u_k - \left( \phi - \frac{1}{2} \right) \sum_{k=1}^{K} u_k \right\} \log N + O_{\mathrm{p}}(1). \quad (16.19)$$

From Eq. (16.15), we obtain

$$Q \leq -\sum_{n=1}^{N} \log\left(p(\boldsymbol{x}^{(n)}|\boldsymbol{w}^*)\exp\left(O_{\mathrm{p}}\left(\frac{1}{N}\right)\right)\right)$$
$$= NS_N(\mathcal{D}) + O_{\mathrm{p}}(1). \tag{16.20}$$

From Eqs. (16.12), (16.19), and (16.20), we have proved that $\widetilde{F}^{\mathrm{VB}}(\mathcal{D})$ is upper-bounded by the right-hand side of Eq. (16.19) for any data set $\mathcal{D}$ and $\{u_k; S_k \leq u_k \leq T_k\}$.

If the number of states such that $\overline{N}^{z}_{(k,i)} = \Theta_{\mathrm{p}}(N)$ is less than $S_k$, i.e., $u_k < S_k$ for some $k$, the predictive distribution $\langle p(\boldsymbol{x}|\boldsymbol{w})\rangle_{r_w(\boldsymbol{w})}$ cannot approach the true distribution $p(\boldsymbol{x}|\boldsymbol{w}^*)$. Then Lemma 15.4 implies that $Q - NS_N(\mathcal{D}) = \Omega_{\mathrm{p}}(N)$. Hence, minimizing the coefficient of the leading term in Eq. (16.19) under the constraints $S_k \leq u_k \leq T_k$ for all $k$, we complete the proof.

## 16.2 Hidden Markov Models

Next we analyze the VB free energy of hidden Markov models (HMMs) (Hosino et al., 2005, 2006b), introduced in Section 4.2.2. Suppose that we observe $N$ sequences, $\mathcal{D} = \{\boldsymbol{X}^{(1)}, \ldots, \boldsymbol{X}^{(N)}\}$, where each sequence $\boldsymbol{X}^{(n)} = (\boldsymbol{x}^{(n,1)}, \ldots, \boldsymbol{x}^{(n,T)})$ has length $T$. We consider the asymptotic analysis for the VB free energy as the number of i.i.d. sample sequences tends to infinity, i.e., $N \to \infty$ while $T$ is a fixed constant.

The model for observed and hidden sequences $\boldsymbol{X} = (\boldsymbol{x}^{(1)}, \ldots, \boldsymbol{x}^{(T)})$ and $\boldsymbol{Z} = (\boldsymbol{z}^{(1)}, \ldots, \boldsymbol{z}^{(T)})$ is given by

$$p(\boldsymbol{X}|\boldsymbol{w}) = \sum_{\boldsymbol{Z}} p(\boldsymbol{X}, \boldsymbol{Z}|\boldsymbol{w}), \tag{16.21}$$

$$p(\boldsymbol{X}, \boldsymbol{Z}|\boldsymbol{w}) = \prod_{m=1}^{M} b_{1,m}^{x_m^{(1)}} \prod_{t=2}^{T}\prod_{k=1}^{K}\prod_{l=1}^{K} a_{k,l}^{z_l^{(t)} z_k^{(t-1)}} \prod_{m=1}^{M} b_{k,m}^{z_k^{(t)} x_m^{(t)}},$$

$$p(\boldsymbol{A}|\phi) = \prod_{k=1}^{K} \mathrm{Dirichlet}_K\left(\widetilde{\boldsymbol{a}}_k; (\phi, \ldots, \phi)^\top\right),$$

$$p(\boldsymbol{B}|\xi) = \prod_{k=1}^{K} \mathrm{Dirichlet}_M\left(\widetilde{\boldsymbol{b}}_k; (\xi, \ldots, \xi)^\top\right),$$

where $\phi > 0$ and $\xi > 0$ are hyperparameters. Let $\mathcal{H} = \{\boldsymbol{Z}^{(1)}, \ldots, \boldsymbol{Z}^{(N)}\}$ be the set of hidden sequences. Under the constraint, $r(\mathcal{H}, \boldsymbol{w}) = r_{\mathcal{H}}(\mathcal{H}) r_w(\boldsymbol{w})$, the VB posteriors are given by

$$
\begin{aligned}
r_w(\boldsymbol{w}) &= r_{\boldsymbol{A}}(\boldsymbol{A}) r_{\boldsymbol{B}}(\boldsymbol{B}), \\
r_{\boldsymbol{A}}(\boldsymbol{A}) &= \prod_{k=1}^{K} \mathrm{Dirichlet}_K \left( \widetilde{\boldsymbol{a}}_k; (\widehat{\phi}_{k,1}, \ldots, \widehat{\phi}_{k,K})^\top \right), \\
r_{\boldsymbol{B}}(\boldsymbol{B}) &= \prod_{k=1}^{K} \mathrm{Dirichlet}_M \left( \widetilde{\boldsymbol{b}}_k; (\widehat{\xi}_{k,1}, \ldots, \widehat{\xi}_{k,M})^\top \right), \\
r_{\mathcal{H}}(\mathcal{H}) &= \prod_{n=1}^{N} r_Z(\boldsymbol{Z}^{(n)}), \\
r_Z(\boldsymbol{Z}^{(n)}) &= \frac{1}{C_{\boldsymbol{Z}^{(n)}}} \exp\left( \sum_{t=2}^{T} \sum_{k=1}^{K} \sum_{l=1}^{K} z_k^{(n,t)} z_l^{(n,t-1)} \left\{ \Psi(\widehat{\phi}_{k,l}) - \Psi\left( \sum_{l'=1}^{K} \widehat{\phi}_{k,l'} \right) \right\} \right. \\
&\quad \left. + \sum_{t=1}^{T} \sum_{k=1}^{K} \sum_{m=1}^{M} z_k^{(n,t)} x_m^{(t)} \left\{ \Psi(\widehat{\xi}_{k,m}) - \Psi\left( \sum_{m'=1}^{M} \widehat{\xi}_{k,m'} \right) \right\} \right),
\end{aligned}
$$

where $C_{\boldsymbol{Z}^{(n)}}$ is the normalizing constant. After the substitution of Eq. (15.5), the free energy is given by

$$
\begin{aligned}
F = \sum_{k=1}^{K} & \left\{ \log\left( \frac{\Gamma(\sum_{l=1}^{K} \widehat{\phi}_{k,l})}{\prod_{l=1}^{K} \Gamma(\widehat{\phi}_{k,l})} \right) + \sum_{l=1}^{K} \left( \widehat{\phi}_{k,l} - \phi \right) \left( \Psi(\widehat{\phi}_{k,l}) - \Psi(\textstyle\sum_{l'=1}^{K} \widehat{\phi}_{k,l'}) \right) \right. \\
& \left. + \log\left( \frac{\Gamma(\sum_{m=1}^{M} \widehat{\xi}_{k,m})}{\prod_{m=1}^{M} \Gamma(\widehat{\xi}_{k,m})} \right) + \sum_{m=1}^{M} \left( \widehat{\xi}_{k,m} - \xi \right) \left( \Psi(\widehat{\xi}_{k,m}) - \Psi(\textstyle\sum_{m'=1}^{M} \widehat{\xi}_{k,m'}) \right) \right\} \\
& - K \log\left( \frac{\Gamma(K\phi)}{(\Gamma(\phi))^K} \right) - K \log\left( \frac{\Gamma(M\xi)}{(\Gamma(\xi))^M} \right) - \sum_{n=1}^{N} \log C_{\boldsymbol{Z}^{(n)}}.
\end{aligned}
$$

The variational parameters satisfy

$$
\begin{aligned}
\widehat{\phi}_{k,l} &= \overline{N}_{k,l}^{[z]} + \phi, \\
\widehat{\xi}_{k,m} &= \overline{N}_{k,m}^{[x]} + \xi,
\end{aligned}
$$

for the expected sufficient statistics defined by

$$
\begin{aligned}
\overline{N}_{k,l}^{[z]} &= \sum_{n=1}^{N} \sum_{t=2}^{T} \left\langle z_l^{(n,t)} z_k^{(n,t-1)} \right\rangle_{r_{\mathcal{H}}(\mathcal{H})}, \\
\overline{N}_{k,m}^{[x]} &= \sum_{n=1}^{N} \sum_{t=1}^{T} \left\langle z_k^{(n,t)} \right\rangle_{r_{\mathcal{H}}(\mathcal{H})} x_m^{(n,t)}.
\end{aligned}
$$

We assume the following condition.

**Assumption 16.2** *The true distribution $q(\boldsymbol{X})$ has $K_0$ hidden states and emits $M$-valued discrete symbols:*

$$q(\boldsymbol{X}) = p(\boldsymbol{X}|\boldsymbol{w}^*) = \sum_{\boldsymbol{Z}} \prod_{m=1}^{M} (b^*_{1m})^{x_m^{(1)}} \prod_{t=2}^{T} \prod_{k=1}^{K_0} \prod_{l=1}^{K_0} (a^*_{kl})^{z_l^{(t)} z_k^{(t-1)}} \prod_{m=1}^{M} (b^*_{km})^{z_k^{(t)} x_m^{(t)}}, \tag{16.22}$$

*where $\sum_{\boldsymbol{Z}}$ is taken over all possible values of the hidden variables. Moreover, the true parameter is defined by*

$$\boldsymbol{w}^* = (\boldsymbol{A}^*, \boldsymbol{B}^*) = ((a^*_{kl}), (b^*_{km})),$$

*where $\boldsymbol{A}^* \in \mathbb{R}^{K_0 \times K_0}$ and $\boldsymbol{B}^* \in \mathbb{R}^{K_0 \times m}$. The number of hidden states $K_0$ of the true distribution is the smallest under this parameterization (Ito et al., 1992) and all parameters $\{a^*_{kl}, b^*_{km}\}$ are strictly positive:*

$$\boldsymbol{w}^* = ((a^*_{kl} > 0), (b^*_{km} > 0)) \quad (1 \leq k, l \leq K_0, 1 \leq m \leq M).$$

*The statistical model given by Eq.* (16.21) *can attain the true distribution, thus the model has $K$ ($\geq K_0$) hidden states.*

Under this assumption, the next theorem evaluates the relative VB free energy. Here $S_N(\mathcal{D}) = -\frac{1}{N}\sum_{n=1}^{N} \log p(\boldsymbol{X}^{(n)}|\boldsymbol{w}^*)$ is the empirical entropy of the true distribution (16.22).

**Theorem 16.4** *The relative VB free energy of HMMs satisfies*

$$\widetilde{F}^{\mathrm{VB}}(\mathcal{D}) = F^{\mathrm{VB}}(\mathcal{D}) - NS_N(\mathcal{D}) = \lambda'^{\mathrm{VB}}_{\mathrm{HMM}} \log N + O_{\mathrm{p}}(1),$$

*where*

$$\lambda'^{\mathrm{VB}}_{\mathrm{HMM}} = \begin{cases} \frac{K_0(K_0-1)+K_0(M-1)}{2} + K_0(K-K_0)\phi & \left(0 < \phi \leq \frac{K_0+K+M-2}{2K_0}\right), \\ \frac{K(K-1)+K(M-1)}{2} & \left(\frac{K_0+K+M-2}{2K_0} < \phi\right). \end{cases} \tag{16.23}$$

*Proof* As in the models discussed in the previous sections, we evaluate the KL divergence from the posterior distribution to the prior distribution of parameters:

$$\begin{aligned} R &= \mathrm{KL}(r_w(\boldsymbol{w})\|p(\boldsymbol{w})) \\ &= \sum_{k=1}^{K} \Bigg[ \log \Gamma(\overline{N}_k + K\phi) - \overline{N}_k \Psi(\overline{N}_k + K\phi) \\ &\quad - \sum_{l=1}^{K} \left\{ \log \Gamma(\overline{N}^{[z]}_{k,l} + \phi) - \overline{N}^{[z]}_{k,l} \Psi(\overline{N}^{[z]}_{k,l} + \phi) \right\} \end{aligned}$$

$$+ \log \Gamma(\overline{N}_k + M\xi) - \overline{N}_k \Psi(\overline{N}_k + M\xi)$$

$$- \sum_{m=1}^{M} \left\{ \log \Gamma(\overline{N}_{k,m}^{[x]} + \xi) - \overline{N}_{k,m}^{[x]} \Psi(\overline{N}_{k,m}^{[x]} + \xi) \right\} \Bigg] + O_{\mathrm{p}}(1). \tag{16.24}$$

Using the inequalities of the di-gamma and the log-gamma functions in Eqs. (15.8) and (15.9), we have

$$R = \sum_{k=1}^{K} \left[ \left( K\phi - \frac{1}{2} \right) \log(\overline{N}_k + K\phi) - \sum_{l=1}^{K} \left( \phi - \frac{1}{2} \right) \log(\overline{N}_{k,l}^{[z]} + \phi) \right.$$

$$\left. + \left( M\xi - \frac{1}{2} \right) \log(\overline{N}_k + M\xi) - \sum_{m=1}^{M} \left( \xi - \frac{1}{2} \right) \log(\overline{N}_{k,m}^{[x]} + \xi) \right] + O_{\mathrm{p}}(1). \tag{16.25}$$

We divide the sum over $k$ and $l$ in Eq. (16.25) to the necessary $K_0$ and redundant $K - K_0$ terms. Moreover, we assume that additional $l$ $(0 \leq l \leq K - K_0)$ hidden states are used, i.e., having $\overline{N}_k = \Theta_{\mathrm{p}}(N)$.

$$\frac{R}{\log N} = \sum_{k=1}^{K_0} \left\{ \left( K\phi + \frac{M}{2} - 1 \right) - \sum_{l=1}^{K_0} \left( \phi - \frac{1}{2} \right) \right\} + g(l) + O_{\mathrm{p}}\left( \frac{1}{\log N} \right), \tag{16.26}$$

where $g(l)$ is given by

$$g(l) = \left( K\phi + \frac{M}{2} - 1 \right) l - \left( \phi - \frac{1}{2} \right) (2K_0 l + l^2).$$

If the number of states with $\overline{N}_k = \Theta_{\mathrm{p}}(N)$ is less than $K_0$, Lemma 15.4 implies that $\widetilde{Q} = -\log C_{\mathcal{H}} - NS_N(\mathcal{D}) = \Omega_{\mathrm{p}}(N)$ for data sequences in the strongly $\varepsilon$-typical set. Otherwise, we can upper-bound $F^{\mathrm{VB}}(\mathcal{D}) - NS_N(\mathcal{D})$ so that $\widetilde{Q} = O_{\mathrm{p}}(1)$ similarly to the models in the previous sections. Hence, minimizing the right-hand side of Eq. (16.26) with respect to $l$, we can evaluate the VB free energy.

The minimum of $g(l)$ is achieved by

$$\begin{cases} l = 0 & \left( 0 < \phi \leq \frac{K_0 + K + M - 2}{2K_0} \right), \\ l = K - K_0 & \left( \frac{K_0 + K + M - 2}{2K_0} < \phi \right). \end{cases}$$

Putting this back into Eq. (16.26), we obtain the theorem. □

Next we consider the simple left-to-right HMMs.

**Assumption 16.3** *In the simple left-to-right HMMs, transition from each hidden state is constrained to itself or the next hidden state:*

$$\{a_{k,l} = 0, l \neq \{k, k+1\}\}. \tag{16.27}$$

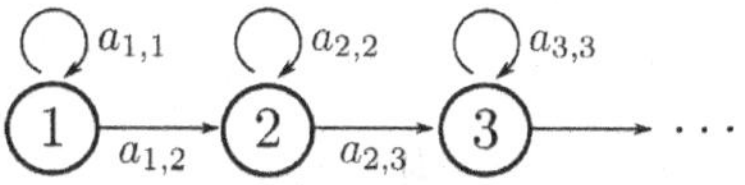


Figure 16.1 State transition diagram of a left-to-right HMM.

*Thus, only* $a_{k,k+1}$ *is a substantial parameter in the transition probability. Figure 16.1 illustrates the state transition diagram of a left-to-right HMM.*

The next theorem evaluates the relative VB free energy of the left-to-right HMM.[1]

**Theorem 16.5** *The relative VB free energy of the left-to-right HMM satisfies*

$$\widetilde{F}^{\mathrm{VB}}(\mathcal{D}) = \lambda'^{\mathrm{VB}}_{\mathrm{LR-HMM}} \log N + O_{\mathrm{p}}(1),$$

*where*

$$\lambda'^{\mathrm{VB}}_{\mathrm{LR-HMM}} = \begin{cases} \frac{(K_0-1)+K_0(M-1)}{2} + \phi & (\phi \leq \frac{M(K-K_0)}{2}), \\ \frac{(K-1)+K(M-1)}{2} & (\phi > \frac{M(K-K_0)}{2}). \end{cases} \tag{16.28}$$

*Proof* From the constraints of the transition probabilities in Eq. (16.27), the asymptotic form of the KL divergence from the VB posterior to the prior is given by

$$\begin{aligned} R &= \mathrm{KL}(r_w(\boldsymbol{w}) \| p(\boldsymbol{w})) \\ &= \sum_{k=1}^{K-1} \left[ \left(2\phi - \frac{1}{2}\right) \log(\overline{N}_k + 2\phi) - \left(\phi - \frac{1}{2}\right) \left\{ \log(\overline{N}^{[z]}_{k,(k+1)} + \phi) + \log(\overline{N}^{[z]}_{k,k} + \phi) \right\} \right] \\ &\quad + \sum_{k=1}^{K} \left[ \left(M\xi - \frac{1}{2}\right) \log(\overline{N}_k + M\xi) - \sum_{m=1}^{M} \left(\xi - \frac{1}{2}\right) \log(\overline{N}^{[x]}_{k,m} + \xi) \right] \\ &\quad + O_{\mathrm{p}}(1). \end{aligned} \tag{16.29}$$

If $K$ hidden states are used, all the variables, $\overline{N}_k$, $\overline{N}^{[z]}_{k,k}$, $\overline{N}^{[z]}_{k,(k+1)}$, and $\overline{N}^{[x]}_{k,m}$ are in the order of $N$, which leads to the asymptotic form in the theorem. If some states are not used, we assume that the $(K_0 + l)$th state is the last state that is effectively used. More specifically, if we consider the case where $\overline{N}_k$, $\overline{N}^{[z]}_{k,k}$, $\overline{N}^{[z]}_{k,(k+1)}$, and $\overline{N}^{[x]}_{k,m}$ are $\Theta_{\mathrm{p}}(N)$ for $K_0+l-1$ states and $\overline{N}^{[z]}_{(K_0+l),(K_0+l+1)} = O_{\mathrm{p}}(1)$ and $\overline{N}^{[z]}_{(K_0+l),(K_0+l)} = \Theta_{\mathrm{p}}(N)$ (and hence, $\overline{N}_{K_0+l} = \Theta_{\mathrm{p}}(N)$), we obtain

$$\frac{R}{\log N} = \frac{K_0 - 1}{2} + \phi + K_0 \frac{M-1}{2} + g(l) + O_{\mathrm{p}}\left(\frac{1}{\log N}\right),$$

[1] This theorem is not obtained as a special case of Theorem 16.4 since some of the transition probabilities are fixed to zero and are no longer parameters.

where

$$g(l) = \frac{M}{2} l$$

for $0 \le l \le K - K_0$. Since the minimum of $g(l)$ is obviously obtained by $l = 0$, we obtain the theorem. □

## 16.3 Probabilistic Context-Free Grammar

In this section, we asymptotically analyze the VB free energy of probabilistic context-free grammar (PCFG), introduced in Section 4.2.3, as the number $N$ of the sequences in the training corpus $\mathcal{D} = \{\boldsymbol{X}^{(1)}, \ldots, \boldsymbol{X}^{(N)}\}$ goes to infinity (Hosino et al., 2006a). The PCFG model is defined by

$$p(\boldsymbol{X}|\boldsymbol{w}) = \sum_{\boldsymbol{Z} \in T(\boldsymbol{X})} p(\boldsymbol{X}, \boldsymbol{Z}|\boldsymbol{w}), \tag{16.30}$$

$$p(\boldsymbol{X}, \boldsymbol{Z}|\boldsymbol{w}) = \prod_{i,j,k=1}^{K} \left(a_{i \to jk}\right)^{c^{\boldsymbol{Z}}_{i \to jk}} \prod_{l=1}^{L} \prod_{i=1}^{K} \prod_{m=1}^{M} (b_{i \to m})^{\widetilde{z}^{(l)}_i x^{(l)}_m},$$

$$\boldsymbol{w} = \{\{\boldsymbol{a}_i\}_{i=1}^{K}, \{\boldsymbol{b}_i\}_{i=1}^{K}\},$$

$$\boldsymbol{a}_i = \{a_{i \to jk}\}_{j,k=1}^{K} \quad (1 \le i \le K),$$

$$\boldsymbol{b}_i = \{b_{i \to m}\}_{m=1}^{M} \quad (1 \le i \le K),$$

$$p(\{\boldsymbol{a}_i\}_{i=1}^{K}|\phi) = \prod_{i=1}^{K} \mathrm{Dirichlet}_{K^2}\left(\boldsymbol{a}_i; (\phi, \ldots, \phi)^\top\right),$$

$$p(\{\boldsymbol{b}_i\}_{i=1}^{K}|\xi) = \prod_{i=1}^{K} \mathrm{Dirichlet}_{M}\left(\boldsymbol{b}_i; (\xi, \ldots, \xi)^\top\right),$$

where $\phi > 0$ and $\xi > 0$ are hyperparameters. Here $T(\boldsymbol{X})$ is the set of derivation sequences that generate $\boldsymbol{X}$, $c^{\boldsymbol{Z}}_{i \to jk}$ is the count of the transition rule from the nonterminal symbol $i$ to the pair of nonterminal symbols $(j, k)$ appearing in the derivation sequence $\boldsymbol{Z}$, and $\widetilde{\boldsymbol{z}}^{(l)} = (\widetilde{z}^{(l)}_1, \ldots, \widetilde{z}^{(l)}_K)$ is the indicator of the (nonterminal) symbol generating the $l$th output symbol of $\boldsymbol{X}$.

Under the constraint, $r(\mathcal{H}, \boldsymbol{w}) = r_{\mathcal{H}}(\mathcal{H}) r_w(\boldsymbol{w})$, the VB posteriors are given by

$$r_w(\boldsymbol{w}) = r_a(\{\boldsymbol{a}_i\}_{i=1}^{K}) r_b(\{\boldsymbol{b}_i\}_{i=1}^{K}),$$

$$r_a(\{\boldsymbol{a}_i\}_{i=1}^{K}) = \prod_{i=1}^{K} \mathrm{Dirichlet}_{K^2}\left(\boldsymbol{a}_i; (\widehat{\phi}_{i \to 11}, \ldots, \widehat{\phi}_{i \to KK})^\top\right),$$

$$
\begin{aligned}
r_b(\{\boldsymbol{b}_i\}_{i=1}^K) &= \prod_{i=1}^K \mathrm{Dirichlet}_M\left(\boldsymbol{b}_i; (\widehat{\xi}_{i\to 1}, \ldots, \widehat{\xi}_{i\to M})^\top\right),\\
r_{\mathcal{H}}(\mathcal{H}) &= \prod_{n=1}^N r_z(\boldsymbol{Z}^{(n)}),\\
r_z(\boldsymbol{Z}^{(n)}) &= \frac{1}{C_{\boldsymbol{Z}^{(n)}}} \exp\left(\gamma_{\boldsymbol{Z}^{(n)}}\right),\\
\gamma_{\boldsymbol{Z}^{(n)}} &= \sum_{i,j,k=1}^K c_{i\to jk}^{\boldsymbol{Z}^{(n)}} \left\{\Psi\left(\widehat{\phi}_{i\to jk}\right) - \Psi\left(\textstyle\sum_{j'=1}^K \sum_{k'=1}^K \widehat{\phi}_{i\to j'k'}\right)\right\}\\
&\quad + \sum_{l=1}^L \sum_{i=1}^K \sum_{m=1}^M \overline{z}_i^{(n,l)} x_m^{(n,l)} \left\{\Psi\left(\widehat{\xi}_{i\to m}\right) - \Psi\left(\textstyle\sum_{m'=1}^M \widehat{\xi}_{i\to m'}\right)\right\},
\end{aligned}
\tag{16.31}
$$

where $C_{\boldsymbol{Z}^{(n)}} = \sum_{\boldsymbol{Z}\in T(\boldsymbol{X}^{(n)})} \exp(\gamma_{\boldsymbol{Z}})$ is the normalizing constant and $T(\boldsymbol{X}^{(n)})$ is the set of derivation sequences that generate $\boldsymbol{X}^{(n)}$. After the substitution of Eq. (15.5), the free energy is given by

$$
\begin{aligned}
F = \sum_{i=1}^K &\left\{\log\left(\frac{\Gamma(\sum_{j,k=1}^K \widehat{\phi}_{i\to jk})}{\prod_{j,k=1}^K \Gamma(\widehat{\phi}_{i\to jk})}\right)\right.\\
&+ \sum_{j,k=1}^K \left(\widehat{\phi}_{i\to jk} - \phi\right)\left(\Psi\left(\widehat{\phi}_{i\to jk}\right) - \Psi\left(\textstyle\sum_{j',k'=1}^K \widehat{\phi}_{i\to j'k'}\right)\right)\\
&\left.+ \log\left(\frac{\Gamma\left(\sum_{m=1}^M \widehat{\xi}_{i\to m}\right)}{\prod_{m=1}^M \Gamma\left(\widehat{\xi}_{i\to m}\right)}\right) + \sum_{m=1}^M \left(\widehat{\xi}_{i\to m} - \xi\right)\left(\Psi\left(\widehat{\xi}_{i\to m}\right) - \Psi\left(\textstyle\sum_{m'=1}^M \widehat{\xi}_{i\to m'}\right)\right)\right\}\\
&- K\log\left(\frac{\Gamma(K^2\phi)}{(\Gamma(\phi))^{K^2}}\right) - K\log\left(\frac{\Gamma(M\xi)}{(\Gamma(\xi))^M}\right) - \sum_{n=1}^N \log C_{\boldsymbol{Z}^{(n)}}.
\end{aligned}
$$

The variational parameters satisfy

$$
\begin{aligned}
\widehat{\phi}_{i\to jk} &= \overline{N}_{i\to jk}^z + \phi,\\
\widehat{\xi}_{i\to m} &= \overline{N}_{i\to m}^x + \xi,
\end{aligned}
$$

where

$$
\begin{aligned}
\overline{N}_{i\to jk}^z &= \sum_{n=1}^N \sum_{l=1}^L \left\langle c_{i\to jk}^{\boldsymbol{Z}^{(n)}}\right\rangle_{r_z(\boldsymbol{Z}^{(n)})},\\
\overline{N}_{i\to m}^x &= \sum_{n=1}^N \sum_{l=1}^L \left\langle \overline{z}_i^{(n,l)}\right\rangle_{r_z(\boldsymbol{Z}^{(n)})} x_m^{(n,l)}.
\end{aligned}
$$

We assume the following condition.

**Assumption 16.4** *The true distribution $q(\boldsymbol{X})$ has $K_0$ nonterminal symbols and $M$ terminal symbols with parameter $\boldsymbol{w}^*$:*

$$q(\boldsymbol{X}) = p(\boldsymbol{X}|\boldsymbol{w}^*) = \sum_{\boldsymbol{Z} \in T(\boldsymbol{X})} p(\boldsymbol{X}, \boldsymbol{Z}|\boldsymbol{w}^*). \tag{16.32}$$

*The true parameters are*

$$\begin{aligned} \boldsymbol{w}^* &= \{\{\boldsymbol{a}_i^*\}_{i=1}^{K_0}, \{\boldsymbol{b}_i^*\}_{i=1}^{K_0}\}, \\ \boldsymbol{a}_i^* &= \{a_{i\to jk}^*\}_{j,k=1}^{K_0} \quad (1 \le i \le K_0), \\ \boldsymbol{b}_i^* &= \{b_{i\to m}^*\}_{m=1}^{M} \quad (1 \le i \le K_0), \end{aligned}$$

*which satisfy the constraints*

$$a_{i\to ii}^* = 1 - \sum_{(j,k)\neq(i,i)} a_{i\to jk}^*, \quad b_{i\to M}^* = 1 - \sum_{m=1}^{M-1} b_{i\to m}^*,$$

*respectively. Since PCFG has nontrivial nonidentifiability as in HMM (Ito et al., 1992), we assume that $K_0$ is the smallest number of nonterminal symbols under this parameterization. The statistical model given by Eq.* (16.30) *includes the true distribution, namely, the number of nonterminal symbols $K$ satisfies the inequality $K_0 \le K$.*

Under this assumption, the next theorem evaluates the relative VB free energy. Here $S_N(\mathcal{D}) = -\frac{1}{N}\sum_{n=1}^{N} \log p(\boldsymbol{X}^{(n)}|\boldsymbol{w}^*)$ is the empirical entropy of the true distribution (16.32).

**Theorem 16.6** *The relative VB free energy of the PCFG model satisfies*

$$\widetilde{F}^{\mathrm{VB}}(\mathcal{D}) = F^{\mathrm{VB}}(\mathcal{D}) - NS_N(\mathcal{D}) = \lambda'^{\mathrm{VB}}_{\mathrm{PCFG}} \log N + O_{\mathrm{p}}(1),$$

*where*

$$\lambda'^{\mathrm{VB}}_{\mathrm{PCFG}} = \begin{cases} \frac{K_0(K_0^2-1)+K_0(M-1)}{2} + K_0(K^2 - K_0^2)\phi & \left(0 < \phi \le \frac{K_0^2+KK_0+K^2+M-2}{2(K_0^2+KK_0)}\right), \\ \frac{K(K^2-1)+K(M-1)}{2} & \left(\frac{K_0^2+KK_0+K^2+M-2}{2(K_0^2+KK_0)} < \phi\right). \end{cases} \tag{16.33}$$

*Proof* Based on Lemma 15.1, similarly to the models in the previous sections, we evaluate $R = \mathrm{KL}(r_w(\boldsymbol{w})\|p(\boldsymbol{w}))$. It is expressed by expected sufficient statistics as

$$R = \sum_{i=1}^{K}\Bigg[\log\Gamma(\overline{N}_i^z + K^2\phi) - \overline{N}_i^z\Psi(\overline{N}_i^z + K^2\phi)$$
$$- \sum_{j,k=1}^{K}\left\{\log\Gamma(\overline{N}_{i\to jk}^z + \phi) - \overline{N}_{i\to jk}^z\Psi(\overline{N}_{i\to jk}^z + \phi)\right\}$$

$$+\log \Gamma(\overline{N}_i^x + M\xi) - \overline{N}_i^x \Psi(\overline{N}_i^x + M\xi)$$
$$\left. - \sum_{m=1}^{M} \left\{ \log \Gamma(\overline{N}_{i\to m}^x + \xi) - \overline{N}_{i\to m}^x \Psi(\overline{N}_{i\to m}^x + \xi) \right\} \right] + O_{\mathrm{p}}(1).$$

Using the inequalities of the di-gamma and the log-gamma functions in Eqs. (15.8) and (15.9), we have

$$R = \sum_{i=1}^{K} \left[ \left( K^2\phi - \frac{1}{2} \right) \log(\overline{N}_i^z + K^2\phi) - \sum_{j,k=1}^{K} \left( \phi - \frac{1}{2} \right) \log(\overline{N}_{i\to jk}^z + \phi) \right.$$
$$\left. + \left( M\xi - \frac{1}{2} \right) \log(\overline{N}_i^x + M\xi) - \sum_{m=1}^{M} \left( \xi - \frac{1}{2} \right) \log(\overline{N}_{i\to m}^x + \xi) \right] + O_{\mathrm{p}}(1). \tag{16.34}$$

We divide the sum over $i$, $j$, and $k$ in Eq. (16.34) to the necessary $K_0$ and redundant $K - K_0$ terms. Moreover, we assume the trained model uses redundant $l$ $(0 \leq l \leq K - K_0)$ nonterminal symbols, i.e., it holds that $\overline{N}_i^z = \Theta_{\mathrm{p}}(N)$:

$$\frac{R}{\log N} = \sum_{i=1}^{K_0} \left\{ \left( K^2\phi + \frac{M}{2} - 1 \right) - \sum_{j,k=1}^{K_0} \left( \phi - \frac{1}{2} \right) \right\}$$
$$+ \sum_{i=1}^{K_0} \left\{ \sum_{j,k=1}^{K_0} \left( \phi - \frac{1}{2} \right) - \sum_{j,k=1}^{K_0+l} \left( \phi - \frac{1}{2} \right) \right\}$$
$$+ \sum_{i=K_0+1}^{K_0+l} \left\{ \left( K^2\phi + \frac{M}{2} - 1 \right) - \sum_{j,k=1}^{K_0+l} \left( \phi - \frac{1}{2} \right) \right\} + O_{\mathrm{p}}\left( \frac{1}{\log N} \right)$$
$$= \left( K^2\phi + \frac{M}{2} - 1 \right) - K_0^2 \left( \phi - \frac{1}{2} \right) + g(l) + O_{\mathrm{p}}\left( \frac{1}{\log N} \right), \tag{16.35}$$

where $g(l)$ is given by

$$g(l) = \left( K^2\phi + \frac{M}{2} - 1 \right) l - \left( \phi - \frac{1}{2} \right) \left\{ (K_0 + l)^3 - K_0^3 \right\}.$$

By Lemma 15.4, similarly to the HMM, we can evaluate the VB free energy by minimizing $g(l)$. The minimum of $g(l)$ is achieved by

$$\begin{cases} l = 0 & \left( 0 < \phi \leq \frac{K_0^2 + KK_0 + K^2 + M - 2}{2(K_0^2 + KK_0)} \right), \\ l = K - K_0 & \left( \frac{K_0^2 + KK_0 + K^2 + M - 2}{2(K_0^2 + KK_0)} < \phi \right). \end{cases}$$

Putting this back into Eq. (16.35), we obtain the theorem. □

## 16.4 Latent Dirichlet Allocation

In this section, we investigate the VB free energy of the latent Dirichlet allocation (LDA) introduced in Section 4.2.4. We also analyze the asymptotic behavior of MAP learning and partially Bayesian learning, which are often used alternatively to VB learning, and discuss similarities and dissimilarities between those learning algorithms.

We consider the following LDA model:

$$p(\boldsymbol{w}^{(n,m)}, \boldsymbol{z}^{(n,m)}|\boldsymbol{\Theta}, \boldsymbol{B}) = p(\boldsymbol{w}^{(n,m)}|\boldsymbol{z}^{(n,m)}, \boldsymbol{B})p(\boldsymbol{z}^{(n,m)}|\boldsymbol{\Theta}),$$

$$p(\boldsymbol{w}^{(n,m)}|\boldsymbol{z}^{(n,m)}, \boldsymbol{B}) = \prod_{l=1}^{L}\prod_{h=1}^{H}(B_{l,h})^{w_l^{(n,m)} z_h^{(n,m)}},$$

$$p(\boldsymbol{z}^{(n,m)}|\boldsymbol{\Theta}) = \prod_{h=1}^{H}(\theta_{m,h})^{z_h^{(n,m)}},$$

$$p(\boldsymbol{\Theta}|\alpha) = \prod_{m=1}^{M}\mathrm{Dirichlet}_H(\widetilde{\boldsymbol{\theta}}_m; (\alpha, \ldots, \alpha)^\top),$$

$$p(\boldsymbol{B}|\eta) = \prod_{h=1}^{H}\mathrm{Dirichlet}_L(\boldsymbol{\beta}_h; (\eta, \ldots, \eta)^\top).$$

Here we have assumed that the priors are symmetric and have hyperparameters $\alpha_1 = \ldots = \alpha_H = \alpha > 0$, $\eta_1 = \ldots = \eta_L = \eta > 0$, respectively.

Under the constraint, $r(\boldsymbol{w}, \mathcal{H}) = r_{\Theta,B}(\boldsymbol{\Theta}, \boldsymbol{B}) r_z\left(\{\{\boldsymbol{z}^{(n,m)}\}_{n=1}^{N^{(m)}}\}_{m=1}^{M}\right)$, the VB posteriors are given by

$$r_z\left(\{\{\boldsymbol{z}^{(n,m)}\}_{n=1}^{N^{(m)}}\}_{m=1}^{M}\right) = \prod_{m=1}^{M}\prod_{n=1}^{N^{(m)}}\mathrm{Multinomial}_{H,1}\left(\boldsymbol{z}^{(n,m)}; \widehat{\boldsymbol{z}}^{(n,m)}\right),$$

$$r_{\Theta,B}(\boldsymbol{\Theta}, \boldsymbol{B}) = r_\Theta(\boldsymbol{\Theta}) r_B(\boldsymbol{B}),$$

$$r_\Theta(\boldsymbol{\Theta}) = \prod_{m=1}^{M}\mathrm{Dirichlet}\left(\widetilde{\boldsymbol{\theta}}_m; \widehat{\boldsymbol{\alpha}}_m\right),$$

$$r_B(\boldsymbol{B}) = \prod_{h=1}^{H}\mathrm{Dirichlet}\left(\boldsymbol{\beta}_h; \widehat{\boldsymbol{\eta}}_h\right).$$

The free energy is given by

$$F = \sum_{m=1}^{M}\left(\log\left(\frac{\Gamma(\sum_{h=1}^{H}\widehat{\alpha}_{m,h})}{\prod_{h=1}^{H}\Gamma(\widehat{\alpha}_{m,h})}\right) - \log\left(\frac{\Gamma(H\alpha)}{\Gamma(\alpha)^H}\right)\right)$$
$$+ \sum_{h=1}^{H}\left(\log\left(\frac{\Gamma(\sum_{l=1}^{L}\widehat{\eta}_{l,h})}{\prod_{l=1}^{L}\Gamma(\widehat{\eta}_{l,h})}\right) - \log\left(\frac{\Gamma(L\eta)}{\Gamma(\eta)^L}\right)\right)$$

$$+\sum_{m=1}^{M}\sum_{h=1}^{H}\left(\widehat{\alpha}_{m,h}-(\overline{N}_h^{(m)}+\alpha)\right)\left(\Psi(\widehat{\alpha}_{m,h})-\Psi(\textstyle\sum_{h'=1}^{H}\widehat{\alpha}_{m,h'})\right)$$
$$+\sum_{h=1}^{H}\sum_{l=1}^{L}\left(\widehat{\eta}_{l,h}-(\overline{W}_{l,h}+\eta)\right)\left(\Psi(\widehat{\eta}_{l,h})-\Psi(\textstyle\sum_{l'=1}^{L}\widehat{\eta}_{l',h})\right)$$
$$+\sum_{m=1}^{M}\sum_{n=1}^{N^{(m)}}\sum_{h=1}^{H}\widehat{z}_h^{(n,m)}\log\widehat{z}_h^{(n,m)},$$

where

$$\overline{N}_h^{(m)}=\sum_{n=1}^{N^{(m)}}\left\langle z_h^{(n,m)}\right\rangle_{r_z\left(\{\{z^{(n,m)}\}_{n=1}^{N^{(m)}}\}_{m=1}^{M}\right)},$$
$$\overline{W}_{l,h}=\sum_{m=1}^{M}\sum_{n=1}^{N^{(m)}}w_l^{(n,m)}\left\langle z_h^{(n,m)}\right\rangle_{r_z\left(\{\{z^{(n,m)}\}_{n=1}^{N^{(m)}}\}_{m=1}^{M}\right)},$$

for the observed data $\mathcal{D}=\{\{\boldsymbol{w}^{(n,m)}\}_{n=1}^{N^{(m)}}\}_{m=1}^{M}$. The variational parameters satisfy

$$\widehat{\alpha}_{m,h}=\overline{N}_h^{(m)}+\alpha, \tag{16.36}$$
$$\widehat{\eta}_{l,h}=\overline{W}_{l,h}+\eta, \tag{16.37}$$
$$\widehat{z}_h^{(n,m)}=\frac{\overline{z}_h^{(n,m)}}{\sum_{h'=1}^{H}\overline{z}_{h'}^{(n,m)}}$$

for

$$\overline{z}_h^{(n,m)}=\exp\Bigg(\left\{\Psi(\widehat{\alpha}_{m,h})-\Psi\left(\textstyle\sum_{h'=1}^{H}\widehat{\alpha}_{m,h'}\right)\right\}$$
$$+\sum_{l=1}^{L}w_l^{(n,m)}\left\{\Psi(\widehat{\eta}_{l,h})-\Psi\left(\textstyle\sum_{l'=1}^{L}\widehat{\eta}_{l',h}\right)\right\}\Bigg).$$

Based on Lemma 15.1, we decompose the free energy as follows:

$$F=R+Q, \tag{16.38}$$

where

$$R=\mathrm{KL}\left(r_\Theta(\boldsymbol{\Theta})r_B(\boldsymbol{B})\|p(\boldsymbol{\Theta}|\alpha)p(\boldsymbol{B}|\eta)\right)$$
$$=\sum_{m=1}^{M}\left(\log\frac{\Gamma(\sum_{h=1}^{H}\widehat{\alpha}_{m,h})}{\prod_{h=1}^{H}\Gamma(\widehat{\alpha}_{m,h})}\frac{\Gamma(\alpha)^H}{\Gamma(H\alpha)}+\sum_{h=1}^{H}(\widehat{\alpha}_{m,h}-\alpha)\left(\Psi(\widehat{\alpha}_{m,h})-\Psi(\textstyle\sum_{h'=1}^{H}\widehat{\alpha}_{m,h'})\right)\right)$$
$$+\sum_{h=1}^{H}\left(\log\frac{\Gamma(\sum_{l=1}^{L}\widehat{\eta}_{l,h})}{\prod_{l=1}^{L}\Gamma(\widehat{\eta}_{l,h})}\frac{\Gamma(\eta_l)^L}{\Gamma(L\eta_l)}+\sum_{l=1}^{L}(\widehat{\eta}_{l,h}-\eta_l)\left(\Psi(\widehat{\eta}_{l,h})-\Psi(\textstyle\sum_{l'=1}^{L}\widehat{\eta}_{l',h})\right)\right), \tag{16.39}$$

$$Q = -\log C_{\mathcal{H}}$$
$$= -\sum_{m=1}^{M} N^{(m)} \sum_{l=1}^{L} V_{l,m} \log\left( \sum_{h=1}^{H} \frac{\exp(\Psi(\widehat{\alpha}_{m,h}))}{\exp\left(\Psi(\sum_{h'=1}^{H} \widehat{\alpha}_{m,h'})\right)} \frac{\exp(\Psi(\widehat{\eta}_{l,h}))}{\exp\left(\Psi(\sum_{l'=1}^{L} \widehat{\eta}_{l',h})\right)} \right). \tag{16.40}$$

Here, $V \in \mathbb{R}^{L \times M}$ is the empirical word distribution matrix with its entries given by $V_{l,m} = \frac{1}{N^{(m)}} \sum_{n=1}^{N^{(m)}} w_l^{(n,m)}$.

### 16.4.1 Asymptotic Analysis of VB Learning

Here we analyze the VB free energy of LDA in the asymptotic limit when $N \equiv \min_m N^{(m)} \to \infty$ (Nakajima et al., 2014). Unlike the analyses for the latent variable models in the previous sections, we do not assume $L, M \ll N$, but $1 \ll L, M, N$ at this point. This amounts to considering the asymptotic limit when $L, M, N \to \infty$ with a fixed mutual ratio, or equivalently, assuming $L, M \sim O(N)$.

We assume the following condition on the true distribution.

**Assumption 16.5** *The word distribution matrix* $V$ *is a sample from the multinomial distribution with the* true *parameter* $U^* \in \mathbb{R}^{L \times M}$ *whose rank is* $H^* \sim O(1)$, *i.e.,* $U^* = B^* \Theta^{*\top}$ *where* $\Theta^* \in \mathbb{R}^{M \times H^*}$ *and* $B^* \in \mathbb{R}^{L \times H^*}$.[2] *The number of topics of the model* $H$ *is set to* $H = \min(L, M)$ *(i.e., the matrix* $B\Theta^\top$ *can express any multinomial distribution).*

The stationary conditions, Eqs. (16.36) and (16.37), lead to the following lemma:

**Lemma 16.7** *Let* $\widehat{B\Theta}^\top = \left\langle B\Theta^\top \right\rangle_{r_{\Theta,B}(\Theta,B)}$. *Then it holds that*

$$\left\langle (B\Theta^\top - \widehat{B\Theta}^\top)_{l,m}^2 \right\rangle_{r_{\Theta,B}(\Theta,B)} = O_p(N^{-2}), \tag{16.41}$$

$$Q = -\sum_{m=1}^{M} N^{(m)} \sum_{l=1}^{L} V_{l,m} \log(\widehat{B\Theta}^\top)_{l,m} + O_p(M). \tag{16.42}$$

*Proof* For the Dirichlet distribution $p(a|\breve{a}) \propto \prod_{h=1}^{H} a_h^{\breve{a}_h - 1}$, the mean and the variance are given as follows:

$$\widehat{a}_h = \langle a_h \rangle_{p(a|\breve{a})} = \frac{\breve{a}_h}{\breve{a}_0}, \qquad \langle (a_h - \widehat{a}_h)^2 \rangle_{p(a|\breve{a})} = \frac{\breve{a}_h(\breve{a}_0 - \breve{a}_h)}{\breve{a}_0^2(\breve{a}_0 + 1)},$$

where $\breve{a}_0 = \sum_{h=1}^{H} \breve{a}_h$.

[2] More precisely, $U^* = B^* \Theta^{*\top} + O(N^{-1})$ is sufficient.

For fixed $N$, $R$, defined by Eq. (16.39), diverges to $+\infty$ if $\widehat{\alpha}_{m,h} \to +0$ for any $(m, h)$ or $\widehat{\eta}_{l,h} \to +0$ for any $(l, h)$. Therefore, the global minimizer of the free energy (16.38) is in the interior of the domain, where the free energy is differentiable. Consequently, the global minimizer is a stationary point. The stationary conditions (16.36) and (16.37) imply that

$$\widehat{\alpha}_{m,h} \geq \alpha, \qquad \widehat{\eta}_{l,h} \geq \eta, \tag{16.43}$$

$$\sum_{h=1}^{H} \widehat{\alpha}_{m,h} = \sum_{h=1}^{H} \alpha + N^{(m)}, \qquad \sum_{l=1}^{L} \widehat{\eta}_{l,h} = \sum_{l=1}^{L} \eta + \sum_{m=1}^{M} (\widehat{\alpha}_{m,h} - \alpha). \tag{16.44}$$

Therefore, we have

$$\left\langle (\Theta_{m,h} - \widehat{\Theta}_{m,h})^2 \right\rangle_{r_\Theta(\boldsymbol{\Theta})} = O_{\mathrm{p}}(N^{-2}) \qquad \text{for all } (m, h), \tag{16.45}$$

$$\left(\max_m \widehat{\Theta}_{m,h}\right)^2 \left\langle (B_{l,h} - \widehat{B}_{l,h})^2 \right\rangle_{r_B(\boldsymbol{B})} = O_{\mathrm{p}}(N^{-2}) \qquad \text{for all } (l, h), \tag{16.46}$$

which leads to Eq. (16.41).

By using Eq. (15.8), $Q$ is bounded as follows:

$$\underline{Q} \leq Q \leq \overline{Q},$$

where

$$\overline{Q} = -\sum_{m=1}^{M} N^{(m)} \sum_{l=1}^{L} V_{l,m} \log\left( \sum_{h=1}^{H} \frac{\widehat{\alpha}_{m,h}}{\sum_{h'=1}^{H} \widehat{\alpha}_{m,h'}} \frac{\widehat{\eta}_{l,h}}{\sum_{l'=1}^{L} \widehat{\eta}_{l',h}} \frac{\exp\left(-\frac{1}{\widehat{\alpha}_{m,h}}\right)}{\exp\left(-\frac{1}{2\sum_{h'=1}^{H} \widehat{\alpha}_{m,h'}}\right)} \frac{\exp\left(-\frac{1}{\widehat{\eta}_{l,h}}\right)}{\exp\left(-\frac{1}{2\sum_{l'=1}^{L} \widehat{\eta}_{l',h}}\right)} \right),$$

$$\underline{Q} = -\sum_{m=1}^{M} N^{(m)} \sum_{l=1}^{L} V_{l,m} \log\left( \sum_{h=1}^{H} \frac{\widehat{\alpha}_{m,h}}{\sum_{h'=1}^{H} \widehat{\alpha}_{m,h'}} \frac{\widehat{\eta}_{l,h}}{\sum_{l'=1}^{L} \widehat{\eta}_{l',h}} \frac{\exp\left(-\frac{1}{2\widehat{\alpha}_{m,h}}\right)}{\exp\left(-\frac{1}{\sum_{h'=1}^{H} \widehat{\alpha}_{m,h'}}\right)} \frac{\exp\left(-\frac{1}{2\widehat{\eta}_{l,h}}\right)}{\exp\left(-\frac{1}{\sum_{l'=1}^{L} \widehat{\eta}_{l',h}}\right)} \right).$$

Using Eqs. (16.45) and (16.46), we have Eq. (16.42), which completes the proof of Lemma 16.7. □

Eq. (16.41) implies the convergence of the posterior. Let $\boldsymbol{u}_m^* = \boldsymbol{B}^* (\widetilde{\boldsymbol{\theta}_m^*})^\top$ be the true probability mass function for the $m$th document and $\widehat{\boldsymbol{u}}_m = \widehat{\boldsymbol{B}}(\widehat{\widehat{\boldsymbol{\theta}}_m})^\top$ be its predictive probability. Define a measure of how far the predictive distributions are from the true distributions by

$$\widehat{J} = \sum_{m=1}^{M} \frac{N^{(m)}}{N} \mathrm{KL}(\boldsymbol{u}_m^* \| \widehat{\boldsymbol{u}}_m). \tag{16.47}$$

Then, by the same arguments as the proof of Lemma 15.4, Eq. (16.42) leads to the following lemma:

**Lemma 16.8** *$Q$ is minimized when $\widehat{J} = O_p(1/N)$, and it holds that*

$$Q = NS_N(\mathcal{D}) + O_p(\widehat{J}N + LM), \qquad \textit{where}$$

$$S_N(\mathcal{D}) = -\frac{1}{N}\log p(\mathcal{D}|\boldsymbol{\Theta}^*, \boldsymbol{B}^*) = -\sum_{m=1}^{M}\frac{N^{(m)}}{N}\sum_{l=1}^{L}V_{l,m}\log(\boldsymbol{B}^*\boldsymbol{\Theta}^*)_{l,m}.$$

Lemma 16.8 simply states that $Q/N$ converges to the empirical entropy $S_N(\mathcal{D})$ of the true distribution if and only if the predictive distribution converges to the true distribution (i.e., $\widehat{J} = O_p(1/N)$).

Let $\widehat{H} = \sum_{h=1}^{H}\theta(\frac{1}{M}\sum_{m=1}^{M}\widehat{\Theta}_{m,h} \sim O_p(1))$ be the number of topics used in the whole corpus, $\widehat{M}^{(h)} = \sum_{m=1}^{M}\theta(\widehat{\Theta}_{m,h} \sim O_p(1))$ be the number of documents that contain the $h$th topic, and $\widehat{L}^{(h)} = \sum_{l=1}^{L}\theta(\widehat{B}_{l,h} \sim O_p(1))$ be the number of words of which the $h$th topic consist. We have the following lemma:

**Lemma 16.9** *$R$ is written as follows:*

$$R = \left\{M\left(H\alpha - \frac{1}{2}\right) + \widehat{H}\left(L\eta - \frac{1}{2}\right) - \sum_{h=1}^{\widehat{H}}\left(\widehat{M}^{(h)}\left(\alpha - \frac{1}{2}\right) + \widehat{L}^{(h)}\left(\eta - \frac{1}{2}\right)\right)\right\}\log N$$
$$+ (H - \widehat{H})\left(L\eta - \frac{1}{2}\right)\log L + O_p(H(M + L)). \tag{16.48}$$

*Proof* By using the bounds (15.8) and (15.9), $R$ can be bounded as

$$\underline{R} \leq R \leq \overline{R}, \tag{16.49}$$

where

$$\underline{R} = -\sum_{m=1}^{M}\log\left(\frac{\Gamma(H\alpha)}{\Gamma(\alpha)^H}\right) - \sum_{h=1}^{H}\log\left(\frac{\Gamma(L\eta)^L}{\Gamma(\eta)}\right) - \frac{M(H-1) + H(L-1)}{2}\log(2\pi)$$
$$+ \sum_{m=1}^{M}\left\{\left(H\alpha - \frac{1}{2}\right)\log\sum_{h=1}^{H}\widehat{\alpha}_{m,h} - \sum_{h=1}^{H}\left(\alpha - \frac{1}{2}\right)\log\widehat{\alpha}_{m,h}\right\}$$
$$+ \sum_{h=1}^{H}\left\{\left(L\eta - \frac{1}{2}\right)\log\sum_{l=1}^{L}\widehat{\eta}_{l,h} - \sum_{l=1}^{L}\left(\eta - \frac{1}{2}\right)\log\widehat{\eta}_{l,h}\right\}$$
$$+ \sum_{m=1}^{M}\left\{-\sum_{h=1}^{H}\frac{1}{12\widehat{\alpha}_{m,h}} - \sum_{h=1}^{H}(\widehat{\alpha}_{m,h} - \alpha)\left(\frac{1}{\widehat{\alpha}_{m,h}} - \frac{1}{2\sum_{h'=1}^{H}\widehat{\alpha}_{m,h'}}\right)\right\}$$
$$+ \sum_{h=1}^{H}\left\{-\sum_{l=1}^{L}\frac{1}{12\widehat{\eta}_{l,h}} - \sum_{l=1}^{L}(\widehat{\eta}_{l,h} - \eta)\left(\frac{1}{\widehat{\eta}_{l,h}} - \frac{1}{2\sum_{l'=1}^{L}\widehat{\eta}_{l',h}}\right)\right\}, \tag{16.50}$$

$$\overline{R} = -\sum_{m=1}^{M} \log\left(\frac{\Gamma(H\alpha)}{\Gamma(\alpha)^{H}}\right) - \sum_{h=1}^{H} \log\left(\frac{\Gamma(L\eta)^{L}}{\Gamma(\eta)}\right) - \frac{M(H-1)+H(L-1)}{2}\log(2\pi)$$
$$+ \sum_{m=1}^{M}\left\{\left(H\alpha - \frac{1}{2}\right)\log\sum_{h=1}^{H}\widehat{\alpha}_{m,h} - \sum_{h=1}^{H}\left(\alpha - \frac{1}{2}\right)\log\widehat{\alpha}_{m,h}\right\}$$
$$+ \sum_{h=1}^{H}\left\{\left(L\eta - \frac{1}{2}\right)\log\sum_{l=1}^{L}\widehat{\eta}_{l,h} - \sum_{l=1}^{L}\left(\eta - \frac{1}{2}\right)\log\widehat{\eta}_{l,h}\right\}$$
$$+ \sum_{m=1}^{M}\left\{\frac{1}{12\sum_{h=1}^{H}\widehat{\alpha}_{m,h}} - \sum_{h=1}^{H}(\widehat{\alpha}_{m,h} - \alpha)\left(\frac{1}{2\widehat{\alpha}_{m,h}} - \frac{1}{\sum_{h'=1}^{H}\widehat{\alpha}_{m,h'}}\right)\right\}$$
$$+ \sum_{h=1}^{H}\left\{\frac{1}{12\sum_{l=1}^{L}\widehat{\eta}_{l,h}} - \sum_{l=1}^{L}(\widehat{\eta}_{l,h} - \eta)\left(\frac{1}{2\widehat{\eta}_{l,h}} - \frac{1}{\sum_{l'=1}^{L}\widehat{\eta}_{l',h}}\right)\right\}. \quad (16.51)$$

Eqs. (16.43) and (16.44) imply that

$$R = \sum_{m=1}^{M}\left\{\left(H\alpha - \frac{1}{2}\right)\log\sum_{h=1}^{H}\widehat{\alpha}_{m,h} - \sum_{h=1}^{H}\left(\alpha - \frac{1}{2}\right)\log\widehat{\alpha}_{m,h}\right\}$$
$$+ \sum_{h=1}^{H}\left\{\left(L\eta - \frac{1}{2}\right)\log\sum_{l=1}^{L}\widehat{\eta}_{l,h} - \sum_{l=1}^{L}\left(\eta - \frac{1}{2}\right)\log\widehat{\eta}_{l,h}\right\} + O_{\mathrm{p}}(H(M+L)),$$

which leads to Eq. (16.48). This completes the proof of Lemma 16.9. □

Since we assumed that the true matrices $\boldsymbol{\Theta}^*$ and $\boldsymbol{B}^*$ are of the rank of $H^*$, $\widehat{H} = H^* \sim O(1)$ is sufficient for the VB posterior to converge to the *true* distribution. However, $\widehat{H}$ can be much larger than $H^*$ with $\left\langle \boldsymbol{B}\boldsymbol{\Theta}^{\top}\right\rangle_{r_{\Theta,B}(\boldsymbol{\Theta},\boldsymbol{B})}$ unchanged because of the nonidentifiability of matrix factorization—duplicating topics with divided weights, for example, does not change the distribution.

Let

$$\widetilde{F}^{\mathrm{VB}}(\mathcal{D}) = F^{\mathrm{VB}}(\mathcal{D}) - NS_N(\mathcal{D}) \quad (16.52)$$

be the relative free energy. Based on Lemmas 16.8 and 16.9, we obtain the following theorem:

**Theorem 16.10** *In the limit when $N \to \infty$ with $L, M \sim O(1)$, it holds that $\widehat{J} = O_{\mathrm{p}}(1/N)$, and*

$$\widetilde{F}^{\mathrm{VB}}(\mathcal{D}) = \lambda'^{\mathrm{VB}}_{\mathrm{LDA}} \log N + O_{\mathrm{p}}(1),$$

*where*

$$\lambda'^{\mathrm{VB}}_{\mathrm{LDA}} = \left\{M\left(H\alpha - \frac{1}{2}\right) + \widehat{H}\left(L\eta - \frac{1}{2}\right) - \sum_{h=1}^{\widehat{H}}\left(\widehat{M}^{(h)}\left(\alpha - \frac{1}{2}\right) + \widehat{L}^{(h)}\left(\eta - \frac{1}{2}\right)\right)\right\}.$$

*In the limit when $N, M \to \infty$ with $\frac{M}{N}, L \sim O(1)$, it holds that $\widehat{J} = o_{\mathrm{p}}(\log N)$, and*

$$\widetilde{F}^{\mathrm{VB}}(\mathcal{D}) = \lambda_{\mathrm{LDA}}^{\prime\mathrm{VB}} \log N + o_{\mathrm{p}}(N \log N),$$

*where*

$$\lambda_{\mathrm{LDA}}^{\prime\mathrm{VB}} = \left\{ M\left(H\alpha - \frac{1}{2}\right) - \sum_{h=1}^{\widehat{H}} \widehat{M}^{(h)}\left(\alpha - \frac{1}{2}\right) \right\}.$$

*In the limit when $N, L \to \infty$ with $\frac{L}{N}, M \sim O(1)$, it holds that $\widehat{J} = o_{\mathrm{p}}(\log N)$, and*

$$\widetilde{F}^{\mathrm{VB}}(\mathcal{D}) = \lambda_{\mathrm{LDA}}^{\prime\mathrm{VB}} \log N + o_{\mathrm{p}}(N \log N),$$

*where*

$$\lambda_{\mathrm{LDA}}^{\prime\mathrm{VB}} = HL\eta.$$

*In the limit when $N, L, M \to \infty$ with $\frac{L}{N}, \frac{M}{N} \sim O(1)$, it holds that $\widehat{J} = o_{\mathrm{p}}(N \log N)$, and*

$$\widetilde{F}^{\mathrm{VB}}(\mathcal{D}) = \lambda_{\mathrm{LDA}}^{\prime\mathrm{VB}} \log N + o_{\mathrm{p}}(N^2 \log N),$$

*where*

$$\lambda_{\mathrm{LDA}}^{\prime\mathrm{VB}} = H(M\alpha + L\eta).$$

*Proof* Lemmas 16.8 and 16.9 imply that the relative free energy can be written as follows:

$$\begin{aligned}
&\widetilde{F}^{\mathrm{VB}}(\mathcal{D}) \\
&= \left\{ M\left(H\alpha - \frac{1}{2}\right) + \widehat{H}\left(L\eta - \frac{1}{2}\right) - \sum_{h=1}^{\widehat{H}} \left( \widehat{M}^{(h)}\left(\alpha - \frac{1}{2}\right) + \widehat{L}^{(h)}\left(\eta - \frac{1}{2}\right) \right) \right\} \log N \\
&\quad + (H - \widehat{H})\left(L\eta - \frac{1}{2}\right) \log L + O_{\mathrm{p}}(\widehat{J}N + LM). \qquad (16.53)
\end{aligned}$$

In the following subsection, we investigate the leading term of the relative free energy (16.53) in different asymptotic limits.

### In the Limit When $N \to \infty$ with $L, M \sim O(1)$

In this case, the minimizer should satisfy

$$\widehat{J} = O_{\mathrm{p}}\left(\frac{1}{N}\right) \qquad (16.54)$$

and the leading term of the relative free energy (16.52) is of the order of $O_p(\log N)$ as follows:

$$\widetilde{F}^{\mathrm{VB}}(\mathcal{D})$$
$$= \left\{ M\left(H\alpha - \frac{1}{2}\right) + \widehat{H}\left(L\eta - \frac{1}{2}\right) - \sum_{h=1}^{\widehat{H}} \left( \widehat{M}^{(h)}\left(\alpha - \frac{1}{2}\right) + \widehat{L}^{(h)}\left(\eta - \frac{1}{2}\right)\right)\right\} \log N$$
$$+ O_p(1).$$

Note that Eq. (16.54) implies the consistency of the VB posterior.

**In the Limit When $N, M \to \infty$ with $\frac{M}{N}, L \sim O(1)$**

In this case,

$$\widehat{J} = o_p(\log N), \qquad (16.55)$$

making the leading term of the relative free energy of the order of $O_p(N \log N)$ as follows:

$$\widetilde{F}^{\mathrm{VB}}(\mathcal{D}) = \left\{ M\left(H\alpha - \frac{1}{2}\right) - \sum_{h=1}^{\widehat{H}} \widehat{M}^{(h)}\left(\alpha - \frac{1}{2}\right)\right\} \log N + o_p(N \log N).$$

Eq. (16.55) implies that the VB posterior is not necessarily consistent.

**In the Limit When $N, L \to \infty$ with $\frac{L}{N}, M \sim O(1)$**

In this case, Eq. (16.55) holds, and the leading term of the relative free energy is of the order of $O_p(N \log N)$ as follows:

$$\widetilde{F}^{\mathrm{VB}}(\mathcal{D}) = HL\eta \log N + o_p(N \log N).$$

**In the Limit When $N, L, M \to \infty$ with $\frac{L}{N}, \frac{M}{N} \sim O(1)$**

In this case,

$$\widehat{J} = o_p(N \log N), \qquad (16.56)$$

and the leading term of the relative free energy is of the order of $O_p(N^2 \log N)$ as follows:

$$\widetilde{F}^{\mathrm{VB}}(\mathcal{D}) = H(M\alpha + L\eta) \log N + o_p(N^2 \log N).$$

This completes the proof of Theorem 16.10. □

Since Eq. (16.41) was shown to hold, the predictive distribution converges to the true distribution if $\widehat{J} = O_p(1/N)$. Accordingly, Theorem 16.10 states that the consistency holds in the limit when $N \to \infty$ with $L, M \sim O(1)$.

Theorem 16.10 also implies that, in the asymptotic limits with small $L \sim O(1)$, the leading term depends on $\widehat{H}$, meaning that it dominates the topic sparsity of the VB solution. We have the following corollary:

**Corollary 16.11** *Let $M^{*(h)} = \sum_{m=1}^{M} \theta(\Theta^*_{m,h} \sim O(1))$ and $L^{*(h)} = \sum_{l=1}^{L} \theta(B^*_{l,h} \sim O(1))$. Consider the limit when $N \to \infty$ with $L, M \sim O(1)$. When $0 < \eta \leq \frac{1}{2L}$, the VB solution is sparse (i.e., $\widehat{H} \ll H = \min(L, M)$) if $\alpha < \frac{1}{2} - \frac{\frac{1}{2} - L\eta}{\min_h M^{*(h)}}$, and dense (i.e., $\widehat{H} \approx H$) if $\alpha > \frac{1}{2} - \frac{\frac{1}{2} - L\eta}{\min_h M^{*(h)}}$. When $\frac{1}{2L} < \eta \leq \frac{1}{2}$, the VB solution is sparse if $\alpha < \frac{1}{2} + \frac{L\eta - \frac{1}{2}}{\max_h M^{*(h)}}$, and dense if $\alpha > \frac{1}{2} + \frac{L\eta - \frac{1}{2}}{\max_h M^{*(h)}}$. When $\eta > \frac{1}{2}$, the VB solution is sparse if $\alpha < \frac{1}{2} + \frac{L-1}{2\max_h M^{*(h)}}$, and dense if $\alpha > \frac{1}{2} + \frac{L\eta - \frac{1}{2}}{\min_h M^{*(h)}}$. In the limit when $N, M \to \infty$ with $\frac{M}{N}, L \sim O(1)$, the VB solution is sparse if $\alpha < \frac{1}{2}$, and dense if $\alpha > \frac{1}{2}$.*

*Proof* From the compact representation when $\widehat{H} = H^*, \widehat{M}^{(h)} = M^{*(h)}$, and $\widehat{L}^{(h)} = L^{*(h)}$, we can decompose a singular component into two, keeping $\widehat{\boldsymbol{B}\boldsymbol{\Theta}^\top}$ unchanged, so that

$$\widehat{H} \to \widehat{H} + 1, \tag{16.57}$$

$$\sum_{h=1}^{H} \widehat{M}^{(h)} \to \sum_{h=1}^{H^*} \widehat{M}^{(h)} + \Delta M \quad \text{for} \quad \min_h M^{*(h)} \leq \Delta M \leq \max_h M^{*(h)}, \tag{16.58}$$

$$\sum_{h=1}^{H} \widehat{L}^{(h)} \to \sum_{h=1}^{H^*} \widehat{L}^{(h)} + \Delta L \quad \text{for} \quad 0 \leq \Delta L \leq \max_h L^{*(h)}. \tag{16.59}$$

Here the lower-bound for $\Delta M$ in Eq. (16.58) corresponds to the case that the least frequent topic is chosen to be decomposed, while the upper-bound to the case that the most frequent topic is chosen. The lower-bound for $\Delta L$ in Eq. (16.59) corresponds to the decomposition such that some of the word-occurrences are moved to a new topic, while the upper-bound to the decomposition such that the topic with the widest vocabulary is copied to a new topic. Note that the bounds both for $\Delta M$ and $\Delta L$ are not always achievable simultaneously, when we choose one topic to decompose.

In the following subsection, we investigate the relation between the sparsity of the solution and the hyperparameter setting in different asymptotic limits.

### In the Limit When $N \to \infty$ with $L, M \sim O(1)$

The coefficient of the leading term of the free energy is

$$\lambda'^{\mathrm{VB}}_{\mathrm{LDA}} = M\left(H\alpha - \frac{1}{2}\right) + \sum_{h=1}^{\widehat{H}} \left(L\eta - \frac{1}{2} - \widehat{M}^{(h)}\left(\alpha - \frac{1}{2}\right) - \widehat{L}^{(h)}\left(\eta - \frac{1}{2}\right)\right). \tag{16.60}$$

Note that the solution is sparse if Eq. (16.60) is increasing with respect to $\widehat{H}$, and dense if it is decreasing. Eqs. (16.57) through (16.59) imply the following:

(I) When $0 < \eta \le \frac{1}{2L}$ and $\alpha \le \frac{1}{2}$, the solution is sparse if

$$L\eta - \frac{1}{2} - \min_h M^{*(h)}\left(\alpha - \frac{1}{2}\right) > 0, \text{ or equivalently,}$$
$$\alpha < \frac{1}{2} - \frac{1}{\min_h M^{*(h)}}\left(\frac{1}{2} - L\eta\right),$$

and dense if

$$\alpha > \frac{1}{2} - \frac{1}{\min_h M^{*(h)}}\left(\frac{1}{2} - L\eta\right).$$

(II) When $0 < \eta \le \frac{1}{2L}$ and $\alpha > \frac{1}{2}$, the solution is sparse if

$$L\eta - \frac{1}{2} - \max_h M^{*(h)}\left(\alpha - \frac{1}{2}\right) > 0, \text{ or equivalently,}$$
$$\alpha < \frac{1}{2} - \frac{1}{\max_h M^{*(h)}}\left(\frac{1}{2} - L\eta\right),$$

and dense if

$$\alpha > \frac{1}{2} - \frac{1}{\max_h M^{*(h)}}\left(\frac{1}{2} - L\eta\right).$$

Therefore, the solution is always dense in this case.

(III) When $\frac{1}{2L} < \eta \le \frac{1}{2}$ and $\alpha < \frac{1}{2}$, the solution is sparse if

$$L\eta - \frac{1}{2} - \min_h M^{*(h)}\left(\alpha - \frac{1}{2}\right) > 0, \text{ or equivalently,}$$
$$\alpha < \frac{1}{2} + \frac{1}{\min_h M^{*(h)}}\left(L\eta - \frac{1}{2}\right),$$

and dense if

$$\alpha > \frac{1}{2} + \frac{1}{\min_h M^{*(h)}}\left(L\eta - \frac{1}{2}\right).$$

Therefore, the solution is always sparse in this case.

(IV) When $\frac{1}{2L} < \eta \le \frac{1}{2}$ and $\alpha \ge \frac{1}{2}$, the solution is sparse if

$$L\eta - \frac{1}{2} - \max_h M^{*(h)}\left(\alpha - \frac{1}{2}\right) > 0, \text{ or equivalently,}$$
$$\alpha < \frac{1}{2} + \frac{1}{\max_h M^{*(h)}}\left(L\eta - \frac{1}{2}\right),$$

and dense if

$$\alpha > \frac{1}{2} + \frac{1}{\max_h M^{*(h)}}\left(L\eta - \frac{1}{2}\right).$$

(V) When $\eta > \frac{1}{2}$ and $\alpha < \frac{1}{2}$, the solution is sparse if

$$L\eta - \frac{1}{2} - \max_h \left(M^{*(h)}\left(\alpha - \frac{1}{2}\right) + L^{*(h)}\left(\eta - \frac{1}{2}\right)\right) > 0, \qquad (16.61)$$

and dense if

$$L\eta - \frac{1}{2} - \max_h \left(M^{*(h)}\left(\alpha - \frac{1}{2}\right) + L^{*(h)}\left(\eta - \frac{1}{2}\right)\right) < 0. \qquad (16.62)$$

Therefore, the solution is sparse if

$$L\eta - \frac{1}{2} - \min_h M^{*(h)}\left(\alpha - \frac{1}{2}\right) - \max_h L^{*(h)}\left(\eta - \frac{1}{2}\right) > 0, \text{ or equivalently,}$$

$$\alpha < \frac{1}{2} + \frac{1}{\min_h M^{*(h)}}\left(L\eta - \frac{1}{2} - \max_h L^{*(h)}\left(\eta - \frac{1}{2}\right)\right),$$

and dense if

$$L\eta - \frac{1}{2} - \max_h M^{*(h)}\left(\alpha - \frac{1}{2}\right) - \max_h L^{*(h)}\left(\eta - \frac{1}{2}\right) < 0, \text{ or equivalently,}$$

$$\alpha > \frac{1}{2} + \frac{1}{\max_h M^{*(h)}}\left(L\eta - \frac{1}{2} - \max_h L^{*(h)}\left(\eta - \frac{1}{2}\right)\right).$$

Therefore, the solution is always sparse in this case.

(VI) When $\eta > \frac{1}{2}$ and $\alpha \geq \frac{1}{2}$, the solution is sparse if Eq. (16.61) holds, and dense if Eq. (16.62) holds. Therefore, the solution is sparse if

$$L\eta - \frac{1}{2} - \max_h M^{*(h)}\left(\alpha - \frac{1}{2}\right) - \max_h L^{*(h)}\left(\eta - \frac{1}{2}\right) > 0, \text{ or equivalently,}$$

$$\alpha < \frac{1}{2} + \frac{1}{\max_h M^{*(h)}}\left(L\eta - \frac{1}{2} - \max_h L^{*(h)}\left(\eta - \frac{1}{2}\right)\right),$$

and dense if

$$L\eta - \frac{1}{2} - \min_h M^{*(h)}\left(\alpha - \frac{1}{2}\right) - \max_h L^{*(h)}\left(\eta - \frac{1}{2}\right) < 0, \text{ or equivalently,}$$

$$\alpha > \frac{1}{2} + \frac{1}{\min_h M^{*(h)}}\left(L\eta - \frac{1}{2} - \max_h L^{*(h)}\left(\eta - \frac{1}{2}\right)\right).$$

Thus, we can conclude that, in this case, the solution is sparse if

$$\alpha < \frac{1}{2} + \frac{L-1}{2\max_h M^{*(h)}},$$

and dense if

$$\alpha > \frac{1}{2} + \frac{L\eta - \frac{1}{2}}{\min_h M^{*(h)}}.$$

Summarizing the preceding, we have the following lemma:

**Lemma 16.12** *When* $0 < \eta \leq \frac{1}{2L}$, *the solution is sparse if* $\alpha < \frac{1}{2} - \frac{\frac{1}{2} - L\eta}{\min_h M^{*(h)}}$, *and dense if* $\alpha > \frac{1}{2} - \frac{\frac{1}{2} - L\eta}{\min_h M^{*(h)}}$. *When* $\frac{1}{2L} < \eta \leq \frac{1}{2}$, *the solution is sparse if* $\alpha < \frac{1}{2} + \frac{L\eta - \frac{1}{2}}{\max_h M^{*(h)}}$, *and dense if* $\alpha > \frac{1}{2} + \frac{L\eta - \frac{1}{2}}{\max_h M^{*(h)}}$. *When* $\eta > \frac{1}{2}$, *the solution is sparse if* $\alpha < \frac{1}{2} + \frac{L-1}{2\max_h M^{*(h)}}$, *and dense if* $\alpha > \frac{1}{2} + \frac{L\eta - \frac{1}{2}}{\min_h M^{*(h)}}$.

**In the Limit When $N, M \to \infty$ with $\frac{M}{N}, L \sim O(1)$**

The coefficient of the leading term of the free energy is given by

$$\lambda'^{\text{VB}}_{\text{LDA}} = M\left(H\alpha - \frac{1}{2}\right) - \sum_{h=1}^{\widehat{H}} \widehat{M}^{(h)}\left(\alpha - \frac{1}{2}\right). \tag{16.63}$$

Although the predictive distribution does not necessarily converge to the true distribution, we can investigate the sparsity of the solution by considering the duplication rules (16.57) through (16.59) that keep $\widehat{\boldsymbol{B}\boldsymbol{\Theta}}^\top$ unchanged. It is clear that Eq. (16.63) is increasing with respect to $\widehat{H}$ if $\alpha < \frac{1}{2}$, and decreasing if $\alpha > \frac{1}{2}$. Combing this result with Lemma 16.12 completes the proof of Corollary 16.11. □

In the case when $L, M \ll N$ and in the case when $L \ll M, N$, Corollary 16.11 provides information on the sparsity of the VB solution, which will be compared with other methods in Section 16.4.2. On the other hand, although we have successfully derived the leading term of the free energy also in the case when $M \ll L, N$ and in the case when $1 \ll L, M, N$, it unfortunately provides no information on sparsity of the solution.

### 16.4.2 Asymptotic Analysis of MAP Learning and Partially Bayesian Learning

For training the LDA model, MAP learning and partially Bayesian (PB) learning (see Section 2.2.2), where $\boldsymbol{\Theta}$ and/or $\boldsymbol{B}$ are point-estimated, are also popular choices. Although the differences in update equations is small, it can affect the asymptotic behavior. In this subsection, we aim to clarify the difference in the asymptotic behavior.

MAP learning, PB-A learning, PB-B learning, and VB learning, respectively, solve the following problem:

$$\min_r F,$$

s.t.

$$\begin{cases} r_{\Theta,B}(\boldsymbol{\Theta}, \boldsymbol{B}) = \delta(\boldsymbol{\Theta}; \widehat{\boldsymbol{\Theta}})\delta(\boldsymbol{B}; \widehat{\boldsymbol{B}}) & \text{(for MAP learning)}, \\ r_{\Theta,B}(\boldsymbol{\Theta}, \boldsymbol{B}) = r_{\Theta}(\boldsymbol{\Theta})\delta(\boldsymbol{B}; \widehat{\boldsymbol{B}}) & \text{(for PB-A learning)}, \\ r_{\Theta,B}(\boldsymbol{\Theta}, \boldsymbol{B}) = \delta(\boldsymbol{\Theta}; \widehat{\boldsymbol{\Theta}})r_B(\boldsymbol{B}) & \text{(for PB-B learning)}, \\ r_{\Theta,B}(\boldsymbol{\Theta}, \boldsymbol{B}) = r_{\Theta}(\boldsymbol{\Theta})r_B(\boldsymbol{B}) & \text{(for VB learning)}, \end{cases}$$

Similar analysis to Section 16.4.1 leads to the following theorem (the proof is given in Section 16.4.5):

**Theorem 16.13** *In the limit when $N \to \infty$ with $L, M \sim O(1)$, the solution is sparse if $\alpha < \underline{\alpha}_{\text{sparse}}$, and dense if $\alpha > \underline{\alpha}_{\text{dense}}$. In the limit when $N, M \to \infty$ with $\frac{M}{N}, L \sim O(1)$, the solution is sparse if $\alpha < \underline{\alpha}_{M\to\infty}$, and dense if $\alpha > \underline{\alpha}_{M\to\infty}$. Here, $\underline{\alpha}_{\text{sparse}}$, $\underline{\alpha}_{\text{dense}}$, and $\underline{\alpha}_{M\to\infty}$ are given in Table 16.1.*

A notable finding from Table 16.1 is that the threshold that determines the topic sparsity of PB-B learning is (most of the case exactly) $\frac{1}{2}$ larger than the threshold of VB learning. The same relation is observed between MAP learning and PB-A learning. From these, we can conclude that point-estimating $\boldsymbol{\Theta}$, instead of integrating it out, increases the threshold by $\frac{1}{2}$ in the LDA model. We will validate this observation by numerical experiments in Section 16.4.4.

Table 16.1 *Sparsity thresholds of VB, PB-A, PB-B, and MAP methods (see Theorem 16.13). The first four columns show the thresholds $(\underline{\alpha}_{\text{sparse}}, \underline{\alpha}_{\text{dense}})$, of which the function forms depend on the range of $\eta$, in the limit when $N \to \infty$ with $L, M \sim O(1)$. A single value is shown if $\underline{\alpha}_{\text{sparse}} = \underline{\alpha}_{\text{dense}}$. The last column shows the threshold $\underline{\alpha}_{M\to\infty}$ in the limit when $N, M \to \infty$ with $\frac{M}{N}, L \sim O(1)$.*

| | $(\underline{\alpha}_{\text{sparse}}, \underline{\alpha}_{\text{dense}})$ | | | | $\underline{\alpha}_{M\to\infty}$ |
|---|---|---|---|---|---|
| $\eta$ range | $0 < \eta \leq \frac{1}{2L}$ | $\frac{1}{2L} < \eta \leq \frac{1}{2}$ | $\frac{1}{2} < \eta < 1$ | $1 \leq \eta < \infty$ | $0 < \eta < \infty$ |
| VB | $\frac{1}{2} - \frac{\frac{1}{2} - L\eta}{\min_h M^{*(h)}}$ | $\frac{1}{2} + \frac{L\eta - \frac{1}{2}}{\max_h M^{*(h)}}$ | $\left(\frac{1}{2} + \frac{L-1}{2\max_h M^{*(h)}}, \frac{1}{2} + \frac{L\eta - \frac{1}{2}}{\min_h M^{*(h)}}\right)$ | | $\frac{1}{2}$ |
| PB-A | | — | | $\left(\frac{1}{2}, \frac{1}{2} + \frac{L(\eta-1)}{\min_h M^{*(h)}}\right)$ | $\frac{1}{2}$ |
| PB-B | 1 | $1 + \frac{L\eta - \frac{1}{2}}{\max_h M^{*(h)}}$ | $\left(1 + \frac{L-1}{2\max_h M^{*(h)}}, 1 + \frac{L\eta - \frac{1}{2}}{\min_h M^{*(h)}}\right)$ | | 1 |
| MAP | | — | | $\left(1, 1 + \frac{L(\eta-1)}{\min_h M^{*(h)}}\right)$ | 1 |

### 16.4.3 Discussion

The preceding theoretical analysis (Thereom 16.13) showed that VB tends to induce weaker sparsity than MAP in the LDA model,[3] i.e., VB requires sparser prior (smaller $\alpha$) than MAP to give a sparse solution (mean of the posterior). This phenomenon is opposite to other models such as mixture models (Chapter 15), Bayesian networks (Section 16.1), hidden Markov models (Section 16.2), and fully observed matrix factorization (Chapter 7), where VB tends to induce stronger sparsity than MAP. This phenomenon might be partly explained as follows: in the case of mixture models, the sparsity threshold depends on the degree of freedom of a single component (Theorem 15.5). This is reasonable because adding a single component increases the model complexity by this amount. Also, in the case of LDA, adding a single topic requires additional $L+1$ parameters. However, the added topic is shared over $M$ documents, which could discount the increased model complexity relative to the increased data fidelity. Corollary 16.11, which implies the dependency of the threshold for $\alpha$ on $L$ and $M$, might support this conjecture. However, the same applies to the matrix factorization, where VB was shown to give a sparser solution than MAP (Chapter 7). Investigation on related models, e.g., Poisson MF (Gopalan et al., 2013), would help us fully explain this phenomenon.

Unlike for the latent variable models in the previous sections, we derived a general form of the asymptotic free energy for LDA, which can be applied to different asymptotic limits and showed that the consistency does not always hold (see Theorem 16.10). Specifically, the standard asymptotic theory requires a large number $N$ of words per document, compared to the number $M$ of documents and the vocabulary size $L$. Assuming such a situation may be reasonable in some collaborative filtering applications, e.g., in the *Last.FM* data which will be used for numerical illustration in Section 16.4.4. However, $L$ and/or $M$ are comparable to or larger than $N$ in many text analysis applications.

The general form of the asymptotic free energy also allowed us to elucidate the behavior of the VB free energy when $L$ and/or $M$ diverges with the same order as $N$. This attempt successfully revealed the sparsity of the solution for the case when $M$ diverges while $L \sim O(1)$. However, when $L$ diverges, we found that the leading term of the free energy does not contain useful information on sparsity of the solution. Higher-order asymptotic analysis will be necessary to further understand the sparsity-inducing mechanism of the LDA model with large vocabulary.

[3] This tendency was pointed out (Asuncion et al., 2009) by using the approximation $\exp(\Psi(n)) \approx n - \frac{1}{2}$ and comparing the stationary condition. The theory here clarified the sparsity behavior of the global solution based on the asymptotic free energy analysis.

### 16.4.4 Numerical Illustration

Here we conduct numerical experiments on artificial and real data for collaborative filtering.

The *artificial* data were created as follows: we first sample the *true* document matrix $\boldsymbol{\Theta}^*$ of size $M \times H^*$ and the *true* topic matrix $\boldsymbol{B}^*$ of size $L \times H^*$. We assume that each row $\widetilde{\boldsymbol{\theta}}_m^*$ of $\boldsymbol{\Theta}^*$ follows the Dirichlet distribution with $\alpha^* = 1/H^*$, while each column $\boldsymbol{\beta}_h^*$ of $\boldsymbol{B}^*$ follows the Dirichlet distribution with $\eta^* = 1/L$. The document length $N^{(m)}$ is sampled from the Poisson distribution with mean $N$. The word histogram $N^{(m)}\boldsymbol{v}_m$ for each document is sampled from the multinomial distribution with the parameter specified by the $m$th row vector of $\boldsymbol{B}^*\boldsymbol{\Theta}^{*\top}$. Thus, we obtain the $L \times M$ matrix $\boldsymbol{V}$, which corresponds to the empirical word distribution over $M$ documents.

As a real-world data set, we used the *Last.FM* data set.[4] *Last.FM* is a well-known social music web site, and the data set includes the triple (“user,” “artist,” “Freq”), which was collected from the playlists of users in

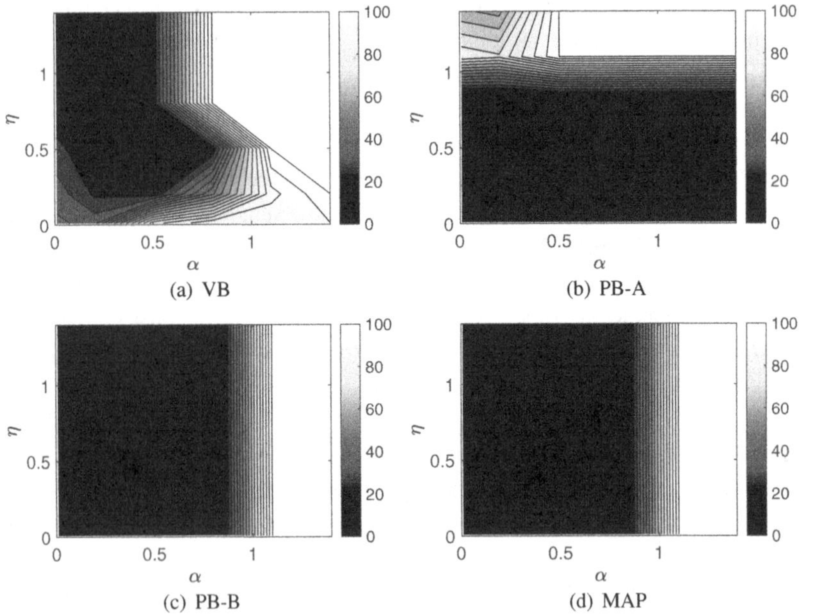


Figure 16.2 Estimated number $\widehat{H}$ of topics by (a) VB learning, (b) PB-A learning, (c) PB-B learning, and (d) MAP learning, on the *artificial* data with $L = 100$, $M = 100$, $H^* = 20$, and $N \sim 10000$.

[4] http://mtg.upf.edu/node/1671

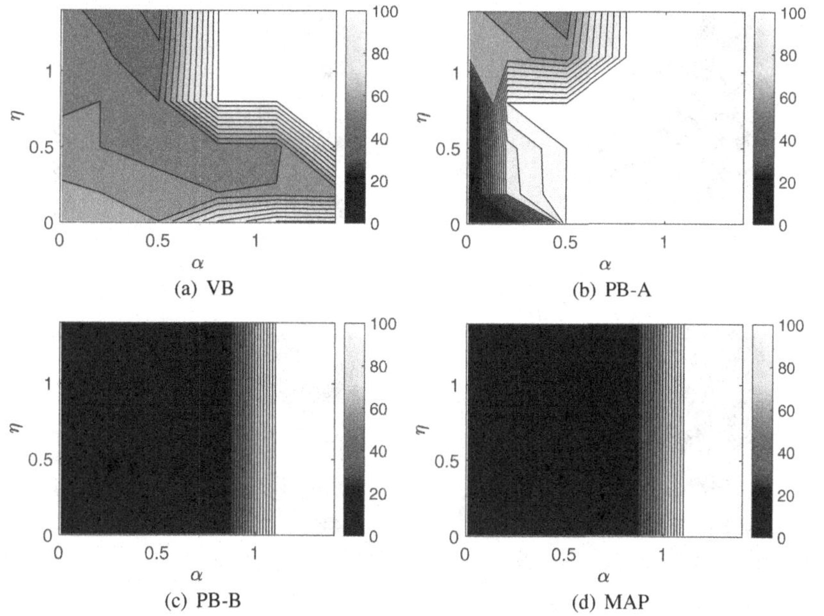


Figure 16.3 Estimated number $\widehat{H}$ of topics on the *Last.FM* data with $L = 100, M = 100$, and $N \sim 700$.

the community by using a plug-in in users' media players. This triple means that "user" played "artist" music "Freq" times, which indicates users' preferred artists. A user and a played artist are analogous to a document and a word, respectively. We randomly chose $L$ artists from the top 1,000 frequent artists, and $M$ users who live in the United States. To find a better local solution (which hopefully is close to the global solution), we adopted a split and merge strategy (Ueda et al., 2000), and chose the local solution giving the lowest free energy among different initialization schemes.

Figure 16.2 shows the estimated number $\widehat{H}$ of topics by different approximate Bayesian methods, i.e., VB, PB-A, PB-B, and MAP learning, on the *artificial* data with $L = 100, M = 100, H^* = 20$, and $N \sim 10000$. We can clearly see that the sparsity threshold in PB-B and MAP learning, where $\boldsymbol{\Theta}$ is point-estimated, is larger than that in VB and PB-A learning, where $\boldsymbol{\Theta}$ is marginalized. This result supports the statement by Theorem 16.13. Figure 16.3 shows results on the *Last.FM* data with $L = 100, M = 100$, and $N \sim 700$. We see a similar tendency to Figure 16.2 except the region where $\eta < 1$ for PB-A learning, in which our theory does not predict the estimated number of topics.

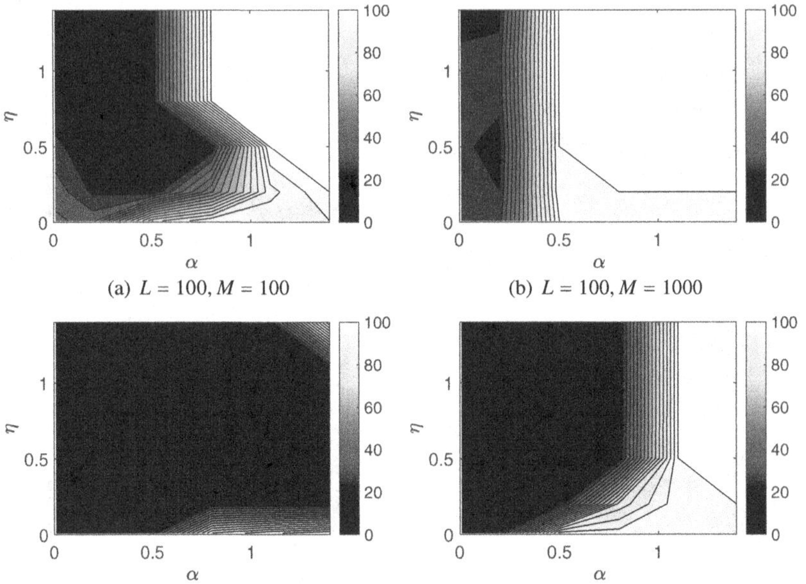


(a) $L = 100, M = 100$ (b) $L = 100, M = 1000$

(c) $L = 500, M = 100$ (d) $L = 500, M = 1000$

Figure 16.4 Estimated number $\widehat{H}$ of topics by VB learning on the *artificial* data with $H^* = 20$ and $N \sim 10000$. For the case when $L = 500, M = 1000$, the maximum estimated rank is limited to 100 for computational reason.

Finally, we investigate how different asymptotic settings affect the topic sparsity. Figure 16.4 shows the sparsity dependence on $L$ and $M$ on the *artificial* data. The graphs correspond to the four cases mentioned in Theorem 16.10, i.e, (a) $L, M \ll N$, (b) $L \ll N, M$, (c) $M \ll N, L$, and (d) $1 \ll N, L, M$. Corollary 16.11 explains the behavior in (a) and (b), and further analysis is required to explain the behavior in (c) and (d).

### 16.4.5 Proof of Theorem 16.13

We analyze PB-A learning, PB-B learning, and MAP learning, and then summarize the results, which proves Theorem 16.13.

**PB-A Learning**

The free energy for PB-A learning is given as follows:

$$F^{\text{PB-A}} = \chi_B + R^{\text{PB-A}} + Q^{\text{PB-A}}, \tag{16.64}$$

where $\chi_B$ is a large constant corresponding to the negative entropy of the delta functions (see Section 2.2.2), and

$$R^{\mathrm{PB-A}} = \left\langle \log \frac{r_{\Theta}(\boldsymbol{\Theta}) r_B(\boldsymbol{B})}{p(\boldsymbol{\Theta}|\alpha) p(\boldsymbol{B}|\eta)} \right\rangle_{r^{\mathrm{PB-A}}(\boldsymbol{\Theta}, \boldsymbol{B})}$$
$$= \sum_{m=1}^{M} \left( \log \frac{\Gamma(\sum_{h=1}^{H} \widehat{\alpha}_{m,h}^{\mathrm{PB-A}})}{\prod_{h=1}^{H} \Gamma(\widehat{\alpha}_{m,h}^{\mathrm{PB-A}})} \frac{\Gamma(\alpha)^H}{\Gamma(H\alpha)} + \sum_{h=1}^{H} \left(\widehat{\alpha}_{m,h}^{\mathrm{PB-A}} - \alpha\right) \left( \Psi(\widehat{\alpha}_{m,h}^{\mathrm{PB-A}}) - \Psi(\sum_{h'=1}^{H} \widehat{\alpha}_{m,h'}^{\mathrm{PB-A}}) \right) \right)$$
$$+ \sum_{h=1}^{H} \left( \log \frac{\Gamma(\eta)^L}{\Gamma(L\eta)} + \sum_{l=1}^{L} (1-\eta) \left( \log(\widehat{\eta}_{l,h}^{\mathrm{PB-A}}) - \log(\textstyle\sum_{l'=1}^{L} \widehat{\eta}_{l',h}^{\mathrm{PB-A}}) \right) \right), \tag{16.65}$$

$$Q^{\mathrm{PB-A}} = \left\langle \log \frac{r_z\left(\{\{\boldsymbol{z}^{(n,m)}\}_{n=1}^{N^{(m)}}\}_{m=1}^{M}\right)}{p(\{\boldsymbol{w}^{(n,m)}\}, \{\boldsymbol{z}^{(n,m)}\}|\boldsymbol{\Theta}, \boldsymbol{B})} \right\rangle_{r^{\mathrm{PB-A}}(\boldsymbol{\Theta}, \boldsymbol{B}, \{\boldsymbol{z}^{(n,m)}\})}$$
$$= -\sum_{m=1}^{M} N^{(m)} \sum_{l=1}^{L} V_{l,m} \log \left( \sum_{h=1}^{H} \frac{\exp\left(\Psi(\widehat{\alpha}_{m,h}^{\mathrm{PB-A}})\right)}{\exp\left(\Psi(\sum_{h'=1}^{H} \widehat{\alpha}_{m,h'}^{\mathrm{PB-A}})\right)} \frac{\widehat{\eta}_{l,h}^{\mathrm{PB-A}}}{\sum_{l'=1}^{L} \widehat{\eta}_{l',h}^{\mathrm{PB-A}}} \right). \tag{16.66}$$

Let us first consider the case when $\eta < 1$. In this case, $F$ diverges to $F \to -\infty$ with fixed $N$, when $\widehat{\eta}_{l,h} = O(1)$ for any $(l, h)$ and $\widehat{\eta}_{l',h} \to +0$ for all other $l' \neq l$. Therefore, the solution is useless.

When $\eta \geq 1$, the solution satisfies the following stationary condition:

$$\widehat{\alpha}_{m,h}^{\mathrm{PB-A}} = \alpha + \sum_{n=1}^{N^{(m)}} \widehat{z}_h^{\mathrm{PB-A}(n,m)}, \qquad \widehat{\eta}_{l,h}^{\mathrm{PB-A}} = \eta - 1 + \sum_{m=1}^{M} \sum_{n=1}^{N^{(m)}} w_l^{(n,m)} \widehat{z}_h^{\mathrm{PB-A}(n,m)}, \tag{16.67}$$

$$\widehat{z}_h^{\mathrm{PB-A}(n,m)} = \frac{\exp\left(\Psi(\widehat{\alpha}_{m,h}^{\mathrm{PB-A}})\right) \prod_{l=1}^{L} (\widehat{\eta}_{l,h}^{\mathrm{PB-A}})^{w_l^{(n,m)}}}{\sum_{h'=1}^{H} \left( \exp\left(\Psi(\widehat{\alpha}_{m,h'}^{\mathrm{PB-A}})\right) \prod_{l=1}^{L} (\widehat{\eta}_{l,h'}^{\mathrm{PB-A}})^{w_l^{(n,m)}} \right)}. \tag{16.68}$$

In the same way as for VB learning, we can obtain the following lemma:

**Lemma 16.14** *Let* $\widehat{\boldsymbol{B}}^{\mathrm{PB-A}} \widehat{\boldsymbol{\Theta}}^{\mathrm{PB-A}\top} = \langle \boldsymbol{B}\boldsymbol{\Theta}^\top \rangle_{r^{\mathrm{PB-A}}(\boldsymbol{\Theta}, \boldsymbol{B})}$. *Then it holds that*

$$\langle (\boldsymbol{B}\boldsymbol{\Theta}^\top - \widehat{\boldsymbol{B}}^{\mathrm{PB-A}} \widehat{\boldsymbol{\Theta}}^{\mathrm{PB-A}\top})_{l,m}^2 \rangle_{r^{\mathrm{PB-A}}(\boldsymbol{\Theta}, \boldsymbol{B})} = O_{\mathrm{p}}(N^{-2}), \tag{16.69}$$

$$Q^{\mathrm{PB-A}} = -\sum_{m=1}^{M} N^{(m)} \sum_{l=1}^{L} V_{l,m} \log(\widehat{\boldsymbol{B}}^{\mathrm{PB-A}} \widehat{\boldsymbol{\Theta}}^{\mathrm{PB-A}\top})_{l,m} + O_{\mathrm{p}}(N^{-1}). \tag{16.70}$$

$Q^{\text{PB-A}}$ *is minimized when* $\widehat{J} = O_{\text{p}}(N^{-1})$, *and it holds that*

$$Q^{\text{PB-A}} = NS_N(\mathcal{D}) + O_{\text{p}}(\widehat{J}N + LM).$$

$R^{\text{PB-A}}$ *is written as follows:*

$$R^{\text{PB-A}} = \left\{ M\left(H\alpha - \frac{1}{2}\right) + \widehat{H}L(\eta - 1) - \sum_{h=1}^{\widehat{H}} \left( \widehat{M}^{(h)} \left(\alpha - \frac{1}{2}\right) + \widehat{L}^{(h)}(\eta - 1) \right) \right\} \log N$$
$$+ (H - \widehat{H})L(\eta - 1)\log L + O_{\text{p}}(H(M + L)). \tag{16.71}$$

Taking the different asymptotic limits, we obtain the following theorem:

**Theorem 16.15** *When* $\eta < 1$, *each column vector of* $\widehat{\boldsymbol{B}}^{\text{PB-A}}$ *has only one nonzero entry. Assume in the following that* $\eta \geq 1$. *In the limit when* $N \to \infty$ *with* $L, M \sim O(1)$, *it holds that* $\widehat{J} = O_{\text{p}}(1/N)$ *and*

$$\widetilde{F}^{\text{PB-A}}(\mathcal{D}) = \lambda'^{\text{PB-A}}_{\text{LDA}} \log N + O_{\text{p}}(1),$$

*where*

$$\lambda'^{\text{PB-A}}_{\text{LDA}} = M\left(H\alpha - \frac{1}{2}\right) + \widehat{H}L(\eta - 1) - \sum_{h=1}^{\widehat{H}} \left( \widehat{M}^{(h)} \left(\alpha - \frac{1}{2}\right) + \widehat{L}^{(h)}(\eta - 1) \right).$$

*In the limit when* $N, M \to \infty$ *with* $\frac{M}{N}, L \sim O(1)$, *it holds that* $\widehat{J} = o_{\text{p}}(\log N)$, *and*

$$\widetilde{F}^{\text{PB-A}}(\mathcal{D}) = \lambda'^{\text{PB-A}}_{\text{LDA}} \log N + o_{\text{p}}(N \log N),$$

*where*

$$\lambda'^{\text{PB-A}}_{\text{LDA}} = M\left(H\alpha - \frac{1}{2}\right) - \sum_{h=1}^{\widehat{H}} \widehat{M}^{(h)} \left(\alpha - \frac{1}{2}\right).$$

*In the limit when* $N, L \to \infty$ *with* $\frac{L}{N}, M \sim O(1)$, *it holds that* $\widehat{J} = o_{\text{p}}(\log N)$, *and*

$$\widetilde{F}^{\text{PB-A}}(\mathcal{D}) = \lambda'^{\text{PB-A}}_{\text{LDA}} \log N + o_{\text{p}}(N \log N),$$

*where*

$$\lambda'^{\text{PB-A}}_{\text{LDA}} = HL(\eta - 1).$$

*In the limit when* $N, L, M \to \infty$ *with* $\frac{L}{N}, \frac{M}{N} \sim O(1)$, *it holds that* $\widehat{J} = o_{\text{p}}(N \log N)$, *and*

$$\widetilde{F}^{\text{PB-A}}(\mathcal{D}) = \lambda'^{\text{PB-A}}_{\text{LDA}} \log N + o_{\text{p}}(N^2 \log N),$$

*where*

$$\lambda_{\mathrm{LDA}}^{\prime \mathrm{PB-A}} = H(M\alpha + L(\eta - 1)).$$

Note that Theorem 16.15 provides no information on the sparsity of the PB-A solution for $\eta < 1$. In the following subsection, we investigate the sparsity of the solution for $\eta \geq 1$.

**In the Limit When $N \to \infty$ with $L, M \sim O(1)$**

The coefficient of the leading term of the free energy is

$$\lambda_{\mathrm{LDA}}^{\prime \mathrm{PB-A}} = M\left(H\alpha - \frac{1}{2}\right) + \sum_{h=1}^{\widehat{H}} \left(L(\eta - 1) - \widehat{M}^{(h)}\left(\alpha - \frac{1}{2}\right) - \widehat{L}^{(h)}(\eta - 1)\right).$$

The solution is sparse if $\lambda_{\mathrm{LDA}}^{\prime \mathrm{PB-A}}$ is increasing with respect to $\widehat{H}$, and dense if it is decreasing. We focus on the case where $\eta \geq 1$. Eqs. (16.57) through (16.59) imply the following:

(I) When $\alpha < \frac{1}{2}$, the solution is sparse if

$$L(\eta - 1) - \max_h \left(M^{*(h)}\left(\alpha - \frac{1}{2}\right) + L^{*(h)}(\eta - 1)\right) > 0, \tag{16.72}$$

and dense if

$$L(\eta - 1) - \max_h \left(M^{*(h)}\left(\alpha - \frac{1}{2}\right) + L^{*(h)}(\eta - 1)\right) < 0. \tag{16.73}$$

Therefore, the solution is sparse if

$$L(\eta - 1) - \min_h M^{*(h)}\left(\alpha - \frac{1}{2}\right) - \max_h L^{*(h)}(\eta - 1) > 0, \text{ or equivalently,}$$

$$\alpha < \frac{1}{2} + \frac{\left(L - \max_h L^{*(h)}\right)(\eta - 1)}{\min_h M^{*(h)}},$$

and dense if

$$L(\eta - 1) - \max_h M^{*(h)}\left(\alpha - \frac{1}{2}\right) - \max_h L^{*(h)}(\eta - 1) < 0, \text{ or equivalently,}$$

$$\alpha > \frac{1}{2} + \frac{\left(L - \max_h L^{*(h)}\right)(\eta - 1)}{\max_h M^{*(h)}}.$$

Therefore, the solution is always sparse in this case.

(II) When $\alpha \geq \frac{1}{2}$, the solution is sparse if Eq. (16.72) holds, and dense if Eq. (16.73) holds. Therefore, the solution is sparse if

$$L(\eta - 1) - \max_h M^{*(h)} \left(\alpha - \frac{1}{2}\right) - \max_h L^{*(h)} (\eta - 1) > 0, \text{ or equivalently,}$$

$$\alpha < \frac{1}{2} + \frac{\left(L - \max_h L^{*(h)}\right)(\eta - 1)}{\max_h M^{*(h)}},$$

and dense if

$$L(\eta - 1) - \min_h M^{*(h)} \left(\alpha - \frac{1}{2}\right) - \max_h L^{*(h)} (\eta - 1) < 0, \text{ or equivalently,}$$

$$\alpha > \frac{1}{2} + \frac{\left(L - \max_h L^{*(h)}\right)(\eta - 1)}{\min_h M^{*(h)}}.$$

Thus, we can conclude that, in this case, the solution is sparse if

$$\alpha < \frac{1}{2},$$

and dense if

$$\alpha > \frac{1}{2} + \frac{L(\eta - 1)}{\min_h M^{*(h)}}.$$

Summarizing the preceding, we have the following lemma:

**Lemma 16.16** *Assume that $\eta \geq 1$. The solution is sparse if $\alpha < \frac{1}{2}$, and dense if $\alpha > \frac{1}{2} + \frac{L(\eta-1)}{\min_h M^{*(h)}}$.*

**In the Limit When $N, M \to \infty$ with $\frac{M}{N}, L \sim O(1)$**

The coefficient of the leading term of the free energy is given by

$$\lambda'^{\text{PB-A}}_{\text{LDA}} = M\left(H\alpha - \frac{1}{2}\right) - \sum_{h=1}^{\widehat{H}} \widehat{M}^{(h)} \left(\alpha - \frac{1}{2}\right). \tag{16.74}$$

Although the predictive distribution does not necessarily converge to the true distribution, we can investigate the sparsity of the solution by considering the duplication rules (16.57) through (16.59) that keep $\widehat{\boldsymbol{B}\boldsymbol{\Theta}^\top}$ unchanged. It is clear that Eq. (16.74) is increasing with respect to $\widehat{H}$ if $\alpha < \frac{1}{2}$, and decreasing if $\alpha > \frac{1}{2}$. Combing this result with Lemma 16.16, we obtain the following corollary:

**Corollary 16.17** *Assume that $\eta \geq 1$. In the limit when $N \to \infty$ with $L$, $M \sim O(1)$, the PB-A solution is sparse if $\alpha < \frac{1}{2}$, and dense if $\alpha > \frac{1}{2} + \frac{L(\eta-1)}{\min_h M^{*(h)}}$.*

*In the limit when $N, M \to \infty$ with $\frac{M}{N}, L \sim O(1)$, the PB-A solution is sparse if $\alpha < \frac{1}{2}$, and dense if $\alpha > \frac{1}{2}$.*

**PB-B Learning**

The free energy for PB-B learning is given as follows:

$$F^{\mathrm{PB-B}} = \chi_{\Theta} + R^{\mathrm{PB-B}} + Q^{\mathrm{PB-B}}, \tag{16.75}$$

where $\chi_{\Theta}$ is a large constant corresponding to the negative entropy of the delta functions, and

$$\begin{aligned} R^{\mathrm{PB-B}} &= \left\langle \log \frac{r_{\Theta}(\boldsymbol{\Theta}) r_B(\boldsymbol{B})}{p(\boldsymbol{\Theta}|\alpha) p(\boldsymbol{B}|\eta)} \right\rangle_{r^{\mathrm{PB-B}}(\boldsymbol{\Theta}, \boldsymbol{B})} \\ &= \sum_{m=1}^{M} \left( \log \frac{\Gamma(\alpha)^H}{\Gamma(H\alpha)} + \sum_{h=1}^{H} (1-\alpha) \left( \log(\widehat{\alpha}_{m,h}^{\mathrm{PB-B}}) - \log(\sum_{h'=1}^{H} \widehat{\alpha}_{m,h'}^{\mathrm{PB-B}}) \right) \right) \\ &\quad + \sum_{h=1}^{H} \left( \log \frac{\Gamma(\sum_{l=1}^{L} \widehat{\eta}_{l,h}^{\mathrm{PB-B}})}{\prod_{l=1}^{L} \Gamma(\widehat{\eta}_{l,h}^{\mathrm{PB-B}})} \frac{\Gamma(\eta)^L}{\Gamma(L\eta)} \right. \\ &\quad \left. + \sum_{l=1}^{L} \left( \widehat{\eta}_{l,h}^{\mathrm{PB-B}} - \eta \right) \left( \Psi(\widehat{\eta}_{l,h}^{\mathrm{PB-B}}) - \Psi(\sum_{l'=1}^{L} \widehat{\eta}_{l',h}^{\mathrm{PB-B}}) \right) \right), \end{aligned} \tag{16.76}$$

$$\begin{aligned} Q^{\mathrm{PB-B}} &= \left\langle \log \frac{r_z\left(\{\{\boldsymbol{z}^{(n,m)}\}_{n=1}^{N^{(m)}}\}_{m=1}^{M}\right)}{p(\{\boldsymbol{w}^{(n,m)}\}, \{\boldsymbol{z}^{(n,m)}\}|\boldsymbol{\Theta}, \boldsymbol{B})} \right\rangle_{r^{\mathrm{PB-B}}(\boldsymbol{\Theta}, \boldsymbol{B}, \{\boldsymbol{z}^{(n,m)}\})} \\ &= -\sum_{m=1}^{M} N^{(m)} \sum_{l=1}^{L} V_{l,m} \log \left( \sum_{h=1}^{H} \frac{\widehat{\alpha}_{m,h}^{\mathrm{PB-B}}}{\sum_{h'=1}^{H} \widehat{\alpha}_{m,h'}^{\mathrm{PB-B}}} \frac{\exp\left(\Psi(\widehat{\eta}_{l,h}^{\mathrm{PB-B}})\right)}{\exp\left(\Psi(\sum_{l'=1}^{L} \widehat{\eta}_{l',h}^{\mathrm{PB-B}})\right)} \right). \end{aligned} \tag{16.77}$$

Let us first consider the case when $\alpha < 1$. In this case, $F$ diverges to $F \to -\infty$ with fixed $N$, when $\widehat{\alpha}_{m,h} = O(1)$ for any $(m, h)$ and $\widehat{\alpha}_{m,h'} \to +0$ for all other $h' \neq h$. Therefore, the solution is sparse (so sparse that the estimator is useless).

When $\alpha \geq 1$, the solution satisfies the following stationary condition:

$$\widehat{\alpha}_{m,h}^{\mathrm{PB-B}} = \alpha - 1 + \sum_{n=1}^{N^{(m)}} \widehat{z}_h^{\mathrm{PB-B}(n,m)}, \qquad \widehat{\eta}_{l,h}^{\mathrm{PB-B}} = \eta + \sum_{m=1}^{M} \sum_{n=1}^{N^{(m)}} w_l^{(n,m)} \widehat{z}_h^{\mathrm{PB-B}(n,m)}, \tag{16.78}$$

$$\widehat{z}_h^{\mathrm{PB-B}(n,m)} = \frac{\widehat{\alpha}_{m,h}^{\mathrm{PB-B}} \exp\left\{\sum_{l=1}^{L} w_l^{(n,m)} \left(\Psi(\widehat{\eta}_{l,h}^{\mathrm{PB-B}}) - \Psi\left(\sum_{l'=1}^{L} \widehat{\eta}_{l',h}^{\mathrm{PB-B}}\right)\right)\right\}}{\sum_{h'=1}^{H} \widehat{\alpha}_{m,h'}^{\mathrm{PB-B}} \exp\left\{\sum_{l=1}^{L} w_l^{(n,m)} \left(\Psi(\widehat{\eta}_{l,h'}^{\mathrm{PB-B}}) - \Psi\left(\sum_{l'=1}^{L} \widehat{\eta}_{l',h'}^{\mathrm{PB-B}}\right)\right)\right\}}. \tag{16.79}$$

In the same way as for VB and PB-A learning, we can obtain the following lemma:

**Lemma 16.18** *Let* $\widehat{\boldsymbol{B}}^{\text{PB-B}}\widehat{\boldsymbol{\Theta}}^{\text{PB-B}\top} = \langle \boldsymbol{B}\boldsymbol{\Theta}^\top \rangle_{r^{\text{PB-B}}(\boldsymbol{\Theta},\boldsymbol{B})}$. *Then it holds that*

$$\langle (\boldsymbol{B}\boldsymbol{\Theta}^\top - \widehat{\boldsymbol{B}}^{\text{PB-B}}\widehat{\boldsymbol{\Theta}}^{\text{PB-B}\top})^2_{l,m} \rangle_{r^{\text{PB-B}}(\boldsymbol{\Theta},\boldsymbol{B})} = O_{\text{p}}(N^{-2}), \tag{16.80}$$

$$Q^{\text{PB-B}} = -\sum_{m=1}^{M} N^{(m)} \sum_{l=1}^{L} V_{l,m} \log(\widehat{\boldsymbol{B}}^{\text{PB-B}}\widehat{\boldsymbol{\Theta}}^{\text{PB-B}\top})_{l,m} + O_{\text{p}}(N^{-1}). \tag{16.81}$$

$Q^{\text{PB-B}}$ *is minimized when* $\widehat{J} = O_{\text{p}}(N^{-1})$*, and it holds that*

$$Q^{\text{PB-B}} = NS_N(\mathcal{D}) + O_{\text{p}}(\widehat{J}N + LM).$$

$R^{\text{PB-B}}$ *is written as follows:*

$$R^{\text{PB-B}} = \left\{ MH(\alpha - 1) + \widehat{H}\left(L\eta - \frac{1}{2}\right) - \sum_{h=1}^{\widehat{H}} \left( \widehat{M}^{(h)}(\alpha - 1) + \widehat{L}^{(h)}\left(\eta - \frac{1}{2}\right)\right)\right\} \log N$$
$$+ (H - \widehat{H})\left(L\eta - \frac{1}{2}\right) \log L + O_{\text{p}}(H(M + L)). \tag{16.82}$$

Taking the different asymptotic limits, we obtain the following theorem:

**Theorem 16.19** *When* $\alpha < 1$*, each row vector of* $\widehat{\boldsymbol{\Theta}}^{\text{PB-B}}$ *has only one nonzero entry, and the PB-B solution is sparse. Assume in the following that* $\alpha \geq 1$*. In the limit when* $N \to \infty$ *with* $L, M \sim O(1)$*, it holds that* $\widehat{J} = O_{\text{p}}(1/N)$ *and*

$$\overline{F}^{\text{PB-B}}(\mathcal{D}) = \lambda'^{\text{PB-B}}_{\text{LDA}} \log N + O_{\text{p}}(1),$$

*where*

$$\lambda'^{\text{PB-B}}_{\text{LDA}} = MH(\alpha - 1) + \widehat{H}\left(L\eta - \frac{1}{2}\right) - \sum_{h=1}^{\widehat{H}} \left( \widehat{M}^{(h)}(\alpha - 1) + \widehat{L}^{(h)}\left(\eta - \frac{1}{2}\right)\right).$$

*In the limit when* $N, M \to \infty$ *with* $\frac{M}{N}, L \sim O(1)$*, it holds that* $\widehat{J} = o_{\text{p}}(\log N)$*, and*

$$\overline{F}^{\text{PB-B}}(\mathcal{D}) = \lambda'^{\text{PB-B}}_{\text{LDA}} \log N + o_{\text{p}}(N \log N),$$

*where*

$$\lambda'^{\text{PB-B}}_{\text{LDA}} = MH(\alpha - 1) - \sum_{h=1}^{\widehat{H}} \widehat{M}^{(h)}(\alpha - 1).$$

*In the limit when $N, L \to \infty$ with $\frac{L}{N}, M \sim O(1)$, it holds that $\widehat{J} = o_{\mathrm{p}}(\log N)$, and*

$$\widetilde{F}^{\mathrm{PB-B}}(\mathcal{D}) = \lambda_{\mathrm{LDA}}^{\prime\mathrm{PB-B}} \log N + o_{\mathrm{p}}(N \log N),$$

*where*

$$\lambda_{\mathrm{LDA}}^{\prime\mathrm{PB-B}} = HL\eta.$$

*In the limit when $N, L, M \to \infty$ with $\frac{L}{N}, \frac{M}{N} \sim O(1)$, it holds that $\widehat{J} = o_{\mathrm{p}}(N \log N)$, and*

$$\widetilde{F}^{\mathrm{PB-B}}(\mathcal{D}) = \lambda_{\mathrm{LDA}}^{\prime\mathrm{PB-B}} \log N + o_{\mathrm{p}}(N^2 \log N),$$

*where*

$$\lambda_{\mathrm{LDA}}^{\prime\mathrm{PB-B}} = H(M(\alpha - 1) + L\eta).$$

Theorem 16.19 states that the PB-B solution is sparse when $\alpha < 1$. In the following subsection, we investigate the sparsity of the solution for $\alpha \geq 1$.

**In the Limit When $N \to \infty$ with $L, M \sim O(1)$**

The coefficient of the leading term of the free energy is

$$\lambda_{\mathrm{LDA}}^{\prime\mathrm{PB-B}} = MH(\alpha - 1) + \sum_{h=1}^{\widehat{H}} \left( L\eta - \frac{1}{2} - \widehat{M}^{(h)}(\alpha - 1) - \widehat{L}^{(h)}\left(\eta - \frac{1}{2}\right)\right).$$

The solution is sparse if $\lambda_{\mathrm{LDA}}^{\prime\mathrm{PB-B}}$ is increasing with respect to $\widehat{H}$, and dense if it is decreasing. We focus on the case where $\alpha \geq 1$. Eqs. (16.57) through (16.59) imply the following:

(I) When $0 < \eta \leq \frac{1}{2L}$, the solution is sparse if

$$L\eta - \frac{1}{2} - \max_h M^{*(h)}(\alpha - 1) > 0, \text{ or equivalently,}$$

$$\alpha < 1 - \frac{1}{\max_h M^{*(h)}}\left(\frac{1}{2} - L\eta\right),$$

and dense if

$$\alpha > 1 - \frac{1}{\max_h M^{*(h)}}\left(\frac{1}{2} - L\eta\right).$$

Therefore, the solution is always dense in this case.

(II) When $\frac{1}{2L} < \eta \leq \frac{1}{2}$, the solution is sparse if

$$L\eta - \frac{1}{2} - \max_h M^{*(h)}(\alpha - 1) > 0, \text{ or equivalently, } \alpha < 1 + \frac{L\eta - \frac{1}{2}}{\max_h M^{*(h)}},$$

and dense if

$$\alpha > 1 + \frac{L\eta - \frac{1}{2}}{\max_h M^{*(h)}}.$$

(III) When $\eta > \frac{1}{2}$, the solution is sparse if

$$L\eta - \frac{1}{2} - \max_h \left( M^{*(h)} (\alpha - 1) + L^{*(h)} \left( \eta - \frac{1}{2} \right) \right) > 0, \tag{16.83}$$

and dense if

$$L\eta - \frac{1}{2} - \max_h \left( M^{*(h)} (\alpha - 1) + L^{*(h)} \left( \eta - \frac{1}{2} \right) \right) < 0. \tag{16.84}$$

Therefore, the solution is sparse if

$$L\eta - \frac{1}{2} - \max_h M^{*(h)} (\alpha - 1) - \max_h L^{*(h)} \left( \eta - \frac{1}{2} \right) > 0, \text{ or equivalently,}$$

$$\alpha < 1 + \frac{1}{\max_h M^{*(h)}} \left( L\eta - \frac{1}{2} - \max_h L^{*(h)} \left( \eta - \frac{1}{2} \right) \right),$$

and dense if

$$L\eta - \frac{1}{2} - \min_h M^{*(h)} (\alpha - 1) - \max_h L^{*(h)} \left( \eta - \frac{1}{2} \right) < 0, \text{ or equivalently,}$$

$$\alpha > 1 + \frac{1}{\min_h M^{*(h)}} \left( L\eta - \frac{1}{2} - \max_h L^{*(h)} \left( \eta - \frac{1}{2} \right) \right).$$

Thus, we can conclude that, in this case, the solution is sparse if

$$\alpha < 1 + \frac{L - 1}{2 \max_h M^{*(h)}},$$

and dense if

$$\alpha > 1 + \frac{L\eta - \frac{1}{2}}{\min_h M^{*(h)}}.$$

Summarizing the preceding, we have the following lemma:

**Lemma 16.20** *Assume that $\alpha \geq 1$. When $0 < \eta \leq \frac{1}{2L}$, the solution is always dense. When $\frac{1}{2L} < \eta \leq \frac{1}{2}$, the solution is sparse if $\alpha < 1 + \frac{L\eta - \frac{1}{2}}{\max_h M^{*(h)}}$, and dense if $\alpha > 1 + \frac{L\eta - \frac{1}{2}}{\max_h M^{*(h)}}$. When $\eta > \frac{1}{2}$, the solution is sparse if $\alpha < 1 + \frac{L-1}{2 \max_h M^{*(h)}}$, and dense if $\alpha > 1 + \frac{L\eta - \frac{1}{2}}{\min_h M^{*(h)}}$.*

**In the Limit When $N, M \to \infty$ with $\frac{M}{N}, L \sim O(1)$**

The coefficient of the leading term of the free energy is given by

$$\lambda'^{\text{PB-B}}_{\text{LDA}} = M(H\alpha - 1) - \sum_{h=1}^{\widehat{H}} \widehat{M}^{(h)}(\alpha - 1). \tag{16.85}$$

Although the predictive distribution does not necessarily converge to the true distribution, we can investigate the sparsity of the solution by considering the duplication rules (16.57) through (16.59) that keep $\widehat{\boldsymbol{B}\boldsymbol{\Theta}}^{\top}$ unchanged. It is clear that Eq. (16.85) is decreasing with respect to $\widehat{H}$ if $\alpha > 1$. Combing this result with Theorem 16.19, which states that the PB-B solution is sparse when $\alpha < 1$, and Lemma 16.20, we obtain the following corollary:

**Corollary 16.21** *Consider the limit when $N \to \infty$ with $L, M \sim O(1)$. When $0 < \eta \leq \frac{1}{2L}$, the PB-B solution is sparse if $\alpha < 1$, and dense if $\alpha > 1$. When $\frac{1}{2L} < \eta \leq \frac{1}{2}$, the PB-B solution is sparse if $\alpha < 1 + \frac{L\eta - \frac{1}{2}}{\max_h M^{*(h)}}$, and dense if $\alpha > 1 + \frac{L\eta - \frac{1}{2}}{\max_h M^{*(h)}}$. When $\eta > \frac{1}{2}$, the PB-B solution is sparse if $\alpha < 1 + \frac{L-1}{2\max_h M^{*(h)}}$, and dense if $\alpha > 1 + \frac{L\eta - \frac{1}{2}}{\min_h M^{*(h)}}$. In the limit when $N, M \to \infty$ with $\frac{M}{N}, L \sim O(1)$, the PB-B solution is sparse if $\alpha < 1$, and dense if $\alpha > 1$.*

**MAP Learning**

The free energy for MAP learning is given as follows:

$$F^{\text{MAP}} = \chi_{\Theta} + \chi_{B} + R^{\text{MAP}} + Q^{\text{MAP}}, \tag{16.86}$$

where $\chi_{\Theta}$ and $\chi_{B}$ are large constants corresponding to the negative entropies of the delta functions, and

$$\begin{aligned} R^{\text{MAP}} &= \left\langle \log \frac{r_{\Theta}(\boldsymbol{\Theta}) r_{B}(\boldsymbol{B})}{p(\boldsymbol{\Theta}|\alpha) p(\boldsymbol{B}|\eta)} \right\rangle_{r^{\text{MAP}}(\boldsymbol{\Theta}, \boldsymbol{B})} \\ &= \sum_{m=1}^{M} \left( \log \frac{\Gamma(\alpha)^H}{\Gamma(H\alpha)} + \sum_{h=1}^{H} (1-\alpha) \left( \log(\widehat{\alpha}^{\text{MAP}}_{m,h}) - \log(\sum_{h'=1}^{H} \widehat{\alpha}^{\text{MAP}}_{m,h'}) \right) \right) \\ &\quad + \sum_{h=1}^{H} \left( \log \frac{\Gamma(\eta)^L}{\Gamma(L\eta)} + \sum_{l=1}^{L} (1-\eta) \left( \log(\widehat{\eta}^{\text{MAP}}_{l,h}) - \log(\textstyle\sum_{l'=1}^{L} \widehat{\eta}^{\text{MAP}}_{l',h}) \right) \right), \end{aligned} \tag{16.87}$$

$$\begin{aligned} Q^{\text{MAP}} &= \left\langle \log \frac{r_z\left(\{\{\boldsymbol{z}^{(n,m)}\}_{n=1}^{N^{(m)}}\}_{m=1}^{M}\right)}{p(\{\boldsymbol{w}^{(n,m)}\}, \{\boldsymbol{z}^{(n,m)}\}|\boldsymbol{\Theta}, \boldsymbol{B})} \right\rangle_{r^{\text{MAP}}(\boldsymbol{\Theta}, \boldsymbol{B}, \{\boldsymbol{z}^{(n,m)}\})} \\ &= -\sum_{m=1}^{M} N^{(m)} \sum_{l=1}^{L} V_{l,m} \log \left( \sum_{h=1}^{H} \frac{\widehat{\alpha}^{\text{MAP}}_{m,h}}{\sum_{h'=1}^{H} \widehat{\alpha}^{\text{MAP}}_{m,h'}} \frac{\widehat{\eta}^{\text{MAP}}_{l,h}}{\sum_{l'=1}^{L} \widehat{\eta}^{\text{MAP}}_{l',h}} \right). \end{aligned} \tag{16.88}$$

Let us first consider the case when $\alpha < 1$. In this case, $F$ diverges to $F \to -\infty$ with fixed $N$, when $\widehat{\alpha}_{m,h} = O(1)$ for any $(h, m)$ and $\widehat{\alpha}_{m,h'} \to +0$ for all other $h' \neq h$. Therefore, the solution is sparse (so sparse that the estimator is useless). Similarly, assume that $\eta < 1$. Then $F$ diverges to $F \to -\infty$ with fixed $N$, when $\widehat{\eta}_{l,h} = O(1)$ for any $(l, h)$ and $\widehat{\eta}_{l',h} \to +0$ for all other $l' \neq l$. Therefore, the solution is useless.

When $\alpha \geq 1$ and $\eta \geq 1$, the solution satisfies the following stationary condition:

$$\widehat{\alpha}_{m,h}^{\mathrm{MAP}} = \alpha - 1 + \sum_{n=1}^{N^{(m)}} \widehat{z}_h^{\mathrm{MAP}(n,m)}, \qquad \widehat{\eta}_{l,h}^{\mathrm{MAP}} = \eta - 1 + \sum_{m=1}^{M} \sum_{n=1}^{N^{(m)}} w_l^{(n,m)} \widehat{z}_h^{\mathrm{MAP}(n,m)}, \tag{16.89}$$

$$\widehat{z}_h^{\mathrm{MAP}(n,m)} = \frac{\widehat{\alpha}_{m,h}^{\mathrm{MAP}} \prod_{l=1}^{L} (\widehat{\eta}_{l,h}^{\mathrm{MAP}})^{w_l^{(n,m)}}}{\sum_{h'=1}^{H} \left( \widehat{\alpha}_{m,h'}^{\mathrm{MAP}} \prod_{l=1}^{L} (\widehat{\eta}_{l,h'}^{\mathrm{MAP}})^{w_l^{(n,m)}} \right)}. \tag{16.90}$$

In the same way as for VB, PB-A, and PB-B learning, we can obtain the following lemma:

**Lemma 16.22** *Let* $\widehat{\boldsymbol{B}}^{\mathrm{MAP}} \widehat{\boldsymbol{\Theta}}^{\mathrm{MAP}\top} = \langle \boldsymbol{B}\boldsymbol{\Theta}^\top \rangle_{r^{\mathrm{MAP}}(\boldsymbol{\Theta},\boldsymbol{B})}$. *Then* $Q^{\mathrm{MAP}}$ *is minimized when* $\widehat{J} = O_{\mathrm{p}}(N^{-1})$, *and it holds that*

$$Q^{\mathrm{MAP}} = NS_N(\mathcal{D}) + O_{\mathrm{p}}(\widehat{J}N + LM).$$

$R^{\mathrm{MAP}}$ *is written as follows:*

$$R^{\mathrm{MAP}} = \left\{ MH(\alpha - 1) + \widehat{H}L(\eta - 1) - \sum_{h=1}^{\widehat{H}} \left( \widehat{M}^{(h)}(\alpha - 1) + \widehat{L}^{(h)}(\eta - 1) \right) \right\} \log N$$
$$+ (H - \widehat{H})L(\eta - 1)\log L + O_{\mathrm{p}}(H(M + L)). \tag{16.91}$$

Taking the different asymptotic limits, we obtain the following theorem:

**Theorem 16.23** *When* $\alpha < 1$, *each row vector of* $\widehat{\boldsymbol{\Theta}}^{\mathrm{MAP}}$ *has only one nonzero entry, and the MAP solution is sparse. When* $\eta < 1$, *each column vector of* $\widehat{\boldsymbol{B}}^{\mathrm{MAP}}$ *has only one nonzero entry. Assume in the following that* $\alpha, \eta \geq 1$. *In the limit when* $N \to \infty$ *with* $L, M \sim O(1)$, *it holds that* $\widehat{J} = O_{\mathrm{p}}(1/N)$ *and*

$$\widetilde{F}^{\mathrm{MAP}}(\mathcal{D}) = \lambda_{\mathrm{LDA}}'^{\mathrm{MAP}} \log N + O_{\mathrm{p}}(1),$$

*where*

$$\lambda_{\mathrm{LDA}}'^{\mathrm{MAP}} = MH(\alpha - 1) + \widehat{H}L(\eta - 1) - \sum_{h=1}^{\widehat{H}} \left( \widehat{M}^{(h)}(\alpha - 1) + \widehat{L}^{(h)}(\eta - 1) \right).$$

*In the limit when* $N, M \to \infty$ *with* $\frac{M}{N}, L \sim O(1)$, *it holds that* $\widehat{J} = o_{\mathrm{p}}(\log N)$, *and*

$$\widetilde{F}^{\mathrm{MAP}}(\mathcal{D}) = \lambda'^{\mathrm{MAP}}_{\mathrm{LDA}} \log N + o_{\mathrm{p}}(N \log N),$$

*where*

$$\lambda'^{\mathrm{MAP}}_{\mathrm{LDA}} = MH(\alpha - 1) - \sum_{h=1}^{\widehat{H}} \widehat{M}^{(h)} (\alpha - 1).$$

*In the limit when* $N, L \to \infty$ *with* $\frac{L}{N}, M \sim O(1)$, *it holds that* $\widehat{J} = o_{\mathrm{p}}(\log N)$, *and*

$$\widetilde{F}^{\mathrm{MAP}}(\mathcal{D}) = \lambda'^{\mathrm{MAP}}_{\mathrm{LDA}} \log N + o_{\mathrm{p}}(N \log N),$$

*where*

$$\lambda'^{\mathrm{MAP}}_{\mathrm{LDA}} = HL(\eta - 1).$$

*In the limit when* $N, L, M \to \infty$ *with* $\frac{L}{N}, \frac{M}{N} \sim O(1)$, *it holds that* $\widehat{J} = o_{\mathrm{p}}(N \log N)$, *and*

$$\widetilde{F}^{\mathrm{MAP}}(\mathcal{D}) = \lambda'^{\mathrm{MAP}}_{\mathrm{LDA}} \log N + o_{\mathrm{p}}(N^2 \log N),$$

*where*

$$\lambda'^{\mathrm{MAP}}_{\mathrm{LDA}} = H(M(\alpha - 1) + L(\eta - 1)).$$

Theorem 16.23 states that the MAP solution is sparse when $\alpha < 1$. However, it provides no information on the sparsity of the MAP solution for $\eta < 1$. In the following, we investigate the sparsity of the solution for $\alpha, \eta \geq 1$.

**In the Limit When $N \to \infty$ with $L, M \sim O(1)$**

The coefficient of the leading term of the free energy is

$$\lambda'^{\mathrm{MAP}}_{\mathrm{LDA}} = MH(\alpha - 1) + \sum_{h=1}^{\widehat{H}} \left( L(\eta - 1) - \widehat{M}^{(h)} (\alpha - 1) - \widehat{L}^{(h)} (\eta - 1) \right).$$

The solution is sparse if $\lambda'^{\mathrm{MAP}}_{\mathrm{LDA}}$ is increasing with respect to $\widehat{H}$, and dense if it is decreasing. We focus on the case where $\alpha, \eta \geq 1$. Eqs. (16.57) through (16.59) imply the following:

The solution is sparse if

$$L(\eta - 1) - \max_h \left( M^{*(h)} (\alpha - 1) + L^{*(h)} (\eta - 1) \right) > 0, \tag{16.92}$$

and dense if

$$L(\eta - 1) - \max_h \left( M^{*(h)} (\alpha - 1) + L^{*(h)} (\eta - 1) \right) < 0. \tag{16.93}$$

Therefore, the solution is sparse if

$$L(\eta - 1) - \max_h M^{*(h)} (\alpha - 1) - \max_h L^{*(h)} (\eta - 1) > 0, \text{ or equivalently,}$$

$$\alpha < 1 + \frac{(L - \max_h L^{*(h)})(\eta - 1)}{\max_h M^{*(h)}},$$

and dense if

$$L(\eta - 1) - \min_h M^{*(h)} (\alpha - 1) - \max_h L^{*(h)} (\eta - 1) < 0, \text{ or equivalently,}$$

$$\alpha > 1 + \frac{(L - \max_h L^{*(h)})(\eta - 1)}{\min_h M^{*(h)}}.$$

Thus, we can conclude that the solution is sparse if

$$\alpha < 1,$$

and dense if

$$\alpha > 1 + \frac{L(\eta - 1)}{\min_h M^{*(h)}}.$$

Summarizing the preceding, we have the following lemma:

**Lemma 16.24** *Assume that $\eta \geq 1$. The solution is sparse if $\alpha < 1$, and dense if $\alpha > 1 + \frac{L(\eta-1)}{\min_h M^{*(h)}}$.*

**In the Limit When $N, M \to \infty$ with $\frac{M}{N}, L \sim O(1)$**

The coefficient of the leading term of the free energy is given by

$$\lambda'^{\text{MAP}}_{\text{LDA}} = MH(\alpha - 1) - \sum_{h=1}^{\widehat{H}} \widehat{M}^{(h)} (\alpha - 1). \tag{16.94}$$

Although the predictive distribution does not necessarily converge to the true distribution, we can investigate the sparsity of the solution by considering the duplication rules (16.57) through (16.59) that keep $\widehat{\boldsymbol{B}\boldsymbol{\Theta}^\top}$ unchanged. It is clear that Eq. (16.94) is decreasing with respect to $\widehat{H}$ if $\alpha > 1$. Combing this result with Theorem 16.23, which states that the MAP solution is sparse if $\alpha < 1$, and Lemma 16.24, we obtain the following corollary:

**Corollary 16.25** *Assume that* $\eta \geq 1$. *In the limit when* $N \to \infty$ *with* $L, M \sim O(1)$, *the MAP solution is sparse if* $\alpha < 1$, *and dense if* $\alpha > 1 + \frac{L(\eta-1)}{\min_h M^{*(h)}}$. *In the limit when* $N, M \to \infty$ *with* $\frac{M}{N}, L \sim O(1)$, *the MAP solution is sparse if* $\alpha < 1$, *and dense if* $\alpha > 1$.

**Summary of Results**

Summarizing Corollaries 16.11, 16.17, 16.21, and 16.25 completes the proof of Theorem 16.13. □

# 17
# Unified Theory for Latent Variable Models

In this chapter, we present a formula for evaluating an asymptotic form of the VB free energy of a general class of latent variable models by relating it to the asymptotic theory of Bayesian learning (Watanabe, 2012). This formula is applicable to all latent variable models discussed in Chapters 15 and 16.[1] It also explains relationships between these asymptotic analyses of VB free energy and several previous works where the asymptotic Bayes free energy has been analyzed for specific latent variable models. We apply this formula to Gaussian mixture models (GMMs) as an example and demonstrate another proof of the upper-bound of the VB free energy given in Section 15.2. Furthermore, this analysis also provides a quantity that is related to the generalization performance of VB learning. Analysis of generalization performance of VB learning has been conducted only for limited cases, as discussed in Chapter 14. We show inequalities that relate the VB free energy to the generalization errors of an approximate predictive distribution (Watanabe, 2012).

## 17.1 Local Latent Variable Model

Consider the joint model

$$p(\boldsymbol{x}, z|\boldsymbol{w}) \tag{17.1}$$

on the observed variable $\boldsymbol{x}$ and the local latent variable $z$ with the parameter $\boldsymbol{w}$. The marginal distribution of the observed variable is[2]

[1] The reduced rank regression (RRR) model discussed in Chapter 14 is not included in this class of latent variable models.

[2] The model is denoted as if the local latent variable is discrete, it can also be continuous. In this case, the sum $\sum_z$ is replaced by the integral $\int dz$. The probabilistic principal component analysis is an example with a continuous local latent variable.

$$p(\boldsymbol{x}|\boldsymbol{w}) = \sum_{z} p(\boldsymbol{x}, z|\boldsymbol{w}). \tag{17.2}$$

For the complete data set $\{\mathcal{D}, \mathcal{H}\} = \{(\boldsymbol{x}^{(n)}, z^{(1)}), \ldots, (\boldsymbol{x}^{(N)}, z^{(N)})\}$, we assume the i.i.d. model

$$p(\mathcal{D}, \mathcal{H}|\boldsymbol{w}) = \prod_{n=1}^{N} p(\boldsymbol{x}^{(n)}, z^{(n)}|\boldsymbol{w}),$$

which implies

$$p(\mathcal{D}|\boldsymbol{w}) = \prod_{n=1}^{N} p(\boldsymbol{x}^{(n)}|\boldsymbol{w}),$$

$$p(\mathcal{H}|\mathcal{D}, \boldsymbol{w}) = \prod_{n=1}^{N} p(z^{(n)}|\boldsymbol{x}^{(n)}, \boldsymbol{w}).$$

We assume that

$$p(\boldsymbol{x}|\boldsymbol{w}^*) = \sum_{z} p(\boldsymbol{x}, z|\boldsymbol{w}^*)$$

with the parameter $\boldsymbol{w}^*$ is the underlying distribution generating data $\mathcal{D} = \{\boldsymbol{x}^{(1)}, \ldots, \boldsymbol{x}^{(N)}\}$. Because of the nonidentifiability of the latent variable model, the set of true parameters,

$$\mathcal{W}^* \equiv \left\{ \widetilde{\boldsymbol{w}}^*; \sum_{z} p(\boldsymbol{x}, z|\widetilde{\boldsymbol{w}}^*) = p(\boldsymbol{x}|\boldsymbol{w}^*) \right\}, \tag{17.3}$$

is not generally a point but can be a union of several manifolds with singularities as demonstrated in Section 13.5.

In the analysis in this chapter, we define and analyze quantities related to generalization performance of a *joint* model, where the local latent variables are treated as observed variables. Although we do not consider the case where the local latent variables are observed, those quantities are useful for relating generalization properties of VB learning to those of Bayesian learning, with which we establish a unified theory connecting VB learning and Bayesian learning of latent variable models.

Thus, consider for a moment the Bayesian learning of the joint model (17.1), where the complete data set $\{\mathcal{D}, \mathcal{H}\}$ is observed. For the prior distribution $p(\boldsymbol{w})$, the posterior distribution is given by

$$p(\boldsymbol{w}|\mathcal{D}, \mathcal{H}) = \frac{p(\mathcal{D}, \mathcal{H}|\boldsymbol{w})p(\boldsymbol{w})}{p(\mathcal{D}, \mathcal{H})}. \tag{17.4}$$

The Bayes free energy of the joint model is defined by

$$F_{\mathrm{Joint}}^{\mathrm{Bayes}}(\mathcal{D},\mathcal{H}) = -\log p(\mathcal{D},\mathcal{H}) = -\log \int p(\mathcal{D},\mathcal{H}|\boldsymbol{w})p(\boldsymbol{w})d\boldsymbol{w}.$$

If $\widetilde{\boldsymbol{w}}^* \in \mathcal{W}^*$ is the true parameter, i.e., the complete data set $\{\mathcal{D},\mathcal{H}\}$ is generated from $q(\boldsymbol{x}, z) = p(\boldsymbol{x}, z|\widetilde{\boldsymbol{w}}^*)$ i.i.d., the relative Bayes free energy is defined by

$$\widetilde{F}_{\mathrm{Joint}}^{\mathrm{Bayes}}(\mathcal{D},\mathcal{H}) = F_{\mathrm{Joint}}^{\mathrm{Bayes}}(\mathcal{D},\mathcal{H}) - NS_N(\mathcal{D},\mathcal{H}), \tag{17.5}$$

where

$$S_N(\mathcal{D},\mathcal{H}) = -\frac{1}{N}\sum_{n=1}^{N}\log p(\boldsymbol{x}^{(n)}, z^{(n)}|\widetilde{\boldsymbol{w}}^*)$$

is the empirical joint entropy. Then the average relative Bayes free energy is defined by

$$\overline{F}_{\mathrm{Joint}}^{\mathrm{Bayes}}(N) = \left\langle \widetilde{F}_{\mathrm{Joint}}^{\mathrm{Bayes}}(\mathcal{D},\mathcal{H}) \right\rangle_{p(\mathcal{D},\mathcal{H}|\widetilde{\boldsymbol{w}}^*)},$$

and the average Bayes generalization error of the predictive distribution for the joint model is defined by

$$\overline{\mathrm{GE}}_{\mathrm{Joint}}^{\mathrm{Bayes}}(N) = \langle \mathrm{KL}(p(\boldsymbol{x}, z|\widetilde{\boldsymbol{w}}^*)\|p(\boldsymbol{x}, z|\mathcal{D},\mathcal{H}))\rangle_{p(\mathcal{D},\mathcal{H}|\widetilde{\boldsymbol{w}}^*)},$$

where

$$p(\boldsymbol{x}, z|\mathcal{D},\mathcal{H}) = \int p(\boldsymbol{x}, z|\boldsymbol{w})p(\boldsymbol{w}|\mathcal{D},\mathcal{H})d\boldsymbol{w}.$$

These two quantities are related to each other as Eq. (13.24):

$$\overline{\mathrm{GE}}_{\mathrm{Joint}}^{\mathrm{Bayes}}(N) = \overline{F}_{\mathrm{Joint}}^{\mathrm{Bayes}}(N+1) - \overline{F}_{\mathrm{Joint}}^{\mathrm{Bayes}}(N). \tag{17.6}$$

Furthermore, the average relative Bayes free energy for the joint model can be approximated as (see Eq. (13.118))

$$\overline{F}_{\mathrm{Joint}}^{\mathrm{Bayes}}(N) \approx -\log \int \exp\left(-N\overline{E}(\boldsymbol{w})\right) \cdot p(\boldsymbol{w})d\boldsymbol{w}, \tag{17.7}$$

where

$$\overline{E}(\boldsymbol{w}) = \mathrm{KL}(p(\boldsymbol{x}, z|\widetilde{\boldsymbol{w}}^*)\|p(\boldsymbol{x}, z|\boldsymbol{w})) = \left\langle \log \frac{p(\boldsymbol{x}, z|\widetilde{\boldsymbol{w}}^*)}{p(\boldsymbol{x}, z|\boldsymbol{w})} \right\rangle_{p(\boldsymbol{x},z|\widetilde{\boldsymbol{w}}^*)}. \tag{17.8}$$

Since the log-sum inequality yields that[3]

$$\sum_z p(\boldsymbol{x}, z|\widetilde{\boldsymbol{w}}^*) \log \frac{p(\boldsymbol{x}, z|\widetilde{\boldsymbol{w}}^*)}{p(\boldsymbol{x}, z|\boldsymbol{w})} \geq p(\boldsymbol{x}|\boldsymbol{w}^*) \log \frac{p(\boldsymbol{x}|\boldsymbol{w}^*)}{p(\boldsymbol{x}|\boldsymbol{w})},$$

we have

$$\overline{E}(\boldsymbol{w}) \geq E(\boldsymbol{w}), \tag{17.10}$$

where

$$E(\boldsymbol{w}) = \mathrm{KL}(p(\boldsymbol{x}|\boldsymbol{w}^*)\|p(\boldsymbol{x}|\boldsymbol{w})) = \left\langle \log \frac{p(\boldsymbol{x}|\boldsymbol{w}^*)}{p(\boldsymbol{x}|\boldsymbol{w})} \right\rangle_{p(\boldsymbol{x}|\boldsymbol{w}^*)}.$$

Hence, it follows from Eq. (13.118) that

$$\begin{aligned}\overline{F}^{\mathrm{Bayes}}(N) &= \left\langle \widetilde{F}^{\mathrm{Bayes}}(\mathcal{D}) \right\rangle_{q(\mathcal{D})} \\ &\approx -\log \int \exp\left(-NE(\boldsymbol{w})\right) \cdot p(\boldsymbol{w}) d\boldsymbol{w} \\ &\leq -\log \int \exp\left(-N\overline{E}(\boldsymbol{w})\right) \cdot p(\boldsymbol{w}) d\boldsymbol{w} \approx \overline{F}^{\mathrm{Bayes}}_{\mathrm{Joint}}(N),\end{aligned} \tag{17.11}$$

where

$$\widetilde{F}^{\mathrm{Bayes}}(\mathcal{D}) = F^{\mathrm{Bayes}}(\mathcal{D}) - NS_N(\mathcal{D})$$

is the relative Bayes free energy defined by the Bayes free energy of the original marginal model,

$$F^{\mathrm{Bayes}}(\mathcal{D}) = -\log p(\mathcal{D}) = -\log \int p(\mathcal{D}|\boldsymbol{w}) p(\boldsymbol{w}) d\boldsymbol{w}$$

and its empirical entropy,

$$S_N(\mathcal{D}) = -\frac{1}{N} \log p(\mathcal{D}|\boldsymbol{w}^*) = -\frac{1}{N} \sum_{n=1}^{N} \log p(\boldsymbol{x}^{(n)}|\boldsymbol{w}^*), \tag{17.12}$$

as in Section 13.3.2.

The asymptotic theory of Bayesian learning (Theorem 13.13) shows that an asymptotic form of $\overline{F}^{\mathrm{Bayes}}_{\mathrm{Joint}}(N)$ is given by

$$\overline{F}^{\mathrm{Bayes}}_{\mathrm{Joint}}(N) = \lambda'^{\mathrm{Bayes}}_{\mathrm{Joint}} \log N - (m'^{\mathrm{Bayes}}_{\mathrm{Joint}} - 1) \log \log N + O(1), \tag{17.13}$$

[3] The log-sum inequality is the following inequality satisfied for nonnegative reals $a_i \geq 0$ and $b_i \geq 0$:

$$\sum_i a_i \log \frac{a_i}{b_i} \geq \left(\sum_i a_i\right) \log \frac{(\sum_i a_i)}{(\sum_i b_i)}. \tag{17.9}$$

This can be proved by subtracting the right-hand side from the left-hand side and applying the nonnegativity of the KL divergence.

where $-\lambda'^{\text{Bayes}}_{\text{Joint}}$ and $m'^{\text{Bayes}}_{\text{Joint}}$ are respectively the largest pole and its order of the zeta function defined for a complex number $z$ by

$$\zeta_{\overline{E}}(z) = \int \overline{E}(\boldsymbol{w})^z p(\boldsymbol{w}) d\boldsymbol{w}. \tag{17.14}$$

This means that the asymptotic behavior of the free energy is characterized by $\overline{E}(\boldsymbol{w})$, while that of the Bayes free energy $F^{\text{Bayes}}$ is characterized by $E(\boldsymbol{w}) = \text{KL}(p(\boldsymbol{x}|\boldsymbol{w}^*)\|p(\boldsymbol{x}|\boldsymbol{w}))$ and the zeta function $\zeta_E(z)$ in Eq. (13.122) as Theorem 13.13. The two functions, $E$ and $\overline{E}$, are related by the log-sum inequality (17.10).

Then Corollary 13.14 implies the following asymptotic expansion of the average generalization error:

$$\overline{\text{GE}}^{\text{Bayes}}_{\text{Joint}}(N) = \frac{\lambda'^{\text{Bayes}}_{\text{Joint}}}{N} - \frac{m'^{\text{Bayes}}_{\text{Joint}} - 1}{N \log N} + o\left(\frac{1}{N \log N}\right). \tag{17.15}$$

With the preceding quantities, we first provide a general upper-bound for the VB free energy (Section 17.2), and then show inequalities relating the VB free energy to the generalization errors of an approximate predictive distribution for the joint model (Section 17.4).

## 17.2 Asymptotic Upper-Bound for VB Free Energy

Given the training data $\mathcal{D} = \{\boldsymbol{x}^{(1)}, \ldots, \boldsymbol{x}^{(N)}\}$, consider VB learning for the latent variable model (17.2) with the prior distribution $p(\boldsymbol{w})$. Under the constraint,

$$r(\boldsymbol{w}, \mathcal{H}) = r_w(\boldsymbol{w}) r_{\mathcal{H}}(\mathcal{H}),$$

the VB free energy is defined by

$$F^{\text{VB}}(\mathcal{D}) = \min_{r_w(\boldsymbol{w}), r_{\mathcal{H}}(\mathcal{H})} F(r),$$

where

$$F(r) = \left\langle \log \frac{r_w(\boldsymbol{w}) r_{\mathcal{H}}(\mathcal{H})}{p(\mathcal{D}, \mathcal{H}|\boldsymbol{w}) p(\boldsymbol{w})} \right\rangle_{r_w(\boldsymbol{w}) r_{\mathcal{H}}(\mathcal{H})} \tag{17.16}$$

$$= F^{\text{Bayes}}(\mathcal{D}) + \text{KL}\left(r_w(\boldsymbol{w}) r_{\mathcal{H}}(\mathcal{H}) \| p(\boldsymbol{w}, \mathcal{H}|\mathcal{D})\right). \tag{17.17}$$

The stationary condition of the free energy yields

$$r_w(\boldsymbol{w}) = \frac{1}{C_w} p(\boldsymbol{w}) \exp \langle \log p(\mathcal{D}, \mathcal{H}|\boldsymbol{w}) \rangle_{r_{\mathcal{H}}(\mathcal{H})}, \tag{17.18}$$

$$r_{\mathcal{H}}(\mathcal{H}) = \frac{1}{C_{\mathcal{H}}} \exp \langle \log p(\mathcal{D}, \mathcal{H}|\boldsymbol{w}) \rangle_{r_w(\boldsymbol{w})}. \tag{17.19}$$

Let us define the relative VB free energy

$$\widetilde{F}^{\mathrm{VB}}(\mathcal{D}) = F^{\mathrm{VB}}(\mathcal{D}) - NS_N(\mathcal{D})$$

by the VB free energy and the empirical entropy (17.12). For arbitrary $\widetilde{w}^* \in \mathcal{W}^*$, substituting Eq. (17.18) into Eq. (17.16), we have

$$\widetilde{F}^{\mathrm{VB}}(\mathcal{D}) = \min_{r_{\mathcal{H}}(\mathcal{H})} \left[ -\log \int p(w) \exp \left\langle \log \frac{p(\mathcal{D}, \mathcal{H}|w)}{r_{\mathcal{H}}(\mathcal{H})} \right\rangle_{r_{\mathcal{H}}(\mathcal{H})} dw \right] + \log p(\mathcal{D}|w^*) \tag{17.20}$$

$$\leq -\log \int \exp \left\{ \sum_{\mathcal{H}} p(\mathcal{H}|\mathcal{D}, \widetilde{w}^*) \log \frac{p(\mathcal{D}, \mathcal{H}|w)}{p(\mathcal{D}, \mathcal{H}|\widetilde{w}^*)} \right\} p(w) dw \tag{17.21}$$

$$\equiv \widetilde{F}^{\mathrm{VB}*}(\mathcal{D}).$$

Here, we have substituted $r_{\mathcal{H}}(\mathcal{H}) \leftarrow p(\mathcal{H}|\mathcal{D}, \widetilde{w}^*) = \frac{p(\mathcal{D},\mathcal{H}|\widetilde{w}^*)}{\sum_{\mathcal{H}} p(\mathcal{D},\mathcal{H}|\widetilde{w}^*)}$ to obtain the upper-bound (17.21). The expression (17.20) of the free energy corresponds to viewing the VB learning as a local variational approximation (Section 5.3.3), where the variational parameter $h(\xi)$ is the vector consisting of the elements $\log p(\mathcal{D}, \mathcal{H}, \xi)$ for all possible $\mathcal{H}$.[4]

By taking the expectation with respect to the distribution of training samples, we define the average relative VB free energy and its upper-bound as

$$\overline{F}^{\mathrm{VB}}(N) = \left\langle \widetilde{F}^{\mathrm{VB}}(\mathcal{D}) \right\rangle_{p(\mathcal{D}|w^*)}, \tag{17.22}$$

$$\overline{F}^{\mathrm{VB}*}(N) = \left\langle \widetilde{F}^{\mathrm{VB}*}(\mathcal{D}) \right\rangle_{p(\mathcal{D}|w^*)}. \tag{17.23}$$

From Eq. (17.7), we have

$$\overline{F}^{\mathrm{Bayes}}_{\mathrm{Joint}}(N) \approx -\log \int e^{-N\overline{E}(w)} p(w) dw \equiv F^{\overline{\mathrm{Bayes}}}_{\mathrm{Joint}}(N),$$

where $\overline{E}(w)$ is defined by Eq. (17.8). Then, the following theorem holds:

**Theorem 17.1** *It holds that*

$$\overline{F}^{\mathrm{Bayes}}(N) \leq \overline{F}^{\mathrm{VB}}(N) \leq \overline{F}^{\mathrm{VB}*}(N) \leq F^{\overline{\mathrm{Bayes}}}_{\mathrm{Joint}}(N). \tag{17.24}$$

*Proof* The left inequality follows from Eq. (17.17). Eq. (17.21) gives

$$\overline{F}^{\mathrm{VB}}(N) \leq \overline{F}^{\mathrm{VB}*}(N)$$
$$= \left\langle \widetilde{F}^{\mathrm{VB}*}(\mathcal{D}) \right\rangle_{p(\mathcal{D}|w^*)}$$

[4] The variational parameter $h(\xi)$ has one-to-one correspondence with $p(\mathcal{H}|\mathcal{D}, \xi)$, and is substituted as $h(\xi) \leftarrow h(\widetilde{w}^*)$ in Eq. (17.21).

$$= -\left\langle \log \int \exp\left\{ \sum_{\mathcal{H}} p(\mathcal{H}|\mathcal{D}, \widetilde{\boldsymbol{w}}^*) \log \frac{p(\mathcal{D}, \mathcal{H}|\boldsymbol{w})}{p(\mathcal{D}, \mathcal{H}|\widetilde{\boldsymbol{w}}^*)} \right\} p(\boldsymbol{w}) d\boldsymbol{w} \right\rangle_{p(\mathcal{D}|\boldsymbol{w}^*)}$$

$$\leq -\log \int \exp\left\{ \left\langle \sum_{\mathcal{H}} p(\mathcal{H}|\mathcal{D}, \widetilde{\boldsymbol{w}}^*) \log \frac{p(\mathcal{D}, \mathcal{H}|\boldsymbol{w})}{p(\mathcal{D}, \mathcal{H}|\widetilde{\boldsymbol{w}}^*)} \right\rangle_{p(\mathcal{D}|\boldsymbol{w}^*)} \right\} p(\boldsymbol{w}) d\boldsymbol{w}$$

$$= -\log \int e^{-N\overline{E}(\boldsymbol{w})} p(\boldsymbol{w}) d\boldsymbol{w} = \overline{F}^{\text{Bayes}}_{\text{Joint}}(N).$$

The first and second equalities are definitions of $\overline{F}^{\text{VB}*}(N)$ and $\widetilde{F}^{\text{VB}*}(\mathcal{D})$. We have applied Jensen's inequality to the convex function $\log \int \exp(\cdot) p(\boldsymbol{w}) d\boldsymbol{w}$ to obtain the last inequality. Finally, the last equality follows from the fact that $p(\mathcal{D}|\boldsymbol{w}^*) p(\mathcal{H}|\mathcal{D}, \widetilde{\boldsymbol{w}}^*) = p(\mathcal{D}, \mathcal{H}|\widetilde{\boldsymbol{w}}^*)$ and the i.i.d. assumption. □

The following corollary is immediately obtained from Theorems 13.13 and 17.1:

**Corollary 17.2** *Let* $0 > -\lambda_1 > -\lambda_2 > \cdots$ *be the sequence of the poles of the zeta function* (17.14) *in the decreasing order, and* $m_1, m_2, \ldots$ *be the corresponding orders of the poles. Then the average relative VB free energy* (17.22) *can be asymptotically upper-bounded as*

$$\overline{F}^{\text{VB}}(N) \leq \lambda_1 \log N - (m_1 - 1) \log \log N + O(1). \tag{17.25}$$

It holds in Eqs. (17.13) and (17.15) that $\underline{\lambda'^{\text{Bayes}}_{\text{Joint}}} = \lambda_1$ and $m'^{\text{Bayes}}_{\text{Joint}} = m_1$ for $\lambda_1$ and $m_1$ defined in Corollary 17.2. Note that $\overline{E}(\boldsymbol{w})$ depends on $\widetilde{\boldsymbol{w}}^* \in \mathcal{W}^*$. For different $\widetilde{\boldsymbol{w}}^*$, we have different values of $\lambda_1$, which is determined by the minimum over different $\widetilde{\boldsymbol{w}}^* \in \mathcal{W}^*$ in Eq. (17.25). Then $m_1$ is determined by the maximum of the order of the pole for the minimum $\lambda_1$. Also note that unlike for Bayesian learning, even if the largest pole of the zeta function is obtained, Eq. (17.25) does not necessarily provide a lower-bound of the VB free energy.

If the joint model $p(\boldsymbol{x}, z|\boldsymbol{w})$, the true distribution $p(\boldsymbol{x}, z|\widetilde{\boldsymbol{w}}^*)$, and the prior $p(\boldsymbol{w})$ satisfy the regularity conditions (Section 13.4.1), it holds that

$$2\lambda'^{\text{Bayes}}_{\text{Joint}} = D,$$

where $D$ is the number of parameters.

If the joint model $p(\boldsymbol{x}, z|\boldsymbol{w})$ is identifiable, even though the true parameter is on the boundary of the parameter space or the prior does not satisfy $0 < p(\boldsymbol{w}) < \infty$, $\lambda'^{\text{Bayes}}_{\text{Joint}}$ can be analyzed similarly to the case of regular models. The GMM with redundant components is an example of such a case, as will be detailed in the next section.

If the joint model $p(\boldsymbol{x}, z|\boldsymbol{w})$ is unidentifiable, we need the algebraic geometrical technique to analyze $\lambda'^{\mathrm{Bayes}}_{\mathrm{Joint}}$ as discussed in Section 13.5.4. This technique is also applicable to identifiable cases as will be demonstrated in a part of the analysis of $\lambda'^{\mathrm{Bayes}}_{\mathrm{Joint}}$ for the GMM in the last part of the next section.

## 17.3 Example: Average VB Free Energy of Gaussian Mixture Model

In this section, we derive an asymptotic upper-bound of the VB free energy of GMMs. Although this upper-bound is immediately obtained from Theorem 15.5 in Section 15.2, it was derived by direct evaluation and minimization of the free energy with respect to the expected sufficient statistics. In this section, we present another derivation through Theorem 17.1 in order to illustrate how the general theory described in Section 17.2 is applied.

Let

$$g(\boldsymbol{x}|\boldsymbol{\mu}) = \mathrm{Gauss}_M(\boldsymbol{x}; \boldsymbol{\mu}, \boldsymbol{I}_M)$$

be the $M$-dimensional uncorrelated Gaussian density and consider the GMM with $K$ components,

$$p(\boldsymbol{x}|\boldsymbol{w}) = \sum_{z} p(\boldsymbol{x}, z|\boldsymbol{w}) = \sum_{k=1}^{K} \alpha_k g(\boldsymbol{x}|\boldsymbol{\mu}_k),$$

where $\boldsymbol{x} \in \mathbb{R}^M$ and $\boldsymbol{w} = (\boldsymbol{\alpha}, \{\boldsymbol{\mu}_k\}_{k=1}^{K})$ denote the parameter vector consisting the mixing weights and the mean vectors, respectively.

Assuming the same prior given by

$$p(\boldsymbol{\alpha}|\phi) = \mathrm{Dirichlet}_K(\boldsymbol{\alpha}; (\phi, \ldots, \phi)^\top), \tag{17.26}$$

$$p(\boldsymbol{\mu}_k|\boldsymbol{\mu}_0, \xi) = \mathrm{Gauss}_M(\boldsymbol{\mu}_k|\boldsymbol{\mu}_0, (1/\xi)\boldsymbol{I}_M), \tag{17.27}$$

and the same true distribution

$$q(\boldsymbol{x}) = p(\boldsymbol{x}|\boldsymbol{w}^*) = \sum_{k=1}^{K_0} \alpha_k^* g(\boldsymbol{x}|\boldsymbol{\mu}_k^*), \tag{17.28}$$

as in Sections 4.1.1 and 15.2, we immediately obtain from the upper-bound in Eq. (15.36) of Theorem 15.5 that

$$\overline{F}^{\mathrm{VB}}(N) \leq \overline{\lambda}'^{\mathrm{VB}}_{\mathrm{MM}} \log N + O(1), \tag{17.29}$$

where

$$\overline{\lambda}'^{\mathrm{VB}}_{\mathrm{MM}} = \begin{cases} (K - K_0)\phi + \frac{MK_0+K_0-1}{2} & (\phi < \frac{M+1}{2}), \\ \frac{MK+K-1}{2} & (\phi \geq \frac{M+1}{2}). \end{cases}$$

In this section, we derive this upper-bound by using Theorem 17.1, which provides an alternative proof to the one presented in Section 15.2. Similar techniques were used for analyzing the Bayes free energy (13.19) in the asymptotic limit (Yamazaki and Watanabe, 2003a,b, 2005). Here, we evaluate the VB free energy and present the details of the proof for the specific choice of the prior distribution.

First, in order to define $p(\boldsymbol{x}, z|\widetilde{\boldsymbol{w}}^*)$ for $z$ with $K$ elements, we extend and redefine the true parameter $\boldsymbol{w}^*$ denoting it as $\widetilde{\boldsymbol{w}}^* = (\widetilde{\boldsymbol{\alpha}}^*, \{\widetilde{\boldsymbol{\mu}}_k^*\}_{k=1}^K)$. Suppose that the true distribution with parameter $\widetilde{\boldsymbol{w}}^*$ has $\widehat{K}$ nonzero mixing weights. For example, we can assume that

$$\widetilde{\alpha}_k^* = \begin{cases} \alpha_k^* & (1 \leq k \leq K_0 - 1), \\ \alpha_{K_0}^* / (\widehat{K} - K_0 + 1) & (K_0 \leq k \leq \widehat{K}), \\ 0 & (\widehat{K} + 1 \leq k \leq K), \end{cases}$$

$$\widetilde{\boldsymbol{\mu}}_k^* = \begin{cases} \boldsymbol{\mu}_k^* & (1 \leq k \leq K_0), \\ \boldsymbol{\mu}_{K_0}^* & (K_0 + 1 \leq k \leq K). \end{cases}$$

Note that the marginal distribution of $p(\boldsymbol{x}, z|\widetilde{\boldsymbol{w}}^*)$ is reduced to Eq. (17.28). Then we have

$$\begin{aligned} \overline{E}(\boldsymbol{w}) &= \int \sum_z p(\boldsymbol{x}, z|\widetilde{\boldsymbol{w}}^*) \log \frac{p(\boldsymbol{x}, z|\widetilde{\boldsymbol{w}}^*)}{p(\boldsymbol{x}, z|\boldsymbol{w})} d\boldsymbol{x} \\ &= \int \sum_{k=1}^K \widetilde{\alpha}_k^* g(\boldsymbol{x}|\widetilde{\boldsymbol{\mu}}_k^*) \log \frac{\widetilde{\alpha}_k^* g(\boldsymbol{x}|\widetilde{\boldsymbol{\mu}}_k^*)}{\alpha_k g(\boldsymbol{x}|\boldsymbol{\mu}_k)} d\boldsymbol{x} \\ &= \sum_{k=1}^K \widetilde{\alpha}_k^* \left\{ \log \frac{\widetilde{\alpha}_k^*}{\alpha_k} + \int g(\boldsymbol{x}|\widetilde{\boldsymbol{\mu}}_k^*) \log \frac{g(\boldsymbol{x}|\widetilde{\boldsymbol{\mu}}_k^*)}{g(\boldsymbol{x}|\boldsymbol{\mu}_k)} d\boldsymbol{x} \right\} \\ &= \sum_{k=1}^{\widehat{K}} \widetilde{\alpha}_k^* \left\{ \log \frac{\widetilde{\alpha}_k^*}{\alpha_k} + \frac{\|\boldsymbol{\mu}_k - \widetilde{\boldsymbol{\mu}}_k^*\|^2}{2} \right\}. \end{aligned}$$

Second, we divide the parameter $\boldsymbol{w}$ into three parts,

$$\boldsymbol{w}_1 = (\alpha_2, \alpha_3, \ldots, \alpha_{\widehat{K}}), \tag{17.30}$$

$$\boldsymbol{w}_2 = (\alpha_{\widehat{K}+1}, \ldots, \alpha_K), \tag{17.31}$$

$$\boldsymbol{w}_3 = (\boldsymbol{\mu}_1, \boldsymbol{\mu}_2, \ldots, \boldsymbol{\mu}_{\widehat{K}}), \tag{17.32}$$

and define

$$\mathcal{W}_1 = \{w_1;\ \ |\alpha_k - \widetilde{\alpha}_k^*| \leq \epsilon, 2 \leq k \leq \widehat{K}\},$$
$$\mathcal{W}_2 = \{w_2;\ \ |\alpha_k| \leq \epsilon, \widehat{K} \leq k \leq K\},$$
$$\mathcal{W}_3 = \{w_3;\ \ \|\mu_k - \widetilde{\mu}_k^*\| \leq \epsilon, 1 \leq k \leq \widehat{K}\},$$

for a sufficiently small constant $\epsilon$. For an arbitrary parameter $w \in \mathcal{W}_1 \times \mathcal{W}_2 \times \mathcal{W}_3 \equiv \mathcal{W}(\epsilon)$, we can decompose $\overline{E}(w)$ as

$$\overline{E}(w) = \overline{E}_1(w_1) + \overline{E}_2(w_2) + \overline{E}_3(w_3), \tag{17.33}$$

where

$$\overline{E}_1(w_1) = \sum_{k=2}^{\widehat{K}} \widetilde{\alpha}_k^* \log \frac{\widetilde{\alpha}_k^*}{\alpha_k} + \left(1 - \sum_{k=2}^{\widehat{K}} \widetilde{\alpha}_k^*\right) \log \frac{1 - \sum_{k=2}^{\widehat{K}} \widetilde{\alpha}_k^*}{1 - \sum_{k=2}^{\widehat{K}} \alpha_k},$$
$$\overline{E}_2(w_2) = \frac{1}{1-c} \frac{1 - \sum_{k=2}^{K_0} \widetilde{\alpha}_k^*}{1 - \sum_{k=2}^{\widehat{K}} \alpha_k} \sum_{k=\widehat{K}+1}^{K} \alpha_k,$$
$$\overline{E}_3(w_3) = \sum_{k=1}^{\widehat{K}} \frac{\widetilde{\alpha}_k^*}{2} \|\mu_k - \widetilde{\mu}_k^*\|^2. \tag{17.34}$$

Here we have used the mean value theorem $-\log(1-t) = \frac{1}{1-c} t$ for some $c$, $0 \leq c \leq t$ with $t = \frac{\sum_{k=\widehat{K}+1}^{K} \alpha_k}{1 - \sum_{k=2}^{\widehat{K}} \alpha_k}$. Furthermore, for $w \in \mathcal{W}(\epsilon)$, there exist positive constants $C_1$, $C_2$, $C_3$, and $C_4$ such that

$$C_1 \sum_{k=2}^{\widehat{K}} (\alpha_k - \widetilde{\alpha}_k^*)^2 \leq \overline{E}_1(w_1) \leq C_2 \sum_{k=2}^{\widehat{K}} (\alpha_k - \widetilde{\alpha}_k^*)^2, \tag{17.35}$$

$$C_3 \sum_{k=\widehat{K}+1}^{K} \alpha_k \leq \overline{E}_2(w_2) \leq C_4 \sum_{k=\widehat{K}+1}^{K} \alpha_k. \tag{17.36}$$

Third, we evaluate the partial free energies defined for $i = 1, 2, 3$ by

$$F_i = -\log \int_{\mathcal{W}_i} \exp(-N\overline{E}_i(w_i)) p(w_i) dw_i, \tag{17.37}$$

where $p(w_i)$ is the product of factors of the prior in Eqs. (17.26) and (17.27), which involve $w_i$ defined in Eqs. (17.30) through (17.32).

It follows from Eqs. (17.24), (17.33), and (17.37) that

$$\overline{F}^{\mathrm{VB}}(N) \leq F_1 + F_2 + F_3 + O(1). \tag{17.38}$$

From Eqs. (17.35) and (17.34), as for $F_1$ and $F_3$, the Gaussian integration yields

$$F_1 = \frac{\widehat{K}-1}{2}\log N + O(1), \quad (17.39)$$

$$F_3 = \frac{M\widehat{K}}{2}\log N + O(1). \quad (17.40)$$

Since

$$N^{\phi}\int_0^{\epsilon} e^{-n\alpha_k}\alpha_k^{\phi-1}d\alpha_k \to \Gamma(\phi) \quad (N\to\infty),$$

for $k = \widehat{K}+1,\ldots,K$, it follows from Eq. (17.36) that

$$F_2 = (K-\widehat{K})\phi\log N + O(1). \quad (17.41)$$

Finally, combining Eqs. (17.38) through (17.41), we obtain

$$\overline{F}^{\mathrm{VB}}(N) \leq \left\{(K-\widehat{K})\phi + \frac{M\widehat{K}+\widehat{K}-1}{2}\right\}\log N + O(1).$$

Minimizing the right-hand side of the preceding expression over $\widehat{K}$ ($K_0 \leq \widehat{K} \leq K$) leads to the upper-bound in Eq. (17.29).

Alternatively, the preceding evaluations of all the partial free energies, $F_1$, $F_2$, and $F_3$, are obtained by using the algebraic geometrical method based on Corollary 17.2. For example, as for $F_2$, the zeta function

$$\zeta_{\overline{E}_2}(z) = \int \overline{E}_2(\boldsymbol{w}_2)^z p(\boldsymbol{w}_2)d\boldsymbol{w}_2$$

has a pole $z = -(K-\widehat{K})\phi$. This can be observed by the change of variables, the so-called blow-up,

$$\alpha_k = \alpha'_k\alpha'_K \quad (k = \widehat{K}+1,\ldots,K-1),$$

$$\alpha_K = \alpha'_K,$$

which yields that $\zeta_{\overline{E}_2}$ has a term

$$\int {\alpha'_K}^{z}{\alpha'_K}^{(K-\widehat{K})\phi-1}\widetilde{\zeta}_{\overline{E}_2}(\widetilde{\boldsymbol{w}}'_2)d\alpha'_K = \frac{\widetilde{\zeta}_{\overline{E}_2}(\widetilde{\boldsymbol{w}}'_2)}{z+(K-\widehat{K})\phi},$$

where $\widetilde{\zeta}_{\overline{E}_2}(\widetilde{\boldsymbol{w}}'_2)$ is a function proportional to

$$\int\left(\sum_{k=\widehat{K}+1}^{K-1}\alpha'_k + 1\right)^z \prod_{k=\widehat{K}+1}^{K-1}{\alpha'_k}^{\phi-1}\prod_{k=\widehat{K}+1}^{K-1}d\alpha'_k.$$

Hence, we can see that $\zeta_{\overline{E}_2}$ has a pole at $z = -(K-\widehat{K})\phi$.

## 17.4 Free Energy and Generalization Error

In this section, we relate the VB free energy to the generalization performance of VB learning. We denote a training data set by $\mathcal{D}^N = \{x^{(1)}, x^{(2)}, \ldots, x^{(N)}\}$ with the number $N$ of training samples as a superscript in this section.

Let $p(x, z|\widetilde{w}^*)$ be the true distribution of the observed variable $x$ and the latent variable $z$, which has the marginal distribution $p(x|w^*)$. We define the generalization error of the predictive distribution for the *joint distribution*,

$$p^{\mathrm{VB}*}(x, z|\mathcal{D}^N) = \langle p(x, z|w) \rangle_{r^*(w;\widetilde{w}^*)} = \int p(x, z|w) r^*(w; \widetilde{w}^*) dw, \tag{17.42}$$

by the Bayes generalization error (13.133)

$$\mathrm{GE}^{\mathrm{VB}*}_{\mathrm{Joint}}(\mathcal{D}^N) = \mathrm{KL}(p(x, z|\widetilde{w}^*) \| p^{\mathrm{VB}*}(x, z|\mathcal{D}^N)), \tag{17.43}$$

and the Gibbs generalization error (13.136) by

$$\mathrm{GGE}^{\mathrm{VB}*}_{\mathrm{Joint}}(\mathcal{D}^N) = \langle \mathrm{KL}(p(x, z|\widetilde{w}^*) \| p(x, z|w)) \rangle_{r^*(w;\widetilde{w}^*)}, \tag{17.44}$$

where

$$r^*(w; \widetilde{w}^*) \propto p(w) \prod_{n=1}^{N} \exp\left( - \sum_{z} p(z|x^{(n)}, \widetilde{w}^*) \log \frac{p(x^{(n)}, z|\widetilde{w}^*)}{p(x^{(n)}, z|w)} \right)$$

is the approximate posterior distribution (17.18) with $p(\mathcal{H}|\mathcal{D}^N, \widetilde{w}^*)$ substituted for $r_{\mathcal{H}}(\mathcal{H})$. We denote their means by

$$\overline{\mathrm{GE}}^{\mathrm{VB}*}_{\mathrm{Joint}}(N) = \left\langle \mathrm{GE}^{\mathrm{VB}*}_{\mathrm{Joint}}(\mathcal{D}^N) \right\rangle_{p(\mathcal{D}^N|w^*)},$$

$$\overline{\mathrm{GGE}}^{\mathrm{VB}*}_{\mathrm{Joint}}(N) = \left\langle \mathrm{GGE}^{\mathrm{VB}*}_{\mathrm{Joint}}(\mathcal{D}^N) \right\rangle_{p(\mathcal{D}^N|w^*)}.$$

Then the following theorem holds:

**Theorem 17.3** *It holds that*

$$\overline{\mathrm{GE}}^{\mathrm{VB}*}_{\mathrm{Joint}}(N) \leq \overline{F}^{\mathrm{VB}*}(N+1) - \overline{F}^{\mathrm{VB}*}(N) \leq \overline{\mathrm{GGE}}^{\mathrm{VB}*}_{\mathrm{Joint}}(N), \tag{17.45}$$

*where $\overline{F}^{\mathrm{VB}*}(N)$ is the upper-bound (17.23) of the average relative VB free energy.*

*Proof* We have

$$\widetilde{F}^{\mathrm{VB}*}(\mathcal{D}^{N+1}) - \widetilde{F}^{\mathrm{VB}*}(\mathcal{D}^N)$$

$$= -\log \frac{\int \prod_{n=1}^{N+1} \exp\left( \sum_z p(z|x^{(n)}, \widetilde{w}^*) \log \frac{p(x^{(n)}, z|w)}{p(x^{(n)}, z|\widetilde{w}^*)} \right) p(w) dw}{\int \prod_{n=1}^{N} \exp\left( \sum_z p(z|x^{(n)}, \widetilde{w}^*) \log \frac{p(x^{(n)}, z|w)}{p(x^{(n)}, z|\widetilde{w}^*)} \right) p(w) dw}$$

$$
\begin{aligned}
&= -\log \int \exp\left( \sum_{z} p(z|\boldsymbol{x}^{(N+1)}, \widetilde{\boldsymbol{w}}^*) \log \frac{p(\boldsymbol{x}^{(N+1)}, z|\boldsymbol{w})}{p(\boldsymbol{x}^{(N+1)}, z|\widetilde{\boldsymbol{w}}^*)} \right) r^*(\boldsymbol{w}; \widetilde{\boldsymbol{w}}^*) d\boldsymbol{w} \\
&= \sum_{z} p(z|\boldsymbol{x}^{(N+1)}, \widetilde{\boldsymbol{w}}^*) \log p(\boldsymbol{x}^{(N+1)}, z|\widetilde{\boldsymbol{w}}^*) \\
&\quad - \log \int \exp\left( \left\langle \log p(\boldsymbol{x}^{(N+1)}, z|\boldsymbol{w}) \right\rangle_{p(z|\boldsymbol{x}^{(N+1)}, \widetilde{\boldsymbol{w}}^*)} \right) r^*(\boldsymbol{w}; \widetilde{\boldsymbol{w}}^*) d\boldsymbol{w} \qquad (17.46) \\
&\geq \sum_{z} p(z|\boldsymbol{x}^{(N+1)}, \widetilde{\boldsymbol{w}}^*) \log \frac{p(\boldsymbol{x}^{(N+1)}, z|\widetilde{\boldsymbol{w}}^*)}{\langle p(\boldsymbol{x}^{(N+1)}, z|\boldsymbol{w}) \rangle_{r^*(\boldsymbol{w}; \widetilde{\boldsymbol{w}}^*)}}.
\end{aligned}
$$

In the last inequality, we have applied Jensen's inequality to the convex function $\log \int \exp(\cdot) p(\boldsymbol{w}) d\boldsymbol{w}$. Taking the expectation with respect to $\prod_{n=1}^{N+1} p(\boldsymbol{x}^{(n)}|\boldsymbol{w}^*)$ in both sides of the preceding inequality yields the left inequality in Eq. (17.45).

By applying Jensen's inequality for the exponential function in Eq. (17.46), and taking the expectation, we have the right inequality in Eq. (17.45). □

The inequalities in Eq. (17.45) are analogous to Eq. (17.6). Let $\lambda'^{\mathrm{VB}*}$ be the free energy coefficient of $\overline{F}^{\mathrm{VB}*}(N)$, i.e.,

$$
\overline{F}^{\mathrm{VB}*}(N) = \lambda'^{\mathrm{VB}*} \log N + o(\log N). \qquad (17.47)
$$

If its difference has the asymptotic form

$$
\overline{F}^{\mathrm{VB}*}(N+1) - \overline{F}^{\mathrm{VB}*}(N) = \frac{\lambda'^{\mathrm{VB}*}}{N} + o\left(\frac{1}{N}\right),
$$

the left inequality in Eq. (17.45) suggests that

$$
\overline{\mathrm{GE}}^{\mathrm{VB}*}_{\mathrm{Joint}}(N) \leq \frac{\lambda'^{\mathrm{VB}*}}{N} + o\left(\frac{1}{N}\right).
$$

This means that the free energy coefficient $\lambda'^{\mathrm{VB}*}$ of $\overline{F}^{\mathrm{VB}*}(N)$ is directly related to the generalization error of VB learning measured by Eq. (17.43). Theorem 17.1 implies that the free energy coefficients satisfy $\lambda'^{\mathrm{VB}*} \leq \lambda'^{\mathrm{Bayes}}_{\mathrm{Joint}}$, which in turn implies from Eq. (17.6) that

$$
\overline{\mathrm{GE}}^{\mathrm{VB}*}_{\mathrm{Joint}}(N) \leq \overline{\mathrm{GE}}^{\mathrm{Bayes}}_{\mathrm{Joint}}(N) \qquad (17.48)
$$

holds asymptotically.

Let

$$
\widehat{r}_w(\boldsymbol{w}) = \operatorname*{argmin}_{r_w(\boldsymbol{w})} \min_{r_{\mathcal{H}}(\mathcal{H})} F(r)
$$

be the optimal VB posterior of the parameter that minimizes the free energy (17.16). The average generalization errors of VB learning are naturally defined by

$$\overline{\mathrm{GE}}^{\mathrm{VB}}_{\mathrm{Joint}}(N) = \left\langle \mathrm{KL}(p(\boldsymbol{x}, z|\widetilde{\boldsymbol{w}}^*) \| p^{\mathrm{VB}}(\boldsymbol{x}, z|\mathcal{D}^N)) \right\rangle_{q(\mathcal{D}^N)}$$

for the joint predictive distribution $p^{\mathrm{VB}}(\boldsymbol{x}, z|\mathcal{D}^N) = \langle p(\boldsymbol{x}, z|\boldsymbol{w}) \rangle_{\widehat{r}_w(\boldsymbol{w})}$, and by

$$\overline{\mathrm{GE}}^{\mathrm{VB}}(N) = \left\langle \mathrm{KL}(p(\boldsymbol{x}|\boldsymbol{w}^*) \| p^{\mathrm{VB}}(\boldsymbol{x}|\mathcal{D}^N)) \right\rangle_{q(\mathcal{D}^N)}$$

for the marginal predictive distribution $p^{\mathrm{VB}}(\boldsymbol{x}|\mathcal{D}^N) = \langle p(\boldsymbol{x}|\boldsymbol{w}) \rangle_{\widehat{r}_w(\boldsymbol{w})}$. It follows from the log-sum inequality (17.9) that

$$\overline{\mathrm{GE}}^{\mathrm{VB}}(N) \leq \overline{\mathrm{GE}}^{\mathrm{VB}}_{\mathrm{Joint}}(N). \tag{17.49}$$

Since the predictive distribution (17.42) is derived from the approximate posterior distribution $r^*(\boldsymbol{w}; \widetilde{\boldsymbol{w}}^*)$ consisting of $p(\mathcal{H}|\mathcal{D}^N, \widetilde{\boldsymbol{w}}^*)$ instead of the minimizer $\widehat{r}_{\mathcal{H}}(\mathcal{H})$ of the free energy, it is conjectured that $\overline{\mathrm{GE}}^{\mathrm{VB}*}_{\mathrm{Joint}}(N)$ provides a lower-bound to $\overline{\mathrm{GE}}^{\mathrm{VB}}_{\mathrm{Joint}}(N)$. At least, the inequalities in Eq. (17.45) imply the affinity of the VB free energy and the generalization error measured by the KL divergence of the joint distributions. The generalization error of the marginal predictive distribution is generally upper-bounded by that of the joint predictive distribution as in Eq. (17.49). Although Eq. (17.48) shows that the average generalization error $\overline{\mathrm{GE}}^{\mathrm{VB}*}_{\mathrm{Joint}}(N)$ of the approximate predictive distribution of VB learning with $r^*(\boldsymbol{w}; \widetilde{\boldsymbol{w}}^*)$ is upper-bounded by that of Bayesian learning in the joint model, the relationship between $\overline{\mathrm{GE}}^{\mathrm{VB}*}_{\mathrm{Joint}}(N)$ and $\overline{\mathrm{GE}}^{\mathrm{VB}}_{\mathrm{Joint}}(N)$ is still unknown.

## 17.5 Relation to Other Analyses

In this section, we discuss the relationships of the asymptotic formulae in Sections 17.2 and 17.4 to the analyses of the Bayes free energy and the generalization error.

### 17.5.1 Asymptotic Analysis of Free Energy Bounds

Asymptotic upper-bounds of the Bayes free energy were obtained for some statistical models, including the GMM, HMM, and the Bayesian network (Yamazaki and Watanabe, 2003a,b, 2005). The upper-bounds are given by the following form:

$$\overline{F}^{\mathrm{Bayes}}(N) \leq \lambda'^{\mathrm{Bayes}}_{\mathrm{Joint}} \log N + O(1), \tag{17.50}$$

where the coefficient $\lambda'^{\mathrm{Bayes}}_{\mathrm{Joint}}$ was identified for each model by analyzing the largest pole of the zeta function $\zeta_{\overline{E}}$ in Eq. (17.14) instead of $\zeta_E$, by using the log-sum inequality (17.10) (Yamazaki and Watanabe, 2003a,b, 2005). Since the largest pole of $\zeta_{\overline{E}}$ provides a lower-bound for that of $\zeta_E$, their analyses provided upper-bounds of $\overline{F}^{\mathrm{Bayes}}(N)$ for the aforementioned models.

On the other hand, the asymptotic forms of the VB free energy were analyzed also for the same models as discussed in Chapters 15 and 16, each of which has the following form:

$$\overline{F}^{\mathrm{VB}}(N) \leq \lambda'^{\mathrm{VB}} \log N + O(1). \tag{17.51}$$

In most cases, asymptotic upper-bounds of $\overline{F}^{\mathrm{Bayes}}(N)$ and $\overline{F}^{\mathrm{VB}}(N)$ coincide, i.e., $\lambda'^{\mathrm{Bayes}}_{\mathrm{Joint}} = \lambda'^{\mathrm{VB}}$ holds while Theorem 17.1 implies that $\lambda'^{\mathrm{VB}} \leq \lambda'^{\mathrm{Bayes}}_{\mathrm{Joint}}$. Hence, it is suggested that this upper-bound is tight in some cases. The zeta function $\zeta_{\overline{E}}$ was also analyzed by the algebraic geometrical technique to evaluate the generalization error for estimating local latent variables (Yamazaki, 2016).

Moreover, the previous analyses of the VB free energy are based on the direct minimization of the free energy over the variational parameters (Chapters 14 through 16). Hence, the analyses are highly dependent on the concrete algorithm for the specific model and the choice of the prior distribution. In other words, it is required to parameterize the free energy explicitly by a finite number of variational parameters in such analyses. Analyzing the right-hand side of Eq. (17.24) is more general and is independent of the concrete algorithm for the specific model. It does not even require that the prior distribution $p(\boldsymbol{w})$ be conjugate since Theorem 17.1 holds for any prior. In such a case, the VB learning algorithm should be implemented with techniques for nonconjugacy such as the local variational approximation and the black-box variational inference (Section 2.1.7). In fact, for mixture models, the upper-bound in Theorem 15.10 averaged over the training samples can be obtained in more general cases. The mixture component $g(\boldsymbol{x}|\boldsymbol{\mu})$ can be generalized to any regular models, while in Chapter 15 it was generalized only to the exponential family.

### 17.5.2 Accuracy of Approximation

For several statistical models, tighter bounds or exact evaluations of the coefficient $\lambda'^{\mathrm{Bayes}}$ of the relative Bayes free energy in Eq. (17.5) have been obtained (Aoyagi and Watanabe, 2005; Yamazaki et al., 2010). If the relative Bayes free energy and VB free energy have the asymptotic forms, $\overline{F}^{\mathrm{Bayes}}(N) = \lambda'^{\mathrm{Bayes}} \log N + o(\log N)$ and $\overline{F}^{\mathrm{VB}}(N) = \lambda'^{\mathrm{VB}} \log N + o(\log N)$, respectively,

$\lambda'^{\mathrm{Bayes}} \leq \lambda'^{\mathrm{VB}}$ holds, and the approximation accuracy of VB learning to Bayesian learning can be evaluated by the gap between them:

$$\overline{F}^{\mathrm{VB}}(N) - \overline{F}^{\mathrm{Bayes}}(N) = (\lambda'^{\mathrm{VB}} - \lambda'^{\mathrm{Bayes}}) \log N + o(\log N).$$

From Eq. (17.17), this turns out to be the KL divergence from the approximate posterior to the true posterior. Such a comparison was first conducted for GMMs (Watanabe and Watanabe, 2004, 2006; Aoyagi and Nagata, 2012). A more detailed comparison was conducted for the Bernoulli mixture model discussed in Section 15.4 (Yamazaki and Kaji, 2013; Kaji et al., 2010). According to the authors' results, $\lambda'^{\mathrm{VB}}$ can be strictly greater than $\lambda'^{\mathrm{Bayes}}$, while $\lambda'^{\mathrm{VB}}$ is not so large as $D/2$, where $D$ is the number of parameters.[5] The arguments in Section 17.2 imply that such a comparison can be extended to general latent variable models by examining the difference between $\min_{\overline{w}^* \in \mathcal{W}^*} \overline{F}^{\mathrm{Bayes}}_{\mathrm{Joint}}(N)$ and $\overline{F}^{\mathrm{Bayes}}(N)$, which is related to the difference between $\lambda'^{\mathrm{Bayes}}_{\mathrm{Joint}}$ and $\lambda'^{\mathrm{Bayes}}$, i.e., the poles of $\zeta_{\overline{E}}$ and $\zeta_E$.

### 17.5.3 Average Generalization Error

Although the generalization performance of the VB learning was fully analyzed in the RRR model as discussed in Chapter 14, little has been known in other models. In Section 17.4, we derived an inequality that implies the relationship between the generalization error and the VB free energy for general latent variable models.

In the exact Bayesian learning, the universal relations (13.138) and (13.139) among the quartet, Bayes and Gibbs generalization losses and Bayes and Gibbs training losses, were proved as discussed in Section 13.5.5 (Watanabe, 2009). It is an important future work to explore such relationships among the quantities introduced in Section 17.4 for VB learning.

[5] For the local latent variable model defined in Section 17.1, Corollary 17.2 combined with Eq. (13.125) implies that

$$2\lambda'^{\mathrm{VB}} \leq D.$$

However, this is not true in general as we discussed for the RRR model in Chapter 14 (see Figure 14.8).

# Appendix A James–Stein Estimator

The James–Stein (JS) estimator (James and Stein, 1961), a shrinkage estimator known to *dominate* the maximum likelihood (ML) estimator, has close relation to Bayesian learning. More specifically, it can be derived as an empirical Bayesian (EBayes) estimator (Efron and Morris, 1973).

Consider an $M$-dimensional Gaussian model with a Gaussian prior for the mean parameter:

$$p(\boldsymbol{x}|\boldsymbol{\mu}) = \mathrm{Gauss}_M(\boldsymbol{x}; \boldsymbol{\mu}, \sigma^2 \boldsymbol{I}_M) = \left(2\pi\sigma^2\right)^{-M/2} \exp\left(-\frac{\|\boldsymbol{x} - \boldsymbol{\mu}\|^2}{2\sigma^2}\right), \tag{A.1}$$

$$p(\boldsymbol{\mu}|c^2) = \mathrm{Gauss}_M(\boldsymbol{\mu}; \boldsymbol{0}, c^2 \boldsymbol{I}_M) = \left(2\pi c^2\right)^{-M/2} \exp\left(-\frac{\|\boldsymbol{\mu}\|^2}{2c^2}\right), \tag{A.2}$$

where the variance $\sigma^2$ of observation noise is assumed to be known. We perform empirical Bayesian learning to estimate the mean parameter $\boldsymbol{\mu}$ and the prior variance $c^2$ from observed samples $\mathcal{D} = \{\boldsymbol{x}^{(1)}, \ldots, \boldsymbol{x}^{(N)}\}$. The joint distribution conditional to the hyperparameter is

$$\begin{aligned}
p(\mathcal{D}, \boldsymbol{\mu}|c^2) &= p(\boldsymbol{\mu}|c^2) \prod_{n=1}^{N} p(\boldsymbol{x}^{(n)}|\boldsymbol{\mu}) \\
&= \frac{1}{(2\pi c^2)^{M/2}(2\pi\sigma^2)^{NM/2}} \exp\left(-\frac{\|\boldsymbol{\mu}\|^2}{2c^2} - \sum_{n=1}^{N} \frac{\left\|\boldsymbol{x}^{(n)} - \boldsymbol{\mu}\right\|^2}{2\sigma^2}\right) \\
&= \frac{1}{(2\pi c^2)^{M/2}(2\pi\sigma^2)^{NM/2}} \exp\left(-\frac{1}{2\sigma^2} \sum_{n=1}^{N} \left\|\boldsymbol{x}^{(n)}\right\|^2 + \frac{N^2 \left\|\overline{\boldsymbol{x}}\right\|^2}{2\sigma^2(N + \sigma^2/c^2)}\right) \\
&\quad \cdot \exp\left(-\frac{N + \sigma^2/c^2}{2\sigma^2} \left\|\boldsymbol{\mu} - \frac{N\overline{\boldsymbol{x}}}{N + \sigma^2/c^2}\right\|^2\right),
\end{aligned}$$

which implies that the posterior is Gaussian,

$$p(\boldsymbol{\mu}|\mathcal{D}, c^2) \propto p(\mathcal{D}, \boldsymbol{\mu}|c^2) \propto \exp\left(-\frac{N + \sigma^2/c^2}{2\sigma^2} \left\|\boldsymbol{\mu} - \frac{N\overline{\boldsymbol{x}}}{N + \sigma^2/c^2}\right\|^2\right),$$

with the mean given by

$$\widehat{\mu} = \frac{N\overline{x}}{N + \sigma^2/c^2} = \left(1 - \frac{\sigma^2}{Nc^2 + \sigma^2}\right)\overline{x}. \tag{A.3}$$

The marginal likelihood is computed as

$$\begin{aligned}
p(\mathcal{D}|c^2) &= \int p(\mathcal{D}, \mu|c^2) d\mu \\
&= \frac{1}{(2\pi c^2)^{M/2}(2\pi\sigma^2)^{NM/2}} \exp\left(-\frac{1}{2\sigma^2}\sum_{n=1}^{N}\left\|x^{(n)}\right\|^2 + \frac{N^2\left\|\overline{x}\right\|^2}{2\sigma^2(N + \sigma^2/c^2)}\right) \\
&\quad \cdot \int \exp\left(-\frac{N + \sigma^2/c^2}{2\sigma^2}\left\|\mu - \frac{N\overline{x}}{N + \sigma^2/c^2}\right\|^2\right) d\mu \\
&= \frac{\exp\left(-\frac{1}{2\sigma^2}\sum_{n=1}^{N}\left\|x^{(n)}\right\|^2 + \frac{N^2\left\|\overline{x}\right\|^2}{2\sigma^2(N+\sigma^2/c^2)}\right)}{(2\pi c^2)^{M/2}(2\pi\sigma^2)^{(N-1)M/2}(N + \sigma^2/c^2)^{M/2}} \\
&= \frac{\exp\left(-\frac{1}{2\sigma^2}\sum_{n=1}^{N}\left\|x^{(n)} - \overline{x}\right\|^2 - \frac{N\left\|\overline{x}\right\|^2}{2\sigma^2} + \frac{N^2\left\|\overline{x}\right\|^2}{2\sigma^2(N+\sigma^2/c^2)}\right)}{(2\pi)^{M/2}(2\pi\sigma^2)^{(N-1)M/2}(Nc^2 + \sigma^2)^{M/2}} \\
&= \frac{\exp\left(-\frac{1}{2\sigma^2}\sum_{n=1}^{N}\left\|x^{(n)} - \overline{x}\right\|^2 - \frac{N\sigma^2/c^2\left\|\overline{x}\right\|^2}{2\sigma^2(N+\sigma^2/c^2)}\right)}{(2\pi)^{M/2}(2\pi\sigma^2)^{(N-1)M/2}(Nc^2 + \sigma^2)^{M/2}} \\
&= \frac{\exp\left(-\frac{1}{2\sigma^2}\sum_{n=1}^{N}\left\|x^{(n)} - \overline{x}\right\|^2 - \frac{N\left\|\overline{x}\right\|^2}{2(Nc^2+\sigma^2)}\right)}{(2\pi)^{M/2}(2\pi\sigma^2)^{(N-1)M/2}(Nc^2 + \sigma^2)^{M/2}} \\
&= \frac{\exp\left(-\frac{1}{2\sigma^2}\sum_{n=1}^{N}\left\|x^{(n)} - \overline{x}\right\|^2\right)}{(2\pi\sigma^2)^{(N-1)M/2}} \cdot \frac{\exp\left(-\frac{N\left\|\overline{x}\right\|^2}{2(Nc^2+\sigma^2)}\right)}{(2\pi(Nc^2 + \sigma^2))^{M/2}}.
\end{aligned} \tag{A.4}$$

This implies that $v = \overline{x}\sqrt{N/(Nc^2 + \sigma^2)}$ is a random variable subject to $\mathrm{Gauss}_M(v; \mathbf{0}, I_M)$. Since its (–2)nd order moment is equal to $\left\langle\|v\|^{-2}\right\rangle_{\mathrm{Gauss}_M(v;\mathbf{0},I_M)} = (M - 2)^{-1}$, we have

$$\left\langle\frac{Nc^2 + \sigma^2}{N\left\|\overline{x}\right\|^2}\right\rangle_{p(\mathcal{D}|c^2)} = \frac{1}{M - 2},$$

and therefore

$$\left\langle\frac{M - 2}{N\left\|\overline{x}\right\|^2}\right\rangle_{p(\mathcal{D}|c^2)} = \frac{1}{Nc^2 + \sigma^2}.$$

Accordingly, $(M - 2)/N\left\|\overline{x}\right\|^2$ is an unbiased estimator of the factor $(Nc^2 + \sigma^2)^{-1}$.

Replacing the factor $(Nc^2 + \sigma^2)^{-1}$ in Eq. (A.3) with its unbiased estimator $(M-2)/N\|\overline{\boldsymbol{x}}\|^2$, we obtain the JS estimator (with degree $M-2$):

$$\widehat{\mu}^{\mathrm{JS}} = \left(1 - \frac{(M-2)\sigma^2}{N\|\overline{\boldsymbol{x}}\|^2}\right)\overline{\boldsymbol{x}}. \tag{A.5}$$

If we estimate $c^2$ by maximizing the marginal likelihood (A.4), we obtain the *positive-part JS estimator* (with degree $M$):

$$\widehat{\mu}^{\mathrm{PJS}} = \max\left(0, 1 - \frac{M\sigma^2}{N\|\overline{\boldsymbol{x}}\|^2}\right)\overline{\boldsymbol{x}}. \tag{A.6}$$

The JS estimator has an interesting property. Let us first introduce terminology. Assume that we observed data $\mathcal{D}$ generated from a distribution $p(\mathcal{D}|\boldsymbol{w})$ with unknown parameter $\boldsymbol{w}$. Consider two estimators $\widehat{\boldsymbol{w}}_1 = \widehat{\boldsymbol{w}}_1(\mathcal{D})$ and $\widehat{\boldsymbol{w}}_2 = \widehat{\boldsymbol{w}}_2(\mathcal{D})$, and measure some error criterion $\mathrm{E}(\widehat{\boldsymbol{w}}, \boldsymbol{w}^*)$ from the true parameter value $\boldsymbol{w}^*$.

**Definition A.1** (*Domination*) We say that the estimator $\widehat{\boldsymbol{w}}_1$ dominates the other estimator $\widehat{\boldsymbol{w}}_2$ if

$$\langle \mathrm{E}(\widehat{\boldsymbol{w}}_1(\mathcal{D}), \boldsymbol{w}^*)\rangle_{p(\mathcal{D}|\boldsymbol{w}^*)} \leq \langle \mathrm{E}(\widehat{\boldsymbol{w}}_2(\mathcal{D}), \boldsymbol{w}^*)\rangle_{p(\mathcal{D}|\boldsymbol{w}^*)} \qquad \text{for arbitrary } \boldsymbol{w}^*,$$

$$\text{and} \qquad \langle \mathrm{E}(\widehat{\boldsymbol{w}}_1(\mathcal{D}), \boldsymbol{w}^*)\rangle_{p(\mathcal{D}|\boldsymbol{w}^*)} < \langle \mathrm{E}(\widehat{\boldsymbol{w}}_2(\mathcal{D}), \boldsymbol{w}^*)\rangle_{p(\mathcal{D}|\boldsymbol{w}^*)} \qquad \text{for a certain } \boldsymbol{w}^*.$$

**Definition A.2** (*Efficiency*) We say that an estimator is efficient if no *unbiased* estimator dominates it.

**Definition A.3** (*Admissibility*) We say that an estimator is admissible if no estimator dominates it.

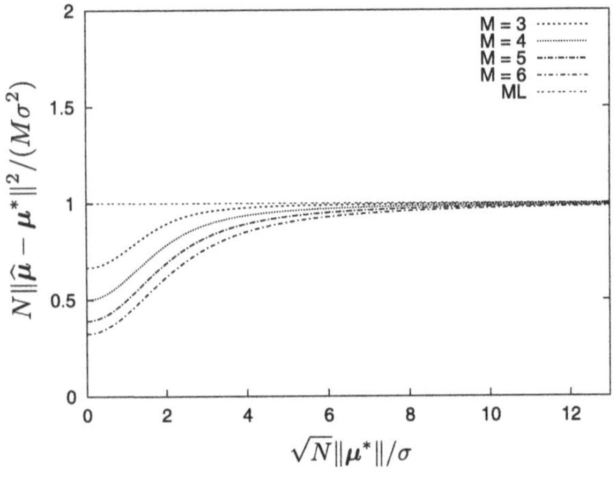


Figure A.1 Generalization error of James–Stein estimator.

Assume that $p(\mathcal{D}|\boldsymbol{w}) = \mathrm{Gauss}_M(\boldsymbol{x}; \boldsymbol{\mu}, \sigma^2 \boldsymbol{I}_M)$. Then the ML estimator,

$$\widehat{\boldsymbol{\mu}}^{\mathrm{ML}} = \overline{\boldsymbol{x}}, \tag{A.7}$$

is known to be *efficient* in terms of the mean squared error

$$\mathrm{E}(\widehat{\boldsymbol{\mu}}, \boldsymbol{\mu}^*) = \left\|\widehat{\boldsymbol{\mu}} - \boldsymbol{\mu}^*\right\|^2 .$$

However, the ML estimator was proven to be inadmissible when $M \geq 3$, i.e., there exists at least one biased estimator that dominates the ML estimator (Stein, 1956). Subsequently, the JS estimator (A.5) was introduced as an estimator dominating the ML estimator (James and Stein, 1961).

Figure A.1 shows the normalized squared loss $N\|\widehat{\boldsymbol{\mu}} - \boldsymbol{\mu}^*\|^2/(M\sigma^2)$ of the ML estimator (A.7) and the JS estimator (A.5) as a function of a scaled true mean $\sqrt{N}\,\|\boldsymbol{\mu}^*\|\,/\sigma$. The ML estimator always gives error equal to one, while the JS estimator gives error dependent on the true value. We can see that the JS estimator dominates the ML estimator for $M \geq 3$. We can easily show that the positive-part JS estimator (A.6) dominates the JS estimator with the same degree.

# Appendix B Metric in Parameter Space

In this appendix, we give a brief summary of the Kullback–Leibler (KL) divergence, the Fisher information, and the Jeffreys prior. The KL divergence is a common (pseudo-)distance measure between distributions, and the corresponding metric in the parameter space is given by the Fisher information. The Jeffreys prior—the uniform prior when the distance between distributions is measured by the KL divergence—is defined so as to reflect the nonuniformity of the density of the volume element in the parameter space.

## B.1 Kullback–Leibler (KL) Divergence

The *KL divergence* between two distributions, $q(\boldsymbol{x})$ and $p(\boldsymbol{x})$, is defined as

$$\begin{aligned}\mathrm{KL}\left(q(\boldsymbol{x})\|p(\boldsymbol{x})\right) &= \int q(\boldsymbol{x})\log\left(\frac{q(\boldsymbol{x})}{p(\boldsymbol{x})}\right)d\boldsymbol{x} \\ &= \int q(\boldsymbol{x})\log\frac{1}{p(\boldsymbol{x})}d\boldsymbol{x} - \int q(\boldsymbol{x})\log\frac{1}{q(\boldsymbol{x})}d\boldsymbol{x} \\ &\geq 0.\end{aligned}$$

If $q(\boldsymbol{x})$ is the true distribution, i.e., $\boldsymbol{x} \sim q(\boldsymbol{x})$, the first term is the average information gain for the one who has (possibly) wrong information (who believes $\boldsymbol{x} \sim p(\boldsymbol{x})$), and the second term is the average information gain, i.e., the *entropy*, for the one who has the correct information (who believes $\boldsymbol{x} \sim q(\boldsymbol{x})$). The KL divergence is not a proper distance metric, since it is not symmetric, i.e., for general $q(\boldsymbol{x})$ and $p(\boldsymbol{x})$,

$$\mathrm{KL}\left(q(\boldsymbol{x})\|p(\boldsymbol{x})\right) \neq \mathrm{KL}\left(p(\boldsymbol{x})\|q(\boldsymbol{x})\right).$$

## B.2 Fisher Information

The *Fisher information* of a parametric distribution $p(\boldsymbol{x}|\boldsymbol{w})$ with its parameter $\boldsymbol{w} \in \mathbb{R}^D$ is defined as

$$\mathbb{S}_{+}^{D} \ni \boldsymbol{F} = \int \frac{\partial \log p(\boldsymbol{x}|\boldsymbol{w})}{\partial \boldsymbol{w}}\left(\frac{\partial \log p(\boldsymbol{x}|\boldsymbol{w})}{\partial \boldsymbol{w}}\right)^{\top} p(\boldsymbol{x}|\boldsymbol{w})d\boldsymbol{x}, \tag{B.1}$$

where $\frac{\partial \log p(\boldsymbol{x}|\boldsymbol{w})}{\partial \boldsymbol{w}} \in \mathbb{R}^D$ is the gradient (column) vector of $\log p(\boldsymbol{x}|\boldsymbol{w})$. Under the regularity conditions (see Section 13.4.1) on the statistical model $p(\boldsymbol{x}|\boldsymbol{w})$, the Fisher information can be written as

$$\boldsymbol{F} = -\int \frac{\partial^2 \log p(\boldsymbol{x}|\boldsymbol{w})}{\partial \boldsymbol{w} \partial \boldsymbol{w}^\top} p(\boldsymbol{x}|\boldsymbol{w}) d\boldsymbol{x}, \tag{B.2}$$

where

$$\left( \frac{\partial^2 \log p(\boldsymbol{x}|\boldsymbol{w})}{\partial \boldsymbol{w} \partial \boldsymbol{w}^\top} \right)_{i,j} = \frac{\partial^2 \log p(\boldsymbol{x}|\boldsymbol{w})}{\partial w_i \partial w_j}.$$

This is because

$$\begin{aligned}
&-\int \frac{\partial^2 \log p(\boldsymbol{x}|\boldsymbol{w})}{\partial w_i \partial w_j} p(\boldsymbol{x}|\boldsymbol{w}) d\boldsymbol{x} \\
&= -\int \frac{\partial}{\partial w_j} \left( \frac{\frac{\partial p(\boldsymbol{x}|\boldsymbol{w})}{\partial w_i}}{p(\boldsymbol{x}|\boldsymbol{w})} \right) p(\boldsymbol{x}|\boldsymbol{w}) d\boldsymbol{x} \\
&= -\int \left( \frac{\frac{\partial^2 p(\boldsymbol{x}|\boldsymbol{w})}{\partial w_i \partial w_j}}{p(\boldsymbol{x}|\boldsymbol{w})} - \frac{\frac{\partial p(\boldsymbol{x}|\boldsymbol{w})}{\partial w_i} \frac{\partial p(\boldsymbol{x}|\boldsymbol{w})}{\partial w_j}}{p^2(\boldsymbol{x}|\boldsymbol{w})} \right) p(\boldsymbol{x}|\boldsymbol{w}) d\boldsymbol{x} \\
&= -\int \frac{\partial^2 p(\boldsymbol{x}|\boldsymbol{w})}{\partial w_i \partial w_j} d\boldsymbol{x} + \int \frac{\frac{\partial p(\boldsymbol{x}|\boldsymbol{w})}{\partial w_i} \frac{\partial p(\boldsymbol{x}|\boldsymbol{w})}{\partial w_j}}{p^2(\boldsymbol{x}|\boldsymbol{w})} p(\boldsymbol{x}|\boldsymbol{w}) d\boldsymbol{x} \\
&= -\frac{\partial^2}{\partial w_i \partial w_j} \int p(\boldsymbol{x}|\boldsymbol{w}) d\boldsymbol{x} + \int \frac{\partial \log p(\boldsymbol{x}|\boldsymbol{w})}{\partial w_i} \frac{\partial \log p(\boldsymbol{x}|\boldsymbol{w})}{\partial w_j} p(\boldsymbol{x}|\boldsymbol{w}) d\boldsymbol{x} \\
&= \int \frac{\partial \log p(\boldsymbol{x}|\boldsymbol{w})}{\partial w_i} \frac{\partial \log p(\boldsymbol{x}|\boldsymbol{w})}{\partial w_j} p(\boldsymbol{x}|\boldsymbol{w}) d\boldsymbol{x}.
\end{aligned}$$

## B.3 Metric and Volume Element

For a small perturbation $\Delta \boldsymbol{w}$ of the parameter, the KL divergence between $p(\boldsymbol{x}|\boldsymbol{w})$ and $p(\boldsymbol{x}|\boldsymbol{w} + \Delta \boldsymbol{w})$ can be written as

$$\begin{aligned}
&\mathrm{KL}\left( p(\boldsymbol{x}|\boldsymbol{w}) \| p(\boldsymbol{x}|\boldsymbol{w} + \Delta \boldsymbol{w}) \right) = \int p(\boldsymbol{x}|\boldsymbol{w}) \log \left( \frac{p(\boldsymbol{x}|\boldsymbol{w})}{p(\boldsymbol{x}|\boldsymbol{w} + \Delta \boldsymbol{w})} \right) d\boldsymbol{x} \\
&= \int p(\boldsymbol{x}|\boldsymbol{w}) \log \left( \frac{p(\boldsymbol{x}|\boldsymbol{w})}{p(\boldsymbol{x}|\boldsymbol{w}) + \frac{\partial p(\boldsymbol{x}|\boldsymbol{w})}{\partial \boldsymbol{w}}^\top \Delta \boldsymbol{w} + \frac{1}{2} \Delta \boldsymbol{w}^\top \frac{\partial^2 p(\boldsymbol{x}|\boldsymbol{w})}{\partial \boldsymbol{w} \partial \boldsymbol{w}} \Delta \boldsymbol{w} + O(\|\Delta \boldsymbol{w}\|^3)} \right) d\boldsymbol{x} \\
&= -\int p(\boldsymbol{x}|\boldsymbol{w}) \log \left( 1 + \frac{\frac{\partial p(\boldsymbol{x}|\boldsymbol{w})}{\partial \boldsymbol{w}}^\top \Delta \boldsymbol{w}}{p(\boldsymbol{x}|\boldsymbol{w})} + \frac{1}{2} \Delta \boldsymbol{w}^\top \frac{\frac{\partial^2 p(\boldsymbol{x}|\boldsymbol{w})}{\partial \boldsymbol{w} \partial \boldsymbol{w}}}{p(\boldsymbol{x}|\boldsymbol{w})} \Delta \boldsymbol{w} + O(\|\Delta \boldsymbol{w}\|^3) \right) d\boldsymbol{x} \\
&= -\int p(\boldsymbol{x}|\boldsymbol{w}) \left( \frac{\frac{\partial p(\boldsymbol{x}|\boldsymbol{w})}{\partial \boldsymbol{w}}^\top}{p(\boldsymbol{x}|\boldsymbol{w})} \Delta \boldsymbol{w} + \frac{1}{2} \Delta \boldsymbol{w}^\top \left( \frac{\frac{\partial^2 p(\boldsymbol{x}|\boldsymbol{w})}{\partial \boldsymbol{w} \partial \boldsymbol{w}}}{p(\boldsymbol{x}|\boldsymbol{w})} - \frac{\frac{\partial p(\boldsymbol{x}|\boldsymbol{w})}{\partial \boldsymbol{w}} \frac{\partial p(\boldsymbol{x}|\boldsymbol{w})}{\partial \boldsymbol{w}}^\top}{p^2(\boldsymbol{x}|\boldsymbol{w})} \right) \Delta \boldsymbol{w} \right) d\boldsymbol{x} \\
&\quad + O(\|\Delta \boldsymbol{w}\|^3)
\end{aligned}$$

$$
\begin{aligned}
&= -\left(\frac{\partial}{\partial w}\int p(x|w)dx\right)^\top \Delta w - \frac{1}{2}\Delta w^\top\left(\frac{\partial^2}{\partial w \partial w}\int p(x|w)dx\right)\Delta w \\
&\quad + \frac{1}{2}\Delta w^\top\left(\int \frac{\partial \log p(x|w)}{\partial w}\frac{\partial \log p(x|w)}{\partial w}^\top p(x|w)dx\right)\Delta w + O(\|\Delta w\|^3) \\
&= \frac{1}{2}\Delta w^\top F \Delta w + O(\|\Delta w\|^3).
\end{aligned}
$$

Therefore, the Fisher information corresponds to the *metric* of the space of distributions when the distance is measured by the KL divergence (Jeffreys, 1946).

When we adopt the Fisher information as the metric, the *volume element* for integrating functions is given by

$$
dV = \frac{1}{\sqrt{2}}\sqrt{\det(F)}dw, \tag{B.3}
$$

where $\frac{1}{\sqrt{2}}\sqrt{\det(F)}$ corresponds to the *density*.

## B.4 Jeffreys Prior

The prior,

$$
p(w) \propto \sqrt{\det(F)}, \tag{B.4}
$$

proportional to the density of the volume element (B.3), is called the *Jeffreys prior* (Jeffreys, 1946). The Jeffreys prior assigns the equal probability to the unit volume element at any point in the parameter space, i.e., it is the *uniform prior* in the distribution space. Since the uniformity is defined not in the parameter space but in the distribution space, the Jeffreys prior is invariant under parameter transformation. Accordingly, the Jeffreys prior is said to be the parameterization invariant *noninformative prior*.

For *singular models*, the Fisher information can have zero eigenvalues, which makes the Jeffreys prior zero. In some models, including the matrix factorization model, zero eigenvalues appear everywhere in the parameter space (see Example B.2). In such cases, we ignore the *common* zero eigenvalues and redefine the (generalized) Jeffrey prior by

$$
p(w) \propto \sqrt{\prod_{d=1}^{\overline{D}} \lambda_d}, \tag{B.5}
$$

where $\lambda_d$ is the $d$th largest eigenvalue of the Fisher information $F$, and $\overline{D}$ is the maximum number of positive eigenvalues over the whole parameter space.

**Example B.1** (Jeffreys prior for one-dimensional Gaussian distribution) The Fisher information of the Gaussian distribution,

$$
p(x|\mu,\sigma^2) = \mathrm{Gauss}_1(x;\mu,\sigma^2) = \left(2\pi\sigma^2\right)^{-1/2}\exp\left(-\frac{(x-\mu)^2}{2\sigma^2}\right),
$$

is calculated as follows. The derivatives of the log likelihood are

$$
\begin{aligned}
\frac{\partial \log p(x|\mu,\sigma^2)}{\partial \mu} &= \frac{\partial}{\partial \mu}\left(-\frac{(x-\mu)^2}{2\sigma^2}\right) \\
&= \frac{x-\mu}{\sigma^2},
\end{aligned}
$$

$$\begin{aligned}
\frac{\partial \log p(x|\mu,\sigma^2)}{\partial \sigma^2} &= \frac{\partial}{\partial \sigma^2}\left(-\frac{1}{2}\log \sigma^2 - \frac{(x-\mu)^2}{2\sigma^2}\right) \\
&= -\frac{1}{2\sigma^2} + \frac{(x-\mu)^2}{2\sigma^4}, \\
\frac{\partial^2 \log p(x|\mu,\sigma^2)}{\partial \mu^2} &= -\frac{1}{\sigma^2}, \\
\frac{\partial^2 \log p(x|\mu,\sigma^2)}{\partial \mu \partial \sigma^2} &= -\frac{x-\mu}{\sigma^4}, \\
\frac{\partial^2 \log p(x|\mu,\sigma^2)}{\partial (\sigma^2)^2} &= \frac{1}{2\sigma^4} - \frac{(x-\mu)^2}{\sigma^6},
\end{aligned}$$

and therefore

$$\begin{aligned}
\boldsymbol{F} &= \left\langle \begin{pmatrix} \frac{1}{\sigma^2} & \frac{x-\mu}{\sigma^4} \\ \frac{x-\mu}{\sigma^4} & \frac{(x-\mu)^2}{\sigma^6} - \frac{1}{2\sigma^4} \end{pmatrix} \right\rangle_{p(x|\mu,\sigma^2)} \\
&= \begin{pmatrix} \frac{1}{\sigma^2} & 0 \\ 0 & \frac{1}{\sigma^4} - \frac{1}{2\sigma^4} \end{pmatrix} \\
&= \begin{pmatrix} \frac{1}{\sigma^2} & 0 \\ 0 & \frac{1}{2\sigma^4} \end{pmatrix}.
\end{aligned}$$

Thus, the Jeffreys priors for $p(x|\mu)$, $p(x|\sigma^2)$, and $p(x|\mu,\sigma^2)$ are

$$\begin{aligned}
p(\mu) &\propto \sqrt{\boldsymbol{F}_{\mu,\mu}} \propto 1, \\
p(\sigma^2) &\propto \sqrt{\boldsymbol{F}_{\sigma^2,\sigma^2}} \propto \frac{1}{\sigma^2}, \\
p(\mu,\sigma^2) &\propto \sqrt{\det(\boldsymbol{F})} \propto \frac{1}{\sigma^3},
\end{aligned}$$

respectively.

**Example B.2** (Jeffreys prior for one-dimensional matrix factorization model) The Fisher information of the one-dimensional matrix factorization (MF) model,

$$p(V|A,B) = \mathrm{Gauss}_1(V; BA, \sigma^2) = \left(2\pi\sigma^2\right)^{-1/2} \exp\left(-\frac{(V-BA)^2}{2\sigma^2}\right), \tag{B.6}$$

is calculated as follows. The derivatives of the log likelihood are

$$\begin{aligned}
\frac{\partial \log p(V|A,B)}{\partial A} &= \sigma^{-2}(V-BA)B, \\
\frac{\partial \log p(V|A,B)}{\partial B} &= \sigma^{-2}(V-BA)A,
\end{aligned}$$

and therefore

$$\begin{aligned}
\boldsymbol{F} &= \frac{1}{\sigma^4}\left\langle \begin{pmatrix} (V-BA)^2B^2 & (V-BA)^2BA \\ (V-BA)^2BA & (V-BA)^2A^2 \end{pmatrix} \right\rangle_{\mathrm{Gauss}_1(V;BA,\sigma^2)} \\
&= \frac{1}{\sigma^2}\begin{pmatrix} B^2 & BA \\ BA & A^2 \end{pmatrix}.
\end{aligned}$$

The Fisher information $\boldsymbol{F}$ has eigenvalues $\lambda_1 = \sigma^{-2}(A^2 + B^2)$ and $\lambda_2 = 0$, since

$$\det\left(\sigma^2 \boldsymbol{F} - \lambda \boldsymbol{I}_2\right) = \det\begin{pmatrix} B^2 - \lambda & BA \\ BA & A^2 - \lambda \end{pmatrix} = (B^2 - \lambda)(A^2 - \lambda) - B^2 A^2$$
$$= \lambda^2 - (A^2 + B^2)\lambda$$
$$= \left(\lambda - (A^2 + B^2)\right)\lambda.$$

The common (over the whole parameter space) zero eigenvalue comes from the invariance of the MF model under the transformation $(A, B) \to (sA, s^{-1}B)$ for any $s \neq 0$. By adopting the generalized definition (B.5) of the Jeffreys prior, the distribution proportional to

$$p(A, B) \propto \sqrt{A^2 + B^2} \tag{B.7}$$

is the parameterization invariant noninformative prior.

The Jeffreys prior is often improper, i.e., the integral of the unnormalized prior over the parameter domain diverges, and therefore the normalization factor cannot be computed, as in Examples B.1 and B.2.

# Appendix C Detailed Description of Overlap Method

Let $V \in \mathbb{R}^{L \times M}$ be the observed matrix, where $L$ and $M$ correspond to the dimensionality $D$ of the observation space and the number $N$ of samples as follows:

$$\begin{aligned} L = D, M = N \quad & \text{if} \quad D \leq N, \\ L = N, M = D \quad & \text{if} \quad D > N. \end{aligned} \tag{C.1}$$

Let

$$V = \sum_{h=1}^{L} \gamma_h \omega_{b_h} \omega_{a_h}^\top \tag{C.2}$$

be the singular value decomposition (SVD) of $V$. The overlap (OL) method (Hoyle, 2008) computes the following approximation to the negative logarithm of the marginal likelihood (8.90) over the hypothetical model rank $H = 1, \ldots, L$:[1]

$$\begin{aligned} 2F^{\mathrm{OL}}(H) &\approx -2 \log p(V) \\ &= (LM - H(L - H - 2)) \log(2\pi) + L \log \pi - 2 \textstyle\sum_{h=1}^{H} \log \left( \frac{\Gamma((M-h+1)/2)}{\Gamma((M-L-h+1)/2)} \right) \\ &\quad + H(M - L)(1 - \log(M - L)) + \textstyle\sum_{h=1}^{H} \sum_{l=H+1}^{L} \log \left( \gamma_h^2 - \gamma_l^2 \right) \\ &\quad + (M - L) \textstyle\sum_{h=1}^{H} \log \gamma_h^2 + (M - H) \sum_{h=1}^{H} \log \left( \frac{1}{\widehat{\sigma}^{2\,\mathrm{OL}}} - \frac{1}{\widehat{\lambda}_h^{\mathrm{OL}}} \right) \\ &\quad - \textstyle\sum_{h=1}^{H} \left( \frac{1}{\widehat{\sigma}^{2\,\mathrm{OL}}} - \frac{1}{\widehat{\lambda}_h^{\mathrm{OL}}} \right) \gamma_h^2 + (L + 2) \left( \sum_{h=1}^{H} \log \widehat{\lambda}_h^{\mathrm{OL}} + (M - H) \log \widehat{\sigma}^{2\,\mathrm{OL}} \right) \\ &\quad + \textstyle\sum_{l=1}^{L} \frac{\gamma_l^2}{\widehat{\sigma}^{2\,\mathrm{OL}}}, \end{aligned} \tag{C.3}$$

where $\Gamma(\cdot)$ denotes the *Gamma function*, and $\{\widehat{\lambda}_h^{\mathrm{OL}}\}_{h=1}^{H}$ and $\widehat{\sigma}^{2\,\mathrm{OL}}$ are estimators for $\{\lambda_h = b_h^2 + \sigma^2\}_{h=1}^{H}$ and $\sigma^2$, respectively, computed by iterating the following updates until convergence:

[1] Our description is slightly different from Hoyle (2008), because the MF model (6.1) does not have the mean parameter shared over the samples.

**Algorithm 23** Overlap method.

1: Prepare the observed matrix $\boldsymbol{V} \in \mathbb{R}^{L\times M}$, following the rule (C.1).
2: Compute the SVD (C.2) of $\boldsymbol{V}$.
3: Compute $F^{\mathrm{OL}}(0)$ by Eq. (C.6).
4: **for** $H = 1$ to $L$ **do**
5: Initialize the noise variance to $\widehat{\sigma}^{2\,\mathrm{OL}} = 10^{-4} \cdot \sum_{h=1}^{L} \gamma_h^2/(LM)$.
6: Iterate Eq. (C.4) for $h = 1, \ldots, H$, and Eq. (C.5) until convergence or any $\widehat{\lambda}_h^{\mathrm{OL}}$ becomes a complex number.
7: Compute $F^{\mathrm{OL}}(H)$ by Eq. (C.3) if all $\{\widehat{\lambda}_h^{\mathrm{OL}}\}_{h=1}^{H}$ are real numbers. Otherwise, set $F^{\mathrm{OL}}(H) = \infty$.
8: **end for**
9: Estimate the rank by $\widehat{H}^{\mathrm{OL}} = \min_{H\in\{0,\ldots,L\}} F^{\mathrm{OL}}(H)$.

$$\widehat{\lambda}_h^{\mathrm{OL}} = \frac{\gamma_h^2}{2(L+2)}\left(1 - \frac{(M-H-(L+2))\widehat{\sigma}^{2\,\mathrm{OL}}}{\gamma_h^2} + \sqrt{\left(1 - \frac{(M-H-(L+2))\widehat{\sigma}^{2\,\mathrm{OL}}}{\gamma_h^2}\right)^2 - \frac{4(L+2)\widehat{\sigma}^{2\,\mathrm{OL}}}{\gamma_h^2}}\right), \tag{C.4}$$

$$\widehat{\sigma}^{2\,\mathrm{OL}} = \frac{1}{(M-H)}\left(\sum_{l=1}^{L} \frac{\gamma_l^2}{L} - \sum_{h=1}^{H} \widehat{\lambda}_h^{\mathrm{OL}}\right). \tag{C.5}$$

When iterating Eqs. (C.4) and (C.5), $\widehat{\lambda}_h^{\mathrm{OL}}$ can become a complex number. In such a case, the hypothetical $H$ is rejected. Otherwise, Eq. (C.3) is evaluated after convergence. For the null hypothesis, i.e., $H = 0$, the negative log likelihood is given by

$$2F^{\mathrm{OL}}(0) = -2\log p(\boldsymbol{V}) = LM\left(\log\left(\frac{2\pi}{LM}\sum_{l=1}^{L}\gamma_l^2\right) + 1\right). \tag{C.6}$$

The estimated rank $\widehat{H}^{\mathrm{OL}}$ is the minimizer of $F^{\mathrm{OL}}(H)$ over $H = 0, \ldots, L$.

Algorithm 23 summarizes the procedure.

# Appendix D Optimality of Bayesian Learning

Bayesian learning is deduced from the basic probability theory, and therefore it is optimal in terms of generalization performance under the assumption that the model and the prior are set reasonably.

Consider a distribution of problems where the true distribution is written as $q(\boldsymbol{x}) = p(\boldsymbol{x}|\boldsymbol{w}^*)$ with the true parameter $\boldsymbol{w}^*$ subject to $q(\boldsymbol{w}^*)$. Although we usually omit the dependency description on the true distribution or the true parameter, the average generalization error, Eq. (13.13), naturally depends on the true distribution, so we here denote the dependence explicitly as $\overline{\mathrm{GE}}(N; \boldsymbol{w}^*)$. Let

$$\overline{\overline{\mathrm{GE}}}(N) = \left\langle \overline{\mathrm{GE}}(N; \boldsymbol{w}^*) \right\rangle_{q(\boldsymbol{w}^*)} \tag{D.1}$$

be the average of the average generalization error over the distribution $q(\boldsymbol{w}^*)$ of the true parameter.

**Theorem D.1** *If we know the distribution $q(\boldsymbol{w}^*)$ of the true parameter and use it as the prior distribution, i.e., $p(\boldsymbol{w}) = q(\boldsymbol{w})$, then Bayesian learning minimizes the average generalization error* (D.1) *over $q(\boldsymbol{w}^*)$, i.e.,*

$$\overline{\overline{\mathrm{GE}}}^{\mathrm{Bayes}}(N) \leq \overline{\overline{\mathrm{GE}}}^{\mathrm{Other}}(N), \tag{D.2}$$

*where $\overline{\overline{\mathrm{GE}}}^{\mathrm{Other}}(N)$ denotes the average generalization error of any (other) learning algorithm.*

*Proof* Let $\boldsymbol{X}^N = (\boldsymbol{x}^{(1)}, \ldots, \boldsymbol{x}^{(N)})$ be the $N$ training samples. Regarding the new test sample as the $(N+1)$th sample, we can write the Bayes predictive distribution as follows:

$$\begin{aligned} p^{\mathrm{Bayes}}(\boldsymbol{x}^{(N+1)}|\boldsymbol{X}^N) &= \int p(\boldsymbol{x}^{(N+1)}|\boldsymbol{w})p(\boldsymbol{w}|\boldsymbol{X}^N)d\boldsymbol{w} \\ &= \frac{\int p(\boldsymbol{w}) \prod_{n=1}^{N+1} p(\boldsymbol{x}^{(n)}|\boldsymbol{w})d\boldsymbol{w}}{\int p(\boldsymbol{w}') \prod_{n=1}^{N} p(\boldsymbol{x}^{(n)}|\boldsymbol{w}')d\boldsymbol{w}'} \\ &= \frac{p(\boldsymbol{X}^{N+1})}{p(\boldsymbol{X}^N)}, \end{aligned} \tag{D.3}$$

where $p(\boldsymbol{X}^N) = \int p(\boldsymbol{w})p(\boldsymbol{X}^N|\boldsymbol{w})d\boldsymbol{w}$ is the marginal likelihood. The average generalization error (D.1) of a learning algorithm with its predictive distribution $r(\boldsymbol{x})$ is given by

$$\overline{\overline{\mathrm{GE}}}(N) = \left\langle \log \frac{p(\boldsymbol{x}^{(N+1)}|\boldsymbol{w}^*)}{r(\boldsymbol{x}^{(N+1)})} \right\rangle_{p(\boldsymbol{X}^{N+1}|\boldsymbol{w}^*)q(\boldsymbol{w}^*)}$$
$$= -\int \left( \int p(\boldsymbol{X}^{N+1}|\boldsymbol{w}^*)q(\boldsymbol{w}^*)d\boldsymbol{w}^* \right) \log r(\boldsymbol{x}^{(N+1)})d\boldsymbol{X}^{N+1} - (N+1)S,$$
$$= -\int q(\boldsymbol{X}^{N+1}) \log r(\boldsymbol{x}^{(N+1)})d\boldsymbol{X}^{N+1} - (N+1)S, \tag{D.4}$$

where

$$q(\boldsymbol{X}^N) = \int p(\boldsymbol{X}^N|\boldsymbol{w}^*)q(\boldsymbol{w}^*)d\boldsymbol{w}^*$$

is the marginal likelihood with the *true* prior distribution $q(\boldsymbol{w}^*)$, and

$$S = -\langle \log p(\boldsymbol{x}|\boldsymbol{w}^*) \rangle_{p(\boldsymbol{x}|\boldsymbol{w}^*)q(\boldsymbol{w}^*)}$$

is the entropy, which does not depend on the predictive distribution $r(\boldsymbol{x})$. Eq. (D.4) can be written as

$$\overline{\overline{\mathrm{GE}}}(N) = -\int q(\boldsymbol{X}^N)\frac{q(\boldsymbol{X}^{N+1})}{q(\boldsymbol{X}^N)} \log r(\boldsymbol{x}^{(N+1)})d\boldsymbol{X}^{N+1} - (N+1)S$$
$$= -\left\langle \int \frac{q(\boldsymbol{X}^{N+1})}{q(\boldsymbol{X}^N)} \log r(\boldsymbol{x}^{(N+1)})d\boldsymbol{x}^{(N+1)} \right\rangle_{q(\boldsymbol{X}^N)} - (N+1)S$$
$$= \left\langle \int \frac{q(\boldsymbol{X}^{N+1})}{q(\boldsymbol{X}^N)} \log \frac{\frac{q(\boldsymbol{X}^{N+1})}{q(\boldsymbol{X}^N)}}{r(\boldsymbol{x}^{(N+1)})} d\boldsymbol{x}^{(N+1)} \right\rangle_{q(\boldsymbol{X}^N)} + \text{const.} \tag{D.5}$$

Since the first term is the KL divergence between $q(\boldsymbol{X}^{N+1})/q(\boldsymbol{X}^N)$ and $r(\boldsymbol{x}^{(N+1)})$, Eq. (D.5) is minimized when

$$r(\boldsymbol{x}^{(N+1)}) = \frac{q(\boldsymbol{X}^{N+1})}{q(\boldsymbol{X}^N)} = q^{\mathrm{Bayes}}(\boldsymbol{x}^{(N+1)}|\boldsymbol{X}^N), \tag{D.6}$$

where $q^{\mathrm{Bayes}}(\boldsymbol{x}^{(N+1)}|\boldsymbol{X}^N)$ is the Bayes predictive distribution (D.3) with the prior distribution set to the distribution of the true parameter, i.e., $p(\boldsymbol{w}) = q(\boldsymbol{w})$. Thus, we have proved that no other learning method can give better generalization error than Bayesian learning with the true prior. □

A remark is that, since we usually do not know the true distribution and the true prior (the distribution of the true parameter), it is not surprising that an approximation method, e.g., variational Bayesian learning, to Bayesian learning provides better generalization performance than Bayesian learning with a nontrue prior in some situations.

# Bibliography

Akaho, S., and Kappen, H. J. 2000. Nonmonotonic Generalization Bias of Gaussian Mixture Models. *Neural Computation*, **12**, 1411–1427.

Akaike, H. 1974. A New Look at Statistical Model. *IEEE Transactions on Automatic Control*, **19**(6), 716–723.

Akaike, H. 1980. Likelihood and Bayes Procedure. Pages 143–166 of: Bernald, J. M. (ed.), *Bayesian Statistics*. Valencia, Italy: University Press.

Alzer, H. 1997. On Some Inequalities for the Gamma and Psi Functions. *Mathematics of Computation*, **66**(217), 373–389.

Amari, S., Park, H., and Ozeki, T. 2002. Geometrical Singularities in the Neuromanifold of Multilayer Perceptrons. Pages 343–350 of: *Advances in NIPS*, vol. 14. Cambridge, MA: MIT Press.

Aoyagi, M., and Nagata, K. 2012. Learning Coefficient of Generalization Error in Bayesian Estimation and Vandermonde Matrix-Type Singularity. *Neural Computation*, **24**(6), 1569–1610.

Aoyagi, M., and Watanabe, S. 2005. Stochastic Complexities of Reduced Rank Regression in Bayesian Estimation. *Neural Networks*, **18**(7), 924–933.

Asuncion, A., and Newman, D.J. 2007. *UCI Machine Learning Repository*. www.ics.uci.edu/~mlearn/MLRepository.html

Asuncion, A., Welling, M., Smyth, P., and Teh, Y. W. 2009. On Smoothing and Inference for Topic Models. Pages 27–34 of: *Proceedings of UAI*. Stockholm, Sweden: Morgan Kaufmann Publishers Inc.

Attias, H. 1999. Inferring Parameters and Structure of Latent Variable Models by Variational Bayes. Pages 21–30 of: *Proceedings of UAI*. Stockholm, Sweden: Morgan Kaufmann Publishers Inc.

Babacan, S. D., Nakajima, S., and Do, M. N. 2012a. Probabilistic Low-Rank Subspace Clustering. Pages 2753–2761 of: *Advances in Neural Information Processing Systems 25*. Lake Tahoe, NV: NIPS Foundation.

Babacan, S. D., Luessi, M., Molina, R., and Katsaggelos, A. K. 2012b. Sparse Bayesian Methods for Low-Rank Matrix Estimation. *IEEE Transactions on Signal Processing*, **60**(8), 3964–3977.

Baik, J., and Silverstein, J. W. 2006. Eigenvalues of Large Sample Covariance Matrices of Spiked Population Models. *Journal of Multivariate Analysis*, **97**(6), 1382–1408.

Baldi, P. F., and Hornik, K. 1995. Learning in Linear Neural Networks: A Survey. *IEEE Transactions on Neural Networks*, **6**(4), 837–858.

Banerjee, A., Merugu, S., Dhillon, I. S., and Ghosh, J. 2005. Clustering with Bregman Divergences. *Journal of Machine Learning Research*, **6**, 1705–1749.

Beal, M. J. 2003. *Variational Algorithms for Approximate Bayesian Inference*. PhD thesis, University College London.

Bicego, M., Lovato, P., Ferrarini, A., and Delledonne, M. 2010. Biclustering of Expression Microarray Data with Topic Models. Pages 2728–2731 of: *Proceedings of ICPR*. Istanbul, Turkey: ICPR.

Bickel, P., and Chernoff, H. 1993. *Asymptotic Distribution of the Likelihood Ratio Statistic in a Prototypical Non Regular Problem*. New Delhi, India: Wiley Eastern Limited.

Bishop, C. M. 1999a. Bayesian Principal Components. Pages 382–388 of: *Advances in NIPS*, vol. 11. Denver, CO: NIPS Foundation.

Bishop, C. M. 1999b. Variational Principal Components. Pages 514–509 of: *Proceedings of International Conference on Artificial Neural Networks*, vol. 1. Edinburgh, UK: Computing and Control Engineering Journal.

Bishop, C. M. 2006. *Pattern Recognition and Machine Learning*. New York: Springer.

Bishop, C. M., and Tipping, M. E. 2000. Variational Relevance Vector Machines. Pages 46–53 of: *Proceedings of the Sixteenth Conference Annual Conference on Uncertainty in Artificial Intelligence*. Stanford, CA: Morgan Kaufmann Publishers Inc.

Blei, D. M., and Jordan, M. I. 2005. Variational Inference for Dirichlet Process Mixtures. *Bayesian Analysis*, **1**, 121–144.

Blei, D. M., Ng, A. Y., and Jordan, M. I. 2003. Latent Dirichlet Allocation. *Journal of Machine Learning Research*, **3**, 993–1022.

Bouchaud, J. P., and Potters, M. 2003. *Theory of Financial Risk and Derivative Pricing—From Statistical Physics to Risk Management*, 2nd edn. Cambridge, UK: University Press.

Brown, L. D. 1986. *Fundamentals of Statistical Exponential Families*. IMS Lecture Notes–Monograph Series 9. Beachwood, OH: Institute of Mathematical Statistics.

Candès, E. J., Li, X., Ma, Y., and Wright, J. 2011. Robust Principal Component Analysis? *Journal of the ACM*, **58**(3), 1–37.

Carroll, J. D., and Chang, J. J. 1970. Analysis of Individual Differences in Multidimensional Scaling via an N-way Generalization of "Eckart–Young" Decomposition. *Psychometrika*, **35**, 283–319.

Chen, X., Hu, X., Shen, X., and Rosen, G. 2010. Probabilistic Topic Modeling for Genomic Data Interpretation. Pages 149–152 of: *2010 IEEE International Conference on Bioinformatics and Biomedicine (BIBM)*.

Chib, S. 1995. Marginal Likelihood from the Gibbs Output. *Journal of the American Statistical Association*, **90**(432), 1313–1321.

Chu, W., and Ghahramani, Z. 2009. Probabilistic Models for Incomplete Multi-dimensional Arrays. Pages 89–96. In: *Proceedings of International Conference on Artificial Intelligence and Statistics*. Clearwater Beach, FL: Proceedings of Machine Learning Research.

Courant, R., and Hilbert, D. 1953. *Methods of Mathematical Physics, Volume 1*. New York: Wiley.

Cramer, H. 1949. *Mathematical Methods of Statistics*. Princeton, NJ: University Press.

Dacunha-Castelle, D., and Gassiat, E. 1997. Testing in Locally Conic Models, and Application to Mixture Models. *Probability and Statistics*, **1**, 285–317.

Dempster, A. P., Laird, N. M., and Rubin, D. B. 1977. Maximum Likelihood for Incomplete Data via the EM Algorithm. *Journal of the Royal Statistical Society*, **39-B**, 1–38.

Dharmadhikari, S., and Joag-Dev, K. 1988. *Unimodality, Convexity, and Applications*. Cambridge, MA: Academic Press.

Ding, X., He, L., and Carin, L. 2011. Bayesian Robust Principal Component Analysis. *IEEE Transactions on Image Processing*, **20**(12), 3419–3430.

Drexler, F. J. 1978. A Homotopy Method for the Calculation of All Zeros of Zero-Dimensional Polynomial Ideals. Pages 69–93 of: Wacker, H. J. (ed.), *Continuation Methods*. New York: Academic Press.

D'Souza, A., Vijayakumar, S., and Schaal, S. 2004. The Bayesian Backfitting Relevance Vector Machine. In: *Proceedings of the 21st International Conference on Machine Learning*. Banff, AB: Association for Computing Machinery.

Durbin, R., Eddy, S., Krogh, A., and Mitchison, G. 1998. *Biological Sequence Analysis: Probabilistic Models of Proteins and Nucleic Acids*. Cambridge: Cambridge University Press.

Efron, B., and Morris, C. 1973. Stein's Estimation Rule and its Competitors—An Empirical Bayes Approach. *Journal of the American Statistical Association*, **68**, 117–130.

Elhamifar, E., and Vidal, R. 2013. Sparse Subspace Clustering: Algorithm, Theory, and Applications. *IEEE Transactions on Pattern Analysis and Machine Intelligence*, **35**(11), 2765–2781.

Felzenszwalb, P. F., and Huttenlocher, D. P. 2004. Efficient Graph-Based Image Segmentation. *International Journal of Computer Vision*, **59**(2), 167–181.

Fukumizu, K. 1999. Generalization Error of Linear Neural Networks in Unidentifiable Cases. Pages 51–62 of: *Proceedings of International Conference on Algorithmic Learning Theory*. Tokyo, Japan: Springer.

Fukumizu, K. 2003. Likelihood Ratio of Unidentifiable Models and Multilayer Neural Networks. *Annals of Statistics*, **31**(3), 833–851.

Garcia, C. B., and Zangwill, W. I. 1979. Determining All Solutions to Certain Systems of Nonlinear Equations. *Mathematics of Operations Research*, **4**, 1–14.

Gershman, S. J., and Blei, D. M. 2012. A Tutorial on Bayesian Nonparametric Models. *Journal of Mathematical Psychology*, **56**(1), 1–12.

Ghahramani, Z., and Beal, M. J. 2001. Graphical Models and Variational Methods. Pages 161–177 of: *Advanced Mean Field Methods*. Cambridge, MA: MIT Press.

Girolami, M. 2001. A Variational Method for Learning Sparse and Overcomplete Representations. *Neural Computation*, **13**(11), 2517–2532.

Girolami, M., and Kaban, A. 2003. On an Equivalence between PLSI and LDA. Pages 433–434 of: *Proceedings of SIGIR*, New York and Toronto, ON: Association for Computing Machinery.

Gopalan, P., Hofman, J. M., and Blei, D. M. 2013. Scalable Recommendation with Poisson Factorization. *arXiv:1311.1704 [cs.IR]*.

Griffiths, T. L., and Steyvers, M. 2004. Finding Scientific Topics. *PNAS*, **101**, 5228–5235.

Gunji, T., Kim, S., Kojima, M., Takeda, A., Fujisawa, K., and Mizutani, T. 2004. PHoM—A Polyhedral Homotopy Continuation Method. *Computing*, **73**, 57–77.

Gupta, A. K., and Nagar, D. K. 1999. *Matrix Variate Distributions*. London, UK: Chapman and Hall/CRC.

Hagiwara, K. 2002. On the Problem in Model Selection of Neural Network Regression in Overrealizable Scenario. *Neural Computation*, **14**, 1979–2002.

Hagiwara, K., and Fukumizu, K. 2008. Relation between Weight Size and Degree of Over-Fitting in Neural Network Regression. *Neural Networks*, **21**(1), 48–58.

Han, T. S., and Kobayashi, K. 2007. *Mathematics of Information and Coding*. Providence, RI: American Mathematical Society.

Harshman, R. A. 1970. Foundations of the PARAFAC Procedure: Models and Conditions for an "Explanatory" Multimodal Factor Analysis. *UCLA Working Papers in Phonetics*, **16**, 1–84.

Hartigan, J. A. 1985. A Failure of Likelihood Ratio Asymptotics for Normal Mixtures. Pages 807–810 of: *Proceedings of the Berkeley Conference in Honor of J. Neyman and J. Kiefer*. Berkeley, CA: Springer.

Hastie, T., and Tibshirani, R. 1986. Generalized Additive Models. *Statistical Science*, **1**(3), 297–318.

Hinton, G. E., and van Camp, D. 1993. Keeping Neural Networks Simple by Minimizing the Description Length of the Weights. Pages 5–13 of: *Proceedings of COLT*. Santa Cruz, CA.

Hoffman, M. D., Blei, D. M., Wang, C., and Paisley, J. 2013. Stochastic Variational Inference. *Journal of Machine Learning Research*, **14**, 1303–1347.

Hofmann, T. 2001. Unsupervised Learning by Probabilistic Latent Semantic Analysis. *Machine Learning*, **42**, 177–196.

Hosino, T., Watanabe, K., and Watanabe, S. 2005. Stochastic Complexity of Variational Bayesian Hidden Markov Models. In: *Proceedings of IJCNN*. Montreal, QC.

Hosino, T., Watanabe, K., and Watanabe, S. 2006a. Free Energy of Stochastic Context Free Grammar on Variational Bayes. Pages 407–416 of: *Proceedings of ICONIP*. Hong Kong, China: Springer.

Hosino, T., Watanabe, K., and Watanabe, S. 2006b. Stochastic Complexity of Hidden Markov Models on the Variational Bayesian Learning (in Japanese). *IEICE Transactions on Information and Systems*, **J89-D**(6), 1279–1287.

Hotelling, H. 1933. Analysis of a Complex of Statistical Variables into Principal Components. *Journal of Educational Psychology*, **24**, 417–441.

Hoyle, D. C. 2008. Automatic PCA Dimension Selection for High Dimensional Data and Small Sample Sizes. *Journal of Machine Learning Research*, **9**, 2733–2759.

Hoyle, D. C., and Rattray, M. 2004. Principal-Component-Analysis Eigenvalue Spectra from Data with Symmetry-Breaking Structure. *Physical Review E*, **69**(026124).

Huynh, T., Mario, F., and Schiele, B. 2008. Discovery of Activity Patterns Using Topic Models. Pages 9–10. In: *International Conference on Ubiquitous Computing (UbiComp)*. New York and Seoul, South Korea: Association for Computer Machinery.

Hyvärinen, A., Karhunen, J., and Oja, E. 2001. *Independent Component Analysis*. New York: Wiley.

Ibragimov, I. A. 1956. On the Composition of Unimodal Distributions. *Theory of Probability and Its Applications*, **1**(2), 255–260.

Ilin, A., and Raiko, T. 2010. Practical Approaches to Principal Component Analysis in the Presence of Missing Values. *Journal of Machine Learning Research*, **11**, 1957–2000.

Ito, H., Amari, S., and Kobayashi, K. 1992. Identifiability of Hidden Markov Information Sources and Their Minimum Degrees of Freedom. *IEEE Transactions on Information Theory*, **38**(2), 324–333.

Jaakkola, T. S., and Jordan, M. I. 2000. Bayesian Parameter Estimation via Variational Methods. *Statistics and Computing*, **10**, 25–37.

James, W., and Stein, C. 1961. Estimation with Quadratic Loss. Pages 361–379 of: *Proceedings of the 4th Berkeley Symposium on Mathematical Statistics and Probability*, vol. 1. Berkeley: University of California Press.

Jeffreys, H. 1946. An Invariant Form for the Prior Probability in Estimation Problems. Pages 453–461 of: *Proceedings of the Royal Society of London. Series A, Mathematical and Physical Sciences*, vol. 186. London, UK: Royal Society.

Jensen, F. V. 2001. *Bayesian Networks and Decision Graphs*. Springer.

Johnstone, I. M. 2001. On the Distribution of the Largest Eigenvalue in Principal Components Analysis. *Annals of Statistics*, **29**, 295–327.

Jordan, M. I., Ghahramani, Z., Jaakkola, T. S., and Saul, L. 1999. Introduction to Variational Methods for Graphical Models. *Machine Learning*, **37**, 183–233.

Kaji, D., Watanabe, K., and Watanabe, S. 2010. Phase Transition of Variational Bayes Learning in Bernoulli Mixture. *Australian Journal of Intelligent Information Processing Systems*, **11**(4), 35–40.

Khan, M. E., Babanezhad, R., Lin, W., Schmidt, M., and Sugiyama, M. 2016. Faster Stochastic Variational Inference Using Proximal-Gradient Methods with General Divergence Functions. Pages 309–318. In: *Proceedings of UAI*. New York: AUAI Press.

Kim, Y. D., and Choi, S. 2014. Scalable Variational Bayesian Matrix Factorization with Side Information. Pages 493–502 of: *Proceedings of AISTATS*. Reykjavik, Iceland: Proceedings of Machine Learning Research.

Kingma, D. P., and Welling, M. 2014. Auto-Encoding Variational Bayes. In: *International Conference on Learning Representations (ICLR)*. *arXiv*:1412.6980

Kolda, T. G., and Bader, B. W. 2009. Tensor Decompositions and Applications. *SIAM Review*, **51**(3), 455–500.

Krestel, R., Fankhauser, P., and Nejdl, W. 2009. Latent Dirichlet Allocation for Tag Recommendation. Pages 61–68 of: *Proceedings of the Third ACM Conference on Recommender Systems*. New York: Association for Computing Machinery.

Krizhevsky, A., Sutskever, I., and Hinton, G. E. 2012. ImageNet Classification with Deep Convolutional Neural Networks. Pages 1097–1105 of: *Advances in NIPS*. Lake Tahoe, NV: NIPS Foundation.

Kurihara, K., and Sato, T. 2004. An application of the variational Bayesian Approach to Probabilistic Context-Free Grammars. In: *Proceedings of IJCNLP*. Banff, AB.

Kurihara, K., Welling, M., and Teh, M. Y. W. 2007. Collapsed Variational Dirichlet Process Mixture Models. In: *Proceedings of IJCAI*. Hyderabad, India.

Kuriki, S., and Takemura, A. 2001. Tail Probabilities of the Maxima of Multilinear Forms and Their Applications. *Annals of Statistics*, **29**(2), 328–371.

Lee, T. L., Li, T. Y., and Tsai, C. H. 2008. HOM4PS-2.0: A Software Package for Solving Polynomial Systems by the Polyhedral Homotopy Continuation Method. *Computing*, **83**, 109–133.

Levin, E., Tishby, N., and Solla, S. A. 1990. A Statistical Approaches to Learning and Generalization in Layered Neural Networks. Pages 1568–1674 of: *Proceedings of IEEE*, vol. 78.

Li, F.-F., and Perona, P. 2005. A Bayesian Hierarchical Model for Learning Natural Scene Categories. Pages 524–531 of: *Proceedings of CVPR*. San Diego, CA.

Lim, Y. J., and Teh, Y. W. 2007. Variational Bayesian Approach to Movie Rating Prediction. In: *Proceedings of KDD Cup and Workshop*. New York and San Jose, CA: Association for Computing Machinery.

Lin, Z., Chen, M., and Ma, Y. 2009. The Augmented Lagrange Multiplier Method for Exact Recovery of Corrupted Low-Rank Matrices. *UIUC Technical Report UILU-ENG-09-2215*.

Liu, G., and Yan, S. 2011. Latent Low-Rank Representation for Subspace Segmentation and Feature Extraction. In: *Proceedings of ICCV*. Barcelona, Spain.

Liu, G., Lin, Z., and Yu, Y. 2010. Robust Subspace Segmentation by Low-Rank Representation. Pages 663–670 of: *Proceedings of ICML*. Haifa, Israel: Omnipress.

Liu, G., Xu, H., and Yan, S. 2012. Exact Subspace Segmentation and Outlier Detection by Low-Rank Representation. In: *Proceedings of AISTATS*. La Palma, Canary Islands: Proceedings of Machine Learning Research.

Liu, X., Pasarica, C., and Shao, Y. 2003. Testing Homogeneity in Gamma Mixture Models. *Scandinavian Journal of Statistics*, **30**, 227–239.

Lloyd, S. P. 1982. Least Square Quantization in PCM. *IEEE Transactions on Information Theory*, **28**(2), 129–137.

MacKay, D. J. C. 1992. Bayesian Interpolation. *Neural Computation*, **4**(2), 415–447.

MacKay, D. J. C. 1995. Developments in Probabilistic Modeling with Neural Networks—Ensemble Learning. Pages 191–198 of: *Proceedings of the 3rd Annual Symposium on Neural Networks*.

Mackay, D. J. C. 2001. Local Minima, Symmetry-Breaking, and Model Pruning in Variational Free Energy Minimization. Available from www.inference.phy.cam.ac.uk/mackay/minima.pdf.

MacKay, D. J. C. 2003. *Information Theory, Inference, and Learning Algorithms*. Cambridge: Cambridge University Press. Available from www.inference.phy.cam.ac.uk/mackay/itila/.

MacQueen, J. B. 1967. Some Methods for Classification and Analysis of Multivariate Observations. Pages 281–297 of: *Proceedings of 5th Berkeley Symposium on Mathematical Statistics and Probability*, vol. 1. Berkeley: University of California Press.

Marčenko, V. A., and Pastur, L. A. 1967. Distribution of Eigenvalues for Some Sets of Random Matrices. *Mathematics of the USSR-Sbornik*, **1**(4), 457–483.

Marshall, A. W., Olkin, I., and Arnold, B. C. 2009. *Inequalities: Theory of Majorization and Its Applications*, 2d ed. Springer.

Minka, T. P. 2001a. Automatic Choice of Dimensionality for PCA. Pages 598–604 of: *Advances in NIPS*, vol. 13. Cambridge, MA: MIT Press.

Minka, T. P. 2001b. Expectation Propagation for Approximate Bayesian Inference. Pages 362–369 of: *Proceedings of UAI*. Seattle, WA: Morgan Kaufmann Publishers Inc.

Mørup, M., and Hansen, L. R. 2009. Automatic Relevance Determination for Multi-Way Models. *Journal of Chemometrics*, **23**, 352–363.

Nakajima, S., and Sugiyama, M. 2011. Theoretical Analysis of Bayesian Matrix Factorization. *Journal of Machine Learning Research*, **12**, 2579–2644.

Nakajima, S., and Sugiyama, M. 2014. Analysis of Empirical MAP and Empirical Partially Bayes: Can They Be Alternatives to Variational Bayes? Pages 20–28 of: *Proceedings of International Conference on Artificial Intelligence and Statistics*, vol. 33. Reykjavik, Iceland: Proceedings of Machine Learning Research.

Nakajima, S., and Watanabe, S. 2007. Variational Bayes Solution of Linear Neural Networks and Its Generalization Performance. *Neural Computation*, **19**(4), 1112–1153.

Nakajima, S., Sugiyama, M., and Babacan, S. D. 2011 (June 28–July 2). On Bayesian PCA: Automatic Dimensionality Selection and Analytic Solution. Pages 497–504 of: *Proceedings of 28th International Conference on Machine Learning (ICML2011)*. Bellevue, WA: Omnipress.

Nakajima, S., Sugiyama, M., Babacan, S. D., and Tomioka, R. 2013a. Global Analytic Solution of Fully-Observed Variational Bayesian Matrix Factorization. *Journal of Machine Learning Research*, **14**, 1–37.

Nakajima, S., Sugiyama, M., and Babacan, S. D. 2013b. Variational Bayesian Sparse Additive Matrix Factorization. *Machine Learning*, **92**, 319–1347.

Nakajima, S., Takeda, A., Babacan, S. D., Sugiyama, M., and Takeuchi, I. 2013c. Global Solver and Its Efficient Approximation for Variational Bayesian Low-Rank Subspace Clustering. In: *Advances in Neural Information Processing Systems 26*. Lake Tahoe, NV: NIPS Foundation.

Nakajima, S., Sato, I., Sugiyama, M., Watanabe, K., and Kobayashi, H. 2014. Analysis of Variational Bayesian Latent Dirichlet Allocation: Weaker Sparsity Than MAP. Pages 1224–1232 of: *Advances in NIPS*, vol. 27. Montreal, Quebec:: NIPS Foundation.

Nakajima, S., Tomioka, R., Sugiyama, M., and Babacan, S. D. 2015. Condition for Perfect Dimensionality Recovery by Variational Bayesian PCA. *Journal of Machine Learning Research*, **16**, 3757–3811.

Nakamura, F., and Watanabe, S. 2014. Asymptotic Behavior of Variational Free Energy for Normal Mixtures Using General Dirichlet Distribution (in Japanese). *IEICE Transactions on Information and Systems*, **J97-D**(5), 1001–1013.

Neal, R. M. 1996. *Bayesian Learning for Neural Networks*. New York: Springer.

Opper, M., and Winther, O. 1996. A Mean Field Algorithm for Bayes Learning in Large Feed-Forward Neural Networks. Pages 225–231 of: *Advances in NIPS*. Denver, CO: NIPS Foundation.

Pearson, K. 1914. *Tables for Statisticians and Biometricians*. Cambridge: Cambridge University Press.

Purushotham, S., Liu, Y., and Kuo, C. C. J. 2012. Collaborative Topic Regression with Social Matrix Factorization for Recommendation Systems. In: *Proceedings of ICML*. Edinburgh, UK: Omnipress.

Rabiner, L. R. 1989. A Tutorial on Hidden Markov Models and Selected Applications in Speech Recognition. Pages 257–286 of: *Proceedings of the IEEE*. Piscataway, NJ: IEEE.

Ranganath, R., Gerrish, S., and Blei, D. M. 2013. Black Box Variational Inference. In: *Proceedings of AISTATS*. Scottsdale, AZ: Proceedings of Machine Learning Research.

Reinsel, G. R., and Velu, R. P. 1998. *Multivariate Reduced-Rank Regression: Theory and Applications*. New York: Springer.

Rissanen, J. 1986. Stochastic Complexity and Modeling. *Annals of Statistics*, **14**(3), 1080–1100.

Robbins, H., and Monro, S. 1951. A Stochastic Approximation Method. *Annals of Mathematical Statistics*, **22**(3), 400–407.

Ruhe, A. 1970. Perturbation Bounds for Means of Eigenvalues and Invariant Subspaces. *BIT Numerical Mathematics*, **10**, 343–354.

Rusakov, D., and Geiger, D. 2005. Asymptotic Model Selection for Naive Bayesian Networks. *Journal of Machine Learning Research*, **6**, 1–35.

Sakamoto, T., Ishiguro, M., and Kitagawa, G. 1986. *Akaike Information Criterion Statistics*. Dordrecht: D. Reidel Publishing Company.

Salakhutdinov, R., and Mnih, A. 2008. Probabilistic Matrix Factorization. Pages 1257–1264 of: Platt, J. C., Koller, D., Singer, Y., and Roweis, S. (eds), *Advances in Neural Information Processing Systems 20*. Cambridge, MA: MIT Press.

Sato, I., Kurihara, K., and Nakagawa, H. 2012. Practical Collapsed Variational Bayes Inference for Hierarchical Dirichlet Process. Pages 105–113. In: *Proceedings of KDD*. New York and Beijing, China: Association for Computing Machinery.

Sato, M., Yoshioka, T., Kajihara, S., et al. 2004. Hierarchical Bayesian Estimation for MEG Inverse Problem. *NeuroImage*, **23**, 806–826.

Schwarz, G. 1978. Estimating the Dimension of a Model. *Annals of Statistics*, **6**(2), 461–464.

Seeger, M. 2008. Bayesian Inference and Optimal Design for the Sparse Linear Model. *Journal of Machine Learning Research*, **9**, 759–813.

Seeger, M. 2009. Sparse Linear Models: Variational Approximate Inference and Bayesian Experimental Design. In: *Journal of Physics: Conference Series*, vol. 197. Bristol, UK: IOP Publishing.

Seeger, M., and Bouchard, G. 2012. Fast Variational Bayesian Inference for Non-Conjugate Matrix Factorization Models. Pages 1012–1018. In: *Proceedings of International Conference on Artificial Intelligence and Statistics*. La Palma, Canary Islands: Proceedings of Machine Learning Research.

Shi, J., and Malik, J. 2000. Normalized Cuts and Image Segmentation. *IEEE Transactions on Pattern Analysis and Machine Intelligence*, **22**(8), 888–905.

Soltanolkotabi, M., and Candès, E. J. 2011. A Geometric Analysis of Subspace Clustering with Outliers. *CoRR*. arXiv:1112.4258 [cs.IT].

Spall, J. 2003. *Introduction to Stochastic Search and Optimization: Estimation, Simulation, and Control*. New York: John Wiley and Sons.

Srebro, N., and Jaakkola, T. 2003. Weighted Low Rank Approximation. In: Fawcett, T., and Mishra, N. (eds), *Proceedings of the Twentieth International Conference on Machine Learning*. Washington, DC: AAAI Press.

Srebro, N., Rennie, J., and Jaakkola, T. 2005. Maximum Margin Matrix Factorization. In: *Advances in Neural Information Processing Systems 17*. Vancouver, BC: NIPS Foundation.

Stein, C. 1956. Inadmissibility of the Usual Estimator for the Mean of a Multivariate Normal Distribution. Pages 197–206 of: *Proceedings of the 3rd Berkeley Symposium on Mathematics Statistics and Probability*. Berkeley: University of California Press.

Takemura, A., and Kuriki, S. 1997. Weights of Chi-Squared Distribution for Smooth or Piecewise Smooth Cone Alternatives. *Annals of Statistics*, **25**(6), 2368–2387.

Teh, Y. W., Newman, D., and Welling, M. 2007. A Collapsed Variational Bayesian Inference Algorithm for Latent Dirichlet Allocation. In: *Advances in NIPS*. Vancouver, BC: NIPS Foundation.

Tipping, M. E. 2001. Sparse Bayesian Learning and the Relevance Vector Machine. *Journal of Machine Learning Research*, **1**, 211–244.

Tipping, M. E., and Bishop, C. M. 1999. Probabilistic Principal Component Analysis. *Journal of the Royal Statistical Society*, **61**, 611–622.

Tomioka, R., Suzuki, T., Sugiyama, M., and Kashima, H. 2010. An Efficient and General Augmented Lagrangian Algorithm for Learning Low-Rank Matrices. In: *Proceedings of International Conference on Machine Learning*. Haifa, Israel: Omnipress.

Tron, R., and Vidal, R. 2007. A Benchmark for the Comparison of 3-D Motion Segmentation Algorithms. Pages 1–8. In: *Proceedings of CVPR*. Minneapolis, MN.

Tucker, L. R. 1996. Some Mathematical Notes on Three-Mode Factor Analysis. *Psychometrika*, **31**, 279–311.

Ueda, N., Nakano, R., Ghahramani, Z., and Hinton, G. E. 2000. SMEM Algorithm for Mixture Models. *Neural Computation*, **12**(9), 2109–2128.

van der Vaart, A. W. 1998. *Asymptotic Statistics*. Cambridge Series in Statistical and Probabilistic Mathematics. Cambridge and New York: Cambridge University Press.

Vidal, R., and Favaro, P. 2014. Low Rank Subspace Clustering. *Pattern Recognition Letters*, **43**(1), 47–61.

Wachter, K. W. 1978. The Strong Limits of Random Matrix Spectra for Sample Matrices of Independent Elements. *Annals of Probability*, **6**, 1–18.

Wainwright, M. J., and Jordan, M. I. 2008. Graphical Models, Exponential Families, and Variational Inference. *Foundations and Trends in Machine Learning*, **1**, 1–305.

Watanabe, K. 2012. An Alternative View of Variational Bayes and Asymptotic Approximations of Free Energy. *Machine Learning*, **86**(2), 273–293.

Watanabe, K., and Watanabe, S. 2004. Lower Bounds of Stochastic Complexities in Variational Bayes Learning of Gaussian Mixture Models. Pages 99–104 of: *Proceedings of IEEE on CIS*.

Watanabe, K., and Watanabe, S. 2005. Variational Bayesian Stochastic Complexity of Mixture Models. Pages 99–104. In: *Advances in NIPS*, vol. 18. Vancouver, BC: NIPS Foundation.

Watanabe, K., and Watanabe, S. 2006. Stochastic Complexities of Gaussian Mixtures in Variational Bayesian Approximation. *Journal of Machine Learning Research*, **7**, 625–644.

Watanabe, K., and Watanabe, S. 2007. Stochastic Complexities of General Mixture Models in Variational Bayesian Learning. *Neural Networks*, **20**(2), 210–219.

Watanabe, K., Shiga, M., and Watanabe, S. 2006. Upper Bounds for Variational Stochastic Complexities of Bayesian Networks. Pages 139–146 of: *Proceedings of IDEAL*. Burgos, Spain: Springer.

Watanabe, K., Shiga, M., and Watanabe, S. 2009. Upper Bound for Variational Free Energy of Bayesian Networks. *Machine Learning*, **75**(2), 199–215.

Watanabe, K., Okada, M., and Ikeda, K. 2011. Divergence Measures and a General Framework for Local Variational Approximation. *Neural Networks*, **24**(10), 1102–1109.

Watanabe, S. 2001a. Algebraic Analysis for Nonidentifiable Learning Machines. *Neural Computation*, **13**(4), 899–933.

Watanabe, S. 2001b. Algebraic Information Geometry for Learning Machines with Singularities. Pages 329–336 of: *Advances in NIPS*, vol. 13. Vancouver, BC: NIPS Foundation.

Watanabe, S. 2009. *Algebraic Geometry and Statistical Learning Theory*. Cambridge: Cambridge University Press.

Watanabe, S. 2010. Asymptotic Equivalence of Bayes Cross Validation and Widely Applicable Information Criterion in Singular Learning Theory. *Journal of Machine Learning Research*, **11**, 3571–3594.

Watanabe, S. 2013. A Widely Applicable Bayesian Information Criterion. *Journal of Machine Learning Research*, **14**, 867–897.

Watanabe, S., and Amari, S. 2003. Learning Coefficients of Layered Models When the True Distribution Mismatches the Singularities. *Neural Computation*, **15**, 1013–1033.

Wei, X., and Croft, W. B. 2006. LDA-Based Document Models for Ad-Hoc Retrieval. Pages 178–185 of: *Proceedings of SIGIR*. Seattle, WA: Association for Computing Machinery New York.

Wingate, D., and Weber, T. 2013. Automated Variational Inference in Probabilistic Programming. *arXiv:1301.1299*.

Yamazaki, K. 2016. Asymptotic Accuracy of Bayes Estimation for Latent Variables with Redundancy. *Machine Learning*, **102**(1), 1–28.

Yamazaki, K., and Kaji, D. 2013. Comparing Two Bayes Methods Based on the Free Energy Functions in Bernoulli Mixtures. *Neural Networks*, **44**, 36–43.

Yamazaki, K., and Watanabe, S. 2003a. Singularities in Mixture Models and Upper Bounds Pages 1–8. of Stochastic Complexity. *Neural Networks*, **16**(7), 1029–1038.

Yamazaki, K., and Watanabe, S. 2003b. Stochastic Complexity of Bayesian Networks. Pages 592–599 of: *Proceedings of the Nineteenth Conference on Uncertainty in Artificial Intelligence*. Acapulco, Mexico: Morgan Kaufmann.

Yamazaki, K., and Watanabe, S. 2004. Newton Diagram and Stochastic Complexity in Mixture of Binomial Distributions. Pages 350–364. In: *Proceedings of ALT*. Padova, Italy: Springer.

Yamazaki, K., and Watanabe, S. 2005. Algebraic Geometry and Stochastic Complexity of Hidden Markov Models. *Neurocomputing*, **69**, 62–84.

Yamazaki, K., Aoyagi, M., and Watanabe, S. 2010. Asymptotic Analysis of Bayesian Generalization Error with Newton Diagram. *Neural Networks*, **23**(1), 35–43.

# Subject Index

Printed in the United States
by Baker & Taylor Publisher Services